The chemistry of
enones

Part 2

THE CHEMISTRY OF FUNCTIONAL GROUPS

A series of advanced treatises under the general editorship of
Professor Saul Patai

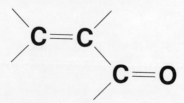

The chemistry of
enones

Part 2

Edited by

SAUL PATAI

and

ZVI RAPPOPORT

The Hebrew University, Jerusalem

1989

JOHN WILEY & SONS

CHICHESTER–NEW YORK–BRISBANE–TORONTO–SINGAPORE

An Interscience® Publication

Library of Congress Cataloging-in-Publication Data:

The Chemistry of enones / edited by Saul Patai and Zvi Rappoport.
    p.   cm. — (The Chemistry of functional groups)
  'An Interscience publication.'
  ISBN 0 471 91563 7 (Part 1)
  ISBN 0 471 92289 7 (Part 2)
  ISBN 0 471 92290 0 (set)
    1. Carbonyl compounds.   2. Olefins.   I. Patai, Saul.
II. Rappoport, Zvi.   III. Series.
QD305.A6C46   1989
547'.036—dc19   88-27713
                                        CIP

British Library Cataloguing in Publication Data:

The Chemistry of Enones.
  1. Enones
  I. Patai, Saul II. Rappoport, Zvi
  III. Series
  547'.036

ISBN 0 471 91563 7 (Part 1)
ISBN 0 471 92289 7 (Part 2)
ISBN 0 471 92290 0 (set)

Printed and bound in Great Britain by
Courier International Ltd, Tiptree, Essex

Contributing authors

M. M. Baizer (deceased) — Department of Chemistry, University of California, Santa Barbara, California 93106, USA

C. L. Bevins — Department of Chemistry, The University of Maryland Baltimore County, Baltimore, Maryland 21228, USA

P. L. Bounds — Department of Chemistry, The University of Maryland Baltimore County, Baltimore, Maryland 21228, USA

G. V. Boyd — Department of Organic Chemistry, The Hebrew University of Jerusalem, Jerusalem 91904, Israel

B. Capon — Department of Chemistry, University of Hong Kong, Pokfulam Road, Hong Kong

M. Dizdaroglu — Center for Chemical Physics, National Bureau of Standards, Gaithersburg, Maryland 20899, USA

D. Duval — Laboratoire de Chimie Physique Organique, Université de Nice, Parc Valrose, 06034 Nice Cédex, France

A. A. Frimer — Department of Chemistry, Bar-Ilan University, Ramat Gan 52100, Israel

J. K. Gawronski — Faculty of Chemistry, Adam Mickiewicz University, Grunwaldzka 6, 60780 Poznań, Poland

S. Géribaldi — Laboratoire de Chimie Physique Organique, Université de Nice, Parc Valrose, 06034 Nice Cédex, France

H. E. Gottlieb — Department of Chemistry, Bar-Ilan University, Ramat-Gan 52100, Israel

N. Greenspoon — Department of Organic Chemistry, The Weizmann Institute of Science, Rehovot, Israel

J. A. S. Howell — Department of Chemistry, University of Keele, Keele, Staffordshire, ST5 5BG, UK

C. R. Johnson — Department of Chemistry, Wayne State University, Detroit, Michigan 48202, USA

E. Keinan — Department of Chemistry, Technion–Israel Institute of Technology, Technion City, Haifa 32000, Israel

J. F. Liebman — Department of Chemistry, The University of Maryland Baltimore County, Baltimore, Maryland 21228, USA

R. D. Little — Department of Chemistry, University of California, Santa Barbara, California 93106, USA

A. Y. Meyer   Department of Organic Chemistry, The Hebrew University of
                     Jerusalem, Jerusalem 91904, Israel

K. Müllen   Department of Organic Chemistry, University of Mainz, J. J.
                     Becher-Weg 18–20, D-6500 Mainz, FRG

P. Neta   Center for Chemical Physics, National Bureau of Standards,
                     Gaithersburg, Maryland 20899, USA

M. R. Peel   Department of Chemistry, Wayne State University, Detroit,
                     Michigan 48202, USA

R. M. Pollack   Department of Chemistry, The University of Maryland Balti-
                     more County, Baltimore, Maryland 21228, USA

G. A. Russell   Department of Chemistry, Iowa State University, Ames, Iowa
                     50011, USA

D. I. Schuster   Department of Chemistry, Faculty of Arts and Science, New
                     York University, 4 Washington Place, Room 514, New York,
                     NY 10003, USA

B. Schweizer   ETH Laboratorium für Organische Chemie, Universitätstrasse
                     16, ETH-Zentrum, CH-8092 Zürich, Switzerland

K. J. Shea   Department of Chemistry, University of California, Irvine,
                     California 92917, USA

C. Thebtaranonth   Department of Chemistry, Faculty of Science, Mahidol Univer-
                     sity, Rama 6 Road, Bangkok 10400, Thailand

Y. Thebtaranonth   Department of Chemistry, Faculty of Science, Mahidol Univer-
                     sity, Rama 6 Road, Bangkok 10400, Thailand

C. R. Theocharis   Department of Chemistry, Brunel University, Uxbridge, Mid-
                     dlesex, UB8 3PH, UK

F. Tureček   The Jaroslav Heyrovsky Institute of Physical Chemistry and
                     Electrochemistry, Machova 7, 12138 Prague 2, Czechoslovakia

P. Wolf   Department of Organic Chemistry, University of Mainz, J. J.
                     Becher-Weg 18–20, D-6500 Mainz, FRG

R. I. Zalewsky   Department of General Chemistry, Academy of Economy,
                     60–967 Poznań, Poland

Foreword

The present volume in 'The chemistry of functional groups' series presents material on ketones and aldehydes containing also a carbon–carbon double bond, i.e. on enones and enals. The two (in the large majority of cases conjugated) functional groups involved, i.e. $C=C$ and $C=O$ influence one another profoundly and their properties and reactions in enones and enals are by no means identical to those which occur alone in simple alkenes or carbonyl compounds. Hence we believed that a separate volume on the $C=C-C=O$ system would be a desirable addition to the series and we are very pleased that we succeeded in securing the collaboration of an international team of authors, scattered widely over three continents.

Two subjects were intended to be covered in this volume, but did not materialize. These were on biochemistry and on enones with strained double bonds. We hope to include these chapters in one of the forthcoming supplementary volumes of the series. A third chapter, on cycloadditions, will be included in Supplement A2, to be published in a few months' time.

Literature coverage in most chapters is up to late 1987 or early 1988.

Jerusalem
December 1988

SAUL PATAI
ZVI RAPPOPORT

The Chemistry of Functional Groups
Preface to the Series

The series 'The Chemistry of Functional Groups' is planned to cover in each volume all aspects of the chemistry of one of the important functional groups in organic chemistry. The emphasis is laid on the functional groups treated and on the effects which it exerts on the chemical and physical properties, primarily in the immediate vicinity of the group in question and secondarily on the behaviour of the whole molecule. For instance, the volume *The Chemistry of the Ether Linkage* deals with reactions in which the C—O—C group is involved, as well as with the effects of the C—O—C group on the reactions of alkyl or aryl groups connected to the ether oxygen. It is the purpose of the volume to give a complete coverage of all properties and reaction of ethers in as far as these depend on the presence of the ether group but the primary subject matter is not the whole molecule, but the C—O—C functional group.

A further restriction in the treatment of the various functional groups in these volumes is that material included in easily and generally available secondary or tertiary sources, such as Chemical Reviews, Quarterly Reviews, Organic Reactions, various 'Advances' and 'Progress' series as well as in textbooks (i.e. in books which are usually found in the chemical libraries of universities and research institutes) should not, as a rule, be repeated in detail, unless it is necessary for the balanced treatment of the subject. Therefore each of the authors is asked not to give an encyclopaedic coverage of his subject, but to concentrate on the most important recent developments and mainly on material that has not been adequately covered by reviews or other secondary sources by the time of writing of the chapter, and to address himself to a reader who is assumed to be at a fairly advanced postgraduate level.

With these restrictions, it is realized that no plan can be devised for a volume that would give a complete coverage of the subject with no overlap between chapters, while at the same time preserving the readability of the text. The Editor set himself the goal of attaining reasonable coverage with moderate overlap, with a minimum of cross-references between the chapters of each volume. In this manner, sufficient freedom is given to each author to produce readable quasi-monographic chapters.

The general plan of each volume includes the following main sections:

(a) An introductory chapter dealing with the general and theoretical aspects of the group.

(b) One or more chapters dealing with the formation of the functional group in question, either from groups present in the molecule, or by introducing the new group directly or indirectly.

(c) Chapters describing the characterization and characteristics of the functional groups, i.e. a chapter dealing with qualitative and quantitative methods of determination including chemical and physical methods, ultraviolet, infrared, nuclear magnetic resonance and mass spectra: a chapter dealing with activating and directive effects exerted by the group and/or a chapter on the basicity, acidity or complex-forming ability of the group (if applicable).

(d) Chapters on the reactions, transformations and rearrangements which the functional groups can undergo, either alone or in conjunction with other reagents.

(e) Special topics which do not fit any of the above sections, such as photochemistry, radiation chemistry, biochemical formations and reactions. Depending on the nature of each functional group treated, these special topics may include short monographs on related functional groups on which no separate volume is planned (e.g. a chapter on 'Thioketones' is included in the volume *The Chemistry of the Carbonyl Group*). In other cases certain compounds, though containing only the functional group of the title, may have special features so as to be best treated in a separate chapter, as e.g. 'Polyethers' in *The Chemistry of the Ether Linkage*, or 'Tetraaminoethylenes' in *The Chemistry of the Amino Group*.

This plan entails that the breadth, depth and thought-provoking nature of each chapter will differ with the views and inclinations of the author and the presentation will necessarily be somewhat uneven. Moreover, a serious problem is caused by authors who deliver their manuscript late or not at all. In order to overcome this problem at least to some extent, it was decided to publish certain volumes in several parts, without giving consideration to the originally planned logical order of the chapters. If after the appearance of the originally planned parts of a volume it is found that either owing to non-delivery of chapters, or to new developments in the subject, sufficient material has accumulated for publication of a supplementary volume, containing material on related functional groups, this will be done as soon as possible.

The overall plan of the volumes in the series 'The Chemistry of Functional Groups' includes the titles listed below:

The chemistry of alkenes (two volumes)
The chemistry of the carbonyl group (two volumes)
The chemistry of the ether linkage
The chemistry of the amino group
The chemistry of the nitro and nitroso groups (two parts)
The chemistry of carboxylic acids and esters
The chemistry of the carbon–nitrogen double bond
The chemistry of the cyano group
The chemistry of amides
The chemistry of the hydroxyl group (two parts)
The chemistry of the azido group
The chemistry of the acyl halides
The chemistry of the carbon–halogen bond (two parts)
The chemistry of the quinonoid compounds (two volumes, four parts)
The chemistry of the thiol group (two parts)
The chemistry of the hydrazo, azo and azoxy groups (two parts)
The chemistry of amidines and imidates
The chemistry of cyanates and their thio derivatives (two parts)
The chemistry of diazonium and diazo groups (two parts)
The chemistry of the carbon–carbon triple bond (two parts)
The chemistry of ketenes, allenes and related compounds (two parts)
The chemistry of the sulphonium group (two parts)
Supplement A: The chemistry of double-bonded functional groups (two parts)
Supplement B: The chemistry of acid derivatives (two parts)
Supplement C: The chemistry of triple-bonded functional groups (two parts)

Supplement D: The chemistry of halides, pseudo-halides and azides (two parts)
*Supplement E: The chemistry of ethers, crown ethers, hydroxyl groups and their sulphur
analogues* (two parts)
Supplement F: The chemistry of amino, nitroso and nitro compounds and their derivatives
(two parts)
The chemistry of the metal–carbon bond (four volumes)
The chemistry of peroxides
The chemistry of organic selenium and tellurium compounds (two volumes)
The chemistry of the cyclopropyl group
The chemistry of sulphones and sulphoxides
The chemistry of organic silicon compounds (two parts)

Titles in press:

Supplement A2: The chemistry of double-bonded functional groups

Titles in preparation:

The chemistry of enols
The chemistry of sulphinic acids, esters and derivatives
The chemistry of sulphenic acids, esters and derivatives

Advice or criticism regarding the plan and execution of this series will be welcomed by
the Editor.

The publication of this series would never have been started, let alone continued,
without the support of many persons. First and foremost among these was the late Dr
Arnold Weissberger, whose reassurance and trust encouraged me to tackle this task. The
efficient and patient co-operation of several staff members of the Publisher also rendered
me invaluable aid (but unfortunately their code of ethics does not allow me to thank them
by name). Many of my friends and colleagues in Israel and overseas helped me in the
solution of various major and minor matters, and my thanks are due to all of them,
especially to Professor Zvi Rappoport. Carrying out such a long range project would be
quite impossible without the non-professional but none the less essential participation and
partnership of my wife.

The Hebrew University   SAUL PATAI
Jerusalem, ISRAEL

Contents

List of abbreviations used

Ac	acetyl (MeCO)
acac	acetylacetone
Ad	adamantyl
All	allyl
An	anisyl
Ar	aryl
Bz	benzoyl (C_6H_5CO)
Bu	butyl (also t-Bu or Bu^t)
CD	circular dichroism
CI	chemical ionization
CIDNP	chemically induced dynamic nuclear polarization
CNDO	complete neglect of differential overlap
Cp	η^5-cyclopentadienyl
DBU	1, 8-diazabicyclo[5.4.0]undec-7-ene
DME	1, 2-dimethoxyethane
DMF	N, N-dimethylformamide
DMSO	dimethyl sulphoxide
ee	enantiomeric excess
EI	electron impact
ESCA	electron spectroscopy for chemical analysis
ESR	electron spin resonance
Et	ethyl
eV	electron volt
Fc	ferrocene
FD	field desorption
FI	field ionization
FT	Fourier transform
Fu	furyl(OC_4H_5)
Hex	hexyl(C_6H_{11})
c-Hex	cyclohexyl(C_6H_{11})
HMPA	hexamethylphosphortriamide
HOMO	highest occupied molecular orbital

xv

i-	iso
Ip	ionization potential
IR	infrared
ICR	ion cyclotron resonance

LCAO	linear combination of atomic orbitals
LDA	lithium diisopropylamide
LUMO	lowest unoccupied molecular orbital

M	metal
M	parent molecule
MCPBA	m-chloroperbenzoic acid
Me	methyl
MNDO	modified neglect of diatomic overlap
MS	mass spectrum

n	normal
Naph	naphthyl
NBS	N-bromosuccinimide
NMR	nuclear magnetic resonance

Pen	pentyl(C_5H_{11})
Pip	piperidyl($C_5H_{10}N$)
Ph	phenyl
ppm	parts per million
Pr	propyl (also i-Pr or Pr^i)
PTC	phase transfer catalysis
Pyr	pyridyl (C_5H_4N)

R	any radical
RT	room temperature

s-	secondary
SET	single electron transfer
SOMO	singly occupied molecular orbital

t-	tertiary
TCNE	tetracyanoethylene
THF	tetrahydrofuran
Thi	thienyl(SC_4H_3)
TMEDA	tetramethylethylene diamine
Tol	tolyl(MeC_6H_4)
Tos	tosyl (p-toluenesulphonyl)
Trityl	triphenylmethyl(Ph_3C)
Xyl	xylyl($Me_2C_6H_3$)

In addition, entries in the 'List of Radical Names' in *IUPAC Nomenclature of Organic Chemistry*, 1979 Edition, Pergamon Press, Oxford, 1979, pp. 305–322, will also be used in their unabbreviated forms, both in the text and in structures.

We are sorry for any inconvenience to our readers. However, the rapidly rising costs of production make it absolutely necessary to use every means to reduce expenses—otherwise the whole existence of our Series would be in jeopardy.

The Chemistry of Enones
Edited by S. Patai and Z. Rappoport
© 1989 John Wiley & Sons Ltd

CHAPTER **14**

Enone electrochemistry

R. DANIEL LITTLE and MANUEL M. BAIZER*

Department of Chemistry, University of California, Santa Barbara, Santa Barbara, CA 93106, USA

I. INTRODUCTION

The α, β-unsaturated ketone (enone) functional group is undoubtedly one of the most useful in organic chemistry. Each atom of the unit can, under appropriate conditions, function as a site at which a reaction can take place.

Enones have often served as substrates for electrochemical investigations[1-3]. For the most part, focus has been upon the generation and study of radical anions rather than

*Deceased July 9, 1988.

radical cations. The reason for this disparity is easy to understand when one realizes that an electronically unperturbed enone possesses both a low-lying highest occupied molecular orbital (HOMO), from which it should be difficult to remove an electron, as well as a low-lying lowest unoccupied molecular orbital (LUMO) into which an electron can easily be added[4].

The terms 'difficult' and 'easily' used above are vague and require refinement. A variety of methods have been used to do so, including molecular orbital calculations[4,5], photoelectron spectroscopy[6-8], electron affinity measurements[9], charge transfer and UV spectroscopy[9], polarography and cyclic voltammetry[10-17].

II. PRODUCTION OF THE RADICAL ANION; REDUCTION POTENTIALS

Several compilations of polarographic reduction potentials, in both protic[10,11] and aprotic[12-14] media, are available. Access to this information is invaluable for the mechanistic insight it can provide (*vide infra*). Generally, the potentials measured under aqueous conditions are considerably less negative than those measured under aprotic conditions[12,13].

A detailed investigation and interpretation of the results obtained from a study of the reduction process as a function of pH has been conducted[15-17] and reviewed[2].

From data collected for a wide range of cyclic and acyclic aldehydes, esters and ketones in anhydrous DMF, it has proven possible to derive a very useful set of empirical rules which allow prediction of reduction potentials within ± 0.1 V as a function of the position and nature of the substituents R^1–R^4[14]. As illustrated in Table 1 substitution of an alkyl group at any one of the available positions shifts the reduction potential by − 0.1 V from a base value of − 1.9 V (vs SCE; $R^1 = R^2 = R^3 = R^4 = H$). An electron-donating alkoxy substituent has a pronounced effect (viz. − 0.3 V) when substituted at either the carbonyl or the β-carbon, but has no effect when placed at the α-carbon. In accord with expectation, substitution of a single phenyl group at any position except the α-carbon makes the enone easier to reduce. Inclusion of a second phenyl group has no significant additional effect.

The presence of a phenyl group at either the carbonyl or the β-carbon makes it possible to observe two, rather than as is the case for many enones, one reduction wave[13]. It has been suggested[13] that the second wave corresponds to conversion of the first formed anion to the dianion. Since the potential associated with the second wave is so negative, even for a highly conjugated system (e.g. $E_{1/2}$ for *trans*-PhCH=CHCOBu-*t* is − 2.23 V and for

TABLE 1. Empirical rules for estimation of reduction potentials

Substituent	Increment in reduction potential for		
	R^1	R^2	R^3 or R^4
Alkyl group	− 0.1	− 0.1	− 0.1
1st alkoxy group	− 0.3	0.0	− 0.3
1st phenyl group	+ 0.4	+ 0.1	+ 0.4

trans-PhCH=CHCOCH$_3$ − 2.61 V), it has been suggested that dianions may rarely, if ever, be involved in the cathodic chemistry of aliphatic enones[13].

In a few cases, the effect upon reduction potential of substituents placed at various positions on the phenyl group of an aryl ketone has been studied and shown to be correlatable with substituent constants using either Hammett or Yukawa–Tsuno relationships[18-20].

III. ELECTRONIC STRUCTURE OF RADICAL ANIONS; ESR STUDIES

Attempts to generate and study radical anions by electron spin resonance (ESR) spectroscopy are thwarted when the compound being studied contains acidic protons located at either end of the enone[13,21]. Replacement of the hydrogens with alkyl or aryl groups allows observation of well-defined ESR spectra and determination of the electron distribution within the radical anion. Independent studies[13,21] show that 40–50% of the unpaired spin density is located at the β-carbon, while the remainder is divided almost equally between the carbonyl carbon and oxygen atoms. Since the unpaired spin density at the α-carbon is nearly zero, one would anticipate and in fact finds (see Section II) that the reduction potential for an enone should be essentially independent of the nature of the substituent attached to that carbon.

From these observations, it is gratifying to recognize that most of the chemistry of enone radical anions is characterized by reactions occurring at the β-carbon (β, β-coupling, protonation), the carbonyl carbon (pinacolization) and on oxygen (protonation). These and other reactions are discussed in Sections VI–XIII.

IV. LIFETIME OF A RADICAL ANION[13,22,23]

The lifetime of an enone radical anion is critically dependent upon several factors including: (a) The nature of the medium in which it is generated. In general, the presence of even low concentrations (e.g. 10^{-1} to 10^{-3} M) of a proton source (e.g. water, ROH, RCO$_2$H) or lithium salts leads to a marked decrease in the lifetime. For example, while the half-life for a 10^{-3} M solution of *trans-t*-BuCH=CHCOBu-*t* in dry DMF was determined by cyclic voltammetry (CV) to be > 10 s at ambient temperature, the addition of 0.03 M CF$_3$CO$_2$H causes a decrease to < 10^{-3}. The reason for this behavior is related to the previously mentioned need to replace acidic hydrogens flanking the enone in order to observe an ESR spectrum. That is, in the presence of a proton donor, the radical anion is protonated, leading to a neutral radical which subsequently dimerizes. Lithium, but interestingly not sodium or quaternary ammonium salts, have a similar effect. (b) The temperature at which the measurement is made. As expected, lower temperatures lead to increased lifetimes. (c) The presence of a functional group with which the radical anion can undergo a reaction intramolecularly (e.g. electrohydrocyclization)[1-3].

V. RADICAL ANION GEOMETRY[13]

Enone radical anions can either lose or, if the temperature is sufficiently low, maintain the geometry of the enone precursor for a time long enough to be discerned. For example, while the experimentally determined difference in free energy between the *cis* and *trans* geometric isomers of *t*-BuCH=CHCOBu-*t* is > 4 kcal mol^{-1} at 27 °C, the difference in their reduction potentials is only 17 mV. Since a potential difference of 1.00 V corresponds to an energy difference of 23.06 kcal mol^{-1}, 17 mV corresponds to only 0.017 × 23.06 kcal mol^{-1}, a value at least ten times less than that expected if each enone was reduced to a common, geometry-equilibrated intermediate. This line of reasoning suggests

If a common intermediate, then

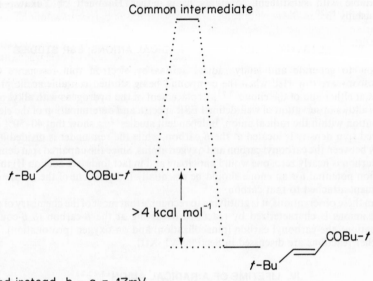

Find instead, b − a = 17mV

SCHEME 1

that each enone affords a geometrically distinct radical anion, as demonstrated in Scheme 1.

It must be noted, however, that these radical anions undergo rapid equilibration on the time scale of an ESR experiment at 25–30 °C, because electrolysis of either the *cis*- or the *trans*-enone in the probe of an ESR spectrometer affords the same well-resolved spectrum, undoubtedly that of the equilibrating forms[13]. That is, the barrier to rotation about the α, β-carbon–carbon bond is significantly lower in the radical anion than in the enone.

The rate of equilibration varies with temperature; it is sufficiently fast at temperatures at or above $- 35$ °C so that no difference in products or product ratio is noted when either the *cis*- or the *trans*-isomer undergoes hydrodimerization[13,24]. However, at $- 78$ °C, the interconversion is slowed to a value where each enone leads to a different mixture of stereoisomeric products. That is, at $T \geqslant - 35$ °C equation 1 applies[13] whereas when $T \leqslant - 78$ °C equation 2 applies, while the *trans*-isomer still affords the *d, l* pair[13].

$$(1)$$

$$(2)$$

It could be synthetically useful if these observations prove general, that is, if enone radical anions can maintain the geometry of the precursor long enough to express that difference in terms of the stereochemical outcome for subsequent coupling processes.

VI. REDUCTIVE DIMERIZATION OF α, β-UNSATURATED KETONES (HYDRODIMERIZATION)

When reduced, α, β-unsaturated ketones can undergo a variety of transformations; most serve to form a new carbon–carbon bond between two or more enone subunits. As illustrated in equation 3, coupling can occur betwee: (a) β-carbon atoms to generate a 1, 6-diketone; (b) two carbonyl carbons, leading to a 1, 2-diol (a pinacol); or (c) the carbonyl carbon of one unit and the β-carbon of the other, creating a γ-hydroxy ketone.

Note that each product corresponds to a dimer of the starting material plus two hydrogens. Consequently, the reduction should be conducted in the presence of a proton donor. The proton donors span a wide range of acidities ranging, as we shall see, from a mineral acid in an aqueous medium to a carbon acid [e.g. $CH_2(CO_2R)_2$] in an organic solvent, often acetonitrile or DMF.

Appropriately, the intermolecular electrochemically initiated hydrodimerization reactions are referred to as electrohydrodimerization (EHD) reactions[2]. The intramolecular

$$
2 \quad \underset{R^4 \quad R^1}{\overset{R^2}{\underset{R^3}{\diagdown}}}\!\!=\!\!O \quad \xrightarrow[+2H^+]{+2e^-} \quad \left[\underset{R^4 \quad R^1}{\overset{R^2}{\underset{R^3}{\diagdown}}}\!\!O \right]_2 \quad \text{and/or} \quad \left[\underset{R^4 \quad R^1}{\overset{R^2}{\underset{R^3}{\diagdown}}}\!\!OH \right]_2
$$

$$
\text{and/or} \quad \underset{R^2 \quad R^4 \quad R^3 \quad R^1}{\overset{R^4 \; HO \quad R^1 \; R^2}{\diagdown}}\!\!=\!\!O \tag{3}
$$

version of the β, β-coupling process is called electrohydrocyclization (EHC)[2]. In both cases, a wide range of electron-withdrawing groups have served as olefin activators (e.g. CN, CHO, CO_2R, $CONR_2$, etc.)[1-3,25,26]. We will focus attention almost exclusively upon α, β-unsaturated ketones and will attempt to provide an overview of the chemistry rather than an exhaustive survey of the very large amount of work which has been published.

In addition to the modes of coupling listed above, reduction of α, β-unsaturated ketones can also lead to: (1) saturation of the C—C π bond, a process which can become important when dimerization is sterically inhibited; and (2) oligomerization or polymerization, processes which are most likely to occur in an aprotic or basic medium.

The scope and limitations of EHD reactions of monoactivated olefins, mixed (or crossed) reductive coupling among them, and EHC reactions have been discussed[27-38]. Two EHD reactions involving enones are illustrated in equations 4 and 5[39,40]. Others are presented in Section VII.

$$
\overset{O}{\bigcirc}\!\!\diagdown \quad \xrightarrow{e^-,\, 95:5\, CH_3CN-H_2O\,(v/v)} \quad \left[\overset{O}{\bigcirc}\!\!\diagdown \right]_2 \tag{4}
$$

95%, d,l + meso

$$
\underset{CH=CHCOCH_3}{\overset{\diagup\!\!\diagdown}{\underset{O}{\diagdown}}} \quad \xrightarrow[H_2SO_4]{\substack{e^-(Hg) \\ 25\%\ aq\ acetone}} \quad \left[\overset{O}{\underset{O}{\diagdown}} \right]_2 \tag{5}
$$

52%

α, β-Unsaturated ketones are suitable coupling partners in mixed couplings for substances which are reduced at more positive cathode voltages[30]. For example, electrolysis at -1.2 to -1.3 V (vs SCE) of a mixture of diethyl fumarate and a tenfold molar excess of methyl vinyl ketone afforded the keto diester shown in ca 85% yield[30] (equation 6).

When the enone is easier to reduce than its partner, then it is desirable to select an enone whose β-carbon is sufficiently encumbered in order to decrease its tendency to undergo self-hydrodimerization, and to couple it with an uncongested acceptor. Examples[30] are

$$\text{EtO}_2\text{C} \diagdown \diagup \diagdown \text{CO}_2\text{Et} \quad + \text{ excess } \text{CH}_2\text{=CHCOCH}_3$$

$$\xrightarrow[\text{Et}_4\text{NOTs, CH}_3\text{CN/H}_2\text{O}]{-1.2 \text{ to } -1.3 \text{ V (SCE)}} \quad \text{EtO}_2\text{C} \diagdown \diagup \diagdown \diagup \diagdown \diagup \diagdown \quad (6)$$

$$85\%$$

given in equations 7 and 8. Attempts to couple 9-benzalfluorene with mesityl oxide, each of which is sterically encumbered, failed[30]. Electrolysis of a mixture of methyl vinyl ketone and 9-benzalfluorene afforded mainly the ketone hydrodimer and no coupled product.

$$\text{PhCH=CHCOCH}_3 + \text{excess CH}_2\text{=CHCN} \xrightarrow[\text{Et}_4\text{NOTs, AcOH}]{-1.4 \text{ to } -1.5 \text{ V (SCE)}}$$

$$\underset{ca\ 32\%}{\text{CH}_3\text{COCH}_2\text{CHPhCHPhCH}_2\text{COCH}_3} + \underset{< 10\%}{\text{CH}_3\text{COCH}_2\text{CHPhCH}_2\text{CH}_2\text{CN}} \quad (7)$$

$$(\text{CH}_3)_2\text{C=CHCOCH}_3 + \text{excess CH}_2\text{=CHCN} \xrightarrow[\text{Et}_4\text{NOTs, CH}_3\text{CN/H}_2\text{O}]{-1.6 \text{ to } -1.7 \text{ V (SCE)}}$$

$$\underset{ca\ 50\%}{\text{CH}_3\text{COCH}_2\text{C(CH}_3)_2\text{CH}_2\text{CH}_2\text{CN}} \quad (8)$$

The EHC reaction can provide a powerful means of constructing polycyclic ring systems. Most often, β, β-coupling occurs[41] (equation 9). However, in another example[42], reduction of the dienoate 1, isomerization of the resulting radical anion, and sigma bond formation between C(α) and C(β') ensues (equation 10). Perhaps β, β'-coupling is simply sterically retarded relative to the alternative pathway. Interestingly, no cleavage of the hydroxyl group was reported.

$$(9)$$

72%

$$(10)$$

A. Mechanism

Given the possibilities outlined above, it may not be surprising to discover that a great deal of time and effort has been expended to determine the mechanism for the reactions[1-3,25,26,39,43-51]. A sufficient amount is known about each so that one can choose remarkably well among a host of conditions those most appropriate to achieve selective and efficient conversion to a single product type. Studies have focused upon the effect that variations in (a) cathode material, (b) catholyte, (c) supporting electrolyte, particularly its cation, (d) concentration of the olefin in the catholyte, and (e) type of cell, have upon the various coupling processes[1-3].

B. Mechanistic Overview; Examples

In general, there exist two schools of thought regarding the mechanism for β, β-coupling under neutral or alkaline conditions. In one, dimerization is thought to occur via the combination of two radical anions (an EC process, i.e. an *e*lectrochemical reaction followed by a *c*hemical reaction)[43,45,51]. The other suggests that the process involves coupling between the initially formed radical anion and the starting enone (the ECE mechanism)[23,46,48-50].

Under acidic conditions, it is generally agreed that dimerization occurs between the neutral radicals formed after preprotonation of the enone on oxygen followed by one-electron reduction to generate an allylic radical[13,24].

A comparatively recent study, illustrating the variation in product composition as a function of the amount of proton donor (in this case, water) present in the reaction medium and the way in which the ratio responds to varying degrees of steric hindrance about the enone subunit, has been published[39]. For the enones **2a–c** illustrated below, the solvent was varied from pure acetonitrile to 5% (v/v) water in acetonitrile; tetra-*n*-butyl-ammonium tetrafluoroborate was used as the supporting electrolyte, a stirred mercury pool as the cathode.

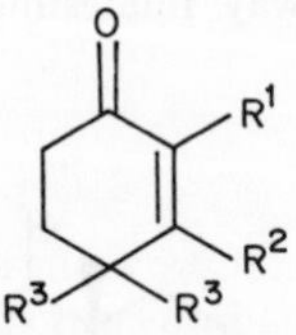

(**a**) $R^1 = R^2 = R^3 = H$

(**b**) $R^1 = R^2 = H$, $R^3 = CH_3$

(**c**) $R^1 = R^2 = R^3 = CH_3$

(**2**)

As illustrated in Table 2, regardless of the water content, coupling between β-carbons is preferred when cyclohexenone is used as the substrate. The addition of methyl groups at C(4) leads to a decrease in the amount of β, β-coupling. As the water content increases, the

TABLE 2. Effect of water content on hydrodimer product ratios

Enone	Water content (% H_2O in CH_3CN)	1, 6-Diketone (%)	γ-Hydroxy ketone (%)	Diol (%)
2a	0	97	—	—
	5	95	—	—
2b	0	52	31	16
	5	28	4	64

amount of 1, 2-diol steadily increases until, at 2% water in acetonitrile, equal quantities of diol and diketone are produced. Eventually, the diol-to-diketone ratio inverts and more diol than diketone is formed.

Severe crowding as in enone **2c** causes a substantial drop in the amount of β, β-coupling and a corresponding increase in the quantities of hydroxy ketone and diol[39] (equation 11). This shift toward more diol as steric hindrance about the β-carbon increases is general and has been noted before[24,52,53].

(2c)	No water:	10%		60%		30%
	5% water:	—		45%		50%

$$(11)$$

These results were interpreted in accord with previous analyses[1-3,13,24-26] to indicate that in the presence of a proton donor, protonation occurs on oxygen to form a neutral allylic radical **3** which subsequently dimerizes by coupling between (a) C(1), leading to diol;

(3)

(b) C(1) and C(3), affording hydroxy ketone; and (c) C(3), providing the 1, 6-diketone. In the absence of the water, the enone plus radical anion pathway was suggested to account for the results.

A similar set of cyclohexenones was subjected to a detailed voltammetric study[23]. Using well-established criteria[51], it was again concluded that electrohydrodimerization proceeds via the radical anion plus enone pathway in the absence of a proton donor, despite evidence accumulated by others based upon voltammetric[43] and chronopotentiometric[45] studies indicating operation of the radical anion dimerization pathway.

The mechanism of electrohydrocyclization has been studied in great detail and with great care[1-3,46]. Here too it was concluded that cyclization of the bisactivated olefins occurs, at least partially, through the attack of a radical anion upon an unreduced double bond[46]. A more recent study dealing with the symmetrical bisenone 4 (equation 12) led to similar conclusions[50]. Note that the final product 5 corresponds formally to one produced by an intramolecular aldol condensation of the product formed in the EHC reaction. Often the products of EHD and EHC reactions undergo well-known 'secondary' processes[2].

$$\xrightarrow[\text{DMF, Bu}_4\text{NBF}_4]{e^- \ (Hg)} \tag{12}$$

(4)   (5)

The potential associated with conversion of 4 to its radical anion is -2.20 to -2.29 V (vs Ag/Ag$^+$) depending upon the scan rate associated with the cyclic voltammetry (CV) experiment. This is nearly the value one would have predicted based upon the use of the set of empirical values listed in Table 1 for estimation of reduction potentials[14] and is nearly the same as that of cyclohexenone. These observations would not be noteworthy except for the way in which they stand in marked contrast to the potentials obtained for bis-α, β-unsaturated esters. For example, the polarographic half-wave potentials for a variety of bisenoates 6 are shifted to a value roughly 200 mV (ca 4.6 kcal) more positive than that associated with the simple model system possessing only one unsaturated ester unit, ethyl crotonate[31]. That is, even though the two unsaturated esters are insulated from one another by a series of methylene units, the presence of the second influences the potential of the first, making the bisenoate easier to reduce. It is generally accepted that this shift to a more positive potential is correlated with a process wherein the polarizable enoates approach one another with the β-carbons sufficiently close to allow the one-electron reduction and sigma-bond formation to occur in concert[1-3].

$$(CH_2)_n \begin{cases} CH{=}CHCO_2C_2H_5 \\ \\ CH{=}CHCO_2C_2H_5 \end{cases}$$

(6)

Based upon precedent of this nature it is curious that a shift to a more positive potential is not observed for the symmetrical bisenone. Perhaps, given the flexibility of the methylene chain linking the α, β-unsaturated ester units to one another, there is a preferred geometry associated with cyclization and attendant potential shift which is unattainable for the comparatively rigid bicyclic enone. It appears as though simply bringing the β-carbons near one another is not sufficient to cause a shift.

With these observations and comments in mind, it is interesting to note the behavior of the rigid bicyclo(3.3.1)enones (7)[54] (equations 13–16). Again, no shift in potential is observed. CV data (Pt, CH$_3$CN, 0.4 M Bu$_4$NBF$_4$, Ag/AgNO$_3$ reference electrode) indicate two one-electron reduction waves, one at -2.0 V, the other at -2.75 V. Preparative scale reduction of 7a and 7b illustrates an important and useful feature of controlled potential electrolysis. That is, different products can sometimes be obtained depending upon whether the reactions are carried out at the first or second wave.

$$\text{(13)}$$

$$\text{(14)}$$

$$\text{(15)}$$

$$\text{(16)}$$

VII. STEREOCHEMISTRY OF β,β-COUPLING

While the stereochemical outcome of several EHD and EHC reactions has been determined (equations 1, 2, 9, 17 and 18)[13,22,41,55,56] and on occasions there exists a high degree of stereoselectivity[13,22,24,41,55,57,58], the factors leading to and controlling the selectivity have, unfortunately, not been thoroughly investigated.

$$\text{(17)}^{56}$$

$$\text{(18)}^{22}$$

In a few cases, such as that of apoverbenone (**8**)[22], it has been suggested that the major product is formed as a result of a least-hindered side approach to the face of the allylic radical **9** opposite the *gem*-methyl group.

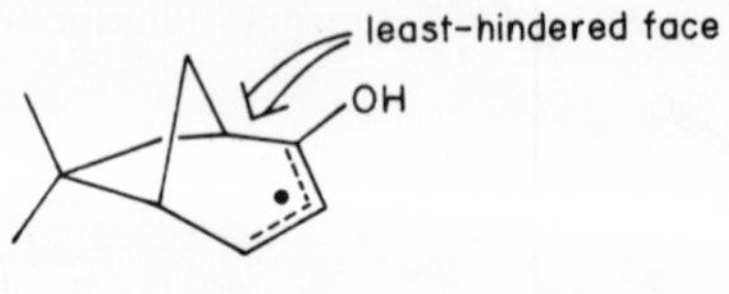

(**9**)

The reaction illustrated in equation 9 is remarkable not only for its stereospecificity, but also for the regiospecificity; only the β, β-coupled hydrodimer is formed[41].

VIII. ELECTROGENERATED BASE (EGB) PROPERTIES OF ENONE RADICAL ANIONS

It was noted previously that the presence of acidic hydrogens at either end of the enone makes it extremely difficult, and in some cases impossible, to obtain an ESR spectrum of the radical anion[13,21]. It was also indicated that even when the hydrogens are replaced by alkyl or aryl groups, the addition of an external proton source greatly diminishes the lifetime of the radical anion[13].

One can use this propensity of radical anions to act as a base, an electrogenerated base (EGB)[59], to affect a variety of transformations. For example, reduction (Hg cathode, -1.90 V, DMF, Pr_4NBF_4) of only a small amount (0.13%) of cyclohexenone leads to the Michael adduct **10**[60] (equation 19).

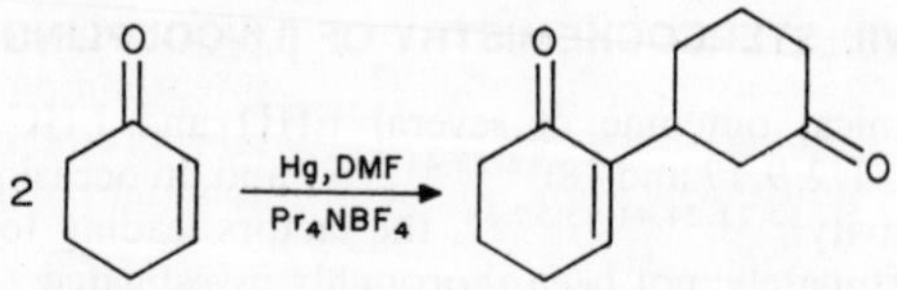

(**10**) 65%

As illustrated in Scheme 2, only a catalytic amount (often 1–10%) of the enone need be reduced to the radical anion, since the latter is used in a catalytic fashion.

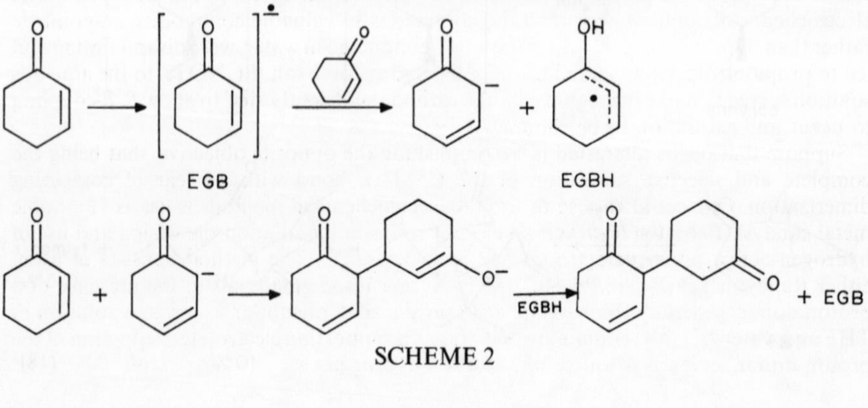

SCHEME 2

Similarly, dimerization of 1-phenyl-1-penten-3-one **11** can be achieved after the passage of less than 0.2 faraday mol^{-1} of electricity. When the reaction is conducted in DMF with lithium perchlorate as the supporting electrolyte, the product is formed stereospecifically and in quantitative yield[61] (equation 20). Interestingly, use of Bu_4NBr in place of lithium perchlorate affords equal amounts of two dimers in addition to polymer[61] (equation 21).

$$2PhCH\!=\!CHCOCH_2CH_3 \xrightarrow{\;e^-\;}$$

$$\text{(11)}$$

quantitative (20)

34%

34% (21)

Occasionally, radical anions are sufficiently long lived so that they can be trapped by added electrophiles such as acetic anhydride[54,62-65] or carbon dioxide[1-3,66,67]. In the absence of a trapping agent and in the absence of a suitable proton donor, radical anions and dianions can undergo trimerization, oligomerization and polymerization[2,66,68].

IX. SATURATION OF THE C—C π BOND

To accomplish the efficient synthesis of any compound requires that one build into as many steps of a sequence as possible a high degree of selectivity or, preferably, specificity. A classic example of the need for such selectivity stems from efforts to reduce acrylonitrile electrochemically and convert it to the commercially valuable commodity adiponitrile rather than to propionitrile[1-3]. Initial studies, conducted in water, were disappointing and led to propionitrile. However, addition of the hydrotropic salt, Et_4NOTs, to the aqueous solution served to make the region near the cathode sufficiently 'dry' to allow β,β-coupling to occur and saturation to be eliminated[27,28].

Suppose that one is interested in accomplishing the opposite objective, that being the complete and selective saturation of the C—C π bond with no fear of competing dimerization. One could choose to use nonelectrochemical methods, such as H_2, noble metal catalyst. Recently, however, an elegant solution based upon the design and use of hydrogen-active powder electrodes has been devised[69]. The method consists of using either Raney nickel (R-Ni), Pd—C or Pt—C as cathode materials in the presence of a proton donor, generally chloroacetic acid, pivalic acid, phenol or water, in a solution of THF and water (9:1, v/v) containing $NaClO_4$ as a supporting electrolyte. Reduction of the proton donor serves as a source of adsorbed hydrogen.

Three substrates were examined: 2-cyclohexen-1-one, 4-methyl-3-penten-2-one and *trans*-3-phenyl-2-propenal. For each substrate, all electrode/proton donor combinations were examined. In general, R-Ni and Pd—C afforded high selectivity (up to 100%) for the conversion to cyclohexanone and to 4-methyl-2-pentanone; Pt—C proved less satisfactory. *trans*-3-Phenyl-2-propenal proved to be a difficult case, affording substantial quantities of *trans*-3-phenyl-2-propen-1-ol in addition to the desired product, 3-phenylpropanal.

Prior to this work researchers attempted to use Raney nickel[70-75] and metal blacks[76-80] as cathode materials. However, it has been noted[69] that electrolytic hydrogenation with hydrogen active powder electrodes has several advantages over direct uncatalyzed electrolysis. For example, the large surface area of the electrode leads to an increase in the rate of hydrogenation. Furthermore, hydrodimerization can most often be avoided entirely, since proton discharge to form atomic hydrogen on the catalyst surface can be accomplished at potentials more positive than those required for generation of an enone radical anion. Finally, reactions are conducted under mild conditions at room temperature and atmospheric pressure.

X. PINACOL FORMATION

If the β-carbon of an enone is sterically hindered and the carbonyl carbon is not, then pinacolization can often be carried out in preference to β, β-coupling. Many examples illustrating this characteristic are known[39,81-84] and several are illustrated in equations 22[83], 23[83] and 24[84].

$$\text{(22)}$$

$$\text{(23)}$$

$$\text{(24)}$$

The reduction shown in equation 24[84] is particularly interesting for it is suggested that Cr^{3+} interacts with the carbonyl oxygen of β-ionone to form a Lewis acid–Lewis base complex which is easier to reduce than the enone in its absence, i.e. the Cr^{3+} behaves like a proton.

It is difficult to convert efficiently retinal 12 to its pinacol, unless diethyl malonate is used as the proton donor[85] (equation 25). The reason(s) for this behavior is (are) not well understood.

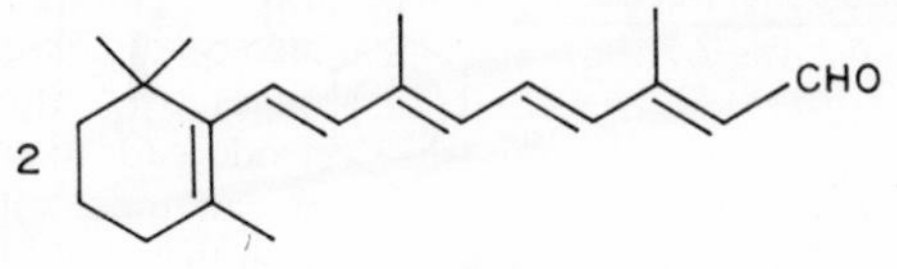

(12)

*

(a) -1.00V, CH_3CN, Bu_4NOAc, AcOH; 11% product.

(b) -1.4V, CH_3CN, $CH_2(CO_2Et)_2$, Bu_4NClO_4; 50% product

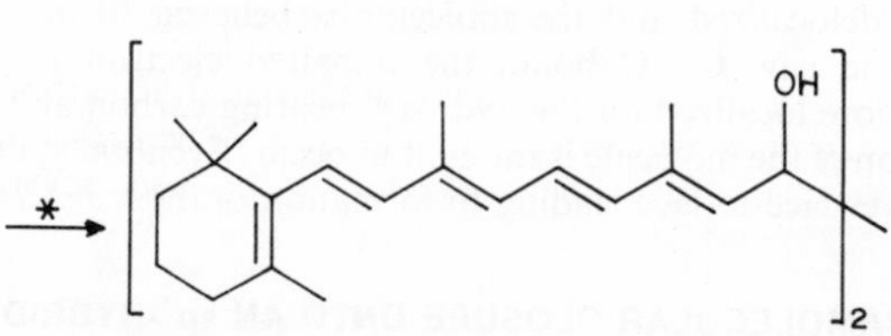

(25)

A. Stereochemistry of Pinacolization

In each of the reactions shown above, a mixture of d, l and *meso* stereoisomers is formed. For example, the 71% pinacol formed in the dimerization of β-ionone corresponds to a 2:1 mixture of *meso* and d, l isomers. While the factors controlling these reactions are reasonably well understood[86,87], stereochemical assignments have rarely been made. A glaring exception to this generalization follows.

From a remarkable study of the stereochemical outcome of the pinacolization of a series of 1, 9, 10, 10a-tetrahydro-3(2H)-phenanthrones 13, it was possible to obtain detailed information concerning the preferred approach of the reacting partners and the importance of the electrode surface during the reaction[88,89]. Furthermore, an expression of chiral recognition was observed. That is, formation of the new sigma bond was shown to occur preferentially between enones of the same chirality [e.g. (+)- with (+)-, or (−)- with (−)-enone was preferred over the combination of (+)- with (−)-enone].

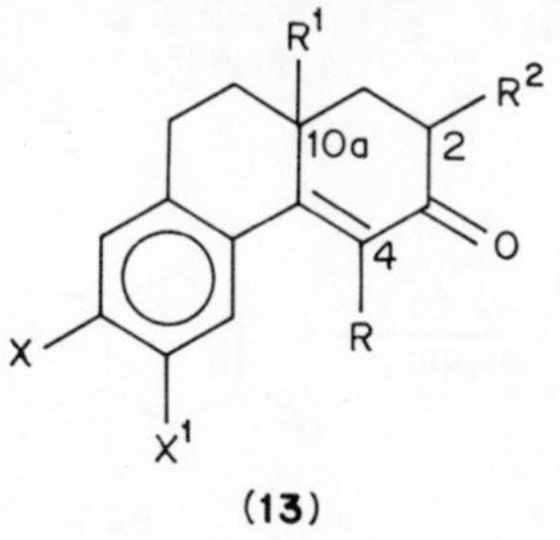

(13)

Consider the stereoselective conversion illustrated in equation 26. Only the product with a *trans* relationship between the hydrogen at C(10a) and the hydroxy group at C(3), a *threo* relationship about the new sigma bond and a *trans*-relationship between the hydrogen at C(10a′) and the hydroxy group at C(3′) was formed[89].

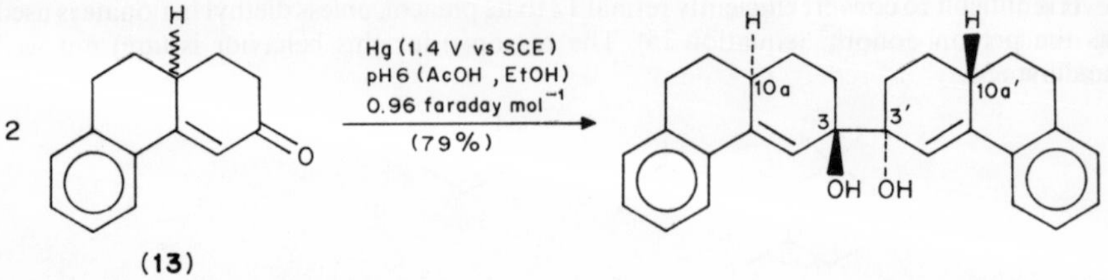

(26)

Arguments are presented[89] which lead to the conclusion that in a neutral medium (pH 6), the two reacting ketones are initially adsorbed selectively so that the least hindered face of each is directed toward the surface of the electrode. The unpaired electron is considered to be completely delocalized and the molecule is believed to lie relatively flat. During formation of the new C—C bond, the unpaired electron is presumed to become progressively more localized on the hydroxyl-bearing carbon and the desorption of the aromatic portion of the molecule is thought to occur. Eventually, the two reactive species orient themselves face-to-face leading to formation of the *trans, threo, trans*-diol.

XI. INTRAMOLECULAR CLOSURE ONTO AN sp³-HYBRIDIZED CARBON

A variety of bicyclic systems can be constructed by capitalizing upon the ability of a suitably positioned radical anion to close onto an sp³-hybridized carbon bearing a tosylate or mesylate as a leaving group. Examples[90] are given in equations 27–29. It is clear that even the presence of a fully substituted β-carbon does not prevent cyclization from occurring and in high yield.

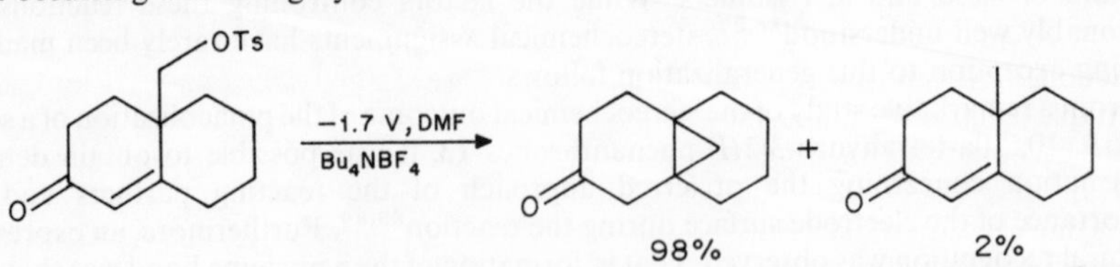

(27)

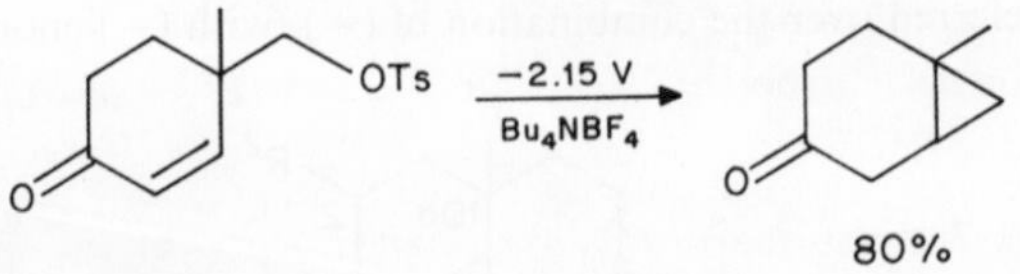

(28)

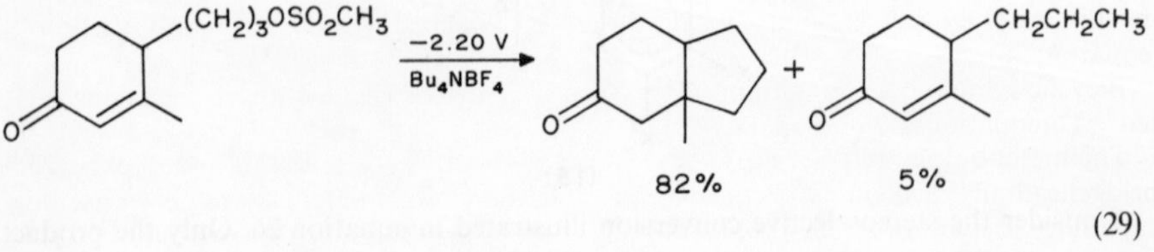

(29)

In these examples, the enone functions as the electrophore, the tosylate or mesylate bearing carbon as the center being attacked (the acceptor). However, when the enone is tethered to an alkyl halide and reduction is carried out in the presence of a cobalt(III) catalyst such as vitamin B_{12} (equation 30)[91], then the role of electrophore and acceptor reverse[91-93]. The initially formed complex between the catalyst and the alkyl halide **14** can be reduced at a potential which is sufficiently negative to cleave the Co^{3+}—C bond but not low enough to reduce the enone[94].

$$ \text{(30)} $$

95%

(1:1 *cis/trans*)

(14)

XII. NONCONJUGATED ENONES

Reduction of a ketone linked to an alkene[95,96], an allene[97,98] or an alkyne[98,99] by a chain of variable length and composition leads to formation of a C—C bond between that unit and the carbonyl carbon. The reactions are often conducted at constant current either in DMF or in a 1:9 (v/v) mixture of methanol and dioxane containing Et_4NOTs as a supporting electrolyte[95,96]. Five- and six-membered rings are formed efficiently, but four- and seven-membered rings are not (equations 31-34)[96].

Cyclization proceeds regioselectively; given the choice between forming a five- or a six-membered ring, five is preferred. However, even formation of a five-membered ring is thwarted when the internal carbon of the olefinic linkage bears an alkyl group as shown in equation 35[96].

Substitution of two alkyl groups on the terminal olefinic carbon apparently slows the rate of closure sufficiently so that formation of an acyclic tertiary alcohol becomes a competitive process. The supporting electrolyte serves as a source of the new alkyl group, in this case an ethyl group, which becomes attached to the carbonyl carbon[96] (equation 36).

Bicyclic compounds containing a bridgehead hydroxyl group can also be constructed[96,98] (equations 37 and 38).

The methodology has been extended to the preparation of both endo- and exocyclic bridgehead allylic alcohols[98] (equations 39 and 40). Unfortunately, attempts to use this capability to synthesize ene-diol-containing natural products such as isoamijiol[100] **(15)**

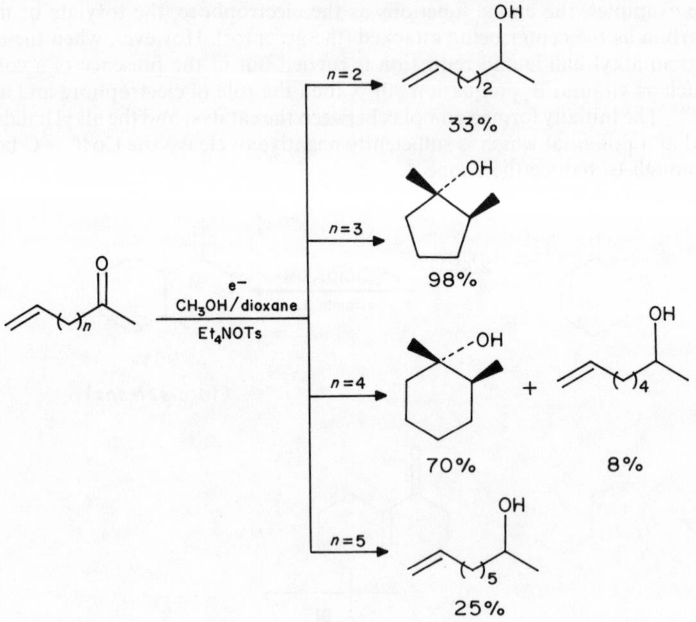

(31)

33%

(32)

98%

(33)

70%   8%

(34)

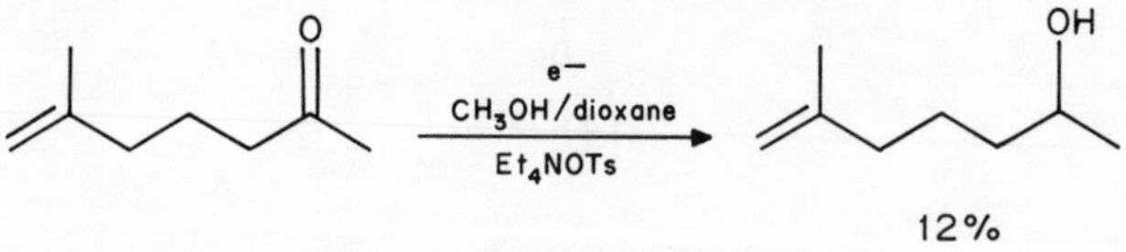

25%

$$\xrightarrow[\text{Et}_4\text{NOTs}]{\substack{e^-\\ \text{CH}_3\text{OH/dioxane}}}$$

(35)

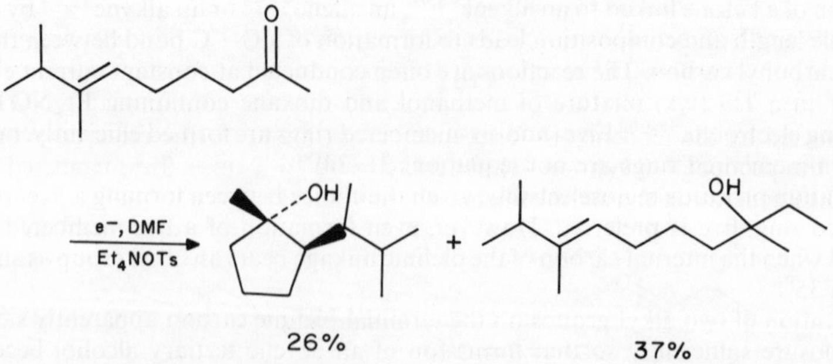

12%

$$\xrightarrow[\text{Et}_4\text{NOTs}]{e^-,\ \text{DMF}}$$

(36)

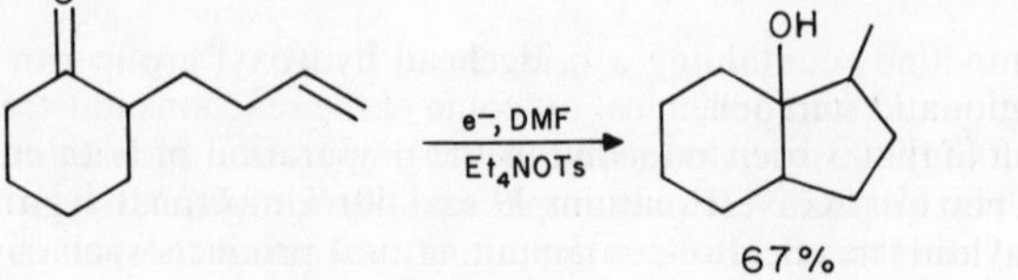

26%   37%

$$\xrightarrow[\text{Et}_4\text{NOTs}]{e^-,\ \text{DMF}}$$

(37)

67%

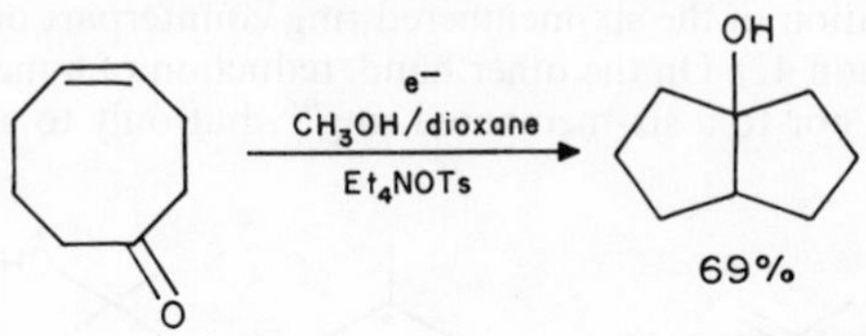

$$ (38) $$

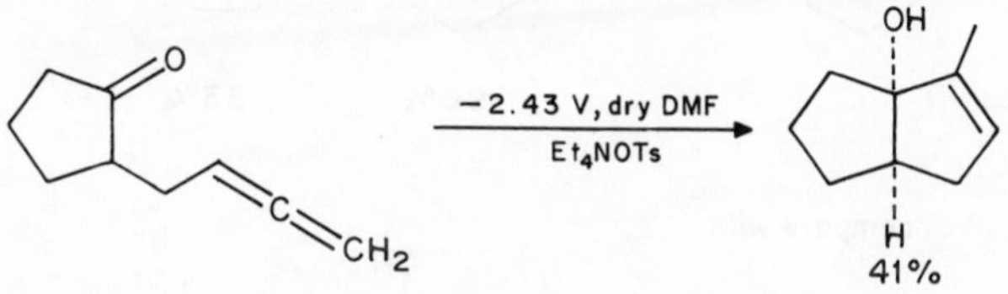

69%

$$ (39) $$

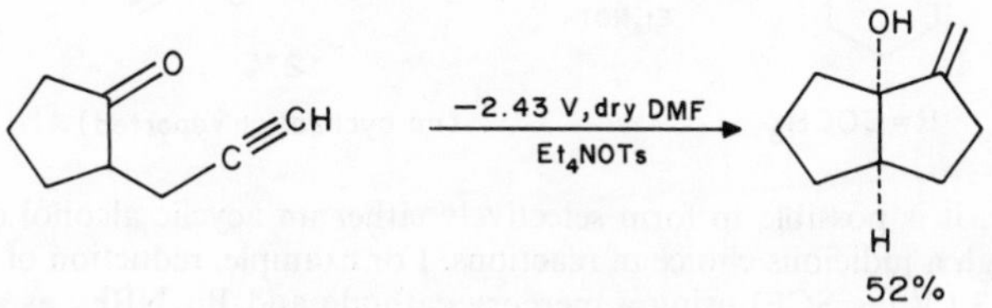

41%

$$ (40) $$

52%

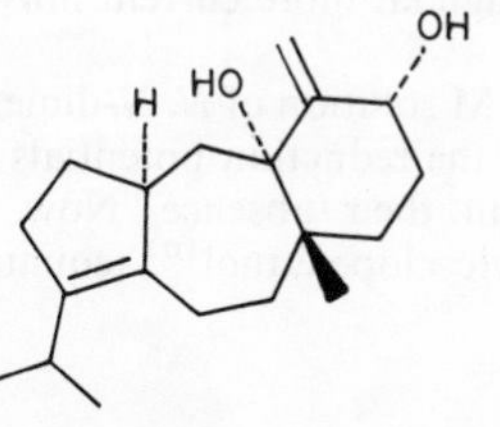

(15)

were thwarted by the tendency to form endo- in preference to the required exocyclic π bond in the product[98] (equation 41). Again, closure to form a five-membered ring is preferred to generating the six-membered alternative.

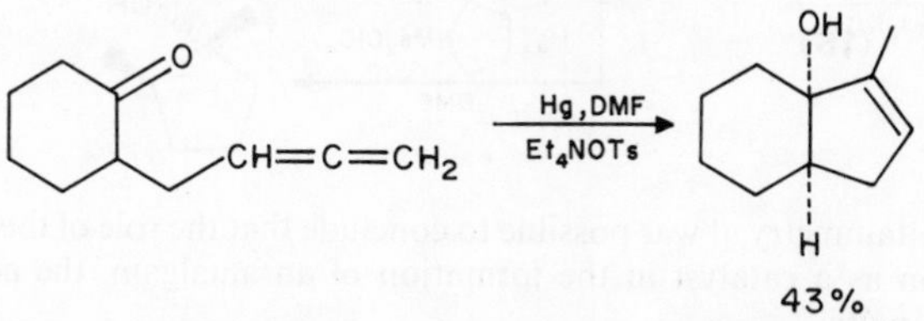

$$ (41) $$

43%

Both the regio- and stereochemical outcome of the reactions illustrated in this section are reminiscent of that associated with 5-hexen-1-yl radical cyclization[101]. However, the similarity is at best qualitative. For example, substitution of an alkyl group at C(5) of the 5-hexenyl radical leads to a decrease in the rate of cyclization to form a five-membered ring to a point where formation of the six-membered ring counterpart occurs at a faster rate and is preferred (equation 42). On the other hand, reduction of 6-methyl-6-hepten-2-one leads neither to a five- nor to a six-membered ring[96], but only to a carbonyl reduction product (equation 43).

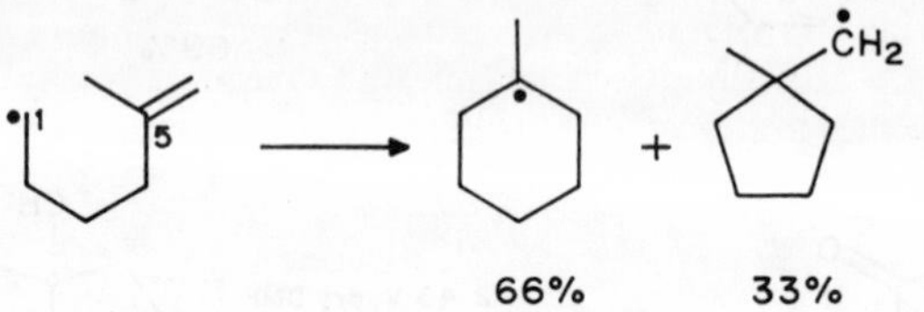

$$\tag{42}$$

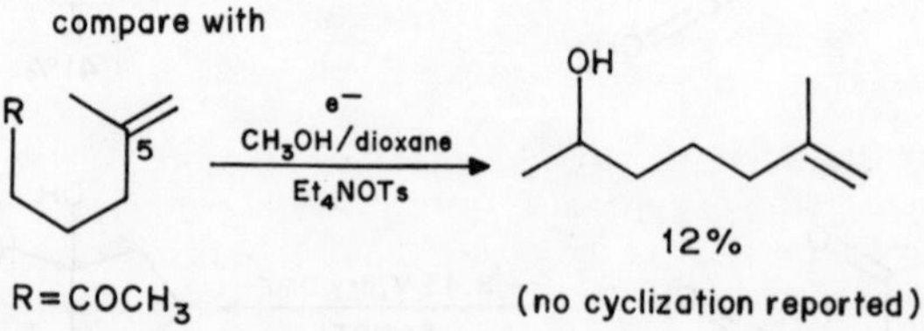

66%   33%

compare with

$$\tag{43}$$

R=COCH$_3$ (no cyclization reported)

Remarkably, it is possible to form selectively either an acyclic alcohol or a cyclized product through a judicious choice of reactions. For example, reduction of hept-6-en-2-one (16) at -3.1 V (vs SCE) using a mercury cathode and Bu$_4$NBF$_4$ as a supporting electrolyte affords hept-6-en-2-ol in 85% yield[102] (equation 44). A similar result is obtained using a graphite electrode, though far more current must be passed to consume starting material[102].

The addition of either a 0.01 M solution of N, N-dimethylpyrrolidinium or tetraethylammonium perchlorate causes the reduction potentials to shift to a value some 300 to 400 mV more positive than in their absence. Now, the major product (90–94%) corresponds to cis-1, 2-dimethylcyclopentanol[102] (equation 45).

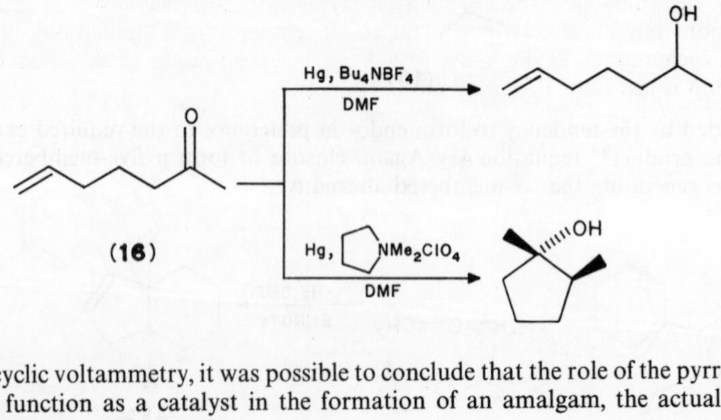

$$\tag{44}$$

(16)

$$\tag{45}$$

From cyclic voltammetry, it was possible to conclude that the role of the pyrrolidinium salt is to function as a catalyst in the formation of an amalgam, the actual reducing agent[102] (equation 46).

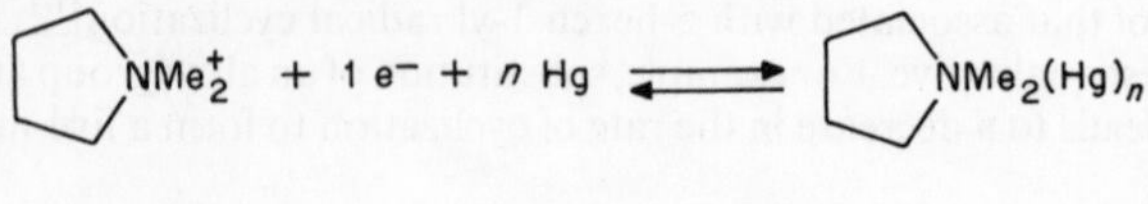

$$\tag{46}$$

XIII. OXIDATION OF ENONES

It was indicated in the introductory portion of this chapter that very little of what is known about the electrochemistry of enones involves, as a primary step, oxidation of the functional group. Once again, the reason for this behavior stems from the fact that most enones have low-lying HOMOs, thereby making it difficult to remove an electron at those potentials which are accessible electrochemically. One noteworthy *apparent* counter-example to these generalizations[103,104] is illustrated in equation 47.

$$(47)$$

It should be noted, however, that the indenone behaves more like an aryl olefin than an enone. That is, the net effect of appending three aromatic groups to the olefin dominates any effect(s) due to the presence of the carbonyl and the chemistry which is observed is much like that of an aryl-substituted olefin as demonstrated for *trans*-stilbene[105] (equation 48).

$$(48)$$

In the absence of a nucleophile, the indenone radical cation can be trapped by an anodically electroinactive species such as styrene. In this way, [4 + 2] and [2 + 2] cycloadditions have been carried out at room temperature[104] (equations 49 and 50). Yields of cycloadduct as high as 70% have been reported, even when electricity consumption is less than 1 faraday mol^{-1}.

$$(49)$$

$$(50)$$

Finally, the interesting and potentially synthetically useful rearrangement pictured in equation 51 is initiated by oxidation at a Pt anode[106]. It is suggested, though it seems unlikely on energetic grounds, that the first step involves a one-electron oxidation of the enone found in ring C. Whatever the case may be, it is likely, and it has been suggested[106], that a carbocation is formed adjacent to C(10) and that it triggers the skeletal rearrangement.

$$\xrightarrow[\text{AcOH, Et}_3\text{N}]{-e^-}$$

(51)

XIV. ACKNOWLEDGEMENTS

RDL is pleased to express his gratitude to the members of the University of British Columbia Chemistry Department for their warm hospitality, a stimulating working atmosphere, access to a fine library and a quiet office for carrying out the chores required to write this chapter.

XV. REFERENCES

1. M. M. Baizer, in *Organic Electrochemistry*, 1st edn. (Ed. M. M. Baizer), Chap. 9, Dekker, New York, 1973, pp. 399–411.
2. M. M. Baizer and L. G. Feoktistov, in *Organic Electrochemistry*, 2nd edn. (Eds. M. M. Baizer and H. Lund), Dekker, New York, 1983, Chap. 10, pp. 359–373; Chap. 20, pp. 658–673.
3. M. M. Baizer, in *Organic Electrochemistry*, 3rd edn. (Eds. M. M. Baizer and H. Lund), Chap. 10 and 26, Dekker, New York, in preparation.
4. I. Fleming, *Frontier Orbitals in Organic Chemical Reactions*, Wiley, New York, 1976.
5. K. N. Houk, *J. Am. Chem. Soc.*, **95**, 4092 (1973).
6. R. Sustmann and R. Schubert, *Angew. Chem., Int. Ed. Engl.*, **11**, 840 (1972).
7. D. W. Turner, C. Baker, A. D. Baker and C. R. Brundle, *Molecular Photoelectron Spectroscopy*, Wiley, London, 1970.
8. R. Sustmann and H. Trill, *Tetrahedron Lett.*, 4271 (1972).
9. G. Breglieb, *Angew. Chem., Int. Ed. Engl.*, **3**, 617 (1964).
10. M. Kotake (Ed.), *Constants of Organic Compounds*, Asakura Publishing Co. Ltd., Tokyo, 1963, pp. 680–693.
11. L. Meites, *Polarographic Techniques*, 2nd edn., Wiley–Interscience, New York, 1965, pp. 671–711.
12. C. K. Mann and K. K. Barnes, *Electrochemical Reactions in Nonaqueous Systems*, Dekker, New York, 1970, pp. 177–180.
13. K. W. Bowers, R. W. Giese, J. Grimshaw, H. O. House, N. H. Kolodny, K. Kronberger and D. K. Roe, *J. Am. Chem. Soc.*, **92**, 2783 (1970).
14. H. O. House, L. E. Huber and M. J. Umen, *J. Am. Chem. Soc.*, **94**, 8471 (1972).
15. P. Zuman, D. Barnes and A. Ryslova-Kejharova, *Discuss Faraday Soc.*, **45**, 202 (1968).
16. V. Toure, M. Levey and P. Zuman, *J. Electroanal. Chem.*, **56**, 285 (1974).

17. P. Zuman and L. Spritzer, *J. Electroanal. Chem.*, **69**, 433 (1976).
18. S. S. Katiyar, M. Lalithambika and G. C. Joshi, *J. Electroanal. Chem.*, **53**, 439 (1974).
19. S. S. Katiyar, M. Lalithambika and D. N. Dhar, *J. Electroanal. Chem.*, **53**, 449 (1974).
20. N. Takano, N. Takeno and Y. Otsuji, in *Recent Advances in Electroorganic Synthesis* (Ed. S. Torii), Elsevier, New York, 1987, pp. 211–214.
21. G. A. Russell and G. R. Stevenson, *J. Am. Chem. Soc.*, **93**, 2432 (1971).
22. J. Grimshaw and H. R. Juneja, *J. Chem. Soc., Perkin Trans. 1*, 2529 (1972).
23. P. Margaretha and P. Tissot, *Nouv. J. Chim.*, **3**, 13 (1979).
24. A. J. Fry, in *Topics in Current Chemistry*, **34**, Springer-Verlag, New York, 1972, pp. 1–46.
25. M. M. Baizer and J. P. Petrovich, in *Prog. Phys. Org. Chem.*, **7**, 189 (1970).
26. F. Beck, *Angew. Chem., Int. Ed. Engl.*, **11**, 760 (1972).
27. M. M. Baizer, *Tetrahedron Lett.*, 973 (1963).
28. M. M. Baizer, *J. Electrochem. Soc.*, **111**, 215 (1964).
29. M. M. Baizer and J. D. Anderson, *J. Electrochem. Soc.*, **111**, 223 (1964).
30. M. M. Baizer and J. D. Anderson, *J. Org. Chem.*, **30**, 3138 (1965).
31. J. P. Petrovich, J. D. Anderson and M. M. Baizer, *J. Org. Chem.*, **31**, 3897 (1966).
32. M. M. Baizer, J. P. Petrovich and D. A. Tyssee, *J. Electrochem. Soc.*, **117**, 173 (1970).
33. J. Andersson and L. Eberson, *Nouv. J. Chim.*, **1**, 413 (1977).
34. J. Andersson and L. Eberson, *J. Chem. Soc., Chem. Commun.*, 565 (1976).
35. M. M. Baizer and J. D. Anderson, *J. Org. Chem.*, **30**, 1357 (1965).
36. J. D. Anderson, M. M. Baizer and E. J. Prill, *J. Org. Chem.*, **30**, 1645 (1965).
37. J. Andersson, L. Eberson and C. Svensson, *Acta Chem. Scand., Ser. B*, **32**, 234 (1978).
38. M. M. Baizer and J. D. Anderson, *J. Org. Chem.*, **30**, 1348 (1965).
39. P. Tissot, J.-P. Surbeck, F. O. Gülaçar and P. Margaretha, *Helv. Chim. Acta*, **64**, 1570 (1981).
40. H. Satonaka, Z. Saito and T. Shimura, *Bull. Chem. Soc. Jpn.*, **46**, 2892 (1973).
41. L. Mandell, R. F. Daley and R. A. Day, *J. Org. Chem.*, **41**, 4087 (1976).
42. B. Terem and J. H. P. Utley, *Electrochim. Acta*, **24**, 1081 (1979).
43. V. J. Puglisi and A. J. Bard, *J. Electrochem. Soc.*, **119**, 829 (1972).
44. C. P. Andrieux, L. Nadjo and J. M. Savéant, *J. Electroanal. Chem.*, **42**, 223 (1973).
45. S. C. Rifkin and D. H. Evans, *J. Electrochem. Soc.*, **121**, 769 (1974).
46. C. P. Andrieux, D. J. Brown and J. M. Savéant, *Nouv. J. Chim.*, **1**, 157 (1977).
47. E. Touboul and G. Dana, *J. Org. Chem.*, **44**, 1397 (1979).
48. V. D. Parker, *Acta Chem. Scand., Ser. B*, **35**, 147 (1981).
49. V. D. Parker, *Acta Chem. Scand., Ser. B*, **35**, 149 (1981).
50. P. Margaretha and P. Tissot, *Helv. Chim. Acta*, **65**, 1949 (1982).
51. E. Lamy, L. Nadjo and J. M. Savéant, *J. Electroanal. Chem.*, **42**, 189 (1973).
52. P. Margaretha and P. Tissot, *Helv. Chim. Acta*, **60**, 1472 (1977).
53. H. O. House, *Modern Synthetic Reactions*, 2nd edn., W. A. Benjamin, Menlow Park, CA, 1972, p. 183.
54. J. M. Mellor, B. S. Pons and J. H. A. Stibbard, *J. Chem. Soc., Perkin Trans. 1*, 3092 (1981).
55. J. Grimshaw and R. J. Haslett, *J. Chem. Soc., Perkin Trans. 1*, 395 (1979).
56. J. Grimshaw and J. Trocha-Grimshaw, *J. Chem. Soc., Perkin Trans. 1*, 2584 (1973).
57. J. P. Morizur, B. Furth and J. Kossanyi, *Bull. Soc. Chim. Fr.*, 1422 (1967).
58. J. Simonet, *Compt. Rend.*, **267C**, 1548 (1968).
59. M. M. Baizer, *Tetrahedron*, **40**, 935 (1984).
60. M. M. Baizer, J. L. Chruma and D. A. White, *Tetrahedron Lett.*, 5209 (1973).
61. F. Fournier, D. Davoust and J.-J. Basselier, *Tetrahedron*, **41**, 5677 (1985).
62. J. P. Coleman, R. J. Kobylecki and J. H. P. Utley, *J. Chem. Soc., Chem. Commun.*, 104 (1971).
63. T. J. Curphey, L. D. Trivedi and T. Layloff, *J. Org. Chem.*, **39**, 3831 (1974).
64. H. Lund and C. Degrand, *Acta Chem. Scand., Ser. B*, **33**, 57 (1979).
65. E. A. H. Hall, G. P. Moss, J. H. P. Utley and B. C. L. Weedon, *J. Chem. Soc., Chem. Commun.*, 586 (1976).
66. S. Wawzonek and A. Gundersen, *J. Electrochem. Soc.*, **111**, 324 (1964).
67. D. A. Tyssee and M. M. Baizer, *J. Org. Chem.*, **39**, 2823 (1974).
68. J. Simonet, *Compt. Rend.*, **263C**, 1546 (1966).
69. T. Osa, T. Matsue, A. Yokozawa and T. Yamada, *Denki Kagaku*, **54**, 484 (1986).
70. B. Sakurai and T. Arai, *Bull. Chem. Soc. Jpn.*, **28**, 93 (1955).
71. T. Chiba, M. Okimoto, H. Nagai and Y. Takata, *Bull. Chem. Soc. Jpn.*, **56**, 719 (1983).

72. M. Fujihira, A. Yokozawa, H. Kinoshita and T. Osa, *Chem. Lett.*, 1089 (1982).
73. T. Osa, T. Matsue, A. Yokozawa and T. Yamada, *Denki Kagaku*, **52**, 629 (1984).
74. T. Osa, T. Matsue, A. Yokozawa and M. Fujihira, *Denki Kagaku*, **53**, 104 (1985).
75. K. Park, P. N. Pintauro, M. M. Baizer and K. Nobe, *J. Electrochem. Soc.*, **132**, 1850 (1985).
76. M. Sakuma, *J. Electrochem. Soc. Jpn.*, **28**, 164 (1960).
77. H. Kita and N. Kubota, *Electrochim. Acta*, **27**, 861 (1982).
78. T. Nonaka, M. Takahashi and T. Fuchigami, *Denki Kagaku*, **51**, 129 (1983).
79. J. Mizuguchi and S. Matsumoto, *J. Pharm. Soc. Jpn.*, **78**, 129 (1953).
80. D. Pletcher and M. Razaq, *Electrochim. Acta*, **26**, 819 (1981).
81. D. Miller, L. Mandell and R. A. Day, *J. Org. Chem.*, **36**, 1683 (1971).
82. H. Lund, *Acta Chem. Scand.*, **11**, 283 (1957).
83. R. E. Sioda, R. Terem, J. H. P. Utley and B. C. L. Weedon, *J. Chem. Soc., Perkin Trans. 1*, 561 (1976).
84. D. W. Sopher and J. H. P. Utley, *J. Chem. Soc., Chem. Commun.*, 1087 (1979).
85. L. A. Powell and R. M. Wightman, *J. Am. Chem. Soc.*, **101**, 4412 (1979).
86. J. H. Stocker and R. M. Jenevein, *J. Org. Chem.*, **33**, 2145 (1968).
87. A. Bewick and H. P. Cleghorn, *J. Chem. Soc., Perkin Trans. 2*, 1410 (1973).
88. E. Touboul and G. Dana, *Tetrahedron*, **31**, 1925 (1975).
89. E. Touboul and G. Dana, *J. Org. Chem.*, **44**, 1397 (1979).
90. P. G. Gassman, O. M. Rasmy, T. O. Murdock and K. Saito, *J. Org. Chem.*, **46**, 5455 (1981).
91. R. Scheffold, M. Dike, S. Dike, T. Herold and L. Walder, *J. Am. Chem. Soc.*, **102**, 3642 (1980).
92. H. Bhandal, G. Pattenden and J. J. Russell, *Tetrahedron Lett.*, **27**, 2299 (1986).
93. S. Torii, T. Inokuchi and T. Yukawa, *J. Org. Chem.*, **50**, 5875 (1985).
94. R. Scheffold, in *Recent Advances in Electroorganic Synthesis* (Ed. S. Torii), Elsevier, New York, 1987, pp. 275–282.
95. T. Shono and M. Mitani, *J. Am. Chem. Soc.*, **93**, 5284 (1971).
96. T. Shono, I. Nishiguchi, H. Ohmizu and M. Mitani, *J. Am. Chem. Soc.*, **100**, 545 (1978).
97. G. Pattenden and G. M. Robertson, *Tetrahedron Lett.*, **24**, 4617 (1983).
98. G. Pattenden and G. M. Robertson, *Tetrahedron*, **41**, 4001 (1985).
99. T. Shono, I. Nishicuchi and H. Ohmizu, *Chem. Lett.*, 1233 (1976).
100. G. Pattenden and G. M. Robertson, *Tetrahedron Lett.*, **27**, 399 (1986).
101. B. Giese, *Radicals in Organic Synthesis: Formation of Carbon–Carbon Bonds*, Pergamon Press, New York, 1986.
102. E. Kariv-Miller and T. J. Mahachi, *J. Org. Chem.*, **51**, 1041 (1986).
103. J. Delaunay, A. M. Orliac and J. Simonet, *Nouv. J. Chim.*, **10**, 133 (1986).
104. J. Simonet, in *Recent Advances in Electroorganic Synthesis* (Ed. S. Torii), Elsevier, New York, 1987, pp. 9–15.
105. F. D. Mango and W. A. Bonner, *J. Org. Chem.*, **29**, 1367 (1964).
106. M. Yoshikawa and I. Kitagawa, in *Recent Advances in Electroorganic Synthesis* (Ed. S. Torii), Elsevier, New York, 1987, pp. 97–104.

The Chemistry of Enones
Edited by S. Patai and Z. Rappoport
© 1989 John Wiley & Sons Ltd

CHAPTER **15**

The photochemistry of enones

DAVID I. SCHUSTER

Department of Chemistry, New York University, New York, NY 10003, USA

David I. Schuster

I. GENERAL INTRODUCTION

Organic compounds containing a ketonic or aldehydic carbonyl group as well as a carbon–carbon double bond undergo a wide variety of reactions on exposure to ultraviolet radiation which are not observed in compounds containing only one of these functional groups. Enones have a very rich photochemistry, depending on the relative proximity of the C=O and C=C moieties. The discussion below will therefore deal in turn with α,β-unsaturated ketones in which the two moieties are conjugated, then with homoconjugated β,γ-unsaturated ketones, and finally with intramolecular interactions between C=O and C=C moieties that are sufficiently separated such that there is no direct chromophoric interaction between them as judged from UV absorption spectroscopy.

Since the photochemistry of enones and their spectroscopy is discussed extensively in textbooks[1-4] and in recent literature reviews[5-9], the discussion below will attempt to summarize and categorize the types of reactions that are observed on UV excitation of enones, with emphasis on recent findings reported in the literature.

For those not familiar with terms and concepts commonly used in photochemistry[10], it is useful to first consider the orbital description of ground and excited states given in Figure 1 and the modified Jablonski diagram given in Figure 2. The ground electronic state of the molecule is designated S_0. Promotion of an electron from the highest occupied molecular orbital (HOMO) of the molecule in its ground electronic state to the lowest unoccupied molecular orbital (LUMO) will occur on absorption of a single photon of UV light of frequency v, light whose energy is $hv = E_{LUMO} - E_{HOMO}$ (h is Planck's constant). The first law of photochemistry is that a substance undergoing photochemical change does so through the absorption of a single quantum of light. In solution, the absorption of light by a molecule at a given wavelength λ or frequency v, where $v = hc/\lambda$, depends directly on the concentration of the absorber c (in mol l^{-1}), the path length l (in cm), and the decadic molar extinction coefficient ε (in units of 1 mol^{-1} cm^{-1}) which is characteristic of the molecule and changes with wavelength. In order for light absorption to occur with high probability (corresponding to a large value of ε and of the related oscillator strength f), there has to be a change in symmetry of the total electronic wave function in proceeding from the ground to the excited state. Thus, certain electronic transitions are highly allowed according to quantum mechanics, while others are strongly forbidden. We will discuss specific types of transitions a little later. Electronic excitation takes place in $ca\,10^{-15}$ s and gives an electronic state of the molecule in which the electron in the HOMO and the remaining electron in the LUMO are still spin-paired, one with spin state $+\frac{1}{2}$ and the other with spin $-\frac{1}{2}$. This singlet excited state is designated S_1. Excitation at shorter wavelengths (higher energy) allows direct population of higher singlet excited states (S_2, S_3, etc.) by promotion of an electron from the HOMO to an MO of higher energy than the LUMO, or from an MO of lower energy than the HOMO to one of the unoccupied MOs. Each such transition corresponds to a different UV absorption band of the molecule, and has its own particular transition probability and corresponding extinction coefficient ε. Each electronic excited state has its own characteristic electron distribution, reactivity and lifetime.

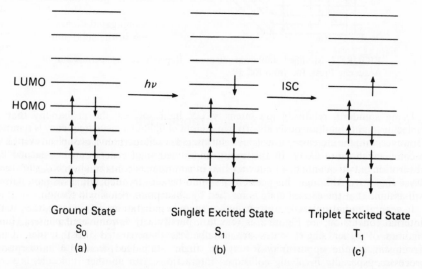

FIGURE 1. Orbital description of ground state and singlet and triplet electronic excited states. Reproduced by permission of Academic Press, Inc. from Ref. 10

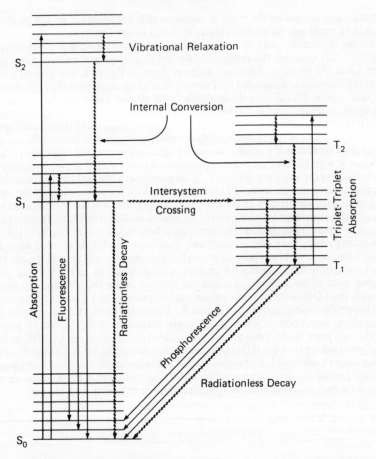

FIGURE 2. Modified Jablonski diagram. Reproduced by permission of Academic Press, Inc. from Ref. 10

Using standard, relatively low-intensity UV light sources, the probability that a molecule might simultaneously absorb two quanta of light of a given frequency is remote. However, with the increasing use of powerful lasers as excitation sources, such an event has become much more likely. In this event, an excited singlet state can be reached by absorption of two quanta of a frequency that otherwise would not be absorbed, such that $2h\nu = E_{exc} - E_0$. Although this possibility should be kept in mind, the discussion below will assume that the excited state is reached by absorption of a single photon.

In general, radiationless decay from higher-energy excited states $S_n (n > 1)$ to the S_1 state (internal conversion; see Figure 2) is very fast, particularly in condensed phases. Thus, lifetimes of upper singlet states are usually less than 10^{-12} s (1 ps). Under these circumstances, the opportunity for upper singlet states to participate in chemical processes, especially involving collisional interactions with another molecule, is very limited. Also, vibrational relaxation within a given excited state is also so fast (rate ca 10^{13} s^{-1}) that population by light absorption of upper vibrational levels of a given electronic excited state, such as S_1, results in rapid radiationless decay to the lowest vibrational level of that state, which therefore is the origin of all the processes which result in depopulation of that electronic state. The lifetime of the S_1 state τ_s is limited primarily by the rate at which a quantum of light is emitted as fluorescence to regenerate the ground state. The greater the transition moment or oscillator strength associated with light absorption, the greater is the probability and rate of fluorescence emission. First-order rate constants for fluorescence emission, measured using pulse techniques (specifically single photon counting) by the exponential decrease in fluorescence intensity following excitation, are of the order of 10^7–$10^{10} s^{-1}$, corresponding to singlet lifetimes of 10^{-7}–10^{-10} s. Thus, for any photochemical change to occur directly from a singlet excited state, the rate must be very fast in order to compete with rapid radiative decay to the ground state. Since fluorescence decay originates almost entirely from the lowest vibrational level of the S_1 state, fluorescence spectra are red-shifted compared to absorption spectra, and the spectra have a mirror-image appearance in cases where the excited state undergoes no appreciable geometric changes prior to light emission.

As can be seen in Figures 1 and 2, an electronic spin flip can occur to generate an excited state in which the two odd electrons (usually one in the formerly HOMO and the other in the formerly LUMO) are no longer spin correlated. This state is a triplet excited state, since the total spin of two unpaired electrons can be either $+1, 0$ or -1. The process in which a triplet excited state is generated from a singlet state is known as intersystem crossing. Radiative (phosphorescence) and nonradiative decay from the triplet manifold to regenerate the ground state S_0 can occur, but since these processes involve coupling of states of different spin parity, they are quantum mechanically spin-forbidden, and have rate constants which are several orders of magnitude less than for corresponding decay from S_1 to S_0. The lifetimes of triplet excited states, particularly the lowest triplet state T_1, are usually much longer than corresponding singlet excited states, often by several orders of magnitude. These triplets are therefore much more likely to undergo chemical reactions than the corresponding singlets particularly bimolecular reactions with an added reagent or the solvent. Thus it is not surprising that most of the photochemical reactions of enones to be discussed later occur via triplet and not singlet excited states. Those in which singlet excited states have been implicated are exclusively unimolecular processes (rearrangements and fragmentations) whose rates can be competitive with those of singlet decay processes.

Mechanisms exist which allow quantum-mechanical coupling of excited singlet and triplet states of ketones, the most important of which is spin–orbit coupling[11], so that intersystem crossing in these systems is generally very rapid (rate constants 10^8–$10^{11} s^{-1}$) and efficient (quantum efficiencies often of the order of unity). Thus, fluorescence of enones is rarely observed. Triplet states of enones as well as other types of systems can also be generated efficiently by transfer of triplet excitation from an electronically excited donor (sensitizer) by the following scheme (equations 1 and 2)[12],

$$Sens_0 + h\nu \longrightarrow {}^1Sens^* \longrightarrow {}^3Sens^* \tag{1}$$

$$^3Sens^* + E \xrightarrow{k_q} Sens_0 + {}^3E^* \tag{2}$$

where E is an enone and k_q is the second-order rate constant for transfer of triplet excitation. The ideal situation is shown schematically in Figure 3, in which the S_1 and T_1 states of the donor (sensitizer) are, respectively, lower and higher in energy than the corresponding states of the acceptor (enone). In this case, use of appropriate excitation wavelengths (controlled by the choice of lamps and filters) allows direct excitation exclusively of the donor, and triplet transfer to the acceptor will occur at or close to a diffusion-controlled rate, depending primarily on the frequency of encounters of excited

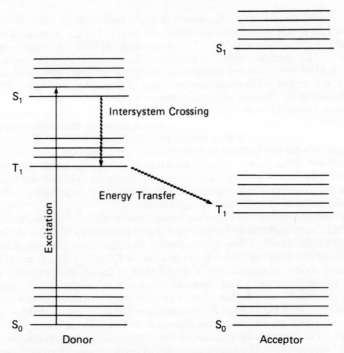

FIGURE 3. Schematic description of triplet excitation transfer, Reproduced by permission of Academic Press, Inc. from Ref. 10

donor and ground-state acceptor. The main advantage of generation of acceptor triplets by the triplet sensitization route is that the acceptor singlets are bypassed; for mechanistic purposes, this permits characterization of the reactivity of triplet states uncontaminated by singlet contributions. In some systems, most notably certain classes of hydrocarbons, triplets can be efficiently generated only by the sensitization route.

Triplet energy transfer can also be used to get information about dynamics of reactive triplet states. Thus, the yield of product derived from chemical reaction of a donor triplet will be reduced in the presence of an appropriate acceptor. In this case, the enone can serve as the donor and any of a series of appropriate triplet quenchers (e.g. naphthalene, conjugated dienes, oxygen, etc.) can be utilized. In the simplest case, the quenching follows the Stern–Volmer relationship given in equation 3

$$\phi_i^0/\phi_i^Q = 1 + k_q\tau_D[Q] \tag{3}$$

where ϕ_i^0 and ϕ_i^Q are the respective quantum efficiencies for process i in the presence and absence of quencher Q, and k_q is the bimolecular rate constant for quenching of the donor triplet excited state whose lifetime is τ_D in the absence of the quencher. The quantum efficiency ϕ_i is defined as the number of molecules undergoing process i divided by the number of quanta of light absorbed in a given period of time. Thus, for a chemical reaction, the relative quantum efficiencies given in equation 3 are equal to the relative yields of the product(s) formed in the presence and absence of the added quencher, which can be conveniently measured using appropriate spectroscopic or chromatographic techniques following co-irradiation of samples with and without known concentrations of quencher Q. Plots of relative product yields vs. quencher concentration should be linear according to equation 3 if there are no kinetic complications, with an intercept of 1.0 and a slope equal to $k_q\tau_D$. If a value for k_q is known or can be estimated (a value equal to the diffusion-controlled rate is often assumed), this technique allows estimation of triplet lifetimes τ_D. If more than one chemical transformation occurs via a common triplet excited state, the Stern–Volmer quenching slopes corresponding to each reaction should have identical slopes. Conversely, if Stern–Volmer quenching plots for formation of different products resulting from excitation of a given compound have experimentally distinguishable slopes, the reactions must occur via different triplet excited states or conceivably via some other quenchable intermediates.

The quenching relationship of equation 3 will be observed when a triplet state is intercepted by any added reagent, and is not limited to triplet energy transfer. As an example, we shall consider later the interaction of triplet states of cyclic conjugated enones with alkenes to give cycloaddition products. Furthermore, sensitizers function not only as agents for transfer of electronic excitation, but also in electron transfer processes in appropriate situations, according to equations 4 and 5[13]:

$$D^* + A_0 \longrightarrow D^{\cdot +} + A^{\cdot -} \tag{4}$$

$$A^* + D_0 \longrightarrow D^{\cdot +} + A^{\cdot -} \tag{5}$$

Thus, either the donor or the acceptor can serve as the excited component, which is usually in a singlet excited state. The free-energy change for a photoinduced electron-transfer process is given by equation 6, known as the Weller equation[14],

$$\Delta G_{et} = E(D/D^+) - E(A^-/A) - E_{0,0} - e_0^2 a \varepsilon \tag{6}$$

where the first term is the oxidation potential of the donor, the second is the reduction potential of the acceptor, the third is the excitation energy of the sensitizer, and the last term is the energy gained by bringing the two radical ions to the encounter distance a in a solvent of dielectric constant ε; in polar solvents the last term is negligibly small, but it can be significant in nonpolar media. We shall encounter cases in which enone radical ions generated by sensitized electron transfer undergo reactions not characteristic of singlet or triplet excited states. Interesting developments in this rapidly growing area of organic photochemistry can be expected in the next few years.

II. ULTRAVIOLET SPECTROSCOPY AND ENERGIES OF ELECTRONIC EXCITED STATES OF ENONES

Before discussing the photochemistry of enones, it is necessary to review the UV spectroscopy of these compounds[15]. The lowest energy electronic transition in formaldehyde and simple aldehydes and ketones is the promotion of an electron from the nonbonding orbital on oxygen into the vacant antibonding π orbital of the carbonyl group ($n \rightarrow \pi^*$). Since these orbitals are formally orthogonal for a planar carbonyl group, this transition is quantum mechanically forbidden; it is observed, but the extinction coefficient ε is very small (10^1–10^2 l mol^{-1} cm^{-1}). The lowest energy excited singlet state, S_1, is therefore a $^1n, \pi^*$ state. For simple aldehydes and ketones and nonconjugated enones, this transition is usually observed in the range of 290–330 nm, corresponding to an excitation

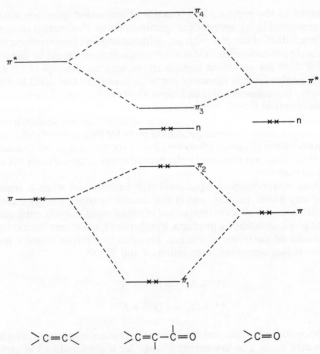

FIGURE 4. Qualitative energy-level diagram for α,β-unsaturated ketones (adapted from Reference 16)

energy of the $^{1}n, \pi^*$ state (the difference in energy of the lowest vibrational levels of the S_1 and S_0 states) of 80–85 kcal mol^{-1}. For formaldehyde, λ_{max} is 304 nm in the vapor phase, and ε_{max} is only 181 mol^{-1} cm^{-1}. The next higher-energy electronic transition is promotion of an electron from the bonding carbonyl π_{CO}-MO to the corresponding antibonding π_{CO}^*-MO. For simple carbonyl compounds, this transition occurs at ca 180–220 nm with ε of the order of 10^4 l mol^{-1} cm^{-1}, corresponding to an excitation energy of the S_2 state ($^{1}\pi, \pi^*$) 140–150 kcal mol^{-1} above the ground state. This energy is similar to that required for excitation of a nonbonding electron on oxygen into the σ^*-MO (the antibonding MO for the C—O sigma bond), and in some cases (formaldehyde in particular) it is not clear whether the $n \rightarrow \sigma^*$ or $\pi \rightarrow \pi^*$ transition is of the lower energy; in most cases, it is generally assumed that the second UV absorption band (going from lower to higher energy) is the $\pi \rightarrow \pi^*$ transition.

For simple alkenes, the lowest energy UV absorption corresponds to a $\pi \rightarrow \pi^*$ transition, and generally occurs between 170 and 210 nm, depending on the substitution pattern on the C=C chromophore, corresponding to an S_1 excitation energy of the order of 140–150 kcal mol^{-1}.

For α,β-unsaturated ketones, interaction of the C=O and C=C molecular orbitals leads to the qualitative energy-level diagram shown in Figure 4[16]. The lowest energy π-MO (π_1) is considerably lower in energy than either the isolated C=C or C=O π—MOs, while the highest occupied π-MO (π_2) is higher in energy than in the isolated chromophores. There is also substantial energy lowering of the LUMO (π_3^*) and a corresponding increase in energy of π_4^*. The energy of the nonbonding (n) orbital on oxygen is not significantly affected by bringing the C=C and C=O moieties into

TABLE 1. UV absorption spectra of selected α,β-unsaturated ketones in ethanol

Ketone	λ_{max} (mm)	ε_{max}
$CH_2{=}C(C_2H_5){-}COCH_3$	221	6450
$CH_3{-}CH{=}CH{-}COCH_3$	224	9750
(cyclohexenyl)—COCH₃	234 / 306 (CH₃CN)	13,000 / 42
(methyl octalone)	237 / 312	15,800 / 56
(cyclohexene)—COCH₃, —CH₃	249	6890
(cyclopentene)—COCH₃	239	13,000
(cyclopentene)—COCH₃, —CH₃	253	10,010
(4,4-dimethylcyclohexenone)	224 / 318	15,600 / 35
(trimethylcyclohexenone)	235 / 321	9500 / 37.6
(octahydrophenanthrenone)	234 (2-PrOH) / 315 (2-PrOH)	18,620 / 62

conjugation. The result of conjugation is that the energies of both the $n \rightarrow \pi^*$ and $\pi \rightarrow \pi^*$ transitions in the $C{=}C{-}C{=}O$ chromophore are lowered in energy relative to the isolated chromophores, i.e. they are shifted to higher wavelength. Typically, the $\pi \rightarrow \pi^*$ absorption band (S_0–S_2) occurs with λ_{max} 220–250 nm and $\varepsilon_{max} > 10^4\,\mathrm{l\,mol^{-1}\,cm^{-1}}$. The location of the absorption maximum for such compounds can be estimated very closely using a set of rules proposed by Woodward, depending on the location of substituents, orientation relative to other carbocyclic rings and ring size (e.g. cyclopentenone absorbs at slightly lower wavelength than cyclohexenone, 218 vs. 225 nm). Table 1 gives values for the $\pi \rightarrow \pi^*$ transitions and the corresponding singlet excitation energies for some typical conjugated enones in ethanol. The corresponding $n \rightarrow \pi^*$ transitions for enones are in the 300–350 nm region, corresponding to S_1 excitation energies of 75–85 kcal mol^{-1} relative to the lowest vibrational level of S_0. The band intensities are slightly higher ($\varepsilon \sim 50$–100 l mol^{-1} cm^{-1}) than for simple aliphatic aldehydes and ketones. The $\pi \rightarrow \pi^*$ and $n \rightarrow \pi^*$ absorption bands of enones (and indeed of simple ketones) are shifted in opposite directions by an increase in solvent polarity. A red (bathochromic) shift is observed for $\pi \rightarrow \pi^*$ absorption bands and a blue (hypsochromic) shift is observed for the $n \rightarrow \pi^*$ absorption. The latter effect is rationalized in terms of greater stabilization (energy lowering) of the n electrons in hydrogen bonding solvents than the antibonding π-MO (π_3^* in Figure 5), which in turn is stabilized (presumably due to greater contributions of structures involving polarization of charge) relative to the bonding MO (π_2 in Figure 4) by an increase in solvent polarity.

When the C=C and C=O chromophores are separated by a single tetrahedral carbon atom in β, γ-unsaturated ketones, interaction of the π systems still occurs, but to a much lesser extent than in α, β-enones because of the restrictions placed by the molecular geometry on the overlap of p orbitals between the chromophores, as shown in Figure 5. The result is that the energies of the MOs are not affected to nearly as great an extent as

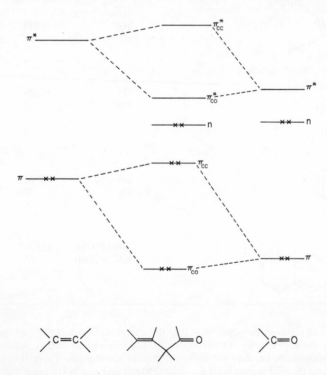

FIGURE 5. Qualitative energy-level diagram for β, γ-unsaturated ketones.

depicted in Figure 4 for the α, β-enones. The $\pi \to \pi^*$ absorption for typical β, γ-enones without conjugating substituents is centred at ca 220 nm and the $n \to \pi^*$ absorption typically has λ_{max} 290–310 nm. What is notable is the intensification of the $n \to \pi^*$ absorption for many (but not all) β, γ-enones, with values of ε 3–10 times as large as for α, β-enones (see Table 2 for representative examples). This effect has received a great deal of attention from spectroscopists, and is discussed at length in a review by Houk[6] on the spectroscopy and photochemistry of β, γ-enones. Labhart and Wagniere[17] suggested that this intensification results from overlap of the n orbital on oxygen with the alkene p orbitals, so that the $n \to \pi^*$ transition in effect borrows intensity from the $\pi \to \pi^*$ transition. That is, in this situation, the $n \to \pi^*$ transition can be viewed as promotion of an electron from an n orbital mixed to some extent with the $\pi_{C=C}$ orbital to a π_{CO}^* orbital which is mixed with the $\pi_{C=C}^*$ orbital, conferring 'allowedness' to this transition, calculated as about 1% of that of a fully allowed transition. It has also been noted that those β, γ-enones which show large intensification of the $n \to \pi^*$ transition also show large optical rotations and Cotton effects, due to the inherent dissymmetry of the chromophore. For β, γ-enones in which the p orbitals of the carbonyl carbon and the C=C bond are not directed at each other, such as 3-cyclopentenone and 3-cyclohexenone, $n \to \pi^*$ intensification is not observed. These spectral properties are of relevance to the photochemical behaviour of β, γ-enones, as is well recognized[6].

The nature of the triplet excited states of enones is of particular significance in understanding the photochemistry of these systems. Triplet states are always of lower energy than the corresponding singlet excited states, but the energy gap is a function of the electronic configuration. Thus, the singlet–triplet energy gap is much larger for π, π^* states than for n, π^* states. The large difference in energy between the S_1 and S_2 ($^1\pi, \pi^*$ and $^1n, \pi^*$)

Table 2. UV absorption spectra of typical β, γ-unsaturated ketones

Compound	λ_{max}	ε_{max}	Solvent
	304	327	EtOH
	309	301	95% EtOH
	307	289	CHCl$_3$
	210 308	3000 290	EtOH

TABLE 2. (*continued*)

Compound	λ_{max}	ε_{max}	Solvent
	202 298	3000 110	95% EtOH
	290	120	EtOH
	277	108	C_6H_{12}
	278	55	Not given
	282	41	MeOH
	292	252	MeOH
	222 295	931 121	t-BuOH
	290	78	C_6H_{14}
	298	118	C_6H_{14}

states of simple carbonyl compounds guarantees that the lowest-energy triplet state T_1 is indeed the $^3n, \pi^*$ state, as is borne out by phosphorescence and $S_0 \rightarrow T_1$ absorption measurements at low temperatures. For conjugated α, β-enones, the energies of the triplet n, π^* and π, π^* states are very similar, so that either one may become the T_1 state, depending on substituents and the solvent. Interesting inversions in the ordering of the states have been observed, since increasing solvent polarity stabilizes $^3\pi, \pi^*$ states and destabilizes $^3n, \pi^*$ states. For β, γ-enones, calculations indicate that in general the T_1 state is a π, π^* state, which is consistent with the observed photochemistry[5,6].

III. TYPICAL PHOTOCHEMISTRY OF COMPARATIVE MODEL SYSTEMS

In order to put the photochemistry of enones into proper perspective, it is useful to summarize the photochemical behaviour of model monochromophoric alkene and carbonyl compounds in order to see what changes in the photochemistry ensue in when both chromophores are present in the same molecule. Since these model reactions are discussed at length in photochemistry texts which can be consulted for details[1-4], they will be presented here only briefly.

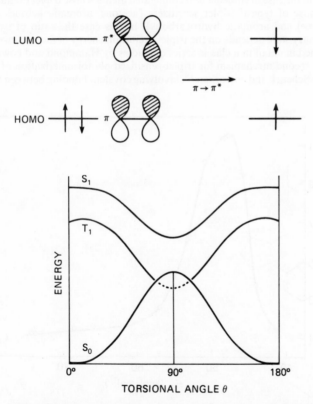

FIGURE 6. Dependence of energies of S_0, S_1 and T_1 states of alkenes on the torsional angle. Reproduced by permission of Academic Press, Inc. from Ref. 10

A. Photochemistry of Alkenes

This brief discussion will concern compounds containing only a single C=C moiety, for comparison with the photochemistry of enones. For discussions of the very rich and interesting photochemistry of dienes, trienes and more extended polyenes, which is not directly relevant to the main subject of this chapter, the reader should consult any of a number of reviews of the literature.

1. Cis-trans isomerization of alkenes

The prototypical reaction which ensues on electronic excitation of acyclic alkenes is isomerization around the C=C bond (*cis–trans* or *Z–E* isomerization). If no other reactions occur which interfere with the isomerization process, a photostationary mixture of isomers results from excitation of either the *Z* or *E* alkene. It is generally agreed that this reaction takes place via singlet excited states, since intersystem crossing is slow in most alkenes compared with the rate of relaxation of the planar excited singlet to a more stable perpendicular geometry, at which point rapid radiationless decay takes place to the ground-state potential surface at or near its energy maximum (see Figure 6).

Triplet-sensitized isomerization of alkenes via alkene triplets is also well known, particularly in the case of stilbenes and conjugated dienes whose triplet excitation energies lie below those of typical triplet sensitizers [acetone, aromatic ketones (particularly benzophenone) and aromatic hydrocarbons]. In this case the ratio of isomers at the photostationary state depends on the triplet excitation energy of the sensitizer, which has been examined in detail in a classic series of studies by Hammond and coworkers[18] (see Figure 7). A second mechanism for triplet-sensitized photoisomerization of alkenes was proposed by Schenck and coworkers[19], involving covalent bonding between the sensitizer

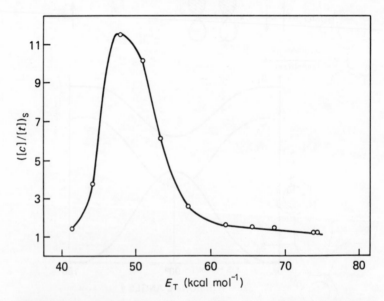

FIGURE 7. Dependence of the ratio of *cis*-stilbene/*trans*-stilbene at the photostationary state on the triplet excitation energy of the sensitizer. Reprinted from Ref. 18 by courtesy of Marcel Dekker, Inc

and the alkene to give a triplet 1,4-biradical, rotation around the former C=C bond and fragmentation (see Scheme 1). Although the Schenck mechanism has been discarded in favor of the Hammond triplet energy-transfer mechanism in the case of stilbenes and dienes, the mechanism has been invoked for photosensitized isomerization of alkenes in cases where the energetics of triplet energy transfer are unfavorable, i.e. with low-energy sensitizers and/or alkenes with high triplet excitation energies, as in sensitized isomerization of 2-butene ($E_T \sim 80 \, \text{kcal} \, \text{mol}^{-1}$).

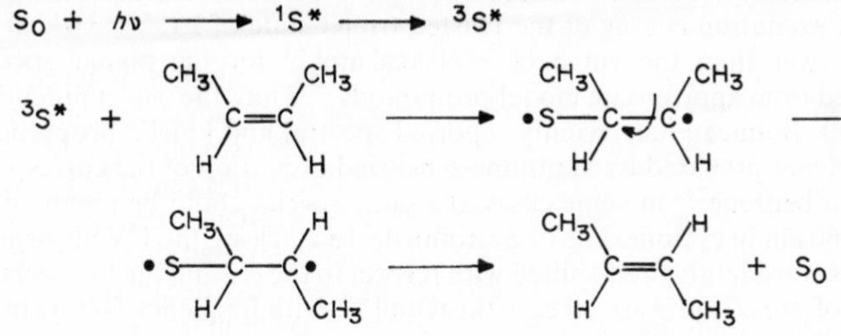

SCHEME 1

Photoisomerization of medium ring cycloalkenes is of particular interest with respect to corresponding reactions of analogous cycloalkenones. Not surprisingly, photosensitized excitation of *cis* cyclooctenes leads to isolable *trans* cyclooctenes[20], and *trans* cycloheptenes have been implicated in sensitized photoaddition reactions (see below) of *cis* cycloheptenes and have been directly detected using nanosecond flash photolysis techniques[21]. Direct or triplet-sensitized excitation of 1-phenylcyclohexene 1 yields *trans*-

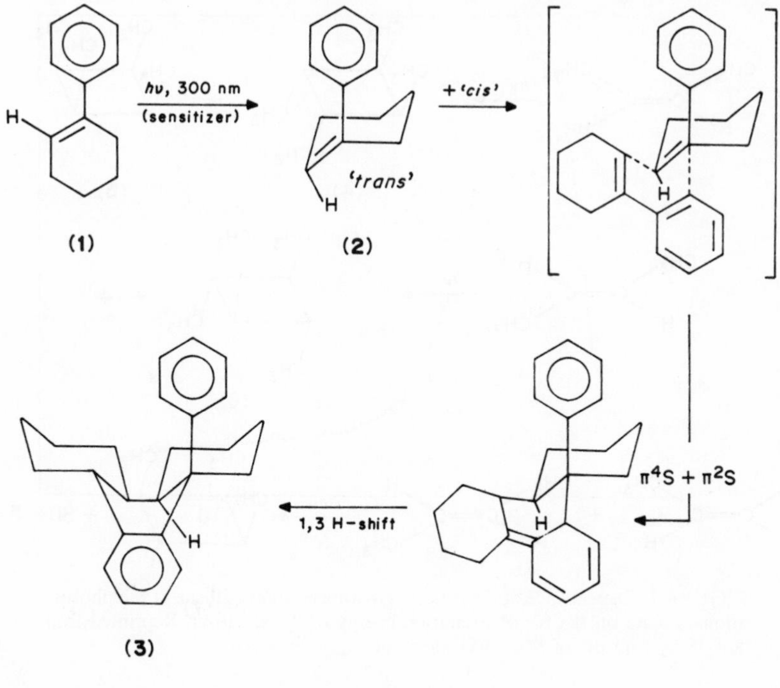

(7)

1-phenylcyclohexene **2**, which has been directly detected as a transient intermediate using nanosecond flash photolysis[22]. It has also been trapped chemically by reaction with acidic methanol[23] and by stereospecific [4 + 2] addition to *cis*-1-phenylcyclohexene to give **3** (equation 7)[24]. The lifetime of 9μs for **2** in methanol obtained by flash techniques has been confirmed using time-resolved photoacoustic calorimetry[25], and the strain energy of **2** vs. **1** is 44.7 ± 5 kcal mol^{-1}. The barriers for thermal reversion of **2** to **1** in methanol and benzene are ∼ 7 kcal mol^{-1} and 10.6 kcal mol^{-1} respectively, which accounts for the relative kinetic stability of **2**. It is of interest that photoacoustic calorimetric data indicate that the triplet excitation energy of the twisted triplet state of **1** (56 ± 3.4 kcal mol^{-1}) is only slightly lower than the value of ∼ 60 kcal mol^{-1} for the planar spectroscopic triplet estimated from appropriate model compounds[25]. Thus, the triplet potential surface of **1** is quite flat. Bonneau has recently reported spectral and kinetic properties of eight '*trans*' cyclohexenes prepared by xanthone-sensitized excitation of the corresponding *cis* cyclohexenes in benzene[26]; in some cases, the same species could be prepared by direct excitation at 266 nm in cyclohexane or acetonitrile. In all cases, the UV absorption of the '*trans*' isomer is considerably red-shifted with respect to the *cis*, and the barriers to thermal isomerization of '*trans*' to *cis* are all *ca* 10 kcal mol^{-1}, with frequency factors in the range 10^{12}–10^{13} s^{-1} [26].

2. Photodimerization of alkenes

Intermolecular photodimerization of alkenes to give cyclobutanes is a well-known reaction that can be brought about on either direct or triplet-sensitized excitation. Thus, direct irradiation of liquid *cis*-2-butene gives dimers **4** and **5** (equation 8) while irradiation of *trans*-2-butene gives **4** and **6** (equation 9); isomerization to 1-butene competed with dimerization[27]. Irradiation of a mixture of *cis*- and *trans*-2-butene gave dimer **7** in addition to **4–6** (equation 10). These stereospecific photodimerizations, necessarily observed only at low alkene conversion because of competitive *cis–trans* isomerization, indicate that the excited alkene undergoing dimerization does not also undergo isomerization, which was rationalized in terms of rapid formation of excited state–ground state complexes en route to dimers. Dilution with neopentane decreased dimer yields drastically. Tetramethylethylene also photodimerizes on direct excitation.

Triplet-sensitized photodimerization of simple alkenes, such as ethylene, can be brought about in the vapour phase using mercury, while small and medium ring cycloalkenes photodimerize in solution using typical organic sensitizers. Thus cyclopropene **8** dimerizes (equation 11) in acetone (in the presence of benzophenone, which may or may not play a role)[28] and cyclopentene also photodimerizes in acetone (equation 12)[29]. An instructive example is provided by norbornene **9**[30]. Dimerization can be sensitized by acetophenone ($E_T = 74\,\text{kcal mol}^{-1}$) but not by benzophenone ($E_T = 69\,\text{kcal mol}^{-1}$); in the latter case cycloaddition occurs to give the oxetane **10** (equation 13). Both reactions occur using xanthone ($E_T = 72\,\text{kcal mol}^{-1}$). Thus, there is a competition between triplet energy transfer and cycloaddition to alkenes, depending on the relative triplet excitation energies of the sensitizer and the alkene.

(**8**)   80%   20%   (11)

(12)

(**9**)   (13)

(**10**)

Copper salts also can be used to catalyze photodimerization of cyclic but not acyclic alkenes, via a Cu(I)–olefin complex[31]. With cyclohexene or cycloheptene, the major products are the *trans*-fused dimers **11** and **12** (equations 14 and 15). It was suggested that these products arise by Cu(I)-catalyzed photoisomerization to the *trans* cycloalkenes, perhaps still complexed to Cu(I), which then undergo stereoelectronically controlled $[\pi^2_s + \pi^2_a]$ addition to the respective *cis* cycloalkenes. The fact that cyclooctene and acyclic dienes do not undergo Cu(I)-catalyzed photodimerization can be ascribed to reduced reactivity of twisted alkene–Cu intermediates in these cases, Cu(I)-catalyzed photodimerization of cyclopentene and norbornene gives exclusively *cis*-fused dimers (equations 16 and 17). In the latter case, quantum yield measurements suggest that the reaction proceeds via a 2:1 norbornene–Cu(I) triflate complex in which the Cu is simultaneously bonded to the π systems of both alkenes.

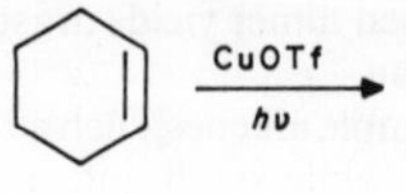

(14)

(11) 49%

(15)

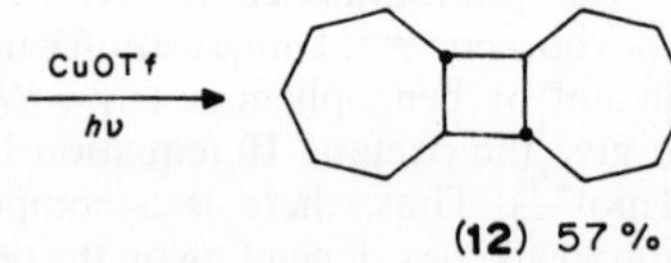

(12) 57%

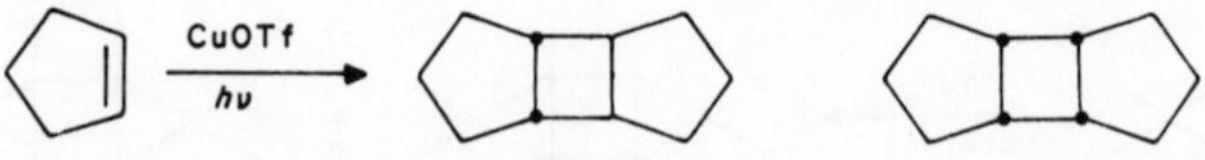

30%   3%

(16)

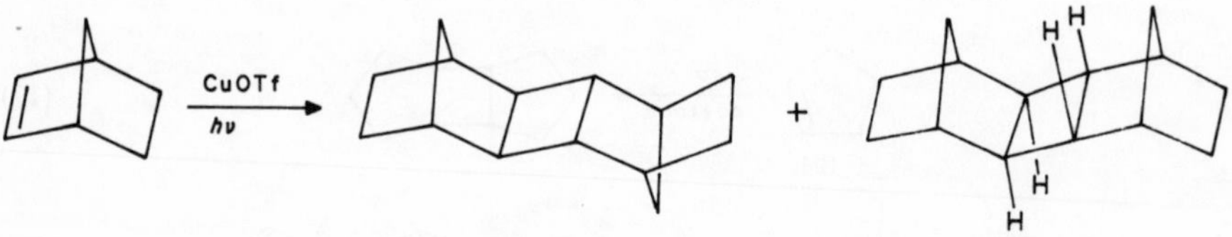

(major)   (minor)

(17)

3. Photoaddition of nucleophiles

Photoaddition of methanol to cyclohexenes and cycloheptenes to give ethers occurs in the presence of high-energy sensitizers, such as benzene, toluene and xylene (the latter is used most frequently)[23a,32,33]. Typical examples are shown in equations 18–20. Such photoaddition reactions are not observed using acyclic or cyclooctenes. The addition reaction is usually accompanied by alkene isomerization. It has been proposed that the reaction involves triplet-sensitized isomerization to a trans cyclohexene or trans cycloheptene, which on protonation gives a carbocation, which is either captured by the nucleophile or loses a proton to give the rearranged alkene (see Scheme 2). It is proposed that unstrained *trans* or acyclic alkenes do not possess sufficient

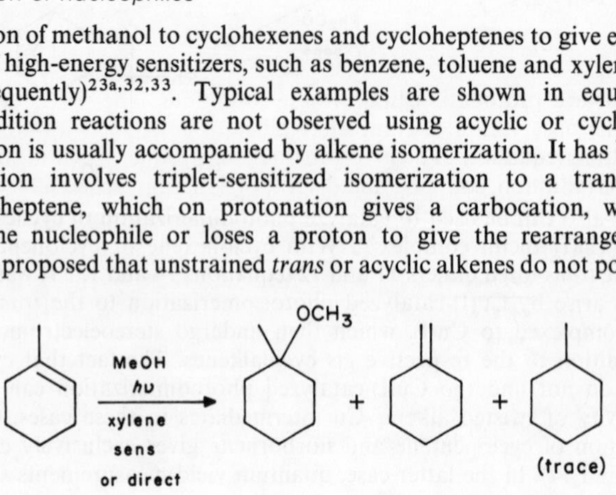

(18)

(trace)

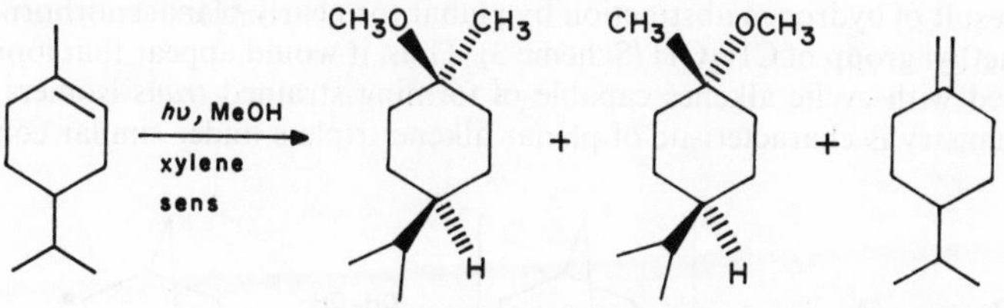

$$(19)$$

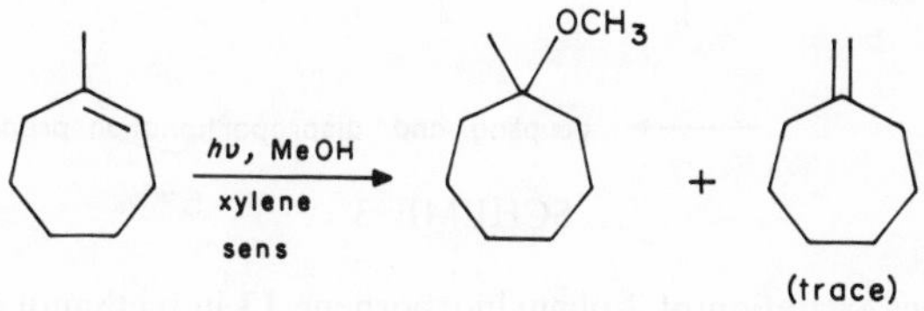

(trace)

$$(20)$$

$n = 6, 7$

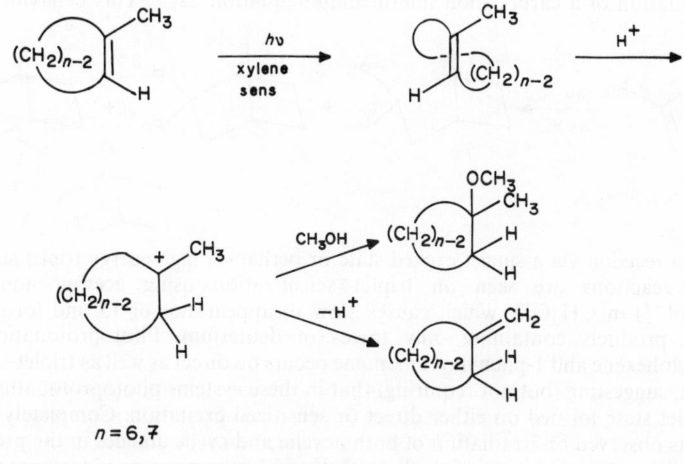

SCHEME 2

driving force toward protonation under these conditions. The proposed mechanism is supported by the observation of completely different behavior with norbornene under identical conditions (equation 21)[23a,32,33]. In this case, products resulting from typical free radical reactions (addition, coupling and disproportionation) are observed, which clearly

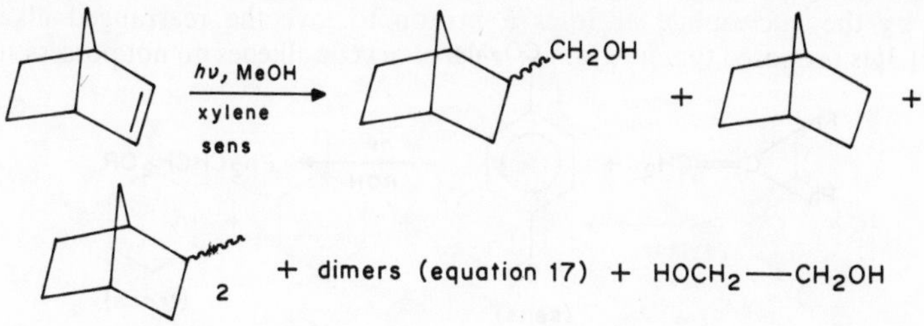

$$2 \qquad + \text{ dimers (equation 17)} + \text{HOCH}_2\text{---CH}_2\text{OH} \qquad (21)$$

arise as a result of hydrogen abstraction by planar (or nearly planar) norbornene triplets from the methyl group of CH_3OH (Scheme 3). Thus, it would appear that ionic reactions are observed with cyclic alkenes capable of forming strained *trans* isomers, while free radical chemistry is characteristic of planar alkene triplets under similar conditions[34].

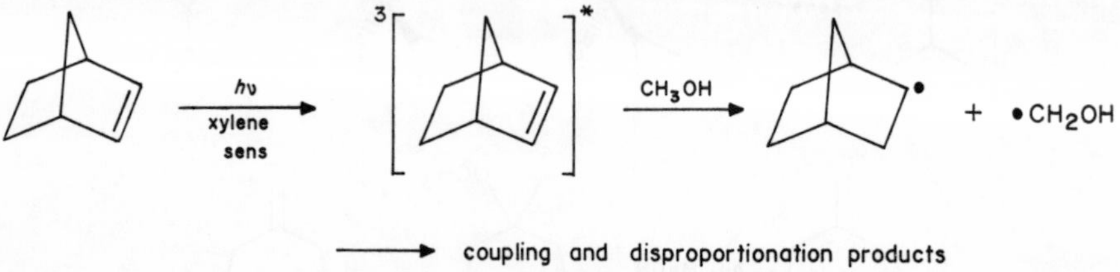

coupling and disproportionation products

SCHEME 3

Curiously, direct excitation of 2-phenylnorbornene **13** in methanol efficiently gives a tertiary ether (with Markownikoff regiochemistry) and other products resulting from initial formation of a carbocation intermediate (equation 22)[35]. This behavior of **13** is

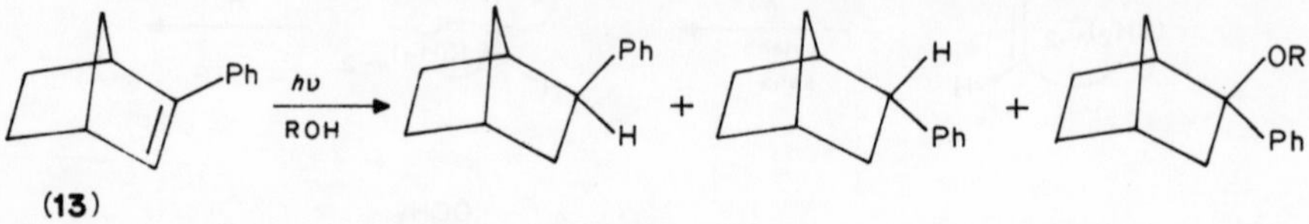

(22)

ascribed to reaction via a singlet excited state or perhaps a high-energy triplet state, since no such reactions are seen on triplet sensitization using acetophenone ($E_T = 74\,\text{kcal mol}^{-1}$) in CH_3OD, which causes slow disappearance of **13** and formation of reduction products containing only traces of deuterium. Photoprotonation of 1-phenylcyclohexene and 1-phenylcycloheptene occurs on direct as well as triplet-sensitized excitation, suggesting (but not requiring) that in these systems photoprotonation occurs via a triplet state formed on either direct or sensitized excitation. Completely different behavior is observed on irradiation of both acyclic and cyclic alkenes in the presence of nucleophilic reagents (a variety of alcohols, acetic acid, potassium cyanide) using methyl *p*-cyanobenzoate, *p*-dicyanobenzene or 1-cyanonaphthalene as sensitizers[36]. As shown in equations 23–25, anti-Markownikoff addition products are formed in moderate to excellent yields to the complete exclusion of Markownikoff addition of the nucleophiles. A general mechanism for these addition reactions is shown in Scheme 4. The key step is electron transfer from the alkene to the sensitizer singlet excited state to give the alkene radical cation (Alk$^{+\cdot}$) and the sensitizer radical anion (Sens$^{-\cdot}$)[13]. Quenching of sensitizer

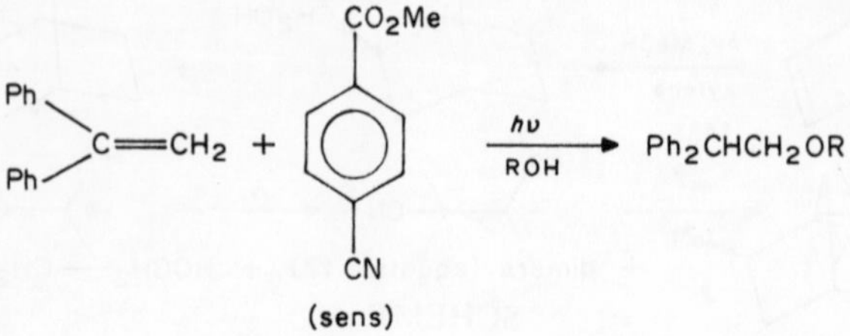

(23)

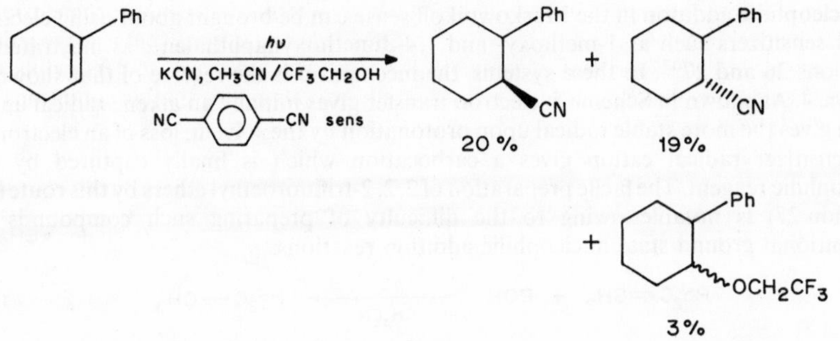

$$20\% \qquad 19\%$$

$$+ \qquad (24)$$

$$3\%$$

$$18\% \qquad 5\%$$

$$54\%$$

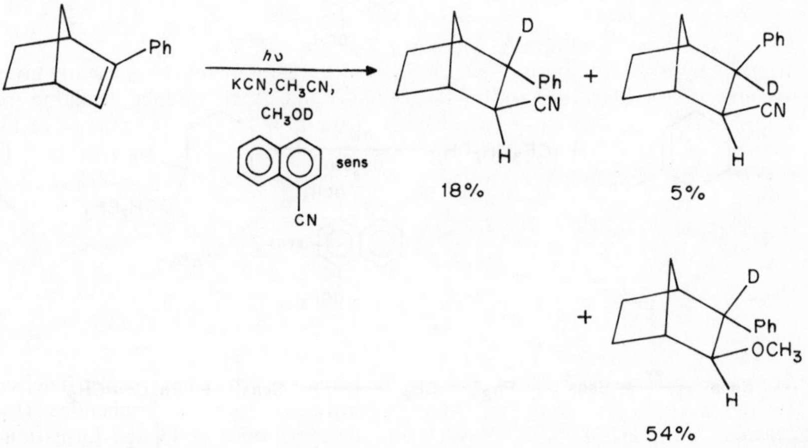

$$(25)$$

fluorescence by alkenes which undergo photoaddition supports an electron transfer mechanism. Nucleophilic addition to $Alk^{+\cdot}$ occurs in an anti-Markownikoff sense to generate the more stable free radical, which is then reduced to an anion by back electron transfer from $Sens^{-\cdot}$, followed finally by protonation. If the reaction is run using a deuteriated solvent ROD, the product incorporates one deuterium at the position predicted by this mechanism.

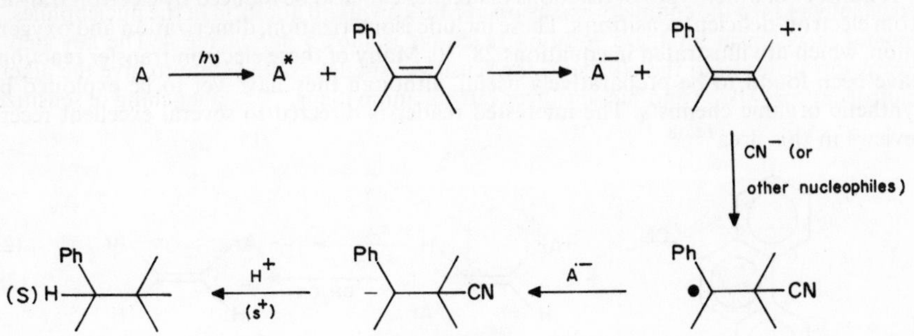

SCHEME 4

Nucleophilic addition in the Markownikoff sense can be brought about using electron donor sensitizers such as 1-methoxy- and 1,4-dimethoxynaphthalene, as illustrated in equations 26 and 27[37]. In these systems, the mechanism is the reverse of that shown in Scheme 4. As shown in Scheme 5, electron transfer gives initially an alkene radical anion which gives the more stable radical upon protonation by the solvent; loss of an electron to the sensitizer radical cation gives a carbocation which is finally captured by the nucleophilic reagent. The facile preparation of 2,2,2-trifluoroethyl ethers by this route (see equation 27) is notable owing to the difficulty of preparing such compounds by conventional ground-state nucleophilic addition reactions.

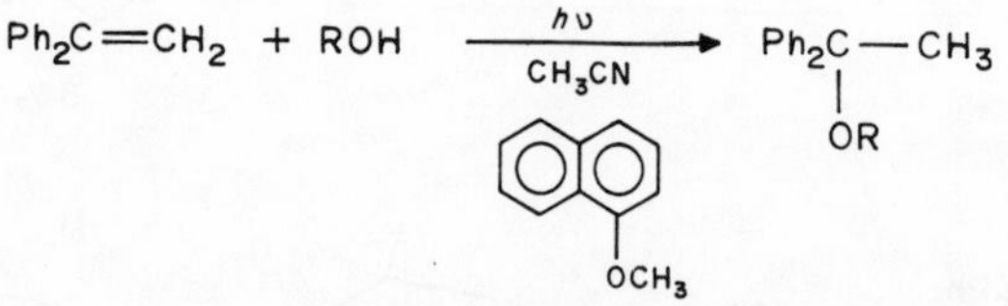

$$Ph_2C{=}CH_2 + ROH \xrightarrow[CH_3CN]{h\upsilon} Ph_2C{-}CH_3 \quad (26)$$

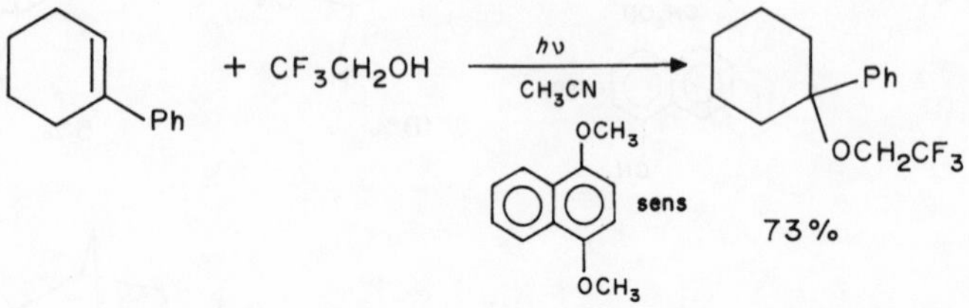

(27)

73%

$$Sens \xrightarrow{h\upsilon} Sens^+ + Ph_2C{=}CH_2 \longrightarrow Sens^{+\cdot} + Ph_2C{=}CH_2^{-}$$

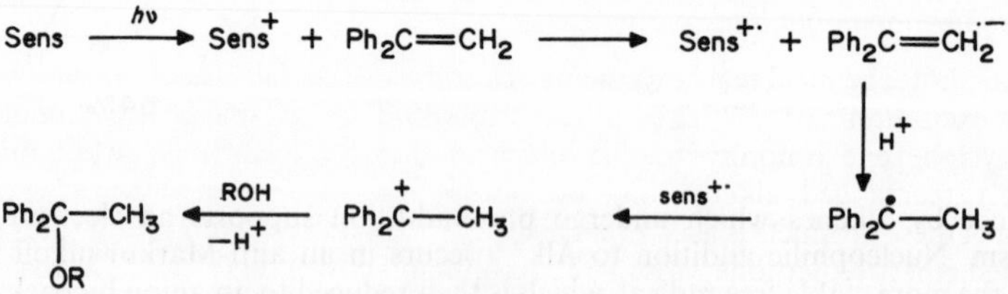

SCHEME 5

A number of other types of reactions of alkenes can also be induced by electron transfer from electron-deficient sensitizers. These include isomerization, dimerization and oxygenation, which are illustrated in equations 28–30. Many of these electron transfer reactions have been found to be preparatively useful, although they have yet to be exploited by synthetic organic chemists. The interested reader is directed to several excellent recent reviews in this area[13,38].

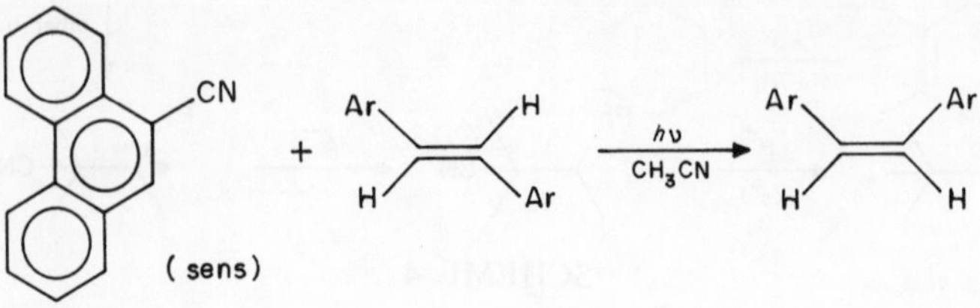

(28)

$$^1A^* + \text{(phenyl vinyl ether)} \longrightarrow A^{\bullet-} + \text{(phenyl vinyl ether radical cation)}^{+\bullet}$$

$$\longrightarrow \text{(cyclobutane-OPh, OPh)} + \text{(cyclobutane-OPh, OPh)} \qquad (29)$$

$$40 : 60$$

$$\underset{\substack{O_2, CH_3CN \\ sens}}{\xrightarrow{h\nu}} \; Ph_2C{=}O \; + \; \text{(epoxide Ph, Ph, Ph, Ph)}$$

$$+ \; Ph_3COH \; + \; Ph_3CC(=O){-}Ph \qquad (30)$$

4. Photorearrangements and related reactions

Unusual photochemical rearrangements of tetrasubstituted alkenes have been observed on direct excitation[23b,34,39]. Thus, direct excitation of tetramethylethylene and 1,2-dimethylcyclohexene in nonhydroxylic solvents (ether, hydrocarbons) gives a mixture of structurally rearranged alkenes and cyclopropanes while, in hydroxylic media, the formation of these products is accompanied by the formation of a mixture of saturated and unsaturated ethers (equations 31 and 32). Kropp and coworkers[23b,34,39] have suggested that these reactions occur by initial formation of a π, R(3s) Rydberg excited state of the

$$\xrightarrow[ROH]{h\nu} \; \text{(—OR)} \; + \; H\text{—}\text{(—OR)} \; + \; \text{(alkene)} \; + \; \text{(cyclopropane—H)} \qquad (31)$$

$$\text{(1,2-dimethylcyclohexene)} \xrightarrow[MeOH]{h\nu} \text{(—OCH}_3) + \text{(—OCH}_3)$$

$$+ \; \text{(—OCH}_3) \; + \; \text{(methylenecyclohexane)} \qquad (32)$$

alkene (R ← N transition in spectroscopic terms), in which the orbital containing the excited electron is much larger than the molecular core which, in effect, becomes positively charged. As Kropp puts it, 'the (excited) electron has been placed in a sort of holding pattern; it has been removed from the core and yet not completely separated from the core's influence'. Structure **14** in Scheme 6 is Kropp's pictorial designation for the π, R(3s) Rydberg state for tetramethylethylene. The energy for this UV transition decreases with the degree of substitution on the C=C bond, from 7.12 eV (174 nm) for CH_2=CH_2 to 5.40 eV (230 nm) for $(CH_3)_2C$=$C(CH_3)_2$; for tetrasubstituted alkenes, the $\pi \to R$ transition may well be the lowest energy transition in solution, but in any event the Rydberg character of the S_1 state will increase with alkyl substitution. This is consistent with the marked changes in photochemistry observed as a function of degree of substitution on C=C. It should be noted that the π, π^* and Rydberg states remain widely separated in the triplet manifold, so that only the $^3\pi, \pi^*$ state need be considered in discussion of triplet reactivity of alkenes.

Kropp proposed that a Rydberg state undergoes two key reactions, as illustrated in Scheme 6: rearrangement to carbenes **15** (path A) and nucleophilic trapping to give alkoxy radical **16** and solvated electrons (path B). Products **17** and **18** arise from carbene **15** by a 1,2-H shift and C—H insertion, respectively, while ethers **19** and **20** arise by disproportionation of radical **16**. The two hydrocarbon products **21** and **22** are proposed to arise by capture of an electron by the starting olefin to give radical anion **23**, and protonation by the solvent to give radical **24** which undergoes disproportionation to give the isolated products. Related observations with a variety of tri- and tetrasubstituted alkenes are presented and discussed in Kropp's excellent review article[34].

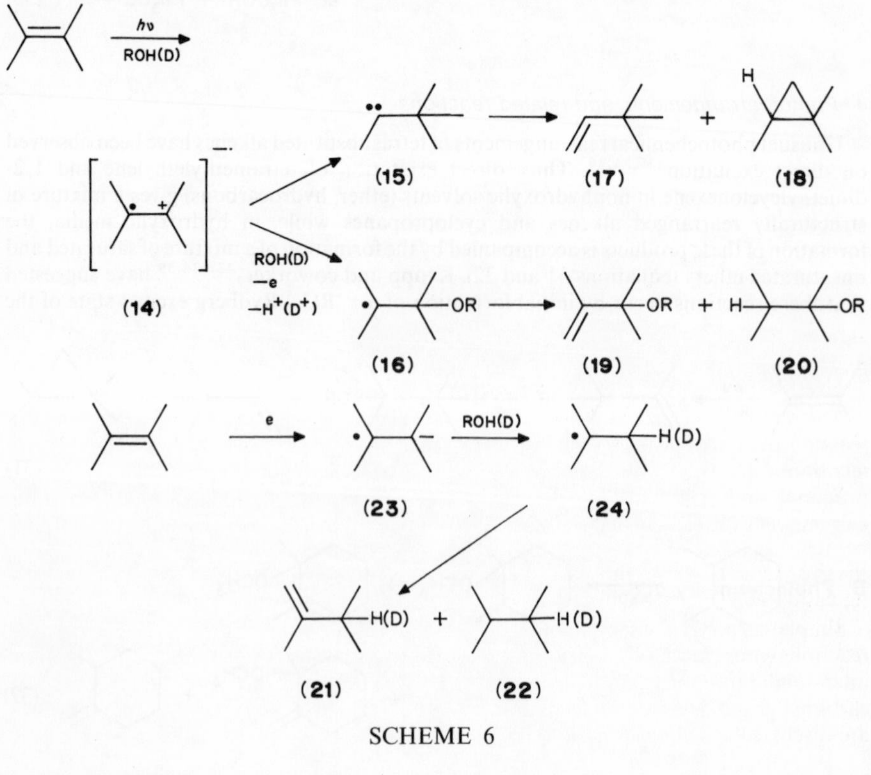

SCHEME 6

Ring opening of cyclobutenes to give dienes and the reverse process are classic electrocyclic reactions, which are predicted by one or another version of orbital symmetry theory to occur photochemically by a disrotatory path[40]. Although photochemical formation of cyclobutenes from 1,3-dienes is well known and indeed occurs stereospecifically in accord with theoretical predictions, the reverse ring opening is not well known. The problem is that 1,3-dienes absorb at longer wavelengths and with greater intensity than cyclobutenes, so that under the conditions required to effect ring opening of cyclobutenes, the reverse photochemical ring closure of dienes should be a facile process. One of the few reported studies of cyclobutene ring opening involves compounds 25 and 26 (equations 33 and 34)[41]. The former indeed affords cis, cis-1, 1'-bicyclohexenyl 27 and the fragmentation product 28, while the latter gives only the isomeric fragmentation product 29; disrotatory ring opening of 26 would afford the highly strained cis, trans isomer of 27 (i.e. 30). Compound 30 was proposed as the intermediate in the photosensitized conversion of 27 to 25 and has indeed been detected and characterized by laser flash techniques[26].

(25)   (27)   (28)   (33)

(26)   (29)

(34)

(30)

5. Hydrogen-atom abstraction

Abstraction of hydrogen from solvents or added reagents is a relatively rare mode of reaction of electronically excited alkenes, since the other types of reactions previously mentioned are usually much faster and therefore dominate. H-atom abstraction from 2-propanol and methanol has been reported for 1,1-diphenylethylene and 1,1-di-t-butylethylene. It is likely that the reactive excited states in these systems are $^3\pi, \pi^*$ states[42].

B. Photochemistry of Ketones

Simple carbonyl compounds (aldehydes and ketones) undergo several prototypical reactions whose mechanisms are reasonably well understood at the present time. These are inter- and intramolecular hydrogen abstraction, cleavage of C—C bonds α- to the carbonyl group, and intermolecular addition to olefins to give oxetanes. These processes are discussed at length in basic texts, so they will be only briefly reviewed here.

1. Photoreduction as a consequence of hydrogen abstraction

Photoreduction of ketones in hydrogen-donor solvents or in the presence of added reagents has been known ever since the pioneering studies of Ciamician and Silber at the turn of the century[43]. Thus, irradiation of benzophenone in 2-propanol or in benzene containing benzhydrol efficiently produces benzpinacol (equations 35 and 36). In classic mechanistic investigations, Hammond and coworkers established that this reaction proceeds via triplet n, π^* excited states according to the mechanism shown in Scheme 7[44].

$$Ph_2C{=\!=}O + (CH_3)_2CHOH \xrightarrow{h\upsilon} \underset{\underset{\text{(major)}}{\overset{\underset{\displaystyle OH \quad OH}{|\quad\;\;|}}{Ph_2C{-}CPh_2}}}{} + (CH_3)_2C{=\!=}O + \underset{\underset{\text{(minor)}}{\overset{\underset{\displaystyle OH \quad OH}{|\quad\;\;|}}{Ph_2C{-}C(CH_3)_2}}}{}$$

$$(35)$$

$$Ph_2C{=\!=}O + Ph_2CHOH \xrightarrow[\text{benzene}]{h\upsilon} \underset{\overset{\underset{\displaystyle OH \quad OH}{|\quad\;\;|}}{Ph_2C{-}CPh_2}}{} \qquad (36)$$

$$^1Ph_2C{=\!=}O \xrightarrow{h\upsilon} {}^1Ph_2CO^* \longrightarrow {}^3Ph_2CO^*$$

$$\downarrow R_2CHOH$$

$$\text{products} \longleftarrow Ph_2\overset{\displaystyle .}{C}OH + R_2\overset{\displaystyle .}{C}OH$$

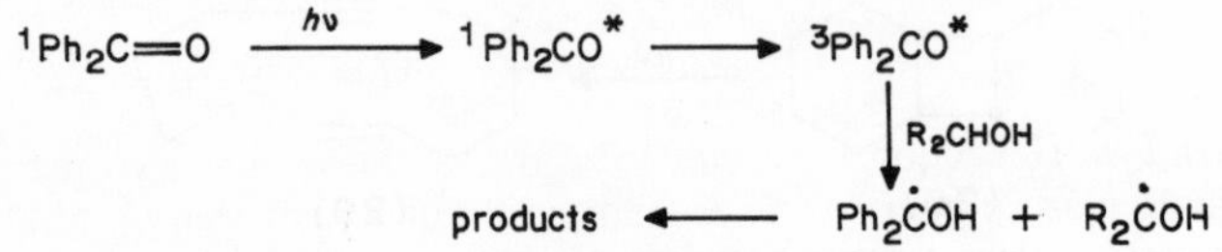

SCHEME 7

The key step is abstraction of the hydrogen attached to the carbinol carbon by benzophenone triplet, for which a kinetic isotope effect k_H/k_D of 2.8 has been determined. The quantum efficiency for disappearance of benzophenone is ca unity using benzhydrol as the reductant, indicating that triplets are formed with 100% efficiency (the rate constant for intersystem crossing has been determined to be ca 10^{11} l mol^{-1} s^{-1}). In 2-propanol, the QE for disappearance of ketone approaches 2.0 at high ketone concentrations because of hydrogen atom transfer from Me$_2\overset{.}{C}$OH to ketone[45]; the rate constant for this process, which Steel and coworkers view as a simultaneous electron/proton transfer, to be differentiated from transfer of a hydrogen atom as such, has recently been determined to be $3.5 \pm 1.5 \times 10^4$ l mol^{-1} s^{-1} [46]. The pinacol product is formed by combination of two Ph$_2\overset{.}{C}$OH radicals. When the aryl group in the ketone is different from that in the corresponding hydrol, as a result of either incorporation of a substituent or an isotopic label, it is found that the initial products are as shown in equation 37, i.e. the pinacol is derived only from the ketone[47]. This result indicates that proton/electron transfer occurs as shown in equation 38, in which the initial ketyl radical is converted to ketone and a

$$Ph_2C{=\!=}O + Ar_2CHOH \xrightarrow{h\upsilon} \underset{\overset{\underset{\displaystyle OH \quad OH}{|\quad\;\;|}}{Ph_2C{-}CPh_2}}{} \qquad (37)$$

$$Ar_2\overset{.}{C}OH + Ph_2C{=\!=}O \longrightarrow Ph_2\overset{.}{C}OH + Ar_2C{=\!=}O \qquad (38)$$

second molecule of ketone is reduced; the rate constant for this reaction has been determined by Steel and coworkers to be $1.3 \pm 0.2 \times 10^4 \, l\,mol^{-1}\,s^{-1}$ [46]. Thus, in the classic Hammond mechanism of Scheme 7 the two initially formed ketyl radicals do not directly combine, undoubtedly because of spin restrictions arising from the fact that they are produced as a triplet radical pair.

Photoreduction is general for ketones whose lowest triplet is an n, π^* state, which includes all aliphatic ketones and aromatic ketones with electron-withdrawing substituents on the aromatic ring. Benzophenones with electron-donor substituents undergo such reaction much less efficiently or not at all, attributed to low-lying π, π^* or charge-transfer triplet states[48]. This unreactive group of ketones also includes carbonyl derivatives of naphthalene and anthracene.

Photoreduction of ketones by amines is a well-known process, illustrated in equation 39 for benzophenone in the presence of triethylamine[49]. In this case, electron transfer occurs from the amine to the ketone triplet to give a radical ion pair, followed by proton transfer to give the ketyl radical $Ph_2\dot{C}OH$, which then either dimerizes to pinacol or abstracts a second hydrogen atom to give the secondary alcohol. The efficiency of the electron transfer process is governed by the factors discussed previously in Section I, the most critical factor being the ionization potential of the amine.

$$3Ph_2C{=\!\!=}O^* + Et_3N: \longrightarrow Ph_2\dot{C}{-\!\!-}^{-} + Et_3N^{+\cdot} \longrightarrow$$

$$Ph_2\dot{C}OH + CH_3\dot{C}H{-\!\!-}NEt_2 \longrightarrow$$

$$\text{coupling and disproportionation products} \qquad (39)$$

The Norrish Type II reaction, illustrated in Scheme 8 for γ-methylvalerophenone (MVP), is the intramolecular counterpart of the intermolecular hydrogen-abstraction process discussed above[50]. This is a general process for ketones possessing a lowest n, π^* triplet with accessible γ-hydrogens on the side-chain. Aromatic ketones undergo this reaction exclusively from triplet states because of rapid intersystem crossing, while aliphatic ketones (which have values of k_{isc} of the order of 10^8–$10^9\,s^{-1}$) generally react from both singlet and triplet n, π^* states[51]. Studies of appropriately substituted compounds show that the singlet component of the reaction is largely stereospecific, while the triplet component gives alkenes with mixed stereochemistry. In cases where the γ-carbon of the side-chain is fully substituted, H abstraction from the next (δ) position is sometimes observed[52]. The Norrish Type II fragmentation to alkenes and ketones is usually accompanied by the formation of low yields of cyclobutanols, as shown in Scheme 8.

2. Norrish Type I cleavage of ketones

Irradiation of aliphatic ketones in the vapor phase usually leads to formation of an acyl–alkyl radical pair by homolytic cleavage of one of the C—C bonds to the carbonyl carbon, illustrated in equation 40[53]. The acyl radical usually loses CO to give a second alkyl radical. The products arise by combination and disproportionation of the various radicals. Many acyclic and alicyclic ketones undergo similar reactions in solution[54]. Intermediate

$$R^1{-\!\!\!\overset{\overset{\displaystyle O}{\|}}{C}\!\!\!-}R^2 \xrightarrow{h\nu} R^1\overset{\overset{\displaystyle O}{\|}}{C}{}^\cdot + R^{2\,\cdot} + R^{1\,\cdot} + R^2\overset{\overset{\displaystyle O}{\|}}{C}{}^\cdot \qquad (40)$$

$$\downarrow$$

$$\text{products}$$

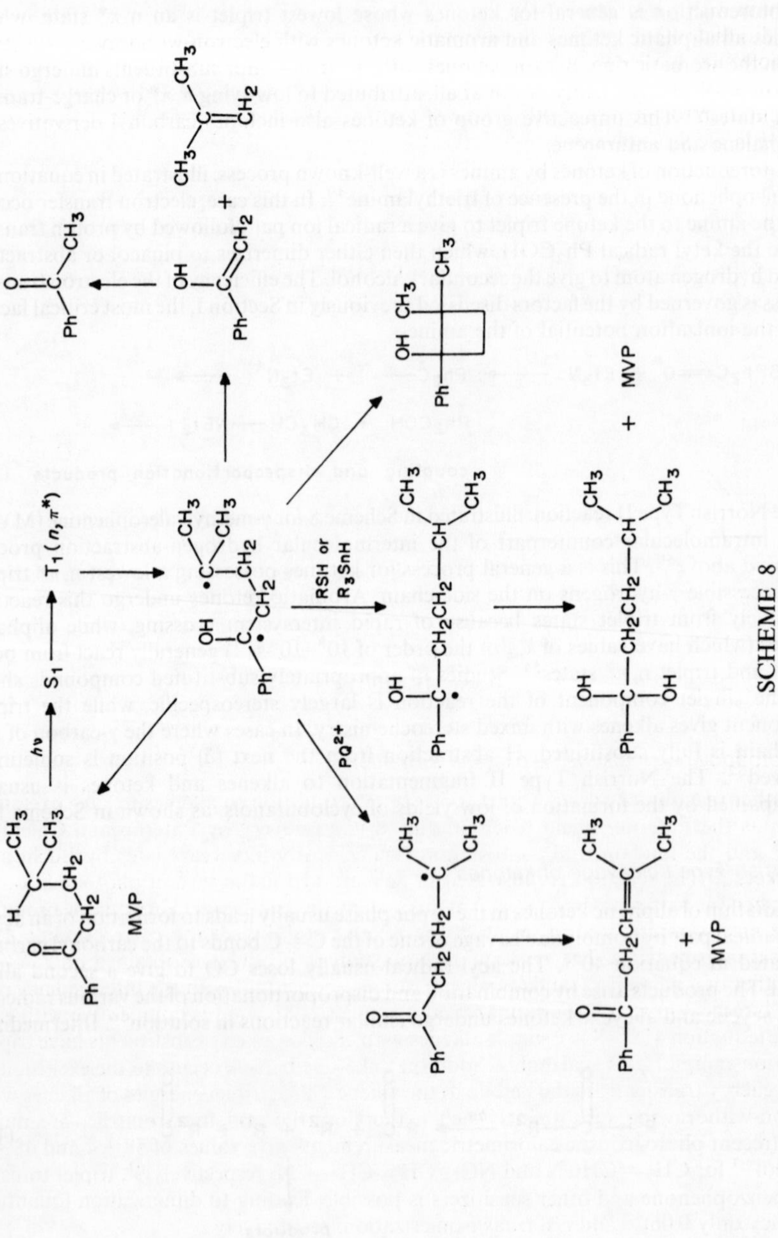

$T_1(n,\pi^*)$

S_1

$h\nu$

MVP

RSH or
R_3SnH

PQ^{2+}

+ MVP

+ MVP

SCHEME 8

radicals have been directly detected by electron spin resonance (ESR) techniques, and observations of nuclear polarization under these conditions also provide evidence for radical intermediates[55].

Norrish I cleavage of aliphatic ketones can occur from both singlet and triplet n, π^* states, and sometimes competes directly with Norrish Type II reactions when there is a side-chain with γ-hydrogens[54]. In the case of cyclohexanone, cleavage affords a 1,6-acyl-alkyl diradical that gives a ketene and an unsaturated aldehyde by competitive intramolecular 1,5-H migrations (see Scheme 9)[56]. In the case of cyclobutanone, the initial 1,4-acyl–alkyl diradical has a choice of (a) cleavage to CO and a trimethylene diradical, (b) fragmentation to ketene and ethylene, or (c) rearrangement to an oxacarbene that can be trapped in alcohol solvents (see Scheme 10)[57].

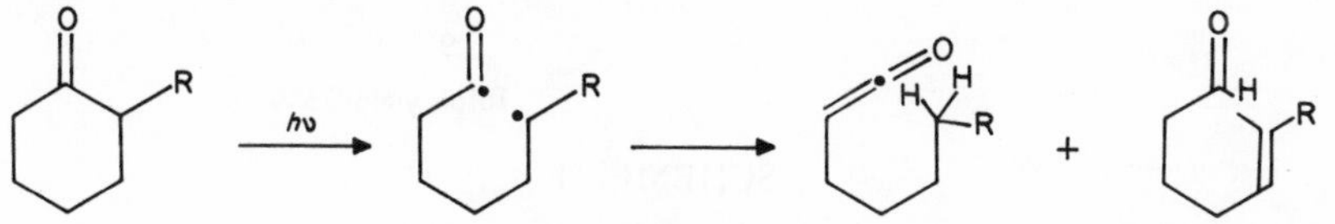

SCHEME 9

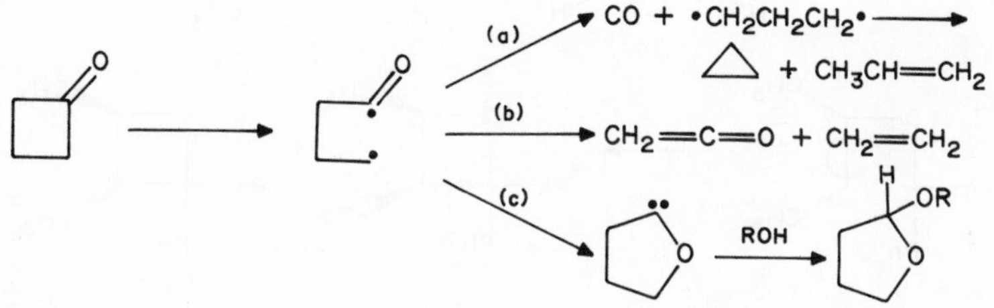

SCHEME 10

3. Photoaddition to alkenes—oxetane formation

Another reaction of ketone triplet n, π^* states is addition to alkenes to give oxetanes, known as the Paterno–Büchi reaction, due to its discovery by Paterno and Chieffi in 1909[58] and the fundamental contributions made nearly fifty years later by Büchi and coworkers[59]. The reaction is illustrated in Scheme 11 for the case of photoaddition of benzophenone to isobutene. As is seen in this system, the relative yield of isomeric products can be nicely rationalized in terms of the relative stability of the corresponding 1,4-diradical intermediates. The same mixture of isomeric products is obtained on reaction with either of a pair of (Z)- and (E)-alkene isomers, consistent with the intermediacy of a triplet biradical in which rotation around a single C—C bond is competitive with ring closure (equation 41)[60]. Since simple alkenes with alkyl or alkoxy substituents have triplet excitation energies $\geqslant 74$ kcal mol^{-1}, addition to ketone triplets occurs to the exclusion of triplet energy transfer from the ketone to the alkene. Since triplet energies of alkenes with electron-withdrawing substituents, such as acrylonitrile and fumaronitrile, are much lower (recent photoacoustic calorimetric measurements give values of 58 ± 4 and 48 ± 3 kcal mol^{-1} for CH_2=CHCN and NC—CH=CH—CN, respectively)[61], triplet transfer from benzophenone and other sensitizers is possible, leading to dimerization (quantum efficiency only 0.06)[62] and cis–trans isomerization, respectively.

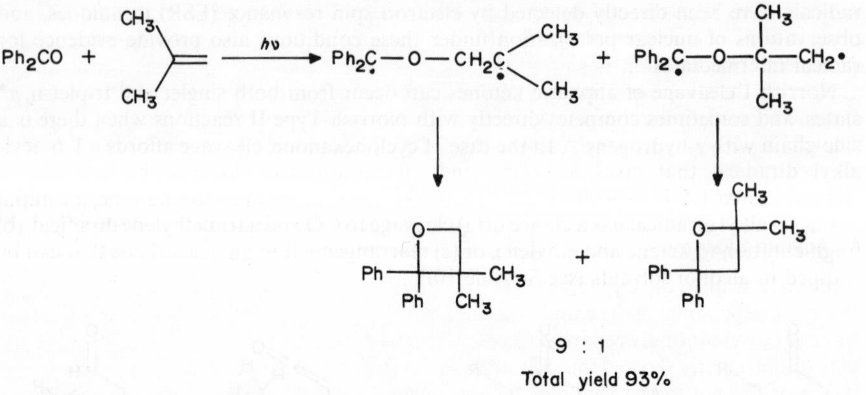
9 : 1
Total yield 93%

SCHEME 11

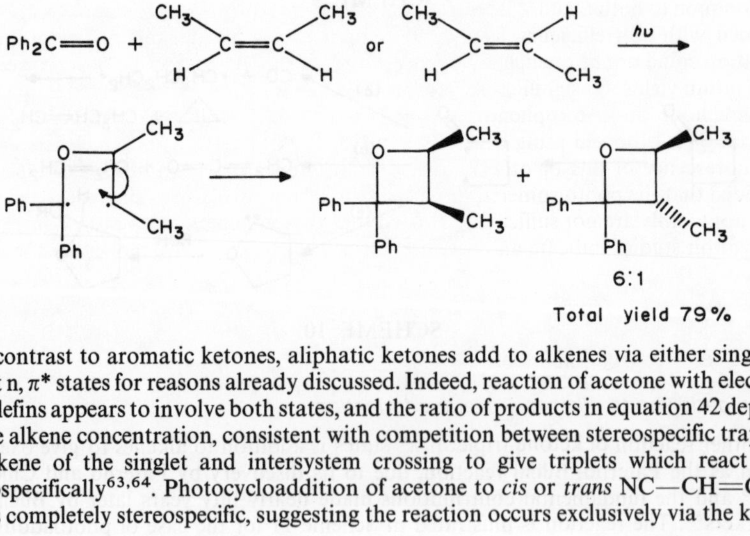
6:1
Total yield 79%

(41)

In contrast to aromatic ketones, aliphatic ketones add to alkenes via either singlet or triplet n, π^* states for reasons already discussed. Indeed, reaction of acetone with electron-rich olefins appears to involve both states, and the ratio of products in equation 42 depends on the alkene concentration, consistent with competition between stereospecific trapping by alkene of the singlet and intersystem crossing to give triplets which react non-stereospecifically[63,64]. Photocycloaddition of acetone to *cis* or *trans* NC—CH=CH—CN is completely stereospecific, suggesting the reaction occurs exclusively via the ketone

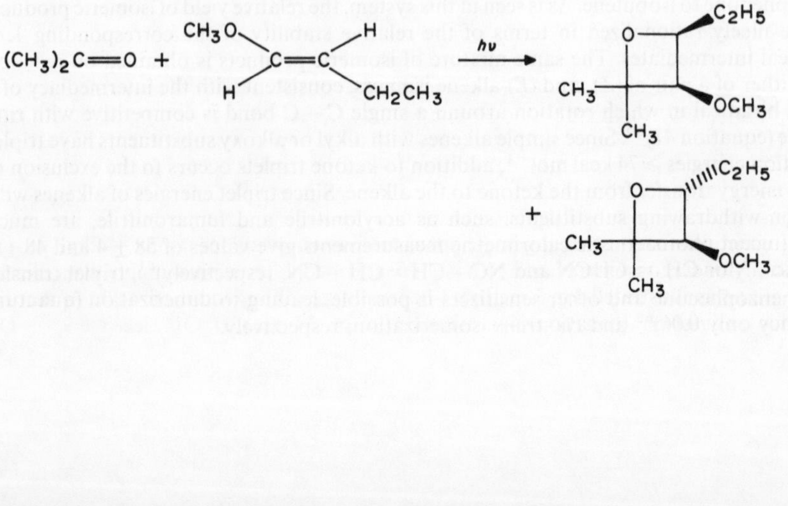

(42)

S_1 state. This is supported by the fact that acetone fluorescence is not quenched by electron-deficient alkenes, and that the cycloaddition is not affected by typical triplet quenchers. This reaction is suggested to involve interaction between the electron-poor π system of the alkene and the electron-rich π system of the ketone S_1 state. The course of oxetane formation has been rationalized in terms of perturbational molecular orbital theory[65].

IV. PHOTOCHEMISTRY OF α, β-UNSATURATED KETONES

A. Acyclic Systems

In early studies of acyclic conjugated enones three general types of behavior were observed, depending on the enone structure. A large group of enones, typified by 3-penten-2-one (31), were initially reported to be resistant to change on UV excitation[66], although later studies clearly showed that 31 undergoes efficient E–Z isomerization when irradiated at 313 or 238 nm in the vapor phase or at 254 or 313 nm in hexane or ether solution (equation 43)[67]. None of the deconjugation product 4-penten-2-one 32 was detected in the solution studies. The sum of the quantum efficiencies for $Z \rightarrow E$ and $E \rightarrow Z$ isomerization for 31 as well as for enone 33 was significantly less than 1.0, indicating that a twisted excited state common to both E and Z isomers cannot be an intermediate in the isomerization, if it is formed with unit efficiency from both isomers[67]. The photoisomerization on direct irradiation could not be quenched by piperylene (1, 3-pentadiene), stilbene or oxygen, and the quantum yields are significantly greater than for sensitized photoisomerization using propiophenone and acetophenone ($E_T = 74.6$ and 73.6 kcal mol^{-1}, respectively); no sensitization is observed using benzophenone ($E_T = 68.5$ kcal mol^{-1}). No fluorescence or phosphorescence of this or other simple acyclic enones has been observed. Thus, it is concluded that the photoisomerization on direct excitation involves singlet excited states which apparently are not sufficiently twisted that they are common to both isomers. The sensitization studies indicate an excitation energy of *ca* 70 ± 1 kcal mol^{-1} for the enone triplet.

Conjugated enones possessing a γ-hydrogen as well as at least one γ-alkyl group additionally undergo isomerization to a β, γ-unsaturated ketone, presumably via a dienol intermediate, as illustrated for 5-methyl-3-hexan-2-one (34) in equation 44[66]. In accord with this suggestion (equation 44), irradiation of 34 in CH_3OD led to 95% D-incorporation at C_3, presumably upon ketonization of dienol **35** (later studies to be discussed below establish this mechanism with virtual certainty). Certain enones such as **36** are not converted to their deconjugated isomers on direct excitation, although they do incorporate deuterium on irradiation in CH_3OD, indicating formation of a dienol isomer which gives exclusively the conjugated enone on ketonization[66].

(44)

Weedon and coworkers have recently reported a series of studies of dienol formation on irradiation of a large number of acyclic conjugated enones[68]. A clear pattern of photochemical reactivity in these systems has emerged from these studies. Thus, virtually all acyclic γ-alkyl-α, β-unsaturated ketones undergo intramolecular H transfer from the γ-carbon to the carbonyl oxygen (analogous to the Norrish Type II reaction); this singlet excited state process proceeds stereoselectively to give a (Z)-dienol (illustrated for enone **36** in Scheme 12). Quantum yields for this process are of the order of 10%. The (Z)-dienols can (a) be trapped as their trimethylsilyl ethers, (b) undergo a noncatalyzed 1,5-sigmatropic H shift to regenerate the starting enone or (c) reketonize under acid or base catalysis to give a β, γ-unsaturated ketone[65]. ^{1}H-NMR spectra taken at $-76\,°C$ of solutions of enone **36** in MeOH-d_4 irradiated in NMR tubes in an acetone/dry ice slurry show a new set of signals belonging to (Z)-dienol **37**; similar results were found for the photoconversion of **38** to **39** (Scheme 12)[69]. Conversion of enone to dienol is generally incomplete under these conditions, probably owing to overlap of UV absorption spectra of the tautomers. The dienols are cleanly reconverted to the starting enones when the solutions are brought from $-76\,°C$ to ambient temperatures; no deconjugation products are formed under these conditions, supporting Weedon's proposal that reketonization occurs via an uncatalyzed 1,5-hydrogen shift. The reversion process follows clean first-order kinetics whose temperature dependence yields the following activation parameters: for **37**, $A = 4 \times 10^8\,s^{-1}$ and $E_a = 15 \pm 1\,kcal\,mol^{-1}$; for **39**, $A = 1 \times 10^6\,s^{-1}$ and $E_a = 11 \pm 1\,kcal\,mol^{-1}$. These parameters should be compared with those for the (Z)-enol **41** derived from o-methylacetophenone **40**; E_a for reversion to **40** is $8.9\,kcal\,mol^{-1}$, very close

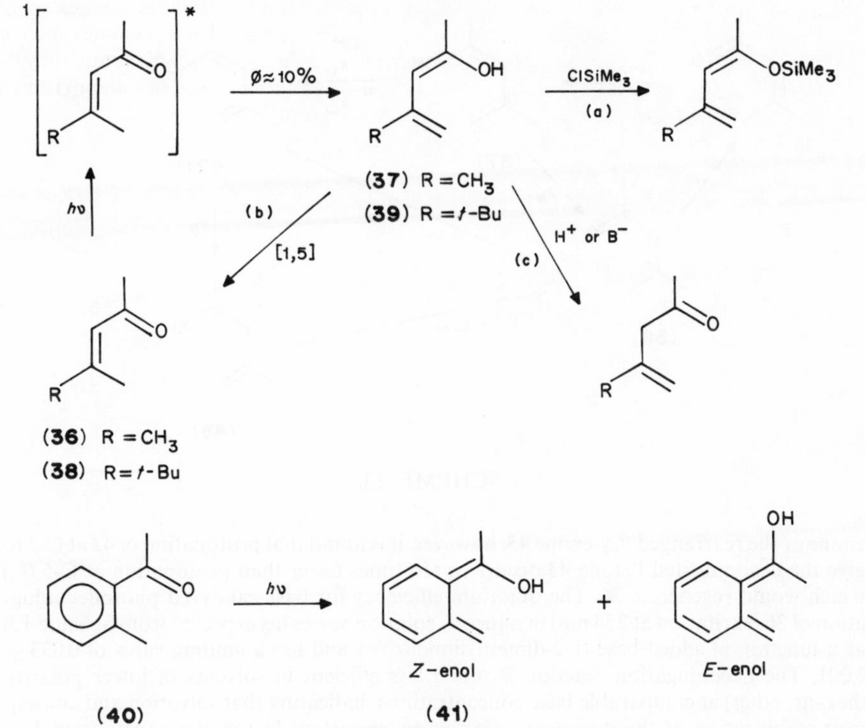

SCHEME 12

to that for **37** and **39**, but the pre-exponential factor in the case of **41** is much greater, $3 \times 10^{12} \, \text{s}^{-1}$. Weedon speculates that this may reflect the constrained cisoid geometry of **41** which optimizes the suprafacial orbital overlap required for the 1,5-hydrogen shift. Thus, the Z-dienols derived from acyclic enones are much longer-lived than their aromatic analogs, or even than simple enols which are relatively stable in the absence of acid and base catalysts. The latter situation can be ascribed to the difficulty of ketonization via a symmetry-allowed antarafacial 1,3-hydrogen shift, which is the only available mechanism in the absence of acids and bases. Attempts to trap dienols using reactive dienophiles in Diels–Alder reactions have thus far been unsuccessful. Such reactions have been successful for the relatively long-lived (E)-dienols derived from o-alkyl aromatic ketones such as **40**[70], but the corresponding (Z)-dienols (e.g. **41**) are too short-lived to permit interception by dienophiles.

When enones such as **36** are excited with a 20-μs UV pulse in aqueous basic solution, transients are produced with UV absorption maxima at ca 290 nm[71]. The transient absorption, which Weedon assigns to dienolate anions (**42** in Scheme 13), decays by clean first-order kinetics with a rate depending on the enone and the pH of the solution. The data indicate that equilibration of the dienol and dienolate is rapid compared with rates of ketonization (k_σ and k_β in Scheme 13) which vary with pH depending on the proportions of dienol and dienolate. Indeed, the variation of the first-order decay rate constant with pH resembles a titration curve, and these data can be used to obtain pK_as of the dienols. Thus, for **37** the pK_a is 10.42 ± 0.01. Protonation of the dienolate can give either the starting α, β-

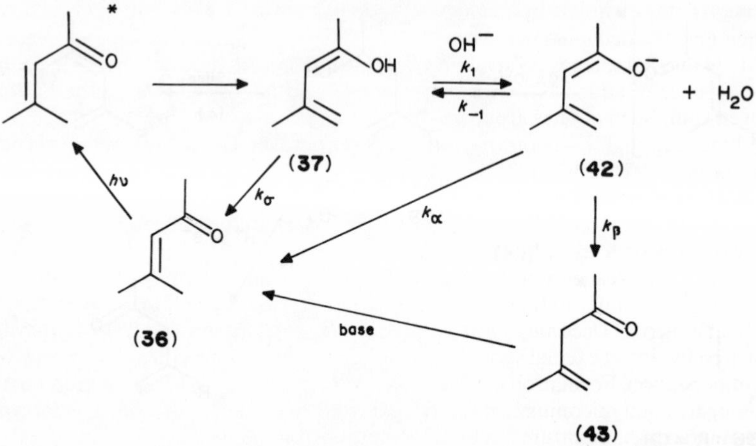

SCHEME 13

enone or the rearranged β, γ-enone 43; however, it is found that protonation of 42 at C-3 to give the deconjugated ketone 43 occurs *ca* ten times faster than protonation at C-5 (k_α) which would regenerate 36. The quantum efficiency for base-catalyzed photodeconjug-ation of 36 (excitation at 254 nm) in aqueous solution varies (as expected from Scheme 13) as a function of added base (1, 2-dimethylimidazole), and has a limiting value of 0.033 ± 0.001. The deconjugation reaction is much less efficient in solvents of lower polarity (hexane, ether) at comparable base concentrations, indicating that solvation and conseq-uent stabilization of the dienolate anion is an important factor; the uncatalyzed 1, 5-hydrogen shift dominates in nonpolar solvents. If the strength of the added base is decreased, deconjugation is inhibited since the equilibrium between dienol and dienolate is shifted toward the dienol, which ketonizes via the 1, 5-hydrogen shift. If the strength of the base is increased too much, the efficiency of photodeconjugation (e.g. in the presence of triethylamine) drops to zero. This is attributed to thermal base-catalyzed conversion of 43 to 36; indeed, rapid reconjugation of 43 occurs in the dark in the presence of triethylamine in methanol at a rate much faster than that of photodeconjugation under comparable conditions[71].

The reactions described for enone 36 are general, as Weedon and coworkers have demonstrated for a large series of acyclic and cyclic enones[68]. Some general conclusions regarding the effect of enone structure on the efficiency of deconjugation can be drawn from the data. Thus, a substituent in the γ-position to the carbonyl (as in 34, 44, 45 and 46) interferes with adoption of the cisoid (skewed or planar) conformation of the dienol required for the suprafacial 1, 5-hydrogen shift, thus increasing the opportunity for

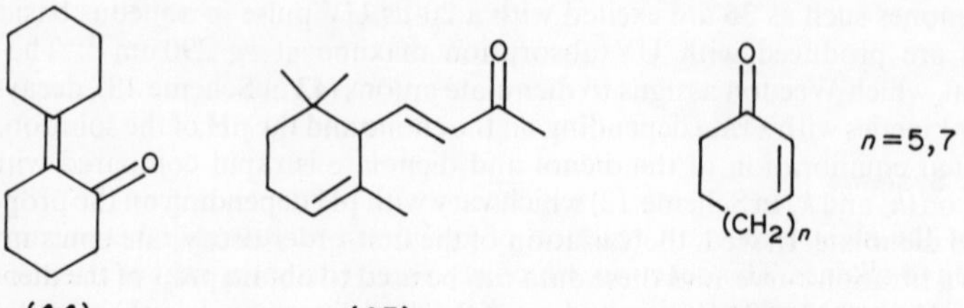

(44)   (45)   (46)

conversion to the dienolate by added base or the solvent itself; protonation of the dienolate can then give the deconjugated enone. When there is no substituent at the γ-carbon (as with **31**, **34** and **36**) there is no structural inhibition for formation of the (Z)-dienol, which (in the absence of added base) reverts exclusively to the conjugated enone by the 1, 5-hydrogen shift. Such enones are therefore inert to photodeconjugation in the absence of added base, although D-incorporation in deuteriated solvents indicates that dienols are indeed formed. The few exceptions to these generalizations can be rationalized on consideration of pertinent structural features in each system[68].

Photodeconjugation of α, β-unsaturated esters on irradiation at 254 nm in the presence of a weak base such as 1, 2-dimethylimidazole has also been reported by Weedon and coworkers[72]. This reaction, although technically outside the scope of this review, shows structural effects similar to those of the α, β-enones discussed above, and the mechanism is entirely analogous. Deconjugation again appears to involve intramolecular hydrogen abstraction by singlet excited states to give the corresponding (Z)-dienol, competitive with Z–E isomerization. Formation of the dienolate followed by protonation gives a mixture of the conjugated and unconjugated esters. For esters which are constrained with respect to Z–E isomerization, quantum yields for deconjugation approach 0.3.

Photodimerization of acyclic α, β-enones generally does not compete with the reactions discussed above, but there are a few exceptions. Ciamician and Silber reported photodimerization of dibenzylideneacetone **47** in solution to give the cyclobutane **48**[73], while later studies showed that uranyl chloride sensitized dimerization gave **49** (equation 45)[74]. There are several reports of photodimerization of chalcones Ar—CH= CH—CO—Ar' in solution as well as in the solid state; the former reactions have been assigned a triplet mechanism in accord with the extensive studies involving photodimerization of cyclic enones, to be discussed later[75].

(45)

B. Cyclic Systems

Perhaps the most important reaction of cyclic α, β-unsaturated ketones is photocycloaddition to alkenes. This reaction, which has received a great deal of attention recently with respect to mechanistic studies and synthetic applications, will be discussed separately below. The following discussion will first focus on other types of photoreactions of cyclic α, β-enones, grouped according to ring size.

1. Cyclopropenones and cyclobutenones

There are very few reports concerning the photochemistry of cyclopropenones and cyclobutenones. As shown in equation 46, cyclopropenones undergo fragmentation to give acetylenes and carbon monoxide[76], while cyclobutenones undergo ring opening to vinyl ketenes (equation 47), which can be detected by infrared spectroscopy when the irradiation is carried out at 77 K or lower temperatures[77].

$$R-C{\equiv}C-R + CO \qquad (46)$$

$$\qquad (47)$$

2. Cyclopentenones

a. Photodimerization. Photodimerization of cyclopentenone (CP) **50** gives a mixture of the *cis*-fused head-to-head and head-to-tail dimers **51** and **52** (equation 48)[78]. The reaction can be quenched by piperylene ($E_T = 57$–59 kcal mol^{-1}) and sensitized by xanthone ($E_T = 74.2$ kcal mol^{-1}) without affecting the ratio of the dimers, indicating they arise from a common triplet state precursor. The product ratio depends on solvent polarity, with the proportion of **51** increasing as solvent polarity or the concentration of enone (which is interpreted as a solvent effect) is increased, although **52** remains the major product under all conditions examined to date. The quantum efficiency for dimerization of 1.0 M **50** in acetonitrile is 0.34, and sensitization studies indicate that the inefficiency arises after triplet formation, i.e. $\phi_{isc} = 1.0$[79]. Wagner and Bucheck[79] argue that the reactive excited state of **50** is more likely a π, π^* than an n, π^* triplet state, and that the inefficiency arises from competitive decay to two ground-state enones from unidentified intermediates (collision complexes, π complexes, triplet 1, 4-biradicals) en route to dimers. This question will be considered later in more detail in connection with enone–alkene photocycloadditions.

$$\qquad (48)$$

(50) **(51)** (minor) **(52)** (major)

From quenching studies, Wagner and Bucheck[79] estimated lifetimes for CP triplets assuming quenching by piperylene is diffusion-controlled (a rate constant of $1.0 \times 10^{10} \, l \, mol^{-1} \, s^{-1}$ was assumed), from which they could obtain values for rate constants for capture of CP triplets by ground state CP ($6.6 \times 10^8 \, l \, mol^{-1} \, s^{-1}$) and for unassisted radiationless decay of the triplet ($4 \times 10^7 \, s^{-1}$), corresponding to a limiting triplet lifetime of 25 ns. Direct measurement of these quantities using nanosecond flash techniques by Heibel and Schuster[80] indicate that Wagner's rate constants are too high. The triplet lifetime (τ_T) of CP (50) is 130 ns in acetonitrile at 0.008 M, a concentration at which dimerization is insignificant, upon excitation at 355 nm using a Nd: YAG laser; the transient triplet decay was monitored at 300 nm (a full discussion of laser flash excitation of cyclic enones will be given below). This directly measured value of τ_T is significantly greater than Wagner's estimates[79] or Bonneau's earlier measurement of 30 ns at higher enone concentrations[81]. The plot of $(\tau_T)^{-1}$ vs. [CP] is linear (see equation 49)[80], where τ_0 is the triplet lifetime of CP at infinite dilution and k_a is the rate constant for interception of the triplet by ground state CP, which is found to be $1.2 \times 10^8 \, l \, mol^{-1} \, s^{-1}$, a factor of five less than Wagner's estimated value. The rate constant for quenching of CP triplets by 1-methylnaphthalene ($E_T = 61 \, kcal \, mol^{-1}$) in acetonitrile is $3.8 \times 10^9 \, l \, mol^{-1} \, s^{-1}$, considerably lower than the diffusion-controlled limit assumed by Wagner[79].

$$(\tau_T)^{-1} = (\tau_0)^{-1} + k_a[CP] \tag{49}$$

b. Photorearrangements. Irradiation of 5-substituted cyclopentenones 53 results in ring contraction to cyclopropylketenes 54, which are usually isolated as the esters 55 (equation 50)[82,83]. This transformation has been observed for a variety of compounds. The ketene can be directly detected by its characteristic IR absorption at $2110 \, cm^{-1}$ when reaction is carried out in pentane; addition of methanol gives the ester. The occurrence of α-cleavage in these systems is to be contrasted with the absence of such a pathway in the photochemistry of structurally analogous 6,6-disubstituted cyclohexenones. It is likely that this is a triplet state reaction.

$R^1 = R^2 = CH_3$ or Ph

$R^1 = H, R^2 = Et, Pr, OEt$

4-Acyl-2,5-di-t-butylcyclopentenones 56 rearrange to bicyclo[2.1.0]pentanones 57 (equation 51) on UV irradiation[84]. Isotopic labelling indicates that the reaction occurs by migration of the acyl group from C-4 to C-3 and formation of a new bond between C-2 and C-4. Mechanistically, this is an oxa-di-π-methane photorearrangement which is characteristic of β,γ-enones[5,6]. A related rearrangement involves acylcyclopentenone 58 which rearranges to the butenolide 59, presumably via the bicyclo[2.1.0]pentanone 60 and ketene 61 (equation 52)[85]. By analogy with the photo-chemistry of β,γ-enones, it is likely that these reactions proceed via triplet excited states, although this has not been demonstrated.

A related photorearrangement occurs with 62. In this case, a phenyl shift would give the bicyclo[2.1.0]pentanone 63, which on ring opening would give ketene 64, the source of the isolated product **65** (equation 53)[86,87]. This reaction is structurally analogous to the chemistry of 4-arylcyclohexenones discussed later. As in that case, triplet intermediates are implicated by sensitization and quenching studies.

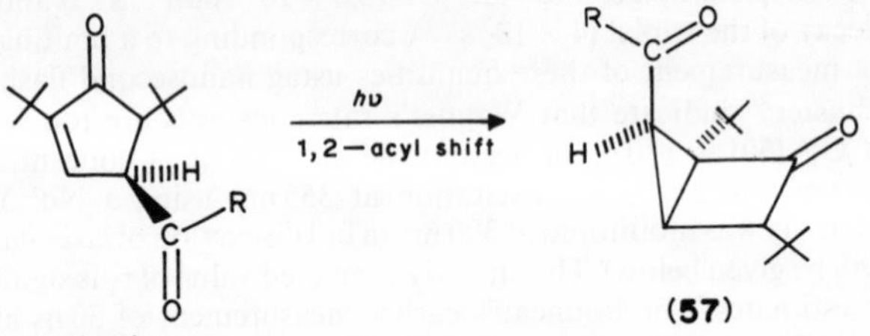

(56) **(57)** (51)

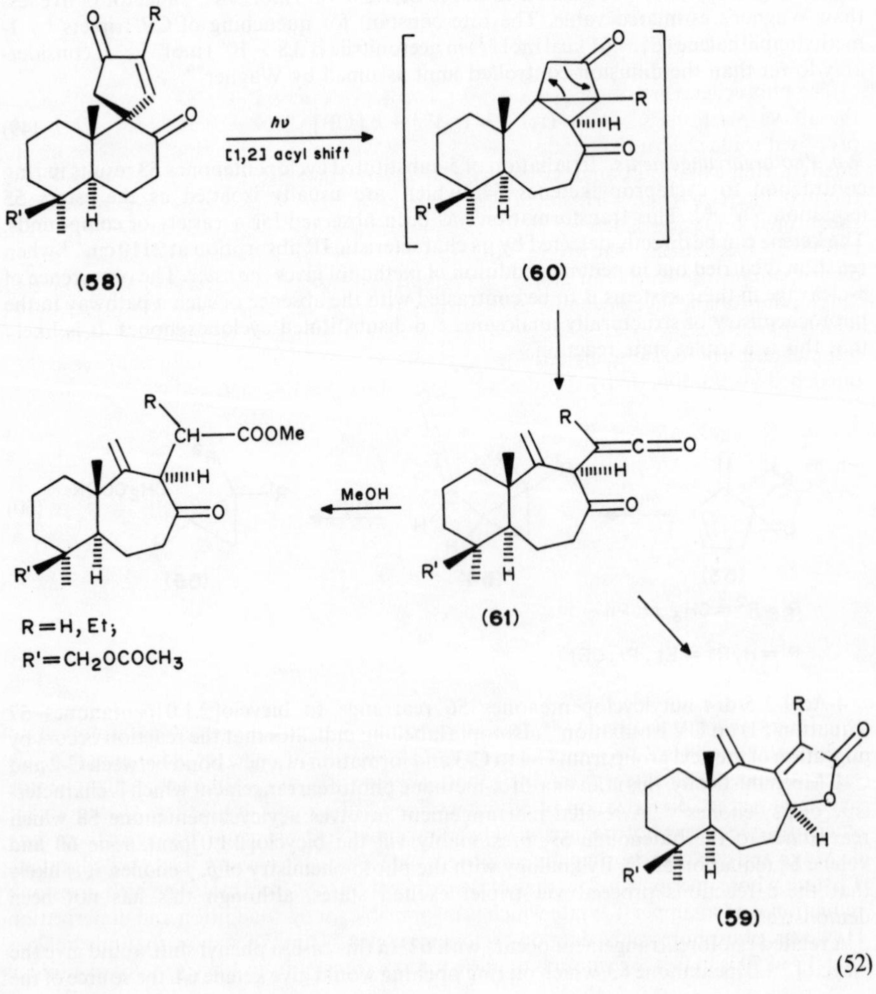

(58) **(60)** **(61)** **(59)**

R=H, Et; R'=CH_2OCOCH_3

(52)

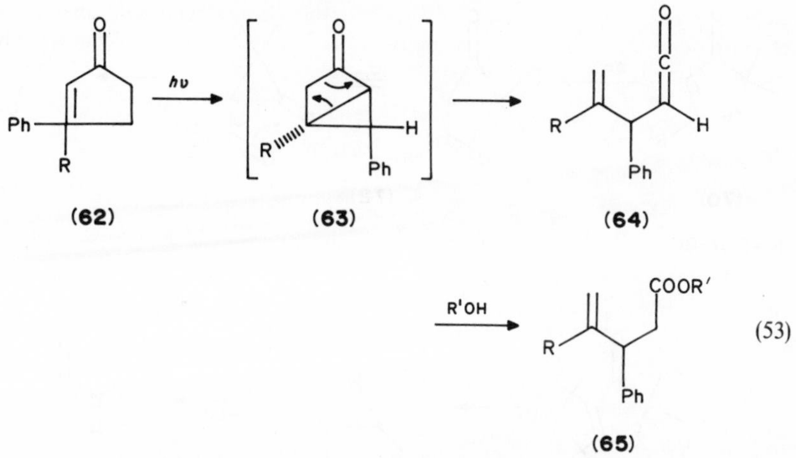

(62)   **(63)**   **(64)**

$$(53)$$

(65)

The photochemistry of simple derivatives of 3(2H)-furanones shows some analogies to the above reactions[88]. Thus, the 2,5-diphenyl derivative **66** rearranges to **67** by the proposed route shown in equation 54, involving α-cleavage to **68**, ring closure to **69** and finally a ring expansion analogous to the well-known vinylcyclopropane–cyclopentene thermal interconversion. However, completely different behavior is seen with alkyl-substituted furanones such as **70** which photorearrange cleanly to lactones **71**. The proposed mechanism, shown in equation 55, involves initial isomerization to cyclopropanone **72**, which can give **71** directly on ring expansion; each step represents a vinylcyclopropane–cyclopentene interconversion. Supporting evidence derives from **73**, which photodecarbonylates to give **74**, a reaction which is believed to result from fragmentation of the sterically crowded cyclopropanone **75**. These reactions are efficiently quenched by 2,3-dimethyl-1,3-butadiene, suggesting that they proceed via enone triplets.

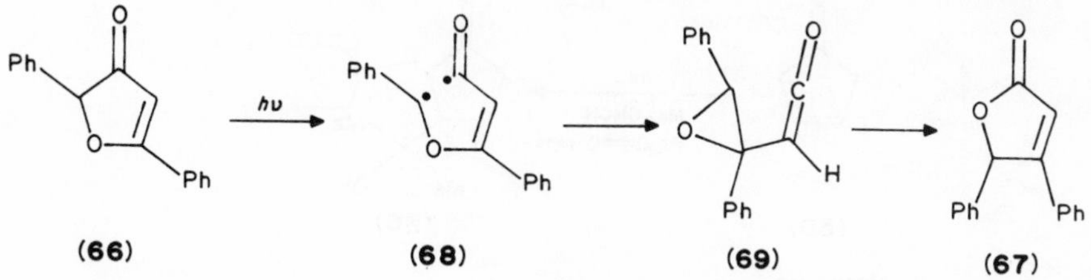

(66)   **(68)**   **(69)**   **(67)**

$$(54)$$

c. Inter- and intramolecular hydrogen abstraction. Irradiation of dilute solutions of cyclopentenone **50** in 2-propanol gives an adduct **76** in addition to the usual dimers[89]. The same adduct **76** is formed on benzophenone sensitization, although benzophenone sensitizes neither dimerization of **50** nor cycloaddition of **50** to alkenes. Moreover, CP efficiently quenches photoreduction of benzophenone in 2-propanol. Based on these data, de Mayo and coworkers originally proposed that CP reacts via two triplet excited states, a T_1 state whose energy is below that of benzophenone (68.5 kcal mol^{-1}) which gives only reduction, and an upper T_2 state which is responsible for cycloaddition and dimerization. However, phosphorescence data on cyclopentenones **77** and **78**[90] show that it is unlikely that T_1 of **50** is in fact low enough to allow energy transfer from Ph$_2$CO to be as rapid as

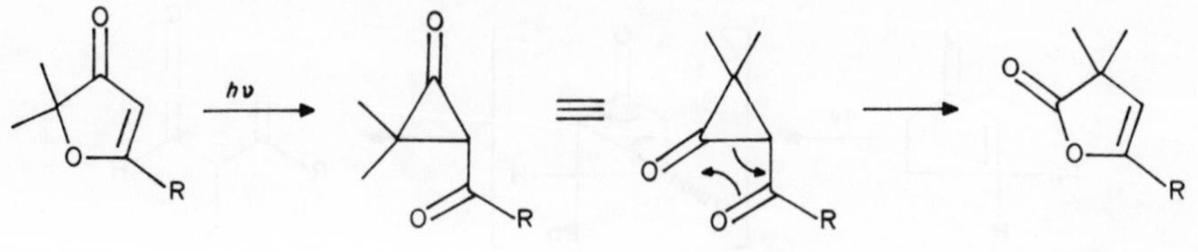

(70)   **(72)**   **(71)**

R=Et, t-Bu

(55)

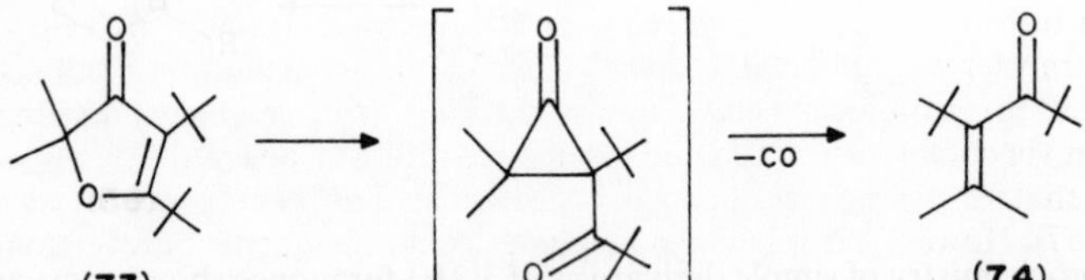

(73)   **(74)**

(75)

required by de Mayo's data[89]. This anomaly was resolved by invoking the mechanism shown in Scheme 14 involving so-called 'chemical sensitization', in which the species quenched by CP is not benzophenone triplet excited state but rather the diphenylketyl radical. As indicated earlier, hydrogen transfer from ketyl radicals to ground-state ketones is a well-documented reaction. This type of 'sensitization' has to be considered in situations where triplet energy transfer is unlikely for energetic reasons.

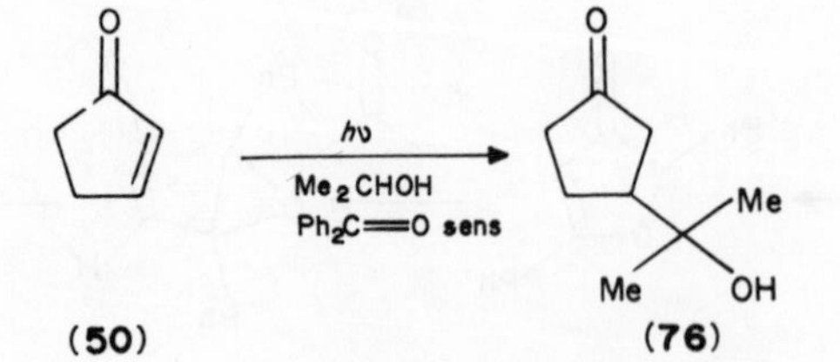

(50)   **(76)**

Mechanism

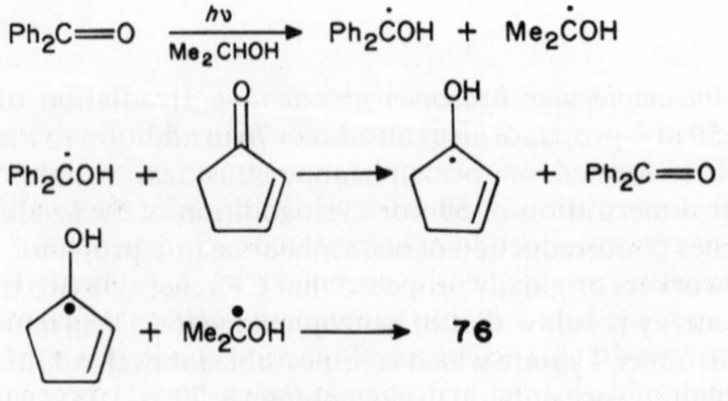

$$Ph_2C{=}O \xrightarrow[Me_2CHOH]{h\nu} Ph_2\overset{\bullet}{C}OH + Me_2\overset{\bullet}{C}OH$$

SCHEME 14

(77) (78)

Irradiation of the 4-substituted cyclopentenone **79** in benzene gives ketones **81a**, **81b** and **81c**, which are logically derived from the diradical **80** formed by hydrogen transfer from the side-chain to the β-carbon of the enone (equation 56)[91]. Agosta and coworkers used deuterium-labeled compounds to demonstrate that 1, 5-hydrogen transfer via a six-membered transition state is preferred over 1, 6-hydrogen transfer; no evidence for 1, 4-hydrogen transfer was obtained. Using **82** in which the diastereotopic methyl groups were distinguished by isotopic labeling, it was possible to discriminate between hydrogen abstraction via conformation **82a** and **82b** to give diradical **83a** and **83b**, respectively[92]. It was found that 92% of the reaction to give the indicated products proceeds via **82a** and **83a** (equation 57). However, it was not possible to assess the degree of reversion to starting materials from the diradicals, nor the extent to which nonvolatile products (totaling 35%) derive from one or the other biradical.

(79) (80)

(81a) (81b) (81c)

(56)

In these and related systems, no hydrogen transfer to the α-carbon is observed, and all the data are consistent with exclusive hydrogen transfer to the β-carbon of the enone. Irradiation of 4, 4-dimethylcyclopentenone in t-butyl alcohol gives exclusively the 2-t-butoxy adduct, again indicating hydrogen abstraction at the β-carbon followed by radical coupling (a Michael-type nucleophilic photoaddition should occur at the β-carbon, as is observed with some substituted cyclohexenones)[82]. The intramolecular hydrogen abstractions are efficiently quenched by 2, 3-dimethyl-1, 3-butadiene ($E_T \sim$ 60 kcal mol^{-1}) and sensitized by propiophenone ($E_T = 74$ kcal mol^{-1}), pointing to a triplet excited intermediate. The course of reaction suggests a π, π^* rather than an n, π^* triplet; the latter should abstract hydrogen at the carbonyl oxygen (see above) leading to radical coupling at the β-carbon, which is not observed.

$$(57)$$

A recent study concerns cyclopentenone **84**, which is converted on UV excitation into the fused tricyclic ketones **85a** and **85b** as shown in equation 57a. Once again, this is consistent with initial hydrogen abstraction at the β-carbon of the enone to give biradical **85** which, on coupling, gives the observed products[93].

$$(57a)$$

That hydrogen abstraction by cyclopentenones does not always occur at the β-carbon is shown by the fact that direct excitation of CP in cyclohexane gives both 2- and 3-cyclohexylcyclopentanones, perhaps due to reaction via both n,π^* and π,π^* triplets which, as mentioned previously, should have similar excitation energies[78b,79,91,92,94].

3. 2-Cycloheptenones and 2-cyclooctenones

Before discussing the complex photochemistry of cyclohexenones, it is useful to first consider the photochemical behavior of medium ring cyclic enones, particularly cycloheptenones and cyclooctenones, which illustrate the possibilities for reaction of electronic excited states in flexible as opposed to rigid ring systems.

a. Cis–trans isomerization. Based upon the fact that simple alkenes, medium and large ring cycloalkenes ($n \geqslant 8$) and acyclic enones all undergo Z–E (*cis–trans*) isomerization, it was reasonable to investigate whether medium ring cyclic α,β-enones also undergo this reaction. Eaton and Lin first reported the conversion of *cis* cyclooctenone **86** to the *trans* isomer **87** on UV (> 300 nm) irradiation in cyclohexane (see equation 58), as detected by loss of the UV absorption of **86** at 223 nm, shift of the $n \rightarrow \pi^*$ absorption λ_{max} from 321 to 283 nm, and appearance of a new IR band for C=O absorption at 1727 cm^{-1} in place of the original band at 1675 cm^{-1} [95]. Other new IR bands observed are similar to those found in *trans* but not *cis* cyclooctene. Since only *ca* 80% conversion of **86** was observed, Eaton and Lin concluded that under the conditions of the experiment photoequilibration of **86** and **87** was achieved. Evidence in support of photochemical formation of **87** was obtained by isolation of *trans*-fused Diels–Alder adducts **89** and **90** upon reaction of the product of irradiation of **86** with diene **88** (equation 58), and the fact that dienes that react sluggishly with **86** (such as cyclopentadiene and furan) react readily with the presumed *trans* enone **87**. Furthermore, the *trans* enone **87** dimerizes in the dark at room temperature, although the structures of the dimers have never been reported.

(58)

Related to the above results is the observation that irradiation of acetylcyclooctene **91** in the presence of cyclopentadiene (CPD) gives the *trans*-fused [4 + 2] adduct **92**, and the fact that the same product is isolated upon addition of CPD (in the dark) to a solution of **91** after UV irradiation [96]. Thus, the adduct **92** would appear to arise from thermal addition of CPD to the *trans* enone **93** (equation 59). Eaton indeed detected a new material assumed to be **93** on excitation of **91** at room temperature as well as dry ice temperatures, but details of this study were never reported.

Shortly thereafter the Corey and Eaton groups both reported the detection of *trans*-2-cycloheptenone **95** from irradiation of the *cis* isomer **94** at low temperatures ($-160\,°$C to $-195\,°$C) using either a thin film of **94** or a dilute solution in 95:5 cyclohexane-isopentane [97, 98]. The main evidence in support of the structure of **95** was the characteristic low-temperature IR spectrum, featuring C=O absorption at 1715 cm^{-1} (vs. 1664 cm^{-1} for **94**) and other spectral shifts consistent with conversion of **94** to **95**. The new absorption bands persisted if the samples were kept at temperatures below $-160\,°$C. However, if the frozen samples were warmed slowly to $-120\,°$C or higher, the IR absorption bands

(59)

assigned to **95** completely disappeared, and the bands characteristic of **94** reappeared with reduced intensity, superimposed on absorption bands of cycloheptenone dimers (see below). If **94** is irradiated in the presence of CPD or furan, *trans*-fused Diels–Alder adducts **96–99** are formed in good yield (equation 60). The adduct **96** was also obtained on irradiation of **94** in glassy methylcyclohexane at − 190 °C followed by treatment in the dark with a cold solution of CPD in pentane, and subsequent warming. These results support the contention that the reactive intermediate produced on irradiation of **94** is indeed a ground-state *trans* cycloheptenone **95** in which conjugation between the C=O and C=C moieties is sharply reduced vis-à-vis the corresponding *cis* isomer **94**.

(60)

The formation of a *trans* cycloheptenone was confirmed using laser flash techniques[99]. Flash photolysis of **94** produced a transient species with λ_{max} 265 nm with a lifetime of 45 s in cyclohexane but much shorter lifetimes in alcoholic solvents (74 ms in EtOH, 33 ms in MeOH). The reduced transient lifetime in alcohol solutions reflects nucleophilic attack by alcohols on **95** (see below), analogous to reaction of alcohols with *trans* cycloalkenes discussed in Section III.A.3. The transient decay in cyclohexane is first order at low excitation energies, but at higher energies corresponding to larger concentrations of the transient the decay is mixed first and second order, which suggests that at least a component of photodimerization (at least at high excitation energies) involves interaction of two *trans* cycloheptenones. In polar and protic solvents, the transient decay is mainly first order, due to reaction with the solvent (see below). From the temperature dependence of the rate of decay of the transient in cyclohexane solution, an activation energy of $15.2 \pm 0.5\, \text{kcal mol}^{-1}$ and a pre-exponential factor of $2 \times 10^9\, \text{s}^{-1}$ were determined by Bonneau and coworkers[99]. The much lower value of the activation energy for thermal isomerization of **95** to **94** determined by Goldfarb[100] was rationalized by Bonneau[99] as a reflection of photoinduced *trans* → *cis* isomerization caused by the analyzing light source. Figure 8 shows the approximate potential surfaces for the ground and excited states of 2-cycloheptenone proposed by Bonneau and coworkers[99].

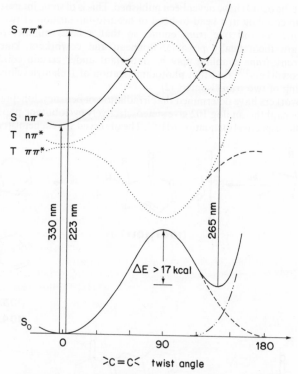

FIGURE 8. Approximate potential surfaces of the ground and excited states of 2-cycloheptenone. Reproduced by permission of Gantier Villars from Ref. 99

By analogy with the mechanism of $Z-E$ photoisomerization of acyclic α, β-enones, it has been assumed that the isomerizations of **86** to **87** and **94** to **95** proceed via triplet excited states. Bonneau[81] observed a very short-lived (11 ns) transient on flash excitation of **94** in cyclohexane at 353 nm, the absorption spectrum of which was similar to that of **95**. The 11 ns transient can be quenched by oxygen but not (at least not efficiently) by piperylene. Bonneau speculates that this species is a highly twisted π, π^* triplet excited state of **94**, represented by the minimum in the $T_{\pi\pi^*}$ potential curve shown in Figure 8, whose very short lifetime can be understood in terms of the small energy difference between the triplet excited state and ground-state potential-energy surfaces at (or close to) a C=C twist angle of $90°$ [81]. The closer the approach of these two surfaces, the better the coupling of the ground and excited states, resulting in more rapid radiationless decay. The dynamics associated with the surface crossing and considerations of momentum of the molecule as it passes through the 'funnel' on the triplet surface[101] suggest that formation of the ground-state *trans* enone may be facilitated over return to the ground-state *cis* enone.

b. Photodimerization. In all the papers on the photochemistry of cycloheptenone **94** from the earliest until the present, formation of enone dimers has been observed under almost all reaction conditions in a wide variety of solvents. In their 1965 paper on *trans* cycloheptenone, Eaton and Lin[98] indicate that the structures of the dimers were determined, but the details have never been published. This is of some interest, since dimers could arise from coupling in a head-to-head or head-to-tail fashion of two *trans* enones and/or from one *cis* and one *trans* enone, so that a large number of regio- and stereoisomers are theoretically possible. Bonneau and coworkers' kinetic studies[99] indicate that *trans–trans* coupling may be important under certain conditions, while Caldwell and coworkers[102] find that photodimerization of 1-phenylcyclohexene mainly involves coupling of two *trans* isomers (2).

Hart and coworkers have determined that irradiation of benzocycloheptadienones **100** and **101** and the naphtho analog **102** give stereoselectively cyclobutane photodimers **103**, **104**, **105** and **106**, respectively (equation 61)[103]. The observed stereochemistry is consistent

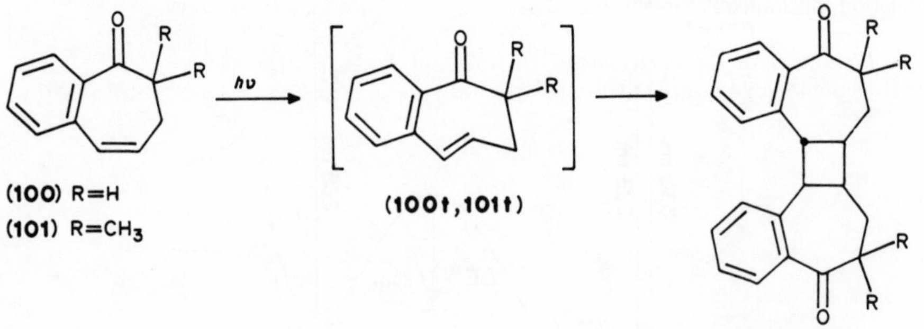

(100) R=H
(101) R=CH₃

(100†,101†)

(103) R=H
(104) R=CH₃

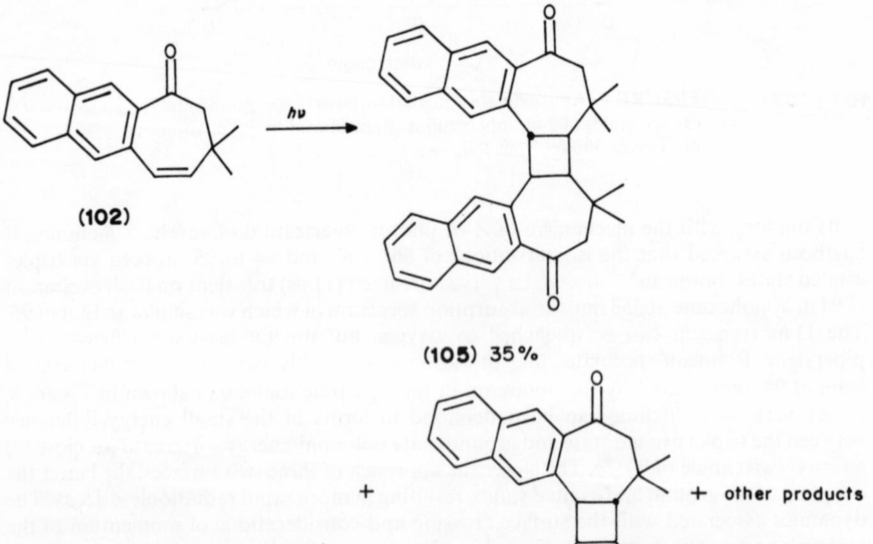

(102)

(105) 35%

+

+ other products

(106) 15%

(61)

with concerted ground-state dimerization of two *trans* cycloalkenones in a symmetry-allowed $_\pi2_s + _\pi2_a$ manner. Support for photogeneration of *trans* cycloalkenones is provided by the formation of *trans*-fused [4 + 2] adducts of these and several other cycloheptadienones upon irradiation in furan. A different mode of photodimerization is seen with **107** which gives **108** and **109**. This reaction course is rationalized as seen in Scheme 15 by addition of the *trans* isomer of **107** (**107 t**) to the styrene moiety of the starting enone, followed by suprafacial 1, 3-acyl shifts to give the isolated products. Hart suggests that even the cyclobutane-type photodimers as in the case of **101** may arise by initial cycloaddition of the *trans* enone (**101t**) to the styryl moiety, followed by a 1, 3-shift (see Scheme 16), in which case it would not be necessary to postulate two completely different reaction mechanisms for photodimerization of structurally similar molecules.

 c. Photoaddition of nucleophiles. Noyori and Kato[104] found that irradiation of cycloheptenone **94** in protic solvents (alcohols, acetic acid, aqueous acetonitrile, diethy-

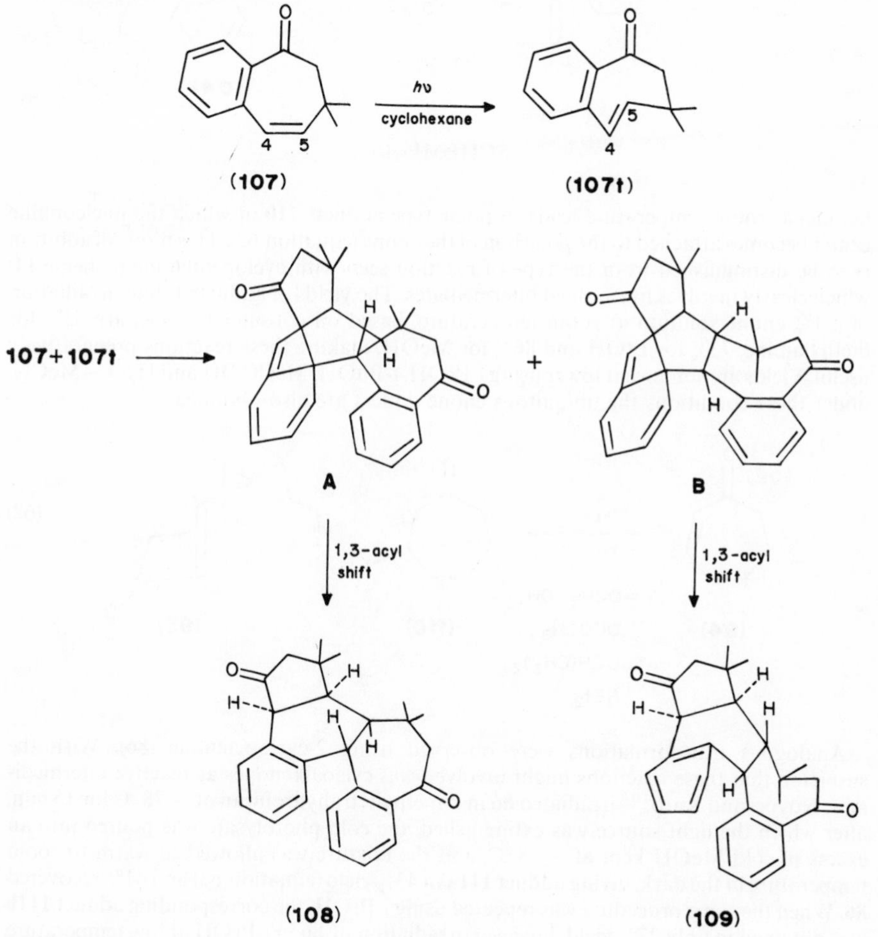

SCHEME 15

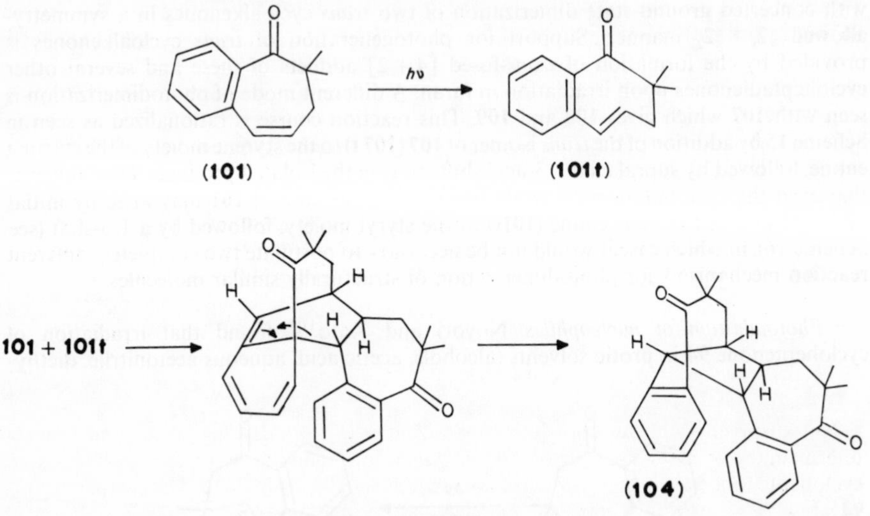

(101) $h\nu$ →   **(101†)**

101 + 101† → → **(104)**

SCHEME 16

lamine) at room temperature leads to polar-type adducts **110** in which the nucleophilic center becomes attached to the β-carbon of the enone (equation 62). This mode of addition is to be distinguished from the type of reaction seen with cyclopentenone (Scheme 14) which clearly involves free radical intermediates. The yields of adducts **110** on irradiation of a 1% enone solution at room temperature, based on consumed enone, are 55% for diethylamine, 73% for EtOH and 86% for MeOH, making these reactions preparatively useful. Yields are somewhat lower using i-PrOH, t-BuOH, MeCOOH and H_2O—MeCN; under these conditions the ubiquitous enone dimers are also obtained.

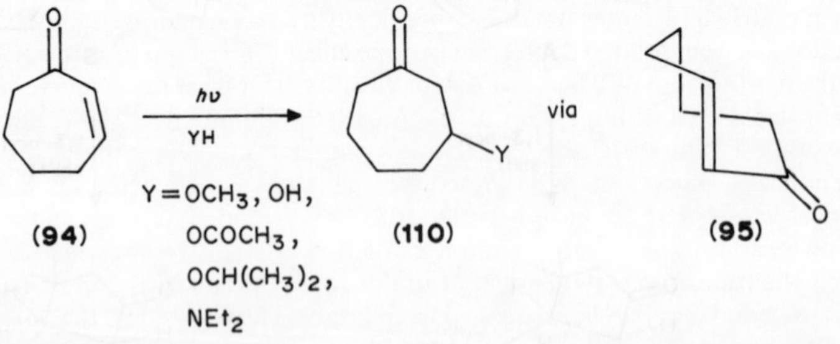

$$\text{(62)}$$

$$Y = OCH_3, OH,$$
$$OCOCH_3,$$
$$OCH(CH_3)_2,$$
$$NEt_2$$

(94)   **(110)**   **(95)**

Analogous transformations were observed using 2-cyclooctenone (**86**). With the suspicion that these reactions might involve *trans* cycloalkenones as reactive intermediates, Noyori and Kato[104] irradiated **86** in 2-methyltetrahydrofuran at $-78\,^\circ$C for 15 min, after which the light source was extinguished, the cold photolysate was poured into an excess of cold MeOH kept at $-78\,^\circ$C, and the mixture was allowed to warm to room temperature in the dark, giving adduct **111a** in 43% yield (equation 63) and 41% recovered **86**. When the same procedure was repeated using i-PrOH, the corresponding adduct **111b** was obtained in only 27% yield; however, irradiation of **86** in i-PrOH at low temperature followed by treatment with a large excess of MeOH gave almost exclusively **111a** and only a trace of **111b**, demonstrating that these alcohols are not reacting with an excited state of **86**, but rather with a long-lived reaction intermediate, probably *trans* cyclooctenone **87**.

(63)

Not surprisingly, using the same approach it was more difficult to demonstrate the intermediacy of *trans* cycloheptenone **95** in the photoadditions of nucleophiles to *cis* cycloheptenone **94**, due to the much shorter lifetime of **95** vis-à-vis **87**. Thus, irradiation of **94** at −78 °C in liquid nitrogen in EPA (ether–pentane–alcohol glass), addition of cold MeOH in the dark and gradual warming to room temperature gave only enone dimers and no MeOH adducts. Irradiation of **94** at −196 °C in MeOH followed by warming also failed to produce MeOH adducts. However, substitution of diethylamine for methanol in the former experiment led to formation of adduct **110** (Y = NEt_2) in 25% yield; no thermal reaction of **94** and Et_2NH was observed under similar conditions[104].

When the ring size is expanded to nine (*cis*-2-cyclononenone), the *trans* enone is stable enough to be isolated and survives treatment with MeOH at 0 °C, although addition occurs when the solution is heated at 100° C. However, neither *cis*- nor *trans*-2-cyclododecenone show any reactivity toward nucleophiles even under these forcing conditions.

Hart and coworkers have determined the stereochemistry of photoinduced addition of methanol to **86**, **94** and a number of fused benzo analogs using CH_3OD[105]. Photoaddition places the methoxy and deuterium stereospecifically *trans*, a reaction course observed with benzo analogs as well. A large deuterium isotope effect is observed in 1:1 MeOH/MeOD, favoring the light solvent by a factor of 4.4 for **94** and 6.0 for **86** at room temperature. Thus, proton transfer is clearly important in the rate-determining step. The results require a regio- as well as stereospecific reaction mechanism involving the respective *trans* cycloalkenones as key reaction intermediates, as shown in Scheme 17. Basically, the authors postulate *syn* addition of MeOH(D) to the ground-state *trans* enone, involving either stepwise addition via the dipolar ion Z or a concerted process in which Z or a similar structure is the transition state. Note that in this highly twisted structure, one face of the twisted C=C bond is completely shielded from attack. Hart considers the possibility that Z might relax conformationally to Z′ to permit charge delocalization prior to protonation, which might be expected to lead to nonstereospecific protonation (or deuteration). It was determined that base-catalyzed Michael addition to these enones in fact also proceeds in a stereospecifically *trans* manner, presumably via an anion analogous to Z′. Thus, reaction via Z′ cannot be ruled out, although the *syn* addition mechanism of Scheme 17 is clearly very attractive.

A somewhat different course of reaction is taken by benzocyclooctadienones such as **112**[106]. Irradiation in methanol results in transannular reaction to give **113** and its dehydration product **114**. Hart again envisages initial formation of a *trans* isomer of **112** (i.e. **112t**), which then reacts as shown in Scheme 18; the formation of only one

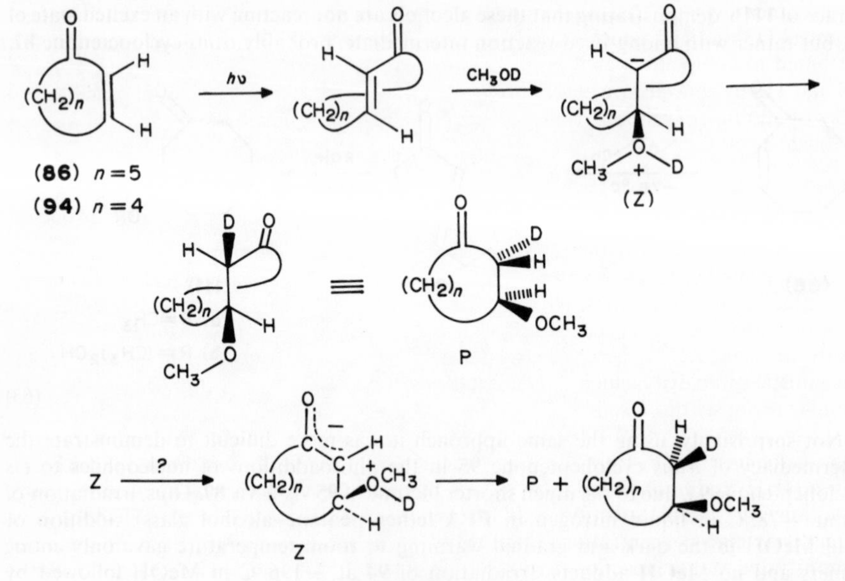

(86) $n = 5$
(94) $n = 4$

SCHEME 17

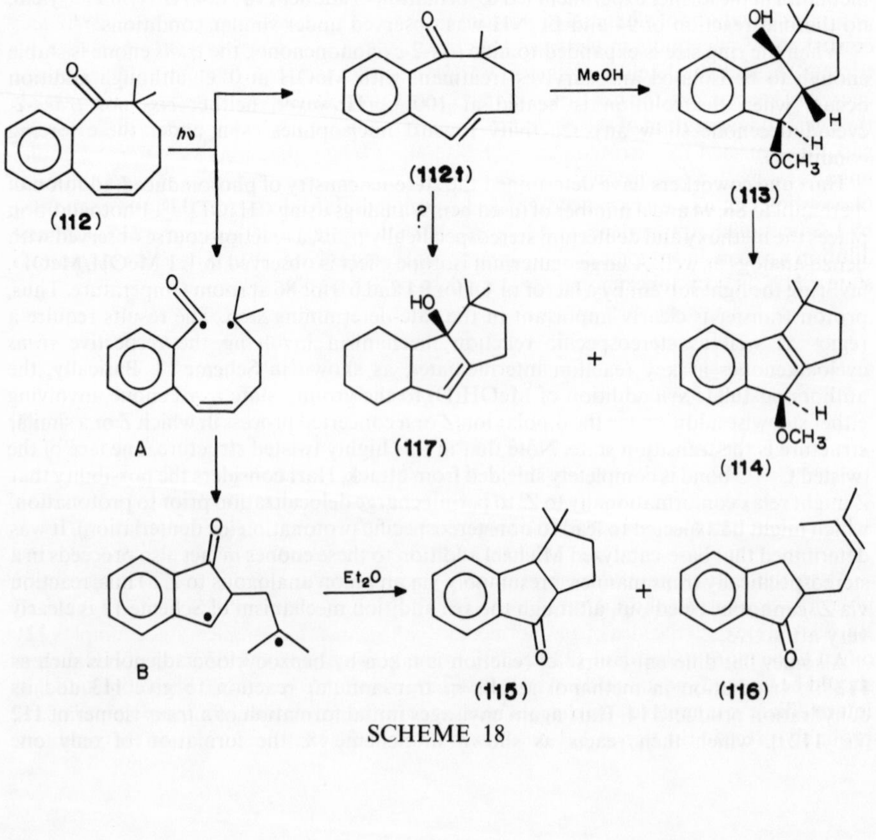

(112)

(112†)

(113)

A

(117)

(114)

B

(115)   **(116)**

SCHEME 18

stereoisomer suggests that nucleophilic attack and ring closure may be synchronous. Products 115 and 116, which are also formed along with 117 on irradiation in ether, are attributed to competitive α-cleavage to biradical A, cyclization to B, and formation of 115 and 116 by ring closure and hydrogen transfer, respectively. Product 117 most likely arises by addition of water to trans-112 and dehydration, although the mechanism was not established.

4. 2-Cyclohexenones

The photochemistry of cyclohexenones, particularly substituted systems, is especially rich and complicated compared with the photochemistry of acyclic enones and cyclic enones with larger and smaller rings. Nonetheless, the similarities as well as the differences can often be understood as effects of ring size as opposed to fundamental differences in the electronic structure of the chromophore itself. Extensive recent investigations reveal mechanistic complexity which does not appear to exist in the photochemistry of the α, β - enones previously discussed.

a. Photodimerization. Photodimerization of cyclohexenone itself (118) to give the head-to-head (HH) and head-to-tail (HT) dimers 119 and 120 has been known for many years (equation 64)[107]. Most substituted cyclohexenones also undergo this reaction in solution at relatively high concentrations ($\geqslant 0.2$ M). Classic sensitization and quenching studies demonstrated that the reaction involves a triplet state of 118 lying *ca* 70 kcal mol^{-1} above the ground state, which was concluded to be the lowest-energy triplet state of the enone[79,107]. The configuration of the triplet was assigned as $^3\pi, \pi^*$ by analogy to photodimerizations of alkenes (see above), on the basis of calculations by Zimmerman and coworkers of differences in electron densities on the C=C bond in n, π^* vis-à-vis π, π^* triplets[108], and the likelihood that twisting around the C=C bond would lower the energy of the π, π^* vs. the n, π^* triplet. Particularly in polar solvents, it was proposed that the energetic separation of the two states would be at least a few kcal mol^{-1}, although the gap was expected to narrow in nonpolar solvents where reactions via the n, π^* triplet might be expected (see below)[79]. From Wagner and Bucheck's studies of the kinetics of photo-dimerization of 118 in acetonitrile, assuming diffusion-controlled triplet quenching by 1, 3-pentadiene (piperylene) and 1, 3-cyclohexadiene, the triplet lifetime of 118 at infinite dilution was concluded to be *ca* 2 ns, and the rate constant for capture of ^{3}CH* by ground-state CH (CH = cyclohexenone) was found to be 1.1×10^8 l mol^{-1} s^{-1} [79].

$$(64)$$

(118)   (119)   (120)

As in the case of photodimerization of cyclopentenone, there is an effect of solvent polarity on the ratio of dimers 119 and 120. The lack of regiospecificity led Wagner and Bucheck[79] to reject the idea of an intermediate charge transfer complex, since complex 121 ought to be more stable than 122, leading to the prediction that formation of HH dimers should be favored substantially over HT dimers, contrary to the facts. They conclude that intermediate π complexes or charge-transfer complexes, with differing dipole moments, probably precede the triplet 1, 4-biradicals which are direct precursors of the products.

The quantum yields (0.20 at 1 M CH in acetonitrile) indicate that significant percentages of these biradicals fragment to regenerate ground-state enone[79].

(121)   (122)

As mentioned above, photodimerization of cyclohexenones is quite general. Isophorone **123** yields three photodimers (equation 65), and once again the ratio of the HH dimer to the two stereoisomeric HT dimers varies as a function of solvent polarity[109]. Mechanistic

(65)

complexities are suggested by the following observations of Chapman and coworkers: (a) plots of $(\phi_{dim})^{-1}$ vs. [isophorone]$^{-1}$ in acetic acid give straight lines with significantly different slopes and intercepts for HH and HT dimerization; (b) identical linear Stern–Volmer plots for quenching of both modes of dimerization by isoprene or ferric acetylacetonate are obtained, but differential quenching is observed using di-t-butyl nitroxide; (c) the ratio of HH vs. HT dimerization is different on benzophenone sensitization (benzophenone absorbed *ca* 32% of the incident light) than on direct irradiation of **123**. The last observation in particular led Chapman to propose that two different triplet states of **123** are responsible for HH vs. HT photodimerization; if only one triplet were involved, the reaction course ought to be the same on direct or triplet-sensitized excitation, unless there was some anomaly associated with benzophenone photosensitization. The latter might be a possibility if the triplet excitation energy of benzophenone ($E_T = 68.5$ kcal mol^{-1}) were less than that of **123**. As indicated above, Wagner concluded that for CH itself E_T is probably > 70 kcal mol^{-1}[79], so that triplet energy transfer from benzophenone to **123** might be uphill, which could introduce other mechanisms for sensitization (e.g. Schenck-type processes as discussed earlier). In other studies of cyclohexenones to be described below, higher-energy triplet sensitizers were used and product ratios were the same as on direct enone excitation. Results (a) and (b) above are compatible with a single triplet precursor for both HH and HT dimers assuming the kinetic scheme given in Scheme 19[109]. The key point is that distinctly different double reciprocal plots of quantum yield vs. enone concentration, as in (a) above, will be observed if there are distinctly different rate constants k_a and k_a' for formation of metastable intermediates (whether they be π complexes or biradicals) en route to HH and HT dimers, and different factors ϕ_p and ϕ_p' for the fractions of these adducts which proceed on to dimers in competition with reversion to enone ground states. If HH and HT dimers arose from a common enone triplet, triplet quenching should alter the yield but not the ratio of the dimers, as indeed seen in (b)[109].

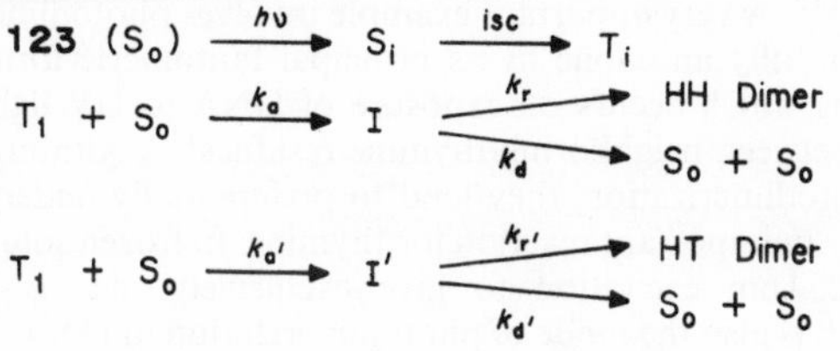

$$123 \; (S_0) \xrightarrow{h\upsilon} S_i \xrightarrow{isc} T_i$$

$$T_1 + S_0 \xrightarrow{k_a} I \begin{array}{c} \xrightarrow{k_r} \text{HH Dimer} \\ \xrightarrow{k_d} S_0 + S_0 \end{array}$$

$$T_1 + S_0 \xrightarrow{k_{a'}} I' \begin{array}{c} \xrightarrow{k_{r'}} \text{HT Dimer} \\ \xrightarrow{k_{d'}} S_0 + S_0 \end{array}$$

$$\varnothing_p = \frac{k_r}{k_r + k_d}$$

$$\varnothing_p' = \frac{k_{r'}}{k_{r'} + k_{d'}}$$

SCHEME 19

Photodimerization of 4,4-dimethylcyclohex-2-en-1-one **124** has been studied by Nuñez and Schuster[110]. Three dimers are formed upon irradiation of neat enone, two of which were formed in sufficient quantity to allow structure determination as the HH dimer **125** and the HT dimer **126**; the third (trace) dimer appeared to isomerize to **126** upon prolonged standing at room temperature and was therefore tentatively assigned structure **127** (equation 66). As with isophorone (**123**), plots of ϕ_{HH}^{-1} and ϕ_{HT}^{-1} vs. [enone]$^{-1}$ were

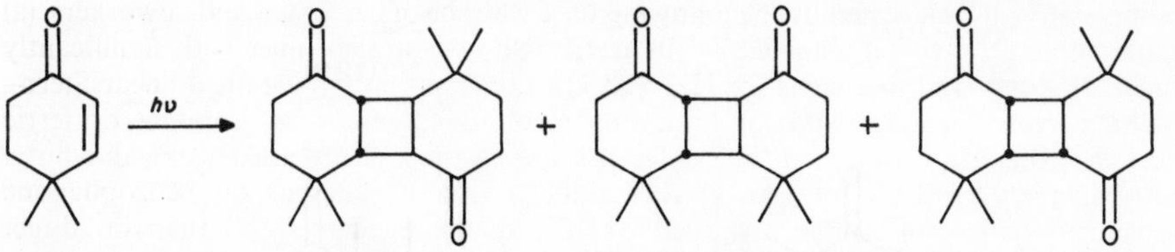

(124)   (126)   (125)   (127)

(66)

linear but with distinctly different slopes and intercepts, consistent with Scheme 19 but also compatible with dimerization via two different triplets. Photosensitized excitation of **124** in 2-propanol was carried out using p-methoxyacetophenone (MAP), not only because of its relatively high triplet energy (71.7 kcal mol^{-1}) but also since photoreduction of MAP in 2-propanol is very inefficient, ϕ_{isc} is high and self-quenching is unimportant. It was found that the yields of all the photoproducts of **124** (concentration 0.5 M) including dimers **125** and **126** were the same as on direct excitation under the same conditions. Tucker[111] later found that formation of the two dimers **125** and **126** from **124** in 2-propanol was quenched to the same extent by 1-methylnaphthalene ($E_T = 60$ kcal mol^{-1}), indicating they indeed arise from a common triplet. On the other hand, Nuñez[110] found that the ratio of **126** to **125** changed as a function of enone concentration in 2-propanol, from 6.4 at 0.10 M to 2.1 at 1.5 M, which could be considered as evidence for their formation from two different triplets. However, CH shows the same behavior in benzene but not in acetonitrile, which was attributed by Hammond and coworkers[107] to changes in the polarity of the medium as a result of increasing enone concentration. It is concluded that the same explanation holds for **124** in 2-propanol[110]; analogous experiments in other solvents were not undertaken.

Even steroidal enones undergo photodimerization, as shown with compounds **128** and **129** in equation 67[112]. A very important example involves photodimerization of thymine **130**, which is technically an enone in its principal tautomeric form. One of the most important reactions which occurs on exposure of DNA to UV light is formation of a dimeric structure between neighboring thymine residues[113]. Although other pyrimidine bases undergo photodimerization, they tend to preferentially undergo photohydration, which is a relatively unimportant reaction for thymine. In frozen solution, thymine reacts on exposure to 254 nm excitation to give exclusively the *cis–syn–cis* dimer **131** (equation 68), which is also the mode of photodimerization in DNA[114]. Using photosensitizers, the other regio- and stereoisomeric *cis*-fused thymine dimers are formed[115]. The dimers can be split by shorter wavelength excitation or by a natural photoreactivating enzyme which serves in nature to repair radiation-damaged DNA. Details of the nature and mechanism of operation of this enzyme can be found in photobiology texts[116].

(67)

$R = C_8H_{17}, COCH_3, OH, H$

(128)

(129)

(68)

(130)   (131)

b. Photoreduction. Photoreduction of enones could involve in principle either the n, π^* or π, π^* triplet states, and in fact both states have been invoked to rationalize the course of reactions of these systems. Irradiation of testosterone acetate **132** in ether gives 2% of cyclobutane dimer **133**, 30% of pinacol **134** and 15% of a mixture of diastereomeric adducts **135** (equation 69)[117]. The latter two products are clearly attributable to initial hydrogen abstraction from the solvent by the oxygen atom of a triplet n, π^* state of the enone. Irradiation of **132** in toluene gives the saturated ketone **136** and the toluene adduct **137**

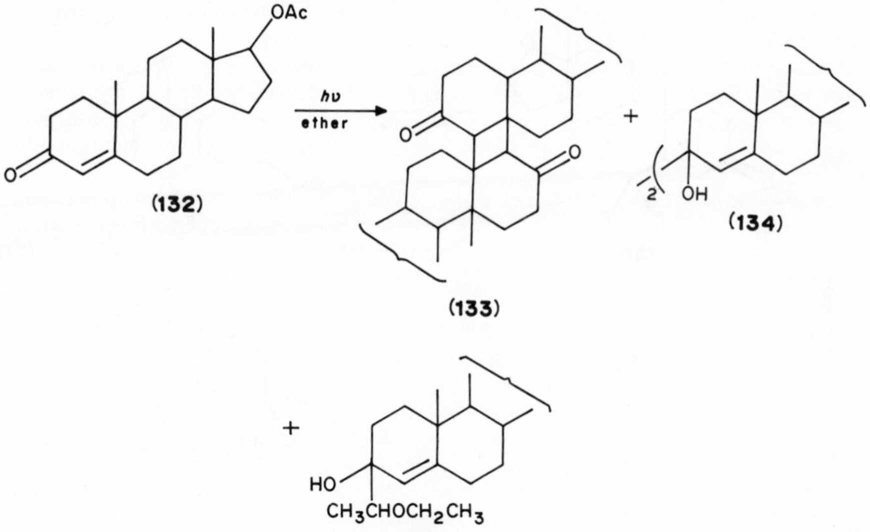

(69)

with an α-benzyl group (equation 70)[118]. In ethanol, **136** was again formed in 20% yield in addition to rearrangement products to be discussed later. Thus, in these solvents the course of photoreduction seems to be most readily rationalized in terms of reaction via $^3\pi, \pi^*$ states. In contrast, as shown in equation 71, the difluoro-substituted steroid enone **138** undergoes reduction to an allylic alcohol in t-BuOH (a solvent in which photoreduction is rarely observed) and to a carbonyl adduct in toluene, again implicating an n, π* triplet[119]. Since γ-fluorine substitution in cyclohexenones has been found to stabilize the n, π* vis-à-vis the π, π* triplet[120], this result is not very surprising.

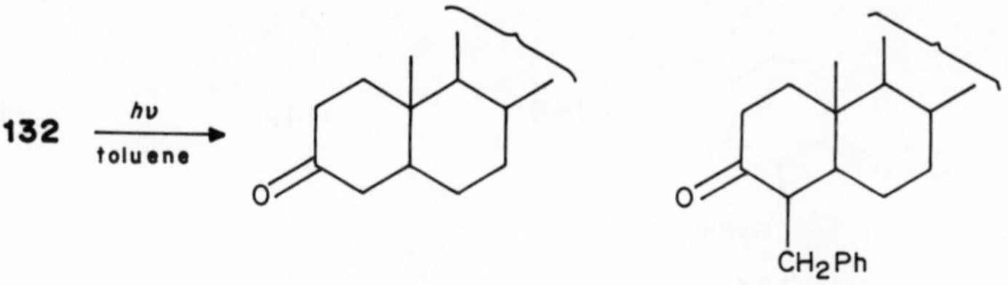

(70)

The octalone **139** upon irradiation in 2-propanol (IPA) was reported[121] to give the saturated ketone **140** (31%), the deconjugated ketone **141** and rearrangement products **142** and **143** to be discussed later (equation 72); no dimers or products of reduction of the C=O group were reported. On irradiation of **139** in toluene, the main products were again **140** and the α-adduct **144** (equation 72). Later studies by Chan and Schuster[122] showed that the original assignment of stereochemistry to the ring junction in **140** was incorrect, as the rings are cis- and not trans-fused, which has mechanistic implications that will be clear shortly. Photoreduction of the C=C and not the C=O bond of isophorone **123** to give **145** was reported to take place in nonpolar solvents such as cyclohexane, but photoreduction did not compete with photodimerization in 2-propanol at the enone concentrations utilized[109]. These reactions of **123** and **139** fit the pattern of reactivity

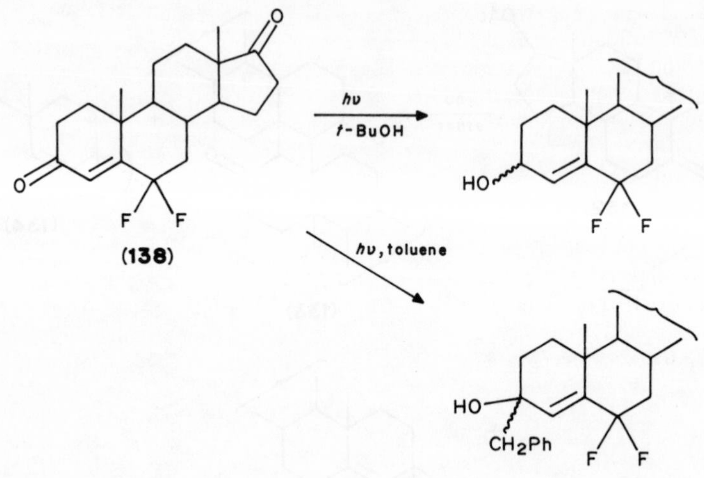

(71)

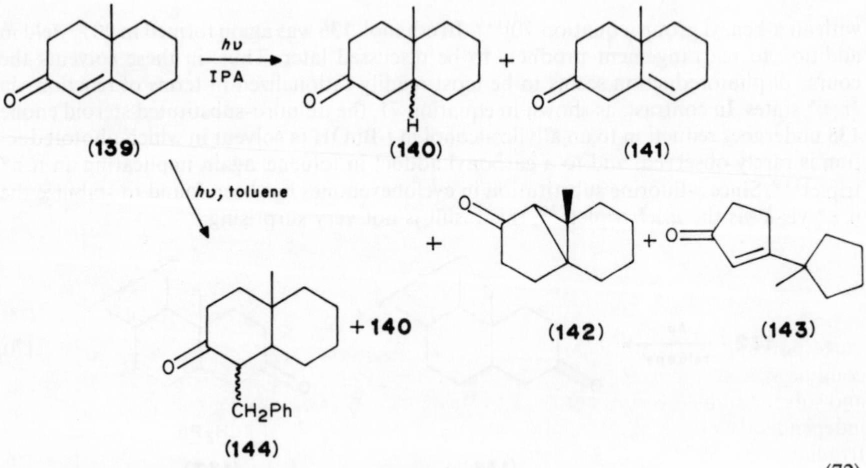

(72)

expected of a $^3\pi, \pi^*$ state in which initial hydrogen abstraction occurs at the β-carbon of the enone, followed by abstraction of a second hydrogen or combination (as in toluene) with solvent-derived radicals.

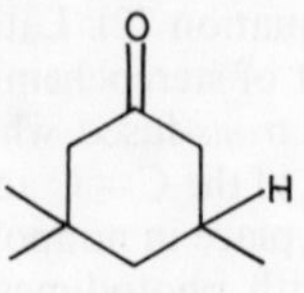

A detailed study of photoreduction of enone **124** in IPA was undertaken by Nuñez and Schuster[110]. Irradiation of a 0.3 M solution gave the saturated ketone **146** (16%), dimers **126** (12%) and **125** (2%), the rearrangement products **147** (36%) and **148** (34%), and traces of 3-isopropylcyclopentan-1-one (**149**) (equation 73). The yields were the same in two runs

corresponding to 16% and 29% conversion of **124**, and the mass balance under these conditions is excellent, indicating that the formation of other products (such as pinacols and solvent adducts) is unimportant under these conditions. The allylic alcohol **150** was independently prepared and shown not to be present in the above photolysis mixture. Irradiations were carried out in IPA-O-d, IPA-d_8 and $(CD)_3)_2$CHOH in the hope of determining the site on the enone of initial H (or D) abstraction from IPA[123]. Neither the starting enone **124** nor the reduction product **147** underwent H–D exchange in these media after 24 h in the dark; a slight reduction in the NMR signal of the α-protons in **146** was detected after the solution was kept for 96 h in the dark. Using IPA-O-d as the solvent, significant D-incorporation into **146** was observed after 24 h irradiation (using mass spectroscopic analysis), but there was no significant incorporation of deuterium into the rearrangement products. Base treatment of the photolysate led to *ca* 50% loss of deuterium in labeled **147**, indicating that the principal (if not exclusive) site of labeling was at C-2. When irradiation of **124** was carried out in $(CD_3)_2$CHOH, there was no significant incorporation of deuterium into any of the products, indicating that hydrogen transfer from methyl groups in the solvent-derived radical $(CH(D)_3)_2$COH to starting enone or radical intermediates (e.g. as occurs with benzophenone; see above) is unimportant. The yield of **146** when **124** was irradiated in IPA-d_8 was sharply reduced compared to the yield in unlabeled IPA, to the point where insufficient quantities of product could be isolated to determine the site of deuterium incorporation. The kinetic isotope effect $k_r(H)/k_r(D)$ was determined to be 9.6 ± 0.8 based upon the yields of **146** produced by simultaneous irradiation of **124** (0.3 M) in t-BuOH solutions containing an equal amount of IPA or IPA-d_8; the yield of the photorearrangement product **147** was the same in the two solutions[110,123]. Finally, irradiation of **124** in toluene gives the reduction product **146** and the α-benzyl adduct **151**, identified by comparison of chromatographic and spectral properties with a sample synthesized independently[110].

The mechanism shown in Scheme 20 accounts for all the experimental observations[123]. Thus, photoreduction of **124** is initiated by hydrogen abstraction at the β-carbon of an enone $^3\pi, \pi^*$ state, as is the case with most (but not all) of the cyclohexenones previously discussed, as well as cyclopentenones (see above). The enoxyl radical **152** can abstract a second hydrogen from the solvent (*not* from the solvent-derived radical) to give either **146** directly or the enol **153**. Deuterium incorporation from IPA-O-d takes place upon ketonization of **153**, suggesting that most of **146** is formed via the enol. The very large kinetic isotope effect (KIE) indicates that hydrogen transfer is well developed at the transition state for hydrogen abstraction, consistent with a symmetric C—H—C transition state; hydrogen transfer from C to O is characterized by a much smaller KIE, indicating an early transition state in which the extent of formation of the O—H bond is much less. The effect of temperature (43–71 °C) on photoreduction vis-à-vis photorearrangement of **124** was measured in IPA, from which a rough estimate of E_{act} for hydrogen abstraction of 5.2 ± 0.3 kcal mol^{-1} could be obtained; there was virtually no effect of temperature on the yields (relative quantum efficiencies) of the photorearrangement products **147** and **148**[110]. This value for E_{act} is also consistent with hydrogen abstraction by a π, π^* triplet excited state[124].

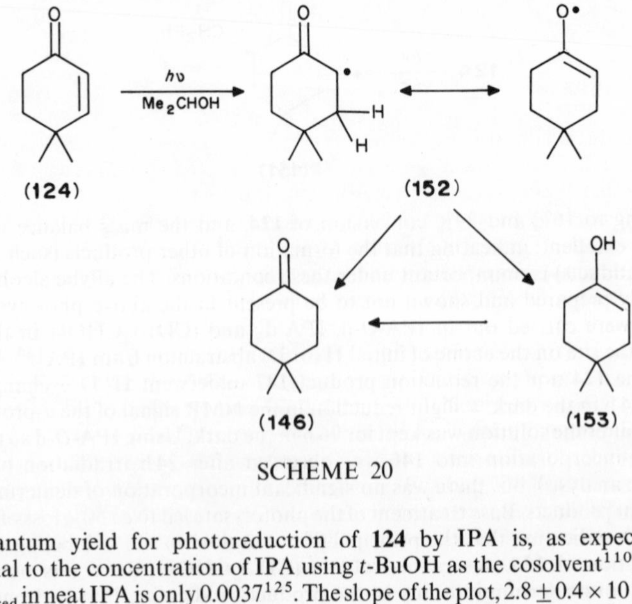

SCHEME 20

The quantum yield for photoreduction of **124** by IPA is, as expected, linearly proportional to the concentration of IPA using t-BuOH as the cosolvent[110]; the limiting value for ϕ_{red} in neat IPA is only 0.0037[125]. The slope of the plot, $2.8 \pm 0.4 \times 10^{-4}$ l mol^{-1}, is equal to $\phi_{isc} k_r \tau_T$, where k_r is the rate constant for hydrogen abstraction and τ_T is the enone triplet lifetime in the absence of IPA. Sensitization experiments indicate that intersystem crossing for **124** is totally efficient (i.e. $\phi_{isc} \sim 1.0$). Stern–Volmer plots for quenching of formation of **124** by naphthalene in neat IPA are linear, with slopes ranging from 11.6–14.9 l mol^{-1} corresponding to an enone triplet lifetime of *ca* 2.6 ns in neat IPA, assuming triplet energy transfer is diffusion-controlled with $k_g = 5 \times 10^9$ l mol^{-1} s^{-1} [110]. Using this value for τ_T, the quantum yield data give a value for k_r of 1.0×10^5 l mol^{-1} s^{-1}. Problems associated with direct determination using laser flash techniques of the lifetime of the triplet state of **124** responsible for photoreduction will be discussed later.

The photochemistry of 4a-methyl-4, 4a, 9, 10-tetrahydro-2(3*H*)-phenanthrone **154** provides a clear example of simultaneous reaction via both $^3n, \pi^*$ and $^3\pi, \pi^*$ triplet states. As shown by Chan and Schuster[122], irradiation of **154** in IPA gives the five products shown in equation 74: the *cis*- and *trans*-fused reduced ketones **155** and **156**, pinacol **157**, allylic alcohol **158** and the rearranged ketone (lumiketone) **159**. Quenching by napthalene shows

(74)

that these products fall into two distinct groups according to the Stern–Volmer plot in Figure 9: **155** and **159** on the one hand, and **156**, **157** and **158** on the other. The data clearly demonstrate that these products arise from two different triplet states of **154** which are quenched differentially by naphthalene. The nature of the products clearly indicates that the latter group arises from an n, π^* triplet, while the former group arises from a π, π^* triplet. The most interesting point is that each of the stereoisomeric dihydroketones **155** and **156** is produced stereospecifically from a different enone triplet, and do not arise from a common triplet precursor by a stereorandom reaction. The selective formation of the *cis* dihydroketone **155** from the same triplet responsible for photorearrangement (see below) is consistent with the proposal that the geometry of the π, π^* triplet is twisted to the point that the hydrogen donor is able to approach the β-carbon only from the same side of the molecule as the angular methyl (see Scheme 21)[122]. In contrast, a more or less planar n, π^* state should undergo hydrogen abstraction on oxygen to give the ketyl radical **160**, which is the precursor for **156**, **157** and **158** (Scheme 21). The stereoselective formation of **156** can be rationalized if the hydrogen donor approaches the planar ketyl radical **160** exclusively on the least-hindered face of the molecule, i.e. opposite to the angular methyl group. On this basis, it seemed surprising that photoreduction of octalone **139** via a twisted $^3\pi, \pi^*$ state should give a *trans*-fused dihydroketone **140**[121]. Restudy of this reaction showed that the structure of the dihydroketone was originally misassigned and that, as predicted, it is actually the *cis*-fused ketone **161**. Mechanistically, this supports the proposal that in

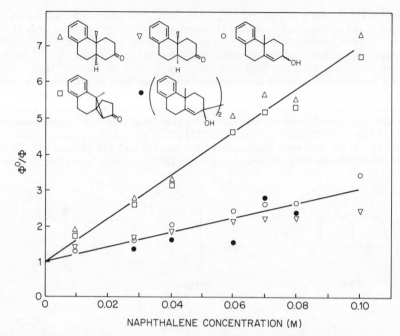

NAPHTHALENE CONCENTRATION (M)

FIGURE 9. Stern–Volmer plots for naphthalene quenching of the photochemistry of 4a-methyl-4, 4a, 9, 10-tetrahydro-2(3H)-phenanthrone **154** in isopropyl alcohol[122]. Reprinted with permission from *J. Am. Chem. Soc.*, **108**, 4561 (1986). Copyright (1986) American Chemical Society

sufficiently flexible cyclohexenones, including compound **139**, the lowest π, π^* triplet assumes a twisted conformation whose geometry controls the course of both photoreduction and photorearrangement processes.

    c. Photorearrangements of cyclohexenones. (i) General considerations. The molecular rearrangements of 4, 4-disubstituted cyclohexenones have been the subject of great deal of attention for almost thirty years, and several reviews on this subject have appeared[109b,126-129], including one by the present author in 1980[5]. The subject has also been well covered in basic texts on organic photochemistry[1-4]. This article will attempt to briefly summarize the basic features of these classic photorearrangements, and then to indicate the important contributions in this area made in the last several years.

    4, 4-Dialkylcyclohex-2-en-1-ones undergo unimolecular photorearrangement to bicyclo[3.1.0]hexan-2-ones (so-called lumiketones) usually accompanied by ring contraction to 3-substituted cyclopent-2-en-1-ones, upon irradiation in a variety of polar and nonpolar solvents. These transformations are illustrated by the photorearrangements of 4, 4-dimethylcyclohexenone **124** and testosterone acetate **132**, two of the first systems investigated, shown in equations 75 and 76[130,131]. As indicated earlier, these reactions are competitive with photodimerization and photoreduction of the enones, depending on the enone concentration and the nature of the solvent. Formation of deconjugated ketones also occurs in some systems, such as octalone **139** (see equation 72)[121]. As will be seen later, this competition between photochemical pathways can be put to advantage in mechanistic studies. Chemical yields of lumiketone are usually optimal in polar solvents such as *t*-BuOH in which photoreduction and deconjugation are minimized. In acetic acid, enone

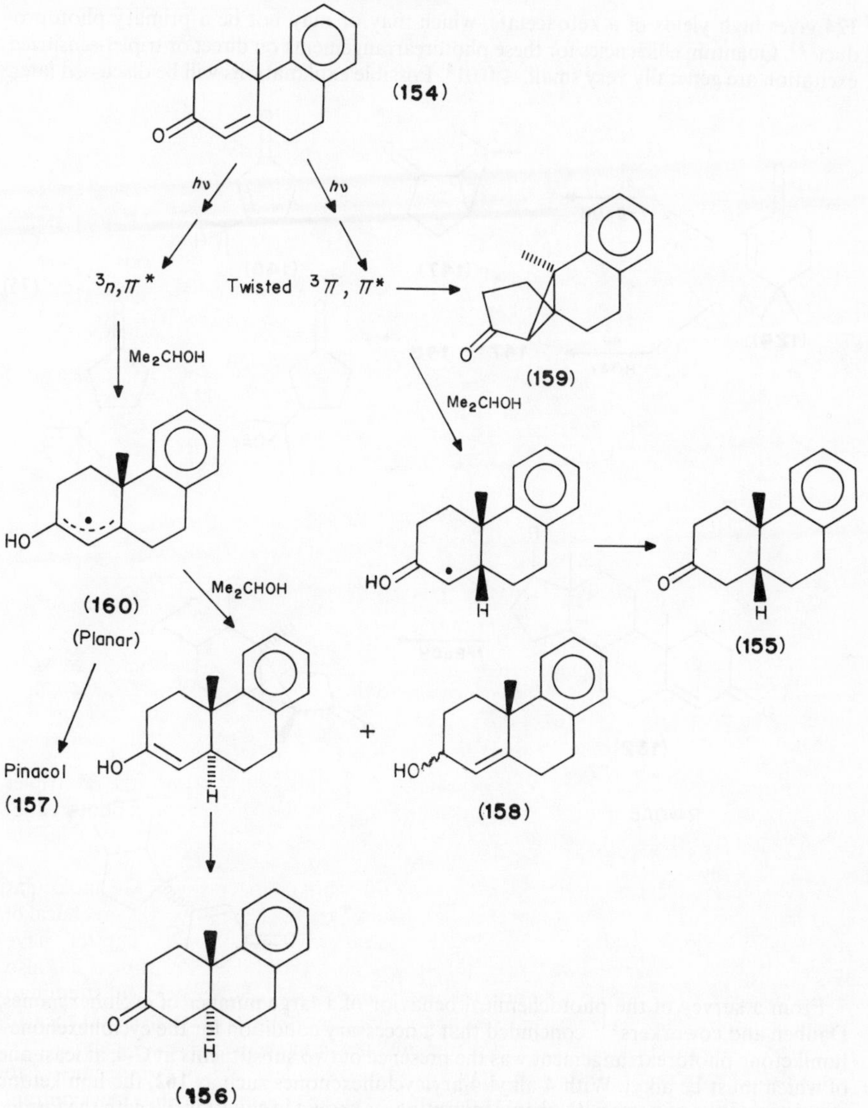

SCHEME 21

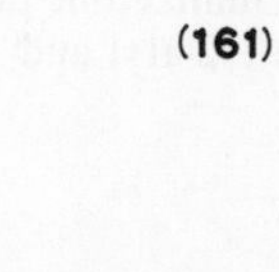

124 gives high yields of a ketoacetate, which may or may not be a primary photoproduct[131]. Quantum efficiencies for these photorearrangements on direct or triplet-sensitized excitation are generally very small, $\leqslant 0.01$[5]. Possible explanations will be discussed later.

From a survey of the photochemical behavior of a large number of cyclohexenones, Dauben and coworkers[132] concluded that a necessary condition for the cyclohexenone–lumiketone photorearrangement was the presence of two substituents at C-4, at least one of which must be alkyl. With 4-alkyl-4-arylcyclohexenones such as **162**, the lumiketone rearrangement competes with phenyl migration, as shown in equation 77, with the former more prominent in more polar protic and aprotic solvents (such as MeCN, DMF, 30% MeOH) while phenyl migration products are the exclusive products in benzene and ether[133]. 4,4-Diarylcyclohexenones such as **163** give only products of phenyl migration on direct or sensitized excitation, as seen in equation 78[134]. Irradiation of 4-alkyl-4-vinylcyclohexenones such as **164** leads to vinyl migration (see equation 79)[135,136]. These reactions are structurally analogous to the well-known di-π-methane photorearrangements[137].

Although they are related, it is useful to separate discussions of the two types of rearrangements of cyclohexenones, the lumiketone photorearrangement (also known as the Type A rearrangement)[126] and the 1,2-aryl and 1,2-vinyl migrations.

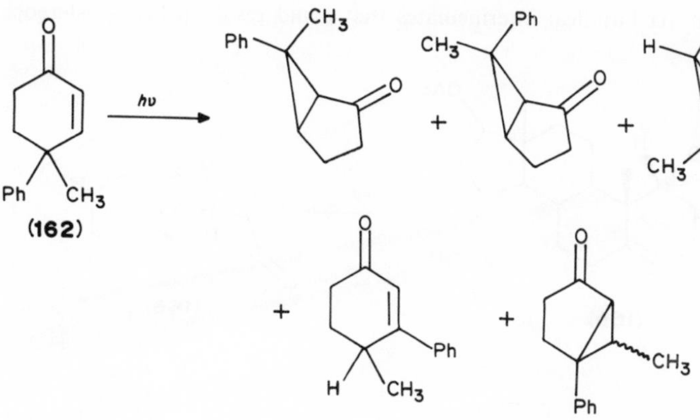

(162)

(77)

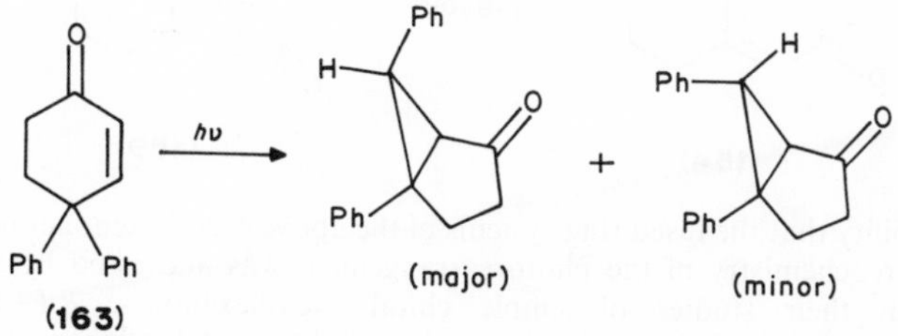

(163)

(major) (minor)

(78)

(164)

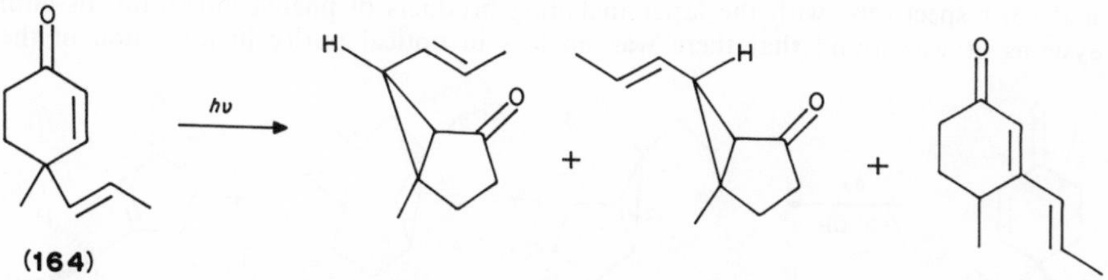

(79)

(ii) *Stereochemistry and mechanism of the lumiketone photorearrangement.* Several key studies have served to define the stereochemistry associated with the rearrangement of cyclohexenones to bicyclo[3.1.0]hexanones (lumiketones). First, Jeger and coworkers established that the stereochemistry of the rearrangement products of testosterone is as shown in equation 76 with H in place of OAc, and that other possible diastereomeric products are not formed[138]. Secondly, Schaffner and coworkers demonstrated that 1α-deuteriotestosterone acetate 165 rearranged stereospecifically to 166 with retention of configuration at C-1 and inversion of configuration at C-10 (analogous respectively to C-5 and C-4 of a simple cyclohexenone), as shown in equation 80[118]. Chapman and coworkers demonstrated that photorearrangement of optically active phenanthrone 154 to its lumiketone 159 proceeded stereospecifically (equation 81) with inversion of configuration at C-10 and loss of less than 5% optical purity (enantiomeric excess)[139]. These results were interpreted in terms of a more or less concerted bond-switching process as opposed to rearrangement via biradical intermediates that could result in loss of stereochemical integrity.

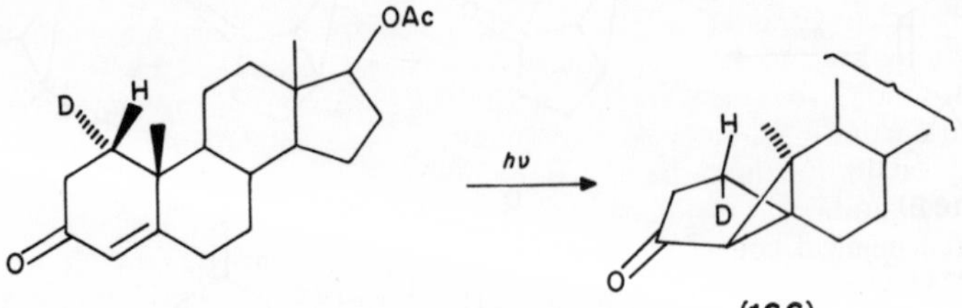

(80)

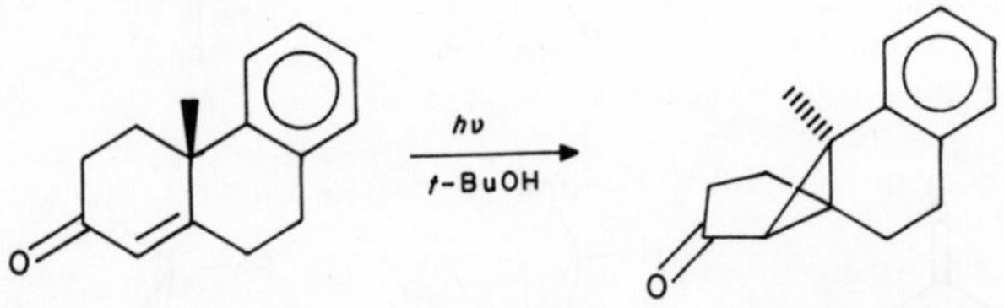

(165)   **(166)**

(81)

(154)   **(159)**

The possibility that the fused ring systems of the above cyclohexenones might obscure the 'true' stereochemistry of the photorearrangement was addressed by Schuster and coworkers in their studies of simple chiral cyclohexenones, R-(+)-4-methyl-4-propylcyclohexenone **167** and R-(+)-4-methyl-4-phenylcyclohexenone **169**[140]. The photoproducts with their stereochemical assignments are shown in equations 82 and 83, respectively, with the latter including products of phenyl migration. In both systems, it was found that there was no loss in optical purity in formation of the

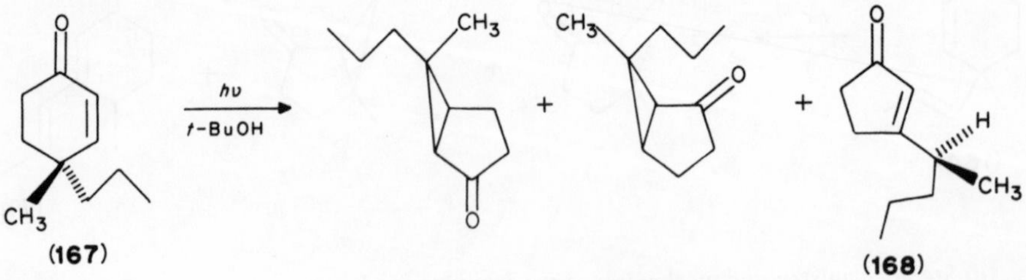

(167)   **(168)**

(82)

(169)

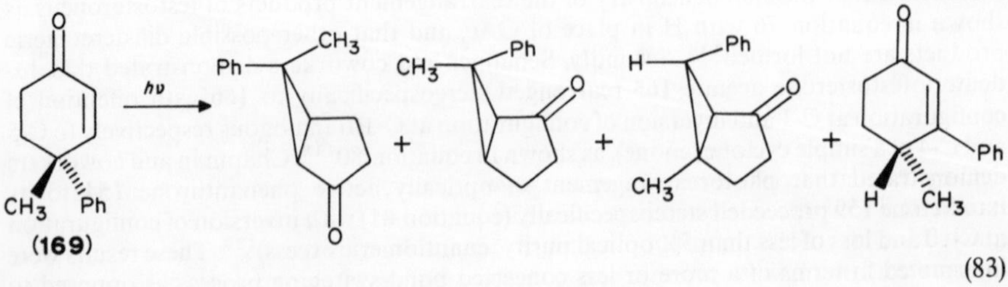

(83)

lumiketones nor in the recovered enones, even after 325 h continuous irradiation in the case of **167**. These data, coupled with those above, establish with certainty that cleavage of the bond between C-4 and C-5 in cyclohexenones must be concerted with formation of the new bond between C-5 and C-3. In other words, no triplet diradical intermediate which is sufficiently long-lived to allow stereorandomization at either radical site as a result of rotations around C—C single bonds can intervene in formation of lumiketone as well as reversion to starting material (recall that quantum efficiencies for photorearrangement are notoriously small). Furthermore, reactions proceed stereospecifically with inversion of absolute configuration at C-4 (C-10 in steroids)[140].

The stereochemical course of reaction in simple cyclohexenones is summarized in Scheme 22. The reaction is stereospecific on each face of the cyclohexenone ring system, with retention of absolute configuration at C-5 and inversion at C-4, leading to formation of diastereomeric lumiketones (with respect to *exo–endo* configuration of the substituents) in which the bicyclo[3.1.0]hexanone ring systems have opposite chirality. Thus, despite the fact that it originates from an enone triplet state (see discussion below), the cyclohexenone–lumiketone photorearrangement has the appearance of a concerted reaction, with a stereochemical course corresponding to a $_\pi 2_a + _\sigma 2_a$ process, in Woodward–Hoffmann terminology[141], involving antarafacial addition to both the C_2—C_3 π bond and the C_4—C_5 σ bond. In steroids and analogous fused-ring enones, such as **132** and **154**, reaction can occur only on one face of the enone because of steric constraints, necessarily affording only a single lumiketone.

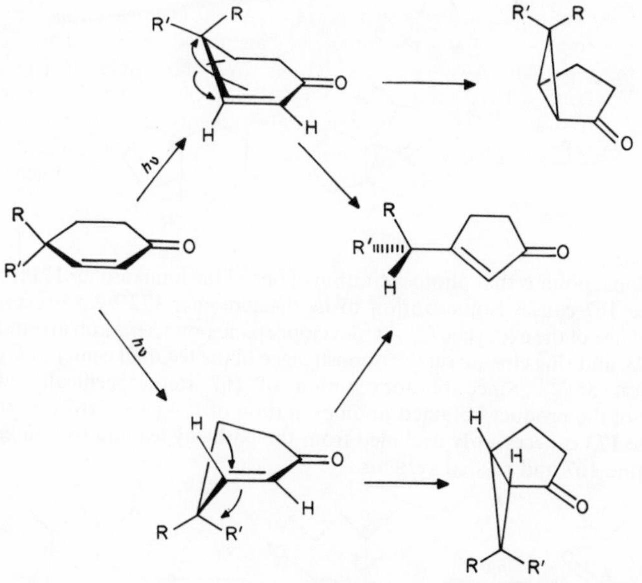

SCHEME 22

The potential inconsistency of a symmetry-allowed process proceeding from a triplet-excited state in which electrons are unpaired has been addressed by Shaik[142], who concludes that in certain situations spin inversion and product formation may occur concomitantly. This is possible when both spin inversion and orbital symmetry requirements are met along the same reaction coordinate, which is precisely the case with the twisting motion required in order in achieve the geometry corresponding to a concerted $_{\pi}2_a + _{\sigma}2_a$ intramolecular cycloaddition, as discussed above. Shaik raises the interesting possibility that such a process might be triplet sublevel specific, i.e. that the x, y and z sublevels of the triplet state might react with differing efficiencies. No studies along these lines have been reported.

It is clear from Scheme 22 that the reactive triplet-excited state of the enone (see below) must undergo substantial twisting around the C=C bond in order for the bond-switching process corresponding to a $_{\pi}2_a + _{\sigma}2_a$ cycloaddition to occur as shown. It was predicted that structurally analogous cyclohexenones whose structures preclude significant twisting around the C=C bond would not undergo the lumiketone photorearrangement[140]. This was verified by Schuster and Hussain[143] with enone **170** which undergoes photoreduction and radical-type solvent photoaddition, but neither rearrangement nor polar-type addition reactions (equation 84).

An additional point is that photoexcitation of one of the lumiketones **171** from optically active enone **167** causes isomerization to its diastereomer **172** by a process that must involve cleavage of the exocyclic C_1—C_6 cyclopropane bond, rotation around C_5—C_6 in biradical **173**, and ring closure on the opposite face of the trigonal center at C_6 to give **172** (see equation 85)[140]. Since photoexcitation of **167** stereospecifically afforded the enantiomer of the product obtained upon excitation of $R - (+) - $ **167** (see equation 82), intermediate **173** is necessarily excluded from the pathway leading to lumiketones from cyclohexenone **167** and related systems.

The Type A photorearrangement of cyclohexenones is formally analogous to the photorearrangement of cyclohexadienones to bicyclo[3.1.0]hex-3-en-2-ones, also called a lumiketone rearrangement, typified by the conversion of **174** to **175** shown in equation

$86^{144,145}$. This reaction, which proceeds via dienone triplets, has been shown in suitable systems to be stereospecific with inversion of configuration at C_4 [146]; thus, in a formal sense it is also an intramolecular $_\pi 2_a + _\sigma 2_a$ cycloaddition. However, it has been demonstrated unequivocally that the photorearrangement of cyclohexadienones proceeds stepwise via zwitterion intermediates $(176)^{126,144,145}$, which can be trapped in certain cases[145,147,148], and is therefore not a concerted intramolecular cycloaddition. Furthermore, the quantum efficiencies (QE) for the cyclohexadienone photorearrangements are quite high (generally 0.8–1.0), indicating that the second C=C bond plays a key mechanistic role[144,145]. Note also that lumiketones are formed in high yield from cyclohexadienones such as 174^{144}, while corresponding 4,4-diphenylcyclohexenones react exclusively by phenyl migration[134].

$$(86)$$

In an attempt to link the cyclohexenone and cyclohexadienone photorearrangements mechanistically, as well as to account for the formation of polar addition products (see equation 75) Chapman[127,128,131] proposed that the cyclohexenone photorearrangements proceed via a 'polar state' **177** (equation 86a) although it was never specified whether **177** represents an excited or ground-state species. Such a species did provide a convenient way of accounting for the formation of bicyclo[3.1.0]hexanones by mechanistic analogy with carbocation rearrangements, as shown in equation 86a, although the subsequently observed stereospecificity would be hard to rationalize on the basis of stepwise reaction of a dipolar intermediate. It is even less obvious how to rationalize the direct formation of ring-contracted cyclopentenones as shown in equations 75 and 76 via a dipolar species without invoking one or more hydride shifts. Irradiation of optically active **167** gives

$$(86a)$$

optically active cyclopentenone **168** (see equation 82) with the absolute configuration as shown, although it is not known whether this rearrangement is totally or only partially stereospecific[140]. The predominant course of reaction is as depicted in Scheme 22 involving a formal $_\sigma 2_a + _\sigma 2_a$ cycloaddition of the C_4—C_5 σ bond to the C_3—H bond, i.e. the hydrogen migration from C_3 to C_4 results in inversion of configuration at C_4. This certainly is not the stereochemical course of reaction expected if **168** arose via ring contraction of a dipolar species such as **177**.

A study of 10-hydroxymethyloctalone **178** was undertaken by Schuster and Brizzolara[149] specifically to test Chapman's 'polar state' theory[127,128]. It was anticipated that irradiation of **178** would produce a CH_2OH fragment, either as a radical or a carbocation, depending on whether the precursor was a dipolar or diradical species. The products and reaction course of **178** are shown in Scheme 23. It is clear that there are two competitive pathways for **178**: (a) rearrangement to lumiketone **179**, and (b) hydrogen abstraction–fragmentation to give hydroxymethyl radical and dienol **180**, which is the precursor to octalones **181** and **182**. Path (a) was the sole reaction course in t-BuOH, while reaction via (b) as well as (a) occurred on irradiation of **178** in $CHCl_3$, toluene, cumene and (curiously) benzene. Triplet quenching experiments showed that both pathways occur from a common triplet excited state of **178** which must have diradical and not dipolar character, in order to account for the nature of the fragmentation products and the effect of solvent on the reaction course[149].

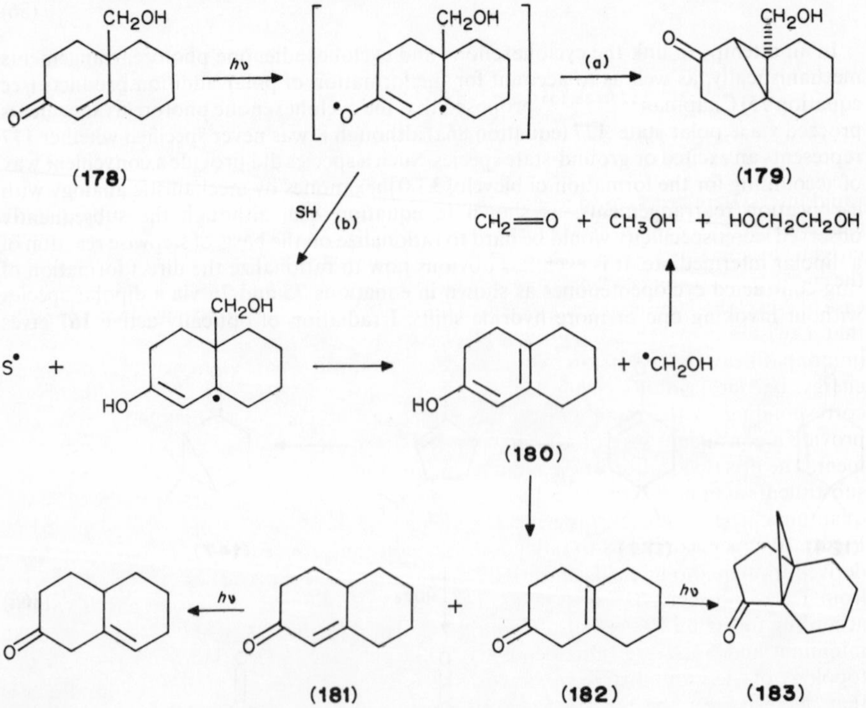

SCHEME 23

Based on the observations summarized above, Schuster and coworkers[140] suggested that the mechanism of the cyclohexenone–lumiketone photorearrangement involves

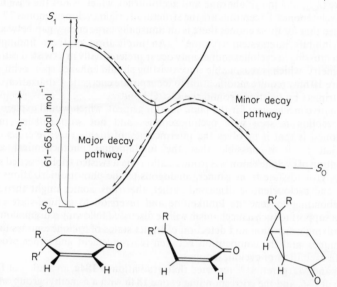

FIGURE 10. Proposed topology of the triplet and ground-state potential surfaces along the reaction coordinate for conversion of 2-cyclohexenones to bicyclo[3.1.0]hexan-2-ones (lumiketones). Reprinted with permission from *J. Am. Chem. Soc.*, **100**, 4504 (1978). Copyright (1978) American Chemical Society

rapid intersystem crossing from the enone S_1 to the T_1 state ($^3\pi, \pi^*$), which then relaxes energetically by twisting around the C=C bond as shown in Figure 10. Intersystem crossing by spin inversion back to S_0 at the twisted geometry should be favored because of the energetic proximity of the S_0 and T_1 surfaces at, or close to, the 90° geometry, as in the case of 2-cycloheptenone (Figure 8). The diagram in Figure 10 is based on the assumption that a twisted cyclohexenone ground state partitions between formation of lumiketone (minor pathway) and reversion to starting enone (major pathway). The existence of a small energy barrier leading to lumiketone on the ground-state surface from the point corresponding to the minimum on the triplet surface, as shown in Figure 10, would provide a convenient way of rationalizing the low quantum efficiency for the rearrangement. The precise location of the minima and maxima in Figure 10 should depend on the substituents at or near the enone chromophore, accounting for structural variations on the quantum efficiency for rearrangement. If the course of reaction is indeed as depicted in Figure 10, it is meaningless to talk about rate constants for triplet decay and reaction as derived from quantum yields and triplet lifetimes, as if these processes competed directly from T_1, as in other types of systems. Thus, the rate of decay of cyclohexenone triplets according to Figure 10 depends only on the energy difference between T_1 at its potential minimum and S_0 at the same geometry, while the reaction efficiency depends on the topology of the ground-state surface, i.e. the fraction of twisted ground-state molecules that make it over the top. However, the possibility that formation of lumiketone is concerted with spin inversion as suggested by Shaik[142] is by no means excluded.

According to the above picture, the efficiency of photorearrangment ought to be temperature dependent, but this has yet to be studied. The observation[143] that enone **170** does not undergo photorearrangement is consistent with this description of the reaction. One of the more intriguing observations is that **170** is weakly fluorescent at room temperature (λ_{max} 385 in cyclohexane and acetonitrile), which is not the case for simple acyclic or cyclic enones. Exceptions are the structurally rigid cyclopentenones **77** and **78**; it was proposed that for these enones there is an unusually large energy gap between S_1 and T_1, which inhibits intersystem crossing[90]. An implication of these findings is that intersystem crossing in cyclohexenones may occur preferentially in a twisted rather than a planar geometry, which is reasonable since twisting should enhance spin–orbit coupling. Thus, Figure 10 may require modification to incorporate energetic stabilization of singlets as well as triplets by twisting around the C=C bond.

An alternative mechanism for the photorearrangement, which would explain why this mode of reaction is seen with cyclohexenones and not with smaller and larger cycloalkenones, is that it involves the intermediacy of highly reactive *trans* cyclohexenones. That is, it is possible that the fundamental photochemical act upon photoexcitation of cyclohexenones is isomerization (via a twisted triplet-excited state) to a high-energy ground-state *trans* isomer, analogous to the photoisomerizations of cycloheptenone and cyclooctenone discussed earlier; the *trans* isomer might then partition between rebonding to generate lumiketone and reversion to ground-state *cis* enone. Evidence in support of such a mechanism will be discussed following a discussion of recent studies involving generation and detection of triplet states of cyclohexenones using laser flash techniques, and the competition between rearrangement and other processes on steady-state excitation of cyclohexenones.

Cruciani and Margaretha[150] reported that irradiation of **184a**, an analog of **124** with a CF_3 group at C-6, and the corresponding enone **184b** with a 6-methyl group, affords the usual rearrangement products, as shown in equation 87; however, in these systems the cyclopentenones are formed in higher yields than the lumiketone, in contrast to the behavior seen with the unsubstituted enone **184c** ($= 124$). They also noted that the

(184)

(a) R=CF$_3$

(b) R=CH$_3$

(c) R=H (=**124**)

R=CF$_3$ only

(87)

reduction of **184a** to the saturated ketone occurred to a significant extent in *t*-BuOH and CH_3CH, which is not the case with analogous enones, which was not explained. The relative quantum yields for rearrangement of these enones at 350 nm are **184a** < **184b** < **184c**. They suggest that the lowering of the quantum yield is probably due to conformational changes in the enone excited states; if so, there ought to be substantial enhancement of triplet lifetimes, as discussed in section IV.B.4.d. The authors interpret the shift of the ratio of rearrangement products toward ring contraction as evidence that these products arise by the route shown in equation 88, i.e. ring opening to a substituted 5-hexenyl radical, ring closure selectively to the *trans*-disubstituted five-membered ring, which then either undergoes ring closure to the lumiketone or a 1,2-hydrogen shift to give the cyclopentenone. Such a photorearrangment mechanism was previously considered and discarded based upon stereochemical data for model cyclohexenones, as discussed earlier. Thus, in the interests of mechanistic simplicity and in accord with Occam's Razor[151], it seems best at present to interpret these findings in terms of the rearrangement mechanisms discussed above, in the absence of compelling reasons to assign a special mechanism to this set of enones. For reasons which are far from clear at this time, the formation of lumiketone from the twisted enone triplet state of **184a** and **184b** by essentially a $_\pi 2_a + _\sigma 2_a$ route is slower than the $_\sigma 2_a + _\sigma 2_a$ route which leads to cyclopentenones.

$$(88)$$

(iii) Di-π-methane rearrangements: 1,2-aryl and 1,2-vinyl migrations. As shown in equation 77, when an aryl group is present at C_4 as in enone **162**, an aryl migration pathway competes with the lumiketone rearrangement[133]. With two aryl groups at C_4, only aryl migration is observed, which gives a mixture of stereoisomers in which the isomer with a 6-endo aryl group dominates. From the work of Zimmerman and coworkers, the mechanism of this transformation is well understood[134,152]. Migratory aptitudes have been determined from studies using enones with two different 4-aryl substituents, and they establish that aryl migration occurs to a carbon center (C_3) with odd electron character. Once again, the results are inconsistent with the 'polar state hypothesis'[127]. Sensitization and quenching studies establish that the rearrangement occurs via triplet-excited states which are formed with close to unit efficiency. From the dependence of product ratios on solvent polarity in the case of enone **162** Dauben and coworkers[133] proposed that it is the enone ^{3}n, π^* state which is the intermediate in the aryl migration pathway, while the $^3\pi, \pi^*$ state is the species responsible for the lumiketone rearrangement, in agreement with the assignments made earlier. Differential quenching of formation of **185** and **186** by naphthalene on irradiation of **162** in ethanol (equation 88a) supports the proposal that these products indeed arise via two different triplet excited states which are not in thermal equilibrium.

Except for the fact that these aryl migrations proceed from ^{3}n, π^* states, the rearrangement is analogous to the di-π-methane rearrangements extensively studied by Zimmerman and his coworkers[136]. The formation of the major rearrangement product with a 6-endo-aryl group in the reaction of **163** can be rationalized in terms of a bridged intermediate **187**[153]. However, the fact that the 6-exo-aryl product is also formed suggests that this reaction is not concerted, and that it occurs at least in part via the open diradical intermediate shown in Scheme 24. Quantum yields for aryl migration as high as 0.18 have

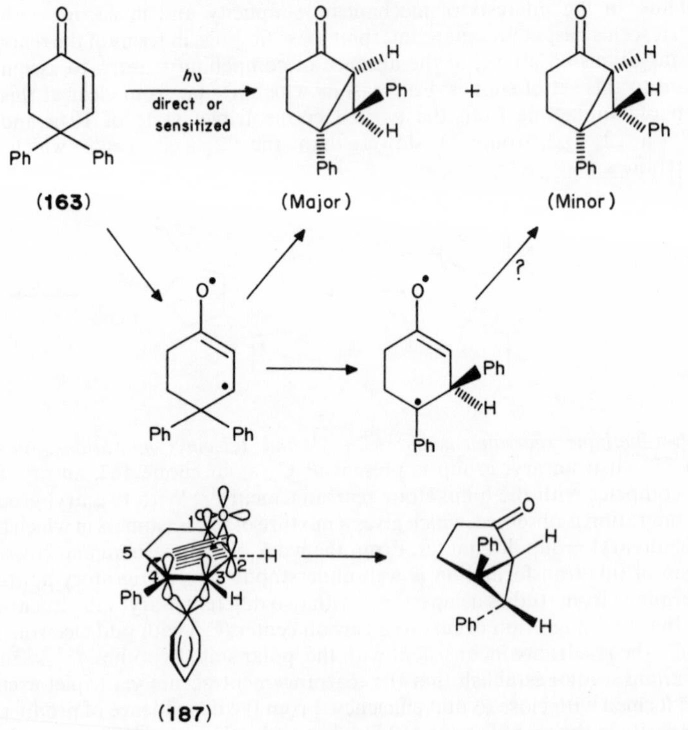

(163)   (Major)   (Minor)

(187)

SCHEME 24

been measured[153], but they vary with the nature of the migrating and nonmigrating groups. Assuming that decay to the ground state and rearrangement are competitive processes of the triplet state, rate constants for these processes (k_d and k_r, respectively) can be determined from quantum yields and triplet lifetimes; the latter are determined from Stern–Volmer triplet quenching plots, assuming that triplet energy transfer is diffusion controlled. (No studies involving direct determination using laser flash techniques of triplet lifetimes for enones 162, 163 or similar enones have been reported, so the validity of this assumption has yet to be tested experimentally.) Values of k_r determined on this basis depend on the nature of the migrating and nonmigrating groups, while k_d values show little variation, and are ca 10^9 s^{-1}. Zimmerman concludes that the 'decay to product seems to have little in common with the decay back to reactant'.

The stereochemistry of the phenyl shift in 162 was determined for the chiral system by Schuster and coworkers[140]. Both 188 and 189 were formed stereospecifically without any loss of optical purity. By relating the absolute configurations of the products and the starting materials, it was shown that both rearrangements occurred as shown in Scheme 25 with complete inversion of configuration at C_4. Thus, it appears that phenyl bridging and ring contraction are synchronous in this system, since the epimer of 189 with a 6-exo-phenyl is not formed; reaction via the open diradical 190, on the other hand, should lead to both epimers.

Zimmerman and coworkers have recently reported interesting studies on 4,4-biphenylylcyclohexenone 191, to determine the effect of incorporating in the molecule a

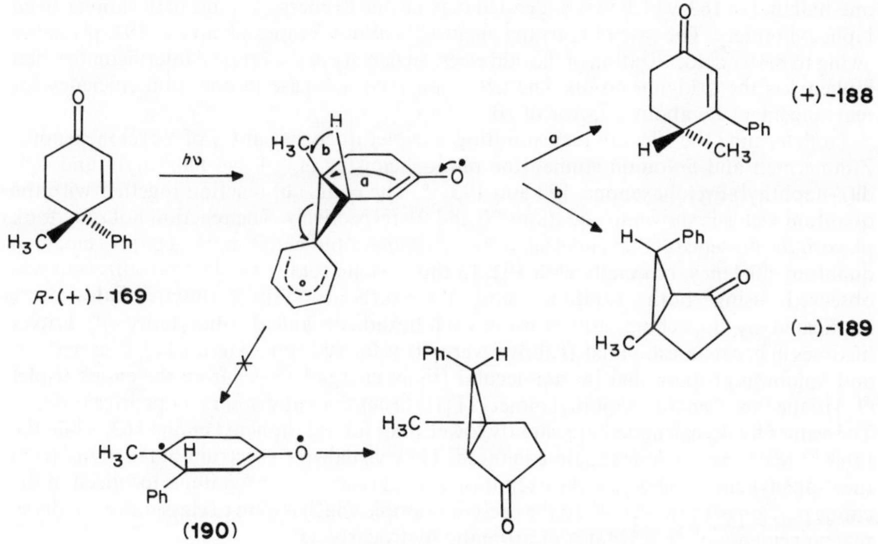

SCHEME 25

moiety whose triplet energy should be approximately the same as that of the enone moiety[154]. The course of reaction of this system, shown in equation 89, is similar to that for the 4,4-diphenylenone, except that the quantum yields shown are considerably larger.

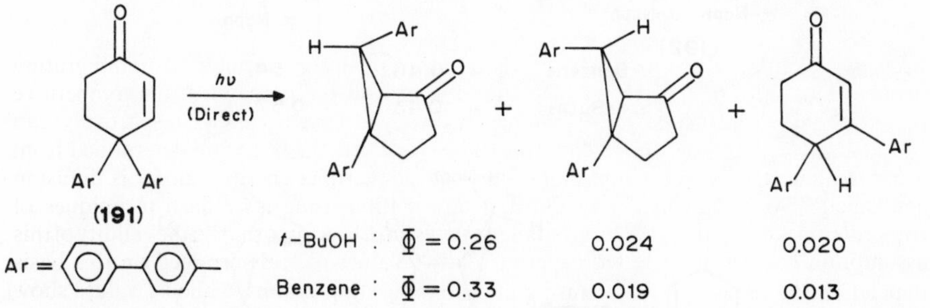

| | t-BuOH : | $\Phi = 0.26$ | 0.024 | 0.020 |
| | Benzene : | $\Phi = 0.33$ | 0.019 | 0.013 |

(89)

Triplet sensitization by either xanthone (E_T 74 kcal mol^{-1}) or benzophenone (E_T 69 kcal mol^{-1}) gave the same products with undiminished quantum yields, while the quantum yields on sensitization by thioxanthone ($E_T = 65$ kcal mol^{-1}) were much lower, indicating uphill triplet energy transfer. They assigned a triplet energy of ca 69 kcal mol^{-1} to enone **191**. The reaction was quenched by 1,3-cyclohexadiene; from the Stern–Volmer slopes, Zimmerman and coworkers calculated a triplet lifetime for **191** of 3.1 ns in t-BuOH and 2.9 ns in benzene. They suggest that equilibration of the triplet excitation between the enone and biphenyl moieties is faster than the rate of rearrangement in this system, with excitation initially localized in the enone moiety on direct excitation and in either moiety on triplet sensitization. From the data, they calculated a k_d value in benzene which is about one-half that of **163**, which they suggested may be due to energy storage in the longer-lived biphenyl moiety. The rate of rearrangement k_r is about 5 times greater in **191**, probably owing to better delocalization of the odd electron density in the bridged intermediate when biphenyl is the bridging group. The net result is an increase in quantum efficiency for rearrangement by about a factor of 10.

To determine the effect of incorporating a triplet quencher at C_4 of a cyclohexenone, Zimmerman and Solomon studied the photochemistry of 4,4-di(α-naphthyl)- and 4,4-di(β-naphthyl)-cyclohexenone, **192** and **193**[155]. The course of reaction together with the quantum yields is shown in equations 90 and 91 respectively. The reaction not only took place in the presence of these internal triplet quenchers, but with a marked improvement in quantum efficiency, especially with **192**. In this system, totally efficient sensitization was observed using both xanthone and thioxanthone. Triplet intermediates were implicated by quenching studies using cyclohexadiene and di-t-butylnitroxyl. Triplet lifetimes in benzene, estimated as above, were 6.0 ns for **192** and 7.3 ns for **193**. Zimmerman and Solomon propose that intramolecular triplet energy transfer from the enone triplet (T_2) to the lower-energy naphthyl moiety (T_1) is faster than any other competitive process. The values for k_d calculated are slightly lower than for the diphenyl enone **163**, while the rates of rearrangement are again enhanced. They visualize a spectrum of reactivity from the diphenyl enone, in which the excitation (in the reactive $^3n, \pi^*$ state) is localized in the enone portion of the molecule, to the present example which is akin to classic di-π-methane rearrangements of $^3\pi, \pi^*$ states of aromatic hydrocarbons[136].

$$(90)$$

| | Benzene | $\bar{\Phi} =$ | 0.46 | 0.54 |
| | t-BuOH | $\bar{\Phi} =$ | 0.43 | 0.57 |

$$(91)$$

| | Benzene | $\bar{\Phi} =$ | 0.38 | 0.02 |
| | t-BuOH | $\bar{\Phi} =$ | 0.40 | 0.02 |

A mechanistically analogous rearrangement of enone **164** has been observed involving migration of a 4-vinyl substituent (equation 79)[135]; E–Z isomerization of the starting material competed with the rearrangement. No lumiketone product was observed in this study. It is likely that the reaction occurs via triplet states, but this was not established. It is also interesting that **164** does not undergo a cyclization reaction analogus to that seen with

194 (equation 92), even on excitation into S_2 (see discussion in Section IV.B.4.c.vi), which the authors attribute to conformational problems in **194** preventing the intramolecular hydrogen abstraction required for the cyclization.

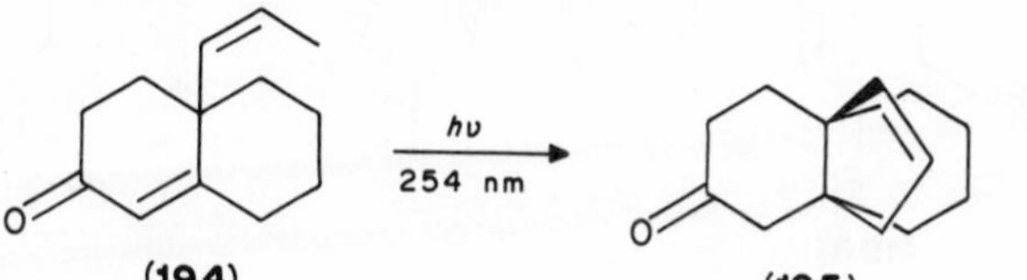

$$ (92) $$

(194)   **(195)**

The possibility that cyclopropyl substituents might undergo 1, 2-shifts analogous to those seen for aryl and vinyl substituents was investigated by Hahn and coworkers[156]. Irradiation of the 4, 4-dicyclopropyl-2-cyclohexenone **196** gave only lumiketones, as shown in equation 93. The reactions seen on irradiation of 4-cyclopropyl-4-phenyl-2-cyclohexenone **197** followed the pattern seen previously with enone **162**, namely 1, 2-phenyl migration in nonpolar solvents, and lumiketone formation in addition in polar solvents. Again, competitive reaction via $^3\pi, \pi^*$ and $^3n, \pi^*$ states was invoked.

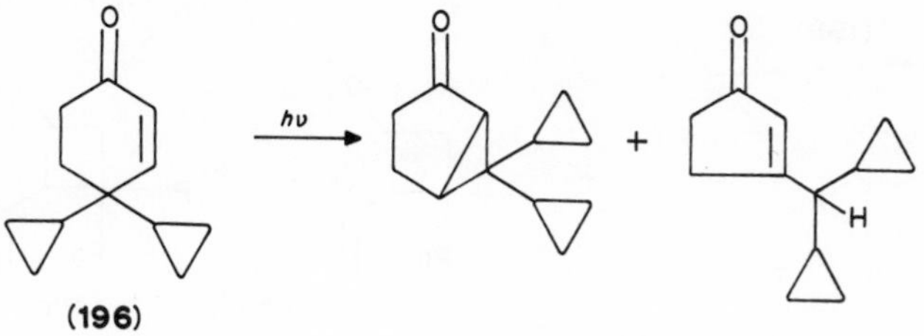

(196)

$$ (93) $$

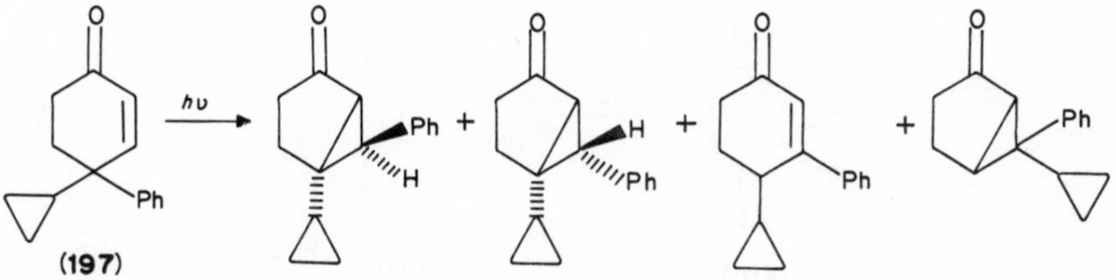

(197)

(iv) Ring contraction to cyclobutanones. Some time ago, Zimmerman and Sam[157,158] reported that ring contraction to a cyclobutanone competed with phenyl migration on irradiation of 4, 5-diphenyl-2-cyclohexenone **198**, as shown in equation 94. Recent studies by Zimmerman and Solomon[159] extend the earlier investigations and go a long way to establishing the mechanism of this interesting rearrangement as well as its relationship to the photochemical rearrangements discussed above. Thus, 4, 5, 5-triphenyl-2-cyclohexenone **199** gives a variety of irradiation products, shown in equation 95, involving both phenyl migration and ring contraction, while 4-methyl-5, 5-diphenyl-2-cyclohexenone **200** gives only ring contraction to **201** (equation 96). On acetophenone sensitization the same products are formed with identical quantum yields as upon direct excitation, and both reactions are quenched by 1, 3-cyclohexadiene. This establishes that these reactions occur via enone triplets, whose lifetimes (assuming diffusion-controlled quenching) are 7.4 and 8.1 ns, respectively.

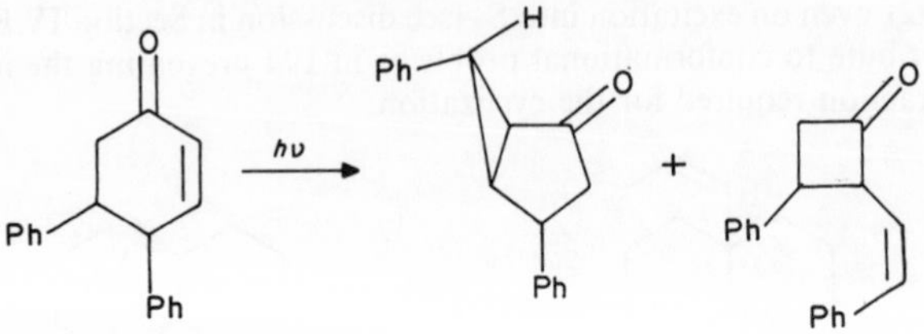

(94)

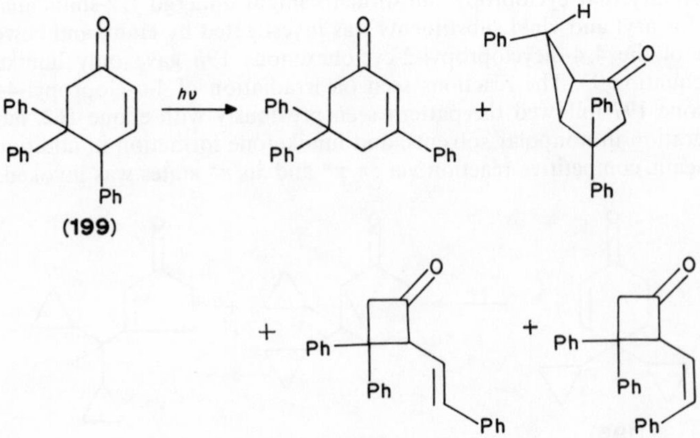

(95)

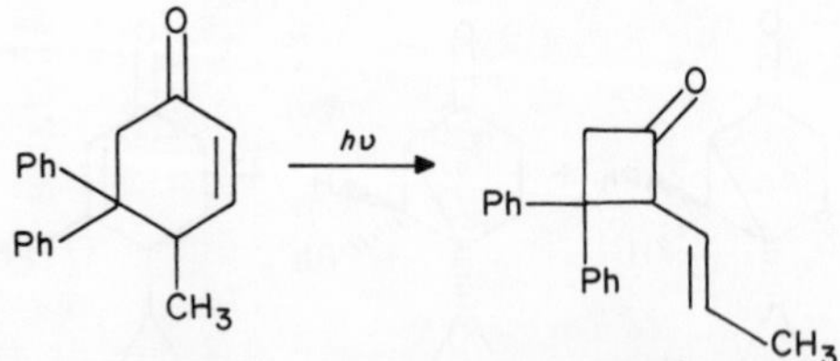

(96)

Since cleavage to ketenes and 1,1-diphenylethylene, followed by recyclization, is a possible route to the cyclobutanones, irradiations were carried out in the presence of potential ketene traps, namely in ethanol and in benzene containing cyclohexylamine. No ester or amide products were detected. A further test was to carry out the irradiation of **200** in the presence of 1,1-di(p-tolyl)ethylene and to look for crossover products; none were observed. These experiments strongly suggest that these ring contractions occur intramolecularly by cleavage of only one C—C bond.

Stereochemical studies clarify the picture. Irradiation of optically active enone **200** gave nearly racemic cyclobutanone **201**, with $6.7 \pm 2.0\%$ residual enantiomeric excess. Recovered starting material had not undergone any racemization. Zimmerman and Solomon[159] discuss these results within the mechanistic framework shown in Scheme 26, using the model for n, π* triplet states originally proposed many years ago[126,144,160]. It is proposed that the key reaction of this triplet is 4,5-bond fission to give diradical **202**. In principle, as shown in Scheme 26, this triplet could cleave to 1,1-diphenylethylene and a

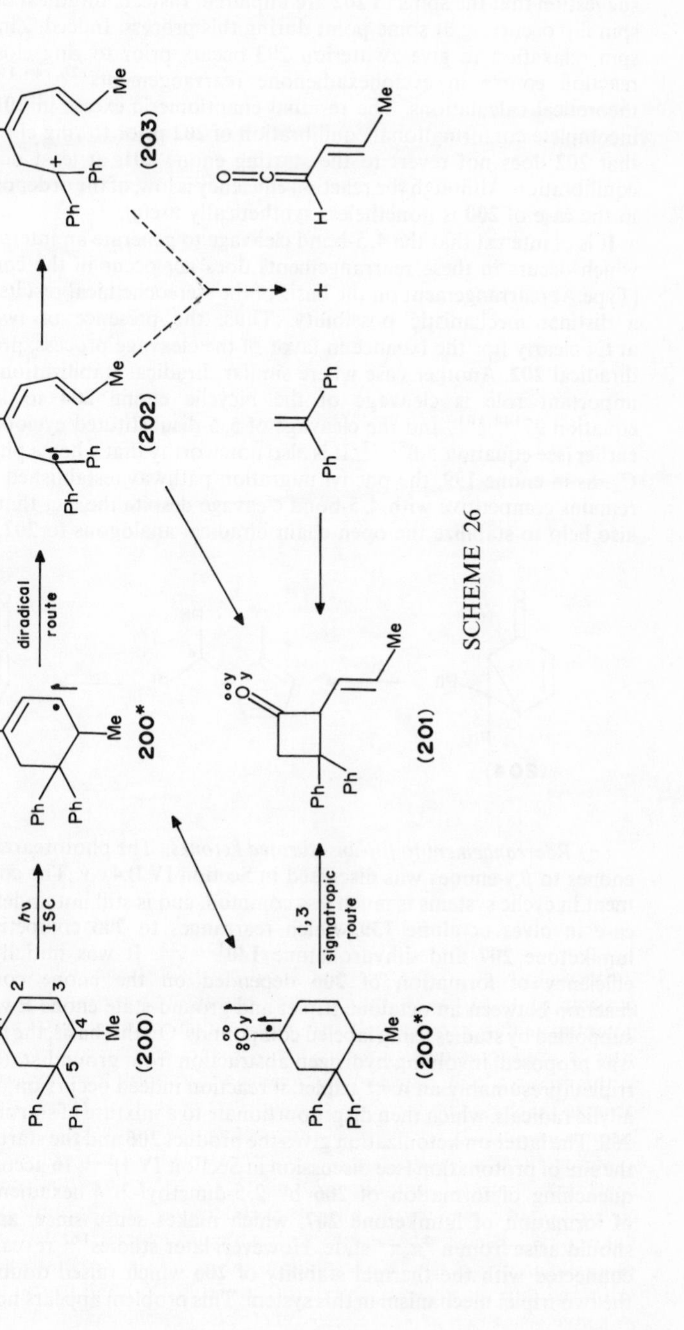

Me

(203)

Me

O

Ph
Ph
+
Ph

O

H
Me

+

Ph
Ph

Me

(202)

Ph
Ph

diradical
route

O
y

Me

200*

Ph
Ph

O
y

Me

(201)

Ph
Ph

hv
ISC

1,3
sigmatropic
route

O
y

1
2
3
4
5

Me

(200)

Ph
Ph

O
y

Me

(200*)

Ph
Ph

SCHEME 26

ketene, but this process does not appear to take place, as indicated above, supporting the suggestion that the spins in **202** are unpaired. Instead, diradical **202** closes to **201**, with a spin flip occurring at some point during this process. Indeed, Zimmerman suggests that spin relaxation to give zwitterion **203** occurs prior to ring closure, analogous to the reaction course in cyclohexadienone rearrangements[126,144,160], and supported by theoretical calculations. The residual enantiomeric excess in **201** probably arises from incomplete conformational equilibration of **202** prior to ring closure. The data establish that **202** does not revert to the starting enone **201**, at least not after conformational equilibration. Although the reaction efficiency is low, of the order of 0.01–0.03, the reaction in the case of **200** is nonetheless synthetically useful.

It is of interest that the 4, 5-bond cleavage to generate an intermediate triplet diradical which occurs in these rearrangements does *not* occur in the course of the lumiketone (Type A) rearrangement on the basis of the stereochemical results[139,140] although it was a distinct mechanistic possibility. Thus, the presence of two phenyl substituents at C_5 clearly tips the balance in favor of the cleavage process, providing stabilization of diradical **202**. Another case where similar diradical stabilization undoubtedly plays an important role is cleavage of the bicyclic enone **204** to ketene **205**, shown in equation 97[144,161], and the cleavage of 5, 5-disubstituted cyclopentenones **53** discussed earlier (see equation 50)[82,83]. It is also noteworthy that when a phenyl group is present at C_4, as in enone **199**, the phenyl migration pathway (established by deuterium labeling) remains competitive with 4, 5-bond cleavage despite the fact that this phenyl group can also help to stabilize the open-chain biradical analogous to **202**.

$$(97)$$

(204)   **(205)**

(v) Rearrangement to β,γ-unsaturated ketones. The photorearrangement of acyclic α,β-enones to β,γ-enones was discussed in Section IV.B.4.c.v. The corresponding rearrangement in cyclic systems is much less common, and is still not understood. The best known case involves octalone **139**, which rearranges to **206** competitive with formation of lumiketone **207** and dihydroketone **140**[121,122]. It was initially reported[121] that the efficiency of formation of **206** depended on the enone concentration, indicating reaction between an octalone triplet and ground-state enone E was involved, which was supported by studies using labeled compounds. On this basis, the mechanism in Scheme 27 was proposed, involving hydrogen abstraction from ground-state enone by the octalone triplet (presumably an n, π* triplet, if reaction indeed occurs on oxygen) to give a pair of allylic radicals, which then disproportionate to a mixture of starting enone **139** and dienol **208**. The latter on ketonization gives the product **206** and the starting enone depending on the site of protonation (see discussion in Section IV.1)[121]. In accord with this mechanism, quenching of formation of **206** by 2, 5-dimethyl-2, 4-hexadiene was much less than of formation of lumiketone **207**, which makes sense since, as discussed earlier, **207** should arise from a $^3\pi, \pi^*$ state. However, later studies[162] revealed analytical problems connected with the thermal stability of **206** which raised doubts as to the validity of the two triplet mechanism in this system. This problem appears not to have been resolved.

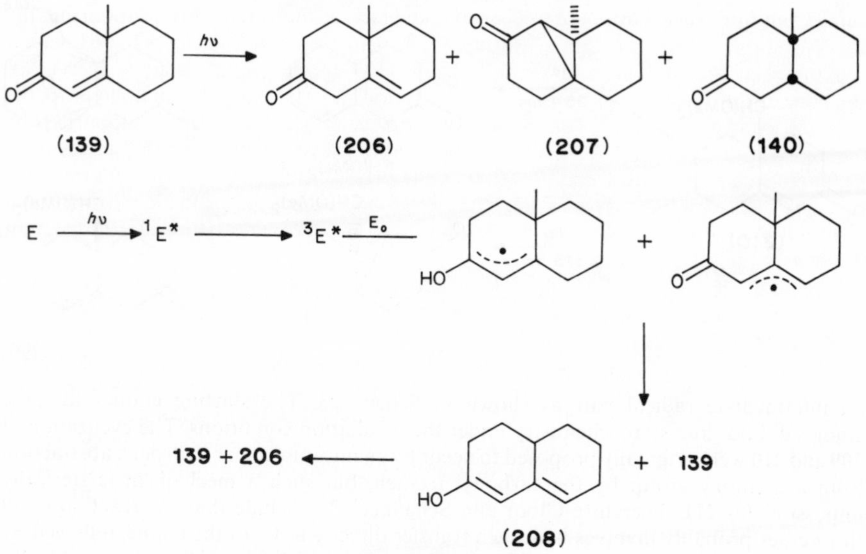

SCHEME 27

(vi) Allylic rearrangements and cyclizations: wavelength-dependent photochemistry. As discussed in the earlier review of enone rearrangements by this author[5], enones **209**, **210** and **211** give different sets of reactions on excitation at 254 nm into the $\pi \to \pi^*$ absorption band, and at $\geqslant 313$ or $\geqslant 340$ nm on excitation into the $n \to \pi^*$ absorption band[163]. These are shown in equations 98 and 99. It can be seen that $n \to \pi^*$ excitation of **209** and **210** leads to deconjugation (see above) and lumiketone formation, while $\pi \to \pi^*$ excitation results in allylic rearrangements ([1,3]-sigmatropic shifts)[164] along with cyclization of the ether moiety to the β-carbon of these enones. Enone **211** is dead on long-wavelength excitation, but at short wavelengths cyclization is observed.

The allylic rearrangements are definitely intramolecular, while stereochemical studies show that the migrating group loses stereochemistry in the course of reaction, suggesting that the [1,3]-shift is not concerted, but involves as the major pathway formation of

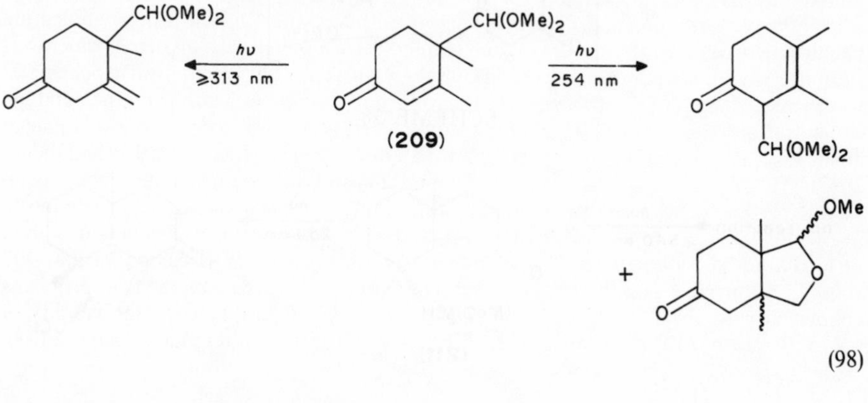

(98)

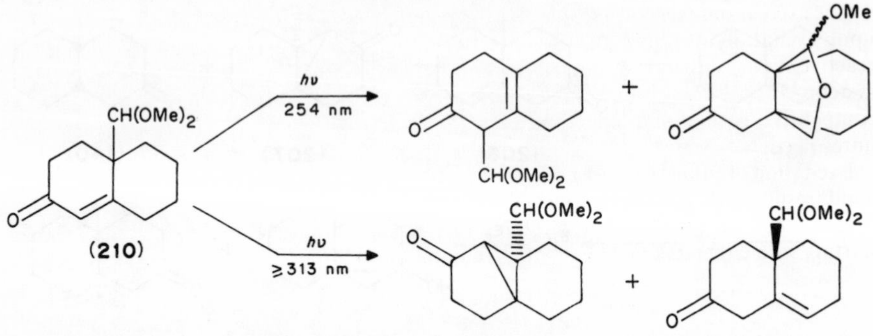

$$(99)$$

an intermediate radical pair as shown in Scheme 28. The starting enones in these studies did not lose stereochemistry under the irradiation conditions. The cyclization of 209 and 210 were originally proposed to occur by an intramolecular hydrogen abstraction from a methoxy group by the carbonyl oxygen, but such a mechanism is sterically impossible for 211. Therefore Gloor and Schaffner[165] conclude that the reaction in all three cases probably involves hydrogen transfer directly to C_α of the enone, followed by radical cyclization at C_β. Note that this is different from the reaction course observed with cyclopentenones by Agosta and coworkers (equation 50)[82,83].

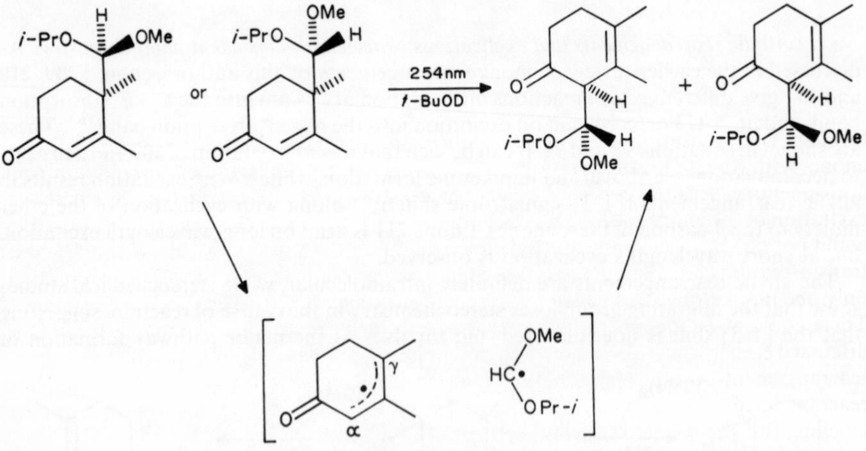

SCHEME 28

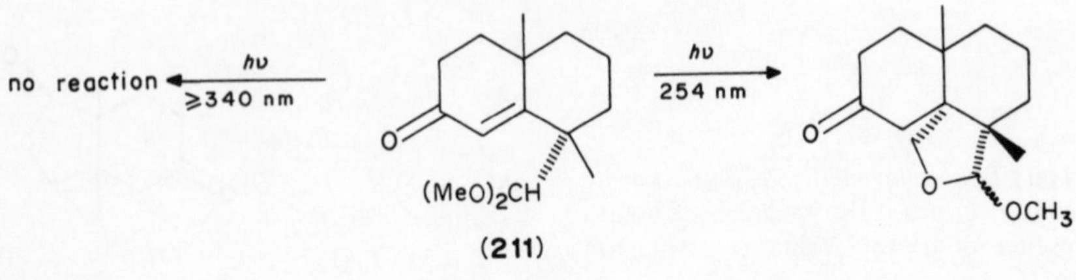

The excited state responsible for the [1, 3]-shifts and radical cyclization is not accessible upon excitation into the enone S_1 state, but rather requires excitation into S_2. Thus, the reactive state is either S_2 or the triplet state T_3, since both T_1 and T_2 lie below S_1 in these systems[163,165]. The authors prefer an interpretation in which S_2 undergoes reaction competitive with radiationless decay to S_1, but this matter remains experimentally unresolved.

Excitation of either the (E)- or (Z)-propenyl enones 212 and 213 in the n → π* absorption bands results in E–Z isomerization, deconjugation and lumiketone formation, as shown in equation 100; only the lumiketone with the (E) configuration in the side-chain was formed starting with either 212 or 213[135]. On irradiation into S_2 the tricyclic ketone 195 was also

(100)

formed. The cyclization efficiency was surprisingly not much higher starting with the enone with the (Z) configuration in the side-chain. The product distributions on acetophenone sensitization in benzene or t-BuOH were similar to those on direct excitation of the enones at long wavelengths, i.e. excitation into S_1. The cyclization to 195 could not be sensitized using acetophenone under any conditions. Mechanistically, these transformations are analogous to those discussed above, except for the observation that only one lumiketone stereoisomer is obtained starting with either of the isomeric enones. There could be factors operating in this system which are different from those in systems discussed earlier which allow isomerization in the side-chain concomitant with the Type A rearrangement. Another possibility is that the lumiketone is formed by a di-π-methane rearrangement via diradical 214. The two routes can be distinguished by appropriate labeling, but the results of such experiments have not been reported.

(214)

(vii) Rearrangements of photogenerated enone radical anions. Givens and Atwater[166] recently reported the reaction of octalones 215 upon irradiation in 2-propanol in the presence of triethylamine (equation 101). The process was shown to involve electron

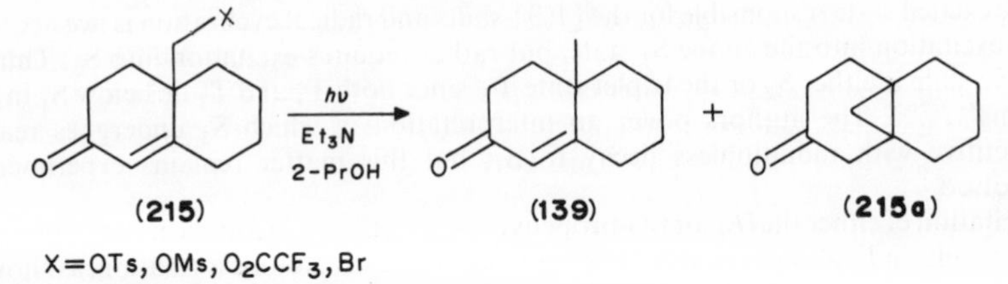

(215)   (139)   (215a)   (101)

X = OTs, OMs, O₂CCF₃, Br

transfer from the amine to the excited octalone by the appropriate inverse dependence of quantum efficiency on the inverse of the amine concentration, and piperylene quenching indicated that it was a triplet process. The surprising finding is that the major product is octalone **139** (> 95%) rather than the cyclopropyl ketone **215a**, which is the major product from reduction of **215** by lithium in liquid ammonia, a process that is also supposed to proceed via a radical anion of **215**[167]. Scheme 29 shows the mechanism proposed by Givens and Atwater, involving internal nucleophilic displacement to give a cyclopropyl-carbinyl radical A which should undergo facile reversible ring opening to the more stable homoallyl-neopentyl radical B. The ratio of products would then depend critically on the nature and concentration of the hydrogen donor, which in this case could be the solvent or $Et_3N^{+\cdot}$. Under reductive conditions, such as Na/NH_3, it is proposed that the intermediate radicals are first reduced to the corresponding anions, which are probably also in equilibrium, to give the two final products on protonation. It is clear that the main pathway in the photochemical system involves free radical ring opening and hydrogen abstraction to give octalone **139**.

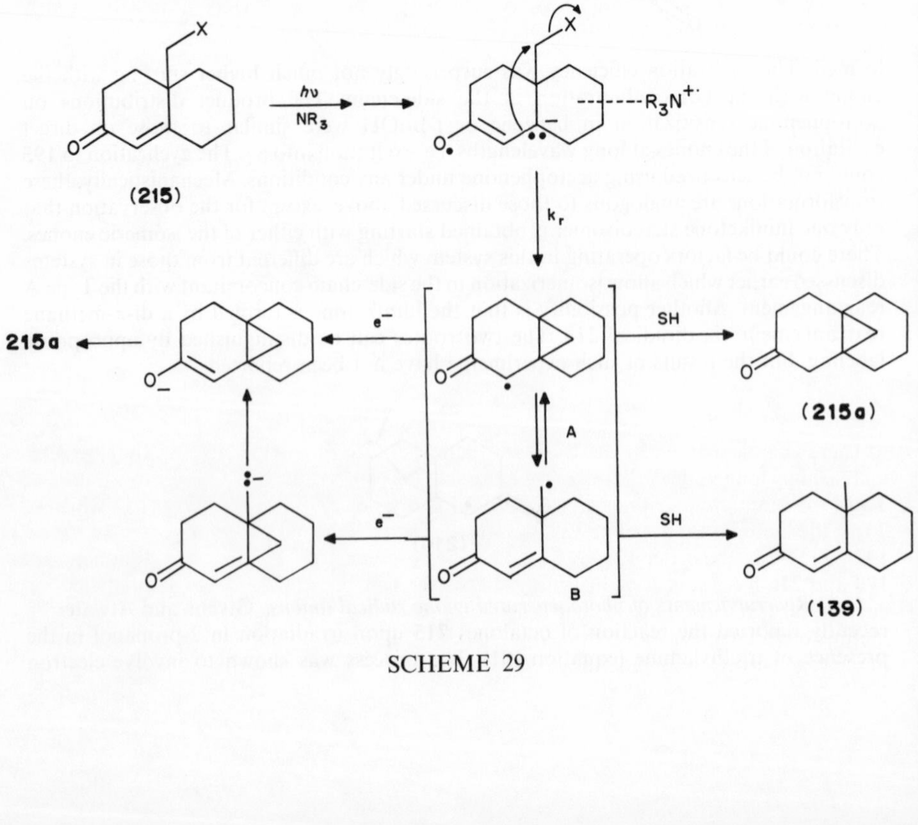

SCHEME 29

d. Direct observation of triplet states of cyclohexenones by nanosecond laser flash techniques. On laser flash excitation of 1-acetylcyclohexene **168** and cycloheptenone **94**[81], two transient intermediates are observed, one very short-lived (16 and 11 ns, respectively, in cyclohexane) and the other relatively long-lived (15 and 45 μs, respectively). The long-lived species B have been identified as the ground-state *trans* enones, as previously discussed. The short-lived transient A from acetylcyclohexene which has a very different UV spectrum (λ_{max} 285 nm) from that of the *trans* enone B (λ_{max} 345 nm), was shown to be a direct precursor of B since the rate of decay of A was equal to the rate of growth of B. Since transient A is quenched by oxygen (see Table 3), Bonneau and Fornier de Violet[168] conclude that A is a twisted (orthogonal) triplet state of acetylcyclohexene. The similarity of the absorption spectrum for transient A from cycloheptenone (λ_{max} 280 nm), coupled with its short lifetime, lead Bonneau to conclude that this species was also a twisted triplet, as previously discussed[81].

Using the same technique, Bonneau[81] observed short-lived transients from cyclohexenone and cyclopentenone (τ 25 and 30 ns, respectively, in cyclohexane) which absorbed in the same region as transients A above, and which were quenchable by oxygen; he concluded that these were also triplet π, π^* states of these enones[81]. In these cases, no transient absorption was observed that could be assigned to a ground-state *trans* enone. Subsequently, a number of other cyclohexenones have been examined using the laser flash technique by Schuster and coworkers in collaboration with Bonneau, Scaiano and Turro[80,169-174] using different laser excitation wavelengths (Nd:Yag laser at 353 or 265 nm; nitrogen laser at 337 nm) in a variety of solvents. The data are summarized in Table 3. In addition to the parent system, the cyclohexenones studied to date include 4, 4-dimethylcyclohexenone **124**, testosterone as well as testosterone acetate **132**, octalone **139**, phenanthrone **154**, bicyclo[4.3.0]nonenones **170** and **216**, 3-methylcyclohexenone **217**, 2, 4, 4-trimethylcyclohexenone **218** and 4,4,6,6-tetramethylcyclohexenone **219**. All of these enones produce transients that show strong UV absorption in the range 270–350 nm with maxima in most cases at *ca* 280 nm; in all cases, these transients are quenched by oxygen. The remote possibility that the '280 nm transients' of the cyclohexenones might be triplets of cyclohexadienols produced by photoenolization is countered by the fact that enone **219**, which cannot photoenolize, produces a transient with similar absorption spectrum and lifetime as **124**. By analogy with the systems reported previously and on the basis of other considerations (see below), it has been concluded that these transients are relaxed twisted cyclohexenone triplet π, π^* states.

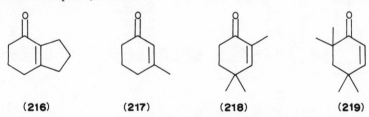

    (216)   **(217)**   **(218)**   **(219)**

There is a clear trend in the transient lifetimes given in Table 3, in that they become increasingly long as the structural constraints to twisting around the C=C bond increase. Thus, the monocyclic enones give transient lifetimes of 25–40 ns except for **217** which is 70 ns; the lifetimes for octalone **139**, the tricyclic phenanthrone **154** and the steroid enone **132** increase in the order **139** < **154** < **132**; the conformationally rigid bicyclononenones **170** and **216** have very long lifetimes, > 1500 ns. This order is precisely as predicted for energetically relaxed twisted enone triplets, based on Bonneau's proposal[81] that the lifetime of such triplets is determined principally by the energy gap between the minimum in the π, π^* triplet surface and the energy of the ground-state enone at that same geometry.

TABLE 3. Triplet lifetimes of enones and quenching rate constants measured by laser flash techniques

Enone	Solvent	τ_{dir}	τ_{ext}	k_q NA(MN) mol^{-1}s^{-1}	k_q Piperylene lmol^{-1}s^{-1}	k_q O$_2$ lmol^{-1}s^{-1}	Ref.
2-Cycloheptenone 94	Cyclohexane (CH)	11			$\leqslant 10^7$		81, 99
1-Acetylcyclohexene 91	CH	16			$\leqslant 10^7$		81, 168
2-Cyclohexenone 118	CH	25(28)	29	7.5×10^8	4×10^7	3.5×10^9	169(174)
	Acetonitrile (AN)	23(40)	24	1.0×10^9	$<10^8$	7×10^9	169, 172(174)
	MeOH	27					174
4,4-Dimethylcyclohexenone 124	t-BuOH	32	29	4.5×10^8	$<10^8$ (nonlinear)		169
	i-PrOH (IPA)	21	28	8.5×10^8			169
	CH	24(28)				7.5×10^9	169, 172
	AN	23	28	1.0×10^9		7.1×10^9	169, 172
4,4,6-tetramethylcyclohexenone 219	MeOH	40					174
	AN	34					174
3-Methylcyclohexenone 217	AN	72	69	(4.6×10^9)			80
	CH	72		(10×10^{10})			80
Octalone 139	AN	57	58	(1.3×10^9)			173
	CH	47					173
Cyclopentenone 50	AN (0.006 M)	125					80
	AN (0.06 M)	43	48	4.1×10^9			80
	CH (0.016 M)	68	80	1.3×10^{10}			80
Phenanthrone 154	CH	30			$<10^8$	5×10^9	81
	AN	131				3.8×10^9	169
	IPA	150	145	4.0×10^8	2×10^8 (nonlinear)	2.5×10^9	169
Testosterone acetate 132	AN	385	413	(4.4×10^9)			173
	AN	380	295	5.0×10^9			172
	CH	292					173
Testosterone	AN	330			10^9	2.2×10^9	81
	CH	440					169
Bicyclononenone 170	IPA	1,400	1,500	4.5×10^9	$\sim 10^9$		169, 172
	t-BuOH	2,300				3.0×10^9	169, 172
Bicyclononenoal 216	AN	1,700		$(\geqslant 10^{10})$			173
	CH	1,850					173

The monocyclic enones have sufficient flexibility to allow close approach of the two surfaces at something near to an orthogonal triplet geometry, as in the case of cycloheptenone (Figure 8), while the T_1–S_0 energy gap for the bicyclononenones should be much larger, corresponding to a planar enone chromophore. Thus, based on the structural dependence of the transient lifetime data, the assignment of the transient absorption centered at ca 280 nm to energetically relaxed π, π^* triplets seems secure. Supporting evidence from quenching data is given below. This assignment for the lowest triplet excited state, at least for cyclohexenone, cyclopentenone and 1-acetylcyclohexene, has recently received confirmation by direct observation of these enone triplets at 77 K using time-resolved electron paramagnetic resonance[174a].

The ability of dienes to quench the '280 nm transients' has been determined in several cases from the dependence of transient lifetime on quencher concentration, with often confusing results. The more rigid enones are quenched linearly by dienes such as piperylene and 1, 3-cyclohexadiene at close to diffusion-controlled rates, suggesting that triplet energy transfer is energetically favorable. However, plots of $(\tau_{obs})^{-1}$ or optical density (ΔOD) vs. [piperylene] curve downward at higher diene concentrations for the more flexible systems, such as cyclohexenone itself, enone **124** and to a lesser extent phenanthrone **154**[81,172]. It is known that cyclohexenones undergo photoaddition to conjugated dienes[110,175], which would gradually deplete the diene concentration, accounting at least in part for the observed nonlinear quenching behavior. Quenching rate constants k_q, estimated from the linear portion of these plots at low diene concentrations, are given in Table 3. It is obvious that these are much below the diffusion-controlled limit, indicating (a) that the triplet excitation energies of these enones is less than or equal to that of piperylene (58–59 kcal mol^{-1}) and/or (b) that there is a geometric inhibition of triplet energy transfer from the nonplanar enone triplets. In any event, the contrast between the linear and efficient quenching by piperylene of transient triplets derived from testosterone and enone **170** and the inefficient nonlinear quenching of transient triplets derived from the more flexible cyclohexenones strongly supports the proposal that the triplet energies of the more rigid enones are in the range of 67–70 kcal mol^{-1}, while the latter are highly twisted species with energies closer to 60 kcal mol^{-1}.

These conclusions are supported by studies using naphthalene (NA, $E_T = 60.9$ kcal mol^{-1}) and 1-methylnaphthalene (MN, $E_T = 59.6$ kcal mol^{-1}) as triplet quenchers[80,169,170,172]. Because ground-state UV absorption by naphthalenes obscures the transient absorption of the enones, it is not possible to directly measure transient quenching by naphthalenes. However, the growth of NA/MN triplet absorption at 413/420 nm can be easily observed, and in general gives an excellent fit to a simple first-order rate law. The rise time for NA or MN triplet absorption depends on quencher concentration as shown in equation 102, where τ_0 is the lifetime of the donor (enone) triplet

$$k_{obs} = 1/\tau_{obs} = 1/\tau_0 + + k_q[Q] \qquad (102)$$

in the absence of quencher and k_q is the quenching rate constant corresponding to transfer of triplet excitation from the enone triplet to NA or MN. Values of k_q and τ_0 obtained from the slope and reciprocal of the intercept of plots of $(\tau_{growth})^{-1}$ vs. [NA] or [MN] are given in Table 3; given the relatively large error in estimating intercepts from these plots, the agreement between these extrapolated values of τ_0 and the triplet lifetimes τ_T determined by measurement of triplet decay at 280–350 nm is excellent. There can be little doubt that the species transferring triplet excitation energy to NA and MN is indeed the species responsible for UV absorption at 280–350 nm[170]. The variation in k_q is again consistent with the argument that the donor is an energetically relaxed enone triplet: triplet transfer is effectively diffusion controlled for the more rigid cyclohexenones, especially **170** and **216**, slightly less so for the steroid enones, even less for octalone **139** and substantially lower for the monocyclic enones. This trend is exactly as expected if the cyclohexenone triplet energy is gradually being reduced by the increasing ability to twist around the C=C bond. Thus, the rigid enones must have triplet excitation energies well above 65 kcal mol^{-1}, probably closer to 70 kcal mol^{-1}, while the energies of the twisted triplets of simple cyclohexenones must be near 60 kcal mol^{-1}, as previously concluded on the basis of other experimental evidence. The energies of several enone triplets have recently been determined by time-resolved photoacoustic calorimetry, and are completely consistent with these proposals[241].

The variation of rate constants for oxygen quenching $k_q(O_2)$ with enone structure also makes sense on this basis. Bonneau and coworkers[176] have observed differing rates of quenching of planar vis-à-vis perpendicular styrene and α-naphthylethylene triplets by oxygen. This was attributed to changes in spin statistics associated with different mechanisms of oxygen quenching, a spin-exchange mechanism ($k_q \sim k_{diff}/3 = 9 \times 10^9$ l mol^{-1} s^{-1}) for quenching of perpendicular triplets, and an energy transfer mechanism ($k_q \sim k_{diff}/9 = 3 \times 10^9$ l mol^{-1} s^{-1}) for quenching of planar triplets[177]. The same trend is seen with the cyclohexenones in Table 3, as higher values of $k_q(O_2)$ are consistently observed for the more twisted cyclohexenone triplets compared with the enone triplets constrained to planarity. This effect should be associated with variations in yields of singlet oxygen with enone structure, but this has yet to be studied.

e. Competition between various reaction pathways of photoexcited cyclohexenones. As indicated above, cyclohexenones can undergo $[_\pi2 + _\pi2]$ dimerization, reduction and rearrangement on direct or triplet-sensitized excitation. In addition, as will be discussed in detail below, they undergo photoaddition to alkenes to give cyclobutanes and (less often) oxetanes. In the presence of amines, photoexcited cyclohexenones give a mixture of dimers, reduction (**146**) and addition (**222**) products, as shown in equation 103 for enone **124** and triethylamine[111,172,178]. The simplest mechanism for this reaction involves electron

(103)

(124)   **(146)**   **(222)**   **(147)**

transfer from amines to enone triplets to generate a radical ion pair which, after proton transfer, gives the pair of free radicals **220** and **221**; radical combination would give the β-adducts **222** as a pair of diastereomers, while a second hydrogen abstraction would give the saturated ketone **146** (Scheme 30). Pienta and McKimmey[178] reported that the ratio of (**222** + **146**) to photodimers was linear with Et$_3$N concentration (from 1 to 7 M), and that the same ratio was independent of the enone concentration (from 0.006 to 1.6 M). On this basis, they proposed that all of these products arose from a dimeric excited species (or excimer). However, this mechanism is inconsistent with a number of observations of Schuster and coworkers, to be described below. An attempt to replicate Pienta's data[178] was unsuccessful. Insogna and Schuster[179] found that with an enone concentration of 0.3 M and amine concentrations above 0.5 M, the [2 + 2] photodimers formed from **124** in the absence of amine (see above) were no longer formed; in addition to adduct **222** and ketone **146**, two new products were observed whose mass indicates they are stereoisomeric dimers of radical **220** or (less likely) the corresponding species with the odd electron at C$_2$. It is possible that these products of reductive dimerization were mistaken for [2 + 2] photodimers in the earlier study[178]. The product distribution is in fact totally consistent with Scheme 30 and it is therefore not necessary to postulate reaction via a triplet excimer.

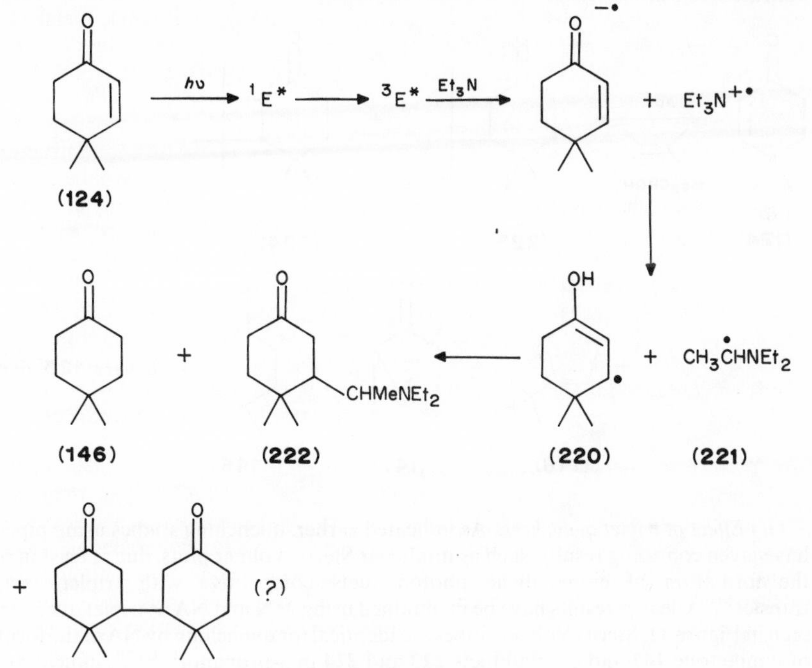

SCHEME 30

A series of studies have been undertaken to determine the competition between the various reaction pathways of photoexcited **124** and, to a lesser extent, other cyclohexenones, as a function of (a) enone concentration, (b) triplet quenchers, (c) amines as quenchers and (d) alkenes as quenchers. These will be summarized below, with details to be given elsewhere.

(i) Variation of enone concentration. Quantum yields of photodimers of cyclohexenones depend on the enone concentration, as originally reported by Wagner and Bucheck[79]. What was surprising was that, upon increasing the concentration of **124** from 0.23 to 1.64 M in 2-propanol, there was no effect within experimental error on the quantum yields of photoreduction product **146** and lumiketone **147**[172]; over this concentration range, the optical densities at ≥ 300 nm were all > 2.0, so the results could not be explained by differential light absorption. Furthermore, a linear double reciprocal plot of $(\phi_{dim})^{-1}$ vs. [enone]$^{-1}$ was observed, with different slopes for the two dimers, as in Wagner's earlier study of cyclohexenone itself[79]. In a second experiment involving **124** in 2-propanol in the presence of tetramethylethylene (TME), it was found that increasing enone concentration from 0.93 to 1.86 M caused only a very slight reduction (overall $< 12\%$) in the yields of the cycloadducts **223** and **224** (equation 104) while there was a very large increase ($> 173\%$) in dimer yields. In the inverse experiment, a negligible effect of increasing the concentration of alkene (cyclohexene) on the yields of photodimers of **124** was observed[111]. These data indicate: (1) photodimerization of **124** occurs from a different

triplet state of **124** than is responsible for rearrangement, reduction or cycloaddition to alkenes. This conclusion is confirmed by both flash and steady-state quenching studies as indicated below.

(*ii*) *Effect of triplet quenchers.* As indicated earlier, quenching studies using piperylene have given confusing results, such as nonlinear Stern–Volmer plots, due at least in part to the formation of enone–diene photoadducts competitive with triplet excitation transfer[110]. Cleaner results have been obtained using MN and NA as triplet quenchers. As seen in Figure 11, Stern–Volmer slopes are identical for quenching by NA of the formation of lumiketone **147** and cycloadducts **223** and **224** in 2-propanol[111,170], indicating these three products arise from one and the same triplet (or, much less likely, two thermally equilibrated triplets). When **124** is irradiated in neat TME, oxetane **225** is obtained in addition to [2 + 2] cycloadducts **223** and **224**[180]. The Stern–Volmer slope for quenching of formation of **225** is experimentally different from that for the other two adducts[111], supporting the proposal that the oxetane arises from a ^{3}n, π^* state and the other adducts from a $^3\pi, \pi^*$ state.

A series of studies of the effect of naphthalene on the products of irradiation of **124** were carried out by Schuster and Nuñez[110]. They observed that Stern–Volmer slopes for formation of the photoreduction product **146** in 2-propanol were consistently 15–20% lower than for the photorearrangement products **147** and **148** (which gave the same slopes within experimental error). The same effect was seen using 1:1 *i*-PrOH benzene. However, in *t*-BuOH–toluene (1:1) NA had virtually no effect on the formation of **146** while it quenched formation of **147** and **148** with efficiency similar to that in the other solvents.

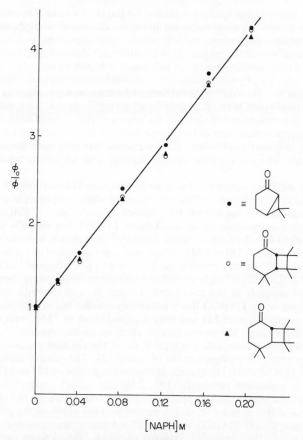

FIGURE 11. Quenching by naphthalene [Naph] of concomitant photorearrangement of enone **124** and its photoaddition to tetramethylethylene in 2-propanol (from Reference 111)

Based upon these data, it is concluded that photoreduction and photorearrangement of enone **124** occur via different triplet states of the enone. This conclusion is supported by similar observations using acenaphthene ($E_T = 59.2$ kcal mol^{-1}) as triplet quencher, and by the finding of very pronounced differential quenching by naphthalene of photoreduction vs. photorearrangement of the related enone **226** in 2-propanol (equation 105). Photorearrangement of **226** is much less efficient than for **124**, and sensitization studies with p-methoxyacetophenone indicate ϕ_{isc} is ~ 0.75 for **226** compared with 1.0 for **124**.

$$\text{(226)} \xrightarrow[\text{Me}_2\text{CHOH}]{h\upsilon} \quad + \quad \tag{105}$$

Otherwise, the course of the reaction is similar for the two systems. The conclusion that different enone triplets are responsible for photorearrangement and photoreduction is supported by results using amines and alkenes as triplet quenchers, as discussed below. The effect of the solvent on the magnitude of the differential quenching effect is tentatively attributed to relative stabilization of the two reactive triplet states.

Differential quenching by naphthalene of the photochemical reactions of phenanthrone 154 in 2-propanol was discussed earlier in Section IV.B.4.6, and was taken as evidence for simultaneous reaction via both $^3n,\pi^*$ and $^3\pi,\pi^*$ states[122]. In that case, unlike 124, the triplet leading to lumiketone 159 also was the source of the cis-fused reduction product 155. By analogy, photoreduction of 124 to give 146 may also occur from both $^3n,\pi^*$ and $^3\pi,\pi^*$ states by different mechanisms and to extents that vary with the nature of the solvent, although there are as yet no data to support such an interpretation.

(iii) Effect of amines as quenchers. The reactions of enone 124 that occur in the presence of triethylamine as well as other amines was discussed above. The effect of amines on the other photoreactions of enone 124 has been studied. Dunn[172] showed that triethylamine strongly quenches rearrangement to lumiketone 147 but has virtually no effect on photoreduction to 146 in 2-propanol, while Tucker[111] showed that both photorearrangement and cycloaddition of 124 to TME in 2-propanol are quenched to the same extent by triethylamine, although neither reaction was quenched by t-butylamine[172]. DABCO (1,4-diazabicyclo[2.2.2]octane) strongly quenches lumiketone formation in 2-propanol and acetonitrile, but uniquely in this case Stern–Volmer plots of the data have distinctly upward curvature; while DABCO has a noticeably smaller but nonetheless significant effect on both dimerization of 124 and photocycloaddition to TME, with nicely linear Stern–Volmer behavior, it has no measurable effect on photoreduction[111].

These data clearly demonstrate the separation of the reaction pathways leading to photorearrangement and photoreduction of enone 124. The relative ability of amines (DABCO > Et$_3$N > t-BuNH$_2$) to quench photorearrangement of 124 to 147 is inversely related to their ionization potentials (7.10, 7.50 and 8.64 eV, respectively), strongly suggesting that the triplet-quenching process involves electron transfer, as has been shown for other ketone–amine systems[181], and in accord with the nature of the products in the case of Et$_3$N (see above). DABCO does not afford enone–amine adducts nor induce photoreduction of the enone[172], consistent with the known reluctance of DABCO$^{+\cdot}$ to lose a proton. The pronounced curvature seen in the Stern–Volmer plots for quenching by DABCO of photorearrangement of 124, and the diversion of these curves from the linear plots for quenching by DABCO of photocycloaddition to TME, are puzzling[111]. The quenching data discussed previously as well as the effect of alkenes on photorearrangement (see below) all strongly indicate that these reactions occur from a common triplet. The relatively long lifetime of the DABCO radical cation compared with radical cations derived from the other amines studied may be a factor. One possibility is that DABCO or DABCO$^{+\cdot}$ may intercept an intermediate, not evident from other studies, on the way to lumiketone from the reactive triplet.

The effect of DABCO on the photochemistry of phenanthrone 154 in 2-propanol was also investigated[172]. A clear distinction between quenching of formation of 156 as opposed to 155 and 159 (which showed the same Stern–Volmer slope) was observed, consistent with earlier evidence that these products arise from different triplets.

(iv) Effect of alkenes as triplet quenchers. Photorearrangement of 124 and photocycloaddition to alkenes apparently take place via a common triplet-excited state, according to the naphthalene quenching data[111,170]. If so, one would expect that alkenes ought to inhibit lumiketone formation. Since the enone triplet implicated in the lumiketone rearrangement is apparently highly twisted, the observation that photoaddition of enones such as 124 to electron-rich alkenes affords *trans*-fused cycloadducts as major products

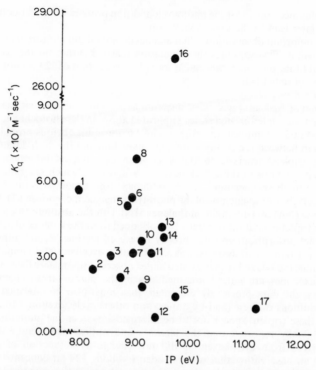

FIGURE 12. Rate constants for quenching of photorearrangement of enone **124** by alkenes and alkynes as a function of alkene ionization potential. Quencher: 1. 2,3-dimethyl-2-octene; 2. tetramethylethylene; 3. 2-methyl-2-butene;   4. norbornene;   5. 2,4,4-trimethyl-1-pentene; 6. cyclohexene;   7. *trans*-3-hexene; 8. cyclopentene; 9. *trans*-2-pentene; 10. *cis*-2-pentene; 11. 2-methyl-1-butene; 12. 4-octyne; 36. 1-heptene; 14. 3,3-dimethyl-1-butene;   15. 3,3-dimethyl-1-butyne;   16. *cis*-1,2-dichloroethylene; 17. maleic anhydride; 18. bicyclo[2.2.2]-2-octene; 19. dimethyl acetylenedicarboxylate

Studies by Schuster and coworkers[182,183] established that alkenes which undergo photocycloaddition to cyclic enones such as **124**, cyclopentenone and cyclohexenone quench the lumiketone rearrangement of **124** in 2-propanol and acetonitrile. The efficiency of this quenching, measured by the slopes ($K_q = k_q \tau$) of the linear Stern–Volmer plots of the quenching data, was studied as a function of the ionization potential (IP) of twenty alkenes and alkynes by Rhodes and Schuster[183], where k_q is the second-order rate constant for interaction of the alkene with the reaction intermediate of lifetime τ. A linear inverse correlation of $\log k_q^{rel}$ vs. alkene IP was anticipated, because it has been widely accepted since the pioneering studies of Corey and coworkers[184] that the initial alkene–enone interaction involves formation of a π complex (exciplex) with the alkene acting as donor and the enone triplet as acceptor. In fact, no such correlation of the quenching data with IP was observed (see Figure 12). As an example, tetramethylethylene (TME) has the second lowest IP of the alkenes used in Rhodes' study, but was one of the poorest quenchers of the photorearrangement[183]. In two cases in which pairs of *cis–trans* alkene isomers were utilized, the *cis* alkene was the better quencher. The data suggest that steric effects dominate over electronic effects in this system, and raise serious questions about the donor–acceptor mechanism for the photocycloaddition process. Recent laser flash data to be discussed later lead to the same conclusion.

One other important observation is that alkenes do not inhibit photoreduction of **124** to **146** in 2-propanol. This supports the conclusion reached earlier on the basis of other quenching data that photorearrangement and photoreduction of **124** do not occur via a common triplet-excited state.

f. Correlation of flash data and quenching data from continuous irradiations. Given the multitude of enone triplet-excited states implicated above in the photochemical reactions of **124** and **154** and, by implication, other cyclohexenones, an obvious problem concerns the relationship between the steady-state quenching data and the flash data. That is, are the transient triplets observed in the laser flash experiments intermediates in the photochemical reactions of these cyclohexenones, and if so, which reactions?

The answer to these questions can be obtained from a comparison of the Stern–Volmer slopes $(k_q \tau_T)$ for quenching of the photorearrangement of enones **124**, **154** and **226** by naphthalene (NA) or 1-methylnaphthalene (MN) with the absolute values of the rate constants for triplet-excitation transfer k_q (as obtained from the kinetics of growth of NA and MN triplet absorption) and the directly measured lifetimes τ_T for transient triplet decay[172]. For a number of systems, this agreement is excellent, and cannot be simply fortuitous. Thus, the relaxed triplet-excited states observed upon laser flash excitation of these cyclohexenones are indeed intermediates in the photorearrangements of these systems. From the data previously presented, they must also be intermediates in the photocycloadditions of these (and by implication other) cyclohexenones to alkenes.

However, these triplets appear not to be intermediates in enone photodimerizations. Bonneau originally reported that the lifetimes and optical densities of the '280 nm' transient derived from cyclohexenone did not change as a function of enone concentration[81], an observation that was extended to enone **124** (at concentrations up to 0.99 M) by Dunn and Schuster[172]. From the dependence of the quantum efficiency of photodimerization on enone concentration, a pronounced reduction in the lifetime of the enone triplet undergoing photodimerization should be seen; however, no such effect on the '280 nm transient' is observed. Also, the triplet lifetimes observed in the flash experiments are much longer than those estimated in Wagner's study of photodimerization[79]. Thus, it appears that photodimerization proceeds via a short-lived higher-energy triplet, perhaps a planar π, π^* triplet, rather than the relaxed (twisted) species observed in the flash studies. The triplet which leads to the enone dimers cannot be a precursor of the triplet observed by laser flash techniques, since the optical density of the latter is not reduced as the enone concentration is increased, up to 0.99 M in the case of **124**. This, perhaps surprising, conclusion requires that the twisted triplet is formed independently, perhaps via a twisted singlet excited state of the enone.

Flash excitation of cyclohexenones in the presence of amines has provided controversial data. New long-lived transient absorption is observed in the region of the triplet absorption (270–350 nm) which quickly obscures the triplet decay and makes lifetime measurements difficult[185,186]. However, Pienta has reported that at low amine concentrations, the enone triplet lifetimes appear to increase, which he associates with a rapid equilibrium between the enone triplet and an enone–amine exciplex, as in Scheme 31. Equilibrium constants for formation of the exciplex could be calculated from these data[186]. However, Dunn, Schuster and Bonneau performed similar experiments, and were unable to see the effect of amines on enone triplet lifetimes reported by Pienta[185]. Recently, Weir, Scaiano and Schuster studied the effect of several amines on the triplet decay of several cyclohexenones[174]. After correction for light emission following the flash (by subtracting the photomultiplier response after the flash with the analyzing beam off), they find that the amines definitely quench the enone triplets. The second-order rate constants determined in their study are given in Table 4. From the dependence of the initial optical

$$EN \longrightarrow {}^1EN^* \longrightarrow {}^3EN^* \overset{K_{eq}}{\underset{\text{Amine}}{\rightleftharpoons}} (EN \cdots \text{Amine})^*$$

$$k_{obs} = \frac{k_1 + k_2 K_{eq}[\text{Amine}]}{1 + K_{eq}[\text{Amine}]}$$

Where $\quad k_1 = (\tau_T)^{-1} \quad$ and $\quad k_2 = (\tau_{EN \cdots \text{Amine}}^*)$

SCHEME 31

density of the long-lived absorption on amine concentration, the lifetimes of the enone triplets intercepted by the amines were determined, which allows their identification as the '280 nm' transients, i.e. as relaxed ${}^3\pi, \pi^*$ enone triplets. The quenching rate constants, taken with values of τ_T, agree with Stern–Volmer slopes for quenching by amines of photorearrangement in the case of 124[111,172].

Dunn, Schuster and Bonneau[185] have observed new transient absorption centered at 450 nm when enones are excited at 353 nm in the presence of DABCO. They established that this absorption is due to DABCO$^{+\cdot}$. From the rate of growth of this absorption, rate constants for the interaction of DABCO with the relevant enone triplet, and the lifetime of the enone triplet which is intercepted by DABCO were determined. It is clear that DABCO is intercepting the '280 nm transient', i.e. the twisted enone π, π^* triplet, in an electron transfer process which generates amine radical cations. Once again, excellent agreement was found between the directly measured quenching rate constants and the initial slopes for quenching by DABCO of enone photorearrangement. There can be little doubt from these data that the twisted enone π, π^* state which leads to lumiketone and alkene addition products is the species which is intercepted by amines. The shorter-lived triplet responsible for photodimerization is also quenched by amines, but with much lower efficiency according to the steady-state data[172].

The effect of alkenes on the rate of decay of the transient enone triplets will be discussed below in the section dealing with the enone–alkene photocycloaddition process.

g. Intermolecular photocycloaddition of cyclic enones to alkenes. (i) Introduction. Eaton and Hurt[78a] originally discovered the photodimerization of cyclopentenone, and shortly afterward Eaton reported that 2-cyclopentenone reacts similarly with cyclopentene to give cycloadduct 227 (equation 106)[187]. Corey and coworkers studied analogous $[_\pi2 + _\pi2]$ photocycloaddition of cyclohexenone to a variety of alkenes, and established many of the basic features of this reaction, as discussed below[184]. Corey first recognized the potential of the enone–alkene photocycloaddition reaction as a key element of a scheme for synthesis of natural products, as demonstrated in his synthesis of caryophyllene[188]. Since

TABLE 4. Triethylamine quenching of cyclohexenone triplets

Compound	Solvent	k_q (l mol^{-1} s^{-1})
118	acetonitrile	$(9.0 \pm 0.8) \times 10^7$
	cyclohexane	$(9.2 \pm 4.6) \times 10^7$
124	methanol	$(1.1 \pm 0.4) \times 10^8$
	acetonitrile	$(3.7 \pm 0.5) \times 10^7$
192	cyclohexane	$(1.3 \pm 0.5) \times 10^8$
154	acetonitrile	$(2.0 \pm 0.6) \times 10^6$

these pioneering studies, inter- and intramolecular photocycloadditions of cyclic enones (cyclopentenones and cyclohexenones for the most part) to alkenes (also called photoannelations) have become probably the most frequently utilized photochemical reaction in the arsenal of synthetic organic chemists. Several excellent reviews of the applications of this methodology have been recently published, so that there is no need here to review this large literature in detail[7-9,189]. This discussion will be concerned with the basic features of the intermolecular reaction, and recent studies relating to its mechanism. The synthetically important intramolecular enone–alkene photocycloaddition possesses several other features which will be discussed separately.

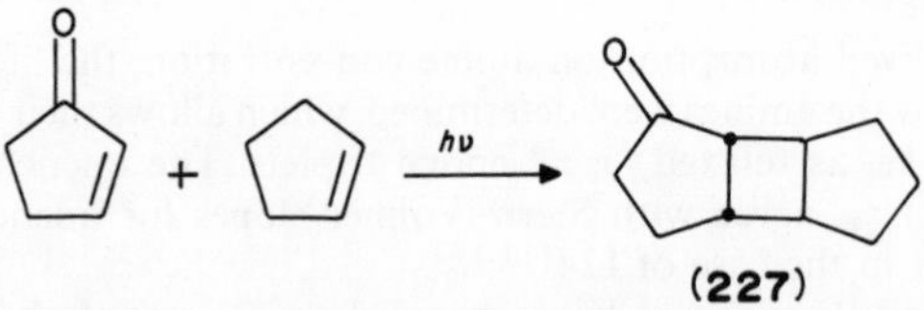

$$(106)$$

(227)

(ii) Scope, regiochemistry and stereochemistry of the [2 + 2] photocycloaddition of cyclic enones to alkenes. Corey and coworkers[184] originally established that cyclohexenone undergoes photocycloaddition to a variety of alkenes, including isobutylene (equation 107), 1,1-dimethoxyethylene (DME) (equation 108), cyclopentene (equation 109), allene (equation 110), vinyl acetate (equation 111), methyl vinyl ether

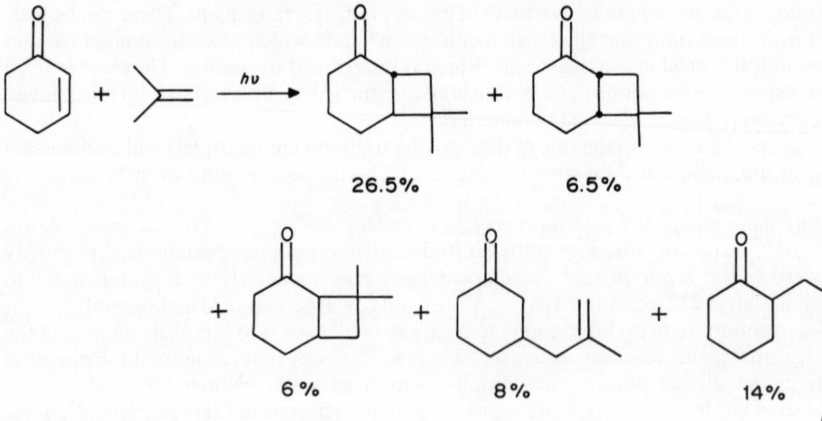

26.5%   6.5%

6%   8%   14%

$$(107)$$

49%   21%

+ 6% other products

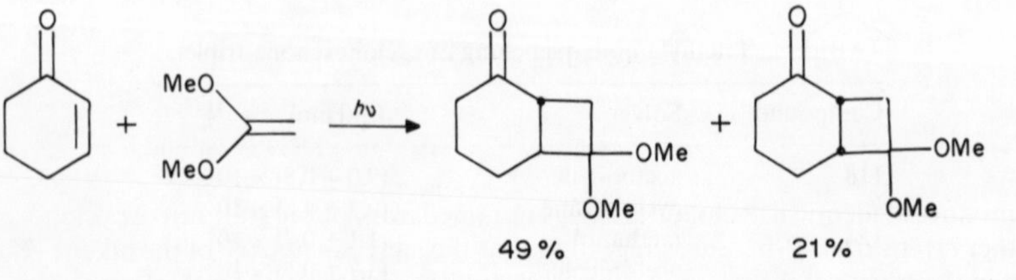

$$(108)$$

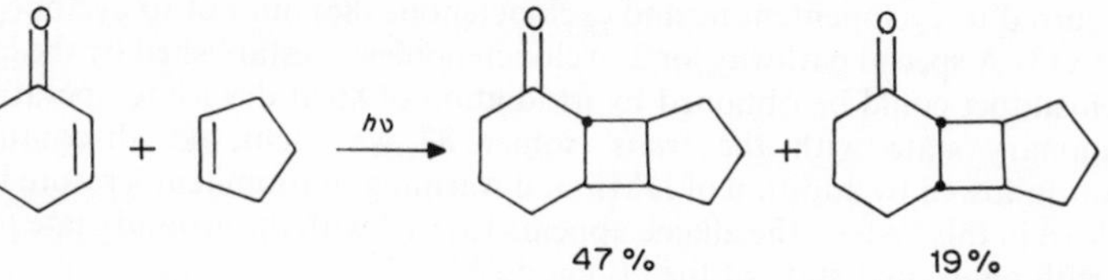

(109)

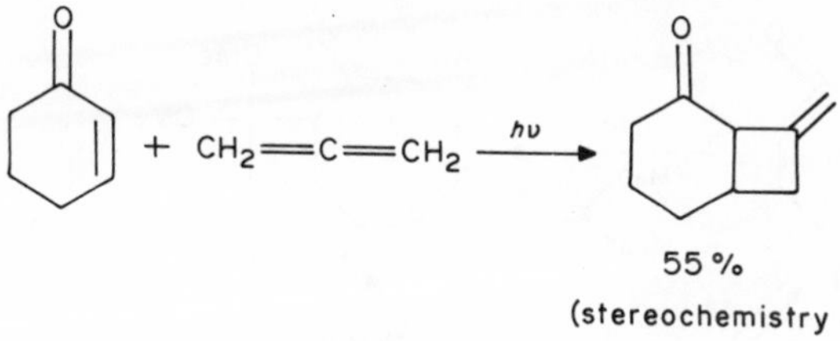

47 %   19 %

(110)

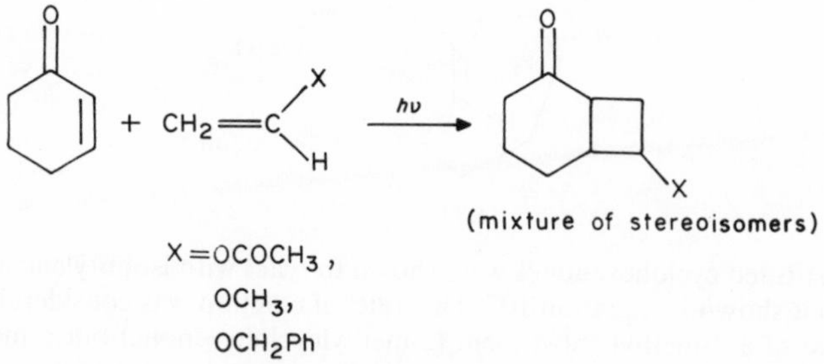

55 %
(stereochemistry probably *cis*)

(111)

(mixture of stereoisomers)

X = OCOCH$_3$,
   OCH$_3$,
   OCH$_2$Ph

(equation 111) and benzyl vinyl ether (equation 111). As can be seen from equations 107–109, the major cycloadduct in these systems has a *trans* fusion of the four- and six-membered rings, which was a feature that Corey immediately recognized as having potential value in the synthesis of complex ring systems. The stereochemistry of the adducts in the other cases was not established. In the case of isobutylene, the cycloadducts were accompanied by olefinic ketones which Corey suggested were formed via disproportionation of 1,4-diradical intermediates (see below). Orientational specificity was clear in all these cases. A much 'slower' reactions was observed between cyclohexenone and acrylonitrile, which gave four adducts whose structures and stereochemistry were not established, although it was suggested that they have the regiochemistry shown in equation 112, opposite to that seen above.

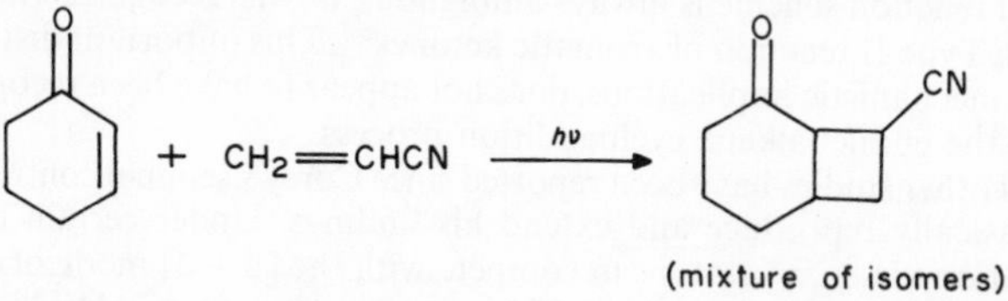

(112)

(mixture of isomers)

A mixture of identical cycloadducts was obtained from photoaddition of cyclohexenone to either *cis*- or *trans*-2-butene, suggesting that the stereochemistry of the alkene reactant is lost in the course of the reaction. Recovery and IR analysis of the starting materials after various reaction times established that < 1% isomerization of the alkene had occurred[184].

Utilizing DME as his model alkene, Corey and coworkers established that photoaddition occurred to cyclopentenone and cyclooctenone (**86**) but not to cycloheptenone **94** (equation 113). A special pathway for 2-cyclooctenone was established by the fact that the same cycloadduct could be obtained by irradiation of **86** at dry ice temperatures until a photostationary state with the *trans* isomer **87** was achieved, discontinuation of irradiation, followed by addition of DME and warming to room temperature in the dark. Thus, at least in this system, the alkene appears to react with the ground-state *trans* enone, and not with an excited state of the *cis* enone[184].

(113)

Methyl-substituted cyclohexenones were shown to react with isobutylene in a manner analgous to that shown in equation 107. The 'rate' of reaction was considerably reduced by the presence of a 2-methyl substituent (2-methylcyclohexenone) but a methyl at C_3 (enone **217**) had no effect on the 'rate'.

Corey and coworkers[184] determined 'relative rate factors' for reaction of five alkenes with cyclohexenone from irradiation of the enone in the presence of pairs of alkenes in large molar excess, with cyclopentene as the reference alkene. The numbers (corrected for statistical factors) were as follows: DME, 4.66; methoxyethylene, 1.57; cyclopentene, 1.00; isobutylene, 0.13–0.40; allene, 0.23. These 'rate factors' were of key importance in Corey's mechanistic proposals, as will be seen shortly.

Since it will be a matter of considerable importance latter in this discussion, it should be pointed out that the relative 'rates' frequently mentioned in Corey's paper[184] are of course not really rates at all but rather relative quantum efficiencies for disappearance of starting material and/or appearance of products. The relationship of relative or even absolute quantum efficiencies of product formation to rates of particular steps in a multistep photochemical reaction scheme is always ambiguous, as was recognized many years ago for the Norrish Type II reaction of aromatic ketones[50]. This important distinction, which has important mechanistic implications, does not appear to have been recognized in prior discussions of the enone–alkene cycloaddition process.

A number of other studies have been reported since Corey's seminal contributions to this area which basically reproduce and extend his findings. Under certain circumstances, oxetane formation via $^3n, \pi^*$ states can compete with the [2 + 2] mode of cycloaddition. Thus, as mentioned earlier, when enone **124** is irradiated in neat TME[180] oxetane **225** is obtained in addition to *trans*- and *cis*-fused cycloadducts and open-chain adducts, but no trace of **225** can be detected when the reaction is carried out in acetonitrile as solvent[111,172]. Earlier observations of differential quenching of formation of the two cycloadducts using di-*t*-butylnitroxyl[180] were interpreted in terms of two different triplet precursors for the stereoisomeric adducts; more recent studies using naphthalene as quencher demonstrate clearly that both $[2+2]$ adducts arise from a common triplet state[111], and that $(t\text{-Bu})_2\text{NO}$ is probably intercepting triplet 1,4-biradical intermediates.

Steroid enones also give a mixture of cis- and trans-fused cycloadducts with simple alkenes. Thus, Rubin and coworkers[190] found that testosterone propionate **228** reacts with cyclopentene to give a 4.5:1 mixture of cis- and trans-fused adducts as shown in equation 114, while the corresponding dienone **229** gives only a single trans-fused adduct (equation 115). Rubin compared the ratio of cis- to trans-fused adducts in equation 114 and in addition of 2-cyclohexenone to cyclopentene in ethyl acetate solvent at room temperature and in dry ice–acetone ($-78\,^\circ$C). The cis–trans ratio of adducts in equation 114 decreased as the temperature was lowered and also varied with the alkene concentration; the product ratio in the cyclohexenone–cyclopentene reaction appeared to be relatively insensitive to temperature changes. No quantum yield data were reported.

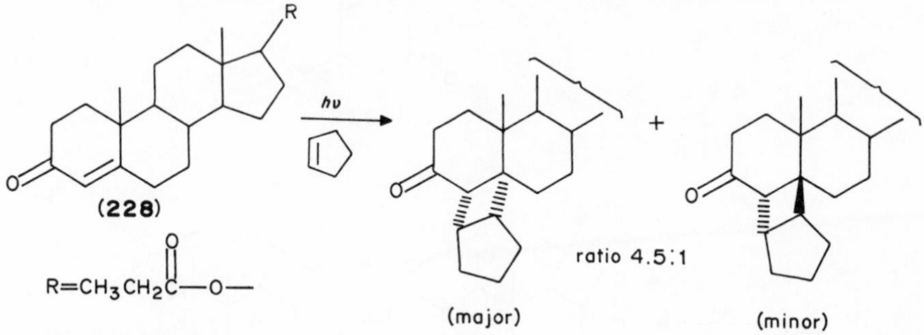

ratio 4.5:1

(major) (minor)

$R = CH_3CH_2\overset{\displaystyle O}{\overset{\|}{C}}\!\!-\!O-$

(114)

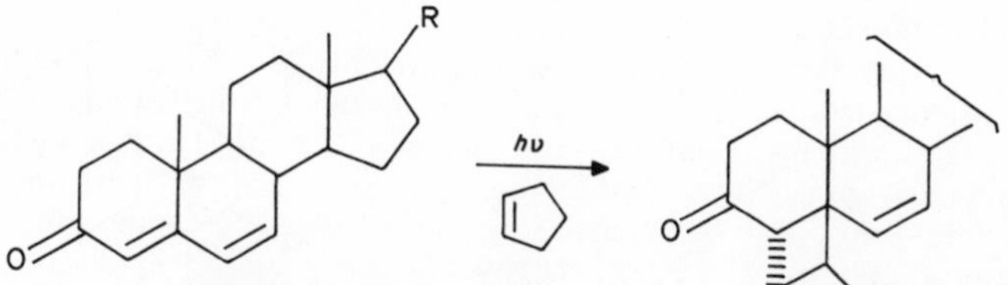

(115)

Lenz[191] has studied photocycloaddition of the Δ^1-steroid enone **230** to cyclopentene, a ketene acetal and isobutene. In all cases, trans-fused adducts are formed as major products. The products in each case are shown in equations 116–118; note the formation of disproportionation product in the last equation as in Corey's original study

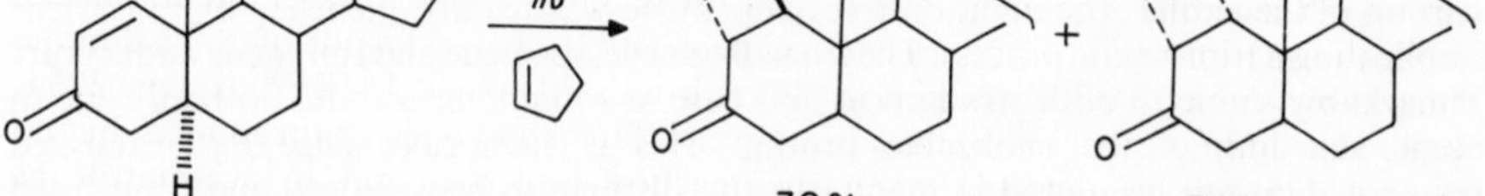

38% 56%

(230)

(116)

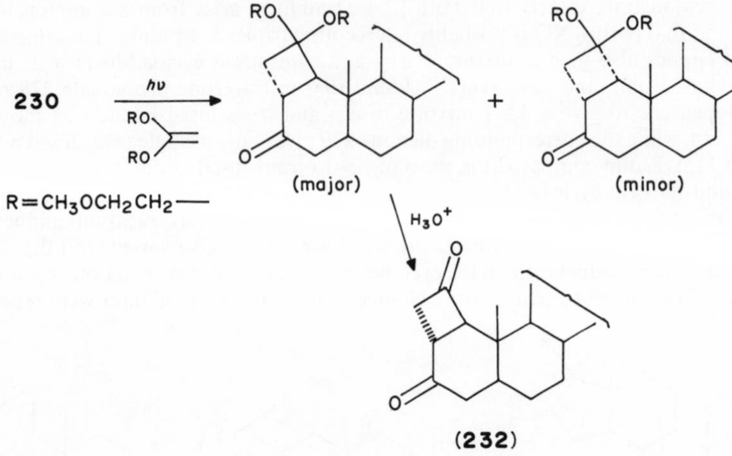

(117)

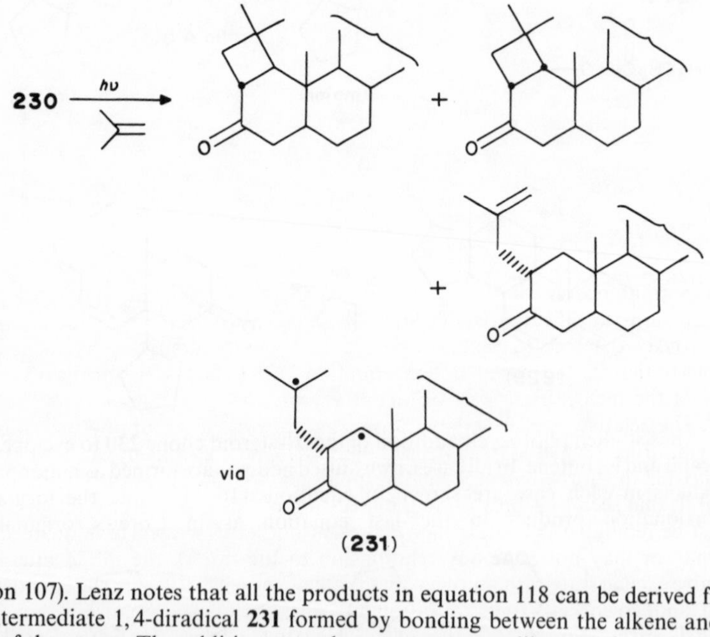

(118)

(equation 107). Lenz notes that all the products in equation 118 can be derived from the same intermediate 1,4-diradical **231** formed by bonding between the alkene and the β-carbon of the enone. The addition to cyclopentene was readily quenched by piperylene implicating a triplet state process. The *trans*-fused cyclopentene and isobutene adducts are remarkably stable to both strong acid and base at room temperature, probably due to steric shielding of the enolizable proton. This is significant since base-catalyzed epimerization has been used in many cases to distinguish between *cis*- and *trans*-fused adducts[7,8,184]; thus, such assignments must be made with care. By hydrolysis of the *trans*-fused adduct in equation 117, Lenz was able to isolate the first *trans*-fused cyclobutanone **232**.

Cantrell and coworkers[192] studied photocycloaddition reactions of 3-R-cyclohexenones, with R = methyl, phenyl and acetoxy, to several alkenes. The course of reaction of the 3-methyl enone (**217**) with DME is shown in equation 119 as an example of the behavior observed. The regiochemistry is similar to that observed by Corey[184] for the parent system, and again a mixture of cis-and trans-fused isomers is obtained. Irradiation of **217** in the presence of cis- or trans-1, 2-dichloroethylene gave the same two major adducts in slightly different yields (equation 120) plus three unidentified minor products. Once again, one sees loss of alkene stereochemistry en route to the photoadducts. The photoadducts of 3-phenylcyclohexenone are all presumed to be cis-fused, since they are stable to base (see below).

26%   38%

(119)

(120)

An important observation[192] was that photocycloaddition of **217** to acrylonitrile 'proceeded surprisingly rapidly'. The structures and stereochemistry of the adducts were not rigorously determined, but it appears that the predominant regiochemistry is analogous to that for the parent system (equation 112). In fact, acrylonitrile was the 'most reactive' of the olefins used with **217**, contrary to Corey's results with cyclohexenone itself[184]. The 'relative rates' found by Cantrell and coworkers[192] for photoaddition to **217** are: acrylonitrile, 7.68; ethoxyethylene, 1.96; DME, 1.27; cyclopentene, 1.00; isobutene, 0.59; trans-1, 2-dichloroethylene, 0.40. Except for acrylonitrile, the trend is similar but not identical to that seen by Corey (note the inversion of DME and ethoxyethylene). Again, one must be reminded that these data represent not 'rates' but relative quantum yields, which may or may not have any relationship to the rate of the initial interaction of the reactive excited state of the enone with the alkene, as will be clear in the later discussion. Cantrell and coworkers tried to rationalize the apparently 'abnormal' reactivity of acrylonitrile by invoking some rather *ad hoc* mechanistic alternatives. It is significant that the other enones studied did not react especially 'fast' with acrylonitrile.

McCullough and coworkers[193] isolated three cycloadducts from photoaddition of 4, 4-dimethylcyclohexenone **124** to cyclopentene (CP), whose structures are given in equation 121. The major adduct has the cis–anti–cis stereochemistry, and the other two adducts have trans-fused cyclobutane rings. Similar behavior is observed on photoaddition of CP to 3-methylcyclohexenone **217** as seen in equation 120, in agreement with findings from Cantrell's laboratory[192]. However, photoaddition of CP to 3-phenylcyclohexenone (**233**) gives only cis-fused cycloadducts, as shown in equation 123.

(121)

(122)

(123)

McCullough and coworkers also showed[193] that the ratio of rearrangement products to CP adducts of enone **124** in methanol was unchanged in the presence of 0.05 M naphthalene, although the efficiency of reaction was reduced by a factor of 3. These data are in agreement with results of the more extensive quenching studies of Tucker discussed earlier, which demonstrated conclusively that photoaddition of **124** to tetramethylethylene and photorearrangement to **147** in 2-propanol are quenched to exactly the same extent by naphthalene[170,172]. McCullough and coworkers[193] propose that formation of *trans*-fused adducts in major amounts from **124** and **217** with CP can be rationalized by attack of a nonplanar triplet state of the enone on ground-state alkene, with initial bonding at C_2 of the enone, and rapid formation of the second bond of the cyclobutane before the enone moiety can relax to its equilibrium configuration. Since it is likely that these reactions proceed via 1,4-biradicals[194] (see below), the second step must be fast, since it is otherwise difficult to understand why an equilibrated biradical would give highly strained adducts with *trans*-fused four- and six-membered rings. The fact that neither Cantrell[192] nor McCullough[193] found evidence for *trans*-fused adducts from 3-phenylcyclohexenone and a variety of alkenes implies that (a) this enone does not twist about the C=C bond and/or (b) the intermediate biradical is stabilized by the phenyl group, enhancing the probability that it will assume a relaxed geometry prior to ring closure. These arguments will be considered later in the detailed discussion of the mechanism of the photocycloaddition reaction.

The effect of incorporating large alkyl groups at C_3 of the enone on the stereochemical course of photoaddition was examined by Singh[195] with carvenone **234** and 3-*tert*-butyl-2-cyclohexenone. From addition of **234** to ethoxyethylene and DME, both *cis*- and *trans*-fused cycloadducts were isolated, the latter as minor products. In all other reactions of these two enones, only *cis*- fused adducts were formed. Photocycloaddition of **234** to dimethyl maleate **235** was sluggish, but one adduct identified as **236** could be isolated in low yield (equation 124). Loss of stereochemical integrity of the alkene moiety is again apparent in this reaction. Photoaddition of **234** to dimethyl acetylenedicarboxylate to give **237** was also observed (equation 125).

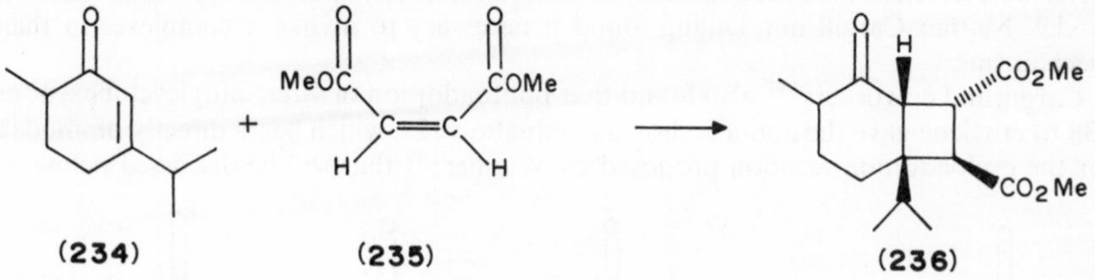

(234) **(235)** **(236)**

(124)

234 + MeOC—C≡C—COMe ⟶

(237)

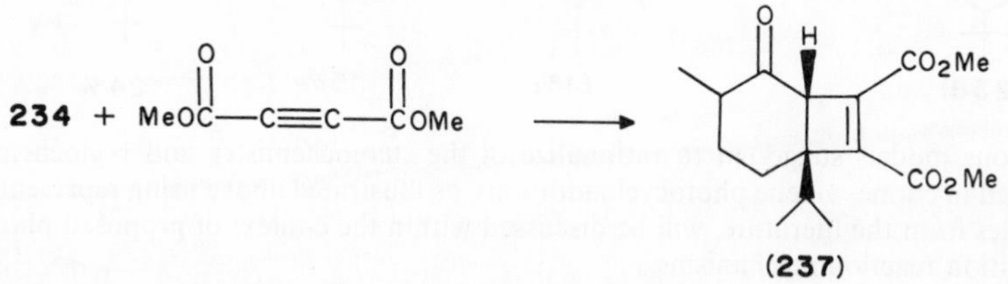

(125)

Three cycloadducts were observed by Cargill and coworkers[196] on photoaddition of $\Delta^{1,6}$-bicyclo[4.3.0]nonen-2-one **216** to either *cis*- or *trans*-2-butene, as shown in equation 126. As seen previously, stereochemical integrity of the alkene is lost. The product distribution from each alkene isomer could be rationalized in terms of preferential

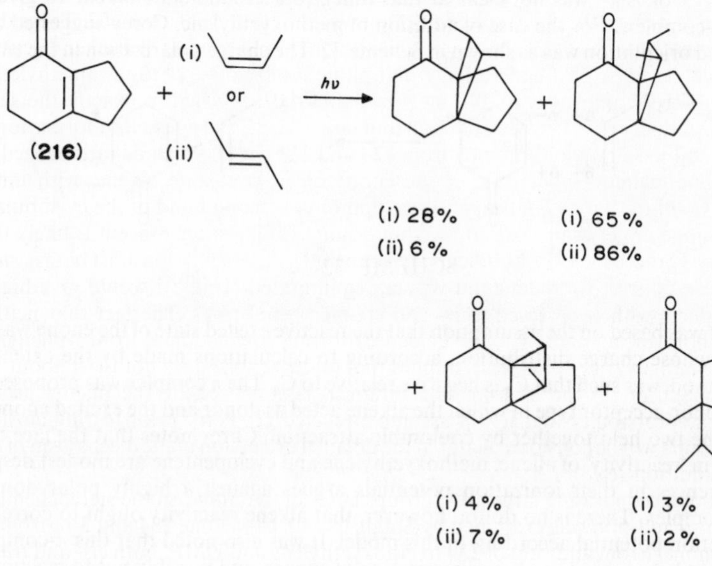

(216) (ii)

(i) 28%   (i) 65%
(ii) 6%   (ii) 86%

(i) 4%   (i) 3%
(ii) 7%   (ii) 2%

(126)

formation of 1, 4-biradicals by bonding to the β-carbon and not the α-carbon of the enone, and by rotational equilibration of the biradicals prior to ring closure. The same conclusion was reached by Dilling and coworkers[197] from studies of addition of cyclopentenone to *cis*- and *trans*-1, 2-dichloroethylene. In both studies, however, no provision is made for possible reversion of biradical intermediates to ground-state enone and alkene (quantum yields were not measured in either study) which can seriously affect this type of mechanistic model. Neither Cargill nor Dilling found it necessary to invoke π complexes in their mechanisms.

Cargill and coworkers[198] also found that photoaddition of 4-*tert*-butylcyclohexenone **238** to ethylene gave the adducts shown in equation 127, which bears directly on models for the cycloaddition reaction proposed by Wiesner[199] that will be discussed below.

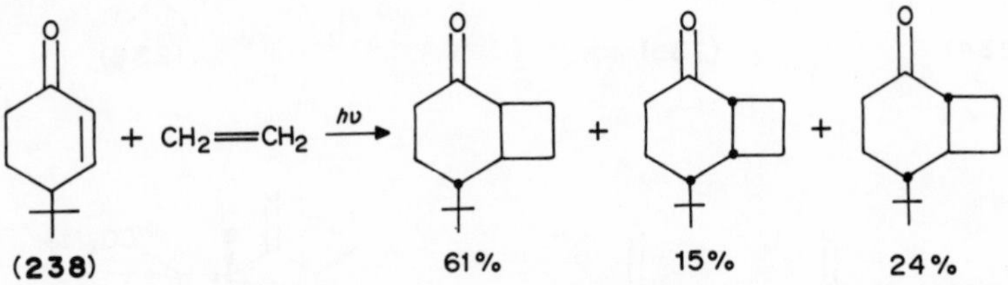

$$(127)$$

(238)   61%   15%   24%

Various models suggested to rationalize of the stereochemistry and regiochemistry observed in enone–alkene photocycloadditions, as illustrated above using representative examples from the literature, will be discussed within the context of proposed photocycloaddition reaction mechanisms.

(iii) The Corey–de Mayo mechanism for photocycloaddition of enones to alkenes. On the basis of the regiochemistry and the 'relative rate factors' associated with addition of alkene to photoexcited cyclohexenone, Corey suggested in 1964[184] that the first step of the reaction involved interaction of an enone excited state, which most likely was a triplet state (whether n, π^* or π, π^* was not clear at that time), to a ground-state alkene to give an 'oriented π-complex'. For the case of addition of methoxyethylene, Corey suggested that the preferred orientation was as shown in Scheme 32. The charge polarization in the enone

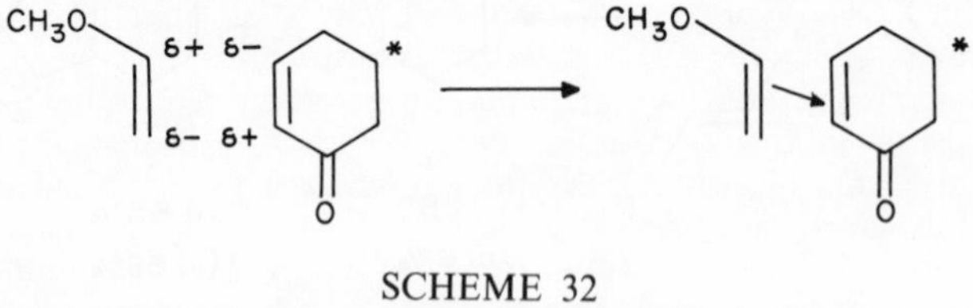

SCHEME 32

component was based on the assumption that the reactive excited state of the enone was an n, π^* state whose charge distribution, according to calculations made by the extended Hückel method, was such that C_β is negative relative to C_α. The π complex was proposed to be of the donor-acceptor type in which the alkene acted as donor and the excited enone as acceptor, the two held together by coulombic attraction. Corey notes that the face that differences in 'reactivity' of allene, methoxyethylene and cyclopentene are modest despite large differences in their ionization potentials argues against a highly polar donor–acceptor complex. There is no doubt, however, that alkene reactivity ought to correlate with ionization potential according to this model. It was also noted that this π-complex model cannot be extended to photodimerization of enones, and possibly not to reaction of enones with olefins possessing electron-withdrawing substituents such as CN or COOR. The importance of steric effects could not be assessed at the time the π-complex model was introduced. No evidence for ground-state π complexation of cyclohexenone and ethoxyethylene was observed.

Corey rejects the alternative hypothesis that the orientation in photocycloaddition is controlled by preferences in diradical formation, since it does not predict the correct regiochemistry in photoaddition of cyclohexenone to DME. Also, it was not in accord with the 'relative rate factors' determined earlier. However, it was necessary to invoke 1, 4-diradicals in order to rationalize the formation of disproportionation products as in equation 107, and the loss of stereochemistry upon photoaddition of cyclohexenone to the 2-butenes. The overall scheme proposed by Corey and coworkers[184] is as shown in Scheme 33.

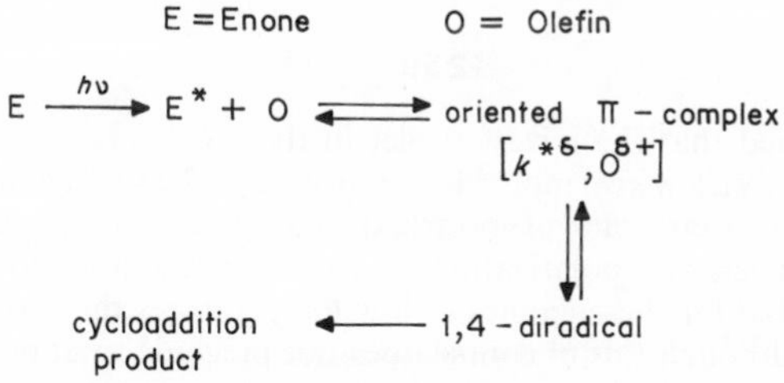

$$E = \text{Enone} \qquad O = \text{Olefin}$$

$$E \xrightarrow{h\nu} E^* + O \rightleftharpoons \text{oriented } \pi - \text{complex}$$
$$\left[_k *^{\delta-}, O^{\delta+} \right]$$

$$\text{cycloaddition} \longleftarrow 1, 4 - \text{diradical}$$
$$\text{product}$$

SCHEME 33

Before citing recent evidence bearing directly on the mechanism of photocycloaddition of enones to alkenes, some criticism of Corey's proposed mechanism[184] can be made in hindsight. First of all, without quantum yield data, it is impossible to say very much about the nature of the intermediates involved in this reaction. Any and all of the intermediates proposed by Corey (enone triplet, π complex, biradicals) can in principle revert to ground-state enone and alkene in competition with progress forward to cycloadduct. The importance of such reversion was not at all apparent until the first quantitative studies of this reaction were made by Loutfy and de Mayo[200] (see below). Without such information, it is not possible to relate the orientational specificity to preferred formation of one or another π complex, since the partitioning factors for progress vs. reversion may be very different for isomeric π complexes and biradicals. That is, the quantum yield ratio (or product yields) obtained in competition experiments, whether involving formation of adducts of two different olefins or isomers from a single olefin, is not related in a simple way to the rate of the initial enone–alkene association. That biradical reversion may well be the most important factor leading to quantum inefficiency in cycloadditions is suggested by McCullough and coworkers' observation[201] that irradiation of 3-phenyl-2-cyclohexenone 233 and cis-2-butene gave a much higher yield of trans-2-butene than of addition products, which he attributes to preferred reversion vs. cyclization of the intermediate 1, 4-biradical (239) in this system (see equation 128). They demonstrated by sensitization and quenching studies that this reaction proceeds via a relatively long-lived (1.59 μs) triplet state of 233; recent flash data[80] demonstrate that indeed the 3-phenylcyclohexenone T_1 state is very long-lived, as is T_1 of 3-phenylcyclopentenone[202].

Secondly, Corey's model[184] for the oriented π complex is based upon the assumption that the reactive excited state of the enone is a triplet n, π^* state. Corey diligently but unsuccessfully searched for phosphorescence from cyclohexenones in order to directly identify the lowest triplet state as n, π^* or π, π^*. Studies by Kearns, Marsh and Schaffner[203] of phosphorescence emission from steroidal enones at 77 K and 4.2 K published only a few

(128)

(233)   (major)   (minor)

via

(239)

years later established that the lowest triplet in these cases is a π, π^* triplet, with the lowest n, π^* triplet only a few kcal mol^{-1} higher in energy. The assignment was based upon the diffuseness of the spectra, lack of spectral overlap with $S_0 \rightarrow T_{n,\pi}$ absorption, lifetime data, heavy atom effects and polarization measurements. Jones and Kearns concluded sometime later[204] that for these enones at low temperatures the enone chromophore is essentially planar, although this of course does not preclude twisting around the C=C bond in these enones at higher temperature, or even at very low temperatures for conformational unconstrained cyclohexenones. The closeness of n, π^* and π, π^* states of 2-cyclopentenones is evident from phosphorescence studies on rigid systems carried out by Cargill, Saltiel and coworkers[90]. Those compounds without substituents on the C=C bond appeared to emit from $^3n, \pi^*$ states whereas those with substituents on the C=C bond showed emission from π, π^* triplets. The emission from the former group could be changed to that of the second group simply by adsorption on silica gel, which stabilizes $^3\pi, \pi^*$ states relative to n, π^* states. They suggest that the lowest relaxed triplet of simple cyclopentenones and cyclohexenones in solution is probably $^3\pi, \pi^*$ due to stabilization by torsion around the C=C bond. It is now known with virtual certainty from the studies summarized in Section IV.B.4.e. that the reactive excited state in enone–alkene photocycloadditions is the lowest π, π^* state of the enone, whose charge distribution is not predicted to be as shown in Scheme 32.

There are therefore very good grounds for challenging Corey's assignment of the nature of the reactive triplet state of cyclohexenone in photocycloaddition to alkenes, and the structure for the 'oriented π-complex' shown in Scheme 32, despite the fact that this model has been very successful in correlating regiochemical data in a large number of examples[7,8,189].

In his review of the 'enone photoannelation' reaction in 1971[205], de Mayo pointed out the kinetic deficiencies of Corey's original mechanism, and explicitly considered reversion of all possible reaction intermediates to ground-state enone and alkene (see Scheme 34). He and his coworkers measured quantum yields for photoaddition of cyclopentenone and in no case were they greater than 0.50 in neat olefin. Representative data (all at 334 nm) are: cyclohexene, 0.50; cyclopentene, 0.32; tetramethylethylene, 0.12; DME, 0.34 (at 313 nm). The effect of triplet quenchers (piperylene, acenaphthene) on the addition of cyclopentenone to cyclohexene in benzene and cyclohexane was determined. From the slopes of Stern–Volmer plots, assuming diffusion-controlled quenching, rate constants k_r for interaction of cyclohexene with cyclopentenone were calculated to be $2.3-5.0 \times 10^8$ l mol^{-1} s^{-1} and the rate constant k_d for unimolecular decay of the enone triplet was found to be $9-46 \times 10^8$ s^{-1}, depending on the quencher and the solvent. De Mayo recognized that these values of k_d were unreasonably large, indicating some deficiency in

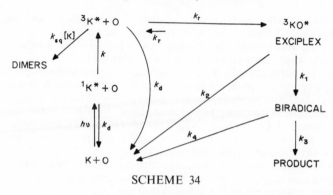

SCHEME 34

the reaction scheme, but it was unclear what the problem was. A temperature dependence of the quantum yield for photoannelation of cyclopentenone with several different olefins was observed; in some systems, these values increase as the temperature is lowered (cyclohexene, cyclopentene, cis-dichloroethylene) while in others (3-hexene) it decreases. For cyclopentenone–cyclohexene, the value increased from 0.46 at 27 °C to 0.72 at − 102 °C, while for cyclopentenone–cyclopentene it more than doubled from 0.23 at 27 °C to 0.61 at − 71 °C. It was concluded that the large changes in ϕ_{add} with temperature result from changes in the fraction of intermediate that gives product rather than from changes in the fraction of enone triplets trapped by alkene.

Loutfy and de Mayo[200] carried out the most extensive quantitative studies of enone photoannelation published to date. They studied the dependence of quantum yields for additions to cyclopentenone and cyclohexenone on temperature and on the alkene concentration at varying temperatures, as well as quenching of photoaddition at various temperatures with 2, 5-dimethyl-2, 4-hexadiene. To get rate constants, they assumed as before that quenching by the diene is diffusion controlled. From their data, values of k_d of $1.1 \pm 0.1 \times 10^8 \, s^{-1}$ and $3.3 \pm 0.3 \times 10^9 \, s^{-1}$ were found for cyclopentenone and cyclohexenone at concentrations of 0.10 and 0.14 M, respectively, corresponding to triplet lifetimes of ca 10 and 3 ns under these conditions. These lifetimes are much shorter than the lifetimes of these triplets measured by flash techniques (see Section IV.B.4.d), indicating that the values for the quenching rate constants assumed by Loutfy and de Mayo are too high by about an order of magnitude. Thus, their values for k_r are also too high by an order of magnitude. However, this problem does not significantly affect the results of their study which were: (a) a triplet exciplex (Corey's π complex) is formed irreversibly and is short-lived; (b) the exciplex collapses to a 1,4-biradical which cyclizes or reverts to starting materials; (c) biradical reversion is the main source of inefficiency in the cycloaddition; (d) there is insufficient evidence from this or prior work to indicate whether the first bond is formed α or β to the carbonyl group. However, their data do not require reaction via an exciplex, since direct formation of a triplet 1,4-biradical would be in accord with their data and other data in the literature.

In Corey's and de Mayo's studies, as well as in subsequent reviews, it is taken for granted that enone dimerization is a special case of enone photoannelation. This assumption, while structurally reasonable, is surprising in the context of the exciplex hypothesis, since the rate constant for self-quenching of the triplet k_{sq} is larger than the rate constant for triplet capture (k_r) by most electron-rich alkenes. However, the kinetic evidence given earlier (Section IV.B.4.e.i.) suggests that at least in the case of simple cyclohexenones, such as the 4,4-dimethyl enone 124, annelation and dimerization occur via different enone triplet excited states. The generality of this finding for other enones remains to be established.

(iv) Recent kinetic studies and alternative mechanisms for enone–alkene photocycloadditions. Several recent studies by Schuster and coworkers cast serious doubt on the Corey–de Mayo exciplex mechanism[200] for photoaddition of enones to alkenes, which has been widely adopted by investigators in this field[7–9].

As discussed earlier (Section IV.B.4.e.iv), an investigation was undertaken of the effect of alkenes on the lumiketone photorearrangement of 4, 4-dimethylcyclohexenone **124**, since it seemed likely that both photorearrangement and photocycloaddition originated from a common triplet state of the enone. Indeed, linear Stern–Volmer plots for quenching by a variety of alkenes of the photorearrangement of **124** to **147** were observed[182,183]. However, as previously mentioned, the slopes of these plots showed absolutely no correlation with the ionization potential of the alkenes (see Figure 12)[183]. For formation of a π complex in which the alkenes were acting as donors, a linear relationship of $\log k_q$ with ionization potential would be expected. The failure to observe such a correlation means that the Corey–de Mayo mechanism is wrong in at least one of two regards: (1) the triplet of **124** reacts with alkene to give 1,4-diradicals directly without the involvement of a discrete exciplex intermediate; (2) the alkenes may intercept some other intermediate on the pathway to lumiketone **147**.

The direct observation of enone triplet-excited states using laser flash techniques made it possible to directly measure the rates of interaction of alkenes with these triplets. Results of such a study involving 4, 4-dimethylcyclohexenone **124** have recently been reported[170]. In this case, the congruity of steady-state quenching data with the lifetime and rate constants for quenching measured by flash techniques made it clear that the triplet observed in the flash was indeed the species responsible for both photorearrangement and photocycloaddition to alkenes. However, it came as a surprise that alkenes (TME, DME, cyclopentene, cyclohexene) which form photocycloadducts with **124** with moderate to good quantum efficiency (up to 0.44) do not appear to directly quench this enone triplet, according to studies of the effect of alkene in high concentrations (e.g. up to 3.8 M in the case of TME) on both the rate of decay of the enone triplet at 280 nm and[170,172], in solutions containing methylnaphthalene (MN), the rate of growth of MN triplet absorption at 420 nm[170]. Thus, the extrapolated enone triplet lifetime in acetonitrile (AN) solution obtained from the MN growth kinetics, monitored at 420 nm, was 26 ns in AN alone and 29 ns in AN containing 30% TME; for comparison, the triplet lifetime for **124** in AN measured by transient decay at 280 nm is 27 ± 2 ns. Thus, TME had no effect on either the rate of radiationless decay of the relaxed enone π, π^* triplet or on triplet transfer to MN. In another comparison, τ_T for **124** in neat cyclopentene (23 ns) is indistinguishable from the value of τ_T in isooctane[170]. These data are in complete agreement with earlier results of Dunn[172] on enone **124** and phenanthrone **154** with cyclohexene and DME obtained using a different laser flash apparatus.

These flash data are in marked contrast with the effect of the same alkenes as quenchers of the rearrangement of **124** to lumiketone **147**[182,183]. There is a clear mismatch between values of $k_q\tau_T$ obtained from the quenching experiments and upper limits to $k_q\tau_T$ calculated from the flash data[170]. These data require that at least in this system these alkenes must be intercepting an intermediate I formed from the relaxed (twisted) enone triplet but not the triplet itself, as indicated in Scheme 35. The nature of this intermediate is not precisely defined by any of the studies carried out to date, but an intriguing possibility is that I is a ground-state *trans* isomer of **124**, that is, a *trans* cyclohexenone[170]. Similar experiments have not yet been carried out using the parent compound, so it is not clear whether these results can be generalized. The consequences of photocycloaddition via a *trans* cyclohexenone will be discussed after first considering cases in which the enone triplet is definitely intercepted by alkenes.

It was anticipated that enones which were structurally constrained from formation of a ground-state *trans* isomer would react directly with alkenes. This indeed is the case for

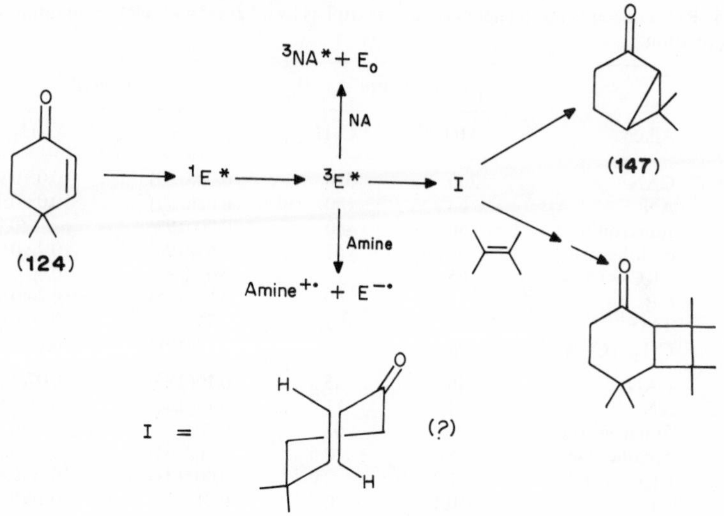

SCHEME 35

cyclopentenone (50)[80], 3-methylcyclohexenone (217)[80], testosterone acetate (132)[173] and $\Delta^{1,6}$-bicyclo[4.3.0]nonen-2-one (216)[173]. In these cases, the rate of decay of the '280 nm' transient triplet was enhanced in the presence of added alkenes in both acetonitrile and cyclohexane solutions. The quenching rate constants given in Table 5, determined from slopes of plots of $(\tau_{obs})^{-1}$ vs. alkene concentration, represent the first absolute values of rate constants determined for interaction of enone triplets with alkenes. For several of these systems absolute or relative quantum yield data have been obtained, which are also given in Table 5.[242]

Two important conclusions can be drawn from these data[242]. One is that there is no correlation between the quantum efficiency for adduct formation and the rate of reaction of the enone triplet with alkenes, which is hardly surprising given the example of the Norrish Type II reaction of aromatic ketones in which triplet 1,4-biradicals also play a crucial role[50]. Secondly, and perhaps more surprising, in all cases studied thus far the rates of interaction of enone triplets with electron-deficient alkenes are much larger than for electron-rich alkenes, which is completely contrary to expectations based on Corey's π-complex hypothesis[184]. Moreover, for some enones (such as 216) photoadducts with electron-deficient olefins are formed in good yields. Thus, Cantrell's observation[192] of enhanced reactivity of acrylonitrile toward photoexcited enone 217 was not anomalous.

The possibility that the primary interaction of enone triplets with electron-deficient alkenes such as acrylonitrile (AN) might involve triplet energy transfer must be considered. Liu and Gale[206] and, independently, Hosaka and Wakamatsu[207] discovered many years ago that dimerization of AN to give cis- and trans-1,2-dicyanocyclobutane can be sensitized by benzophenone and a number of other triplet sensitizers. The triplet energy of AN was estimated to be ca 62 kcal mol^{-1}, and a recent measurement by photoacoustic calorimetry indeed places it at 58 ± 4 kcal mol^{-1} and that of fumaronitrile at 48 ± 3 kcal mol^{-1}[61]. It was therefore necessary to determine if enones could also sensitize dimerization of AN and α-chloroacrylonitrile (CAN). Authentic AN and CAN dimers were first prepared by benzophenone sensitization as per the literature[206,207]. Using cyclopentenone, 3-methylcyclohexenone 217 and bicyclononenone 216, whose triplets had

TABLE 5. Rate constants for quenching of enone triplets by alkenes[a] and relative quantum yield for adduct formation[b]

Ketone	Alkene[c]	$k_q \times 10^{-7}$ (l mol^{-1} s^{-1})		$\Phi^d(\phi_{tc})^g$	
		MeCN	C$_6$H$_{12}$	MeCN	C$_6$H$_{12}$
50	CAN	200	520	0.04(0.99)	0.05(0.99)
	AN	63	180	0.08(0.97)	0.03(0.99)
	fumaronitrile	160	460	0.00(0.99)	(0.99)
	cyclohexene	33	42	0.64(0.94)	0.42(0.96)
	Cl$_2$C=CCl$_2$	65		0.00(0.97)	0.00
	CP	15	40	0.56(0.85)	0.26(0.95)
	TME		99	0.71	0.29(0.99)
	CH$_2$=CCl$_2$	78		0.18(0.98)	0.15
217	CAN	46	35	0.10(0.95)	0.07(0.91)
	AN	15	11	0.14(0.84)	0.08(0.87)
	fumaronitrile		67	0.20	(0.95)
	cyclohexene	5.2	0.5	0.16(0.66)	0.07(0.12)
	Cl$_2$C=CCl$_2$	1.2	2.0	0.00(0.31)	0.00(0.41)
	CP	<0.1	0.5	0.21	0.10(0.14)
	TME	<0.1		0.08	0.03
	DME	0.7			
	CH$_2$=CCl$_2$			0.07	0.04
132	AN	24			
	CP	6		0.21[e]	
216	AN	130		1.60[f]	
	cyclohexene	27		4.16[f]	
	CP	3.8		1.00[f]	(0.048)[e](0.91)
	DME	26		0.62[f]	

[a] Determined from lifetimes of enone triplets of flash excitation of 355 nm as a function of alkene concentration.
[b] Adducts determined by GC/MS. Conversion <10%.
[c] CAN = α-chloroacrylonitrile. AN = acrylonitrile. CP = cyclopentene. TME = tetramethylethylene. DME = 1,1-dimethylethylene.
[d] Quantum yield at 313 nm at 0.50 M alkene.
[e] Quantum yield at 313 nm in neat cyclopentene.
[f] Relative quantum yield at 0.75 M alkene.
[g] Quantum yield for triplet capture (see text).

been shown to be highly reactive toward both AN and CAN, it was found that only trace quantities of AN or CAN dimers could be detected upon irradiation of these enones in neat alkene. It was also possible that enone–AN adducts could arise by triplet transfer to AN followed by attack of AN triplets on ground-state enone. However, when a mixture of benzophenone (1.0 M) and cyclopentenone (0.2 M) was irradiated in neat AN under conditions where more than 97% of the light was absorbed by benzophenone, with the enone at a concentration greater than that needed to furnish AN adducts in good yield, only AN dimers were produced and no enone–AN adducts could be detected[80,242]. Thus, it is concluded that triplet transfer from cyclopentenone to AN is very inefficient compared to formation of triplet 1,4-biradicals en route to cycloadducts.

On the basis of these new data, and the criticisms of the π-complex hypothesis made earlier, one can speculate that these enone triplets may react with alkenes to give 1,4-biradicals directly without the intervention of exciplexes as discrete intermediates.

Inefficiency in adduct formation would then result from a combination of two factors: decay of the enone triplet to ground-state enone competitive with formation of triplet 1,4-biradicals (^{3}BIR), and reversion to enone and alkene ground states from each of the sequentially formed biradicals ^{3}BIR I and ^{3}BIR II competitive with cyclization and disproportionation (Scheme 36); the importance of each process will vary with each specific enone–alkene system depending on the rate constants of the competitive processes. The analogy to other photochemical reactions proceeding via triplet 1,4-biradicals, most significantly the Norrish II reaction[50], should be obvious.

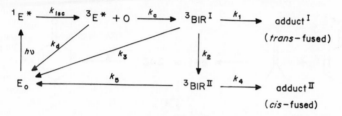

SCHEME 36

The quantum efficiency for triplet capture ϕ_{tc} by the alkene can be calculated from the flash data according to the expression $\phi_{te} = 1 - \tau_{obs}/\tau_0 = k_q[\text{alkene}]\tau_{obs}$. Comparison with quantum yields for adduct formation reveals the extent to which reversion to ground state enone occurs before and after enone triplet interception by alkene[242] (see Table 5). Thus, ϕ_{tc} for capture of enone triplets in neat cyclopentene at room temperature is 0.82 and 0.91 for testosterone acetate and bicyclononenone **216**, respectively, while ϕ_{prod} for these systems is 0.21 and 0.049[173]. Thus, most (but not all) of the reaction inefficiency in these systems is due to biradical reversion, which is especially important in the latter case, probably for steric resons. In general, enone triplet decay will play a more important role for shorter- than for longer-lived triplets.

The photocycloaddition of cyclic enones to electron-deficient alkenes has received little attention from organic photochemists and synthetic chemists probably owing to the strong influence of Corey's 'oriented π-complex' mechanism[184], despite its problematic basis. There will undoubtedly be important applications of such photoaddition reactions in organic synthesis in the future. For example, Stoute, Shimonov and Schuster[209] have found that electron-deficient alkenes such as AN, CAN, maleic anhydride and chloroalkenes form adducts with cycloheptenone **94** at the expense of formation of photodimers of **94**. In contrast, electron-rich alkenes such as DME and TME do not form adducts, as originally observed by Corey and coworkers[184]. The structures of these new adducts and the mechanism of their formation are currently being elucidated. Thus, it is not yet known whether such alkenes react with triplets of **94** or with the *trans* enone. The reactions of electron-deficient alkenes with other cyclic enones is currently under study.

(v) Regiochemistry and stereochemistry of photocycloadditions to cyclohexenones— alternative explanations. Bauslaugh[210] proposed many years ago that the regiochemistry observed by Corey and coworkers[184] could be explained without invoking exciplexes, as a consequence of the competition between cyclization and reversion to ketone and olefin ground states from intermediate biradicals. On the basis of the arguments and data given above, this explanation seems to be sufficient to rationalize the experimental facts. Thus, Bauslaugh[210] analyzes the addition of cyclohexenone to isobutylene in terms of the formation of the four 1,4-biradicals **240–243** shown in Scheme 37. On the basis of radical stabilization, the rate (and efficiency) of formation of **240** should be the greatest, and of **243**

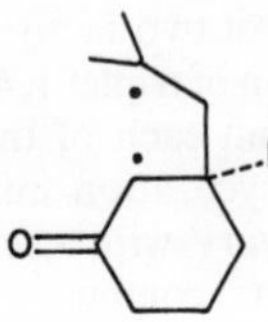

(240) 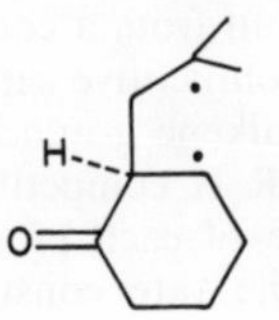   (241) 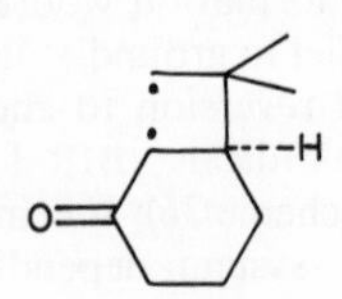   (242) 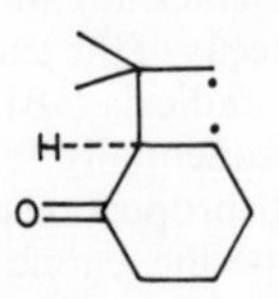   (243)

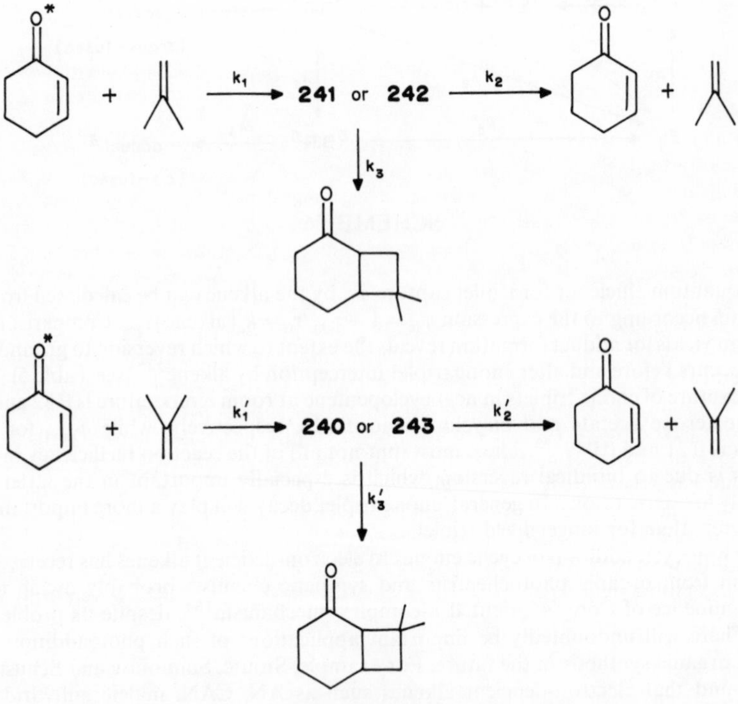

SCHEME 37

the least. The latter probably plays little role in the reaction. Since **240** is the most stable, it is not unreasonable that it would show the most reversion to ground states of starting materials (k_2 in Scheme 37) of any of the biradicals. Thus, if cyclization occurred mainly from **241** and **242** ($k_3/k_2 > k_3'/k_2'$) the predominant product would be the head-to-tail adduct shown in equation 107. As Eaton noted in his 1968 review[96], this preference is in any event 'really rather small'. The fact that analogous addition of cyclopentenone to propene gives about an even distribution of head-to-head and head-to-tail adducts (see equation 129) is rationalized by Bauslaugh in terms of the reduced importance of the k_2 process in this case because of reduced tendency to regenerate a cyclopentenone vs. a cyclohexenone system due to strain; the result would be an increase in the formation of head-to-head adducts via the k_3 path. This experimental result is inconsistent in any event with the exciplex hypothesis.

$$45\% \qquad 55\% \qquad (129)$$

As far as the regiospecificity of the addition of cyclohexenone to DME (equation 108) is concerned, Bauslaugh[210] proposes that the diradical which would lead to head-to-head adducts would require superimposing the polar groups before ring closure can be achieved. It is therefore not surprising that this mode of addition is not observed relative to the alternative process.

Admittedly, the above explanations of Bauslaugh[210] have a definite *ad hoc* flavor, and are not subject to a precise kinetic analysis. The same argument could in fact be made about Corey's original exciplex hypothesis[184]. In order to really assess the validity of this type of analysis, it would be necessary to know the quantum yields of formation of the isomeric biradicals (i.e. the magnitude of k_1 vis-à-vis k_1' in Scheme 37 and the efficiency of triplet capture ϕ_{tc}) and the quantum yields for formation of the regioisomeric products. Since no one has yet devised a way of obtaining all these data, Bauslaugh's type of analysis in terms of biradical reversion vs. cyclization/disproportionation is as good an approach as any for discussing the regiospecificity of these photocycloadditions.

Bauslaugh[210] also proposes a simple steric argument to account for the formation of *trans*-fused cycloadducts from cyclohexenones as major products. Considering again the addition of cyclohexenone to isobutylene, he proposes three staggered conformations for the diradical **241**, the principal if not exclusive source of the head-to-tail adduct which is the predominant product in equation 107. These are shown in structures **241a, 241b** and **241c**. Diradical **241a** would give the *trans*-fused adduct with a diequatorial linkage, **241c** would give the *cis*-fused adduct (axial–equatorial linkage) while **241b** is unable to give either adduct. According to Bauslaugh, examination of models suggests that reaction via **241a** is more favored than via **241c**, since **241a** is conformationally more stable than **241c**, and because **241c** encounters more severe steric problems when it closes to form a cyclobutane than does **241a**. Similarly, the head-to-head adducts formed from cyclohexenone and isobutylene arise from three likely conformations **240a, b** and **240c** of the

(241a)   (241b)   (241c)

(240a)   (240b)   (240c)

diradical **240**; of these, **240a** is clearly the one best suited for ring closure. In this case, Bauslaugh sees no compelling preference for closure of conformer **240a** in a *trans* vs. a *cis* fashion, so that the amount of *trans*-fused adduct should diminish. In fact, Corey and coworkers observed only *cis*-fused head-to-head adduct in this reaction (equation 107). Other reactions which also give head-to-head adducts, as in addition to allene and acrylonitrile, apparently give only *cis*-fused adducts, consistent with this analysis.

Wiesner[199] proposed a model to explain the facial stereoselectivity associated with photoaddition of steroid enones to allene, which he suggested might have generality. Application of principles of conformational analysis, along with the suggestion that the β-carbon of the enone excited state is pyramidal, allowed Wiesner to rationalize why allene adds to one or the other face of the steroid molecule. His argument was basically that the configuration at C_β of the excited enone will be the one which is preferred thermodynamically on the basis of ring strain and nonbonded interactions. However, this model gives the wrong prediction with respect to the direction of addition of ethylene to enone **238** studied by Cargill and coworkers[198] as demonstrated by the two *cis*-fused isomers in equation 127. Moreover, since this model does not take into account competition between reversion to starting materials and ring closure of diradical intermediates, it will not be considered further.

Returning to the Bauslaugh stereochemical analysis[210], it is not obvious how this theory can explain why addition of unsubstituted ethylene to enone **238** gives 24% of a *trans*-fused isomer (equation 127), and why addition of cyclopentene to Δ^1- as well as Δ^4-steroidal enones gives major amounts of *trans*-fused adducts. It seems clear from the kinetic studies discussed earlier that the excited state of cyclohexenones which lead to cycloadducts is a highly twisted $^3\pi,\pi^*$ state. If such a species were to interact directly with alkenes, *trans*-fused adducts would be produced if the twisted geometry could be preserved, as pointed out many years ago by McCullough and coworkers[193]. One could then envisage a scheme such as that shown in Scheme 38 in which the interaction of enone triplet with alkene leads to a geometrically distorted 1, 4-biradical **244** which would give *trans*-fused cycloadducts if cyclization occurred competitively with relaxation of the biradical to the more stable geometry shown in structure **245**, which would be expected to give only *cis*-fused cycloadducts on ring closure[182]. According to this scheme, the biradicals leading to *trans*- and *cis*-fused cycloadducts are formed sequentially rather than concomitantly as in Bauslaugh's scheme[210]. One would therefore expect that any structural feature which would prolong the lifetime of the first-formed biradical or inhibit ring closure would enhance the probability of forming *cis*-fused adducts via the conformationally relaxed biradical **245** as well as return to ground-state enone and alkene. Some of the conformational effects discussed by Bauslaugh might indeed play a role in this regard. Again, it is not possible to discuss this scheme in quantitative terms unless one knew (a) the quantum yield for formation of the first-formed biradical, and (b) the extent of reversion from both **244** and **245**. While the first parameter can be obtained from flash data (see equation 128), there is still no good way of obtaining (b); the best one can do at present is to calculate the total extent of reversion from the difference between ϕ_{tc} and ϕ_{prod}. In order to obtain rate constants for cyclization and reversion, and to understand the dependence of these kinetic parameters on structural features of the reactants, one would need to determine lifetimes of the triplet biradicals; at present these are unknown, but in principle they could be determined by methods analogous to those used by Wagner and Scaiano in their studies of the triplet 1, 4-biradicals involved in the Norrish Type II reaction[211,212].

The observation that steroid enones give good yields of *trans*-fused cycloadducts, under conditions where the enone triplet is directly quenched by alkenes, suggests that the mechanism of Scheme 38 is operative. Thus, testosterone acetate reacts with cyclopentene to give two adducts in *ca* 1:1 ratio, one *cis*-fused and one *trans*-fused[173]. In neat

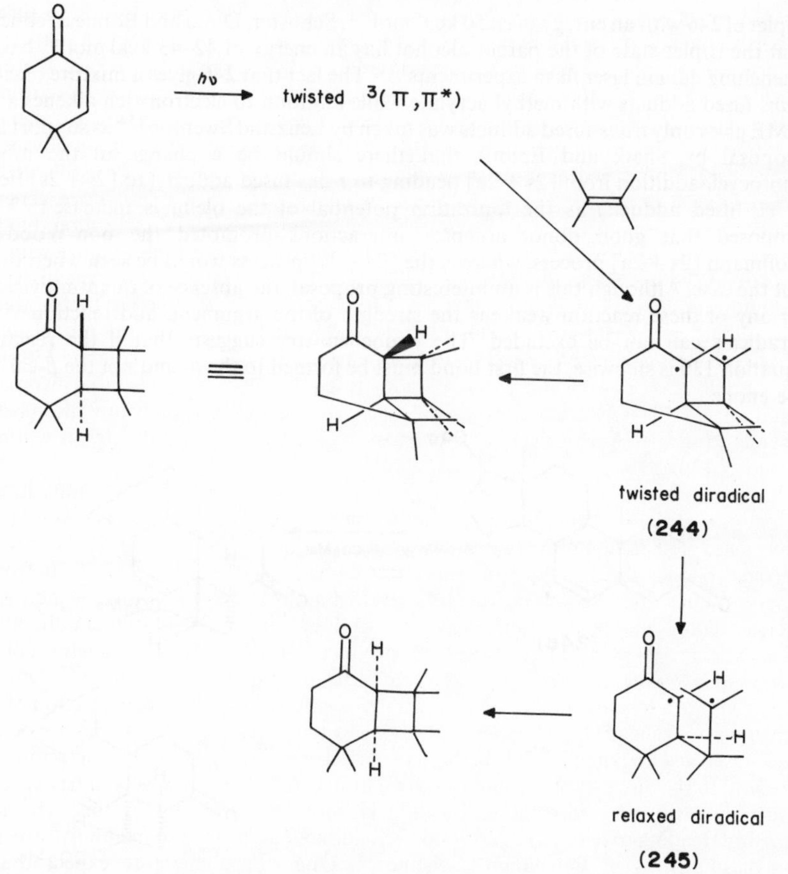

twisted $^3(\pi,\pi^*)$

twisted diradical

(244)

relaxed diradical

(245)

SCHEME 38

cyclopentene, the quantum efficiency of adduct formation is 0.21 while flash data show that 82% of the enone triplets are captured by cyclopentene. Thus, 75% of the initially formed radicals revert to starting materials. However, no conclusion can be drawn as to the extent of reversion to starting materials from twisted biradicals of type **244** vis-à-vis relaxed biradicals of type **245**, although it is likely that reversion from the latter is more important. These data, as well as the findings of Lenz on Δ^1-steroid enones discussed earlier[191], suggest that even in these relatively rigid systems the enone chromophore is significantly twisted in the excited state.

Lenz has extensively investigated photocycloaddition reactions of linear steroid dienones[213] which, in general, are beyond the scope of this review. However, the results of the studies of Lenz and Swenton on photoadditions of dienone **246** to electron-deficient alkenes[213] are of direct relevance to present considerations. Photocycloaddition of **246** to methyl acrylate gave a mixture of cis-and trans-fused adducts, as shown in equation 130; this represents the first example of isolation of trans-fused cycloadducts using an electron-deficient alkene. The cycloaddition could be quenched by a low-energy triplet quencher, 3, 3, 4, 4-tetramethyldiazetidine 1, 2-dioxide, suggesting that reaction occurs via a π,π^*

triplet of **246** with an energy of *ca* 50 kcal mol^{-1}. Schuster, Dunn and Bonneau concluded that the triplet state of the parent alcohol has an energy of 42–43 kcal mol^{-1} based on quenching data in laser flash experiments[171]. The fact that **246** gives a mixture of *cis*- and *trans*-fused adducts with methyl acrylate while addition to electron-rich alkenes such as DME gives only *trans*-fused adducts was taken by Lenz and Swenton[214] as support for the proposal by Shaik and Epiotis that there should be a change in the mode of photocycloaddition from [2s + 2a] (leading to *trans*-fused adducts) to [2s + 2s] (leading to *cis*-fused adducts) as the ionization potential of the olefin is increased[215]. They proposed that good donor–acceptor interactions promoted the non-Woodward–Hoffmann [2s + 2a] process, whereas the [2s + 2s] process would be seen when this was not the case. Although this is an interesting proposal, the absence of quantum yield data for any of these reaction weakens the strength of the argument, and reaction via 1,4-biradicals can not be excluded. The regiochemistry suggests that if the reaction in equation 130 is stepwise, the first bond must be formed to the α- and not the β-carbon of the enone.

(130)

One of the more unusual observations in the photocycloaddition literature is the report by Tobe and coworkers that photoaddition of enone **247** to cyclohexene gives the *cis–anti–trans* adduct **248** in 84% yield and a quantum yield of 0.69 (equation 131)[216].

(131)

Quenching studies with piperylene implicated a triplet state of the enone as the reactive excited state. Analogous systems with smaller ($n = 5$) and larger ($n = 7$ or 8) cycloalkene rings fused to cyclopentenone give mixtures of stereoisomeric adducts. The formation of only one product in equation 131 is rationalized by the authors in terms of more severe nonbonded interactions of the hydrogens in the other possible adducts vis-à-vis **248**, i.e. the mode of ring closure of the intermediate triplet biradical is governed by conformational energetics, as in Bauslaugh's original proposal[210]. This same effect is seen in addition of cyclopentene to enone **216** (equation 132), where the efficiency of triplet capture in neat alkene is 91% but the efficiency of formation of the adduct **248** is only 0.05[173]; nonbonded interactions between hydrogens in **248** are severe, whether the cyclopentane ring is oriented above either the five- or six-membered ring of the enone, which is not yet known. Thus, 95% of the intermediate biradicals in this case revert to starting materials. It would be of interest in this connection to see if the quantum efficiency of adduct formation increases as the ring size of the olefinic reactant is systematically enlarged, as predicted by this mechanism.

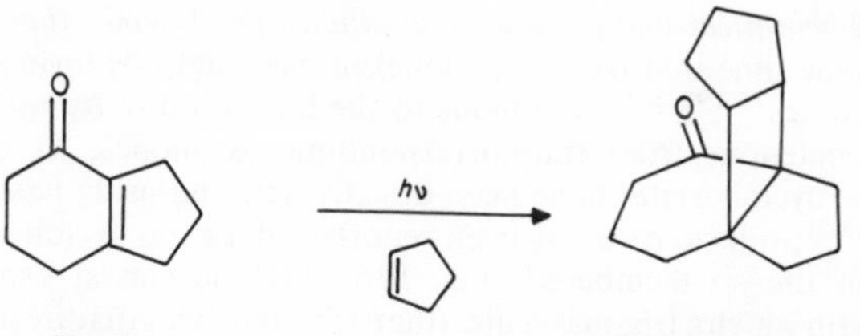

$$(132)$$

(216) (stereochemistry not determined)

It was suggested earlier on the basis of the incompatibility of the flash and steady-state kinetic data on enone **124** that in this system the alkene does not directly intercept the triplet state but rather reacts with an intermediate derived from the triplet, perhaps a *trans* cyclohexenone[170]. If this is indeed the correct mechanism, which is by no means certain, one would have to provide an alternative mechanism for formation of *trans*- and *cis*-fused cycloadducts in this system and other systems which show similar kinetic behavior. (As mentioned earlier, corresponding studies of cyclohexenone itself have yet to be done.) One might anticipate that addition of ground-state *trans* cyclohexenones to alkenes ought to be a rapid process, due to the great strain and consequent high reactivity of the enones. If it were concerted, orbital symmetry rules predict it should be a $_\pi 2_a + _\pi 2_s$ process[40]. Addition to the *trans* enone is expected to occur only suprafacially since one face of the enone moiety is shielded by the ring atoms. Therefore, addition to acyclic alkenes should give only cycloadducts in which the cyclobutane and cyclohexanone rings are *trans* fused (Scheme 39). Similarly concerted photocycloaddition of a *trans* cyclohexenone to

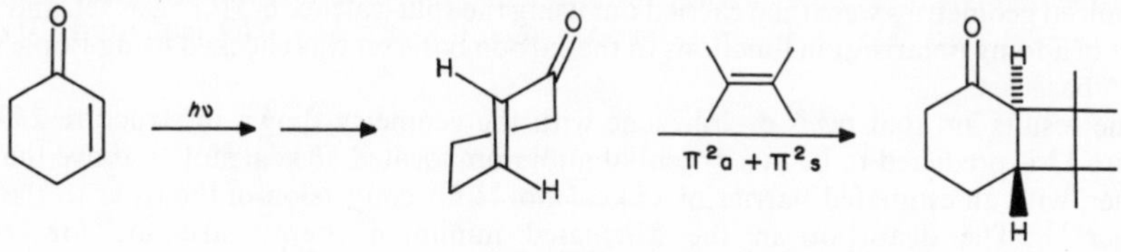

SCHEME 39

cyclopentene should give adducts in which both the five- and six-membered rings are *trans* fused to the cyclobutane ring, which is not observed. Similarly, concerted formation of *cis*-fused cycloadducts on photoaddition of **124** to acyclic alkenes is difficult to rationalize on the basis of a *trans*-cyclohexenone intermediate, since it would require antarafacial addition to the enone component. Therefore, it seems likely that photoadditions of **124** to alkenes are nonconcerted and may proceed via triplet biradical intermediates, although there is no definitive evidence in this connection (e.g. reactions with *cis–trans* pairs of alkenes have not yet been investigated). Since the ground-state and triplet potential surfaces are energetically close at the geometry corresponding to the *trans* cyclohexenone (see below), it is conceivable that intersystem crossing back to a triplet surface may take place when the *trans* enone reacts with alkenes. Although this discussion must be considered to be highly speculative due to the lack of conclusive supporting data, alternative mechanisms should be seriously considered in the case of cyclohexenones which are capable of undergoing severe molecular distortion by twisting around the C=C bonds.

h. trans-2-Cyclohexenones as intermediates in photochemical reactions of cis-2-*cyclohexenones. Theoretical and experimental studies (i) Introduction.* There has been speculation for some time that *trans*-2-cyclohexenones might be formed on photoexcitation of the *cis* enones[96,193,205], analogous to the formation of *trans*-2-cycloheptenone and *trans*-2-cyclooctenone from the corresponding *cis* enones. As discussed earlier, ground-state *trans* cyclohexenes have been directly detected using flash techniques in a number of cases[25,26], but no case has been reported of a *trans* cyclohexene with a third trigonal center in the six-membered ring. Probably the closest example is *trans*-1-acetylcyclohexene in which a trigonal center (the carbonyl carbon) is directly attached to the twisted C=C bond[168].

Schuster, Scaiano and coworkers have reported kinetic data which require that in photocycloaddition of enone **124** to electron rich alkenes the reaction intermediate intercepted by the alkenes is not the enone triplet, which is directly observable in flash experiments, but some species I derived from that triplet (see Scheme 35, Section IV.B.4.g.iv)[170]. The identity of I is by no means established, but one possibility that must be considered is that I is a *trans* cyclohexenone. In the following discussion, theoretical predictions concerning the viability of *trans* cyclohexenones as photochemical reaction intermediates will be discussed followed by experimental findings which bear directly on this question.

(ii) Theoretical treatments of trans-2-cyclohexenone. Verbeek and coworkers have published the results of theoretical *ab initio* calculations relating to the existence of *trans* cyclohexene[217]. To obtain a zeroth-order description of this system, at least a two-configuration wave function is required. They used an equivalent GVB formalism, in which the geometries of *cis* and *trans* cyclohexene were optimized using a minimal STO-3G basis set, assuming C_2 symmetry throughout. Single-point GVB calculations at the optimized geometries were then carried out using the split-valence 6-31G basis set, and the effect of adding polarization functions to the carbon basis set was checked using Pople's 6-31G* basis set.

The results are that *trans* cyclohexene with the geometry shown in structure **249** in Figure 13 is predicted to lie in a potential minimum located 56 kcal mol^{-1} above the *cis* isomer, with an estimated barrier of 15 kcal mol^{-1} for conversion of the *trans* to the *cis* isomer[217]. The distortion in the calculated minimum energy structure for *trans* cyclohexene lies mainly in the C_1—C_2 'double bond'. The π overlap in this compound is poor, reflected in the long C_1—C_2 bond of 1.421 Å and the C_3—C_2—C_1—C_6 torsional angle of 81°. The dihedral angle between the p orbitals is estimated to be about 46°, corresponding to considerable diradical character in **249**, *ca* 30% compared to *ca* 10% for *cis* cyclohexene. The strain in the molecule is also reflected by unusually long C—C single bonds, e.g. 1.564 Å for C_4—C_5 in **249** compared to 1.542 Å for the corresponding bond in *cis* cyclohexene. The transition state for conversion of *trans* to *cis* cyclohexene is nearly a pure (*ca* 90%) biradical, with perpendicular p orbitals[217].

The authors conclude that *trans* cyclohexene corresponds to a local minimum, and that it might be possible to generate and observe it in an inert matrix, as had indeed been

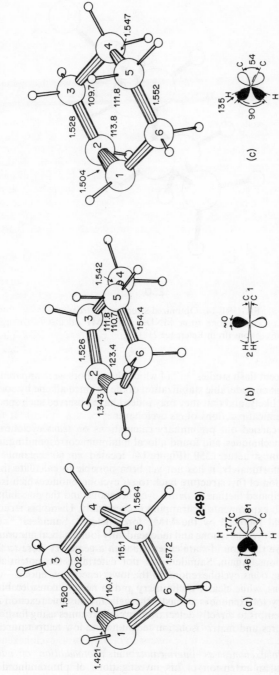

FIGURE 13. Optimized structures for (a) *trans* cyclohexene (**249**) (b) *cis* cyclohexene and (c) the transition state for isomerization of (a) to (b). Reprinted with permission from *J. Org. Chem.*, **52**, 2955 (1987). Copyright (1987) American Chemical Society

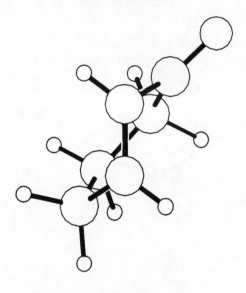

(250)

FIGURE 14. Optimized structure for *trans* cyclohexenone from MNDO and AM1 calculations (from Reference 218)

accomplished in recent flash studies[25,26] of which the authors were apparently unaware. They also noted that considerable stabilization of the strained alkene by coordination to transition metals is likely, and that they may indeed have observed such species in metal-catalyzed photochemical reactions of *cis* cyclohexene[217].

Johnson[218] has carried out preliminary calculations on *trans* cyclohexenone using MNDO and AM1 techniques, and found a local minimum corresponding to the twisted geometry shown in structure **250** (Figure 14) located *ca* 60 kcal mol^{-1} above *cis* cyclohexenone. Unfortunately, it has not yet been possible to calculate the barrier for thermal isomerization of this structure back to *cis* cyclohexenone, which is critical with respect to the anticipated lifetime of *trrans* cyclohexenone and the possibility of directly observing it in flash experiments or trapping it chemically. Using his recently reported two-body force field followed by the MM2 force field[219], Saunders[220] explored the potential surface of 2-cyclohexenone and independently found a local minimum located *ca* 60 kcal mol^{-1} above the ground-state *cis* enone with a geometry close to that found in Johnson's calculations. Again, Saunders did not determine the potential barrier for isomerization of the *trans* cyclohexenone to the lower-energy *cis* enone.

These calculations, while clearly preliminary and in need of considerable refinement, suggest that *trans* cyclohexenone is at least a theoretically possible reaction intermediate, and should fuel attempts to directly detect *trans* cyclohexenones using flash techniques at ambient temperatures and matrix isolation techniques at low temperatures.

(iii) trans-*Cyclohexenones as intermediates in photoaddition of nucleophiles to cyclohexenones.* As an extension of his investigations of photoinduced addition of methanol to cycloheptenones and cyclooctenones, which appeared to occur via ground-

state *trans*-cycloalkenone intermediates[105], Hart and coworkers studied analogous additions to cyclohexenones[221]. Noyori and Kato[104] had previously reported that irradiation of 2-cyclohexenone in methanol solvent gave only a 0.7% yield of 3-methoxycyclohexanone, while other simple cyclohexenones also gave disappointingly low yields of alcohol or water adducts. However, Matsuura and Ogura had reported that a crystalline methanol adduct was formed from Pummerer's ketone, **251**[222]. To obtain further information about the mechanism of this unique photoreaction, Hart and coworkers studied the stereochemistry of the photoaddition using CH$_3$OD[221]. Using NMR spectroscopy, they determined that the product had the structure shown in equation 133, indicating that the reaction had proceeded in a stereospecific *trans* manner,

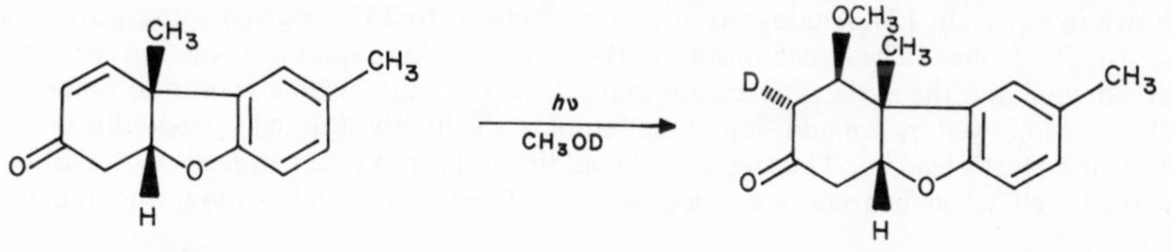

$$(133)$$

completely analogous to the findings on additions of MeOH(D) to seven- and eight-membered cyclic enones[105]. Hart proposed the mechanism shown in Scheme 40 in which irradiation of **251** 'results in an excited state or intermediate (depicted as **252**) in which the carbon–carbon double bond is twisted more than 90°'[221]. Only *syn* addition of methanol to the '*trans*' double bond is possible, since one face is completely blocked by the ring itself. Therefore, the methoxyl group attached to the β-carbon of the enone must end up *cis* to the angular methyl group, as was shown in Matsuura and Ogura's original study[222] and

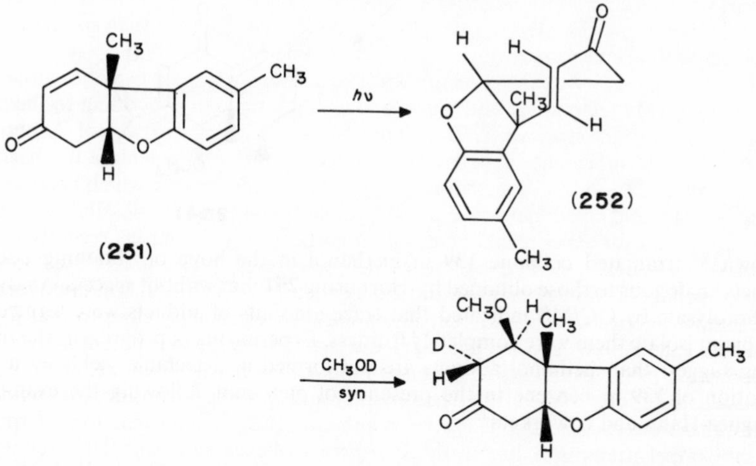

SCHEME 40

confirmed by Hart and coworkers[221]. After the former enone ring untwists, the deuterium ends up *trans* to the methoxy group, i.e. *trans* stereochemistry arises from *syn* addition to a twisted *trans* enone. The isotope effect of 4.3 ± 0.5 found using mixtures of MeOH/MeOD is comparable to that found in additions of methanol to *trans* cycloheptenone and *trans* cyclooctenone[105], although a smaller effect was anticipated for addition of MeOH to the much more reactive and hence less selective *trans* cyclohexenone 252.

Very few other examples of photoadditions of methanol to cyclohexenones have been reported. Thus, Rodriguez-Hahn and coworkers[223] reported that irradiation of decompostin 253 in methanol gave only the 6-epi-methoxy substitution product 254; however, irradiation of 253 in benzene in the presence of methanol gave the methyl ether 255 whose structure was determined by X-ray crystallography. These transformations are shown in equation 134. Analogous addition reactions to 253 occurred using water and isopropyl alcohol. The mechanism of the MeOH photoaddition reaction was not investigated, but the authors speculate that it involves 'initial isomerization to the *trans*-alkenone followed by *syn* addition of methanol to the highly strained 1, 10 double bond', following Hart's lead[221]. The fact that the addition did not occur in neat methanol but worked well when benzene was used as the solvent is very interesting, and remains unexplained.

(134)

Brown[173] irradiated octalone 139 in methanol in the hope of obtaining addition products analogous to those obtained by Hart using 251, but without success. Analysis of the photolysate by GC/MS indicated that trace amounts of adducts were formed, but attempts to isolate them were completely fruitless. Experiments in progress at the time of writing suggest that methanol adducts are not formed in detectable yields even upon irradiation of 139 in benzene in the presence of methanol, following the example of Rodriguez-Hahn and coworkers[223].

(iv) Photoaddition of cyclohexenones to conjugated dienes. Cantrell[175] originally reported that cyclohexenone and cyclopentenone undergo both [2 + 2] and [4 + 2] addition to conjugated dienes. Using acyclic dienes, such as 1,3-butadiene and 2,4-hexadiene, as well as cyclic dienes such as 1,3-cyclohexadiene and spiro[2,4]hepta-2,5-diene, only [2 + 2] adducts were formed, as illustrated in equation 135. However, both

types of adducts were formed from cyclohexenone and cyclopentadiene (equation 136) and from both enones with furan (equations 137 and 138). The quantum efficiencies for addition were somewhat larger with cyclohexenone than with cyclopentenone. Unfortunately, the nature of the ring fusion in the [4 + 2] adducts in equations 136 and 137, which relates to the possible capture of *trans* cyclohexenone by the cyclic dienes, was not established. An interesting mechanistic observation was made with the cyclohexenone–butadiene system[175]. Along with the adducts shown in equation 135, dimers of the diene are formed; the latter were attributed to triplet energy transfer from the enone to the diene based on studies by Hammond and coworkers[224]. Curiously, while the efficiencies of both processes increased as the diene concentration was increased, the ratio of diene dimers to adducts increased as a function of diene concentration. No explanation was offered, but one possibility is that the triplet excitation transfer may be taking place from a different (higher-energy) enone triplet than the triplet state (twisted π, π^*) that is implicated in the cycloaddition process.

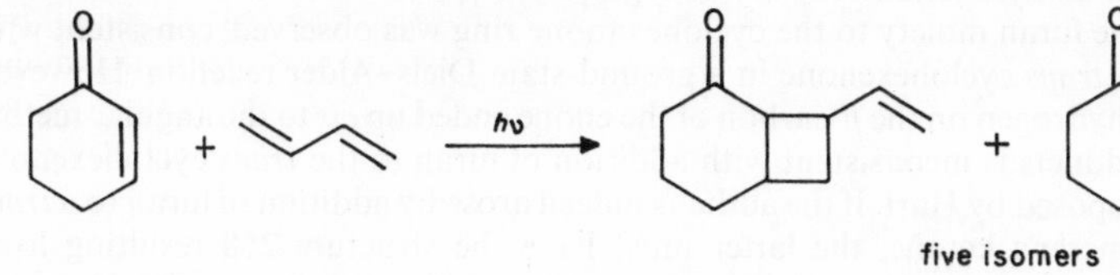

five isomers

(135)

mixture of
stereo- and
regioisomers

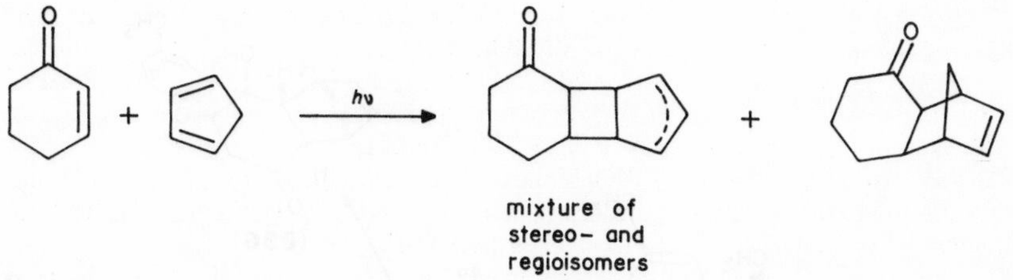

(136)

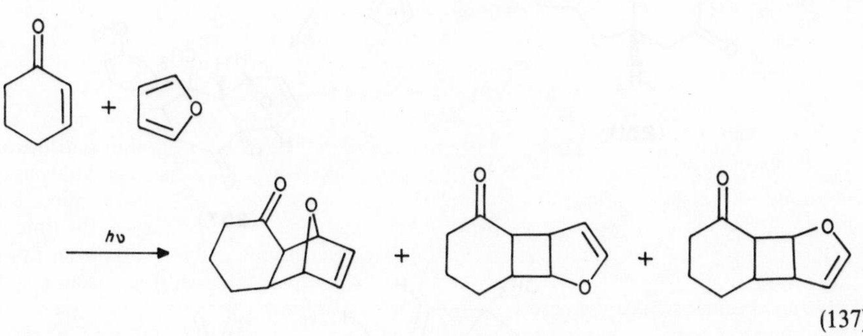

(137)

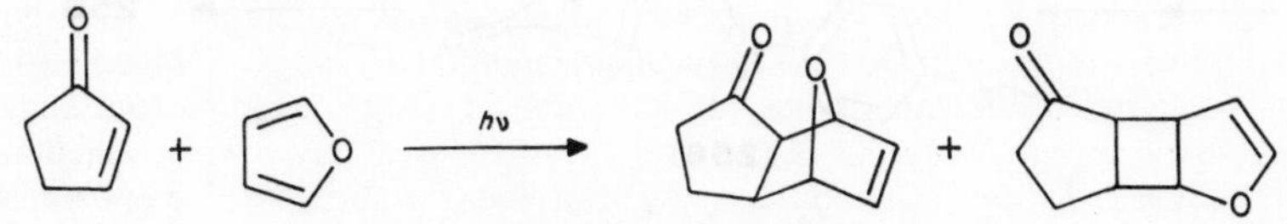

(138)

At the very least, as pointed out by Cantrell[175], these findings indicate why erratic results are often observed in kinetic studies in which dienes are used as potential quenchers of enone triplets, both in flash and steady-state studies. Enone triplets whose energies are substantially reduced by twisting around the C=C bond may be quenched very inefficiently by dienes by triplet energy transfer, which was not appreciated until recently[81]. Under these conditions, cycloaddition processes may dominate. However, it is interesting that even 2-cyclopentenone, whose triplet is not anticipated to undergo substantial distortion due to twisting, forms [2 + 2] adducts relatively efficiently with cyclopentadiene, although absolute quantum yields for this process have not been reported[175].

Since dienes were used effectively as reagents for trapping of *trans* cycloheptenone and *trans* cyclooctenone, the possibility that *trans* cyclohexenones as generated from Pummerer's ketone **251** (see above) might also be capturable using cyclic dienes was investigated by Mintas, Schuster and Williard[225,229]. Indeed, irradiation of **251** in neat furan led to the isolation of two furan adducts assigned structures **256** and **257** on the basis of NMR spectral analysis and X-ray crystallography (equation 139). In both adducts, a *trans* fusion of the furan moiety to the cyclohexanone ring was observed, consistent with interception of a *trans* cyclohexenone in a ground-state Diels–Alder reaction. However, the fact that the hydrogen on the β-carbon of the enone ended up *cis* to the angular methyl group in both adducts is inconsistent with addition of furan to the *trans* cyclohexenone structure **252** proposed by Hart. If the adducts indeed arose by addition of furan to a *trans* isomer of Pummerer's ketone, the latter must have the structure **258** resulting from twisting in the opposite direction, as shown in equation 140, a structure which Hart and coworkers had originally dismissed as untenable because of nonbonded interactions[221].

(139)

(256)

(251)

(257)

(258)

(140)

However, further experiments raise doubts about formation of adducts **256** and **257** via a *trans* cyclohexenone. Methylnaphthalene ($E_T = 61$ kcal mol^{-1}) does not quench adduct formation, which is consistent with the finding that the triplet of lifetime of **251** in acetonitrile is only *ca* 15 ns on laser flash excitation at 308 nm[226]; this finding, in turn, suggests that the π, π^* triplet state of **251** is indeed highly twisted, according to the data and interpretation given in Section IV.B.4.d. However, the free radical tetramethyl-1-piperidinyloxy (TEMPO) as well as oxygen quench formation of the adducts, but to very different extents, demonstrating that these reagents are not intercepting a common precursor of **256** and **257**. This finding was interpreted in terms of the mechanism indicated in Scheme 41, in which it is proposed that a highly twisted triplet state of **251** reacts with furan to give stereoisomeric triplet biradical intermediates I and I', which are the species intercepted by the paramagnetic reagents TEMPO and O_2[225]. Since a ground-state Diels–Alder reaction between **258** and furan should be concerted, even if not entirely synchronous[227], quenching of such a process by TEMPO or O_2 would be unprecedented. The observed stereochemistry suggests that reaction occurs on only one face of the twisted triplet of **251**, but indiscriminately with respect to the oxy bridge in furan. Furthermore, the low quantum yields for formation of **256** (0.062) and **257** (0.065) suggest that reversion to ground state reactants probably occurs predominantly from the relatively long-lived triplet biradicals I and I' rather than from the short-lived triplet state of **251**. These observations raise doubts about the role of highly strained *trans* ground states in other cases where *trans*-fused Diels–Alder adducts have been isolated[96,228].

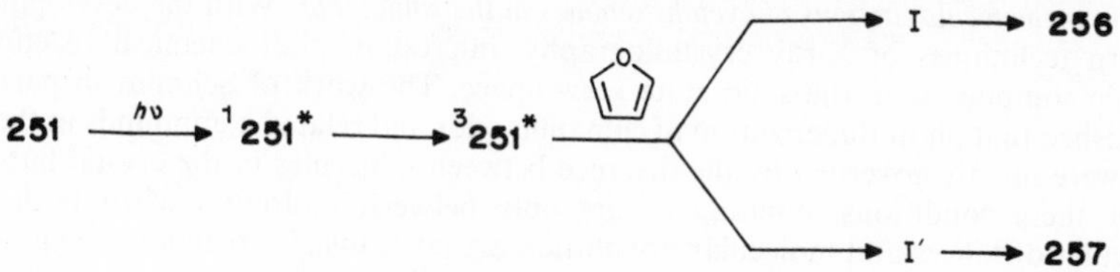

$$\textbf{251} \xrightarrow{h\nu} {}^1\textbf{251}^* \longrightarrow {}^3\textbf{251}^* \xrightarrow{\text{furan}} \begin{cases} \text{I} \longrightarrow \textbf{256} \\ \text{I}' \longrightarrow \textbf{257} \end{cases}$$

SCHEME 41

Photocycloaddition of Pummerer's ketone **251** to several alkenes has also been investigated by Mintas, Schuster and Williard[229]. The predominant cycloadduct formed from **251** and tetramethylethylene (TME) has the *trans*-fused structure **259** (*cis*-fused adducts are formed in at best trace amounts) (equation 141) reminiscent of the course of reaction of enone **124** with TME, while addition to 1,1-dimethoxyethylene (DME) gives a mixture of *cis*- and *trans*-fused adducts (equation 142). The structures of these adducts were determined by X-ray crystallography. It is worth noting that the *cis*-fused DME adducts are formed by attack on opposite faces of the reactive intermediate derived from **251**, whether it be a triplet-state or a ground-state *trans* enone. The short triplet lifetime of **251** precludes studies of triplet quenching by alkenes or dienes using nanosecond flash photolysis; such studies will require the use of picosecond flash techniques.

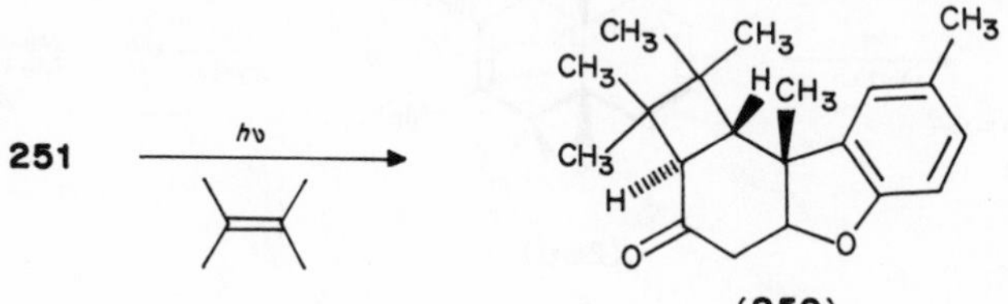

$$\textbf{251} \xrightarrow{h\nu} \textbf{(259)} \tag{141}$$

(142)

In summary, there is as yet no compelling evidence for the formation of ground state *trans* cyclohexenones on photoirradiation of *cis* cyclohexenones, although such intermediates provide an attractive way of rationalizing a number of experimental observations. Further studies directed toward observation and trapping of such species will be awaited with keen interest.

i. Photochemical reactions of cyclohexenones in the solid state. With the development of modern techniques of X-ray crystallography, interest in photochemical reactions of organic compounds in the solid state grew apace. The work of Schmidt in particular established that photodimerization of cinnamic acids and related compounds in the solid state were mainly governed by the distance between molecules in the crystal lattice[230]. Under these conditions, bonding occurs only between molecules when both intermolecular distances and molecular orientation are favorable, i.e. reactions are governed basically by the principle of least motion. However, different forces govern the course of unimolecular reactions in the solid state, as shown by the beautiful work of Scheffer and his colleagues in recent years[231]. Here, the reaction course is determined mainly by molecular conformation. In the solid state, only one molecular conformation is involved, and that is nearly always the lowest-energy conformation of the molecule[232]. In contrast, reactions in solution may proceed via minor populations of more reactive higher-energy conformations if the rate constant for reaction is sufficiently large. Thus, Scheffer and his coworkers have observed many cases in which different products are formed upon irradiation of organic compounds in solution vs. the solid state[231]. The understanding of such differences depends critically on knowledge of the X-ray crystal structures of the systems of interest, which will be assumed in the discussion below.

Some of the most interesting findings in Scheffer's studies concern cyclohexenones. The reactions shown in equations 143–145 illustrate the differences between solution and

(143)

(144)

(145)

solid-state photochemistry of three representative cyclohexenones, **260, 261** and **262**[233]. In solution, the only reaction observed is intramolecular [2 + 2] photocycloaddition, while irradiation of crystals of these systems leads to rearrangements and only traces of the cage compounds. In the crystalline state, these compounds adopt one of the twist conformations A or B shown in Scheme 42, depending on the nature of the substituents at C_4 of the enone moiety; the bulkier substituent prefers to adopt the pseudoequatorial position. In solution, the two conformations are in rapid equilibrium at room temperature.

Thus, as shown in Scheme 42 enone **261** crystallizes in conformation A with an equatorial methyl group. Irradiation leads to intramolecular hydrogen transfer from the allylic position at C_5 to the β-carbon of the enone moiety, establishing the stereochemistry at this center. The resulting biradical cyclizes to give the observed product **263**. Enone **260**, however, with an H in place of methyl at C_4, adopts conformation B in the solid state with an equatorial hydroxyl group. In this case, intramolecular hydrogen transfer to the β-carbon of the enone can occur only from the other allylic position C_8, giving a biradical which closes to ketone **264**; this then cyclizes to the hemiacetal **265**, the isolated product[233].

Other features are illustrated by the photochemistry of enones **266** (R = H or CH$_3$) shown in Scheme 43[234], in which methyl groups on the enone double bond are missing. In solution, as above, only intramolecular [2 + 2] cycloaddition to give a cage compound is observed. In the solid state, Irradiation gives both **267** and **268**, in a ratio that is temperature dependent (2.25:1 at 13 °C, 0.5:1 at − 40 °C), corresponding to a difference in activation energy of 4 kcal mol⁻¹. Path a involves allylic hydrogen transfer to the carbonyl oxygen followed by bonding between the radical centers at C_1 and C_6, while path b involves hydrogen transfer from the other allylic carbon to the β-carbon of the enone, and bonding between C_2 and C_5 of the intermediate diradical. Scheffer and coworkers argue that path a involves reaction of a ³n, π* state and path b reaction of a ³π, π* state, with the former having the larger activation energy. Substitution of methyl

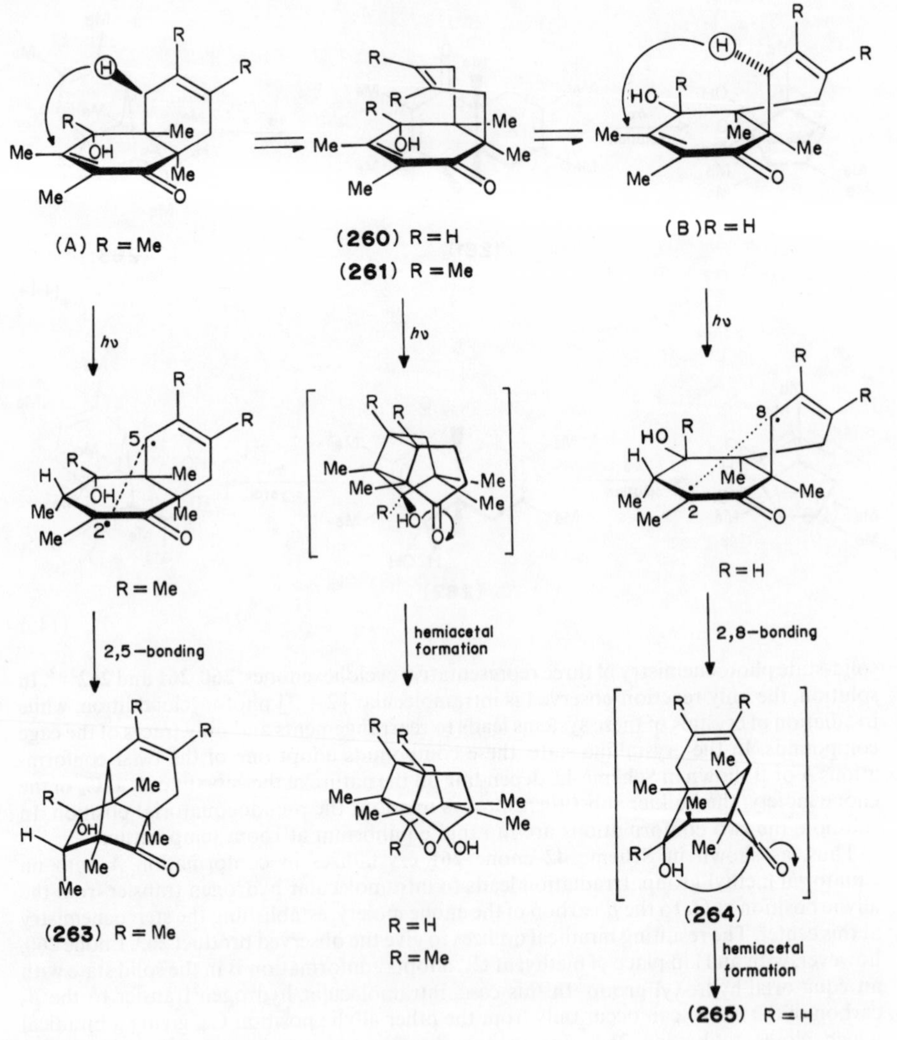

SCHEME 42

groups on the enone C=C bond should stabilize the π, π^* triplet, so it is not surprising that carbonyl abstraction reactions are not seen with enones such as **260–262**[234].

Unusual reactivity was observed for enone **269** in the solid state, in that hydrogen transfer in this case occurs exclusively to the α-carbon of the enone moiety, as shown in Scheme 44, the reverse of the selectivity usually observed. In solution, once again caged products as a result of [2 + 2] cycloadditions are observed. The explanation proposed by Scheffer and coworkers[235,236] involves a crystal-lattice steric effect, or what he calls 'steric compression control'. Since pyramidalization occurs at the carbon that is the migration terminus, the methyl group at this position is forced downward, into close contact with atoms on neighboring molecules in the crystal lattice. In most examples studied

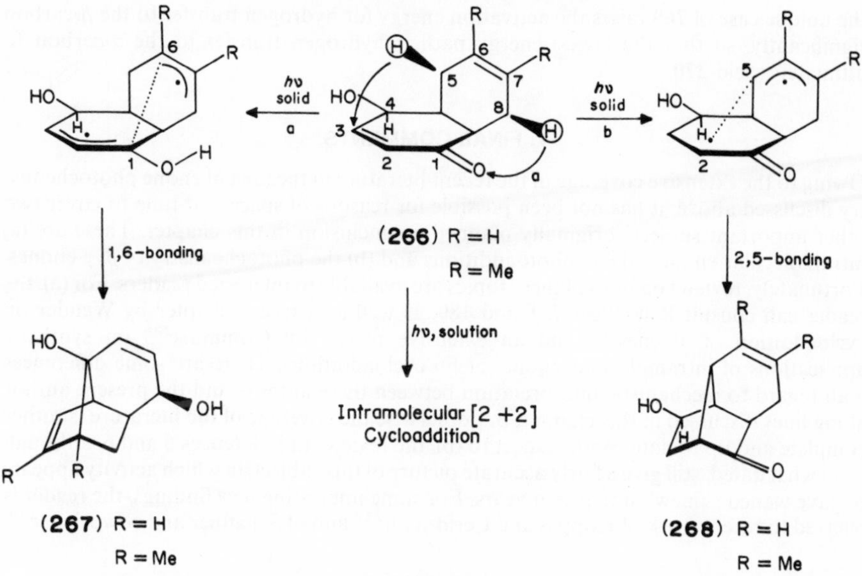
hν solid a
hν solid b
1,6-bonding
(266) R = H
R = Me
hν, solution
Intramolecular [2 +2] Cycloaddition
2,5-bonding
(267) R = H
R = Me
(268) R = H
R = Me

SCHEME 43

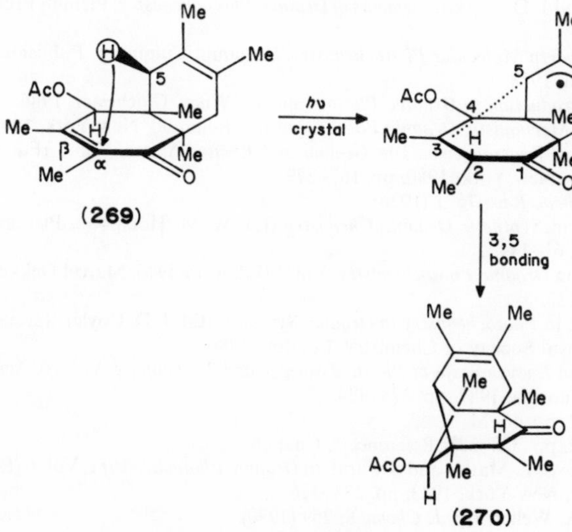
AcO
(269)
hν crystal
3,5 bonding
(270)

SCHEME 44

previously, this effect was comparable at both the α- and β-carbons of the enone moiety, affording no special selectivity. However, in the case of **269** computer simulation studies reveal that steric compression results only from hydrogen transfer to the β-carbon, since a void space surrounds the α-carbon. In the absence of such effects, hydrogen transfer to the α-carbon of the π,π^* triplet is preferred electronically, and is the lower-energy pathway, in solution as well as in the solid state. Scheffer argues that steric compression in the unique case of **269** raises the activation energy for hydrogen transfer to the β-carbon significantly, so that the lowest-energy path is hydrogen transfer to the α-carbon to ultimately yield **270**.

V. FINAL COMMENTS

Owing to the extensive coverage of the recent literature in the area of enone photochemistry discussed above, it has not been possible for reasons of space and time to cover two other important subjects originally planned for inclusion in this chapter. These are (a) intramolecular enone–alkene photoadditions and (b) the photochemistry of β, γ-enones. Fortunately, reviews on both of these topics are available to interested readers. For (a), the reader can consult References 7, 8 and 189, as well as a recent chapter by Wender on cycloaddition of alkenes[237] and an extensive review by Crimmins[238] on synthetic applications of intramolecular enone–olefin cycloadditions. There are some differences with regard to mechanistic interpretation between these authors and the present author along lines discussed in this chapter, but otherwise the coverage of the literature is rather complete and up to date. With respect to (b), the reviews in References 5 and 6, although somewhat dated, still give a fairly accurate picture of this subject, in which activity appears to have waned somewhat in recent years. For some interesting new findings, the reader is referred to recent work of Koppes and Cerfontain[239] and of Schaffner and coworkers[240].

VI. REFERENCES

1. D. O. Cowan and J. D. Drisko, *Elements of Organic Photochemistry*, Plenum Press, New York, 1976.
2. N. J. Turro, *Modern Molecular Photochemistry*, Benjamin/Cummings Publishing Co., Menlo Park, California, 1978.
3. J. D. Coyle, *Introduction to Organic Photochemistry*, Wiley, Chichester, 1986.
4. D. C. Neckers, *Mechanistic Organic Photochemistry*, Reinhold, New York, 1967.
5. D. I. Schuster, in *Rearrangements in Ground and Excited States*, Vol. 3 (Ed. P. de Mayo), Academic Press, New York, 1980, pp. 167–279.
6. K. N. Houk, *Chem. Rev.*, **76**, 1 (1976).
7. A. C. Weedon, in *Synthetic Organic Chemistry* (Ed. W. M. Horspool), Plenum Press, New York, 1984, pp. 61–144.
8. S. W. Baldwin, in *Organic Photochemistry*, Vol. 5 (Ed. A. Padwa), Marcel Dekker, New York, 1981, pp. 123–225.
9. H. A. J. Carless, in *Photochemistry in Organic Synthesis* (Ed. J. D. Coyle), Special Publication No. 57, The Royal Society of Chemistry, London, 1986.
10. D. I. Schuster, in *Encyclopedia of Physical Science and Technology*, Vol. 10, Academic Press, San Diego, California, 1987, pp. 375–424.
11. See Reference 2, pp. 46–51.
12. See Reference 2, pp. 328–338; Reference 1, Chap. 6.
13. For a review, see S. L. Mattes and S. Farid, in *Organic Photochemistry*, Vol. 6 (Ed. A. Padwa), Marcel Dekker, New York, 1983, pp. 233–326.
14. D. Rehm and A. Weller, *Isr. J. Chem.*, **8**, 259 (1970).
15. H. H. Jaffe and M. Orchin, *Theory and Applications of Ultraviolet Spectroscopy*, Wiley, New York, 1962, pp. 204–217.
16. Reference 15, p. 205.
17. H. Labhart and G. Wagniere, *Helv. Chim. Acta*, **42**, 2219 (1959); see also J. N. Murrell, *The Theory of the Electronic Spectra of Organic Molecules*, Wiley, New York, 1963 pp. 164–168, and A. Moscowitz, K. Mislow, M. A. W. Glass and C. Djerassi, *J. Am. Chem. Soc.*, **84**, 1945 (1962).
18. G. S. Hammond, J. Saltiel, A. A. Lamola, N. J. Turro, J. S. Bradshaw, D. O. Cowan, R. C. Counsell, V. Vogt and C. Dalton, *J. Am. Chem. Soc.*, **86**, 3197 (1964); for a recent review, see J. Saltiel and J. L. Charlton, in Reference 5, pp. 25–89.

19. K. Gollnick and G. O. Schenck, *Pure Appl. Chem.*, **9**, 507 (1964); G. O. Schenck and R. Steinmetz, *Tetrahedron Lett.*, 1 (1960).
20. Y. Inoue, S. Takamuku, Y. Kunitomi and H. Sakurai, *J. Chem. Soc., Perkin Trans 1*, 1672 (1980).
21. Y. Inoue, T. Ueoka, T. Kuroda and T. Hagushi, *J. Chem. Soc., Perkin Trans.* 2, 983 (1983), and references cited therein.
22. R. Bonneau, J. Joussot-Dubien, L. Salem and A. J. Yarwood, *J. Am. Chem. Soc.*, **98**, 4329 (1976).
23. See Reference 22 and (a) P. J. Kropp, *J. Am. Chem. Soc.*, **91**, 5783 (1969); (b) P. J. Kropp, E. J. Reardon, Jr., Z. L. F. Gaibel, K. F. Williard and J. N. Hattaway, Jr., *J. Am. Chem. Soc.*, **95**, 7058 (1973).
24. W. G. Dauben, H. C. H. A. van Riel, C. Hauw, F. Leroy, J. Joussot-Dubien and R. Bonneau, *J. Am. Chem. Soc.*, **101**, 1901 (1979).
25. J. L. Goodman, K. S. Peters, H. Misawa and R. A. Caldwell, *J. Am. Chem. Soc.*, **108**, 6803 (1986).
26. R. Bonneau, *J. Photochem.*, **36**, 311 (1987).
27. H. Yamazaki and R. J. Cvetanovic, *J. Am. Chem. Soc.*, **91**, 520 (1969).
28. H. H. Stechl, *Angew. Chem.*, **75**, 1176 (1963).
29. H. D. Scharf and F. Korte, *Chem. Ber.*, **97**, 2425 (1964).
30. D. R. Arnold, R. L. Hinman and A. H. Glick, *Tetrahedron Lett.*, 1425 (1964); D. R. Arnold, D. J. Trecker and E. Whipple, *J. Am. Chem. Soc.*, **87**, 2596 (1965).
31. R. G. Salomon and J. K. Kochi, *J. Am. Chem. Soc.*, **96**, 1137 (1974); R. G. Salomon, K. Folting, W. E. Streib and J. K. Kochi, *J. Am. Chem. Soc.*, **96**, 1145 (1974).
32. P. J. Kropp, *J. Am. Chem. Soc.*, **88**, 4091 (1966); P. J. Kropp and H. J. Krauss, **89**, 5199 (1967).
33. J. A. Marshall and R. D. Carroll, *J. Am. Chem. Soc.*, **88**, 4092 (1966).
34. P. J. Kropp, in *Organic Photochemistry*, Vol. 4 (Ed. A. Padwa), Marcel Dekker, New York, 1979, pp. 1–142.
35. P. J. Kropp, *J. Am. Chem. Soc.*, **95**, 4611 (1973).
36. R. A. Neunteufel and D. R. Arnold, *J. Am. Chem. Soc.*, **95**, 4080 (1973); A. J. Maroulis, Y. Shigemitsu and D. R. Arnold, *J. Am. Chem. Soc.*, **100**, 535 (1978); Y. Shigemitsu and D. R. Arnold, *J. Chem. Soc., Chem. Commun.*, 407 (1975).
37. D. R. Arnold and A. J. Maroullis, *J. Am. Chem. Soc.*, **99**, 7355 (1977).
38. P. S. Mariano and J. L. Stavinoha, in Reference 7, pp. 145–257.
39. T. R. Fields and P. J. Kropp, *J. Am. Chem. Soc.*, **96**, 7559 (1974); P. J. Kropp, E. J. Reardon, Jr., Z. L. F. Gaibel, K. F. Williard and J. N. Hattaway, Jr., *J. Am. Chem. Soc.*, **95**, 7058 (1973). P. J. Kropp, *Mol. Photochem.*, **9**, 9 (1978).
40. R. B. Woodward and R. Hoffmann, *The Conservation of Orbital Symmetry*, Verlag Chemie/Academic Press, Weinheim, 1970.
41. J. Saltiel and L. S. Nghim, *J. Am. Chem. Soc.*, **91**, 5404 (1969); W. G. Dauben and J. E. Haubrich, *J. Org. Chem.*, **53**, 600 (1988). See also W. G. Dauben, R. L. Caigill, R. M. Coates, and J. Saltiel, *J. Am. Chem. Soc.*, **88**, 2742 (1966).
42. H. M. Rosenberg and P. Serve, *J. Am. Chem. Soc.*, **92**, 4746 (1970); see also Reference 34, p. 111.
43. G. Ciamician and P. Silber, *Chem. Ber.*, **33**, 2911 (1900); see also W. D. Cohen, *Recl. Trav. Chim. Pays-Bas.*, **39**, 243 (1920).
44. G. S. Hammond and W. M. Moore, *J. Am. Chem. Soc.*, **81**, 6334 (1959); W. M. Moore, G. S. Hammond and R. P. Foss, *J. Am. Chem. Soc.*, **83**, 2789 (1961); see also H. L. J. Backstrom, *Acta Chem. Scand.*, **14**, 48 (1960).
45. J. N. Pitts, Jr., R. Letsinger, R. Taylor, S. Patterson, G. Recktenwald and R. Martin, *J. Am. Chem. Soc.*, **81**, 1068 (1959); A. Beckett and G. Porter, *Trans. Faraday Soc.*, **59**, 2039 (1963).
46. Y. M. A. Naguib, C. Steel and S. G. Cohen, private communication of results submitted for publication.
47. D. I. Schuster and P. B. Karp, *J. Photochem.*, **12**, 333 (1980); V. Franzen, *Justus Liebigs Ann. Chem.*, **633**, 1 (1960); G. O. Schenck, G. Koltzenberg and E. Roselius, *Z. Naturforsch.*, **24**, 222 (1969).
48. G. Porter and P. Suppan, *Pure Appl. Chem.*, **9**, 499 (1964); G. Porter and P. Suppan, *Trans. Faraday Soc.*, **61**, 1664 (1965).
49. S. Inbar, H. Linschitz and S. G. Cohen, *J. Am. Chem. Soc.*, **102**, 1419 (1980), and references cited therein; *J. Am. Chem. Soc.*, **103**, 1048 (1981).
50. P. J. Wagner, *Acc. Chem. Res.*, **4**, 168 (1971), and primary sources cited.

51. P. J. Wagner and G. S. Hammond, *J. Am. Chem. Soc.*, **87**, 4009 (1965); N. C. Yang, S. P. Elliott and B. Kim, *J. Am. Chem. Soc.*, **91**, 7551 (1969).
52. P. J. Wagner, B. P. Giri, J. C. Scaiano, D. L. Ward, E. Gabe and F. L. Lee, *J. Am. Chem. Soc.*, **107**, 5483 (1985).
53. C. R. Masson, V. Boekelheide and W. A. Noyes, Jr., in *Techniques of Organic Chemistry*, Vol. II (Ed. A. Weissberger), Interscience/Wiley, New York, 1956.
54. For a review and primary references, see Reference 1, pp. 135–181.
55. H. Schuh, E. J. Hamilton, H. Paul and H. Fischer, *Helv. Chim. Acta*, **57**, 2011 (1974); G. P. Laroff and H. Fischer, *Helv. Chim. Acta*, **56**, 2011 (1973); B. Blank, A. Henne, G. P. Laroff and H. Fischer, *Pure Appl. Chem.*, **41**, 475 (1975); K. Muller and G. L. Closs, *J. Am. Chem. Soc.*, **94**, 1002 (1972).
56. P. J. Wagner and R. W. Spoerke, *J. Am. Chem. Soc.*, **91**, 4437 (1969).
57. For a review, see D. R. Morton and N. J. Turro, in *Advances in Photochemistry*, Vol. 9 (Eds. J. N. Pitts, Jr., G. S. Hammond and K. Gollnick), Interscience/Wiley, New York, 1974, pp. 197–309.
58. E. Paterno and C. Chieffi, *Gazz. Chim. Ital.*, **39**, 341 (1909).
59. G. Buchi, C. G. Inman and E. S. Lipinsky, *J. Am. Chem. Soc.*, **76**, 4327 (1954).
60. For a review, see D. R. Arnold, in *Advances in Photochemistry*, Vol. 6 (Eds. W. A. Noyes, Jr., G. S. Hammond and J. N. Pitts, Jr.), Interscience/Wiley, New York, 1968, pp. 301–423.
61. J. A. Lavilla and J. L. Goodman, *Chem. Phys. Lett.*, **141**, 149 (1987).
62. R. S. H. Liu and D. M. Gale, *J. Am. Chem. Soc.*, **90**, 1897 (1968); S. Hosaka and S. Wakamatsu, *Tetrahedron Lett.*, 219 (1968).
63. N. J. Turro and P. A. Wriede, *J. Am. Chem. Soc.*, **92**, 320 (1970); N. J. Turro, J. C. Dalton, K. Dawes, G. Farrington, R. Hautala, D. Morton, M. Niemczyk and W. Schore, *Acc. Chem. Res.*, **5**, 92 (1972).
64. J. A. Barltrop and H. A. J. Carless, *J. Am. Chem. Soc.*, **94**, 1951 (1972).
65. W. C. Herndon, *Tetrahedron Lett.*, 125 (1971); W. C. Herndon and W. B. Giles, *Mol. Photochem.*, **2**, 277 (1970).
66. N. C. Yang and M. J. Jorgenson, *Tetrahedron Lett.*, 1203 (1964).
67. J. F. Graf and C. P. Lillya, *Mol. Photochem.*, **9**, 227 (1979).
68. R. Ricard, P. Sauvage, C. S. K. Wan, A. C. Weedon and D. F. Wong, *J. Org. Chem.*, **51**, 62 (1986).
69. R. M. Duhaime and A. C. Weedon, *Can. J. Chem.*, **65**, 1867 (1987).
70. P. G. Sammes, *Tetrahedron*, **32**, 405 (1976); M. Pfau, J. E. Rowe and N. D. Heindel, *Tetrahedron*, **34**, 3469 (1978).
71. R. M. Duhaime and A. C. Weedon, *J. Am. Chem. Soc.*, **109**, 2479 (1987).
72. R. M. Duhaime, D. A. Lombardo, I. A. Skinner and A. C. Weedon, *J. Org. Chem.*, **50**, 873 (1985).
73. G. Ciamician and P. Silber, *Ber. Dtsch. Chem. Ges.*, **42**, 1386 (1909).
74. G. W. Recktenwald, J. N. Pitts, Jr. and R. L. Letsinger, *J. Am. Chem. Soc.*, **75**, 3028 (1953).
75. G. Montaudo and S. Caccamese, *J. Org. Chem.*, **38**, 710 (1973); S. Caccamese, J. A. McMillan and G. Montaudo, *J. Org. Chem.*, **43**, 2703 (1978).
76. J. Ciabattoni and E. C. Nathan, III, *J. Am. Chem. Soc.*, **91**, 4766 (1969).
77. J. E. Baldwin and M. C. McDaniel, *J. Am. Chem. Soc.*, **89**, 1537 (1967); **90**, 6118 (1968); O. L. Chapman and J. D. Lassilla, *J. Am. Chem. Soc.*, **90**, 2449 (1968).
78. (a) P. E. Eaton and W. S. Hurt, *J. Am. Chem. Soc.*, **88**, 5038 (1966).
    (b) J. L. Ruhlen and P. A. Leermakers, *J. Am. Chem. Soc.*, **88**, 5671 (1966); **89**, 4944 (1967).
79. P. J. Wagner and D. J. Bucheck, *J. Am. Chem. Soc.*, **91**, 5090 (1969).
80. G. E. Heibel and D. I. Schuster, unpublished results.
81. R. Bonneau, *J. Am. Chem. Soc.*, **102**, 3816 (1980).
82. W. C. Agosta and A. B. Smith, III, *J. Am. Chem. Soc.*, **93**, 5513 (1971).
83. W. C. Agosta, A. B. Smith, III, A. S. Kende, R. G. Eilerman and J. Benham, *Tetrahedron Lett.*, 4517 (1969).
84. T. Matsuura and K. Ogura, *J. Am. Chem. Soc.*, **89**, 3850 (1967); *Bull. Chem. Soc. Jpn.*, **43**, 3187 (1970).
85. F. G. Burkinshaw, B. R. Davis and P. D. Woodgate, *J. Chem. Soc. (C)*, 1607 (1970).
86. H. E. Zimmerman and R. D. Little, *J. Am. Chem. Soc.*, **96**, 4623 (1974).

87. S. Wolff and W. C. Agosta, *J. Chem. Soc., Chem. Commun.*, 226 (1972).
88. S. Wolff and W. C. Agosta, *J. Org. Chem.*, **50**, 4707 (1985).
89. P. de Mayo, J-P. Pete and M. Tchir, *Can. J. Chem.*, **46**, 2535 (1968); see also M. Pfau, R. Dulou and M. Vilkas, *Compt. Rend.*, 1817 (1962).
90. R. L. Cargill, A. C. Miller, D. M. Pond, P. de Mayo, M. F. Tchir, K. R. Neuberger and J. Saltiel, *Mol. Photochem.*, **1**, 301 (1969).
91. S. Wolff, W. L. Schreiber, A. B. Smith III and W. C. Agosta, *J. Am. Chem. Soc.*, **94**, 7797 (1972).
92. S. Ayral-Kaloustian, S. Wolff and W. C. Agosta, *J. Am. Chem. Soc.*, **99**, 5984 (1977).
93. Y. Tobe, T. Iseki, K. Kakiuchi and Y. Odaira, *Tetrahedron Lett.*, **25**, 3895 (1984).
94. P. de Mayo, J.-P. Pete and M. Tchir, *J. Am. Chem. Soc.*, **89**, 5712 (1967).
95. P. E. Eaton and K. Lin, *J. Am. Chem. Soc.*, **86**, 2087 (1964).
96. P. E. Eaton, *Acc. Chem. Res.*, **1**, 50 (1968).
97. E. J. Corey, M. Tada, R. LeMahieu and L. Libit, *J. Am. Chem. Soc.*, **87**, 2051 (1965).
98. P. E. Eaton and K. Lin, *J. Am. Chem. Soc.*, **87**, 2052 (1965).
99. R. Bonneau, P. Fornier de Violet and J. Joussot-Dubien, *Nouv. J. Chim.*, **1**, 31 (1977).
100. T. Goldfarb, *J. Photochem.*, **8**, 29 (1978).
101. J. Michl, *Mol. Photochem.*, **4**, 243, 257 (1972); see also Reference 2, Chap. 4.
102. R. A. Caldwell, H. Misawa, E. F. Healy and M. J. S. Dewer, *J. Am. Chem. Soc.*, **109**, 6869 (1987).
103. H. Hart, T. Miyashi, D. N. Buchanan and S. Sasson, *J. Am. Chem. Soc.*, **96**, 4857 (1974); E. Dunkelblum, H. Hart and M. Suzuki, *J. Am. Chem. Soc.*, **99**, 5074 (1977).
104. R. Noyori and M. Kato, *Bull. Chem. Soc. Jpn.*, **47**, 1460 (1974).
105. E. Dunkelblum and H. Hart, *J. Am. Chem. Soc.*, **99**, 644 (1977); H. Hart and E. Dunkelblum, *J. Am. Chem. Soc.*, **100**, 5141 (1978).
106. M. Suzuki, H. Hart, E. Dunkelblum and W. Li, *J. Am. Chem. Soc.*, **99**, 5083 (1977).
107. E. Y. Y. Lam, D. Valentine and G. S. Hammond, *J. Am. Chem. Soc.*, **89**, 3482 (1967).
108. H. E. Zimmerman, R. W. Binkley, J. J. McCullough and G. A. Zimmerman, *J. Am. Chem. Soc.*, **89**, 6589 (1967).
109. (a) O. L. Chapman, P. J. Nelson, R. W. King, D. J. Trecker and A. A. Griswold, *Rec. Chem. Prog.*, **28**, 167 (1967).
   (b) See also O. L. Chapman and D. S. Weiss, in *Organic Photochemistry*, Vol. 3 (Ed. O. L. Chapman), Marcel Dekker, New York, 1973, pp. 197–288.
110. D. I. Schuster and I. M. Nunez, unpublished results; I. M. Nunez, Ph.D. Dissertation, New York University, 1982.
111. P. C. Tucker, Ph.D. Dissertation, New York University, 1988.
112. A. Butenandt, L. Karlson-Poschmann, G. Failer, U. Schiedt and E. Biekert, *Justus Liebigs Ann. Chem.*, **575**, 123 (1952).
113. For a review see J. G. Burr, in *Advances in Photochemistry*, Vol. 6 (Eds. W. A. Noyes, Jr., G. S. Hammond and J. N. Pitts, Jr.), Interscience/Wiley, 1968, pp. 193–299.
114. R. Beukers and W. Berends, *Biochim. Biophys. Acta*, **41**, 550 (1960).
115. A. A. Lamola, *Photochem. Photobiol.*, **7**, 619 (1968).
116. C. S. Rupert, in *Photophysiology*, Vol. 2 (Ed. A. C. Giese), Academic Press, New York, 1964, pp. 283–327.
117. B. Nann, D. Gravel, R. Schorta, H. Wehrli, K. Schaffner and O. Jeger, *Helv. Chim. Acta*, **46**, 2473 (1963).
118. D. Bellus, D. R. Kearns and K. Schaffner, *Helv. Chim. Acta*, **52**, 971 (1969).
119. K. Schaffner, *Tetrahedron*, **30**, 1891 (1974).
120. See K. Schaffner, 23rd Int. Congr. Pure Appl. Chem., 1971 p. 405.
121. See Reference 118, and P. Margaretha and K. Schaffner, *Helv. Chim. Acta*, **56**, 2884 (1973).
122. A. C. Chan and D. I. Schuster, *J. Am. Chem. Soc.*, **108**, 4561 (1986).
123. D. I. Schuster, I. M. Nunez and C. B. Chan, *Tetrahedron Lett.*, **22**, 1187 (1981).
124. See, for example, the kinetic isotope effect in A. Padwa and C. S. Chou, *J. Am. Chem. Soc.*, **102**, 3619 (1980); for energies of activation for hydrogen abstraction from 2-propanol by n, π^* triplet states, see M. Berger, E. McAlpine and C. Steel, *J. Am. Chem. Soc.*, **100**, 5147 (1978).
125. See Reference 109b, and G. Wampfler, Ph.D. Dissertation, Iowa State University, 1970.
126. H. E. Zimmerman, in *Advances in Photochemistry*, Vol. 1 (Eds. W. A. Noyes, Jr., G. S. Hammond and J. N. Pitts, Jr.), Interscience/Wiley, 1963, pp. 183–208.

127. O. L. Chapman, in Reference 126, pp. 323–420.
128. K. Schaffner in *Advances in Photochemistry*, Vol. 4 (Ed. W. A. Noyes, Jr., G. S. Hammond and J. N. Pitts, Jr.), Interscience-Wiley, 1966, pp. 81–112.
129. P. J. Kropp, in *Organic Photochemistry*, Vol. 1 (Ed. O. L. Chapman), Wiley/Interscience, 1967, pp. 1–90.
130. W. W. Kwie, B. A. Shoulders and P. D. Gardner, *J. Am. Chem. Soc.*, **84**, 2268 (1962); B. A. Shoulders, W. W. Kwie, W. Klyne and P. D. Gardner, *Tetrahedron*, **21**, 2973 (1965).
131. O. L. Chapman, T. A. Rettig, A. A. Griswold, A. I. Dutton and P. Fitton, *Tetrahedron Lett.*, 2049 (1963).
132. W. G. Dauben, G. W. Shaffer and N. D. Vietmeyer, *J. Org. Chem.*, **33**, 4060 (1968).
133. W. G. Dauben, W. A. Spitzer and M. S. Kellogg, *J. Am. Chem. Soc.*, **93**, 3674 (1971).
134. H. E. Zimmerman and J. W. Wilson, *J. Am. Chem. Soc.*, **86**, 4036 (1964); H. E. Zimmerman, R. D. Rieke and J. R. Scheffer, *J. Am. Chem. Soc.*, **89**, 2033 (1967); H. E. Zimmerman and R. L. Morse, *J. Am. Chem. Soc.*, **90**, 954 (1968); H. E. Zimmerman and K. G. Hancock, *J. Am. Chem. Soc.*, **90**, 3749 (1968); H. E. Zimmerman and W. R. Elser, *J. Am. Chem. Soc.*, **91**, 887 (1969); H. E. Zimmerman and D. J. Sam, *J. Am. Chem. Soc.*, **88**, 4114, 4905 (1966).
135. F. Nobs, U. Burger and K. Schaffner, *Helv. Chim. Acta*, **60**, 1607 (1977).
136. J. S. Swenton, R. M. Blankenship and R. Sanitra, *J. Am. Chem. Soc.*, **97**, 4941 (1975).
137. For a review, see S. S. Hixson, P. S. Mariano and H. E. Zimmerman, *Chem. Rev.*, **73**, 531 (1973).
138. B. Nann, D. Gravel, R. Schorta, H. Wehrli, K. Schaffner and O. Jeger, *Helv. Chim. Acta*, **46**, 2473 (1963).
139. O. L. Chapman, J. B. Sieja and W. J. Welstead, Jr., *J. Am. Chem. Soc.*, **88**, 161 (1966).
140. D. I. Schuster, R. H. Brown and B. M. Resnick, *J. Am. Chem. Soc.*, **100**, 4504 (1978).
141. Reference 40, pp. 89–100.
142. S. S. Shaik, *J. Am. Chem. Soc.*, **101**, 2736 (1979); see also S. Shaik and N. D. Epiotis, *J. Am. Chem. Soc.*, **100**, 18 (1978).
143. D. I. Schuster and S. Hussain, *J. Am. Chem. Soc.*, **102**, 409 (1980); S. Hussain, Ph.D. Dissertation, New York University, 1979.
144. H. E. Zimmerman and D. I. Schuster, *J. Am. Chem. Soc.*, **84**, 4527 (1962); H. E. Zimmerman and J. S. Swenton, *J. Am. Chem. Soc.*, **89**, 906 (1967).
145. For a recent review of the literature, see K. Schaffner and M. Demuth, in *Rearrangements in Ground and Excited States*, Vol. 3 (Ed. P. de Mayo), Academic Press, New York, 1980, pp. 281–348.
146. B. Frei, C. Ganter, K. Kagi, K. Kocsis, M. Miljkovic, A. Siewinski, R. Wenger, K. Schaffner and O. Jeger, *Helv. Chim. Acta*, **49**, 1049 (1966); D. I. Schuster and K. V. Prabhu, *J. Am. Chem. Soc.*, **96**, 3511 (1974).
147. D. I. Schuster, *Acc. Chem. Res.*, **11**, 65 (1978); D. I. Schuster and K. Liu, *Tetrahedron*, **37**, 3329 (1981).
148. C. J. Samuel, *J. Chem. Soc., Perkin Trans. 2*, 736 (1981); A. G. Schultz, M. Macielag and M. Plummer, *J. Org. Chem.*, **53**, 391 (1988).
149. D. I. Schuster and D. F. Brizzolara, *J. Am. Chem. Soc.*, **92**, 4357 (1970).
150. G. Cruciani and P. Margaretha, *J. Fluorine Chem.*, **37**, 95 (1987).
151. William of Occam (1300–1349): "Essentia non sunt multiplicanda praeter necessitatem", *Encyclopaedia Brittanica*, Vol. 16, 1955 edition, pp. 680–681.
152. H. E. Zimmerman and N. Lewin, *J. Am. Chem. Soc.*, **91**, 879 (1969).
153. H. E. Zimmerman, *Tetrahedron*, **30**, 1617 (1974).
154. H. E. Zimmerman, X. Jian-hua, R. K. King and C. E. Caufield, *J. Am. Chem. Soc.*, **107**, 7724 (1985).
155. H. E. Zimmerman, C. E. Caufield and R. K. King, *J. Am. Chem. Soc.*, **107**, 7732 (1985).
156. R. C. Hahn and G. W. Jones, *J. Am. Chem. Soc.*, **93**, 4232 (1971); R. C. Hahn and D. W. Kurtz, *J. Am. Chem. Soc.*, **95**, 6723 (1973).
157. H. E. Zimmerman and D. J. Sam, *J. Am. Chem. Soc.*, **88**, 4905 (1966).
158. H. E. Zimmerman and R. L. Morse, *J. Am. Chem. Soc.*, **90**, 954 (1968).
159. H. E. Zimmerman and R. D. Solomon, *J. Am. Chem. Soc.*, **108**, 6276 (1986).
160. H. E. Zimmerman, *Tetrahedron*, **19**, **Supp 2**, 393 (1963), *Pure Appl. Chem.*, **9**, 493 (1964).
161. H. E. Zimmerman, J. Nasielski, R. Keese and J. S. Swenton, *J. Am. Chem. Soc.*, **88**, 4895 (1966).
162. P. Margaretha and K. Schaffner, *Helv. Chim. Acta*, **56**, 2884 (1973).

163. J. Gloor, K. Schaffner and O. Jeger, *Helv. Chim. Acta*, **54**, 1864 (1971); K. Schaffner, *Pure Appl. Chem.*, **33**, 329 (1973); J. Gloor, G. Bernardinelli, R. Gerdil and K. Schaffner, *Helv. Chim. Acta*, **56**, 2520 (1973).
164. Reference 40, pp. 114–140.
165. J. Gloor and K. Schaffner, *Helv. Chim. Acta*, **57**, 1815 (1974).
166. R. S. Givens and B. W. Atwater, *J. Am. Chem. Soc.*, **108**, 5028 (1986).
167. G. Stork and J. Tsuji, *J. Am. Chem. Soc.*, **83**, 2783 (1961); G. Stork, P. Rosen, N. Goldman, R. V. Coombs and J. Tsuji, *J. Am. Chem. Soc.*, **87**, 275 (1965).
168. R. Bonneau and P. Fornier de Violet, *C. R. Acad. Sci. Paris, Ser. C*, **284**, 631 (1977).
169. D. I. Schuster, R. Bonneau, D. A. Dunn, J. M. Rao and J. Joussot-Dubien, *J. Am. Chem. Soc.*, **106**, 2706 (1984).
170. D. I. Schuster, P. B. Brown, L. Capponi, C. A. Rhodes, J. C. Scaiano and D. Weir, *J. Am. Chem. Soc.*, **109**, 2533 (1987).
171. D. I. Schuster, D. A. Dunn and R. Bonneau, *J. Photochem.*, **28**, 413 (1985).
172. D. A. Dunn, Ph.D. Dissertation, New York University, 1985; D. A. Dunn and D. I. Schuster, unpublished results.
173. P. B. Brown, Ph.D. Dissertation, New York University, 1988.
174. D. Weir, J. C. Scaiano and D. I. Schuster, *Can. J. Chem.*, (1988), in press.
174a. S. Yamauchi, N. Hirota and J. Higuchi, *J. Phys. Chem.*, **92**, 2129 (1988).
175. T. S. Cantrell, *J. Org. Chem.*, **39**, 3063 (1974).
176. S. Lazare, R. Bonneau and R. Lapouyade, *J. Phys. Chem.*, **88**, 18 (1984); S. Lazare, R. Lapouyade and R. Bonneau, *J. Am. Chem. Soc.*, **107**, 6604 (1985).
177. J. Saltiel and B. Thomas, *Chem. Phys. Lett.*, **37**, 147 (1976); J. Saltiel and B. W. Atwater, in *Advances in Photochemistry*, Vol. 14 (Eds. D. H. Volman, G. S. Hammond and K. Gollnick), Wiley/Interscience, New York, 1988, pp. 6–38.
178. N. J. Pienta and J. E. McKimmey, *J. Am. Chem. Soc.*, **104**, 5501 (1982).
179. A. Insogna and D. I. Schuster, unpublished results.
180. O. L. Chapman, D. Ostren, J. Lasilla and P. Nelson, *J. Org. Chem.*, **34**, 811 (1969).
181. J. B. Guttenplan and S. G. Cohen, *J. Am. Chem. Soc.*, **94**, 4040 (1972); *Tetrahedron Lett.*, 2163 (1972); A. H. Parola, A. W. Rosa and S. G. Cohen, *J. Am. Chem. Soc.*, **97**, 6202 (1975); S. Inbar, H. Linschitz and S. G. Cohen, *J. Am. Chem. Soc.*, **102**, 1419 (1980); K. S. Peters, S. C. Freilich and C. G. Schaeffner, *J. Am. Chem. Soc.*, **102**, 5701 (1980); J. D. Simon and K. S. Peters, *J. Am. Chem. Soc.*, **103**, 6403 (1981).
182. D. I. Schuster, M. M. Greenberg, I. M. Nuñez and P. C. Tucker, *J. Org. Chem.*, **48**, 2615 (1983).
183. C. A. Rhodes and D. I. Schuster, unpublished results.
184. E. J. Corey, J. D. Bass, R. LeMahieu and R. B. Mitra, *J. Am. Chem. Soc.*, **86**, 5570 (1964).
185. D. A. Dunn, D. I. Schuster and R. Bonneau, *J. Am. Chem. Soc.*, **107**, 2802 (1985).
186. N. J. Pienta, *J. Am. Chem. Soc.*, **106**, 2704 (1984).
187. P. E. Eaton, *J. Am. Chem. Soc.*, **84**, 2454 (1962).
188. E. J. Corey, R. B. Mitra and H. Uda, *J. Am. Chem. Soc.*, **86**, 485 (1964); E. J. Corey and S. Nozoe, *J. Am. Chem. Soc.*, **86**, 1652 (1964).
189. W. Oppolzer, *Acc. Chem. Res.*, **15**, 135 (1982).
190. M. B. Rubin, T. Maymon and D. Glover, *Isr. J. Chem.*, **8**, 717 (1970).
191. G. R. Lenz, *Rev. Chem. Intermed.*, **4**, 369 (1981); G. R. Lenz, *J. Chem. Soc., Chem. Commun.*, 803 (1982); G. R. Lenz, *J. Chem. Soc., Perkin Trans. 1*, 2397 (1984).
192. T. S. Cantrell, W. S. Haller and J. C. Williams, *J. Org. Chem.*, **34**, 509 (1969).
193. R. M. Bowman, C. Calvo, J. J. McCullough, P. W. Rasmussen and F. F. Snyder, *J. Org. Chem.*, **37**, 2084 (1972).
194. J. J. McCullough, J. M. Kelly and P. W. Rasmussen, *J. Org. Chem.*, **34**, 2933 (1969).
195. P. Singh, *J. Org. Chem.*, **36**, 3334 (1971).
196. N. P. Peet, R. L. Cargill and D. F. Bushey, *J. Org. Chem.*, **38**, 1218 (1973).
197. W. L. Dilling, T. E. Tabor, F. P. Boer and P. P. North, *J. Am. Chem. Soc.*, **92**, 1399 (1970).
198. R. L. Cargill, G. H. Morton and J. Bordner, *J. Org. Chem.*, **45**, 3929 (1980).
199. K. Wiesner, *Tetrahedron*, **31**, 1655 (1975).
200. R. O. Loutfy and P. de Mayo, *J. Am. Chem. Soc.*, **99**, 3559 (1977).
201. J. J. McCullough, B. R. Ramachandran, F. F. Snyder and G. N. Taylor, *J. Am. Chem. Soc.*, **97**, 6767 (1975).

756 David I. Schuster

202. J. M. Kelly, T. B. H. McMurry and T. H. Work, *J. Chem. Soc., Chem. Commun.*, 280 (1987).
203. D. R. Kearns, G. Marsh and K. Schaffner, *J. Chem. Phys.*, **49**, 3316 (1968); G. Marsh, D. R. Kearns and K. Schaffner, *Helv. Chim. Acta*, **51**, 1890 (1968); *J. Am. Chem. Soc.*, **93**, 3129 (1971).
204. C. R. Jones and D. R. Kearns, *J. Am. Chem. Soc.*, **99**, 344 (1977).
205. P. de Mayo, *Acc. Chem. Res.*, **4**, 41 (1971).
206. R. S. H. Liu and D. M. Gale, *J. Am. Chem. Soc.*, **90**, 1897 (1968).
207. S. Hosaka and S. Wakamatsu, *Tetrahedron Lett.*, 219 (1968).
208. For rare examples of photocycloadditions using electron deficient alkenes see M. T. Crimmins and J. A. DeLoach, *J. Am. Chem. Soc.*, **108**, 800 (1986) and B. D. Challand, H. Hikino, G. Kornis, G. Lange and P. de Mayo, *J. Org. Chem.*, **34**, 794 (1969).
209. V. A. Stoute, J. Shimonov and D. I. Schuster, unpublished results.
210. P. G. Bauslaugh, *Synthesis*, 287 (1970).
211. J. C. Scaiano, *Acc. Chem. Res.*, **15**, 252 (1982) and references cited therein; J. C. Scaiano, C. W. B. Lee, Y. L. Chow and B. Marciniak, *J. Phys. Chem.*, **86**, 2452 (1982).
212. M. V. Encinas, P. J. Wagner and J. C. Scaiano, *J. Am. Chem. Soc.*, **102**, 1357 (1980).
213. G. R. Lenz, *Tetrahedron*, **28**, 2211 (1972).
214. G. R. Lenz and L. Swenton, *J. Chem. Soc., Chem. Commun.*, 444 (1979).
215. N. D. Epiotis and S. Shaik, *J. Am. Chem. Soc.*, **100**, 9 (1978); S. Shaik, *J. Am. Chem. Soc.*, **101**, 3184 (1979).
216. Y. Tobe, A. Doi, A. Kunai, K. Kimura and Y. Odaira, *J. Org. Chem.*, **42**, 2523 (1977).
217. J. Verbeek, J. H. van Lenthe, P. J. J. A. Timmermans, A. Mackor and P. H. M. Budzelaar, *J. Org. Chem.*, **52**, 2955 (1987).
218. R. S. Johnson, University of New Hampshire, private communication of unpublished results.
219. M. Saunders, *J. Am. Chem. Soc.*, **109**, 3150 (1987).
220. M. Saunders, Yale University, private communication of unpublished results.
221. E. Dunkelblum, H. Hart and M. Jeffares, *J. Org. Chem.*, **43**, 3409 (1978).
222. T. Matsuura and K. Ogura, *Bull. Chem. Soc. Jpn.*, **40**, 945 (1967).
223. L. Rodriguez-Hahn, B. Esquivel, A. Ortega, J. Garcia, E. Diaz, J. Cardena, M. Soriano-Garcia and A. Toscano, *J. Org. Chem.*, **50**, 2865 (1985).
224. G. S. Hammond, N. J. Turro and R. S. H. Liu, *J. Org. Chem.*, **28**, 3297 (1963); R. S. H. Liu, N. J. Turro and G. S. Hammond, *J. Am. Chem. Soc.*, **87**, 3406 (1965); W. G. Herkstroeter, A. A. Lamola and G. S. Hammond, *J. Am. Chem. Soc.*, **86**, 4537 (1964).
225. M. Mintas, D. I. Schuster and P. G. Williard, *J. Am. Chem. Soc.*, **110**, 2305 (1988).
226. J. C. Scaiano, data obtained at NRC Laboratories, Ottawa.
227. M. J. S. Dewar, S. Olivella and J. J. P. Stewart, *J. Am. Chem. Soc.*, **108**, 5771 (1986).
228. H. Shinozaki, S. Arai and M. Tada, *Bull. Chem. Soc. Jpn.*, **49**, 821 (1976).
229. M. Mintas, D. I. Schuster and P. G. Williard, *Tetrahedron*, **44**, 6001 (1988).
230. G. M. J. Schmidt, *Solid State Photochemistry* (Ed. D. Ginsburg), Verlag Chemie, New York, 1976.
231. J. Scheffer, M. Garcia-Garibay and O. Nalamasu, in *Organic Photochemistry*, Vol. 8 (Ed. A. Padwa), Marcel Dekker, New York, 1987.
232. J. D. Dunitz, *X-Ray Analysis and the Structure of Organic Molecules*, Cornell University Press, Ithaca, New York, 1979, pp. 312–318.
233. W. K. Appel, Z. Q. Jiang, J. R. Scheffer and L. Walsh, *J. Am. Chem. Soc.*, **105**, 5354 (1983).
234. T. J. Greenhough, J. R. Scheffer, A. S. Secco, J. Trotter and L. Walsh, *Isr. J. Chem.*, **25**, 297 (1985).
235. S. Ariel, S. Askari, J. R. Scheffer, J. Trotter and L. Walsh, *J. Am. Chem. Soc.*, **106**, 5726 (1984).
236. S. Ariel, S. Askari, J. R. Scheffer, J. Trotter and L. Walsh, in *Organic Phototransformations in Nonhomogeneous Media* (Ed. M. A. Fox), American Chemical Society, Washington, D.C., 1985. Chap. 15.
237. P. Wender, in *Photochemistry in Organic Synthesis* (Ed. J. D. Coyle), Chap. 9, Royal Society of Chemistry, London, 1986.
238. M. T. Crimmins, *Chem. Rev.*, in press.
239. M. J. C. M. Koppes and H. Cerfontain, *Recl. Trav. Chim. Pays-Bas*, **107**, 412, 549 (1988).
240. B. Reiman, D. E. Sadler and K. Schaffner, *J. Am. Chem. Soc.*, **108**, 5527 (1986).
241. D. Schuster, G. E. Heibel, R. A. Caldwell, L. A. Melton and W. Tang, unpublished data.
242. D. I. Schuster, G. E. Heibel, P. B. Brown, N. J. Turro and C. V. Kumar, *J. Am. Chem. Soc.*, **110**, 826 (1988).

The Chemistry of Enones
Edited by S. Patai and Z. Rappoport
© 1989 John Wiley & Sons Ltd

CHAPTER **16**

Radiation chemistry of enones

P. NETA AND M. DIZDAROGLU

Center for Chemical Physics, National Bureau of Standards, Gaithersburg, Maryland 20899, USA

I. INTRODUCTION

Radiation chemistry deals with the chemical effects of ionizing radiation, such as X-rays, gamma rays, high energy electrons, or other energetic particles. Ionizing radiation is absorbed in organic materials somewhat indiscriminately and causes ionizations and excitations which may result in bond scission. In discussing the radiation chemistry of an organic compound, we should distinguish between the radiation chemistry of the neat compound, where the energy is absorbed totally by the compound itself, and the radiation chemistry of its solutions, where the energy is absorbed predominantly by the solvent. In the latter case, the solute undergoes chemical changes only via reactions with the primary radicals formed from the solvent. The radiation chemistry of enones was studied mainly in solution, as will become clear from this review, and most often it involved aqueous solutions.

Radiolytic studies have been carried out with only a limited number of enones. Several studies have dealt with the simple enones such as acrolein or crotonaldehyde and with the polyene retinal. A number of papers have been published on ascorbic acid and related compounds. Among the heterocyclic enones, we find a study on pyridones but a very large number of papers on pyrimidine and purine bases. In fact, the amount of research carried out on these bases is orders of magnitude higher than that on all other enones, obviously because of the importance of understanding the basic radiation chemistry of DNA. As a result, many reviews and books dealing with the radiation chemistry of DNA components have been published. In order to keep this chapter on enones somewhat balanced, we shall discuss the DNA bases only briefly and refer the interested reader to the main literature on the topic.

To facilitate discussion of the radiation chemistry of individual compounds in solution, we shall describe here briefly the primary reactions that take place in typical irradiated solvents. The most important and best understood of the solvents is water.

Radiolysis of water results in the production of hydrated electrons, hydrogen atoms, hydroxyl radicals and molecular products (hydrogen and hydrogen peroxide). The yields (G values) of these species in neutral water are approximately $2.8\,e_{aq}^-$, 2.8 OH, 0.6 H, $0.8\,H_2O_2$ and $0.4\,H_2$ (molecules per 100 eV absorbed in solution). In most cases, the molecular products do not interfere with the reactions of the radicals.

Hydrated electrons react with aldehydes and ketones and with conjugated double bonds very rapidly ($k = 10^9 - 10^{10}\,M^{-1}\,s^{-1}$) to form radical anions, which subsequently may protonate to yield neutral radicals. Hydroxyl radicals react with enones very rapidly ($k = 10^9 - 10^{10}\,M^{-1}\,s^{-1}$) by addition to the double bond and more slowly ($k = 10^8 - 10^9\,M^{-1}\,s^{-1}$) by hydrogen abstraction from C—H bonds. Hydrogen atoms also add to double bonds rapidly but they abstract hydrogen much more slowly ($k = 10^5 - 10^7\,M^{-1}\,s^{-1}$) and also may add slowly to the carbonyl group. These reactions will be discussed in more detail in conjunction with each group of compounds. It is clear, however, that if all the primary radicals are allowed to react with the solute, the system will be very complex and the ensuing chemistry may not be meaningful. To simplify the system under study and to direct the reaction toward a desired product one has to manipulate the primary radicals by addition of proper scavengers.

To study one-electron reduction without interference by OH and H one may add a scavenger for these radicals, commonly an alcohol or formate ions, which react with H and OH rapidly and thus prevent their reaction with the solute under study. Moreover, the radicals produced by reactions of H and OH with alcohols and formate may be reducing in nature and thus the net result is one-electron reduction of the solute by e_{aq}^- and by the organic radical, i.e. a system with one radical produced from the solute under study, with no other side-reactions, e.g.

$$\dot{O}H + HCO_2^- \rightarrow H_2O + \dot{C}O_2^- \qquad (k = 3 \times 10^9\,M^{-1}\,s^{-1}) \qquad (1)$$

$$\dot{H} + HCO_2^- \rightarrow H_2 + \dot{C}O_2^- \qquad (k = 2 \times 10^8\,M^{-1}\,s^{-1}) \qquad (2)$$

$$e_{aq}^- + S \rightarrow S^{\cdot -} \qquad (3)$$

$$\dot{C}O_2^- + S \rightarrow CO_2 + S^{\cdot -} \qquad (4)$$

To study the reactions of OH without interference by e_{aq}^- the solution is saturated with N_2O, which converts the hydrated electron into OH radical.

$$e_{aq}^- + N_2O \rightarrow N_2 + OH^- + \dot{O}H \qquad (k = 9 \times 10^9\,M^{-1}\,s^{-1}) \qquad (5)$$

In this case the yield of H atoms amounts only to 10% of that of OH radicals and thus no significant interference by H is experienced. Moreover, H and OH often react with a solute by the same mechanism, i.e. hydrogen abstraction to give the same radical or addition to a double bond to give similar radicals.

To study specifically the reactions of H atoms one uses acidic solutions where the hydrated electron is protonated to give H.

$$e_{aq}^- + H^+ \rightarrow \dot{H} \qquad (k = 2.3 \times 10^{10}\,M^{-1}\,s^{-1}) \qquad (6)$$

The interfering OH reaction may be eliminated by using t-butyl alcohol as a scavenger. This alcohol reacts rapidly with OH ($k = 5 \times 10^8\,M^{-1}\,s^{-1}$) but much more slowly with H ($k = 1.7 \times 10^5\,M^{-1}\,s^{-1}$) and, furthermore, the radical produced by its reaction with OH is relatively unreactive and is not likely to interfere with the study of the reaction of H with the solute.

$$\dot{O}H + (CH_3)_3COH \rightarrow H_2O + \dot{C}H_2C(CH_3)_2OH \tag{7}$$

The above manipulations allow the study of each of the three primary radicals with little interference by the others. Further, they allow the study of one-electron reduction of solutes. To carry out one-electron oxidation of a solute one may attempt to use OH radicals in N_2O saturated solutions. The OH radicals, however, although they are strong oxidants, generally react by addition or abstraction rather than by a one-electron transfer mechanism. Addition of OH may be followed by water elimination to result in a net oxidation process, but for many compounds this is not the case. Therefore, to carry out one-electron oxidation it is advantageous to convert the OH into strict one-electron oxidizing radicals by the intermediacy of halides, thiocyanate, azide or ethylene glycol. For example, bromide ions form Br_2^- radicals

$$Br^- + \dot{O}H \rightarrow \dot{B}r + OH^- \qquad (k = 1.1 \times 10^{10}\,M^{-1}\,s^{-1}) \tag{8}$$

$$\dot{B}r + Br^- \leftrightharpoons \dot{B}r_2^- \qquad (K = 2 \times 10^5\,M^{-1}) \tag{9}$$

azide ions form the azidyl radical

$$N_3^- + \dot{O}H \rightarrow \dot{N}_3 + OH^- \qquad (k = 1.2 \times 10^{10}\,M^{-1}\,s^{-1}) \tag{10}$$

and ethylene glycol undergoes hydrogen abstraction followed by acid- or base-catalyzed water elimination to yield the oxidizing formylmethyl radical.

$$HOCH_2CH_2OH + \dot{O}H \rightarrow HOCH_2\dot{C}HOH + H_2O \qquad (k = 1.8 \times 10^9\,M^{-1}\,s^{-1}) \tag{11}$$

$$HOCH_2\dot{C}HOH \rightarrow \dot{C}H_2CHO + H_2O \tag{12}$$

The radicals Cl_2^{-}, Br_2^{-}, I_2^{-}, $(SCN)_2^{-}$, N_3^{-}, and $\dot{C}H_2CHO$ are strict one-electron oxidants of different redox potentials and may serve to oxidize a variety of enones. Other oxidants may be produced from metal ions or from organic compounds to serve the same purpose.

Radiolysis of a solute in non-aqueous solutions also may lead to oxidation or reduction products and certain solvents are sufficiently well understood to be useful for specific purposes. For example, radiolysis of a solute dissolved in alcohols or ethers results in the formation of its radical anion, and radiolysis in carbon tetrachloride or methylene chloride results in the formation of the radical cation. In both cases the radiolysis produces initially an electron and a positive hole. However, in alcohols the hole is converted into a reducing radical while in halogenated hydrocarbons the hole oxidizes the solute and the electron reacts with the solvent to form an inert halide ion. Further details on the various solvents and the experimental techniques are found in a number of reviews[1,2] and books[3-6].

II. SIMPLE UNSATURATED KETONES AND ALDEHYDES

Irradiation of neat enones, like the irradiation of many olefins, may result in polymerization. Thus gamma radiolysis of frozen acrolein produces a polymer. The rate of polymerization and the structure of the resulting polymer were determined as a function of the irradiation temperature[7], and the results suggested that the polymerization was anionic.

Irradiation of acrolein, methyl vinyl ketone, crotonaldehyde, 3-hexene-2, 5-dione and 2, 4, 6-octatrienal in aqueous solutions containing an alcohol as OH scavenger and deoxygenated by bubbling with Ar led to the formation of the radical anions of these enones[8].

$$RCH{=}CHCO{-}R + e_{aq} \rightarrow RCH{=}CH\dot{C}O^-{-}R \tag{13}$$

The rate constants for reaction 13 must be close to the diffusion-controlled limit ($\sim 10^{10}\,\text{M}^{-1}\,\text{s}^{-1}$) since acetaldehyde and acetone react with e_{aq}^- very rapidly ($k = 3.5 \times 10^9$ and $6 \times 10^9\,\text{M}^{-1}\,\text{s}^{-1}$, respectively)[9]. The optical absorption spectra of the radical anions formed in reaction 13 are in the UV range, the maxima are at 270–280 nm for the radicals with one carbonyl and one double bond, but shift to 350–370 nm when a second carbonyl or a second double bond is conjugated with the basic enone[8]. These ketyl radical anions undergo protonation in neutral or acid solution to form the ketyl radicals.

$$RCH{=}CH\dot{C}O^- + H^+ \leftrightarrows RCH{=}CH\dot{C}OH{-}R \qquad (14)$$

The absorption spectra of the neutral forms are shifted to lower wavelengths, about 250 nm for the simple enones and 320 nm for the more highly conjugated ones. The difference in spectra permits determination of the pK_a values for these radicals. They were found to be 9.6–10.1 for the simple enone radicals, 9.0 for the radical derived from 2, 4, 6-octatrienal and 5.2 for the radical derived from 3-hexene-2, 5-dione[8]. Clearly, an additional conjugated double bond lowers the pK_a somewhat but an additional carbonyl group exerts a very strong effect by withdrawing electrons from the radical site. In the case of the latter radical (from 3-hexene-2, 5-dione) the spin density and the negative charge are divided between the two carbonyl groups equally, so that protonation is greatly facilitated. Both the pK_a values of the radicals and the wave numbers of their absorption maxima gave linear correlation with the transition energies calculated by LCAO methods[8].

Acrolein, crotonaldehyde and methyl vinyl ketone also react with OH radicals. This reaction was studied in N_2O saturated solutions and found to take place with very high rate constants, 3.5–$5.1 \times 10^9\,\text{M}^{-1}\,\text{s}^{-1}$, and to form the corresponding OH adducts[10].

$$RCH{=}CHCOR + \dot{O}H \rightarrow RCH(OH){-}\dot{C}HCOR \qquad (15)$$

These radicals absorb light in the UV region and the absorption maxima are below 240 nm. The spectra were found to change with time due to the second-order decay of the radicals, and the product spectra were pH dependent. The decay may take place by radical–radical combination or disproportionation.

$$2RCH(OH){-}\dot{C}HCOR \nearrow RCH(OH)CH(COR)CH(COR)CH(OH)R \qquad (16)$$
$$\searrow RC(OH){=}CHCOR + RCH(OH)CH_2COR \qquad (17)$$

The contribution of disproportionation was determined to be 30% for acrolein and methyl vinyl ketone and 86% for crotonaldehyde[10]. The reaction of OH with the hydrated form of crotonaldehyde follows a similar mechanism and produces the hydrated enol $CH_3C(OH){=}CHCH(OH)_2$, which undergoes spontaneous dehydration ($k = 54\,\text{s}^{-1}$) to $CH_3C(OH){=}CHCHO$. It also dehydrates in a base catalyzed process by deprotonating to $CH_3C(O^-){=}CHCH(OH)_2$ ($pK_a = 11.6$) and then losing water very rapidly ($k = 1 \times 10^4\,\text{s}^{-1}$)[10].

2, 3-Dihydroxy-2-propenal (triose reductone, TR) reacts with OH radicals ($k = 10^{10}\,\text{M}^{-1}\,\text{s}^{-1}$) to form adducts, which absorb mainly in the UV region[11].

$$CH(OH){=}C(OH)CH{=}O + \dot{O}H \nearrow CH(OH)_2{-}\dot{C}(OH)CH{=}O \qquad (18)$$
$$\searrow \dot{C}H(OH)C(OH)_2CH{=}O \qquad (19)$$

When the substrate is present in its anionic forms ($pK_1 = 5$, $pK_2 = 13$),

$$CH(OH){=}C(OH)CHO \leftrightarrows CH(O^-){=}C(OH)CHO + H^+ \leftrightarrows CH(O^-){=}C(O^-)CHO + 2H^+$$
$$\text{(TRH}_2\text{)} \qquad\qquad \text{(TRH}^-\text{)} \qquad\qquad\qquad \text{(TR}^{2-}\text{)}$$
$$(20)$$

the reaction of OH leads partially to the oxidation product, $(OCHCOCHO)^{\cdot -}$, which exhibits intense absorption at 398 nm due to its highly delocalized π system[11].

$$\left[\begin{array}{ccc} H\underset{\displaystyle O}{C}\!=\!=\!=\!\underset{\displaystyle O}{C}\!=\!=\!=\!\underset{\displaystyle O}{CH} \end{array} \right]^{-}$$

This radical is formed quantitatively by one-electron oxidation of 2,3-dihydroxy-2-propenal with Cl_2^- [$k(TRH_2) = 1.1 \times 10^9$], Br_2^- [$k(TRH_2) = 2.2 \times 10^8$, $k(TRH^-) = 1.8 \times 10^9$], $(SCN)_2^-$ [$k(TRH_2) = 2.7 \times 10^7$, $k(TRH^-) = 9 \times 10^8$], I_2^- [$k(TRH_2) < 10^6$, $k(TRH^-) = 3.4 \times 10^8$] and N_3 [$k(TRH^-) = 4 \times 10^9$ $M^{-1} s^{-1}$][12]. Its absorption at 398 nm is perfectly symmetric with 71 nm width at half maximum and with molar absorptivity of $5500 \, M^{-1} cm^{-1}$. This radical resembles that formed by oxidation of ascorbate in that both are conjugated tricarbonyl anions which absorb at similar wavelengths and protonate only at very low pH. The pK_a for protonation of $TR^{\cdot-}$ was found to be 1.4. Both $TR^{\cdot-}$ and $TRH^{\cdot}$ decay by second-order processes, the neutral form more rapidly than the anion, to yield TRH_2 and TR.

Radical anions of enones were formed also by radiolysis in frozen (77 K) methyltetrahydrofuran glasses and their absorption spectra reported[13].

III. RETINAL AND RELATED COMPOUNDS

Retinal is the chromophore of rhodopsin, the visual pigment, and of bacteriorhodopsin. Therefore, many studies have been carried out on the excited state of retinal, including some by pulse radiolysis. The latter technique was used also to investigate the properties of the radical anions and radical cations of retinal and other related polyenes.

all-trans-retinal

Das and Becker[14] studied the photophysical properties of the triplet state of retinal and of shorter and longer homologues having 3–7 conjugated double bonds next to the aldehyde group. In this series, the peak of the triplet–triplet absorption band was found to change from ca 400 to 500 nm with increase in chain length, and the molar absorptivity increases in the same series by about a factor of four. Some solvent effects were observed on both of the above parameters as well as on the rate of decay of the triplet. The nature of solvent also affected the quantum yield of the lowest triplet state; the effect was minimal for the short homologues, moderate for retinal and considerably higher for the longer homologues, where a decrease by a factor of 5–18 was found on changing from cyclohexane to acetonitrile, benzene and methanol.

Wilbrandt and Jensen[15] produced the lowest triplet state of retinal by pulse radiolysis in benzene or toluene solutions containing naphthalene as a sensitizer. Similarly, Bensasson and coworkers[16] prepared the triplet states of retinal homologues by radiolysis in hexane solutions containing biphenyl. Radiolysis of these solvents results in the formation of the triplet states of naphthalene or biphenyl which then transfer the energy to retinal and its homologues very rapidly (for naphthalene triplet reacting with retinal $k = 5.5 \times 10^9 \, M^{-1} s^{-1}$). The resulting triplet retinal is short-lived and was found to decay with a second-order rate constant of $2k = 6 \times 10^9 \, M^{-1} s^{-1}$. Using time-resolved resonance Raman spectroscopy, they recorded the Raman spectrum of the triplet state of retinal and found strong bands at 1550 and 1186 cm^{-1} and weaker bands at 1137, 1212, 1253, 1305 and 1339 cm^{-1} [15]. By comparing these bands with those of retinal in the ground state they concluded that the triplet state has a higher delocalization of π electrons. They also found similarity between the Raman spectrum of triplet retinal and that of an intermediate observed in the photochemical cycle of bacteriorhodopsin. Later, they compared the absorption and Raman spectra of the triplet state of *all-trans*-retinal with those of the 9-*cis*-, 11-*cis*- and 13-*cis*-isomers[17]. They concluded that each isomer forms a different triplet state or a different mixture of triplet states.

Land and collaborators[18] investigated the radical anions and radical cations of retinal and of shorter (2–3 double bonds) and longer (9 and 13 double bonds) homologues. Pulse radiolysis in deoxygenated hexane solutions produced a mixture of the radical anion and cation, but in the presence of N_2O the anion was absent. The absorption spectra of the radical-anion and -cation of the same polyenal were found to be similar, the maxima differing usually by only 10 nm. However, for the different homologues the maxima changed from 380 nm (for the compound with 3 conjugated double bonds) to 1130 nm (14 double bonds). Pulse radiolysis in methanol solutions gave only the radical anions. These underwent rapid protonation to form species which absorb at much lower wavelengths. Raghavan and colleagues[19] repeated the experiments in methanol and determined the rate of protonation of the radical anion by the solvent to be $7 \times 10^5 \, s^{-1}$. The rate of protonation in 2-propanol was found to be considerably lower, $8.1 \times 10^3 \, s^{-1}$.

$$Ret^{\cdot -} + ROH \rightleftharpoons RetH^{\cdot} + RO^- \tag{21}$$

For homologues of retinal, these rates were dependent on the chain length[20]. They increased by a factor of > 25 in going from a C_{30} to a C_{10} polyene, i.e. an increase in the number of conjugated double bonds stabilizes the radical anion against protonation.

Bobrowski and Das[21] utilized the radiolysis of 2-propanol/acetone/CCl_4 as a source of protons in order to measure the rate of protonation of the radical anions of retinal and other polyenes. For the retinal anion they considered the equilibrium

$$Ret^{\cdot -} + ROH_2^+ \rightleftharpoons RetH^{\cdot} + ROH \tag{22}$$

where ROH_2^+ represents the proton in 2-propanol, and determined k(forward) = $9.6 \times 10^6 \, M^{-1} \, s^{-1}$ and k(reverse) = $3.5 \times 10^4 \, s^{-1}$, and hence an equilibrium constant of $270 \, M^{-1}$.

The same authors also determined the spectra of the radical anion and radical cation of retinal in a wide variety of solvents[22]. They found only small shifts for the radical cation, the peak being between 580 and 600 nm, but large shifts for the radical anion, between 440 and 580 nm. The shifts are hypsochromic on going from non-polar to polar solvents and from aprotic to protic solvents.

The retinal radical cation, but not the anion, forms a complex with a molecule of retinal.

$$Ret^{\cdot +} + Ret \rightleftharpoons (Ret)_2^{\cdot +} \tag{23}$$

This complexation results in a small change in the spectrum which permits determination of the rate and equilibrium constants[23]. The results were solvent dependent. In acetone k(forward) = $1 \times 10^9 \, M^{-1} \, s^{-1}$ and k(reverse) = $2.4 \times 10^6 \, s^{-1}$, hence the equilibrium constant is $K = 430 \, M^{-1}$. In 1,2-dichloroethane $k_f = 1.3 \times 10^9$, $k_r = 8 \times 10^5$ and $K = 1600$, in the same units.

The radical cations of polyenals also react with nucleophiles such as water, triethylamine or Br^- [20]. The rate constants for these reactions increase with decreasing chain length due to increased stabilization of the more highly conjugated radicals. The rate constants were found to be in the range of 10^8–$10^{10} \, M^{-1} \, s^{-1}$ for Br^-, 10^6–$10^9 \, M^{-1} \, s^{-1}$ for triethylamine and 10^3–$10^5 \, M^{-1} \, s^{-1}$ for water.

IV. ASCORBIC ACID AND RELATED COMPOUNDS

Ascorbic acid is an important component of many biological systems and there are indications that some of its biochemical reactions lead to the formation of the ascorbate radical. Radiolytic techniques, in conjunction with optical and ESR detection, have been used extensively to study the properties of ascorbic acid and its radical and thus shed some light on their biochemical role. Although the radiolysis of ascorbic acid in aqueous solutions has been suggested[24] to produce the ascorbate radical, Bielski and Allen[25] were the first to provide conclusive evidence by recording the optical absorption spectrum of the radical. Subsequent studies dealt with the rate and mechanism of reaction of ascorbic acid with various radicals and with the properties of the ascorbate radical.

The reaction of OH radicals with ascorbic acid or ascorbate ion leads to the formation of a mixture of radicals[26,27] because OH may add to the double bond on either end or abstract hydrogen and some of the OH adducts may lose a water molecule. To avoid these complications, the ascorbate radical can be produced by one-electron oxidation of ascorbate with a wide variety of oxidizing radicals.

$$\text{(24)}$$

The rate constants for oxidation increase in going from ascorbic acid to its monoanion ($pK_1 = 4.2$) and dianion ($pK_2 = 11.5$). The values for various radicals are summarized in Table 1. It is seen from the Table that the rate constants vary over many orders of magnitude, depending on the oxidation potential of the radical and its self-exchange rate (i.e., the rate of electron transfer between the radical and its reduction product, e.g., $CO^{3-\bullet} + CO_3^{2-} = CO_3^{2-} + CO_3^{-\bullet}$). Nevertheless, all these radicals produce the ascorbate radical as formulated in reaction 24 with no side-reactions.

Ascorbic acid (AH_2) and ascorbate ions (AH^-) are also oxidized by HO_2 and O_2^- radicals, but the pH dependence of the rate constant is somewhat complex because both the compound and the radical undergo acid–base equilibria[38]. The rate constant for $AH_2 + HO_2$ is 1.6×10^4 and for $AH^- + O_2^-$ it is 5×10^4 $M^{-1} s^{-1}$. However, the reaction of AH^- with HO_2 is much faster and the pH profile shows a maximum at pH 4.5, where the overall rate constant is 1×10^7 $M^{-1} s^{-1}$.

Ascorbate ions react also by hydrogen abstraction with carbon-centered radicals such as $(CH_3)_2\dot{C}OH$ ($k = 1.2 \times 10^6$) and $\dot{C}H(CO_2^-)_2$ ($k = 1.3 \times 10^7$ $M^{-1} s^{-1}$) and these reactions lead to formation of the same ascorbate radical[39].

The ascorbate radical exhibits an optical absorption spectrum with a maximum at 360 nm and molar absorptivity of 3300 $M^{-1} cm^{-1}$[30]. The radical is very long lived in alkaline solutions but decays more rapidly in acidic solutions. The mechanism of decay was suggested to involve an equilibrium between the radical and its dimer, where the dimer may undergo protonation to yield the disproportionation products, ascorbate (HA^-) and dehydroascorbic acid (A)[40].

$$A^{\bullet -} + A^{\bullet -} \leftrightarrows (A)_2^{2-} \tag{25}$$

$$(A)_2^{2-} + H^+ \rightarrow HA^- + A \tag{26}$$

The ascorbate radical is unreactive toward O_2 and most simple organic compounds but can reduce cytochrome c (Fe^{3+}) slowly ($k = 6.6 \times 10^3$ $M^{-1} s^{-1}$)[41]. The low reactivity of the radical is an important factor in the antioxidant activity of ascorbate. As seen from Table 1, ascorbate reduces peroxyl radicals and thus serves as an antioxidant. But ascorbate reduces also the radicals from other antioxidants, such as phenols and tocopherol, and thus may serve as the ultimate antioxidant. One of the factors that

TABLE 1. Rate constants for one-electron oxidation of ascorbic acid

Radical	pH	k^*_{acid} (M^{-1}s^{-1})	pH	k^{**}_{ions} (M^{-1}s^{-1})	Reference
CO$_3^{\cdot-}$			11	1.1×10^9	28
O$_3$	2	6.9×10^5	4.8	5.6×10^7	28
N$_3^{\cdot}$			7	2.9×10^9	28
$\dot{N}H_2$			11.3	7.3×10^8	28
$\dot{N}O_2$			6.5	1.8×10^7	28
$\dot{S}O_3^{-}$	<3	$<1 \times 10^6$	5–10	9×10^6	28
			>12	3×10^8	28
SO$_5^{\cdot-}$	2	2×10^6	7	1×10^8	28
(SCN)$_2^{\cdot-}$	1.8	1×10^7	7	5×10^8	28
Cl$_2^{\cdot-}$	2	6×10^8			28
Br$_2^{\cdot-}$	2	1.1×10^8	7	1×10^9	28
I$_2^{\cdot-}$	2	5×10^6	7	1.4×10^8	28
$\dot{C}H_2CHO$			7	8.8×10^7	29
C$_6$H$_5$O$\cdot$			11	6.9×10^8	30
4-CNC$_6$H$_4\dot{O}$			11	2×10^9	30
4-NH$_2$C$_6$H$_4\dot{O}$			11	5×10^7	30
3-$^-$OC$_6$H$_4\dot{O}$			11	1.1×10^8	30
2-$^-$OC$_6$H$_4\dot{O}$			11	5×10^5	31
Tryptophanyl radical			7	7.3×10^7	32
α-Tocopheryl (Vit. E radical)			7	1.6×10^6	33
CH$_3$O$_2^{\cdot}$			7	2×10^6	34, 35
HOCH$_2$O$_2^{\cdot}$			7	4.7×10^6	34
$^-$O$_2$CCH$_2$O$_2^{\cdot}$			7	2.2×10^6	34
(CH$_3$)$_2$C(OH)CH$_2$O$_2^{\cdot}$			7	2.1×10^6	35
CH$_2$ClO$_2^{\cdot}$			7	9.2×10^7	35
CHCl$_2$O$_2^{\cdot}$			7	2×10^8	35
CCl$_3$O$_2^{\cdot}$			7	2×10^8	35
CBr$_3$O$_2^{\cdot}$			7	2×10^8	36
CF$_3$O$_2^{\cdot}$			7	7×10^8	36
$\dot{S}CH_2CH_2NH_2$			6.5	1.3×10^9	37
$\dot{S}CH_2CH(NH_3^+)CO_2^-$			6.5	1.2×10^9	37
G$\dot{S}$ (Glutathione radical)			6.5	6.0×10^8	37

*This is the rate constant for ascorbic acid (AH$_2$).
**This is the rate constant for the ascorbate ions, AH$^-$ and A^{2-} (depending on pH).

determine the activity of an antioxidant is the potential for its one-electron oxidation to the corresponding radical. For ascorbate and many phenols, these potentials were determined by pulse radiolysis by establishing equilibrium between radicals before they decay and measuring the equilibrium constant[31]. The potential for ascorbate was determined from equilibrium against catechol:

$$A^{2-} + {}^-OC_6H_4O^{\cdot} \leftrightarrows A^{\cdot-} + {}^-OC_6H_4O^- \qquad (27)$$

This electron-transfer equilibrium was established at high pH where the reaction is relatively rapid and the radicals more stable[31]. The one-electron oxidation potential of ascorbate was calculated from the equilibrium constant based on the value for catechol. The potential for neutral solutions was then calculated using the known pK_a values of the compound and the radical. The value was found to be 0.30 V vs NHE, (normal hydrogen electrode) indicating that ascorbate is a stronger one-electron reductant at pH 7 than hydroquinone or catechol[31].

The long lifetime and low reactivity of the ascorbate radical as well as its intense absorption spectrum are ascribed to the highly conjugated system of the tricarbonyl anion, which is inferred from its ESR spectrum. *In situ* radiolysis ESR experiments have demonstrated that the ascorbate radical is present in the anionic form throughout most of the pH range and that it protonates only in strongly acidic solutions ($pK_a = -0.45$)[42]. In the anionic form the three CO bonds form a conjugated system such that the unpaired electron is distributed among all of them. This conclusion was supported also by studies on model compounds such as reductic acid [$CH_2CH_2C(OH)=C(OH)C=O$] and hydroxy-tetronic acid [$OCH_2C(OH)=C(OH)C=O$]. Other models for the ascorbate radical were that derived from γ-methyl-α-hydroxytetronic acid[43], and the nitrogen analogue 2,3,4-trioxopyrrolidine radical anion[44]. In all the above cases the radicals were long lived and accumulated in the *in situ* radiolysis experiments in sufficiently high concentrations to permit determination of the ^{13}C hyperfine constants at the natural abundance level. These parameters provided further insight into the electronic structure of the radicals, beyond that obtained from the easily determined proton hyperfine constants. They provided an estimate of the spin density on the various carbon atoms and suggested that a considerable portion of the unpaired spin density is on the three carbonyl oxygens. The ring oxygen was suggested to have a very small portion of the spin density as well.

V. PYRIDONES

Although pyridones are the tautomeric forms of hydroxypyridines, they exist mainly as the enone forms. In aqueous solutions the enone form predominates by a factor of 340 for 2-pyridone and 2200 for 4-pyridone. This justifies the inclusion of their radiation chemistry in this chapter, although in some respects they may behave in parallel with phenols.

The reaction of OH radicals with 4-pyridone takes place via addition to the 3-position but the adduct undergoes rapid keto-enol tautomerization to the hydroxypyridine form[45].

$$(28)$$

This radical undergoes acid- and base-catalyzed water elimination to form the pyridine-4-oxyl radical[45]. Acid catalysis is by protonation of the radical on the ring nitrogen ($pK_a = 2.5$).

$$(29)$$

Base catalysis is by deprotonation of the 4-OH group ($pK_a = 10$) followed by loss of OH$^-$ (reaction 30). The rate of the latter process is $1.8 \times 10^4 s^{-1}$[45], at least two orders of magnitude lower than the parallel reaction with phenol, due to electron-withdrawing by the ring nitrogen.

The reaction of OH with 2-pyridone yields two isomeric adducts; the OH adds to the 3- and 5-positions, where the electron density is the highest (reactions 31, 32). These adducts also revert to the hydroxypyridine tautomer, but in contrast with the case of the 4-pyridone, they do not eliminate water[45].

$$\text{(30)}$$

$$\text{(31)}$$

$$\text{(32)}$$

The OH adduct of 2, 6-dicarboxypyridone-4 was suggested to remain in the pyridone form and to isomerize to the pyridol tautomer only in alkaline solutions, i.e. after deprotonation of the NH$^+$ group[45].

VI. PYRIMIDINE AND PURINE BASES

The radiation chemistry of pyrimidine and purine bases has been studied very extensively because of its importance in understanding the mechanism of radiation damage to DNA and all living cells. Radiation damage occurs by two pathways, i.e. by direct effect of radiation on the DNA molecule and by the indirect effect resulting from the reaction of DNA with radicals produced in the radiolysis of water. Therefore, the radiation chemistry of the bases was investigated both in the solid phase and in solution. Studies in the solid phase involved single crystals and powders, as well as glasses and frozen solutions, and concentrated on identifying the radicals by ESR spectroscopy. Studies in aqueous solutions applied ESR to determine the structure of transient species, optical pulse radiolysis to determine their kinetic behavior and product analysis to learn their ultimate fate.

The direct effect of radiation on DNA is likely to result in ionization of one of the bases to produce a radical cation (B$^{\cdot+}$) and an electron. The electron is captured by another base moiety to yield a radical anion (B$^{\cdot-}$). This may protonate, most likely on carbon 6, to yield a neutral 5-yl radical equivalent to an H atom adduct. The radical cation may deprotonate, probably losing the NH proton, to form a neutral radical. Alternatively, it may hydroxylate at carbon 6 to give the neutral 5-yl radical. All the above species have been identified by ESR in the solid phase. One of the main conclusions of those studies is that the effect of radiation on DNA is likely to result in the oxidation of guanine and reduction of thymine. It is beyond the scope of this chapter to review all the literature on this subject, but the interested reader is referred to several representative examples[46-54].

When the radiolysis was carried out in frozen alkaline solutions, an additional reaction was observed, that of O$^{\cdot-}$ radicals with thymine leading to hydrogen abstraction from the methyl group[50]. The formation of this radical was confirmed also by irradiating polycrystalline thymine and then dissolving it in an aqueous solution of a spin trapping material, 2-methyl-2-nitrosopropane[55]. The radicals formed in the solid were trapped and identified by ESR; they were found to include also the C5 and C6 H-adducts.

Spin-trapping ESR studies have been carried out also with aqueous solutions irradiated at room temperature. Here the irradiation is done in the presence of the spin trap so that the short-lived radicals are trapped to form very long lived or persistent radicals which are easily identified by ESR[56-58]. Thus the OH adducts to C5 and C6 of several pyrimidine bases and the radicals formed by oxidation of these bases with SO_4^- have been observed. In certain cases, the spin-trap radicals are sufficiently persistent to be separated by chromatography before ESR analysis[59]. This method identified the 5-OH adduct of uracil and distinguished between the *cis* and *trans* isomers. However, the ESR spectra of spin-trapped radicals do not provide as much information on the structure of the initial radicals as is obtained by direct observation of these radicals.

Direct ESR measurements on short-lived radicals in irradiated solutions was achieved by *in situ* radiolysis within the ESR spectrometer. Studies were carried out using pulse or continuous radiolysis. In the pulse radiolysis experiments the radicals produced by the reactions of OH and of e_{aq}^- with pyrimidine bases have been identified and the rate constants for their reactions with oxygen and with thiols were measured[60]. The sensitivity of this technique, however, was lower than that of the steady-state method and thus only the major hyperfine constants were determined with accuracy. The steady-state method provided detailed hyperfine constants but, because of the lack of the time resolution, secondary radicals were observed along with, or instead of, the primary ones[61-65]. These studies identified the radical formed upon H-abstraction from the methyl group of thymine by O^- radicals, but the C5 and C6 OH-adducts were not observed in their initial form, only the products of dehydration or secondary oxidation were identified. In the case of 5-halo- and 5-nitrouracil the OH-adducts underwent rapid loss of HX or HNO_2 to form the 5-oxo-6-yl radicals[62,63].

The main reactions occurring in the irradiated solutions discussed above are shown in equations 33–37. Further details on these reactions were obtained from pulse radiolysis experiments utilizing optical and conductometric detection.

$$(6-yl) \quad (33)$$

$$(5-yl) \quad (34)$$

$$(35)$$

The rate constants for the reactions of representative bases with the primary radicals of water radiolysis and with certain secondary radicals are summarized in Table 2. The

$$(36)$$

$$(37)$$

purine and pyrimidine bases react with OH radicals at nearly diffusion-controlled rates, k approaching 10^{10} M^{-1} s^{-1}. They react with H atoms somewhat more slowly, $k \sim 10^8$–10^9 M^{-1} s^{-1}. Both reactions lead mainly to addition to the 5, 6-double bond as in reactions 33 and 34. In purine bases addition to the 8-position is also possible[66,67]. These adduct radicals have similar absorption spectra[68-72] but can be distinguished through differences in their redox behavior. In general, the 6-yl radicals are reducing while the 5-yl are oxidizing. Table 3 shows that the 5-OH adducts, i.e. 6-yl radicals, reduce tetranitro-methane, quinones, riboflavin and hemin very rapidly while the 6-OH adducts oxidize N, N, N', N'-tetramethyl-p-phenylenediamine (TMPD), also rapidly. This difference per-mitted determination of the relative yields of the two types of radicals. In all pyrimidine bases, the OH addition was found to take place preferentially at C5 (uracil—82%,

TABLE 2. Rate constants for selected reactions of DNA bases with radicals

Base	Radical	pH	$k(M^{-1}s^{-1})$	Reference
Uracil	$\dot{O}H$	7	6×10^9	73
Thymine	$\dot{O}H$	7	6×10^9	73
Thymine	$\dot{O}^-$	> 13	4×10^8	73
Uracil	H	1, 7	3×10^8	73
Thymine	H	1	7×10^8	73
Uracil	e_{aq}^-	7	$\sim 1 \times 10^{10}$	73
Thymine	e_{aq}^-	7	1.7×10^{10}	73
Uracil	$CO_3^{\cdot-}$	7	$< 1 \times 10^4$	28
Uracil	$HPO_4^{\cdot-}$	9	1×10^8	28
Uracil	$H_2PO_4^{\cdot}$	4.5	6×10^8	28
Uracil	$SO_4^{\cdot-}$	7	1×10^9	28
Thymine	$(SCN)_2^{\cdot-}$	7	1×10^6	28
Thymine	$(SCN)_2^{\cdot-}$	12	3×10^7	28
Uracil	$Cl_2^{\cdot-}$	2, 6	4×10^7	28
Uracil	$Br_2^{\cdot-}$	7	$< 1 \times 10^7$	28
Uracil	$Br_2^{\cdot-}$	12	2×10^8	28
Thymine	$CO_2^{\cdot-}$	7	$\sim 5 \times 10^4$	28
5-Bromouracil	$(CH_3)_2\dot{C}OH$	7	2×10^7	74
Adenosine	$(CH_3)_2\dot{C}OH$	7	$< 10^6$	74
Adenosine	$(CH_3)_2\dot{C}OH$	2	5×10^7	74
Isobarbiturate	$\dot{C}H_2CHO$	13.5	1.6×10^9	75

TABLE 3. Rate constants for representative reactions of DNA base radicals[84]

Radical	Reactant	$k \, (M^{-1} s^{-1})$
Thymine-5-OH adduct	O_2	2×10^9
	menaquinone	4×10^9
	tetranitromethane	1.5×10^9
	cysteine	$< 1 \times 10^6$
Thymine-6-OH adduct	TMPD	1.3×10^9
Cytosine-5-OH adduct	riboflavin	1.6×10^9
	hemin c	1.1×10^9
	tetranitromethane	1.1×10^9
Cytosine-6-OH adduct	TMPD	1.1×10^9
Deoxyguanosine-OH adduct	ascorbate ion	1.4×10^9
	NADH	4.0×10^8
	cysteine	8×10^7
Thymine radical anion	O_2	6×10^9
	menaquinone	4×10^9
	orotic acid	1.5×10^9
Uracil radical (ox., pH 13)	adenine	9×10^7
	guanine	7×10^8
	xanthine	8×10^8
	tryptophan	1.4×10^9

thymine—60%, cytosine—87%)[76-79]. Hydrogen abstraction by OH from the methyl group of thymine amounts to 10% contribution. However, at high pH, when OH is converted to O^-, hydrogen abstraction becomes predominant (reaction 35). In the case of 5-halouracils, the contribution of OH addition to C5 was estimated from the extent of dehalogenation (reaction 37)[80].

The 6-OH adduct eliminates a water molecule at high pH (reaction 36)[76,77]. The radical formed in this reaction can be produced directly by one-electron oxidation of the pyrimidine base with an oxidizing radical such as $SO_4^{\bullet-}$. Table 2 shows that the rate constant for oxidation with $SO_4^{\bullet-}$ is very high but weaker oxidants, such as the dihalide radicals, react much more slowly. The rate of oxidation depends also on pH; the ionized forms of the pyrimidine bases are oxidized more rapidly than the neutral forms. The radicals produced in these reactions may behave as oxidants toward other molecules. Table 3 lists several examples for the radical of uracil. In general, such radicals oxidize pyrimidine or purine bases of lower redox potential as well as tryptophan, 5-hydroxy-tryptophan, vitamin E and certain phenols. These reactions were observed in alkaline solutions but are slow in neutral solutions[81].

The reactions of pyrimidine and purine bases with hydrated electrons take place with diffusion-controlled rate constants[82] but electron transfer to these bases from other reducing radicals is a slow process (Table 2). The initial electron adducts have pK_a values near 7[83]. Both forms are strongly reducing and may transfer an electron to oxygen, quinones, nitro compounds, and to other pyrimidines with higher electron affinity, such as orotic acid (Table 3)[85,86]. The electron adducts also protonate slowly on carbon to yield 6-H-adducts, which are oxidizing radicals[87].

$$(38)$$

$$+ \ H^+ \longrightarrow \qquad\qquad (39)$$

The electron adducts of 5-halouracils undergo rapid dehalogenation[88-90]. This process predominates in the case of bromo and iodo derivatives but with chlorouracil it is in competition with the protonation reaction[91,92]. The uracilyl radical produced in reaction 40 is very reactive and can add to another molecule of halouracil or abstract hydrogen from 2-propanol, in the latter case propagating a chain reaction[88-90].

$$\xrightarrow{e_{aq}^-} \qquad\qquad \longrightarrow \qquad\qquad + \ X^- \qquad (40)$$

Radicals produced by addition of H and OH to pyrimidine and purine bases undergo dimerization or disproportionation reactions to form the final products. When oxygen is present in solution, they react with it very rapidly to give peroxyl radicals, which then decay to stable products. These products have been determined over the past three decades by various analytical techniques and under different experimental conditions[79]. Before we discuss the mechanism of formation of final products we briefly summarize the main findings.

Thymine. Monomeric products of thymine radicals have been identified by a number of laboratories[93-97]. Table 4 presents a list of thymine products and their yields from three different sources. The main product is thymine glycol (*cis*- and *trans*-5, 6-dihydro-5, 6-dihydroxythymine). Other OH radical-induced products of thymine are 5-hydroxy-5, 6-dihydrothymine, 6-hydroxy-5, 6-dihydrothymine and 5-hydroxymethyluracil. 5, 6-Dihydrothymine results from H-atom reactions with thymine. The nature of OH radical-induced dimers of thymine has recently been elucidated[97]. The structures of the dimers have been obtained from mass spectral data, which suggested that combination reactions of OH-adduct radicals of thymine lead to dimers. Some of the dimers have been shown to dehydrate presumably during the derivatization process prior to analysis. In the case of

TABLE 4. Products and their yields from γ-radiolysis of thymine in N_2O-saturated aqueous solution

Product	G value		
	Ref. 95	Ref. 96	Ref. 97
Thymine consumption	3.9	2.7	5.5
cis- and *trans*-Thymine glycol	2.26	0.32	1.4
5-Hydroxy-5, 6-dihydrothymine	—	0.08 ⎫	0.5
6-Hydroxy-5, 6-dihydrothymine	0.13	0.1 ⎬	
5, 6-Dihydrothymine	0.1	0.17	0.04
5-Hydroxymethyluracil	0.22	0.27	0.2
Dimers	0.26	—	3.1

TABLE 5. Products and their yields from γ-radiolysis of thymine in aerated aqueous solution[101]

Product	G value
Thymine consumption	2.6
cis- and trans-Thymine glycol	0.248
5-Hydroxy-5, 6-dihydrothymine	0.016
6-Hydroxy-5, 6-dihydrothymine	0.008
5-Hydroxymethyluracil	0.017
5-Hydroxy-5-methylbarbituric acid	} 0.149
5-Hydroxy-5-methylhydantoin	
Formylurea	0.065
Formylpyruvylurea	0.460
cis- and trans-5-Hydroperoxy-6-hydroxydihydrothymine	1.144
cis-6-Hydroperoxy-5-hydroxydihydrothymine	0.081
5-Hydroperoxymethyluracil	0.047
5-Hydroperoxydihydrothymine	0.062
cis- and trans-Hydroperoxydihydrothymine	0.055
5-Hydroperoxy-5-methylhydantoin	0.005
5-Hydroperoxy-5-methylbarbituric acid	0.011
trans-5, 6-Dihydroperoxydihydrothymine	0.009

thymine oligo- and polynucleotides, the dimers resulting from combination reactions of OH-adduct and H-adduct radicals of thymine have also been identified[98,99]. The combination of electron-adduct radicals of thymidine has been demonstrated to yield a dihydrodimer in deareated aqueous solutions of thymidine in the presence of formate ions[100]. The same dimer has also been found to be formed by reaction of formate ions with thymine in N_2O-saturated solution.

In the presence of oxygen, a large number of thymine products have been observed[101-106]. The major products observed in aerated aqueous solution and their yields are listed in Table 5. Some of these products might have been secondary products because large radiation doses have been used. The dominant products are the hydroperoxides, which are formed from the interaction of the HO_2/O_2^- radicals with thymine peroxyl radicals. No dimers have been found in the presence of oxygen.

Uracil. The main monomeric products of uracil in the absence of oxygen are *cis*- and *trans*-uracil glycols and isobarbituric acid[107-111]. The latter one presumably results from dehydration of uracil glycols, and thus it is not a primary product. The yields of the products have been measured under different conditions and pH values. Table 6 shows the

TABLE 6. Products and their yields from γ-radiolysis of uracil in deoxygenated aqueous solution at neutral pH[108]

Product	G value
Uracil consumption	3.75
cis-Uracil glycol	0.81
trans-Uracil glycol	0.92
Isobarbituric acid	0.38
6-Hydroxy-5, 6-dihydrouracil	0.13
Formylurea	0.34
Alloxan	0.25
Dialuric acid	0.15
Dimers	0.21

TABLE 7. Products and yields from the γ-radiolysis of 1,3-dimethyluracil in N$_2$O-saturated aqueous solution[113]

Product	G value		
	pH 3	pH 6.5	pH 10.4
1,3-Dimethyluracil consumption	3.9	5.7	5.1
1,3-Dimethyluracil glycol	1.5	0.85	0.8
1,3-Dimethylisobarbituric acid	0.15	0.1	0.1
5-Hydroxy-5,6-dihydro-1,3-dimethyluracil	0.4	0.75	0.6
6-Hydroxy-5,6-dihydro-1,3-dimethyluracil	0.1	0.2	0.1
Dimers	1.7	3.6	3.2

uracil products and their yields in deoxygenated aqueous solution from one source. There appears to be a disagreement on the yields of products among different laboratories[108–111]. Dimers have also been observed; however, no definite structure could be assigned to dimeric products[109,111,112]. In the case of 1,3-dimethyluracil, structure of dimeric products of OH-adduct radicals in N$_2$O-saturated aqueous solution could be assigned with certainty because this compound was more suitable to analysis than uracil[113]. Mass spectral data suggested that dimers were formed exclusively by combination of a C(5)-OH adduct radical with an identical radical or with a C(5)-H adduct radical. Monomeric products and their yields have also been determined at different pH values (Table 7).

The formation of unstable hydroxyhydroperoxides of uracil in aerated aqueous solution has been observed in earlier studies[114,115]. A number of products have been isolated and identified[108,116,117]. Recently, the radiolysis of uracil has been reinvestigated in N$_2$O/O$_2$-saturated aqueous solution by product analysis and pulse radiolysis[118]. The yields of the products observed (Table 8) are strongly pH dependent and the mechanisms of product formation have been discussed in detail[118]. No dimers have been observed in the presence of oxygen.

Cytosine. Products of cytosine radicals in the absence and presence of oxygen have been identified, and their yields have been measured[119–125]. Tables 9 and 10 summarize the findings in the presence and absence of oxygen, respectively. 5-Hydroxycytosine, which

TABLE 8. Products and their yields from γ-radiolysis of uracil in N$_2$O/O$_2$-saturated aqueous solution[118]

Product	G value		
	pH 3.0	pH 6.5	pH 10.0
Uracil consumption	4.9	5.3	5.2
cis-Uracil glycol	0.6	0.9	1.4
trans-Uracil glycol	0.5	1.1	1.0
Isobarbituric acid	0	0.2	1.2
Formylhydroxyhydantoin	1.6	1.4	0.2
Dialuric acid	0.9	0.4	0.2
Isodialuric acid	0.1	0.2	0.1
5-Hydroxyhydantoin	0.4	0.4	0.3
Unidentified products	0.9	0.6	0.9

TABLE 9. Products and their yields from the γ-radiolysis of cytosine in aerated aqueous solution[120]

Product	G value
Cytosine consumption	2.5
trans-1-Carbamoylimidazolidone-4, 5-diol	0.6
4-Amino-1-formyl-5-hydroxy-2-oxo-3-imidazoline	0.2
cis-Uracil glycol	0.03
trans-Uracil glycol	0.1
5-Hydroxyhydantoin	0.1
Oxaluric acid and ureides	0.2
Parabanic acid	0.03
Biuret	0.06
Formylurea	0.06

TABLE 10. Products and their yields from the γ-radiolysis of cytosine in N_2O-saturated aqueous solution[125]

Product	G value
Cytosine consumption	5.6
Uracil	0.02
Uracil glycol	0.15
5-Hydroxycytosine	1.4
6-Hydroxycytosine	0.07
Cytosine glycol	0.05
5, 6-Dihydroxycytosine	0.20
Dimers	3.2

was observed with a high yield in the absence of oxygen (Table 10), is not a primary product and results from dehydration of cytosine glycol. Dimers have been found only in the absence of oxygen and their structures have been elucidated from mass spectral data[125].

Adenine and Guanine. The radiation chemistry of purine bases is less well understood than that of pyrimidines. The site of attack of the species from the water radiolysis has not been determined definitely. The yield of oxidizing and reducing radicals of some purine nucleotides, which were produced upon OH radical attack, has been determined recently and three sites of OH radical attack for guanine derivatives have been proposed[66]. Several products of adenine have been identified in earlier studies[126-129]. Table 11 lists the products and their yields. The formation of these products may be accounted for by the

TABLE 11. Products and their yields from the γ-radiolysis of adenine in N_2O-saturated aqueous solution[129]

Product	G value
Adenine consumption	1.0
8-Hydroxyadenine	0.35
4, 6-Diamino-5-formamidopyrimidine	0.2
6-Amino-8-hydroxy-7, 8-dihydropurine	0.1

TABLE 12. Products and their yields from the γ-radiolysis of 2'-deoxyguanosine in N_2 and N_2O-saturated aqueous solutions[130]

Product	G value	
	N_2	N_2O
2'-Deoxyguanosine consumption	0.81	1.50
9-(2-Deoxy-β-D-erythropentopyranosyl)-2,4-diamino-5-formamidopyrimid-6-one	0.08	0.09
9-(2-Deoxy-α-D-erythropentopyranosyl)-2,4-diamino-5-formamidopyrimid-6-one	0.26	0.25
9-(2-Deoxy-α-D-erythropentopyranosyl)guanine	0.02	0.03
9-(2-Deoxy-β-D-erythropentopyranosyl)guanine	0.01	0.02
9-(2-Deoxy-α-D-erythropentofuranosyl)guanine	0.02	0.02
9-(2-Deoxy-α-L-threopentofuranosyl)guanine	0.02	0.03
9-(2-Deoxy-β-D-erythropento-1,5-dialdo-1,4-furanosyl)guanine	0.07	0.08
5',8-Cyclo-2',5'-dideoxyguanosine	0.05	0.06
8-Hydroxy-2'-deoxyguanosine	—	0.24
Guanine	0.19	0.38

OH radical attack at the C(8)-position of adenine. The low product yields and the low adenine consumption in N_2O-saturated aqueous solution have been suggested to result from reconstitution reactions of adenine radicals[129].

The radiation chemistry of guanine has been investigated using guanine nucleosides or nucleotides because of the insufficient solubility of guanine in water. Table 12 summarizes the products identified in deareated and N_2O-saturated aqueous solutions of 2'-deoxyguanosine[130]. Similar to the adenine system, the yields of the products and the consumption of 2'-deoxyguanosine have been found to be low.

In the presence of oxygen, addition of the OH radical to the C(4)-position of adenine and peroxidation of the resultant radical has been suggested to account for the degradation of adenine; however, no peroxides have been detected[131]. In aerated aqueous solution, 8-hydroxyadenine has been found as the major product of adenine[127,129]. Some other degradation products of adenine have also been identified; however, their formation has been suggested to result from the decomposition of 8-hydroxyadenine[132]. The knowledge of the radiation chemistry of purines in the presence of oxygen is very limited at present.

Mechanistic Aspects. In the absence of oxygen, disproportionation and combination reactions of the adduct radicals of pyrimidines lead to final products. Some of the major pathways are illustrated in reactions 41–44 using the 1,3-dimethyluracil system as an example[113]. Combination reactions take place between the C(5)-OH adduct radicals and another C(5)-OH adduct or an H-adduct to give the observed dimers (reactions 43 and 44)[113]. Analogous mechanisms for dimer formation have been described for thymine, its oligo- and polynucleotides, and cytosine[97–99,125].

The electron adducts of thymine also undergo dimerization after protonation as illustrated in reaction 45[100].

In the presence of oxygen, peroxyl radicals, which are formed by addition of oxygen to the adduct radicals of pyrimidines, disproportionate to give the final products. The major pathways are illustrated in reactions 46–48 using the uracil system[118].

The pathways in equation 49 have been suggested for the formation of the products of adenine[129]. In the case of guanylic acid (dGMP), reaction with OH radicals was suggested to lead to formation of a radical cation or a protonated OH-adduct (reactions 50–52)[133,134]. Recently, three OH-adduct radicals of guanylic acid have been postulated[66].

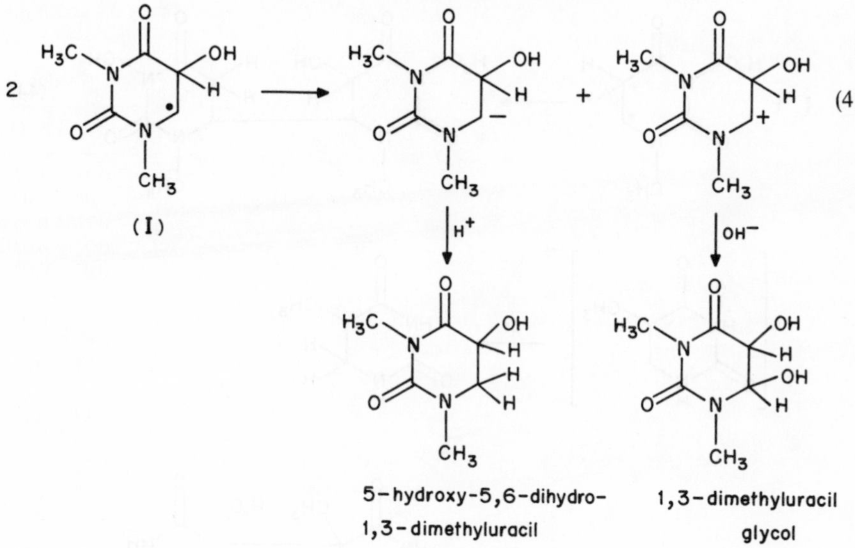
(I)
5-hydroxy-5,6-dihydro-1,3-dimethyluracil
1,3-dimethyluracil glycol

(41)

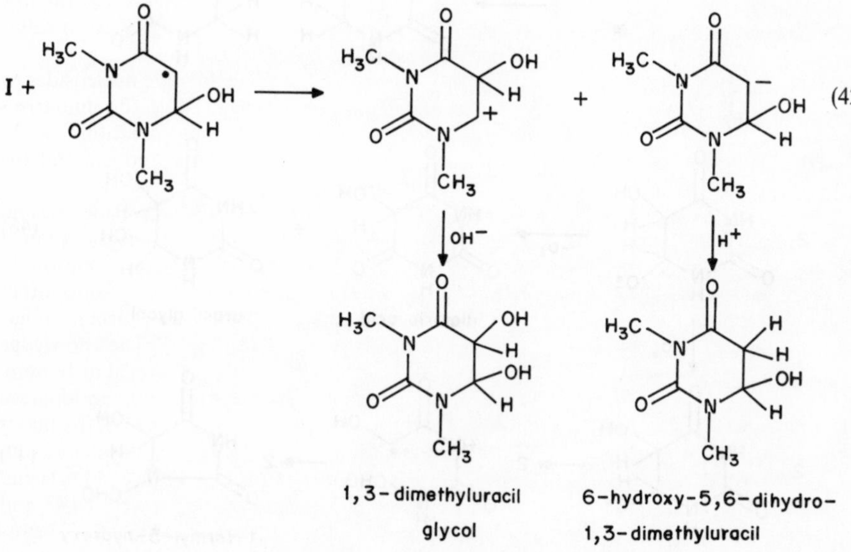
I +
1,3-dimethyluracil glycol
6-hydroxy-5,6-dihydro-1,3-dimethyluracil

(42)

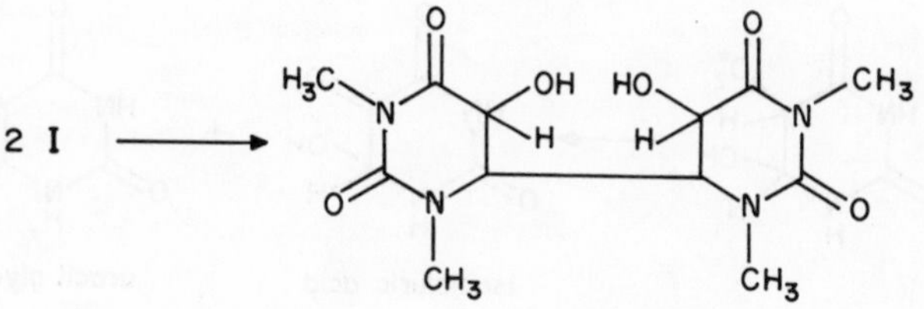
2 I

(43)

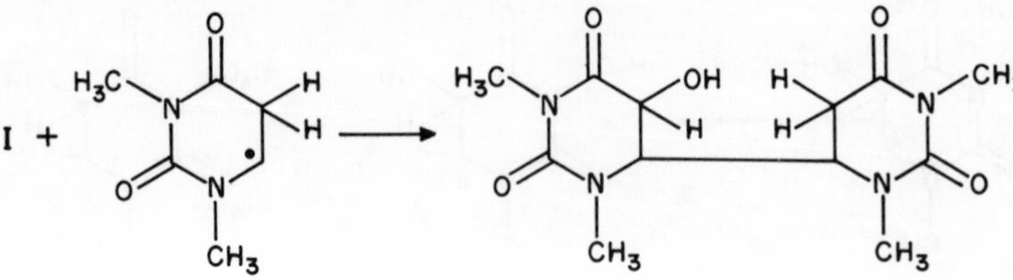

(44)

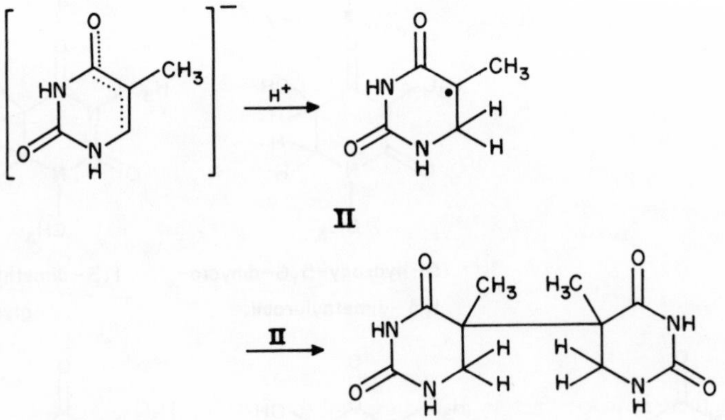

II

(45)

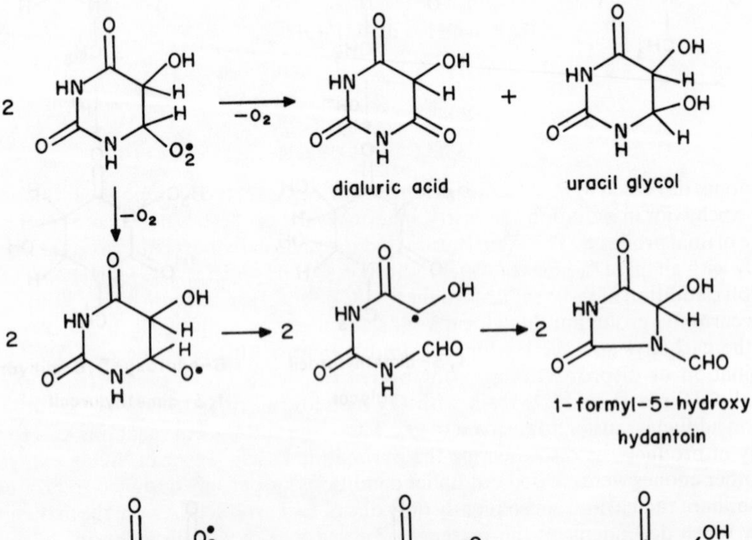

dialuric acid   uracil glycol

(46)

1-formyl-5-hydroxy
hydantoin

(47)

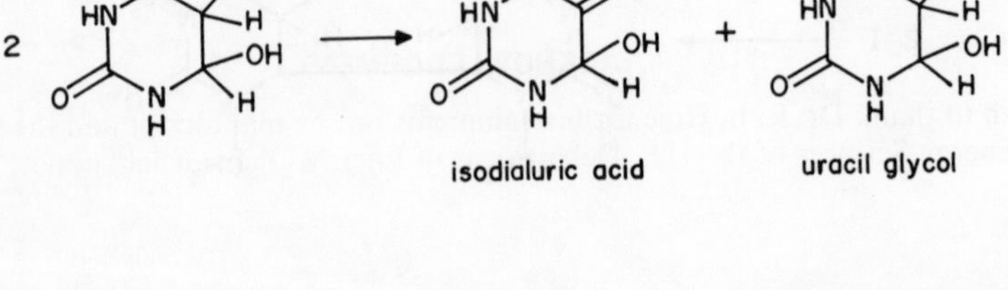

isodialuric acid   uracil glycol

(48)

8-hydroxyadenine   6-amino-8-hydroxy-
7,8-dihydropurine   (49)

4,6-diamino-5-formamido
pyrimidine

At present, there is no satisfactory mechanism for the formation of the products of adenine and guanine.

$$dGMP + OH \rightarrow dGMP^{\cdot +} + OH^- \tag{50}$$

$$dGMP + OH \rightarrow dGMP(OH)^{\cdot} \tag{51}$$

$$dGMP(OH)^{\cdot} + H^+ \rightarrow dGMP(OH)H^{\cdot +} \tag{52}$$

VII. CONCLUSION

The enones discussed in this chapter belong to various groups of compounds which exhibit diverse behavior in radiation chemistry, in terms of the properties of transient radicals and nature of final products. The main feature that is common to most enones is that they react rapidly with all three radicals of water radiolysis, OH, H and e_{aq}^-. The reactions of OH and H involve addition of these radicals to the C=C double bond. On the other hand, e_{aq}^- adds to the carbonyl group and may form a radical anion in which the electron is delocalized over the carbonyl and the conjugated double bonds. All the above radicals decay by combination or disproportionation or by reaction with O_2, if present in solution. In general, OH and H adducts react with O_2 by addition to form peroxyl radicals while electron adducts transfer an electron to O_2. These and subsequent reactions lead to a wide variety of products, as discussed for the pyrimidine bases. However, many experiments with other enones were carried out under conditions specifically designed to produce one predominant radical and subsequently only one or two products. Again, the final outcome is very much dependent on the presence or absence of oxygen in solution.

VIII. ACKNOWLEDGEMENT

We wish to thank Dr. R. E. Huie for his comments on the manuscript and the Office of Basic Energy Sciences of the U.S. Department of Energy for financial support.

IX. REFERENCES

1. A. J. Swallow, *Prog. React. Kinet.*, **9**, 195 (1978).
2. P. Neta, *Adv. Phys. Org. Chem.*, **12**, 223 (1976).
3. I. G. Draganic and Z. D. Draganic, *The Radiation Chemistry of Water*, Academic Press, New York, 1971.
4. A. J. Swallow, *Introduction to Radiation Chemistry*, Wiley, New York, 1973.
5. J. W. T. Spinks and R. J. Woods, *An Introduction to Radiation Chemistry*, Wiley, New York, 1964.
6. M. S. Matheson and L. M. Dorfman, *Pulse Radiolysis*, MIT Press, Cambridge, MA, 1969.
7. E. I. Finkelshtein and A. D. Abkin, *High Energy Chem.*, **3**, 403, 404 (1969).
8. J. Lilie and A. Henglein, *Ber. Bunsenges. Phys. Chem.*, **73**, 170 (1969).
9. M. Anbar, M. Bambenek and A. B. Ross, *Natl. Stand. Ref. Data Ser., Natl. Bur. Stand.*, Report No. 43 (1973).
10. J. Lilie and A. Henglein, *Ber. Bunsenges. Phys. Chem.*, **74**, 388 (1970).
11. H. Horii, Y. Abe and S. Taniguchi, *Bull. Chem. Soc. Jpn.*, **59**, 721 (1986).
12. H. Horii, Y. Abe and S. Taniguchi, *Bull. Chem. Soc. Jpn.*, **58**, 2751 (1985).
13. T. Shida, S. Iwata and M. Imamura, *J. Phys. Chem.*, **78**, 741 (1974).
14. P. K. Das and R. S. Becker, *J. Am. Chem. Soc.*, **101**, 6348 (1979).
15. R. Wilbrandt and N.-H. Jensen, *J. Am. Chem. Soc.*, **103**, 1036 (1981).
16. R. V. Bensasson, E. J. Land, R. S. H. Liu, K. K. N. Lo and T. G. Truscott, *Photochem. Photobiol.*, **39**, 263 (1984).
17. R. Wilbrandt, N.-H. Jensen and C. Houee-Levin, *Photochem. Photobiol.*, **41**, 175 (1985).
18. E. J. Land, J. Lafferty, R. S. Sinclair and T. G. Truscott, *J. Chem. Soc., Faraday Trans. 1*, **74**, 538 (1978).
19. N. V. Raghavan, P. K. Das and K. Bobrowski, *J. Am. Chem. Soc.*, **103**, 4569 (1981).
20. K. Bobrowski and P. K. Das, *J. Phys. Chem.*, **91**, 1210 (1987).
21. K. Bobrowski and P. K. Das, *J. Am. Chem. Soc.*, **104**, 1704 (1982).
22. K. Bobrowski and P. K. Das, *J. Phys. Chem.*, **89**, 5733 (1985).
23. K. Bobrowski and P. K. Das, *J. Phys. Chem.*, **90**, 927 (1986).
24. N. F. Barr and C. G. King, *J. Am. Chem. Soc.*, **78**, 303 (1956).
25. B. H. J. Bielski and A. O. Allen, *J. Am. Chem. Soc.*, **92**, 3793 (1970).
26. B. H. J. Bielski, D. A. Comstock and R. A. Bowen, *J. Am. Chem. Soc.*, **93**, 5624 (1971).
27. M. Schoneshofer, *Z. Naturforsch. B*, **27B**, 649 (1972).
28. P. Neta, R. E. Huie and A. B. Ross, *J. Phys. Chem. Ref. Data*, **17**, 1027 (1988).
29. S. Steenken, *J. Phys. Chem.*, **83**, 595 (1979).
30. R. H. Schuler, *Radiat. Res.*, **69**, 417 (1977).
31. S. Steenken and P. Neta, *J. Phys. Chem.*, **83**, 1134 (1979).
32. B. M. Hoey and J. Butler, *Biochim. Biophys. Acta*, **791**, 212 (1984).
33. J. E. Packer, T. F. Slater and R. L. Willson, *Nature (London)*, **278**, 737 (1979).
34. R. E. Huie and P. Neta, *Int. J. Chem. Kinet.*, **18**, 1185 (1986).
35. J. E. Packer, R. L. Willson, D. Bahnemann and K.-D. Asmus, *J. Chem. Soc., Perkin Trans. 2*, 296 (1980).
36. R. E. Huie, D. Brault and P. Neta, *Chem.-Biol. Interact.*, **62**, 227 (1987).
37. L. G. Forni, J. Monig, V. O. Mora-Arellano and R. L. Willson, *J. Chem. Soc., Perkin Trans. 2*, 961 (1983).
38. D. E. Cabelli and B. H. J. Bielski, *J. Phys. Chem.*, **87**, 1809 (1983).
39. J. L. Redpath and R. L. Willson, *Int. J. Radiat. Biol.*, **23**, 51 (1973).
40. B. H. J. Bielski, A. O. Allen and H. A. Schwarz, *J. Am. Chem. Soc.*, **103**, 3516 (1981).
41. B. H. J. Bielski, H. W. Richter and P. C. Chan, *Ann. N.Y. Acad. Sci.*, **258**, 231 (1975).
42. G. P. Laroff, R. W. Fessenden and R. H. Schuler, *J. Am. Chem. Soc.*, **94**, 9062 (1972).
43. Y. Kirino and R. H. Schuler, *J. Am. Chem. Soc.*, **95**, 6926 (1973).
44. Y. Kirino, P. L. Southwick and R. H. Schuler, *J. Am. Chem. Soc.*, **96**, 673 (1974).
45. S. Steenken and P. O'Neill, *J. Phys. Chem.*, **83**, 2407 (1979).
46. T. Henriksen and W. Snipes, *Radiat. Res.*, **42**, 255 (1970).
47. J. Huttermann, *Int. J. Radiat. Biol.*, **17**, 249 (1970).
48. J. Huttermann, J. F. Ward and L. S. Myers, Jr., *J. Phys. Chem.*, **74**, 4022 (1970); *Int. J. Radiat. Phys. Chem.*, **3**, 117 (1971).

49. R. A. Holroyd and J. W. Glass, *Int. J. Radiat. Biol.*, **14**, 445 (1968).
50. N. B. Nazhat and J. J. Weiss, *Trans. Faraday Soc.*, **66**, 1302 (1970).
51. M. D. Sevilla, in *Excited States in Organic Chemistry and Biochemistry* (Eds. B. Pullman and N. Goldblum), Reidel Publ. Co., Dordrecht, 1977, p. 15.
52. M. D. Sevilla, D. Suryanarayana and K. M. Morehouse, *J. Phys. Chem.*, **85**, 1027 (1981).
53. M. D. Sevilla and S. Swarts, *J. Phys. Chem.*, **86**, 1751 (1982).
54. M. D. Sevilla, S. Swarts, H. Riederer and J. Huttermann, *J. Phys. Chem.*, **88**, 1601 (1984).
55. M. Kuwabara, Y. Lion and P. Riesz, *Int. J. Radiat. Biol.*, **39**, 465 (1981).
56. A. Joshi, H. Moss and P. Riesz, *Int. J. Radiat. Biol.*, **34**, 165 (1978).
57. P. Riesz and S. Rustgi, *Radiat. Phys. Chem.*, **13**, 21 (1979).
58. M. Kuwabara, Y. Lion and P. Riesz, *Int. J. Radiat. Biol.*, **39**, 491 (1981).
59. K. Makino, M. Mossoba and P. Riesz, *J. Phys. Chem.*, **87**, 1074 (1983).
60. G. Nucifora, B. Smaller, R. Remko and E. C. Avery, *Radiat. Res.*, **49**, 96 (1972).
61. P. Neta, *Radiat. Res.*, **49**, 1 (1972); **56**, 201 (1973).
62. P. Neta, *J. Phys. Chem.*, **76**, 2399 (1972).
63. P. Neta and C. L. Greenstock, *Radiat. Res.*, **54**, 35 (1973).
64. K. M. Bansal and R. W. Fessenden, *Radiat. Res.*, **75**, 497 (1978).
65. J. Planinic, *Int. J. Radiat. Biol.*, **38**, 651 (1980).
66. P. O'Neill, *Radiat. Res.*, **96**, 198 (1983).
67. A. J. S. C. Vieira and S. Steenken, *J. Phys. Chem.*, **91**, 4138 (1987).
68. R. M. Danziger, E. Hayon and M. E. Langmuir, *J. Phys. Chem.*, **72**, 3842 (1968).
69. L. S. Myers, Jr. and L. M. Theard, *J. Am. Chem. Soc.*, **92**, 2868 (1970); L. S. Myers, Jr., M. L. Hollis, L. M. Theard, F. C. Peterson and A. Warnick, *J. Am. Chem. Soc.*, **92**, 2875 (1970); L. M. Theard, F. C. Peterson and L. S. Myers, Jr., *J. Phys. Chem.*, **75**, 3815 (1971).
70. C. L. Greenstock, J. W. Hunt and M. Ng, *Trans. Faraday Soc.*, **65**, 3279 (1969); C. L. Greenstock, *Trans. Faraday Soc.*, **66**, 2541 (1970); P. C. Shragge and J. W. Hunt, *Radiat. Res.*, **60**, 233 (1974).
71. A. Hissung and C. von Sonntag, *Z. Naturforsch. B*, **33**, 321 (1978).
72. D. J. Deeble and C. von Sonntag, *Z. Naturforsch. C*, **40**, 925 (1985).
73. G. V. Buxton, C. L. Greenstock, W. P. Helman, and A. B. Ross, *J. Phys. Chem. Ref. Data*, **17**, 513 (1988).
74. A. B. Ross and P. Neta, *Natl. Stand. Ref. Data Ser., Natl. Bur. Stand.*, Report No. 70 (1982).
75. S. Steenken and P. Neta, *J. Phys. Chem.*, **86**, 3661 (1982).
76. S. Fujita and S. Steenken, *J. Am. Chem. Soc.*, **103**, 2540 (1981).
77. D. K. Hazra and S. Steenken, *J. Am. Chem. Soc.*, **105**, 4380 (1983).
78. S. Steenken, *J. Chem. Soc., Faraday Trans. 1*, **83**, 113 (1987).
79. For a review see: C. von Sonntag, *The Chemical Basis of Radiation Biology*, Taylor and Francis, London, 1987.
80. K. M. Bansal, L. K. Patterson and R. H. Schuler, *J. Phys. Chem.*, **76**, 2386 (1972); L. K. Patterson and K. M. Bansal, *J. Phys. Chem.*, **76**, 2392 (1972).
81. S. V. Jovanovic and M. G. Simic, *J. Phys. Chem.*, **90**, 974 (1986).
82. C. L. Greenstock, M. Ng and J. W. Hunt, *Adv. Chem. Ser.*, **81**, 397 (1968).
83. E. Hayon, *J. Chem. Phys.*, **51**, 4881 (1969).
84. P. Neta and A. B. Ross, in *Chemical Kinetics of Small Organic Radicals* (Ed. Z. B. Alfassi), CRC Press, Boca Raton, FL Vol. IV, p. 187 (1988).
85. H. Loman and M. Ebert, *Int. J. Radiat. Biol.*, **18**, 369 (1970).
86. G. E. Adams, C. L. Greenstock, J. J. van Hemmen and R. L. Willson, *Radiat. Res.*, **49**, 85 (1972).
87. S. Das, D. J. Deeble, M. N. Schuchmann and C. von Sonntag, *Int. J. Radiat. Biol.*, **46**, 7 (1984).
88. J. D. Zimbrick, J. F. Ward and L. S. Myers, Jr., *Int. J. Radiat. Biol.*, **16**, 505 (1969).
89. G. E. Adams and R. L. Willson, *Int. J. Radiat. Biol.*, **22**, 589 (1972).
90. K. Bhatia and R. H. Schuler, *J. Phys. Chem.*, **77**, 1888 (1973).
91. B. O. Wagner and D. Schulte-Frohlinde, *Ber. Bunsenges. Phys. Chem.*, **79**, 589 (1975).
92. E. Rivera and R. H. Schuler, *J. Phys. Chem.*, **87**, 3966 (1983).
93. C. Nofre and A. Cier, *Bull. Soc. Chim. France*, 1326 (1966).
94. J. Cadet and R. Teoule, *Int. J. Appl. Radiat. Isot.*, **22**, 273 (1971).
95. G. A. Infante, P. Jirathana, E. J. Fendler and J. H. Fendler, *J. Chem. Soc., Faraday Trans. 1*, **69**, 1586 (1973).
96. S. Nishimoto, H. Ide, T. Wada and T. Kagiya, *Int. J. Radiat. Biol.*, **44**, 585 (1983).
97. M. Dizdaroglu and M. G. Simic, *Int. J. Radiat. Biol.*, **46**, 241 (1984).

98. M. Dizdaroglu and M. G. Simic, *Radiat. Phys. Chem.*, **26**, 309 (1985).
99. L. R. Karam, M. G. Simic and M. Dizdaroglu, *Int. J. Radiat. Biol.*, **49**, 67 (1986).
100. S. Nishimoto, H. Ide, K. Nakamichi and T. Kagiya, *J. Am. Chem. Soc.*, **105**, 6740 (1983).
101. R. Teoule and J. Cadet, *J. Chem. Soc., Chem. Commun.*, 1269 (1971).
102. B. Ekert and R. Monier, *Nature (London)*, **184**, 58 (1959).
103. G. Scholes and J. Weiss, *Nature (London)*, **185**, 305 (1960).
104. R. Teoule and J. Cadet, *Z. Naturforsch.*, **29c**, 645 (1974).
105. G. Scholes, J. Weiss and C. M. Wheeler, *Nature (London)*, **178**, 157 (1956).
106. J. Cadet and R. Teoule, *C.R. Acad. Sci. Paris, Ser. C*, **276**, 1743 (1973).
107. M. N. Khattak and J. H. Green, *Aust. J. Chem.*, **18**, 1847 (1965).
108. G. A. Infante, P. Jirathana, E. J. Fendler and J. H. Fendler, *J. Chem. Soc., Faraday Trans. 1*, **70**, 1162 (1974).
109. P. C. Shragge, A. J. Varghese, J. W. Hunt and C. L. Greenstock, *Radiat. Res.*, **60**, 250 (1974).
110. P. C. Shragge, A. J. Varghese, J. W. Hunt and C. L. Greenstock, *J. Chem. Soc., Chem. Commun.*, 736 (1974).
111. K. M. Idriss Ali and G. Scholes, *J. Chem. Soc., Faraday Trans. 1*, **76**, 449 (1980).
112. K. M. Idriss Ali, *J. Radiat. Res.*, **20**, 84 (1979).
113. M. Al-Sheikhly and C. von Sonntag, *Z. Naturforsch.*, **38b**, 1622 (1983).
114. G. Scholes, J. F. Ward and J. Weiss, *J. Mol. Biol.*, **2**, 379 (1960).
115. D. Barszcz and D. Sugar, *Acta Biochim. Pol.*, **19**, 25 (1961).
116. K. C. Smith and J. E. Hays, *Radiat. Res.*, **33**, 129 (1968).
117. R. Ducolomb, J. Cadet and R. Teoule, *Bull. Soc. Chim. France*, 1167 (1973).
118. M. N. Schuchmann and C. von Sonntag, *J. Chem. Soc., Perkin Trans. 2*, 1525 (1983).
119. B. Ekert and R. Monier, *Nature (London)*, **188**, 309 (1960).
120. M. Polverelli and R. Teoule, *Z. Naturforsch.*, **29c**, 16 (1974).
121. M. Polverelli, J. Ulrich and R. Teoule, *Z. Naturforsch.*, **39c**, 64 (1983).
122. G. P. Zhizhina and K. E. Kruglyakova, *Doklady Chem.*, **180**, 469 (1968).
123. B. S. Hahn, S. Y. Wang, J. L. Flippen and I. L. Karle, *J. Am. Chem. Soc.*, **95**, 2711 (1973).
124. M. N. Khattak and J. H. Green, *Int. J. Radiat. Biol.*, **11**, 113 (1966).
125. M. Dizdaroglu and M. G. Simic, *Radiat. Res.*, **100**, 41 (1984).
126. G. Hems, *Radiat. Res.*, **13**, 777 (1960).
127. J. J. Conlay, *Nature (London)*, **197**, 555 (1963).
128. C. Ponnamperuma, R. M. Lemmon and M. Calvin, *Radiat. Res.*, **18**, 540 (1963).
129. J. J. van Hemmen and J. F. Bleichrodt, *Radiat. Res.*, **46**, 444 (1971).
130. M. Berger and J. Cadet, *Z. Naturforsch.*, **40b**, 1519 (1985).
131. G. Scholes, *Prog. Biophys. Mol. Biol.*, **13**, 59 (1963).
132. N. Mariaggi and R. Teoule, *C.R. Acad. Sci. Paris, Ser. C*, **279**, 1005 (1974).
133. R. L. Willson, P. Wardman and K. D. Asmus, *Nature (London)*, **252**, 323 (1974).
134. K. D. Asmus, D. J. Deeble, A. Garner, K. M. Idriss Ali and G. Scholes, *Br. J. Cancer, Suppl. III*, **37**, 46 (1978).

The Chemistry of Enones
Edited by S. Patai and Z. Rappoport
© John Wiley & Sons Ltd

CHAPTER **17**

The oxygenation of enones

ARYEH A. FRIMER

The Ethel and David Resnick Chair in Active Oxygen Chemistry, Department of Chemistry, Bar Ilan University, Ramat Gan 52100, Israel

I. INTRODUCTION

The discovery of oxygen over 200 years ago can be attributed to three people: Lavoisier, Priestly, and Scheele[1]. The continuing fascination of the scientific community with this element stems from the complicated role molecular oxygen (dioxygen) and its derivatives play not only in the 'breath of life' but more interestingly in oxygen toxicity[2,3]—what might be poetically called 'the breath of death'[4]. This review will focus on the interaction of various active oxygen species with enones, one of the most fascinating and useful organic moieties.

For the purpose of this review, we have surveyed the literature through January 1988 and have discussed variously substituted α, β- and β, γ-unsaturated carbonyl compounds (including keto enols and aci-reductones) as well as ketenes. While we have tried to present a complete picture, no attempt has been made to be encyclopedic and exhaustive.

II. THEORETICAL DESCRIPTION OF ACTIVE OXYGEN SPECIES

Ever since the discovery of oxygen over two centuries ago, mankind has invested a good deal of time and resources in attempting to understand the exact role this life-supporting molecule plays in autoxidative, photooxidative and metabolic processes. Since the electronic makeup of a molecule determines its reactivity, it was to molecular orbital theory and electronic excitation spectroscopy that scientists turned in order to get an exact description of the configuration of the various electronic states of molecular oxygen[5]. We shall limit our discussion to the structure of the lowest three electronic states of dioxygen (O_2) which differ primarily in the manner in which the two electrons of highest energy occupy the two degenerate π_{2p}^* orbitals. Following Hund's rule, in the ground state of O_2, these two electrons will have parallel spins and be located one each in the two degenerate π_{2p}^* orbitals (Figure 1). Such an electronic configuration corresponds to a triplet $^3\Sigma_g^-$ state and we shall henceforth refer to ground-state molecular oxygen as triplet oxygen, $^3O_2$.

This triplet character is responsible for the paramagnetism and diradical-like properties of $^3O_2$. More importantly, this triplet electronic configuration only permits reactions involving one-electron steps. Thus, despite the exothermicity of oxygenation reactions, a spin barrier prevents $^3O_2$ from reacting indiscriminately with the plethora of singlet ground-state organic compounds surrounding it. One could well argue that it is this spin barrier that permits life to be maintained.

The two lowest excited states are both singlets in which the two highest-energy electrons have antiparallel spins. Thus, no spin barrier should exist for their reaction with organic substrates. In the first $(^1\Delta_g)$ state, which lies 22.5 kcal mol^{-1} above the ground state, both of

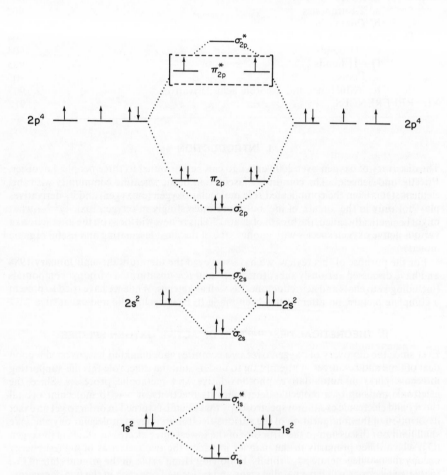

FIGURE 1. Schematic energy-level diagram showing how the atomic orbitals (A.O.) of two atoms of elemental oxygen interact to form the molecular orbitals (M.O.) of molecular oxygen. The electron distribution is, according to Hund's rule, yielding ground-state molecular oxygen ($^3\Sigma_g^-$)

the highest-energy electrons occupy the same π_{2p}^* orbital. In the second, a $^1\Sigma_g^+$ state lying 15 kcal mol^{-1} higher, each of the π_{2p}^* orbitals is half full (Table 1).

In the gas phase the lifetimes of $^1\Delta$ and $^1\Sigma$ oxygen are 45 min and 7 s, respectively[6]. However, in solution these lifetimes are dramatically reduced through collisional deactivation to approximately 10^{-3} and 10^{-9} s, respectively[6,7]. Because the reactions that concern us are generally carried out in solution, it is the longer-lived $^1\Delta O_2$ that is involved

TABLE 1. The three lowest electronic states of molecular oxygen and selected properties

Electronic state	Configuration of π^*_{2p}	Relative energy (kcal mol⁻¹)	Lifetime (s)[6,7] Gas phase	Liquid phase	Valence bond representative
$^1\Sigma^+_g$	↿ ⇂	37.5	7–12	10^{-7}	
$^1\Delta_g$	⇅ __	22.5	2700	10^{-3}	
$^3\Sigma^-_g$	↿ ↿	0	∞	∞	

as the active species. We shall henceforth refer to this longer-lived species as singlet oxygen, $^1O_2$.

A simplified picture of the three lowest electronic states of molecular oxygen and a comparison of some of their properties is presented in Table 1.

The one-electron reduction product of molecular oxygen is the superoxide anion radical. $O_2^-{}^\bullet$ differs from $^3O_2$ and $^1O_2$ in that the former has three—not two—electrons in its π^*_{2p} orbitals. This leads to a situation in which one of the two degenerate π^*_{2p} orbitals is totally occupied while the second is only half full, as outlined in equation 1. It should be noted that no Jahn–Teller splitting can occur with diatomic molecules; hence, all three of the π^*_{2p} electrons in $O_2^-{}^\bullet$ are of equal energy.

$$^3O_2 \uparrow\uparrow \xrightarrow{+e^-} \uparrow\downarrow \uparrow\, O_2^-{}^\bullet \tag{1}$$

III. TRIPLET MOLECULAR OXYGEN

A. Radical Initiated Autoxidation[8–15]

Autoxidation is a general term used to describe the reaction of a substance with molecular oxygen at a temperature generally below 150–200 °C and in the absence of a flame. We will limit our discussion to the oxidation of labile C—H bonds of hydrocarbons, in which case the primary product is the corresponding hydroperoxide. A wide variety of organic compounds will undergo autoxidative hydroperoxidation with the rate being highly dependent on steric and electronic factors.

Autoxidation has been shown to be a free-radical process consisting of the traditional chain-mechanism elements: initiation, propagation and termination, as outlined in equation 2–7.

Initiation

$$RH \xrightarrow[\text{or enzymatically}]{h\nu, \Delta, R^{1\bullet} Mn^+} R^\bullet \tag{2}$$

Propagation

$$R^\bullet + O_2 \longrightarrow ROO^\bullet \tag{3}$$

$$ROO^\bullet + RH \longrightarrow ROOH + R^\bullet \tag{4}$$

Termination

$$R^\bullet + R^\bullet \longrightarrow R—R \tag{5}$$

$$ROO^\bullet + R^\bullet \longrightarrow ROOR \tag{6}$$

$$ROO^\bullet + {}^\bullet OOR \longrightarrow ROOOOR \tag{7}$$

$$\longrightarrow \text{Nonradical products}$$

As summarized in equation 2, initiation requires the generation of free radicals via the homolytic cleavage of bonds. This can be accomplished either thermally (hot point), photochemically (in the absence or presence of photosensitizers), chemically (by reacting with another radical generated from peroxides, azo compounds, etc.) enzymatically or via metal ion catalysis. Although the name '*autoxidation*' suggests that this process can occur without the addition of any outside initiators, truly spontaneous processes are extremely rare[16].

Since ground-state molecular oxygen can be considered a triplet biradical, it is not surprising that its coupling with most carbon centered radicals is essentially a diffusion-controlled process. This coupling is in fact the essence of the first step in the propagation (equation 3) which is 10^6 to 10^8 times faster than the rate-determining hydrogen-abstraction step (equation 4). It follows that those steric and electronic factors which weaken the R—H bond will accelerate the rate of autoxidation. Furthermore, the point of autoxidative attack in a molecule, RH, is generally that which leads to the most stable radical, R·, upon cleavage of the C—H bond.

There is, however, one further factor which will be relevant to our discussion of enones, namely, polar effects[17,18]. Electron-donating substituents α to the C—H bond to be broken accelerate autoxidation, while electron-withdrawing groups decrease the rate of this process. Polar effects of this kind are well known in free radical processes and in the case of autoxidation result from the electrophilicity of the hydrogen-abstracting peroxy radical[13,17,18]. It is believed that a dipolar structure plays an important role in the transition state for this reaction (equation 8).

$$R'OO^\bullet \; H\!:\!\overset{\diagup}{\underset{\diagdown}{C}}\!-\!X \longrightarrow [R'OO\!:^{\delta-} \; \cdots \; H\!\cdot\!\overset{\delta+}{\overset{\diagup}{\underset{\diagdown}{C}}}\!-\!X] \longrightarrow R'OOH \; \cdot\overset{\diagup}{\underset{\diagdown}{C}}\!-\!X \qquad (8)$$

B. Base-catalyzed Autoxidation (BCA)[19]

Organic compounds with acidic hydrogens attached to carbon undergo facile reactions with oxygen in basic media. For example, the *t*-butoxide mediated oxygenation of di- and triphenylmethane generates benzophenone and triphenylmethanol respectively, rapidly and in high yield. Russell and coworkers[19] have proposed that these autoxidations are chain reactions (equations 9–11), generally involving a rate-determining deprotonation of the substrate RH which produces a carbanion $R:^-$ (equation 9). The latter is then oxygenated to the corresponding peroxy anion $ROO:^-$ (equation 10) which deprotonates another molecule of starting material (equation 11), thereby initiating another cycle.

Initiation $\qquad\qquad\qquad RH + B: \longrightarrow R:^- + BH^+ \qquad\qquad\qquad (9)$

Propagation $\qquad\qquad\qquad R:^- + O_2 \longrightarrow ROO:^- \qquad\qquad\qquad\quad (10)$

$\qquad\qquad\qquad ROO:^- + RH \longrightarrow ROOH + R:^- \qquad\qquad (11)$

While the mechanism as written is consistent with the experimental data, the direct combination of a carbanion with triplet dioxygen to yield a peroxy anion violates the Wigner spin-conservation principle (see equation 12)[19a-c,20-26].

$$R\uparrow\downarrow + \; \widehat{O}\underset{\uparrow}{\overset{\uparrow}{|}}\widehat{O} \; \xrightarrow{\;\times\;} R\uparrow\downarrow\bar{O}\!-\!\bar{O}\uparrow\downarrow \qquad\qquad (12)$$

$$(R^-) \qquad (^3O_2) \qquad\qquad (R\!-\!O\!-\!O^-)$$

Russell and coworkers[19a-c,24] suggest that the carbanion R^- may be converted to a free radical by donating an electron to an acceptor before combining with triplet biradical dioxygen. The acceptor is most commonly dioxygen itself, though peroxy radicals or trace metals may be involved as well. A plausible mechanism could then be the sequence outlined below:

$$RH + B^- \longrightarrow R^- + BH \qquad (9)$$

$$R^- + O_2 \rightleftharpoons R^{\cdot} + O_2^{-\cdot} \qquad (13)$$

$$R^{\cdot} + O_2 \longrightarrow ROO^{\cdot} \qquad (14)$$

$$ROO^{\cdot} + O_2^{-\cdot} \longrightarrow ROO^- + O_2 \qquad (15)$$

$$ROO^{\cdot} + R^- \longrightarrow ROO^- + R^{\cdot} \qquad (16)$$

$$ROO^- + RH \longrightarrow ROOH + R^- \qquad (11)$$

We have not included in this sequence a radical coupling between $R^{\cdot}$ and superoxide anion radical, $O_2^{-\cdot}$ (equation 17a), a process with the same outcome as equations 14 and 15 combined. This is simply because electron transfer (see equations 13 and 15) rather than radical coupling is generally observed with superoxide[27] (equation 17b).

$$R^{\cdot} + O_2^{-\cdot} \longrightarrow RO_2^- \qquad (17a)$$
$$\phantom{R^{\cdot} + O_2^{-\cdot}} \longrightarrow R^- + O_2 \qquad (17b)$$

An alternate proposal[21,23,25,26] to that of Russell's is that a change of multiplicity occurs via a carbanion–oxygen complex (equation 18).

$$R^- + O_2 \longrightarrow [R^{\cdot} \cdots O_2^{-\cdot}] \longrightarrow [R^{\cdot} \cdots O_2^{-\cdot}] \longrightarrow R{-}O_2^- \qquad (18)$$

Finally, we should note that although the primary products of BCA processes are generally hydroperoxides, these are rarely isolated under the basic reaction conditions. Instead, the corresponding ketones, alcohols, carboxylic acids or related oxidative cleavage products are obtained. The mechanism for some of these transformations are described in the next section.

C. Reactions of Hydroperoxides[28]

We have noted above that hydroperoxides are the major primary autoxidation products. They are, however, generally quite labile and the reaction product(s) actually isolated depends greatly on the reaction and/or workup conditions (solvent, temperature, pH, etc.). To aid in their handling, hydroperoxides are often reduced to the corresponding alcohols by a variety of reagents including Ph_3P, $(PhO)_3P$, $LiAlH_4$, $NaBH_4$, Na_2SO_3 and Me_2S. In many instances, however, the hydroperoxide product rearranges before it can be treated. It will be of value, therefore, to acquaint ourselves with some of these transformations before we delve into a discussion of enone oxygenations.

1. Homolysis of the peroxy linkage

Owing to the relative weakness of the peroxide bond, its homolysis to alkoxy radicals at room temperature or above (e.g. GLC injector port) is a prevalent phenomenon. In many cases this reaction is to be considered a metal-catalyzed process, particularly since precautions are rarely taken to eliminate the trace amount (10^{-8} mol) of metal ions which suffice to catalyze the homolytic decomposition of hydroperoxides[29] (equation 19).

$$ROOH + M^{+n} \longrightarrow RO^{\cdot} + HO^- + M^{+(n+1)} \qquad (19)$$

Several reaction pathways are available to the alkoxy radical thus generated (equation 20)[30-33]. First, an alcohol can be formed via hydrogen abstraction. Alternatively, β cleavage of a neighboring β hydrogen, alkyl or alkoxy group would lead to a carbonyl compound. In the case of primary and secondary hydroperoxides, loss of a hydrogen atom is quite prevalent. In sum total, this corresponds to the elimination of the elements of water from the hydroperoxide, a process commonly called 'Hock dehydration' (not to be confused with Hock cleavage; Section III.C.3). For tertiary hydroperoxides, carbonyl formation requires carbon–carbon bond scission, while for α-hydroperoxy ethers or esters carbon–oxygen cleavage often results.

(20)

Whenever the expected product is an aldehyde, it may undergo rapid oxidation to the corresponding acid (via the labile peracid).

2. Kornblum–DeLaMare reaction

In the presence of bases (even as weak as dilute aqueous hydroxide, pyridine or basic alumina), peroxides (including hydroperoxides) possessing α-hydrogens can undergo the Kornblum–DeLaMare reaction[34–36]. In this process, which can be viewed as an oxygen analog of an E2 elimination, primary and secondary hydroperoxides are dehydrated to aldehydes or ketones, respectively (equation 21). As might be expected, the reaction is particularly preferred when the resulting ketone is conjugated.

(21)

It should be noted, however, that most alkali bases contain substantial amounts of metal ions which may catalyze competing homolytic decomposition. Hence, Kornblum–DeLaMare dehydrations may well be accompanied by alcohol formation. In some cases, the metal-catalyzed homolysis can be inhibited by the addition of EDTA.

3. Hock cleavage[28]

In principle, the heterolysis of the peroxide bond should generate both a negative and a positive oxygen fragment. The instability of the latter with respect to a carbocation would then initiate skeletal changes in the carbon framework resulting from migration of groups to the electron-deficient oxygen. Such heterolysis and ensuing rearrangements have indeed been observed with hydroperoxides and are generally acid-catalyzed. One classic example is the acid-catalyzed cleavage of a hydroperoxide to an alcoholic and a ketonic fragment, for which the accepted mechanism, first suggested by Criegée[37–40], is outlined in Scheme 1. Relative migratory aptitudes have been determined for this reaction and their qualitative order is as follows[28c].

cyclobutyl > aryl > vinyl > hydrogen > cyclopentyl $\approx$ cyclohexyl $\gg$ alkyl

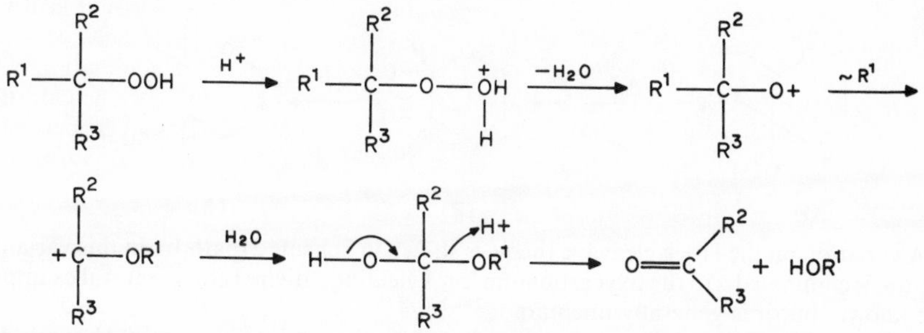

SCHEME 1. Criegée mechanism for the acid-catalyzed cleavage of hydroperoxides

In the particular case of allylic hydroperoxides the migrating group is generally vinylic. In such cases the resulting fragments will both be ketonic (Scheme 2, path a). Because of this fundamental difference in the make-up of the products, this transformation of allylic hydroperoxides to *two* carbonyl fragments, called Hock cleavage[41-45], has for a long time been classified separately. While such cleavages are generally acid-catalyzed[28a,46], several have been reported to occur in the absence of any added acid[28a] and even under basic conditions[47].

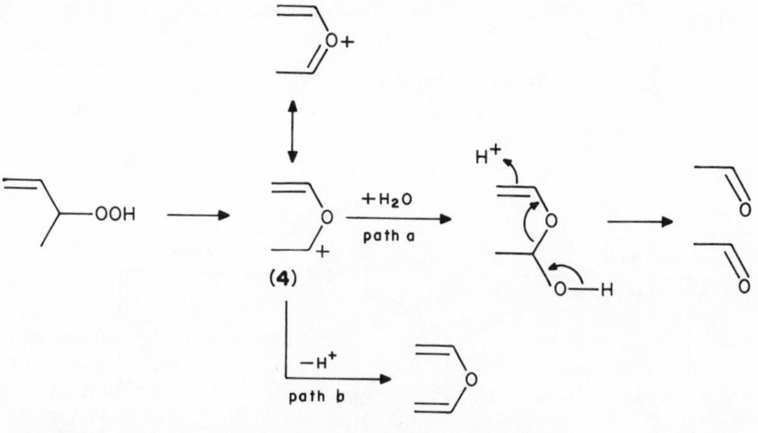

SCHEME 2. Acid-catalyzed cleavage of allylic hydroperoxides: path a: Hock cleavage; path b: divinyl ether formation

Recently, there has been growing experimental evidence[28a,46,48,49] that in some substrates, Hock cleavage can proceed not only via a Criegée mechanism but also through a cyclic dioxetane mechanism, first proposed by Farmer and Sundralingam[28b,50] (equation 22). For example, hydroperoxide **1** has been shown to rearrange to dioxetane **2**, which cleaves slowly in turn to acetoxy keto ester **3** (equation 23)[48,49].

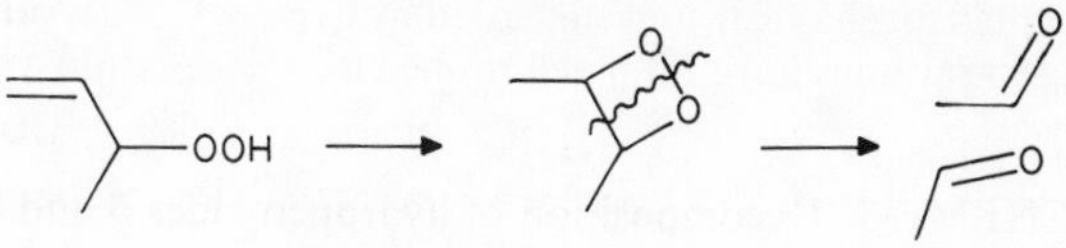

$$(22)$$

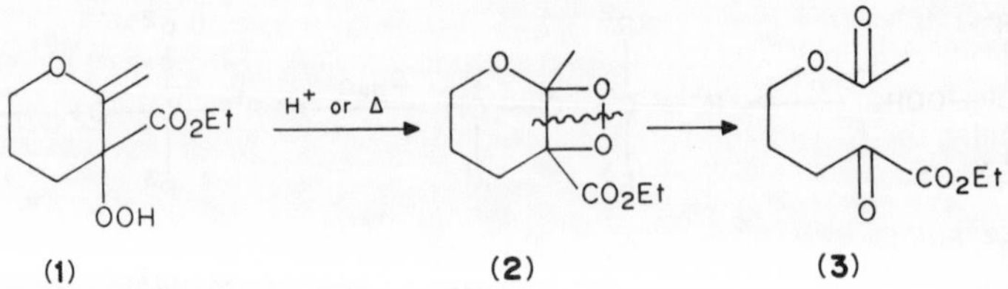

$$(23)$$

A variation on the Hock cleavage theme is shown in Scheme 2 (path b). In this variant a proton is eliminated α to the oxycarbonium ion 4 yielding a divinyl ether. Several examples are known, but it is generally uncommon[28a,b,d].

An interesting example[51,52] of some of the transformations discussed above is the decomposition of isomeric hydroperoxides 5 and 6 (Scheme 3). The former yields 2-

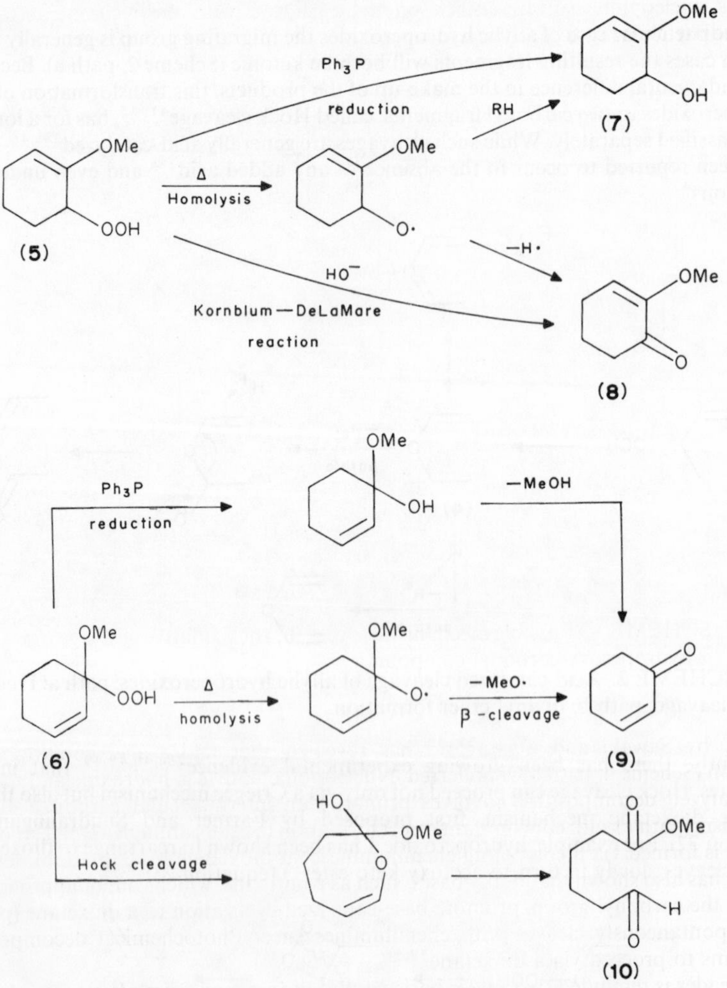

SCHEME 3. Decomposition of hydroperoxides 5 and 6.

methoxy-2-cyclohexen-1-ol (**7**) and 1-one (**8**). Peracetal **6** loses the elements of methyl hydroperoxide yielding **9**, while Hock cleavage generates aldehydo ester **10**. Compounds **7** and **9** can be formed directly upon reduction of **5** and **6** respectively with triphenylphosphine. Finally, the Kornblum–DeLaMare dehydration of **5** yields cyclohexenone **8**.

4. Transformations of hydroperoxy carbonyl compounds

α-Hydroperoxy carbonyl compounds undergo oxidative cleavage catalyzed either thermally, photochemically, by acids or by bases, yielding the corresponding carbonyl fragments. Three mechanisms have been considered (Scheme 4). The first involves acyl group migration (path a) which corresponds to the Criegée hydroperoxide cleavage mechanism (Scheme 1) where the migrating group R′ is RCO. The second mechanism (path b) involves a cyclic α-hydroxy dioxetane intermediate. The third mechanism (path c) involves nucleophilic solvent attack on the carbonyl, with the resulting tetrahedral intermediate cleaving to products.

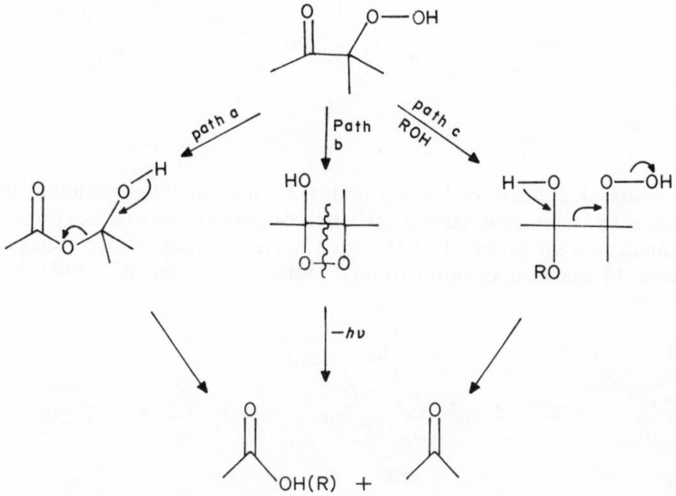

SCHEME 4. Possible mechanistic routes for the oxidative cleavage of α-hydroperoxy carbonyl compounds

Work by Sawaki and Ogata[53-55] has revealed that under acid conditions acyl migration (Scheme 4, path a) is preferred. Nucleophilic base (e.g. hydroxide and methoxide) catalyzed decomposition involves primarily an intermolecular carbonyl addition mechanism (path c) with concomitant direct formation of esters, though a small amount of product is formed via a competing chemiluminescent dioxetane route (path b). Jefford's group[56] has also shown that bulky bases, such as t-butoxide, which cannot approach and bond to the carbonyl group, promote base-catalyzed cyclization to a dioxetane (path b) which spontaneously cleaves with chemiluminescence. Photochemical decomposition also seems to proceed via a dioxetane[57,58].

The reader is reminded that under basic conditions aldehydes are often autoxidized to acids. Furthermore, 1° and 2° hydroperoxides can undergo Kornblum–DeLaMare dehydration to the corresponding ketone (Section III.C.2 above), and this is true for α-hydroperoxy carbonyl compounds as well. The exact mode of decomposition of the latter under basic conditions is quite sensitive to the structure of the substrate. The predominant reaction in the case of unsaturated α-hydroperoxy ketones such as 11 (equation 24) and steroidal α-keto hydroperoxides is dehydration to diketones, while simple saturated α-hydroperoxy ketones generally cleave to diacids[59,60].

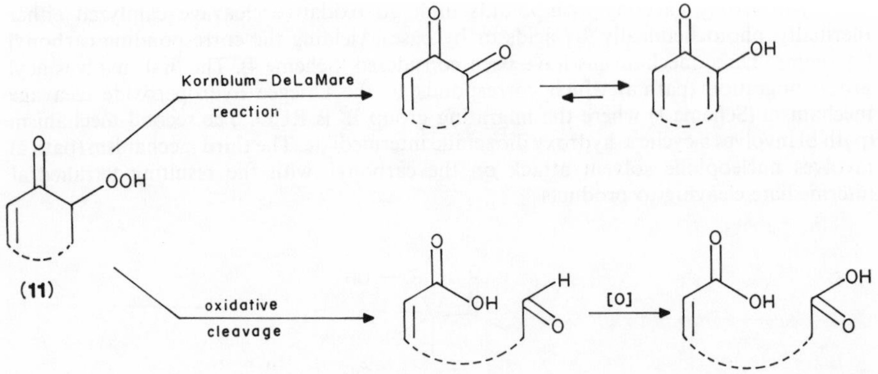

(24)

3-Hydroperoxy-1, 2-dicarbonyl compounds (12, equation 25) are generally quite labile and decompose to carbon monoxide, a carbonyl compound, and a carboxylic acid[61-76]. A likely intermediate is the perlactol 13, though Mayers and Kagan[77] have suggested a role for perlactone 14 when the distant carbonyl moiety is an ester ($R^1 = OR$).

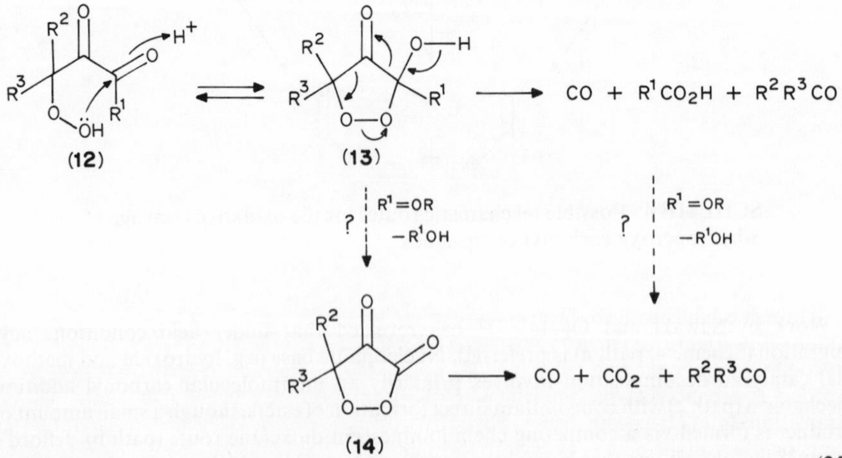

(25)

A simple example of these transformations is the decomposition of 3-hydroperoxycyclohexane-1, 2-dione which cleaves primarily to aldehydo acid 16 via peroxy lactol 15. A small amount of aldehydo keto acid 18 is also produced, presumably through dioxetane 17 (equation 26).

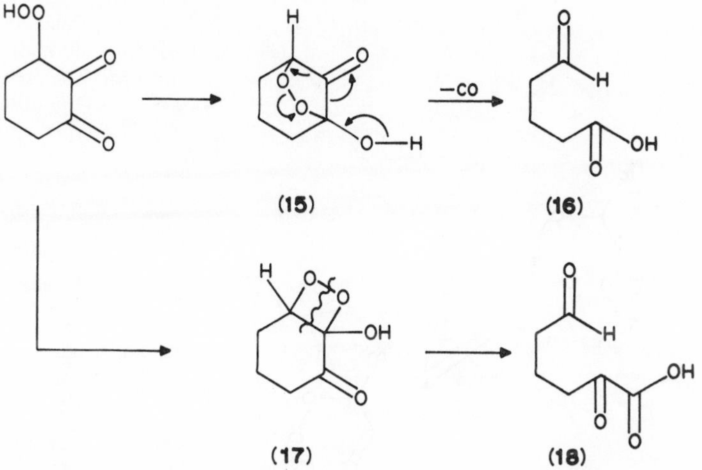

(15)   **(16)**

  (26)

(17)   **(18)**

5. 1,3-Allylic hydroperoxide rearrangement

Before closing this section, we should mention the 1,3-allylic hydroperoxide rearrangement (equation 27, $n = 1$), for which an analogous 1,5-pentadienyl hydroperoxide shift (equation 27, $n = 2$) is also known[28b,d]. The driving force for this transformation seems to be the greater stability of the olefinic linkage in the final product. Thus, allylic hydroperoxide **19**, in which the double bond is trisubstituted, rearranges to isomer **20** in which the olefinic linkage is now tetrasubstituted (equation 28)[48,49].

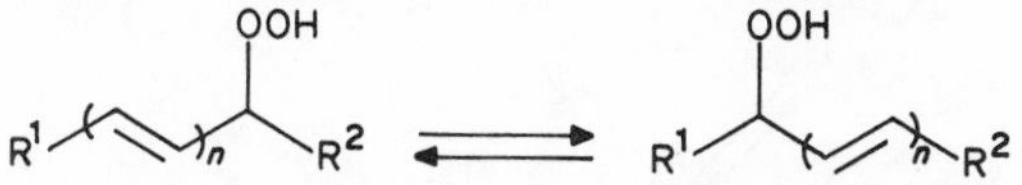

  (27)

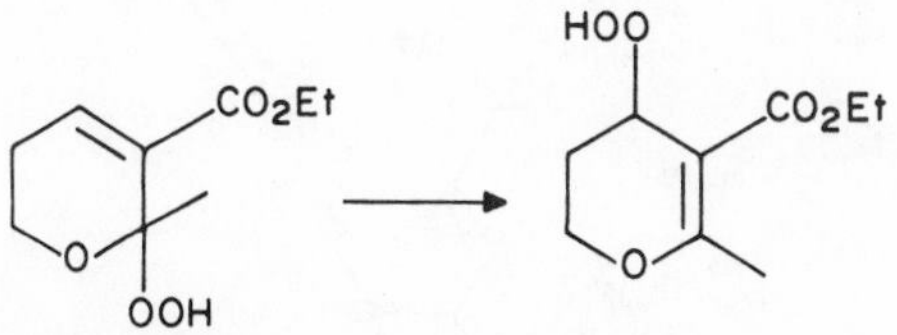

  (28)

(19)   **(20)**

Three mechanisms have been suggested for these[28b,28d,78–82] processes, and they are outlined in Figure 2. The first is a stepwise mechanism involving the intermediacy of a cyclic five-membered ring peroxide **(21)** possessing a free radical at the position 4. The second is a concerted mechanism with the formation of a cyclic five-membered ring transition state **(22)** linking the two allylic hydroperoxy radicals. The final possibility is a β-scission of an allylic peroxy radical to form molecular oxygen and an allyl carbon radical **23**.

The intermediacy of **21** in this transformation can be ruled out because no oxygen entrapment of this radical was observed[80], although authentic **21** do undergo facile oxygenation[81]. Oxygen-18 labeled hydroperoxides rearrange without loss of the label, suggesting the involvement of the concerted mechanism via transition state **22**[82].

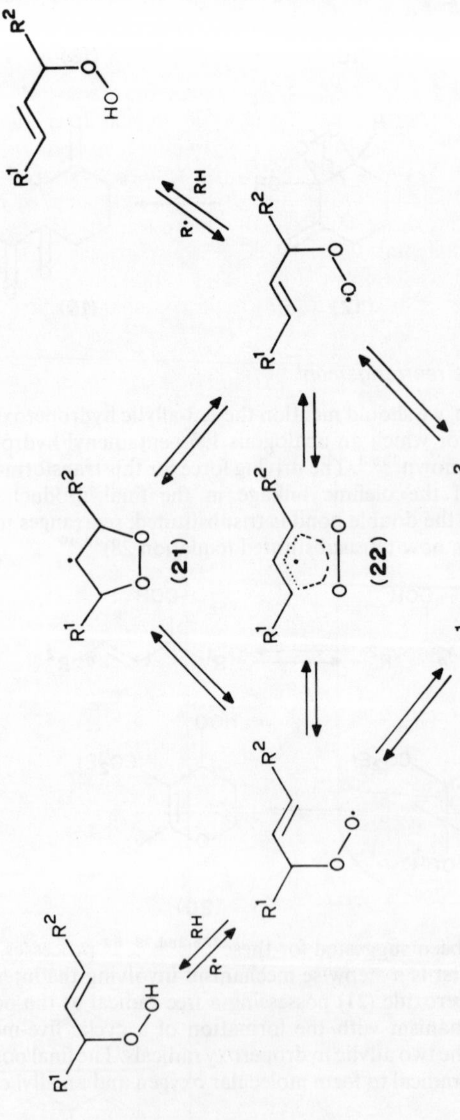

FIGURE 2. Possible mechanisms for the 1,3-allylic hydroperoxide rearrangement

It should be noted that in the corresponding pentadienyl case, the label is lost[78], indicating that in this case the rearrangement proceeds via β-scission yielding a pentadienyl radical, the vinylog of **23**.

D. Autoxidation of Enones

1. General considerations

α, β-Unsaturated carbonyl compounds (**24**, equation 29) are generally quite stable towards autoxidation, despite the fact that the resulting radical **25** is stabilized by the extended conjugation. The inhibition of the rate-determining hydrogen-abstraction step of the propagation (equation 4) can be attributed to the aforementioned 'polar effect' (Section III.A) resulting from the electron-withdrawing carbonyl group. By contrast, β, γ-unsaturated systems (**26**) autoxidize substantially more rapidly—even though the resulting radical (**25**) is the same and the polar effect is also at play (equation 29). The explanation here is that the lower stability of the β, γ system presumably results in a lower activation energy for hydrogen abstraction leading to the conjugated radical **25** (Figure 3).

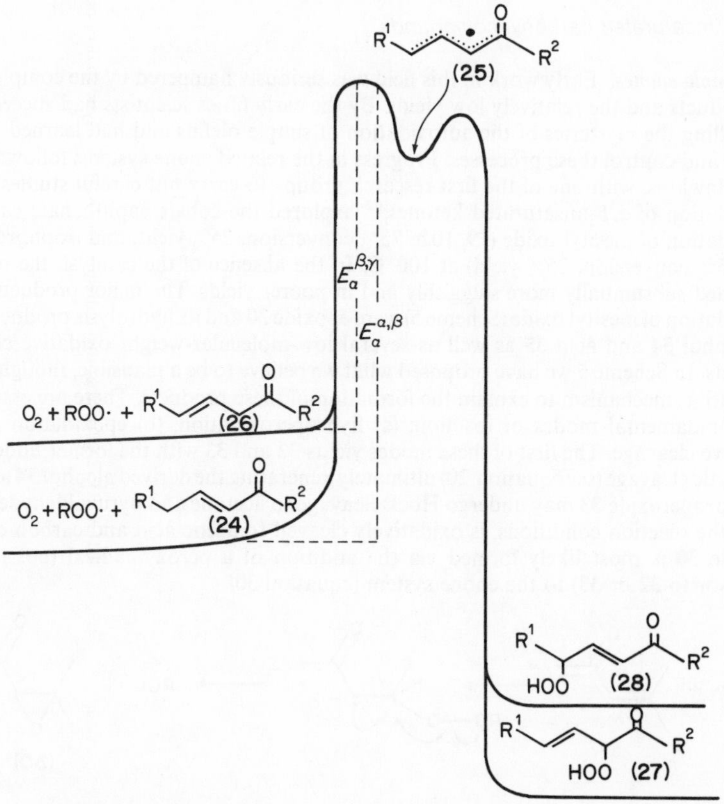

FIGURE 3. Energy profiles for the free radical autoxidation of α, β-enones **24** versus the β, γ-enones **26**

Of the two possible isomeric hydroperoxides **27** and **28**, the conjugated **27** is generally preferred for thermodynamic reasons.

(29)

2. α,β-Unsaturated carbonyl compounds

a. Simple enones. Early work in this field was seriously hampered by the complexity of the products and the relatively low yields. By the early fifties scientists had succeeded in unravelling the mysteries of the autoxidation of simple olefins and had learned how to initiate and control these processes. Progress in the related enone systems followed soon after. Hawkins, with one of the first research groups to carry out careful studies on the autoxidation of α, β-unsaturated ketones,[83] explored the cobalt naphthenate catalyzed autoxidation of mesityl oxide (**29**, 10 h, 75% conversion, 25% yield) and isophorone (**36**, 24 h, 35% conversion, 25% yield) at 100 °C. In the absence of the catalyst, the reaction proceeded substantially more sluggishly and in poorer yields. The major products in the autoxidation of mesityl oxide (Scheme 5) were epoxide **30** and its hydrolysis product glycol **31**, alcohol **34** and acid **35** as well as several low-molecular-weight oxidative cleavage products. In Scheme 5 we have proposed what we believe to be a plausible, though purely speculative, mechanism to explain the formation of these products. There are essentially three fundamental modes of reaction: (a) hydroperoxidation, (b) epoxidation and (c) oxidative cleavage. The first of these modes yields **32** and **33** with the former undergoing homolytic cleavage (see equation 20) ultimately generating the derived alcohol **34** and acid **35**. Hydroperoxide **33** may undergo Hock cleavage to acetone and pyruvaldehyde which, under the reaction conditions, is oxidatively cleaved to acetic acid and carbon dioxide. Epoxide **30** is most likely formed via the addition of a peroxy radical (possibly the precursor to **32** or **33**) to the enone system (equation 30).

(30)

It should be noted at this juncture that there are authors[84] who have suggested that the epoxy ketones result from the rearrangement of the α-oxygenation product (e.g. **33**), as

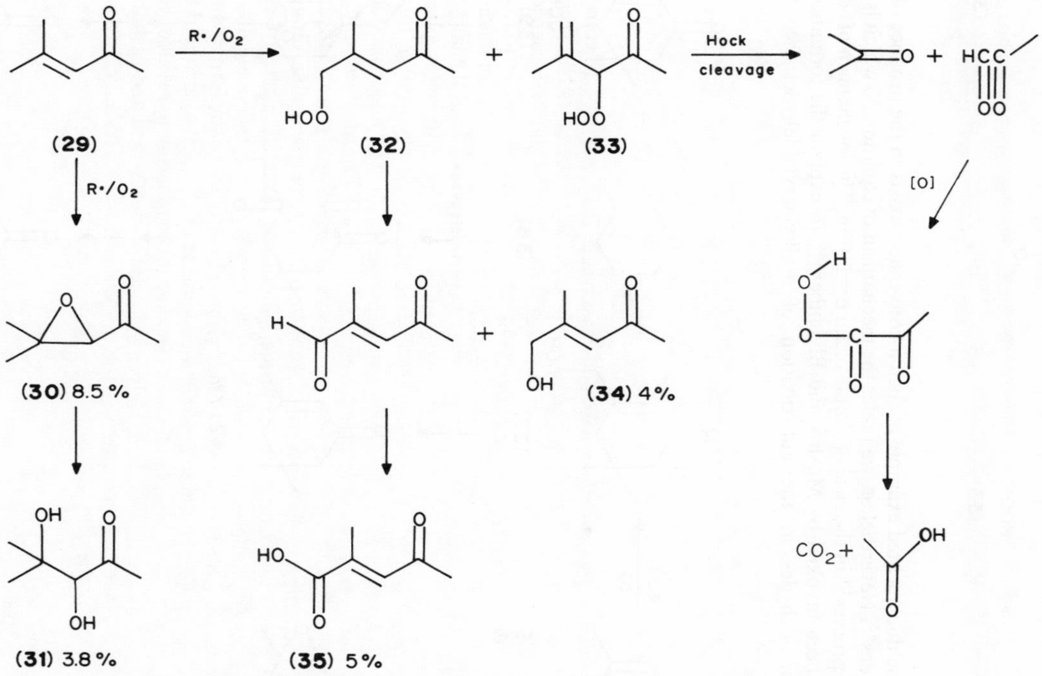

SCHEME 5. Mechanism proposed for the autoxidation of mesityl oxide

outlined in equation 31.

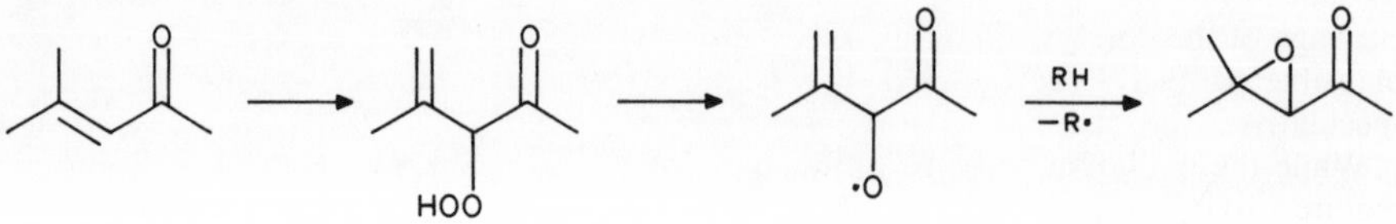

(29) (33) (30)

(31)

Indeed, one documented example of such a conversion exists in the instance of allylic hydroperoxides[85]. Nevertheless, we prefer the mechanism of equation 30 which is a well-precedented process[14] in the case of simple olefin oxidation. The mechanism of equation 30 has also been invoked by Moslov and Blyumberg[86] to explain the formation of α-epoxypropionaldehydes in the autoxidation of α-alkylacrylaldehydes (see Section III.D.2.c).

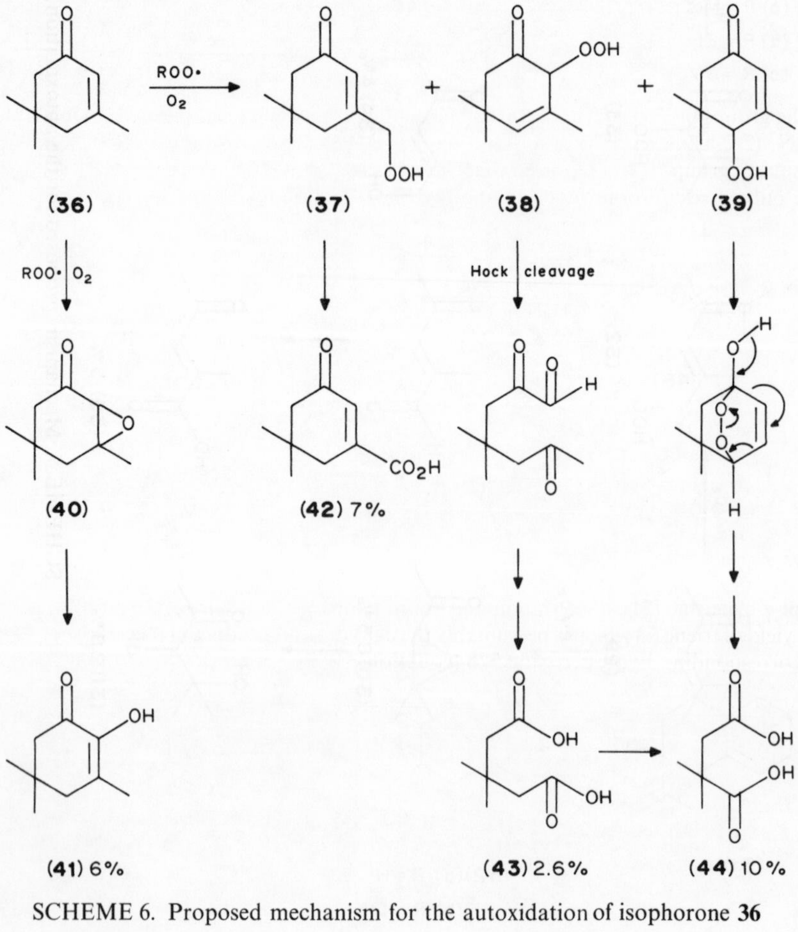

(36) (37) (38) (39)

(40) (42) 7%

(41) 6% (43) 2.6% (44) 10%

SCHEME 6. Proposed mechanism for the autoxidation of isophorone **36**

In the case of isophorone 36, enol 41 and acids 42–44 are the major products. The enol is ostensibly formed via epoxide 40, while the acids presumably result from the oxidative cleavage of the corresponding hydroperoxides 37, 38 and 39, respectively. The various plausible pathways are outlined in Scheme 6, but again the mechanisms are purely speculative.

While the oxidative cleavage products reported in the work of Hawkins[83] seem to require α-oxygenation and the formation of the unconjugated hydroperoxy enones 33 and 38, most subsequent reports involve the γ-hydroperoxide exclusively. Thus Tischenko and Stanishevskii[87,88] have reported that a series of homologous β-isopropyl enones 45 were converted to the corresponding alcohols 47 in relatively high yields by oxygenation and subsequent catalytic reduction (equation 32).

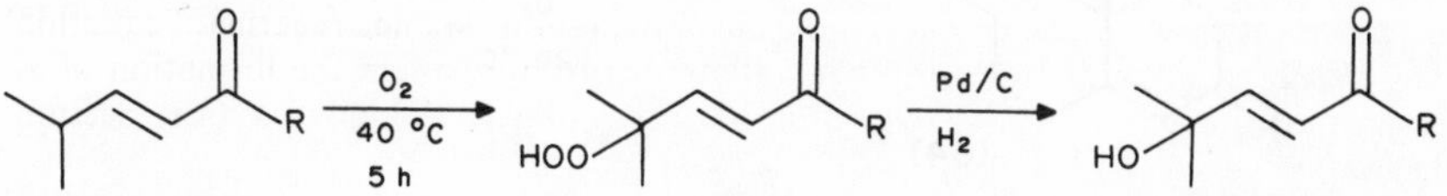

(45) (a) R = Me
　　 (b) R = Et
　　 (c) R = Bu

(46)

(47) 85–95%

(32)

The groups of Volger[89] and Watt[90a] have found that such reactions can be initiated by AIBN (2, 2-azoisobutyronitrile) and/or t-butyl hydroperoxide at 60 °C. Similarly, Gersman's group[90b] has reported that ester 48 (as well as its β, γ-unsaturated analog, 50) gave only γ-oxidation product 49 upon AIBN initiation (equation 33).

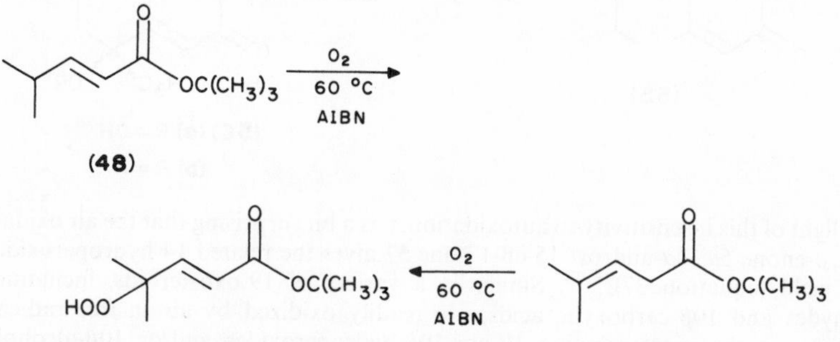

(48)

(49)

(50)

(33)

Epi-α-cyperone (51), after standing at room temperature in air for 15 months, gave a 50% yield of triene 53 which is presumably the dehydration product of the γ-alcohol 52a or the corresponding hydroperoxide 52b (equation 34)[91].

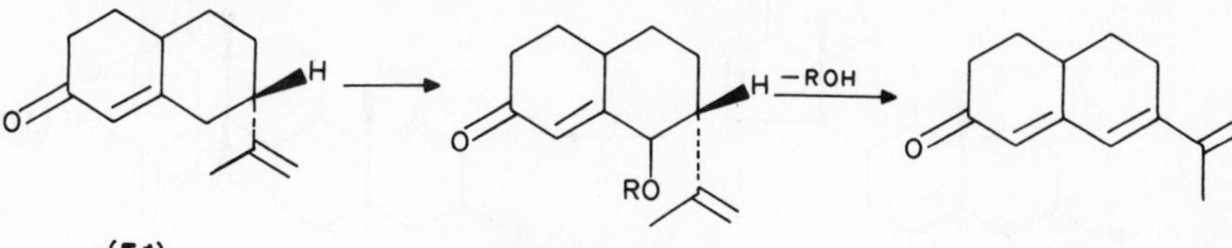

(51)

(52)(a) R = H
　　 (b) R = OH

(53)

(34)

Most conjugated steroids, such as cholest-4-en-3-one (**54**), are not particularly sensitive to autoxidation[15,92,93] (equation 35). Dimethisterone (**55**), too, is stable when exposed to air at 55 °C for as long as 16 h[94]. However, when it is subjected to higher temperatures (65–70 °C) for similar periods, TLC reveals the formation of trace amounts of the corresponding epimeric 6α- and 6β-hydroperoxides **56a**, as well as the derived epimeric alcohols **56b** (equation 36).

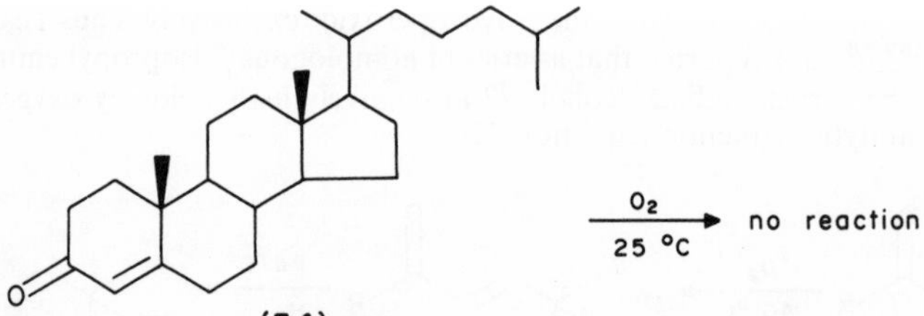

$$\xrightarrow[\text{25 °C}]{O_2} \text{ no reaction} \qquad (35)$$

(54)

$$\xrightarrow[\text{16 h}]{O_2 / 70 \text{ °C}} \qquad\qquad (36)$$

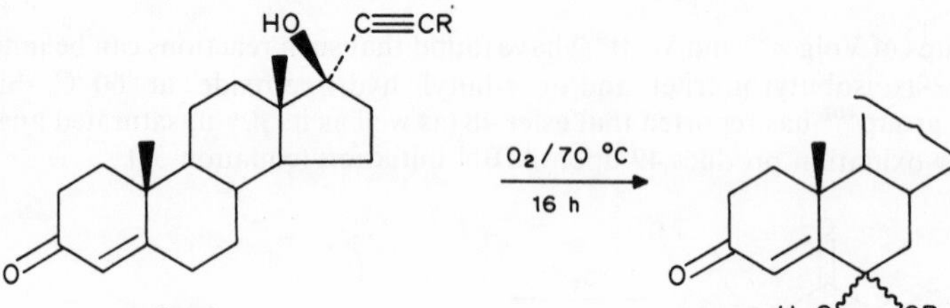

(55)

(56) (a) R=OH
(b) R=H

In light of this insensitivity to autoxidation, it is a bit surprising that the air oxidation of the α, β-enone 5α,14α-androst-15-en-17-one **57** gives the related 14-hydroperoxide **58** in high yield (equation 37)[95,96]. Similarly, a variety of 19-oxosteroids, including 10β-aldehydes and 10β-carboxylic acids, are readily oxidized by air in free radical type reactions to the corresponding 19-nor-10β-hydroperoxides and/or 10β-alcohols[15,97]. Thus androstenal **59** is converted to hydroperoxide **62** with the evolution of carbon monoxide after 3 days of aeration at 50° in the presence of the radical initiator AIBN[97]. A possible mechanism is outlined below (equation 38).

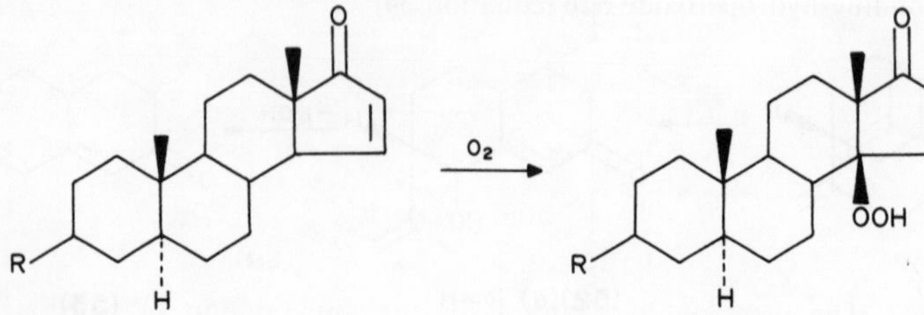

$$\xrightarrow{O_2} \qquad\qquad (37)$$

(57)   **(58)**

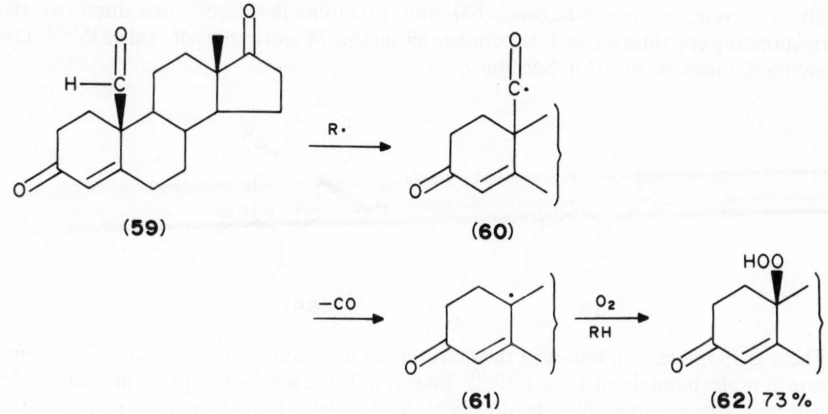

(59) (60)

(61) (62) 73% (38)

(63) R^1=CH$_3$; R^2=R^3=H (70)

(64) R^1,R^2=—(CH$_2$)$_4$—, R^3=CH$_3$

(71) (72)

(66) (65)

(73) (74)

(67) (68) (69)

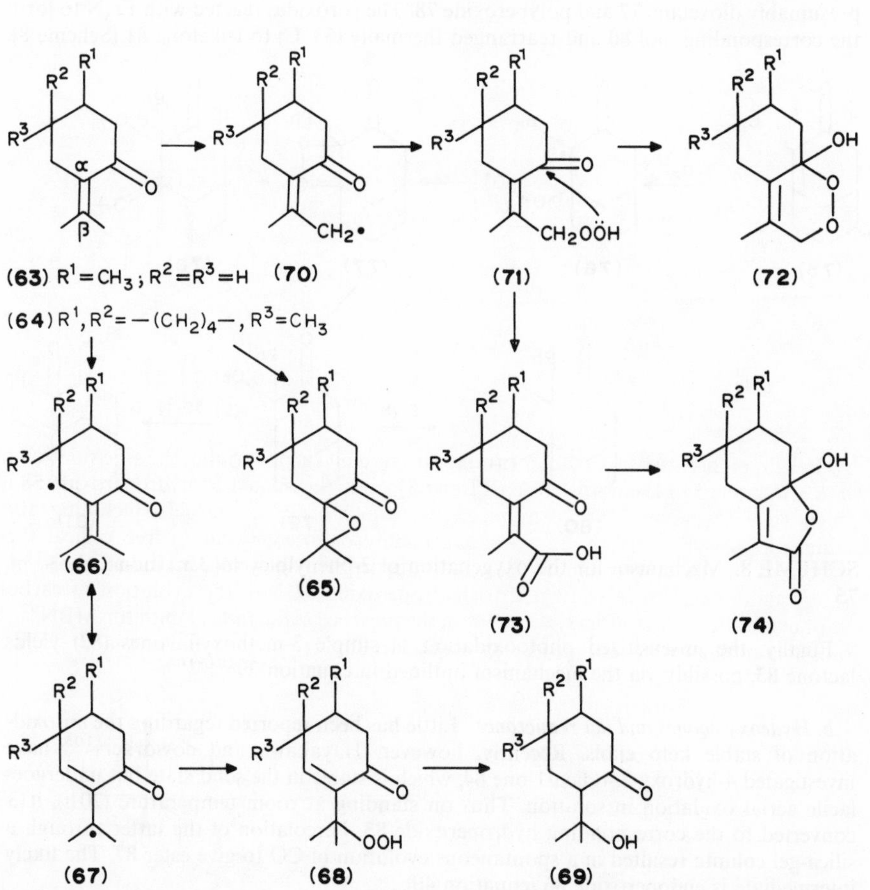

SCHEME 7. The proposed mechanism for the autoxidation of pulegone (63) and fukinone (64)

The exocyclic enones pulegone (63) and fukinone (64) are autoxidized to the corresponding epoxides 65, hydroxy enones 69, lactols 74 and cyclic peroxide 72[84,98]. The likely mechanism is shown in Scheme 7.

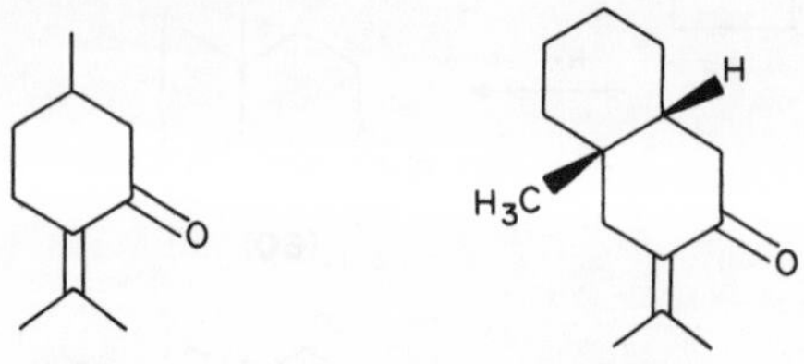

(63)   **(64)**

There is an interesting report in the literature of a spontaneous oxidation of an enone whose double bond is distorted[99,100]. Phenyl-substituted bicyclo[3.3.1]nonenone 75 reacts with oxygen (possibly via diradical 76) to yield a solid mixture of peroxides, presumably dioxetane 77 and polyperoxide 78. The peroxides reacted with Et_3N to form the corresponding diol 80 and rearranged thermally (53 °C) to triketone 81 (Scheme 8).

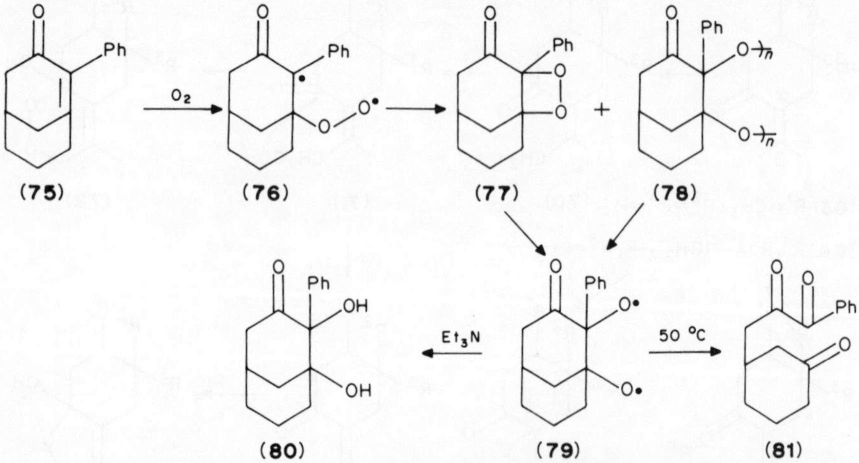

SCHEME 8. Mechanism for the oxygenation of 2-phenylbicyclo[3.3.1]non-1-en-3-one 75

Finally, the unsensitized photooxidation of simple 3-methoxyflavones (82) yields lactone 83, possibly via the mechanism outlined in equation 39[101-104].

b. Hydroxy enones and aci-reductones. Little has been reported regarding the autoxidation of stable keto enols. Recently, however, Hayakawa and coworkers[105] have investigated 4-hydroxy-2,4-dien-1-one 84, which is stable in the solid state but undergoes facile aerial oxidation in solution. Thus on standing at room temperature (20 h), it is converted to the corresponding hydroperoxide 85. Percolation of the latter through a silica-gel column resulted in a spontaneous evolution of CO to give ester 87. The likely intermediate is endoperoxide 86 (equation 40).

Other examples in this category are β-diketones, which exist essentially in their 3-hydroxy-2-en-1-one (keto enol) form[106]. Interestingly, Bredereck and Bauer[107] report that

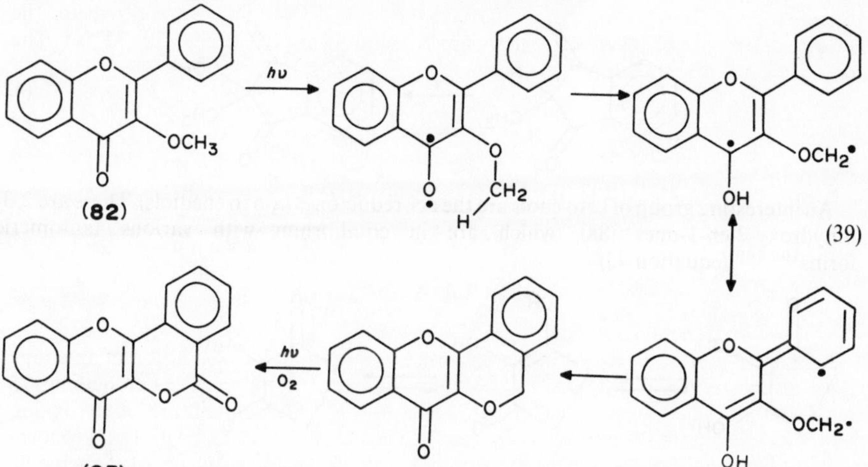

(82)

(39)

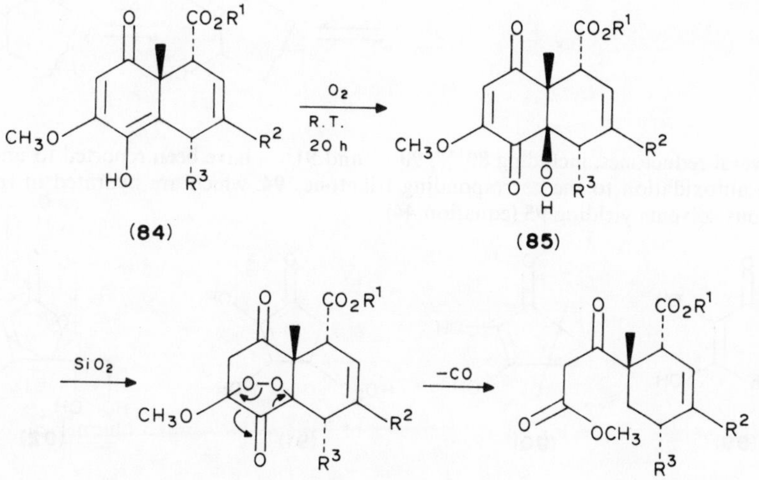

(83)

(84) (85)

(40)

(86) (87)

autoxidation of cyclic 1, 3-diketones with a tertiary C_2 carbon yields the corresponding 2-hydroperoxy-1, 3-diones (equations 41 and 42).

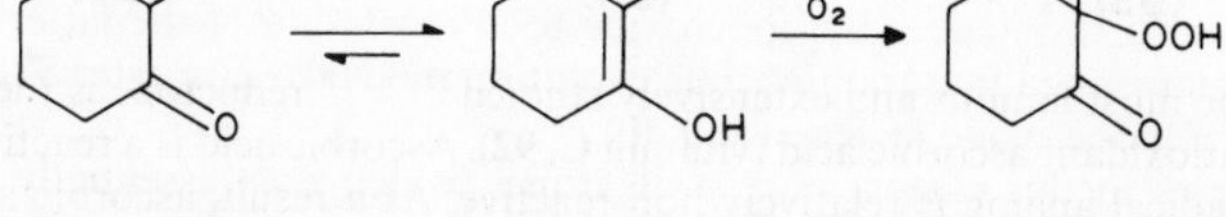

(41)

$R = Me, Et, i-Pr, t-Bu, CH_2Ph_2, Ph$

$$\text{(42)}$$

An interesting group of keto enols are the aci-reductones (α-oxo enediols). These are 2, 3-dihydroxy-2-en-1-ones (88), which are in equilibrium with various tautomeric forms[108,109] (equation 43).

(88)

$$\text{(43)}$$

Several reductones, including 89[110], 90[111] and 91[112], have been reported to undergo facile autoxidation to the corresponding triketones 94, which are hydrated in turn in aqueous solvents yielding 95 (equation 44).

(89)   (90)   (91)   (92)

(93)   (94)   (95)

$$\text{(44)}$$

Perhaps the most famous and extensively studied[113-115] reductone is the biologically important antioxidant ascorbic acid (vitamin C, 92). Ascorbic acid is a reactive reductant, but its free radical analog is relatively non-reactive. As a result, ascorbic acid does not undergo rapid autoxidation[116,117] and is quite stable in the solid state. There are, however, several reports of successful oxygenations of this reductone to the triketone (dehydroascorbic acid) carried out in protic media in the presence of either charcoal[118,119] or palladium carbon catalyst[120].

The mechanistic details for the autoxidative conversion of reductones to triketones has only been explored in the case of Vitamin C. It has been shown that $O_2^{-\cdot}$ is formed in this process[121,122] and, furthermore, that the oxidation rate for the neutral non-dissociated form of ascorbic acid is close to zero[123]. All this suggests that oxidation occurs from the ionized form and that the role of oxygen is not to oxygenate the radical intermediates but to function as an electron acceptor.

A plausible mechanism for the formation of the triketone dehydroascorbic acid is shown in equation 45[124].

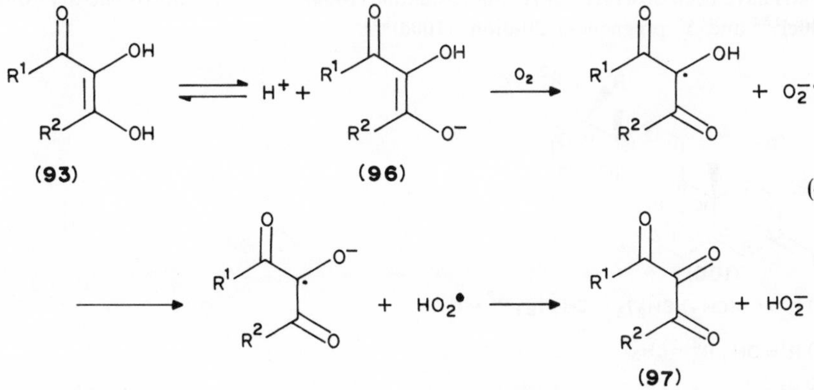

(45)

c. α, β-*Unsaturated aldehydes*. These compounds are oxidized to the related carboxylic acids several orders of magnitude more slowly than the corresponding saturated analogs[19e,86,125]. In addition, Moslov and Blyumberg[86] report the formation of α-epoxypropionaldehyde **99** as a side-product in the autoxidation of α-alkylacrylaldehydes. The mechanism for this process is outlined in equation 46.

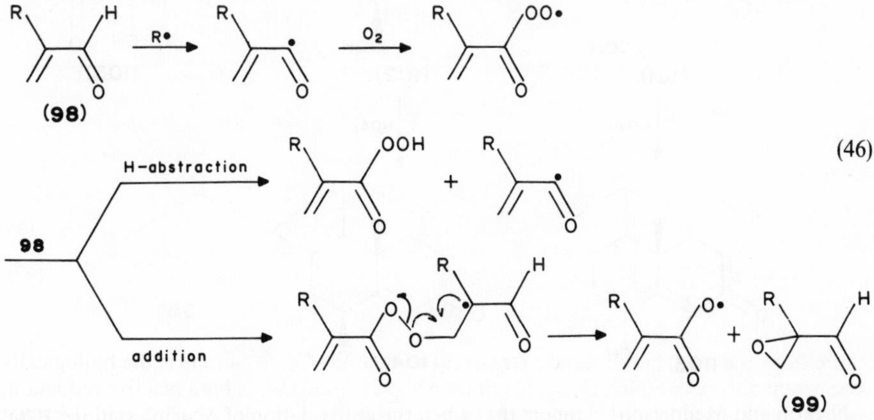

(46)

3. β,γ-Unsaturated carbonyl compounds

It has long been known that β,γ-enones are labile compounds which rearrange readily to their α,β-conjugated analogs and also undergo facile air oxidation at room temperature. The exact nature of these oxidation products was studied by Fieser and colleagues[126–129] who reported that Δ^5-cholesten-3-one (**100a**) combines with molecular oxygen in hexane at 25 °C to yield a 1:1 mixture of 6α- and 6β-hydroperoxy-Δ^4-cholesten-3-one (**101a** and **102a**). Best results (82% yield) are obtained by overnight aeration in the dark of a cyclohexane solution (at 40–50 °C) of the Δ^5-steroid containing a little benzoyl peroxide. The two hydroperoxides are quite stable and are separable by crystallization. Upon reduction with sodium iodide in acetic acid, each of these hydroperoxides is converted to their respective 6-hydroxy compounds **103a** and **104a** (equation 47). Similar results have been observed for Δ^5-androstenone (**100b**)[130,131], Δ^5-androstene-3, 17-dione (**100c**)[132] and Δ^5-pregnene-3, 20-dione (**100d**)[132].

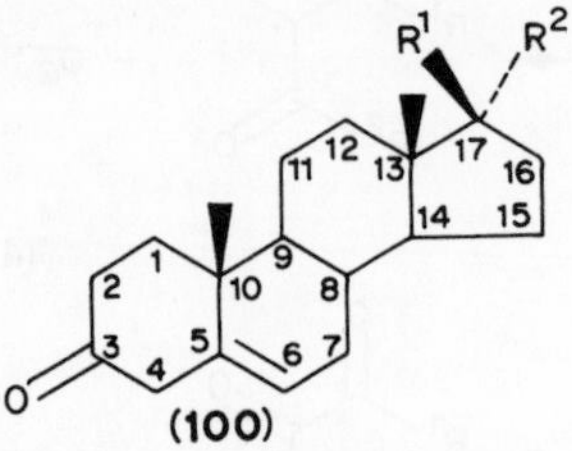

(**a**) R^1=CH(CH$_3$)(CH$_2$)$_3$ C(CH$_3$)$_2$; R^2=H

(**b**) R^1=OH; R^2=CH$_3$

(**c**) R^1,R^2==O

(**d**) R^1=COCH$_3$; R^2=H

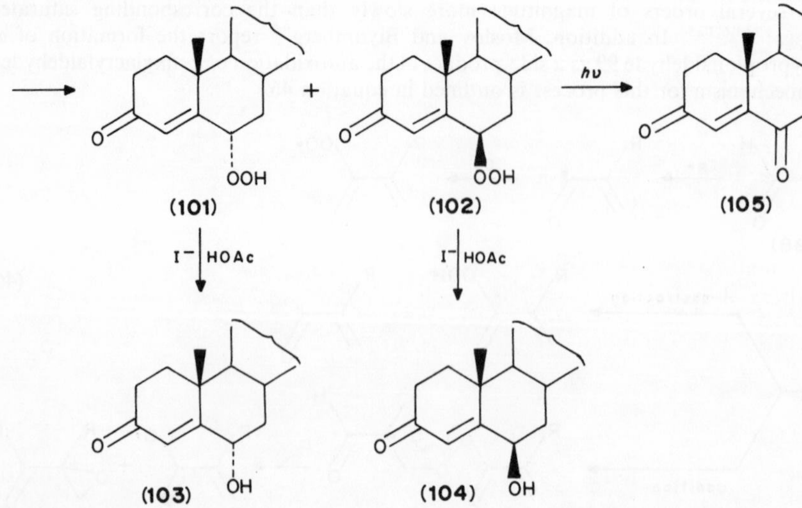

(47)

Nickon and Mendelson[133] report that when the autoxidation of Δ^5-cholestenone **100a** is initiated photochemically (in the absence of sensitizers), after 42 hours of irradiation and subsequent reduction, a 50% yield of a mixture of **103a** and **104a** as well as a 3% yield of diketone **105a** are isolated (equation 47).

de la Mare and Wilson[134] have carried out kinetic studies on these reactions and found that the oxidation of cholest-5-en-3-one (**100a**) with air in CCl_4 at 20 °C is slow, autocatalytic, catalyzed by dibenzoyl peroxide and inhibited by 3, 5-di-t-butylanisole. The products are **101a** and **102a** as in the corresponding reaction in cyclohexane reported by Cox[128]. The same reaction in ethanol was seven times faster. The products were entirely those of oxidation, namely **101a–105a**, and no rearrangement to the Δ^4 analog **54** was observed.

Shapiro and colleagues[130,131] studied the related oxidation of the 19-nor systems **106a–c** and again obtained the corresponding γ-oxidation products, 10β-hydroperoxy compounds **107** (40% yield). The latter are reduced to alcohols **108** with iodide (equation 48). These oxidations occur under a variety of conditions, i.e. with or without fluorescent light irradiation, with or without radical initiators (benzoyl peroxide or AIBN), or it may occur on a suitable substrate such as silica gel. Kirdani and Layne[135] found that, as compared to organic media, the oxidation of norethynodrel **106a** occurs quite slowly in aqueous solution with the initial products being **107a** and **108a**. The oxidation is rapidly catalyzed by horseradish peroxidase in the presence of hydrogen peroxide and manganese ion or by hemoglobin.

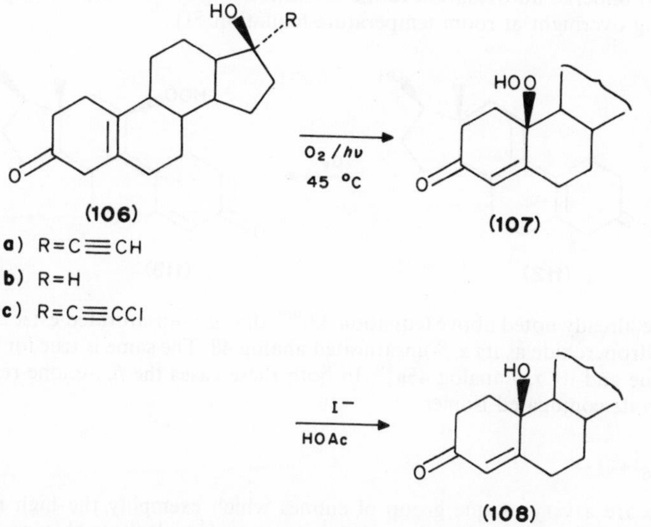

(106)

(**a**) R = C ≡ CH

(**b**) R = H

(**c**) R = C ≡ CCl

(107)

(48)

(108)

β, γ-Unsaturated 17-ketones are also sensitive to air oxidation. Thus, androstenone **109** gives the related 14-hydroperoxide **58** in high yield (equation 49; cf. equation 37)[95,96,136].

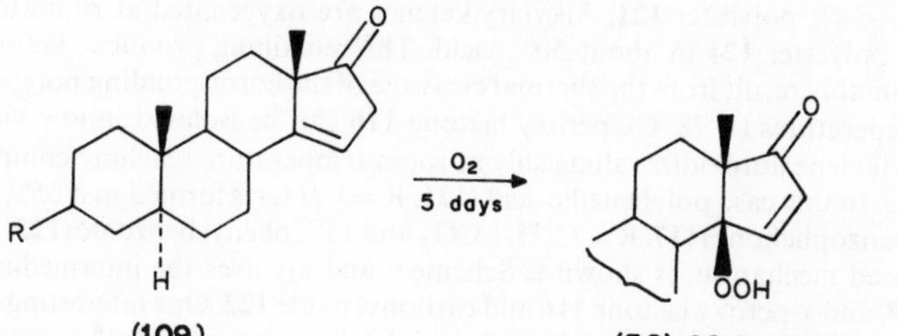

(109)

(49)

(58) 82 % yields

In a related study[137], γ-hydroperoxides **111** can be produced in fair yields by merely allowing the corresponding β,γ-unsaturated podocarpenones **110** to stand under oxygen in ether solution for several days (equation 50). Oxygen bubbling as well as fluorescent lamp irradiation hastens the process.

$$\text{(50)}$$

R=CO$_2$H, CO$_2$CH$_3$ or CH$_2$OH

In the case of $\beta,\gamma,\delta,\varepsilon$-dienones, oxygenation occurs at the ε position with the double bond shifting, in tandem, into conjugation[138-141]. Thus, solid $\Delta^{5(10),9(11)}$-3-ketone **112** is reported to undergo autoxidation to the corresponding $\Delta^{4,9(10)}$-10β-hydroperoxide **113** on standing overnight at room temperature (equation 51).

$$\text{(51)}$$

We have already noted above (equation 33)[90b] that β,γ-unsaturated ester **50** yields the same γ-hydroperoxide as its α,β-unsaturated analog **48**. The same is true for 5-methyl-4-hexen-2-one and its α,β-analog **45a**[89]. In both these cases the β,γ-enone reacted much faster than its conjugated isomer.

4. Ketenes[142,143]

Ketenes are a very unique group of enones which exemplify the high reactivity of cumulenes as well as substituent-dependent behaviour. Unsubstituted ketenes (**114**,R = H, Scheme 9) do not autoxidized readily. On the other hand, dialkylketenes react to completion even at $-20\,^\circ\text{C}$ after several hours, producing polyperester **118** in a 96% yield along with <4% polyester **121**. Alkylarylketenes are oxygenated at room temperature generating polyester **121** in about 50% yield. The remaining products, ketone **117** and CO$_2$, presumably result from the thermal cleavage of the corresponding polyperester **118**. At low temperatures ($-78\,^\circ\text{C}$), peroxy lactone **116** can be isolated in low yields.

Diphenylketene autoxidizes sluggishly at room temperature reaching completion only after 3 days. In this case polybenzilic acid (**121**, R = C$_6$H$_5$) is formed in a 65% yield along with 20% benzophenone (**117**, R = C$_6$H$_5$), CO$_2$ and 15% phenyl benzoate (**123**, R = C$_6$H$_5$). The proposed mechanism is shown is Scheme 9 and involves the intermediacy of an α-lactone **120**, and α-peroxy lactone **116** and carbonyl oxide **122**. One interesting facet of this reaction is that it appears to be initiated completely spontaneously[16].

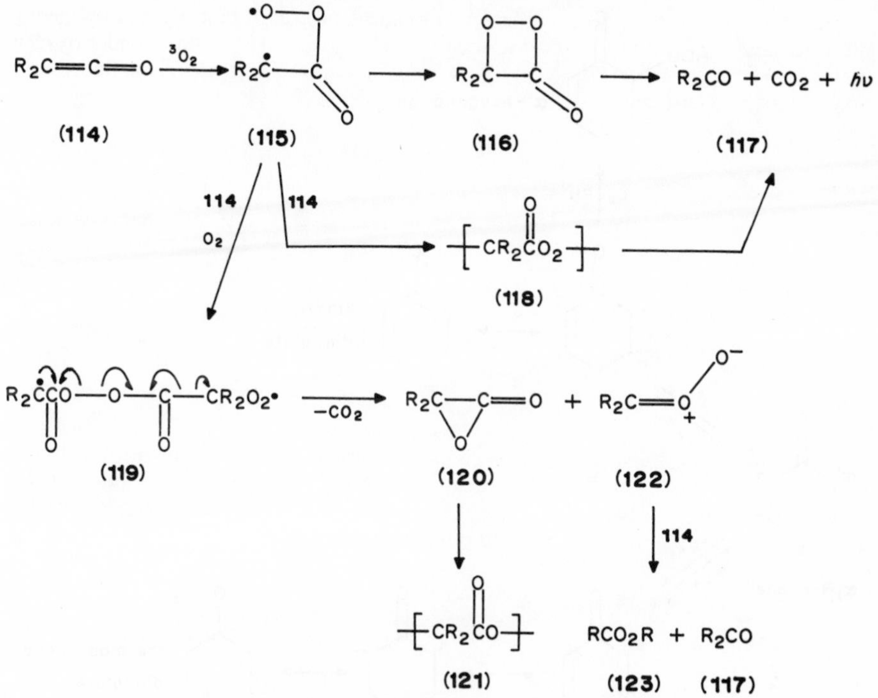

SCHEME 9. Autoxidation of ketenes

E. Base-catalyzed Autoxidation of Enones

1. General mechanism

In α, β-unsaturated carbonyl systems, two different acidic protons are often present, positioned at the α' and γ carbons. Of the two, the α'-hydrogen is the more acidic, presumably for inductive reasons. Nevertheless, abstraction of the γ-hydrogen is thermodynamically preferred since the completely conjugated dienolate anion formed is more stable than its cross-conjugated isomer[144] (Scheme 10).

As a result, enones can give dienolate anion mixtures of various composition, depending on whether the enolates are formed under circumstances in which the composition was determined by the relative rates of proton abstraction (kinetic control) or via equilibration of the various enolate anions (equilibrium or thermodynamic control). A rapid equilibrium between the enolates is achieved only when some proton donor, such as a protic solvent (e.g. t-butoxide in t-butanol) or excess unionized ketone, is present in the reaction mixture. Consequently, a kinetically controlled mixture of enolates is obtained by slowly adding a ketone to excess of strong base in an aprotic solvent at low temperature. On the other hand, protic solvents and elevated temperatures, the slow addition of a strong base to a ketone, or the presence of excess ketone in a solution of enolate anions, all favor the formation of the thermodynamic dienolate[144-148]. Recent research has further shown that high selectivity in the formation of either the linear-conjugated or the cross-conjugated dienolates can be obtained by choosing the correct base–solvent combination[149].

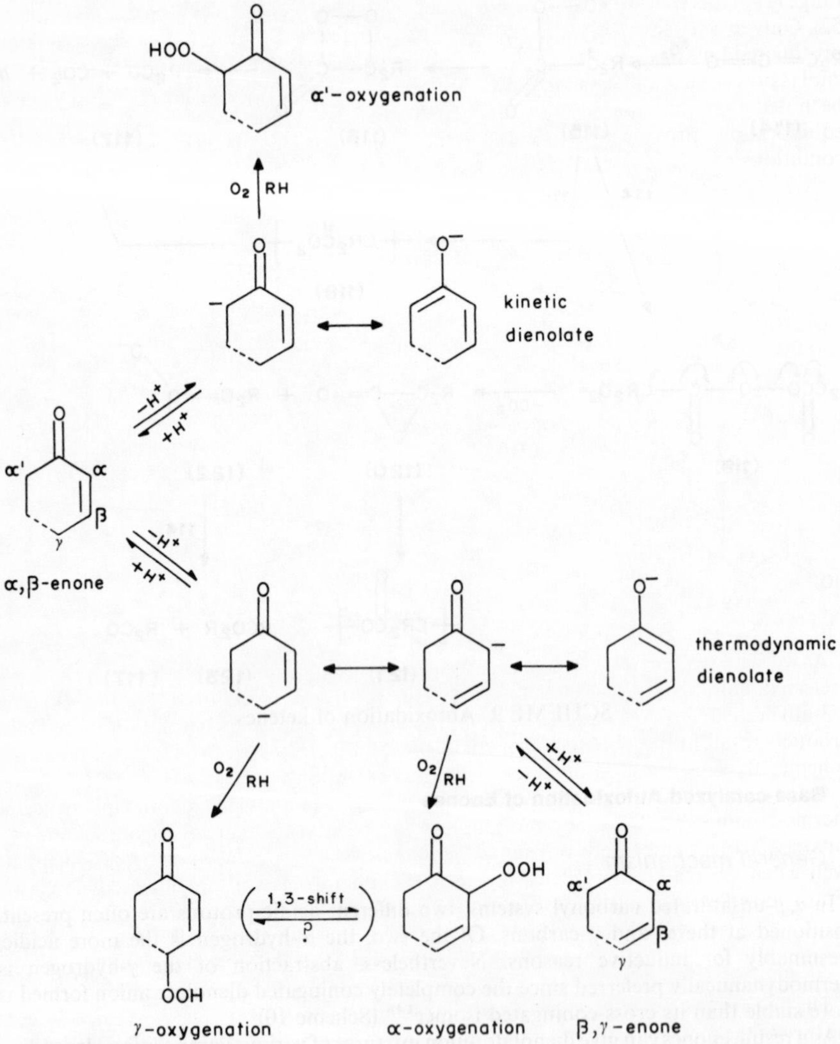

SCHEME 10. Scheme for the deprotonation and oxygenation of enones

The kinetic dienolate reacts with electrophiles (alkyl halide, protons, molecular dioxygen, etc.) at the α′ position, while the thermodynamic dienolate theoretically provides opportunities for electrophilic attack at either the α or γ positions. In fact, however, the thermodynamic dienolates invariably undergo intermolecular alkylation and protonation at the α position, even when that site is sterically quite congested[148,150-152].

We have spoken thus far only about enolate formation and have essentially neglected the intermediacy of the corresponding enol. This is because at basic pH, it is the enolate alone which is the predominant reactive species. In studies[152] on the tautomerization of the conjugated enol **124** of cholest-4-en-3-one (**54**), it has been shown that over a broad pH range (2–8) it is the enolate anion **125** which is protonated during ketonization (equation 52). Only at very low pH is the enol itself protonated. Furthermore, while the enolate is protonated kinetically at the center of the conjugated system (i.e. at C_α or C_4 in **125**), the enol is protonated at the end (i.e. at C_γ or C_6 in **124**). For our purposes, however, it should be noted that the enolate is so much more reactive than the enol that enol–enolate equilibration provides sufficient enolate to favor C_α protonation under most conditions[152].

$$(52)$$

As we shall see shortly, in the case of the *oxygenation* of enolates, while α oxygenation is preferred, both α and γ products are known. Nevertheless, in light of the aforementioned 1,3-allylic hydroperoxide shift (Section III.C.5), it is quite possible that γ-oxygenation products result from the rearrangement of the initially formed α products (see bottom of Scheme 10). This question deserves further investigation.

In the case of β,γ-enones, abstraction of the α-hydrogen is preferred both kinetically and thermodynamically. Thus, deprotonation of β,γ-unsaturated carbonyl compounds permits easy access to the 'thermodynamic dienolate' of the α,β-enone system (see Scheme 10) even when the reaction is carried out at low temperatures and aprotic media. We will return to this point a bit later (Section III.E.4).

2. Epoxidation of α, β-enones

The first studies on the base-catalyzed autoxidation (BCA) of enones were carried out at the turn of the century by Harries[153] and Stahler[154], but it was not until three decades later that systematic research was begun by Treibs[155–161]. The early reactions were carried out in aqueous methanol, above room temperature, and for lengthy reaction times. The yields isolated were generally quite low (<15%). Hydroperoxides formed via α'- or γ-proton abstraction were undoubtedly the primary products, but neither these nor the corresponding ketones or alcohols were isolated. Undoubtedly, these underwent further oxidation and cleavage and unidentified acidic compounds represented the bulk of the products. The major isolated product was the corresponding epoxide **127** or its derivatives formed in a variety of subsequent hydrolytic and/or oxidative rearrangement steps (equation 53).

For example[161], 3-methylcyclohex-2-en-1-one (**132a**), as well as its 5-methyl and 5,5-dimethyl analogs (**132b** and **132c**), yield the corresponding diosphenol methyl ethers **135a–c** (equation 54). Similarly, diosphenol **137** was the main product in the autoxidation of verbenone (**136**, equation 55).

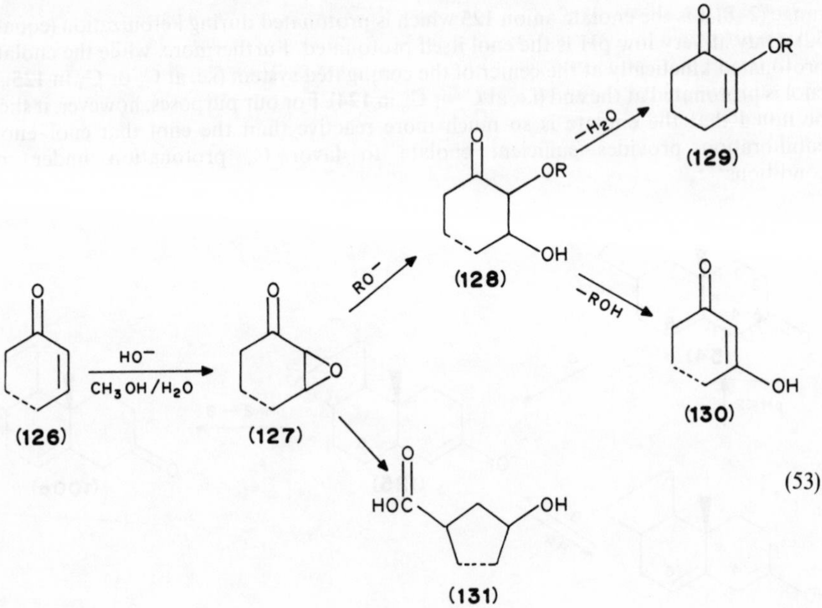

(53)

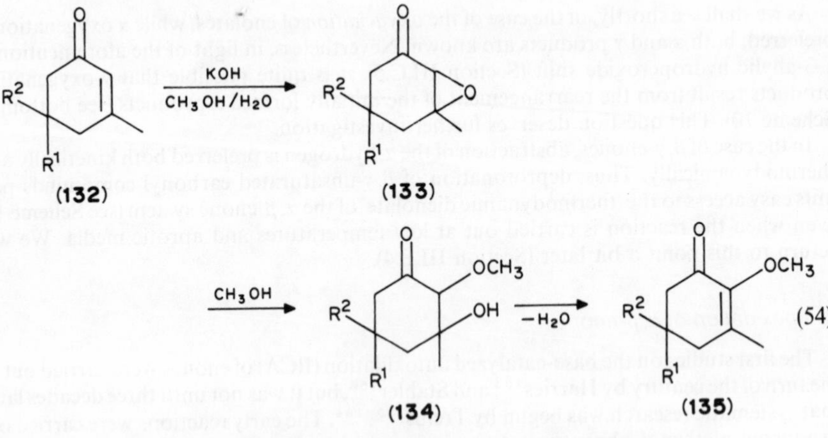

(a) $R^1, R^2 = H$; (b) $R^1 = H, R^2 = CH_3$; (c) $R^1 = R^2 = CH_3$

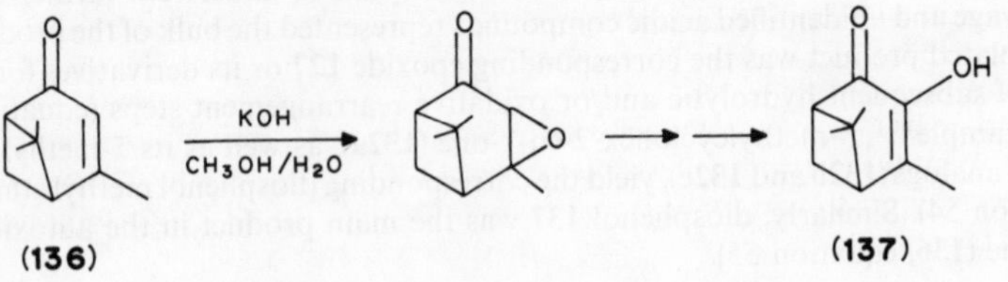

(54)

(55)

On the other hand, carvone[153-156,159,161] (**138**) and carvotanacetone[161] (**142**), which lack a hydrogen α to the carbonyl, form 3-hydroxy enones **141** and **144** (equations 56 and 57). In the case of enone **142**, addition product **143** was also isolated (equation 57), while the analogous addition product **146** was the sole compound isolated from the autoxidation of eucarvone (**145**, equation 58).

Finally, α-hydroxy acids **148** are the primary products from the autoxidation of piperitone (**147a**)[155,157,158,160,161] and carvenone (**147b**)[155,158] as outlined in equation 59.

The mechanism suggested by Treibs[157] (Scheme 11, path a) and quoted by Sosnovsky and Zaret[19d] for the formation of β-hydroxy acids **148** is unnecessarily complicated and in many aspects unprecedented. We prefer the intermediacy of a Favorskii rearrangement (Scheme 11, path b) which is well precedented for α,β-epoxy ketones[162,163].

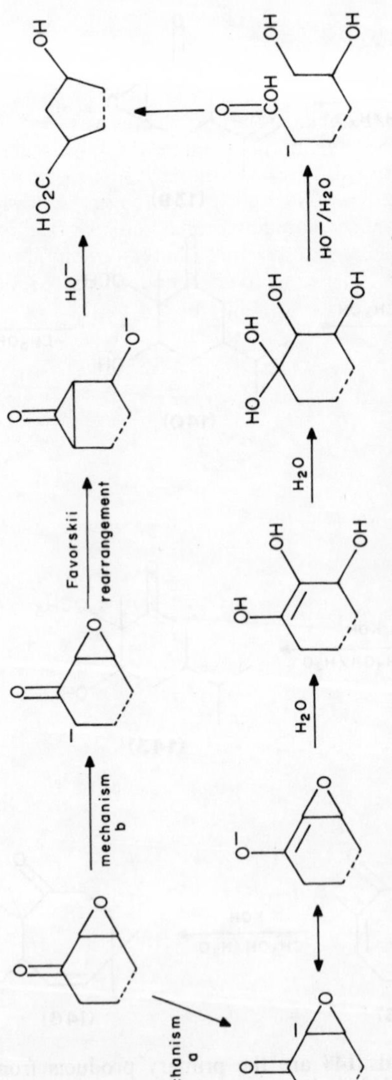

On the other hand, ... (138) and cyclopentanone ... (141), which lack a hydrogen on the carbonyl atom. ... of enone 142, addition produces 143 ... the double bond; addition product 144 was the ... compound isolated from the saturated solution (125, equation 55).

HO₂C — OH

Favorskii rearrangement

HO⁻

mechanism b

mechanism a

H₂O

H₂O

HO⁻/H₂O

SCHEME 11. Possible mechanisms for the formation of β-hydroxyacids from enones

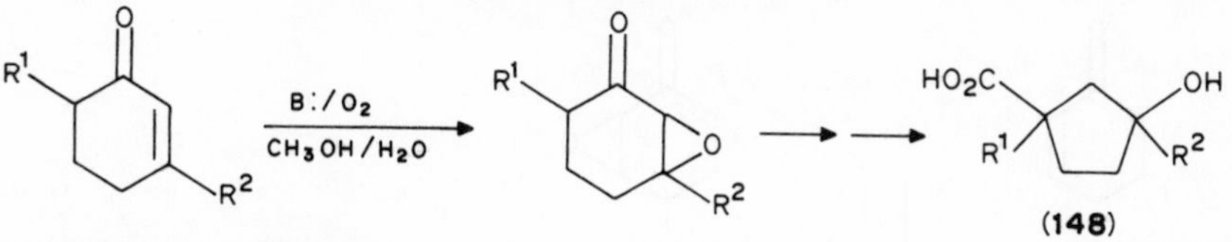

$$(147)(a) \quad R^1 = i\text{-}Pr \; ; \; R^2 = Me$$
$$(b) \quad R^1 = Me \; ; \; R^2 = i\text{-}Pr$$

(59)

The intermediacy of epoxides in all the above cases was verified by demonstrating that pure epoxides generate the same products under the same reaction conditions. Various condensation products were also formed in some instances; however, the vast majority of the product components were unidentified as noted in the beginning of this section.

Related systems have been explored by Frimer and his students[164,165] in aprotic media using potassium hydroxide, superoxide and t-butoxide solubilized in toluene or benzene with 18-crown-6-polyether. These researchers obtained low to moderate yields of epoxides in the BCA of cyclohex-2-en-1-ones **149** and **151** (equations 60 and 61). (For further discussion of this reaction see Section III.E.3.b.)

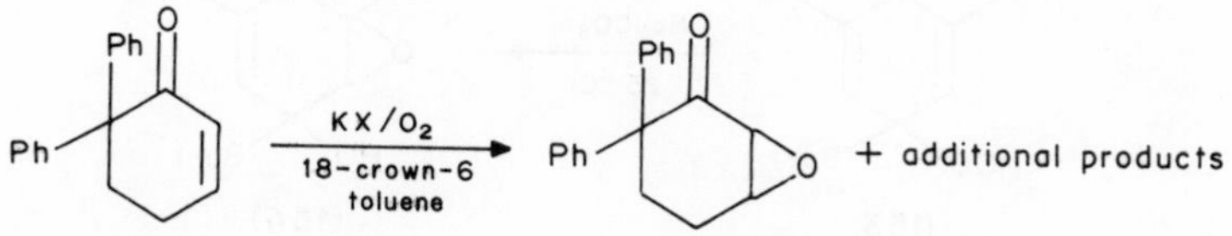

(60)

KOH/25 °C	36% yield
KO$_2$/25 °C	50%
t-C$_4$H$_9$OK/−40 °C	40%

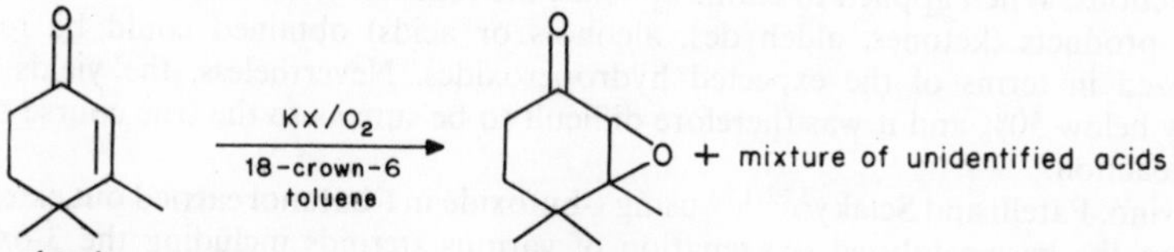

(61)

KOH/25 °C	33% yield
t-C$_4$H$_9$OK/−40 °C	50%

The issue that remains to be resolved is the mechanism of epoxidation in all these cases. Karnojitzky[19e] suggests that 'hydrogen peroxide, formed by the hydrolysis of the allylic hydroperoxide produced initially, can serve as the epoxidizing agent'. Since epoxides are obtained in aprotic media as well, we believe it much more likely that these allylic hydroperoxides themselves are the active agents (equation 62)[164], a suggestion that has been confirmed by recent work of Sugawara and Baizer[166].

In the same vein, Jensen and Foote[167] recently reported that hydroperoxide **153** is converted to epoxide **154** upon treatment with Na$_2$CO$_3$ (equation 63).

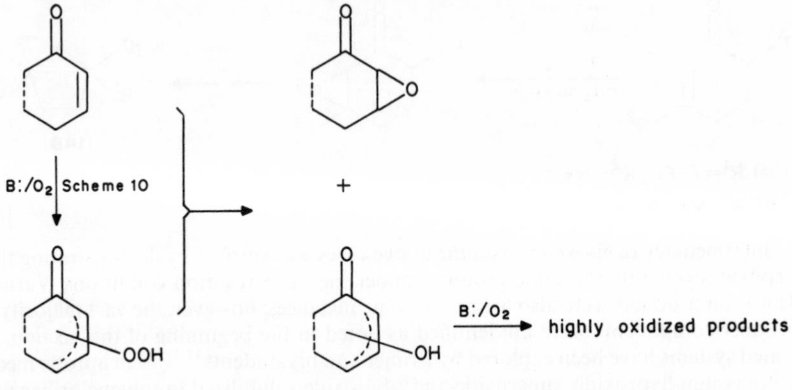

(62)

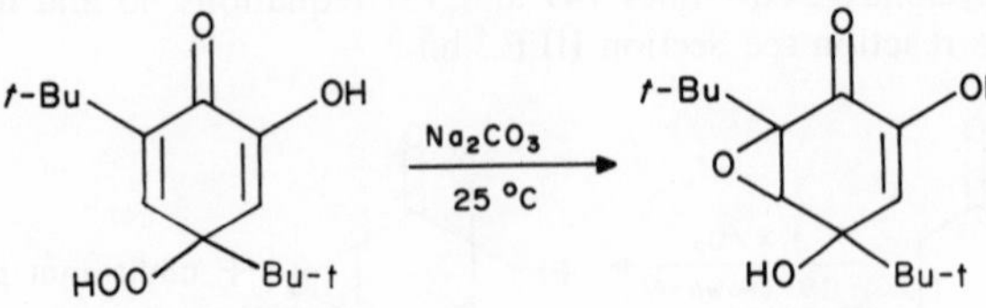

(63)

3. Hydroperoxidation of α, β-enones

a. Protic media. For nearly three decades following the work of Treibs[155-161], further work on the BCA of enones was essentially abandoned. The obvious reasons were the low yields and the complicated reaction mixtures. In the mid 1950s and early 1960s, the research groups of Doering and Barton[168-171] reported on the utility of the non-nucleophilic strong base t-butoxide (commonly dissolved in t-butanol) for carrying out BCA reactions. When applied to enone systems, the reaction yields improved somewhat and the products (ketones, aldehydes, alcohols or acids) obtained could be readily rationalized in terms of the expected hydroperoxides. Nevertheless, the yields were generally below 50% and it was therefore difficult to be sure as to the true course of the overall reaction.

Camerino, Patelli and Sciaky[172,173] using t-butoxide in t-butanol carried out extensive studies on the base-catalyzed oxygenation of various steroids including the 3-oxo-Δ^4 system (155, equation 64). They reported low yields of 4-hydroxy dienone 158 and enedione 105 (see Table 2) which clearly result from the oxygenations of the α(C-4) and γ(C-6) carbons of the thermodynamic enolate 125, followed by Kornblum–DeLaMare dehydration of the resulting hydroperoxides 156 and 101, 102 (equation 64). Other groups have found similar results under slightly different BCA conditions[174,175] (see Table 2).

Camerino and coworkers[172,173] also found that $\Delta^{1,4}$-3-oxosteroids react in a similar fashion. $\Delta^{1,4}$-Pregnadien-11β-ol-3-one-BMD yielded the corresponding 4-hydroxy-1,4,6-triene-3-one in a 35% yield. Holland's group[176,177], on the other hand, found that $\Delta^{1,4}$- and $\Delta^{4,6}$-dien-3-ones were unreactive towards oxygen when the BCA is mediated by Na_2O_2 in aqueous ethanol. Various 4-chloro-3-oxo-Δ^4 steroids (159) also undergo BCA oxidation with t-butoxide in t-butanol yielding 158. A probable mechanism is shown in equation 65.

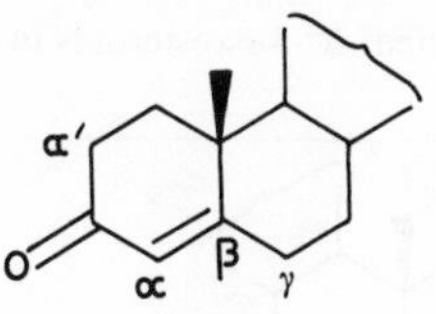

(155)

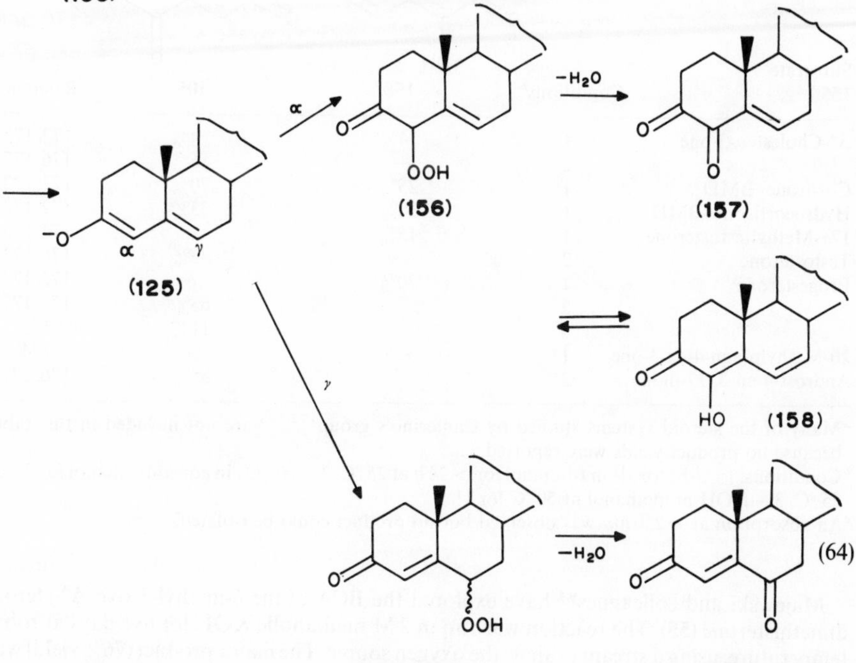

(125)

α → (156)

−H₂O → (157)

(158)

γ → (101, 102)

−H₂O → (105) (64)

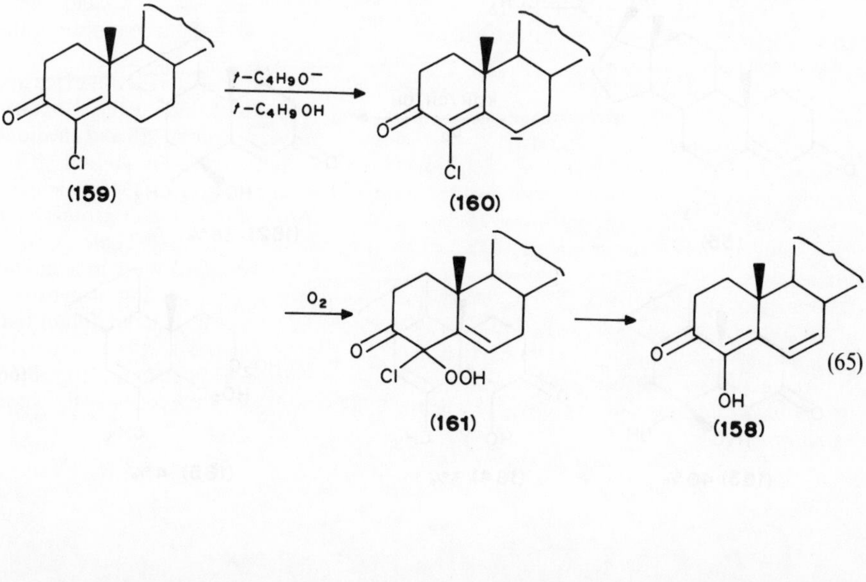

(159)

t−C₄H₉O⁻
t−C₄H₉OH

→ (160)

O₂ → (161)

→ (158) (65)

TABLE 2. Product distribution in the base-catalyzed autoxidation of selected[a] Δ^4-3-oxosteroids in protic media

Substrate 155	Conditions[b]	158	105	Reference
Δ^4-Cholesten-3-one	1	14%	4%	172, 173
	2	—	56%	176, 177
Cortisone–BMD	1	25%	70%	172, 173
Hydrocortisone–BMD	1	30%	33%	172, 173
17α-Methyltestosterone	1	15%	c	172, 173
Testosterone	2	—	43%	176, 177
Progesterone	1	20%	c	172, 173
	2	—	63%	176, 177
	3	—	11%	175
20-Methylpregn-4-en-3-one	1	30%	—	174
Androst-4-en-3, 17-dione	2	—	55%	176, 177

[a]Many of the steroid systems studied by Camerino's group[172,173] are not included in this table because no product yields were reported.
[b]Conditions: 1—t-butoxide in t-butanol for > 24 h at 25 °C. 2—Na$_2$O$_2$ in aqueous ethanol for 2 h at 25 °C. 3—KOH in methanol at 50 °C for $\frac{1}{2}$ h.
[c]An absorption at ∼250 mμ was observed but no product could be isolated.

Majewski and colleagues[94] have explored the BCA of the 6-methyl-3-oxo Δ^4-steroid dimethisterone (**55**). The reaction was run in 2 M methanolic KOH for five days at room temperature, using a stream of air as the oxygen source. The major product (76% yield) was a 1:1 mixture of the epimeric 6-hydroxy analogs **162** and **163**. The 4-hydroxy dienone **164** and diacid **165** were also isolated in low yields (equation 66).

$$\text{(66)}$$

As in the case of the unsubstituted 3-oxo-Δ^4 steroids (equation 64), the primary products under protic conditions are the 4- and 6-hydroperoxides. In the present case, however, Kornblum–DeLaMare dehydration of the latter is precluded; hence homolytic cleavage (Section III.C.1) leading to alcohols **162** and **163** is observed. The 4-hydroperoxide yields **164** (via Kornblum–DeLaMare dehydration and enolization) and diacid **165**, via an oxidative cleavage typical of α-hydroperoxy ketones (Section III.C.4).

In a non-steroidal system, Gersman and colleagues[90b] found that BCA of α, β-unsaturated ester **48** (as well as its β, γ-unsaturated analog **50**) results in a 25% yield of the α-product (**166**, equation 67). This is in contradistinction to free radical autoxidation where only γ-oxidation product **49** is isolated (equation 33).

$$(67)$$

α-*Oxidation* is strongly preferred in many, if not most, cases over γ-oxidation products. This is true, for example, for pulegone (**63**)[178], fukinone (**64**)[178] and dialkylmaleic anhydrides (**169**)[179] (equations 68–70).

$$(68)$$

$$(64) \quad\quad 7.5\% \quad\quad 75\% \quad\quad (69)$$

$$(169) \quad\quad\quad\quad\quad\quad (70)$$

γ-*Oxidation products*, on the other hand, are preferred for epi-α-cyperone (**51**), its dihydro analog **170** (equations 71 and 72)[91], and for butenolide **172** (equation 73)[89]. In the latter case even the weak base triethylamine works efficiently.

$$51 \xrightarrow{\text{KOH/O}_2} 52b \longrightarrow 52a \quad\quad (71)$$

$$(170) \quad\quad\quad\quad (171)\ 45\%\ \text{yield} \quad\quad (72)$$

$$(172)$$

$$(174)\ 50\%$$

$$(173)\ 45\% \quad\quad (73)$$

β-Oxidation products are also observed in the case of diosphenols, i.e. 2-hydroxy-2-en-1-ones[31,61-66,180-185]. Deprotonation of the acidic enol followed by oxygenation at the enolate carbanion results in the formation of a 3-hydroperoxy-1, 2-diketone system (Scheme 12, path a). The latter decomposes as discussed previously (Section III C.4) to CO, a carboxylic acid and a carbonyl group. In the case of cyclic systems, the acid and carbonyl moieties often cyclize to a lactol[31,61-66,180-185]. In addition, since 2-hydroxy-2-en-1-ones are merely the enolic form of α- diketones, it should not be surprising that a benzil–benzilic acid rearrangement (yielding an α-hydroxy acid) often competes in these base-catalyzed processes (Scheme 12, path b)[170,186-191]. As a general rule, nucleophilic bases (HO⁻, CH₃O⁻, C₂H₅O⁻) favor hydroxy acid or ester formation, while the stronger base t-butoxide favors lactol formation.

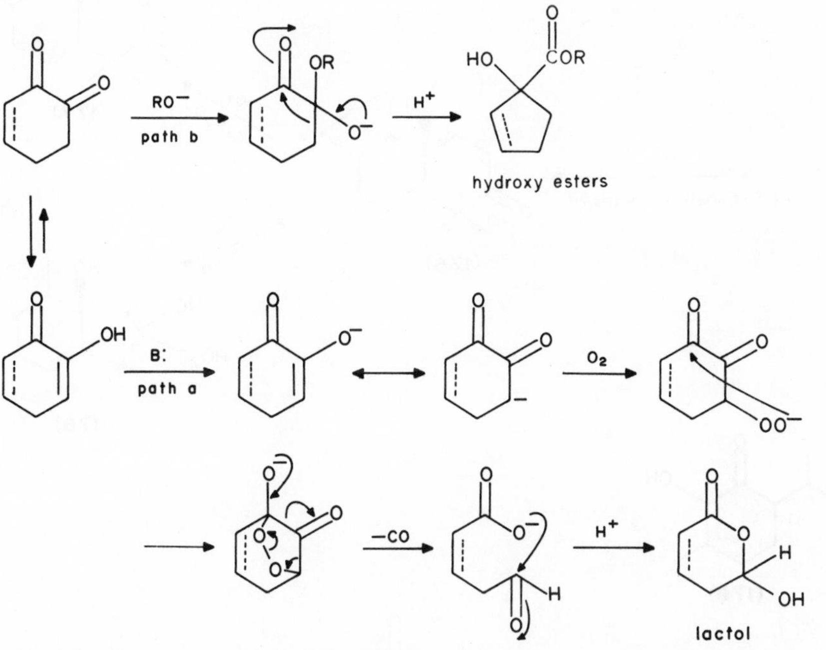

SCHEME 12. *β*-Oxidation and benzilic acid rearrangement of 2-hydroxy-2-en-1-ones

An example of these transformations was reported by Hanna and Ourisson[61a,181], who studied the t-butoxide mediated autoxidation of 4, 4-dimethyl-Δ⁵-cholestenone (175) which yields lactol 177 via the corresponding enol 176 (equation 74). The latter can be isolated and, when treated with ethoxide in ethanol, yields α-hydroxy acid 178 rather than lactol 177.

A more recent example[64] is the BCA of 2-hydroxypiperitone (179) which, under micellar catalysis, yields an acyclic keto acid as the major product (equation 75, path a). In the absence of micellar material, several acidic by-products are formed, some of which presumably involve benzil–benzilic acid rearrangements (equation 75, path b). We will see several more examples in Section III.E.3.b.

At this juncture, we should discuss briefly the BCA of ascorbic acid (92, AH₂). We have already noted (Section II.D.2.b) that the autoxidation of 92 at neutral pH proceeds via the

(74)

(75)

ascorbate ion (AH^-, equation 45) and that the primary role of O_2 is that of an electron acceptor. In general, there is a more rapid uptake of oxygen at basic pH values than at neutral or acidic value with the oxidation product being the triketone dehydroascorbic acid (A). Recently, Afanas'ev and his colleagues[124] reported that the rate of ascorbate anion oxidation in aqueous solution is independent of pH (at pH 6–10) and is completely inhibited by EDTA. This suggests, then, that metal (Fe^{+3}) catalyzed oxidation is the primary mode of reaction in aqueous solution (equation 76).

$$AH^- + Fe^{3+} \longrightarrow AH^{\cdot} + Fe^{2+} \xrightarrow{O_2} A + HO_2^- + Fe^{3+} \tag{76}$$

Presumably the highly hydrated ascorbate is not able to transfer an electron directly to molecular oxygen. In acetonitrile, on the other hand, the solvent apparently forms an unreactive complex with Fe^{+3} ion and inhibits the catalytic process. As a result, only an uncatalyzed direct electron transfer to dioxygen occurs (equation 45).

α'-*Oxidation* of an α, β-unsaturated enone occurs when one of the following conditions is fulfilled: (1) when there are no abstractable γ-hydrogens; (2) when the α' carbon is already partially oxidized; (3) when the reaction is under kinetic control.

An example of the first category is the oxidation of the carotenoids canthaxanthin (**180a**)[192] and astaxanthin (**180b**)[193] which proceeds under base catalysis to yield astacene (**181**, equation 77).

(**180**)

(**a**) R = H

(**b**) R = OH

$$O_2 \Big| \begin{array}{l} t\text{-}C_4H_9O^- \\ t\text{-}C_4H_9OH \end{array} \tag{77}$$

(**181**)

Similarly, Kreiser and Ulrich[194] report that lanosterols **182a–c**, which lack γ hydrogens, are readily converted in 80–100% yields to the corresponding diosphenol **183** (equation 78).

In the second category, we can include the oxidation of a series of α- and β-2-hydroxy and 2-acetoxytestosterones (**184**, equation 79) to the corresponding enols **185**[195,196]. Aqueous alcohol media and various bases (KOH, NaOH, $KHCO_3$, K_2CO_3) have been used to effect this transformation which proceeds in high yield at the α'-carbon, despite an abstractable γ-hydrogen. This is undoubtedly due to the fact that the electron-withdrawing hydroxy group stabilizes the adjacent carbanion[197].

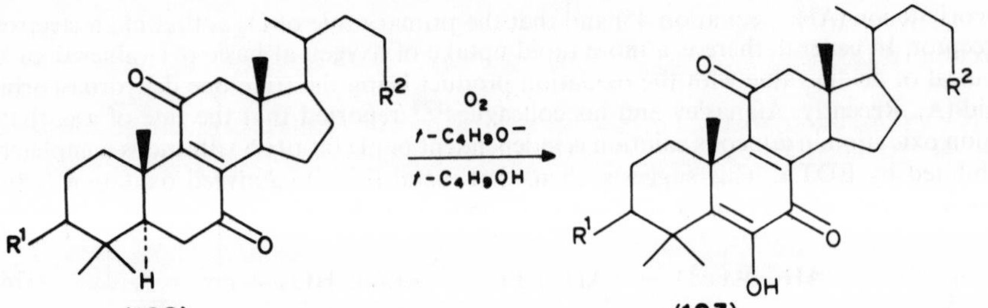

(78)

(a) $R^1 = OH$, $R^2 = CH_2CH(CH_3)_2$

(b) $R^1 = =O$, $R^2 = CH_2CH(CH_3)_2$

(c) $R^1 = OH$, $R^2 = CO_2H$

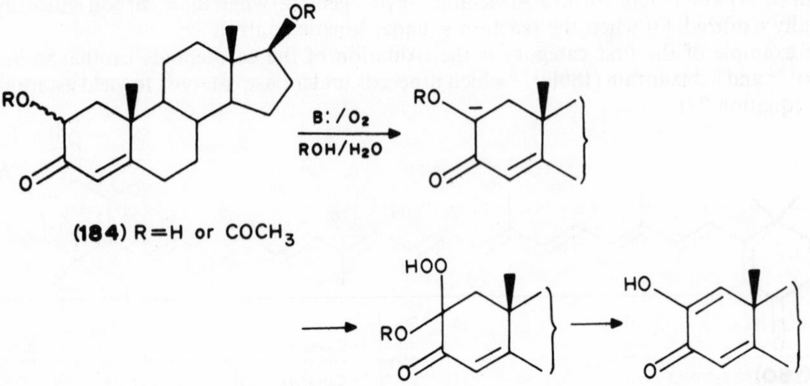

(184) R=H or COCH$_3$

(79)

(185) 70—85%

Kinetic control as a factor in directing oxidation towards the α′ carbon will be discussed in Section III.E.3.b.

We have thus far reviewed α, β, γ and α′ oxidation in the enone system. Gardner and coworkers[198] report that in the case of progesterone, oxidation of these positions competes with oxidation at C-17 resulting in a 'gummy product'. However, in the $\Delta^{1,4}$-analog 186, neither α nor γ oxidation is observed; the major product results from C-17 hydroperoxidation, yielding sterol 187 upon triethylphosphite reduction (equation 80).

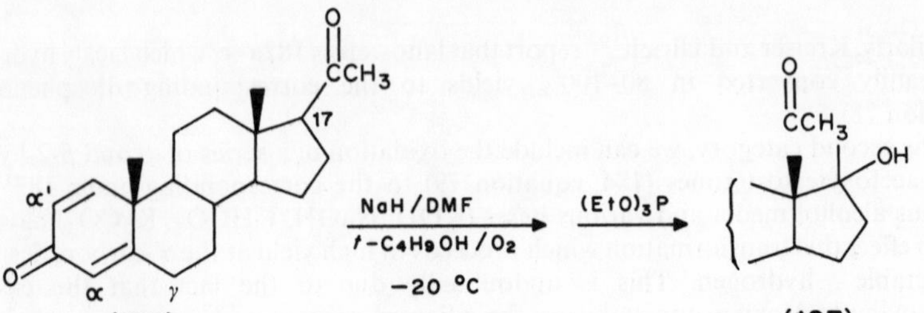

(80)

The oxygenation of 3-hydroxyflavones in protic media will be discussed at the end of Section III.E.3.b.

b. Aprotic media. By the 1970s, chemists had discovered that crown ethers and phase transfer agents would enable them to solubilize a whole variety of inorganic bases even in non-polar aprotic media such as benzene. The BCA reactions of enones carried out in aprotic media proved to give cleaner reaction mixtures in higher yields; what is more, they were easier to control. One of the new bases explored was superoxide anion radical $[O_2^{-\cdot}]$, commonly generated from potassium superoxide $[KO_2]$ and 18-crown-6 polyethers. We will discuss this and some of superoxide's other properties in Section V; meanwhile, let us simply note that the base strength of $O_2^{-\cdot}$ in aprotic media is qualitatively less than *t*-butoxide but greater than hydroxide[27b,d].

Frimer and coworkers[27b,68,164,165] studied the superoxide, *t*-butoxide and hydroxide mediated oxidation of variously substituted cyclohex-2-en-1-ones. 4,4,6,6-Tetrasubstituted cyclohexenones **188** are totally inert to hydroxide and superoxide even after prolonged reaction times (equation 81). This is not surprising, of course, since **188** lacks abstractable acidic hydrogens.

$$\text{(188)} \xrightarrow[>72 \text{ h}]{\text{HO}^- \text{ or } O_2^-} \text{no reaction} \qquad (81)$$

In the case of 6,6-disubstituted cyclohexenones **189**, epoxides **190**, acids **191**, aldehydes **192**, dimers **193** and ketones **194** are the isolated products, with the product distribution depending on the nature of the substituents (equation 82). When the BCA of **189a** is mediated by *t*-butoxide in toluene at $-40\,^\circ\text{C}$, the two major products are epoxide **190a** and hydroxyacid **195** (equation 83).

$$\text{(189)} \xrightarrow[\text{toluene}]{\begin{array}{c} O_2^- \text{ or HO}^- \\ O_2/25\,^\circ\text{C} \end{array}} \text{(190)} + \text{(191)} + \text{(192)}$$

(189)
(a) R=Ph
(b) R=CH$_3$
(c) R=H

(190)   (191)   (192)

$$+ \quad \text{(193)} \quad + \quad R_2CO \quad + \quad \text{trimer} \qquad (82)$$

(194)

$$\text{(189a)} \xrightarrow[\substack{O_2/-40\ ^\circ C \\ \text{toluene}}]{t\text{-}C_4H_9O^-} \text{(190a)} + \text{(195)} \qquad (83)$$

The mechanism proposed for both these transformations is outlined in Scheme 13. Following initial γ-proton removal, condensation of the resulting anion with starting material ultimately produces dimer **193**, while oxygenation generates hydroperoxide, **196**. The latter can epoxidize the substrate, yielding **190**, or decompose to enol **198**. As noted previously, α-ketoenol **198** can undergo either benzil–benzilic acid rearrangement to α-hydroxyacid **195** or oxidation to the lactol **199**. We speculate that this lactol loses CO_2 generating α-hydroperoxy ketone **200** which cleaves to aldehyde **192**. Oxidation of the latter to the corresponding acid **191** is a facile process.

Both 4,4- and 5,5-disubstituted cyclohexenones (**201** and **202** respectively) yield the corresponding enols **204** in generally high yields (equation 84). In the case of **201** it is the α' hydrogen that is removed, since the γ position is blocked. Oxygenation ultimately yields the diketone **203**, which in turn enolizes to **204**. In the case of **202**, the γ hydrogen is preferentially removed generating the thermodynamic enolate (see Scheme 10). The latter is oxygenated α to the carbonyl, leading again to diketone **203** and enol **204**. It should be noted that these enols can be further oxidized under the reaction conditions to the corresponding lactols **205** which, upon $NaBH_4$ reduction, yield lactones **206** (equation 85). Indeed, Frimer and Gilinsky[68] have been able to convert enones to lactols in a one-pot reaction followed by reduction of the lactols to the corresponding δ-valerolactones (**206**) in overall yields approaching 85%.

$$\underset{\text{(201a—d)}}{} \quad \text{or} \quad \underset{\text{(202 a,c,e,f)}}{} \xrightarrow[\text{or } t\text{-}C_4H_9O^-]{O_2^{-\cdot},\ HO^-} \underset{\text{(203 a—f)}}{} \rightleftharpoons \underset{\substack{\text{(204 a—f)} \\ 75\text{—}95\%}}{} \qquad (84)$$

(a) $R^1 = Me,\ R^2 = H$

(b) $R^1 = Ph,\ R^2 = H$

(c) $R^1 = Me,\ R^2 = OEt$

(d) $R^1 = Me,\ R^2 = OMe$

(e) $R^1 = R^2 = Me$

(f) $R^1, R^1 = -(CH_2)_5-,\ R^2 = OMe$

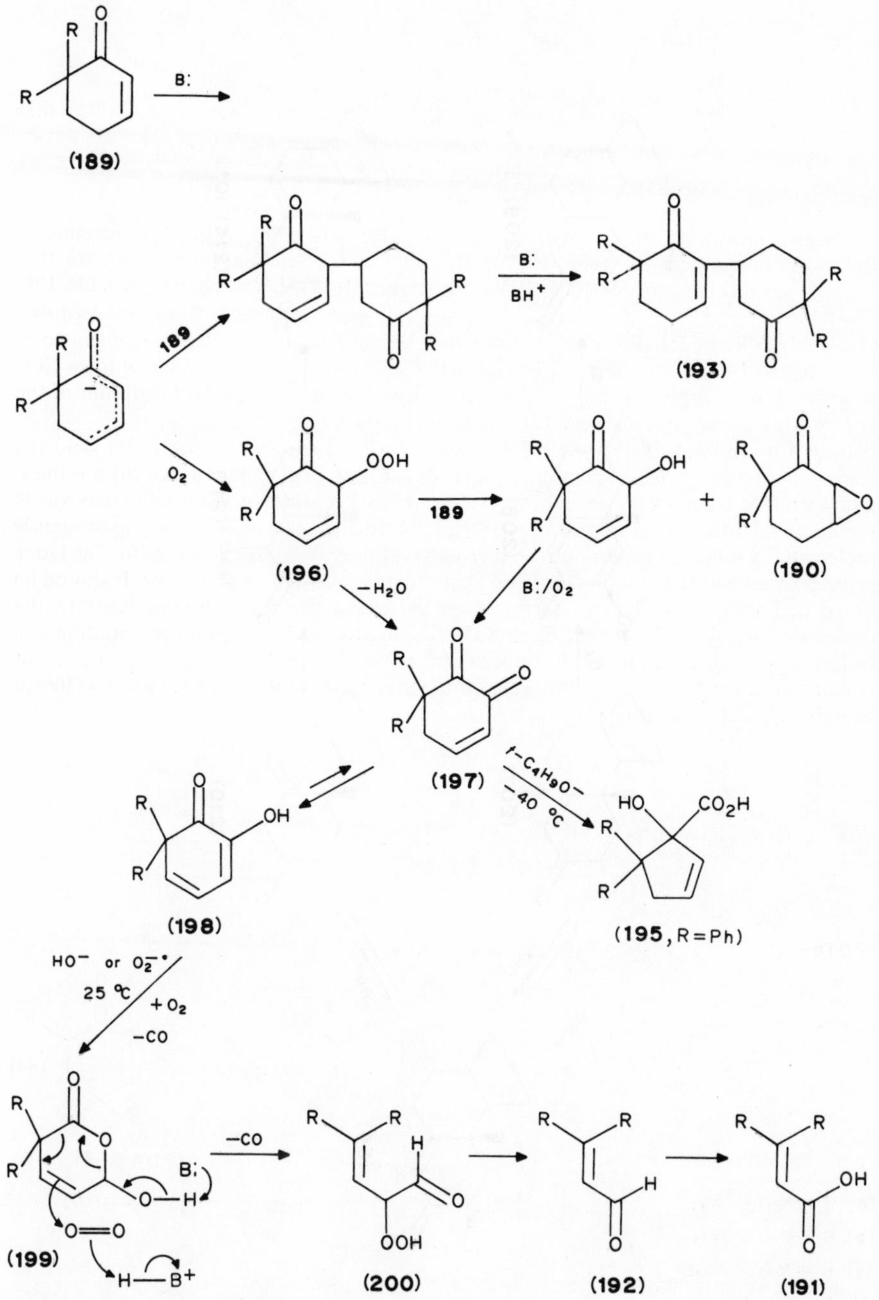

SCHEME 13. Base-catalyzed autoxidation of cyclohexenones with $t\text{-}C_4H_4O^-$ (at $-40\,°C$) and $O_2^{-\cdot}$ or HO^- (at $25\,°C$)

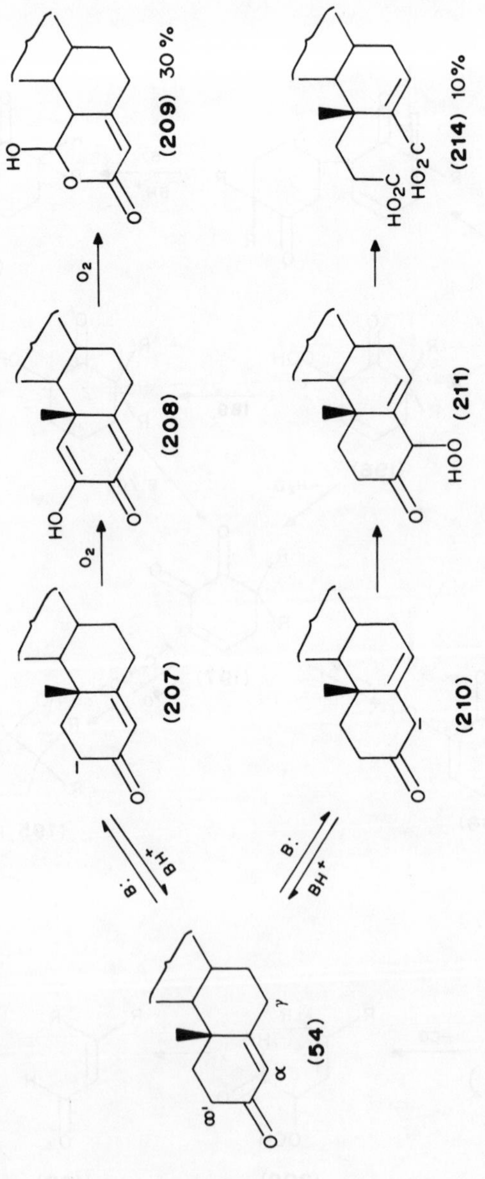

SCHEME 17 Base-catalysed autoxidation of cyclohexenones with molecular O_2 (or with KO₂ and O₂; see ref. 200).

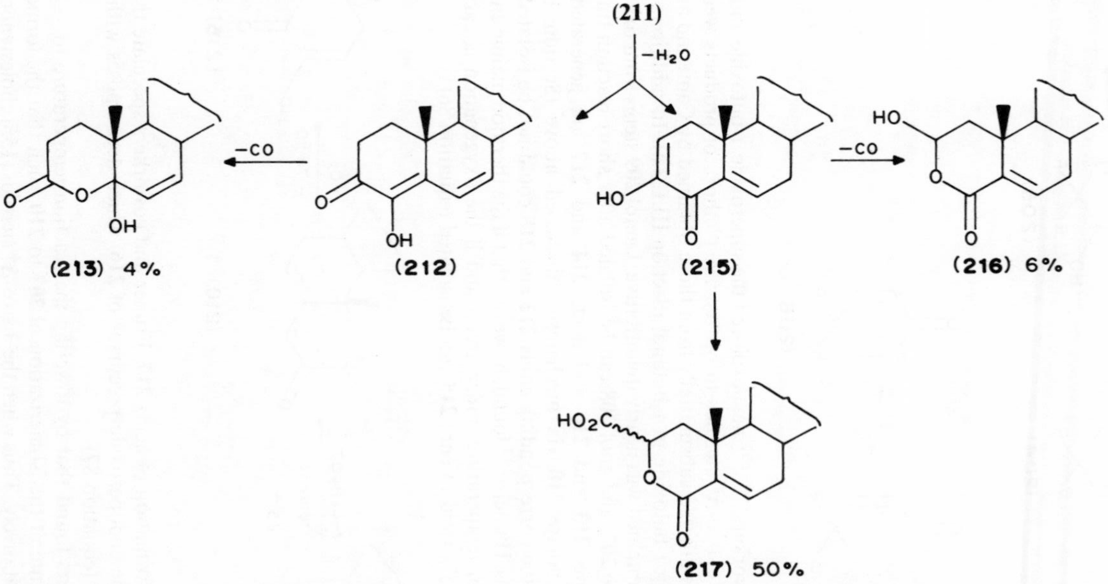

SCHEME 14. Superoxide-, hydroxide- and *t*-butoxide-catalyzed autoxidation of Δ^4-cholestenone[54] in benzene

(85)

Frimer and coworkers[199-202] next explored the superoxide, *t*-butoxide and hydroxide mediated BCA of 3-oxo-Δ^4 steroids in benzene. A plethora of products were obtained (Scheme 14) which differed substantially from those obtained by Camerino and collaborators[172,173] using *t*-butoxide in *t*-butanol (Section III.E.3.a). In addition, the overall product yield so obtained was nearly quantitative. Lactol **209** stems from oxygenation of the kinetic enolate **207**, and enol **208** can be isolated after short reaction times. On the other hand, lactols **213** and **216** and acids **214** and **217** are generated from the thermodynamic enolate **210** via mechanisms discussed above (Sections III.C.4 and III.E.3.a) though the corresponding enols **212** and **215** could not be isolated under the reaction conditions. Hameiri[202] found, however, that if the thermodynamic enolate **210** is generated at room temperature under argon and if the oxygenation is carried out at −78 °C, then a 15% yield of enol **215** can be isolated (equation 86).

(86)

Regarding the formation of acids **217**, Frimer and coworkers speculate that they are generated from the endoperoxide precursor of **216**, which decomposes without loss of carbon monoxide (equation 87).

These researchers found that by lowering the reaction temperature to −20 °C, they could essentially inhibit the isomerization of **207** to **210**, such that the former could be oxygenated quantitatively. Thus when the 3-oxo-Δ^4 steroids (**155**) cholestenone, testosterone, 17α-methyltestosterone, 17α-hydroxyprogesterone, progesterone, cortisone–BMD and cortisolone–BMD are autoxidized with *t*-butoxide in toluene at −25 °C for 1.5–4 h, enol **208** can be isolated in yields of 85–95%. If instead of quenching the reaction to isolate the enol, the reaction is allowed to continue at room temperature for 1–3 days, lactol **209** can be obtained in similar yields. NaBH₄ reduction of lactols **209** yields the corresponding therapeutically active 2-oxa-3-oxo-Δ^4 steroid lactones **218** (equation 88)[201].

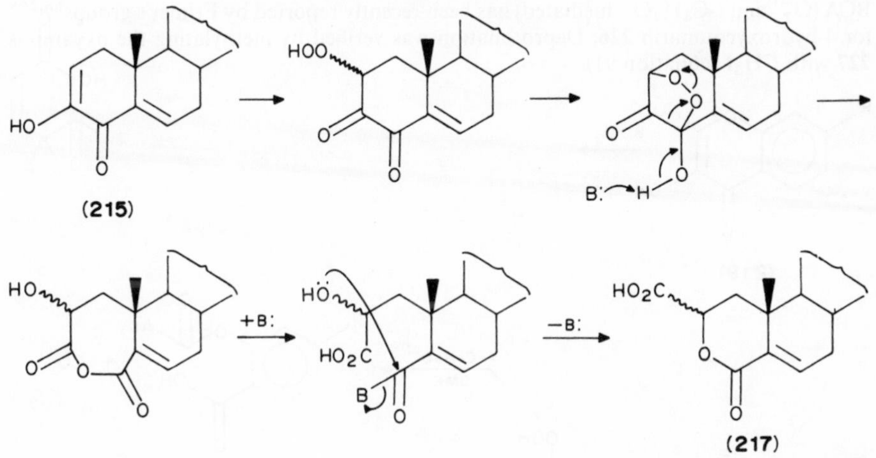

(215)

(217)

(87)

(155)

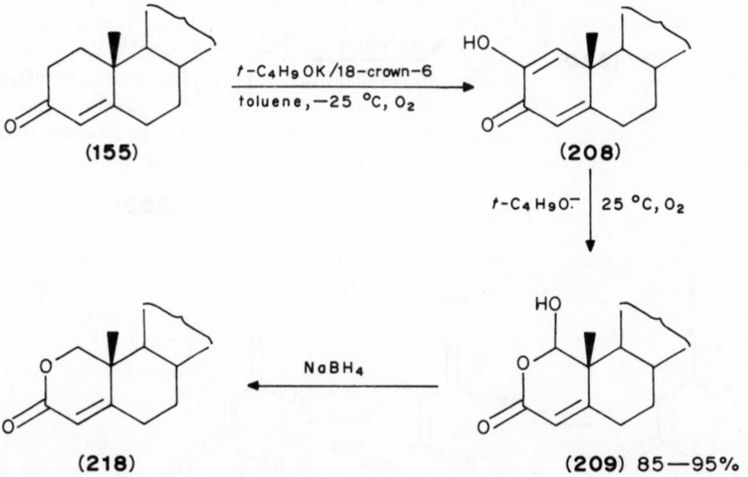

$t\text{-}C_4H_9OK/18\text{-crown-6}$
$\text{toluene},\ -25\ ^\circ C,\ O_2$

(208)

$t\text{-}C_4H_9O^-$ | 25 °C, O_2 (88)

(218)

NaBH₄

(209) 85—95%

Related to the steroidal diosphenols are the 3-hydroxyflavones **219** whose biological role and reactions will be discussed in Section III.F.2. This class of compounds undergoes rapid *t*-butoxide mediated BCA in DMF or DMSO yielding depside **221** and carbon monoxide[62,63,65]. In protic media (H_2O–NaOH or MeOH–MeONa) or in the case of superoxide mediated BCA[180,184,185] the oxidation proceeds slowly to give a mixture of the depside **221** and its solvolysis products **222** and **223** (equation 89).

c. *Miscellaneous*. We have cited above (Section III.D.2.b) that the radical autoxidation of acyclic 1,3-diketones with a tertiary C_2-carbon yields the corresponding 2-hydroxyperoxy-1,3-diones[107]. Interestingly, although dibenzoylmethane (**224a**), 1,3-cyclohexadione (**224b**), ethyl acetoacetate (**224c**) and diethyl malonate (**224d**) are all easily deprotonated by a variety of bases (including superoxide anion), the resulting diketo carbanions **225** are stable to oxygenation[171,203–207] (equation 90). A similar resistance to

BCA ($O_2^{-\cdot}$ and t-$C_4H_7O^-$ mediated) has been recently reported by Frimer's groups[68,208] for 4-hydroxycoumarin **226**. Deprotonation was verified by methylating the oxyanions **227** with CH_3I (equation 91).

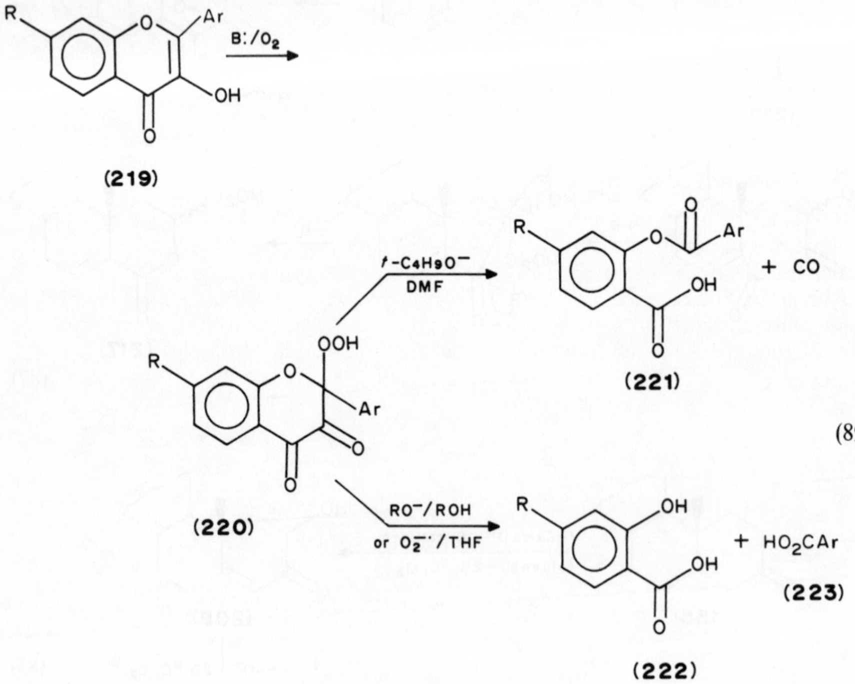

(89)

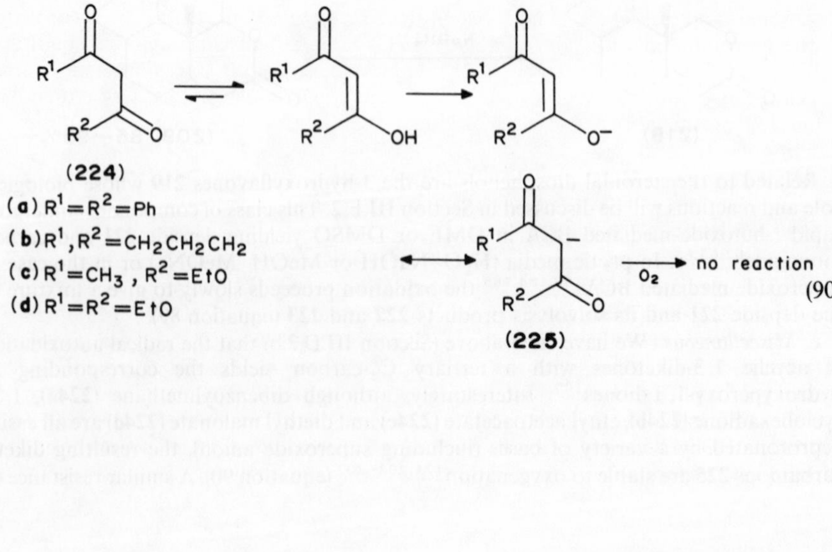

(90)

(224)
(a) $R^1 = R^2 = Ph$
(b) $R^1, R^2 = CH_2CH_2CH_2$
(c) $R^1 = CH_3$, $R^2 = EtO$
(d) $R^1 = R^2 = EtO$

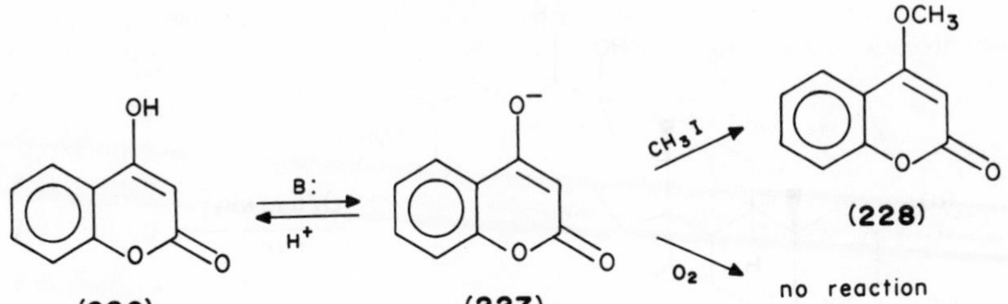

(91)

Nevertheless, as in the radical autoxidation case, once the C_2-atom is alkylated or arylated it is susceptible to oxygenation[103,206,207]. Thus, diethyl 2-methyl-, 2-ethyl- and 2-phenylmalonate all yield products resulting from initial hydroperoxidation at C_2. Furthermore, Young reports[209] that the unsensitized photooxidation of dimedone **229** in basic solution leads to a mixture of products from which one can isolate the monomethyl ester of glutamic acid **230**. A likely mechanism is outlined in equation 92.

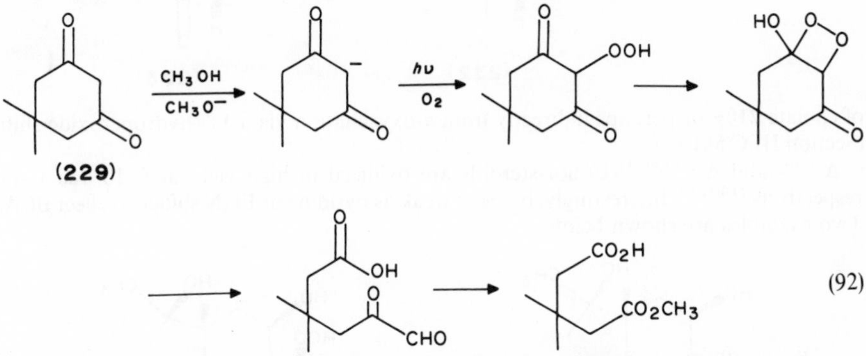

(92)

Canonica and colleagues[210] find that 14-hydroxy-7-en-6-keto steroid **231** is converted to the corresponding 14α-hydroperoxide **233** under reductive elimination conditions (lithium metal in liquid ammonia–THF) without the rigorous exclusion of O_2 during workup. These authors suggest that oxygenation proceeds via the BCA of a dienolate anion **232** resulting from the elimination of the C-4 alkoxide group, as outlined in equation 93.

4. Hydroperoxidation of β,γ-enones

In contradistinction to the sluggish reaction of α,β-enones, the BCA of β,γ-enones is a very facile process. Of the latter group, the Δ^5-3-ketosteroidal system has been the most actively investigated (Table 3 and Scheme 15). In aqueous ethanol, Na_2O_2-mediated BCA of Δ^5-cholestenone (**100a**)[211] yields Δ^4-3, 6-dione **105**. In t-butanol[174], on the other hand, the t-butoxide mediated oxidation of Δ^5-cholesten-3-one yields dienol **212** in a 10% yield. Stern[212,213] studied this same BCA in toluene using t-butoxide at $-78\,°C$ and superoxide at $0\,°C$. At the lower temperature, α-oxygenation product **215** is formed exclusively, while γ-products **103** and **104** predominate for the latter conditions. [We have already had the opportunity to speculate whether γ-oxidation products result directly from γ-oxygenation

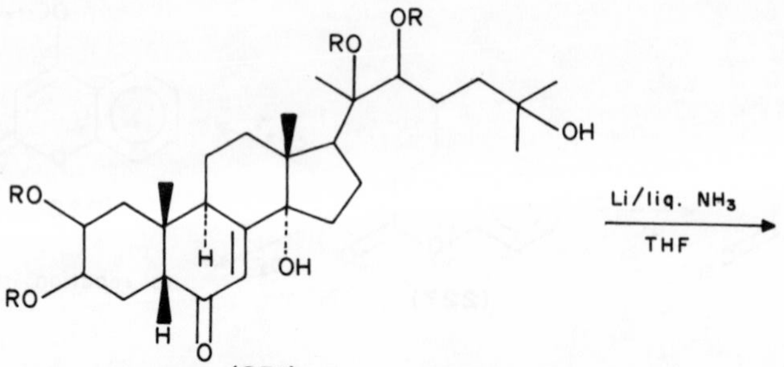

(231)

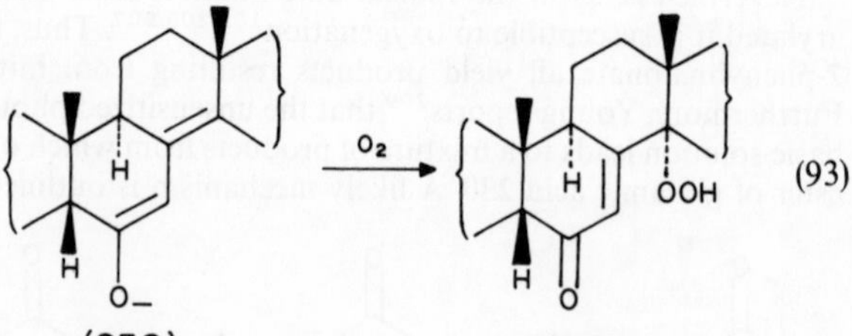

$$\xrightarrow[\text{THF}]{\text{Li/liq. NH}_3}$$

(232) $\xrightarrow{O_2}$ **(233)**   (93)

of enolate **210γ** or perhaps indirectly from α-oxygenation via a 1,3-hydroperoxide shift (Section III.C.5).]

$\Delta^{5(10)}$- and $\Delta^{5(10),9(11)}$-19-nor-steroids are oxidized in high yield at C-10 and C-11 respectively[214-216]. Interestingly, bases as weak as pyridine or Et_3N suffice to effect BCA. Two examples are shown below.

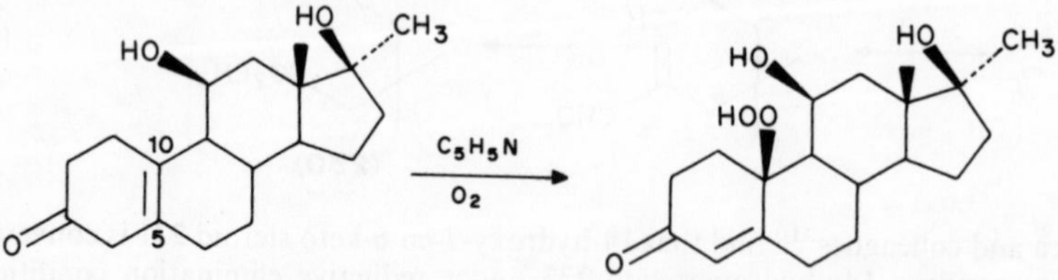

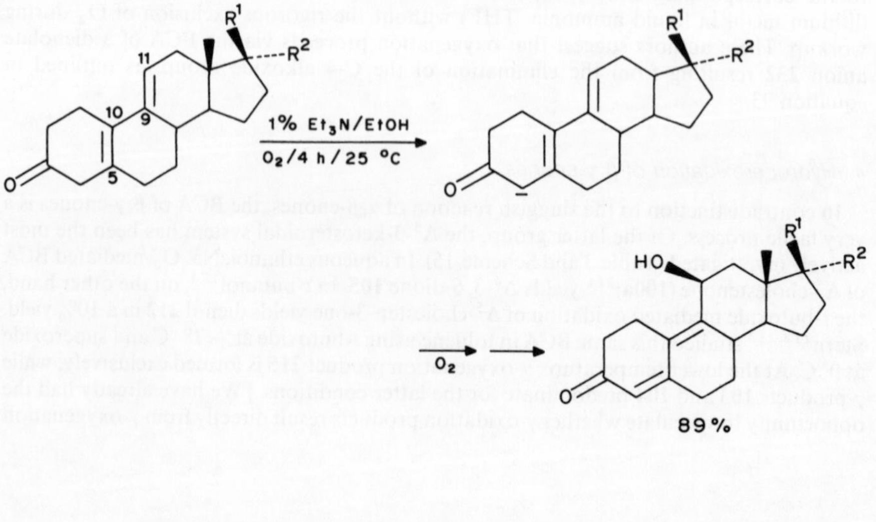

$$\xrightarrow[O_2]{C_5H_5N}$$

$$\xrightarrow[O_2/4\,h/25\,°C]{1\%\ Et_3N/EtOH}$$

$$\xrightarrow{O_2}$$

89%

TABLE 3. Product yields in the base-catalyzed autoxidation of Δ^5-cholestenone

Base:	Na_2O_2	t-butoxide	t-butoxide	superoxide
Solvent:	aq. ethanol	t-butanol	toluene	toluene
Temperature (°C):	25	25	−78	0
Time (h):	2	1.5	1	0.75
215	—	—	100%	31%
104	—	—	—	36%
103	—	—	—	25%
212	—	10%	—	—
105	20%	—	—	—
Reference:	211	174	212	212

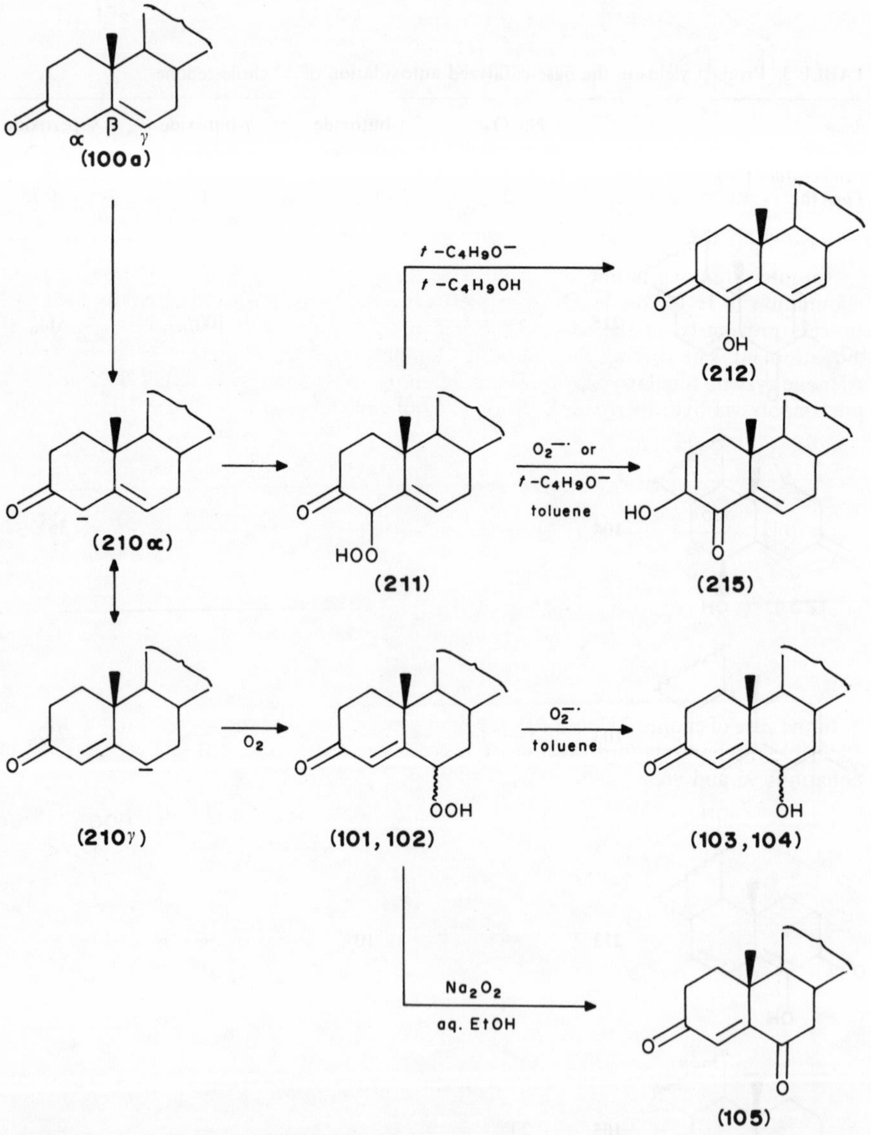

SCHEME 15. Mechanism for product formation in the BCA of Δ^5-cholesten-3-one (100a)

We have already noted above that β,γ-unsaturated ester 50 yields the corresponding γ-oxidation product (equation 67). Similarly, α-safranate undergoes facile t-butoxide mediated BCA in glyme to yield a divinyl methylhydroperoxide and its corresponding dehydration product[217].

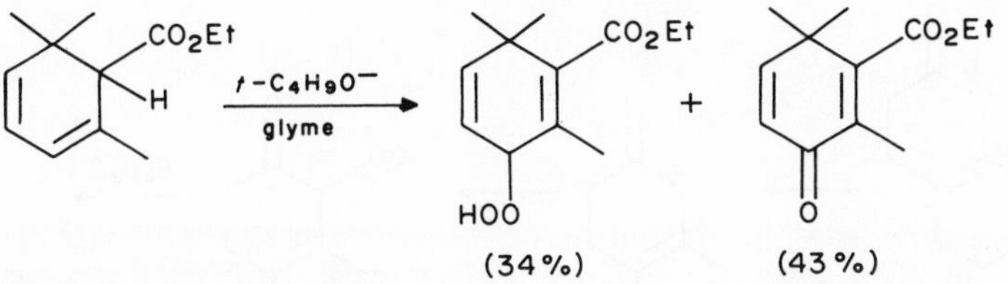

(34%) (43%)

5. Double-bond formation and aromatization

An interesting variation on the theme of hydroperoxidation is the subsequent elimination of H_2O_2 (or H_2O) from the oxidized product. We have already seen this process previously, in the case of the free radical autoxidation of epi-α-cyperone **51** (equation 34). The driving force for the elimination is the formation of a conjugated trienone system. Similarly, 2,4-cycloheptadienone (**234**) upon BCA yields tropone (**236**), presumably via hydroperoxide **235a** or alcohol **235b** (equation 94)[218].

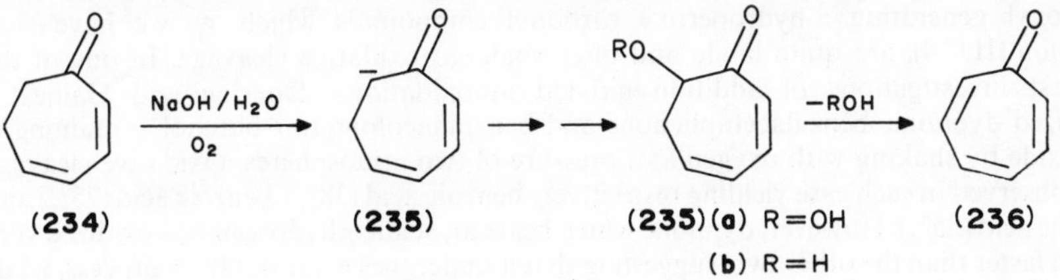

(234) (235) (235)(a) R=OH (236)
(b) R=H

In the case of enones **237** and **238** the final products are the corresponding phenols[219]. Plausible mechanisms (not necessarily those suggested by the authors) are outlined in equations 95 and 96.

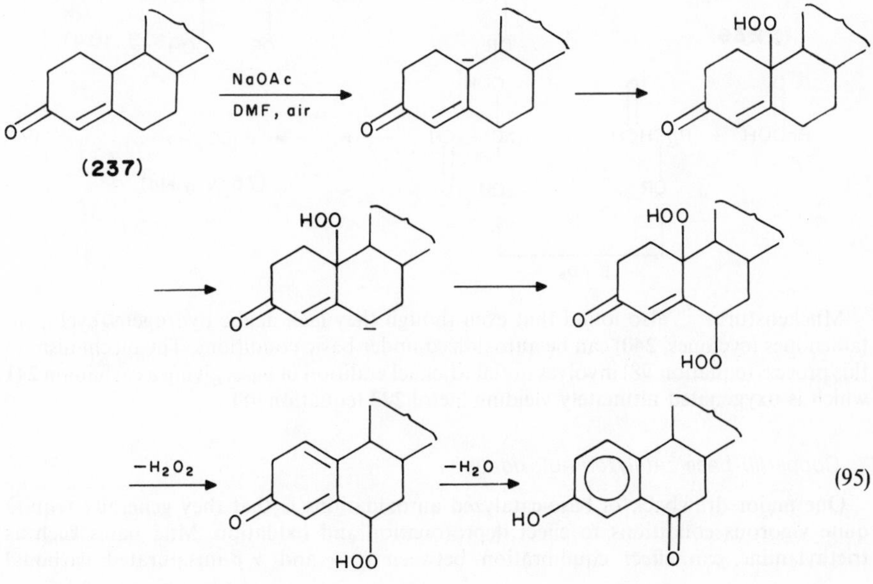

(237)

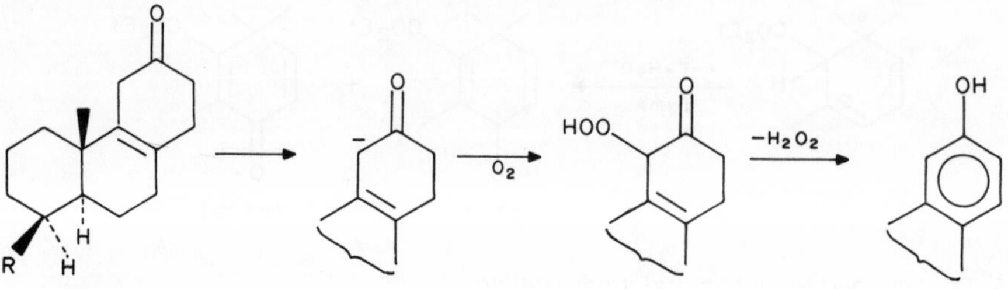

(96)

6. Addition-initiated oxidation

There are several examples in the literature where a BCA process is initiated by Michael addition of the base to an enone system. The resulting enolates are then oxidized α to the carbonyl generating α-hydroperoxy carbonyl compounds which, as we have seen (Section III.C.4), are quite labile and often undergo oxidative cleavage. In one of the earliest investigations of addition-initiated autoxidations, Doering and Haines[171] oxidized dypnone, benzalacetophenone and benzpinacolone in t-butanol containing t-butoxide by shaking with oxygen at a pressure of two atmospheres. Oxidative cleavage was observed in each case yielding respectively benzoic acid (38%), benzoic acid (75%) and pivalic acid (55%). However, dypnone which bears an acidic γ hydrogen was oxidized very much faster than the other two, suggesting that it undergoes a normal BCA process, while the others follow a different autoxidative pathway. The addition-initiated process suggested in the case of benzalacetophenone (239) is outlined in equation 97.

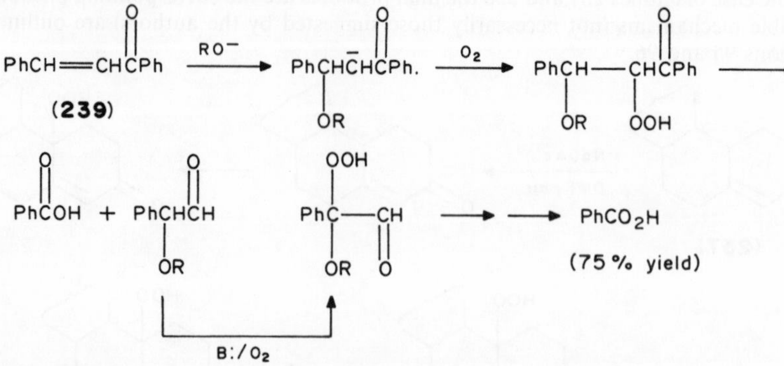

(97)

Muckensturm[220] also found that even though they lack acidic hydrogens, cyclopentadienones (cyclones, 240) can be autoxidized under basic conditions. The mechanism of this process (equation 98) involves initial Michael addition of base, giving a carbanion 241 which is oxygenated ultimately yielding lactol 242 (equation 98).

7. Copper(II)-base catalyzed autoxidations

One major drawback of base-catalyzed autoxidations is that they generally require quite vigorous conditions to effect deprotonation and oxidation. Mild bases such as triethylamine can effect equilibration between β,γ- and α,β-unsaturated carbonyl

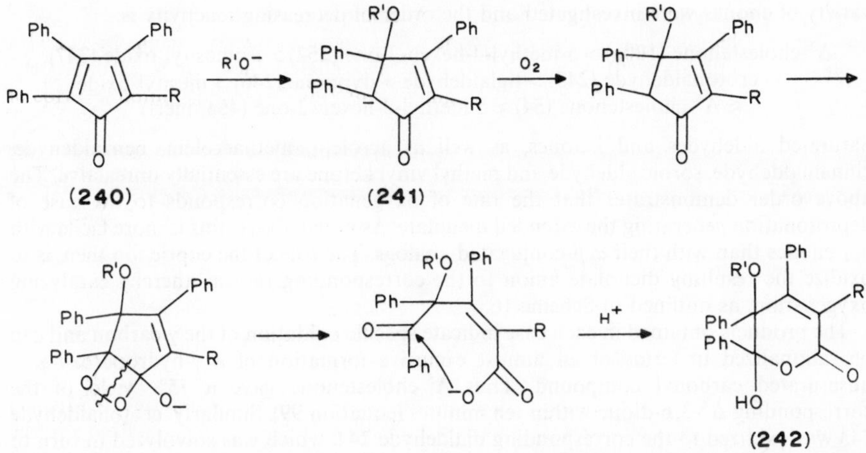

$$(98)$$

compounds[221], but little if any oxidation is observed. Volger and coworkers[222-224] have found, however, that the oxidation of α, β- and β, γ-unsaturated aldehydes and ketones, capable of forming a conjugated dienol, can be effected in mildly alkaline methanolic solutions containing triethylamine and catalytic amounts of cupric-pyridine complexes. A

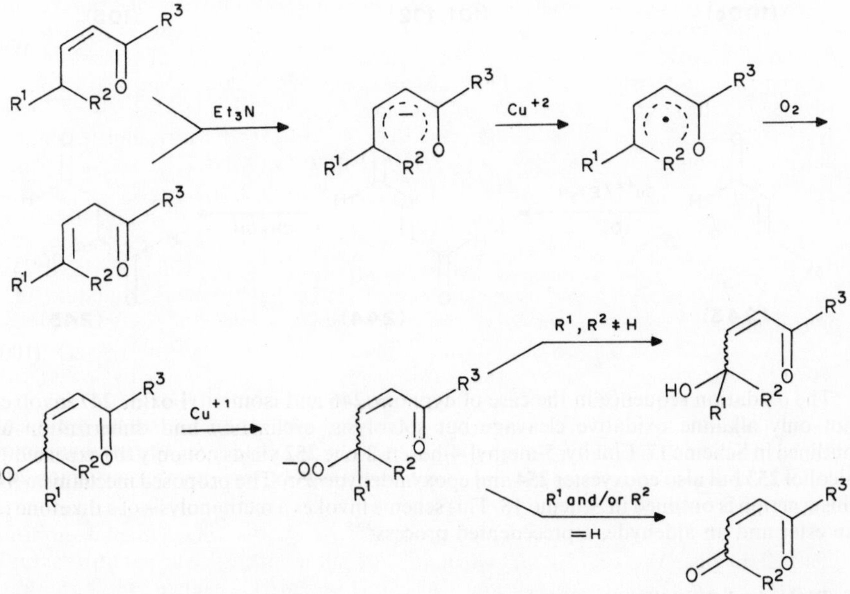

SCHEME 16. General mechanism for autoxidation in the cupric–pyridine–triethylamine–methanol system

variety of enones were investigated and the order of decreasing reactivity is:

Δ^5-cholestenone (**100a**) > 5-methyl-4-hexen-2-one (**252**) > isomesityl oxide (**247**)
> crotonaldehyde (**243**) > tiglaldehyde > dypnone (**246**) > mesityl oxide
$\approx \Delta^4$-cholestenone (**54**) $\approx$ 5-methyl-3-hexen-2-one (**45a**, inert).

Saturated aldehydes and ketones, as well as acrolein, methacrolein, benzaldehyde, cinnamaldehyde, sorbic aldehyde and methyl vinyl ketone are essentially unreactive. The above order demonstrates that the rate of oxygenation corresponds to the ease of deprotonation generating the extended dienolate. As noted above, this is more facile with β,γ-enones than with their α,β-conjugated analogs. The role of the cupric ion then, is to oxidize the resulting dienolate anion to the corresponding radical, thereby catalyzing oxygenation, as outlined in Scheme 16.

The products obtained in each case indicate specific oxidation of the γ-carbon and can be rationalized in terms of an almost exclusive formation of a γ-hydroperoxy-α,β-unsaturated carbonyl compound. Thus Δ^5-cholestenone gave a 75% yield of the corresponding Δ^4-3,6-dione within ten minutes (equation 99). Similarly, crotonaldehyde **243** was oxidized to the corresponding dialdehyde **244**, which was solvolyzed in turn to methoxysuccinaldehyde **245** (62% yield) (equation 100).

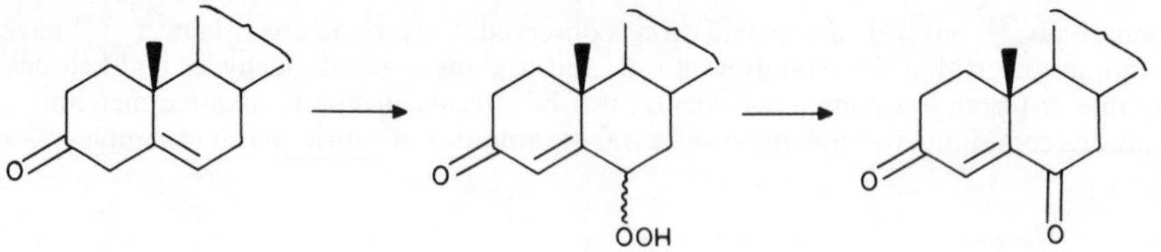

(**100a**)   (**101, 102**)   (**105**)

$$(99)$$

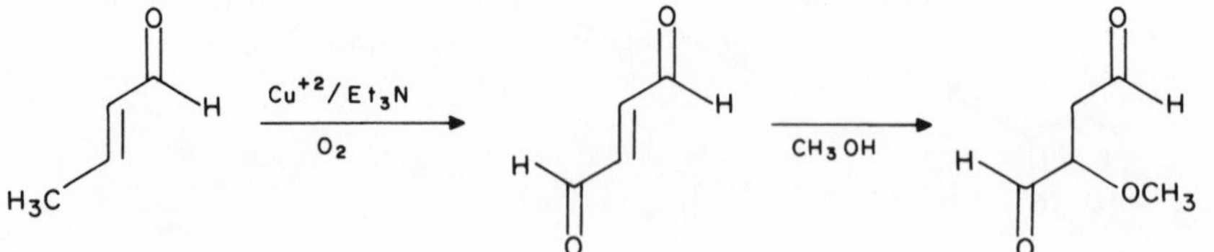

(**243**)   (**244**)   (**245**)

$$(100)$$

The oxidation sequence in the case of dypnone **246** and isomesityl oxide **247** involves not only alkaline oxidative cleavage but solvolysis, cyclization and dimerization as outlined in Scheme 17. Finally, 5-methyl-4-hexen-2-one **252** yields not only the expected γ-alcohol **253** but also epoxyester **254** and epoxyaldehyde **255**. The proposed mechanism for this reaction is outlined in Scheme 18. This scheme invokes a methanolysis of a diketone to an ester and an aldehyde, a precedented process[225].

F. Biological Oxidations

The field of biological oxidations has been encyclopedically reviewed recently[226-230] with the major emphasis on steroids and polyunsaturated fatty acids. A survey of the

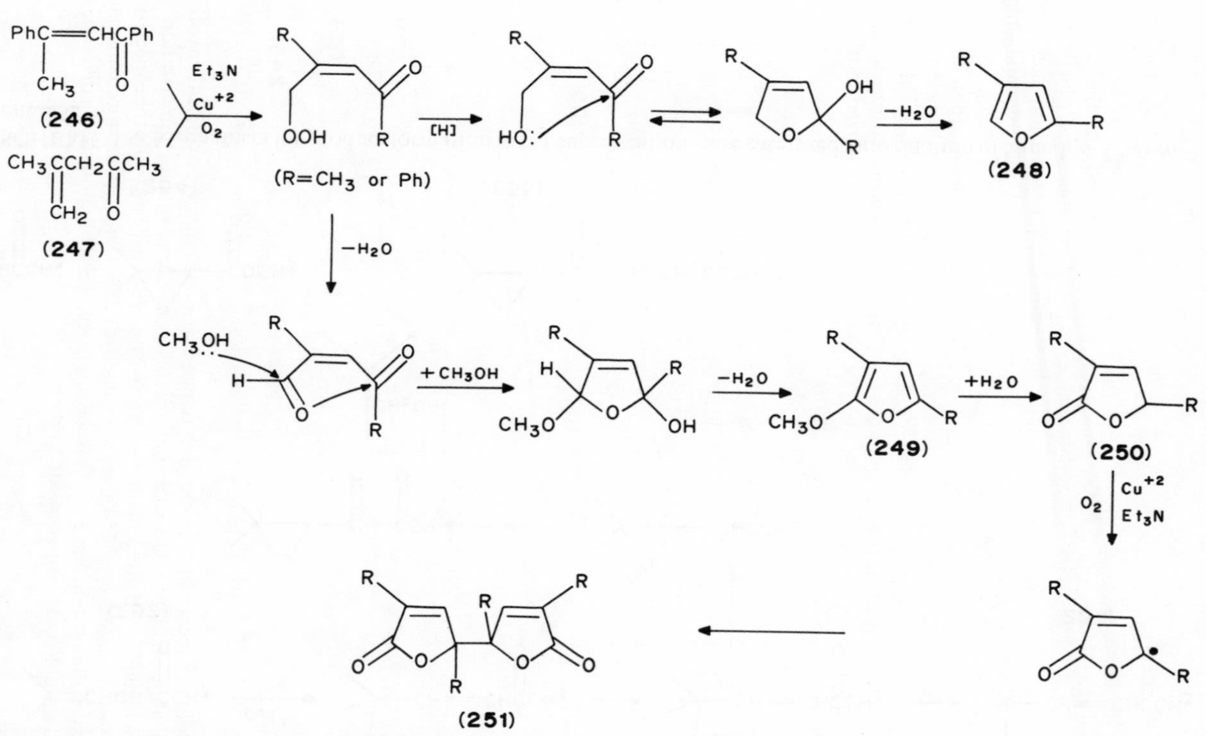

SCHEME 17. Proposed mechanism for the formation of products **248–251** in the Cu^{+2}-catalyzed BCA of dypnone **246** and of isomesityl oxide **247**

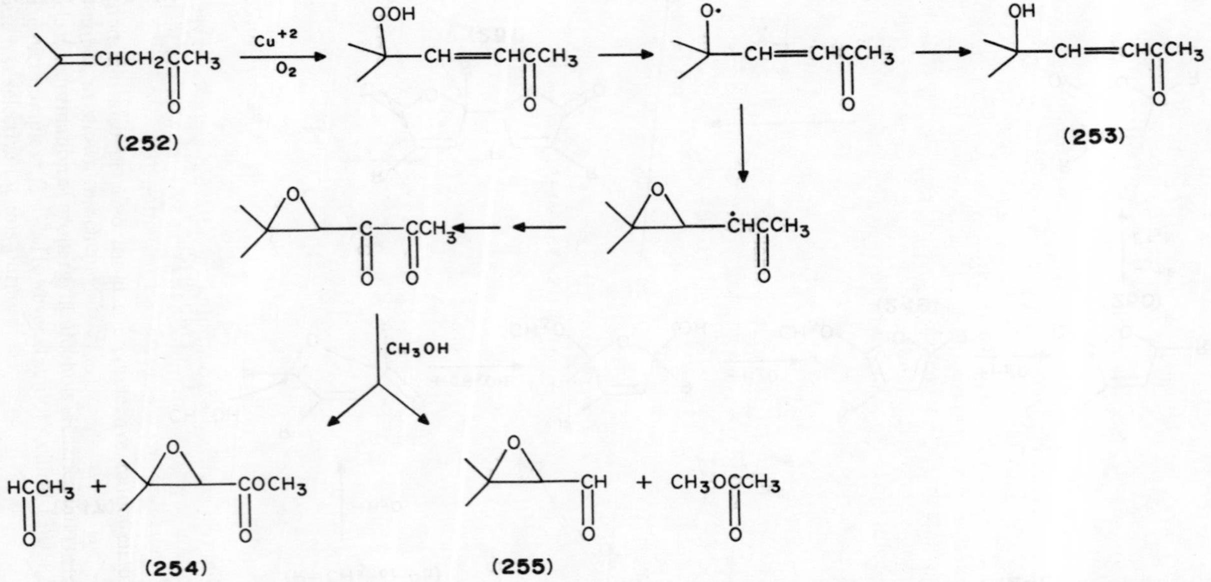

SCHEME 18. Mechanism for product formation in the cupric-cation-base catalyzed autoxidation of 5-methyl-4-hexen-2-one 252

plethora of enone systems investigated is beyond the scope of this review. For the purpose of comparison we will highlight several of the systems discussed elsewhere in this review, namely steroids, cyclohexenones, flavones and chalcones.

1. 3-Oxosteroids and cyclohexenones

a. Microbial hydroxylation. The microbial oxygenation of Δ^4-3-oxosteroids[226-236] results in hydroxylation at C-2β, C-6β, C-11α, C-16α (carbonyl at C-17), C-17α (carbonyl at C-20) and C-21 (carbonyl at C-20). It has been clearly established that these are reactions which involve the direct incorporation of molecular oxygen, but are generally not simple free radical autoxidations. On the contrary, they seem to involve the electrophilic attack of a positively charged hydroxylating species (perhaps HO$^+$) upon an enol or enolate species. Several representative examples follow.

Aspergillus niger effects the C-21 hydroxylation of progesterone[234], while the oxidation of 17-methyltestosterone, testosterone and 4-androstene-3, 17-dione by various species of the fungi *Rhizopus* yields the 11α- and the 6β-hydroxy analogs in a ratio of approximately 4:1 (equation 101).

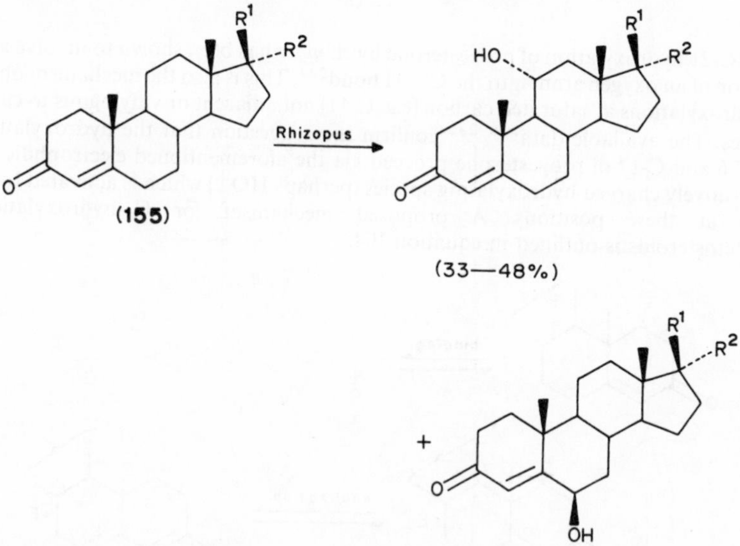

(155)

(33—48%)

+ (101)

(6—10%)

By comparison, the related hexahydronaphthalenone **256** undergoes oxygenation[235] solely at the C-6 carbon (using the steroidal numering system) with the β-epimer **257** predominating over the α (**258**) by a ratio of 13:1 (equation 102).

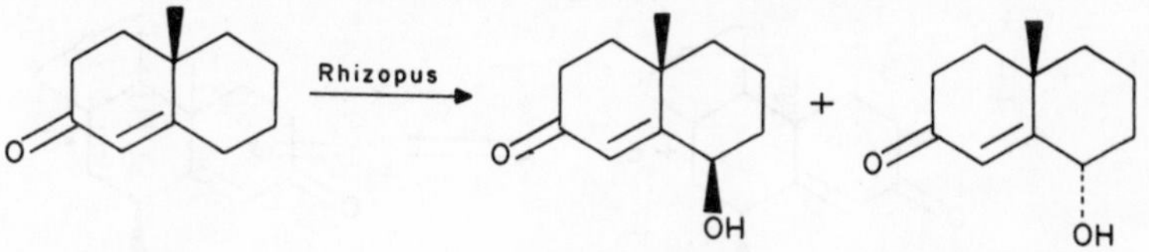

(256)　　　　(257) 22%　　　　(258) 1.7%

In the case of 19-nor-Δ^4-3-ketosteroids (**259**) microbial hydroxylation generally occurs at C-6β, C-10β and C-11α (equation 103)[235].

$$(103)$$

The C-21 hydroxylation of progesterone by *A. niger* has been shown to involve a direct insertion of an oxygen atom into the C—H bond[234]. This is also the mechanism observed for hydroxylations at saturated carbon (e.g. C-11) not adjacent or vinylogous to carbonyl moieties. The available data[226-236] confirm the suggestion that the hydroxylations at C-2, C-6 and C-17 of progesterone proceed via the aforementioned electrophilic attack of a positively charged hydroxylating species (perhaps HO$^+$) which is activated by enolization at these positions. A proposed mechanism for C$_6$-hydroxylation of Δ^4-3-ketosteroids is outlined in equation 104.

$$(104)$$

$\Delta^{5(6)}$-3-ketosteroids have also been reacted with *Rhizopus* species to yield the rearranged Δ^4-analog; hydroxylation at C-6β, C-6α and C-11α, as well as ketone formation at C-6[236]. The formation of the C-6α hydroxylated and ketonic products, unknown in other microbial oxidations but observed in the absence of fungus, as well as other evidence, suggests[236] that in this instance enzymic and non-enzymic processes are competing. Furthermore, the first step in the enzymic process involves isomerization of the Δ^5- to the isomeric Δ^4-steroid.

Interestingly, 19-nor-$\Delta^{5(10)}$-3-ketosteroid **260** as well as enones **261** and **262** are unreactive when incubated with *Rhizopus* species[235].

(260)   **(261)**   **(262)**

b. Lipoxygenase oxidation. Teng and Smith[237] report that soybean lipoxygenase oxidation of Δ^4-cholesten-3-one (**54**) yields a mixture of the corresponding 6α- and 6β-hydroperoxides **101** and **102**, 6α- and 6β-alcohols **103** and **104** and the 3,6-dione **105** (equation 105). The ratio of **101**:**102**:(**103** + **104**):**105** at pH 9.0 is 10:20:3:1.

lipoxygenase
37 °C, 4 h

(54)   **(101a)**   **(102a)**

(103a)   **(104a)**   **(105a)**

(105)

The evidence indicates that hydroperoxides **101** and **102** are the primary products, which are then thermally decomposed to alcohol and ketone derivatives **103**–**105**.

Interestingly, these authors further report that the interconversion of **101** and **102** occurred on storage of the solid sample and in organic solvent solutions. Epimerization of the quasiaxial **102** to the quasiequatorial **101** was favored over the reverse epimerization, which also occurred but to a lesser extent. This epimerization undoubtedly proceeds via the aforementioned β-cleavage process described above (Section III.C.5). For **101** the prominent mode of transformation is dehydration to **105**.

c. Horseradish peroxidase. 6β-Hydroperoxyprogesterone (**102d**) and 6β-hydro-peroxyandrostenedione (**102c**) were biochemically synthesized from the correspond-ing Δ^5-3-ketosteroids (**100**) by using horseradish peroxidase or bovine adrenal mito-chondria as the enzyme source[132] (see equation 47). Hydroperoxide **102d** is further metabolized in the adrenals to 6-keto (**105d**) and 6β-hydroxyprogesterone (**103d**).

2. 3-Hydroxyflavones

The flavonol quercetin (5, 7, 3′, 4′-tetrahydroxyflavone, **263** R = H) is present in the leaves and flowers of higher plants as the 3-O-glycoside rutin which contributes a cream pigmentation. Rutin is aerobically degraded to carbon monoxide and water-soluble products by extracellular enzymes from *Aspergillus* and *Pullularia* species[65,238,239]. Rutin is first hydrolyzed to rutinose and quercetin and the latter is then oxidatively decarbonylated by the action of the dioxygenase quercetinase to give carbon monoxide and depside **266**. In the last step, the depside is hydrolyzed to 2, 4, 6-trihydroxybenzoic acid **267** and protocatechuic acid **268**. The likely mechanism is outlined in Scheme 19, path a (cf. end of Section III.E.3.b and Section IV.C.2). This mechanism involves endoperoxides **265** and is supported by tracer experiments which reveal that the carbon monoxide expelled stems from C-3, and that an oxygen molecule is incorporated into depside **266** and its hydrolyzed products but not into carbon monoxide. These data rule out the intermediacy of dioxetane **269** (Scheme 19, path b)[65,238,239].

3. Chalcones

The peroxidase-catalyzed oxidation of 4, 2′, 4′-trihydroxychalcone **271** has also been explored extensively[240–245]. The major primary product is the corresponding dioxetane **272** which is transformed under the reaction conditions to flavonol **273** and benzoxepinone-spiro-cyclohexadienone **274** or reduced, depending on the contaminants present, to hydrated aurone **275** and dihydroflavonol **276**. The mechanism suggested for these processes is outlined in Scheme 20. Wilson and Wong[243] have demonstrated that the peroxidase-catalyzed oxidation of chalcone **271** to dioxetane **272** utilizes molecular oxygen in equimolar amounts. Although the reaction requires the presence of H_2O_2, only a catalytic net consumption occurs. Thus, the role of the peroxidase is simply to initiate the radical autoxidation of the chalcone.

4. Tetracyclone

The soybean lipoxidase-mediated oxygenation of tetraphenylcyclopentadienone (tetra-cyclone)[246], presumably via the oxidative cleavage outlined in equation 106, path a, and not *cis*-dibenzoylstilbene as earlier suggested (equation 106, path b)[247].

G. Miscellaneous Oxygenations

1. ^{60}Co initiated

Δ^4-cholesten-3-one **54** remains unaffected by ^{60}Co gamma irradiation[92]. We have, however, seen that this steroid is susceptible to lipoxygenase-mediated oxidation (Section III.F.1.b)[237]. The differential behaviour of the ketone in the two systems may be understood if we assume that the enolization of **54** to 3, 5-cholestadien-3-ol (**124**; see

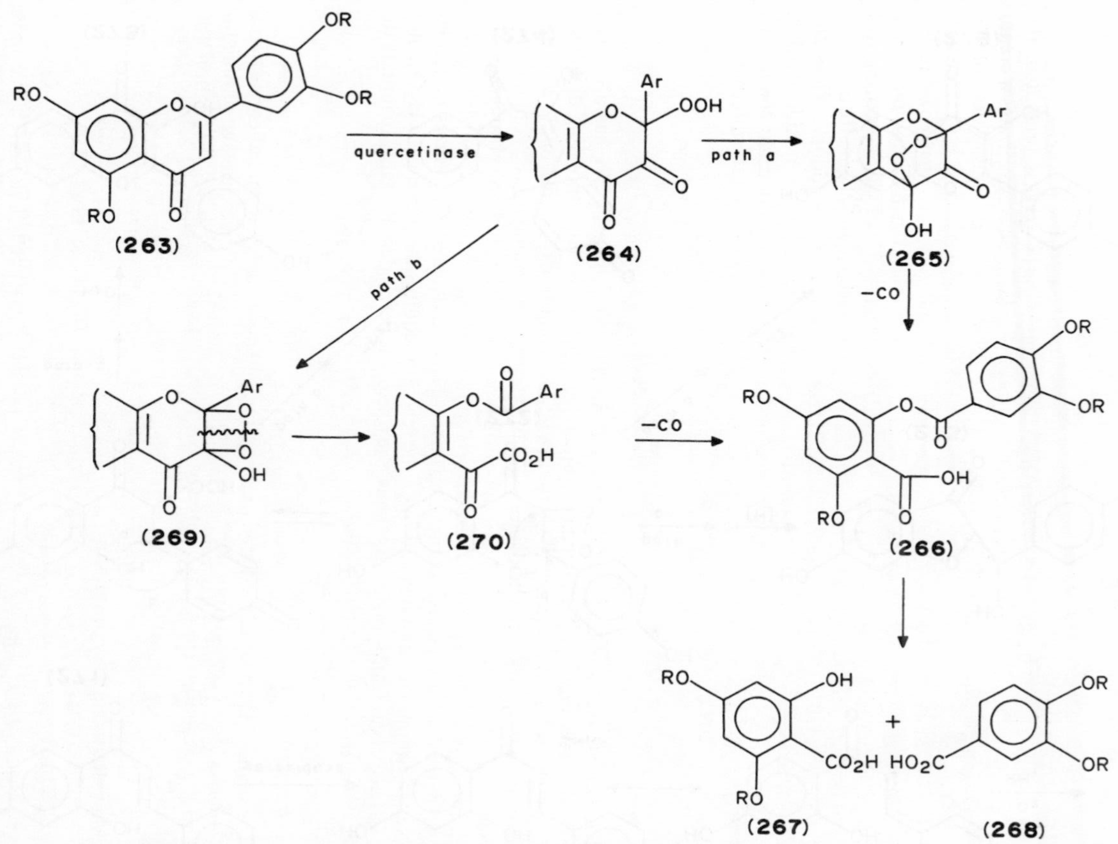

SCHEME 19. Proposed mechanism for the biological oxygenation of 3-hydroxyflavones

848

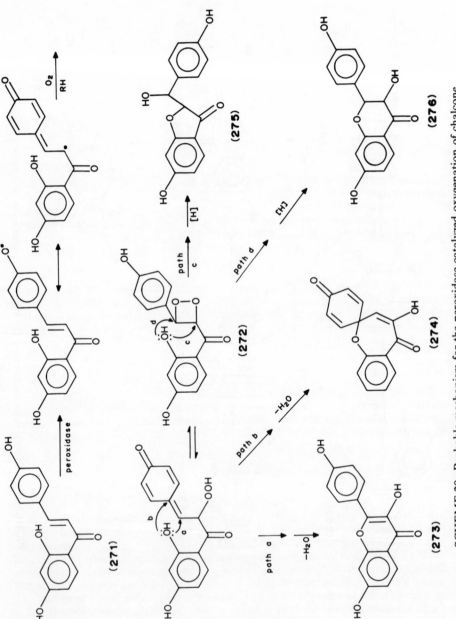

SCHEME 20. Probable mechanism for the peroxidase-catalyzed oxygenation of chalcone

$$(106)$$

equation 52), the likely active steroid intermediate, is more facile in aqueous-buffered enzyme systems than in the solid state[237].

The autoxidation of Δ^5 cholesten-3-one **100a** is initiated by ^{60}Co gamma irradiation. Δ^4-3,6-Dione **105** is the only product formed, presumably as a dehydration product of isomeric hydroperoxides **101** and **102**[92] (see equation 47).

2. Pt catalyzed

While Δ^4-cholestenone **54** resists platinum-catalyzed oxidation, Δ^5-analog **100a** is converted to **54** (2%), alcohols **103** (18%) and **104** (20%) as well as dione **105** (29%)[248]. As in other oxidations of **100**, **103**–**105** are presumed to result from hydroperoxides **101** and **102**. This process is assumed to be a free radical type autoxidation initiated by the platinum catalyst[248].

3. Cu catalyzed

We have already mentioned above the catalytic role copper(II) ions play in the base-catalyzed autoxidation of various enones (Section III.E.7). In such cases the metal ion serves both as an electron acceptor, facilitating the oxygenation of the carbanion, and later as an electron donor to convert the peroxy radical to a peroxyanion (equation 107).

$$RH \xrightarrow{\text{B:}} R^- \xrightarrow{\text{Cu}^{+2}} R^\cdot + O_2 \longrightarrow RO_2^\cdot \xrightarrow{\text{Cu}^{+1}} RO_2^- + RH \longrightarrow ROOH$$

$$(107)$$

a. 2-Hydroxy-2-en-1-ones. Cu(II) salts are also effective in mediating the oxidation of 2-hydroxy-2-en-1-ones under neutral conditions[71,249]. Thus 1,2-cyclohexanediones **277**, which exist primarily in the keto enol form **278**, were rapidly (~ 1 h) oxygenated with the aid of $CuCl_2 \cdot H_2O$ in methanol affording (Scheme 21) the corresponding 1,5- keto acid **284** as the major product along with carbon monoxide, as well as smaller amounts of methyl α-hydroxyadipate (**285**), oxidative cleavage product **286** and coupling product **287** ($R = R^1 = H$). The mechanism outlined in Scheme 21 invokes the initial formation of a Cu(II) complex **279**. The latter is in fact a copper enolate. However, as noted above (Section III.E.3.b), were this a simple BCA, the oxygenation in methanol would have taken 24 h not 1 h[180,185]. Thus, as in the case of the copper-catalyzed BCA, an electron transfer to the copper ion initiates a radical process. Oxygenation at the β-carbon, reduction of the

850

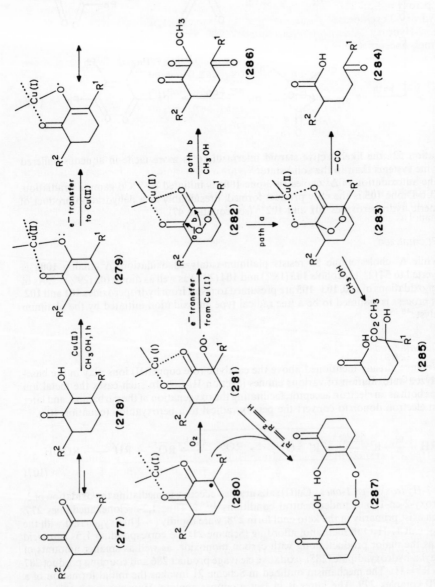

SCHEME 21. Copper(II)-catalyzed oxidation of 2-hydroxy-2-en-1-ones (for $R^1=CH_3$, $R^2 = i\text{-}C_3H_7$ the yields are **284**; 73%; **285**; 80%; **286**; 9%; and **287**; 5%)

peroxy radical by Cu(I) and cyclization of **282** (Scheme 21, path a) yields endo-peroxide **283**. Loss of CO generates keto acid **284**, while methanolysis yields ester **285**. The α-keto peroxy anion **282** can also undergo oxidative cleavage (Section III.C.4) yielding **286** (Scheme 21, path b). Peroxide **287** results from the radical coupling of carbon radical **280** and peroxy radical **281**. These authors[71,249] further report that 1, 2-cyclopentanediones (2-hydroxy-2-cyclopenten-1-ones) undergo copper ion catalyzed oxygenation in a similar way. 2-Hydroxy-2-methyl-4-pyrone (maltol, **288**, equation 108) is unreactive. This inertness appears strange since, as we shall see below, 3-hydroxyflavones which share the pyrone structure are reactive, though via a mechanism different from that of the aforementioned diones.

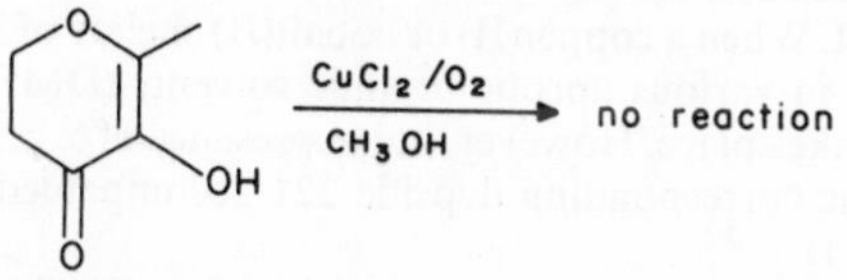

$$\underset{\textbf{(288)}}{} \quad \xrightarrow[\text{CH}_3\text{OH}]{\text{CuCl}_2 / \text{O}_2} \quad \text{no reaction} \qquad (108)$$

b. 3-Hydroxyflavones. In the case of 3-hydroxyflavones **289**, oxygenation occurs quite slowly to yield the corresponding 2-alkoxyflavan-3, 4-diones **291** which are isolated in methanol as hemiacetals or hydrates (equation 109)[250]. One mole equivalent of oxygen is

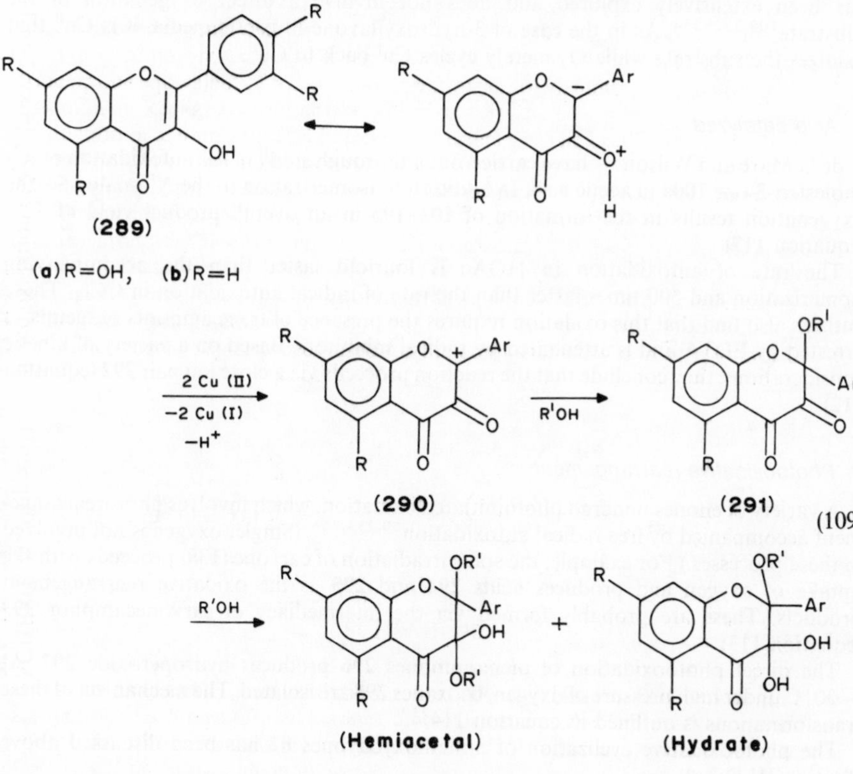

(289)

(a) R=OH,   (b) R=H

$$\xrightarrow[\begin{array}{c}-2\,\text{Cu (I)}\\ -\text{H}^+\end{array}]{2\,\text{Cu (II)}} \quad \textbf{(290)} \quad \xrightarrow{\text{R'OH}} \quad \textbf{(291)}$$

(109)

$$\xrightarrow{\text{R'OH}}$$

(Hemiacetal)   +   (Hydrate)

absorbed but no carbon monoxide is expelled. The rate of reaction is extremely sensitive to the electron density of C-2; thus, for **289a** the reaction reaches completion after 10 h at 20 °C, while for **289b** 26 h at 50 °C is required. The authors invoke the intermediacy of cation **290** generated by the abstraction of two electrons from enol **289** by copper(II). The oxygen does not oxygenate the substrate directly, but rather serves to drive the reaction by reoxidizing Cu^I back to Cu^{II}, as outlined in equation 110.

$$2Cu^I + \frac{1}{2}O_2 + 2H^+ \rightarrow 2Cu^{II} + H_2O \qquad (110)$$

Interestingly, Matsuura and colleagues[239,251] report that in aprotic media, the situation is substantially different. When a copper(II) or cobalt(II) chelate of 3-hydroxyflavone **289b** is treated with oxygen in various aprotic organic solvents (DMF, DMSO, pyridine or CH_2Cl_2), no reaction takes place. However, in the presence of excess flavone oxygenation does occur, yielding the corresponding depside **221** accompanied by carbon monoxide evolution (equation 111).

The depside **221** is also obtained in a 37% yield when the flavone is oxidized using catalytic amounts of Cu(II) acetate. Utaka and Takeda[250] report that the remaining 60% in this latter case is a lactol, formed without carbon monoxide generation. The corresponding lactol is the sole product when **289a** is the substrate (equation 111).

c. Ascorbic acid. The copper-catalyzed oxidation of vitamin C to dehydroascorbic acid has been extensively explored and does not involve a direct oxygenation of the substrate[109,252-254]. As in the case of 3-hydroxyflavone in protic media, it is Cu^{II} that oxidizes the substrate while O_2 merely cycles Cu^I back to Cu^{II}.

4. Acid catalyzed

de la Mare and Wilson[134] have carried out a thorough study of the autoxidation of Δ^5-cholesten-3-one **100a** in acetic acid. In addition to isomerization to the Δ^4-analog **54**, the oxygenation results in the formation of **103–105** in an overall product yield of 72% (equation 112).

The rate of autoxidation in HOAc is fourfold faster than the accompanying isomerization and 500 times faster than the rate of radical autoxidation in CCl_4. These authors also find that this oxidation requires the presence of trace amounts of metals, is arrested by EDTA and is attenuated by radical inhibitors. Based on a variety of kinetic considerations, they conclude that the reaction proceeds via a close ion pair **292** (equation 112).

5. Photooxidative rearrangement

A variety of enones undergo photoinitiated oxidation, which involves photorearrangement accompanied by free radical autoxidation[60,255,256]. (Singlet oxygen is not involved in these processes.) For example, the solar irradiation of carvone (**138**) proceeds with the uptake of oxygen and produces acids **294** and **295** as the oxidative rearrangement products. These are probably formed via the intermediacy of carvonecamphor **293** (equation 113).

The direct photooxidation of menaquinones **296** produces hydroperoxide **297**. At − 30 °C under high pressure of oxygen, trioxanes **298** are isolated. The mechanism of these transformations is outlined in equation 114[257].

The photooxidative cyclization of 3-methoxyflavones **82** has been discussed above (Section III.D.2.a).

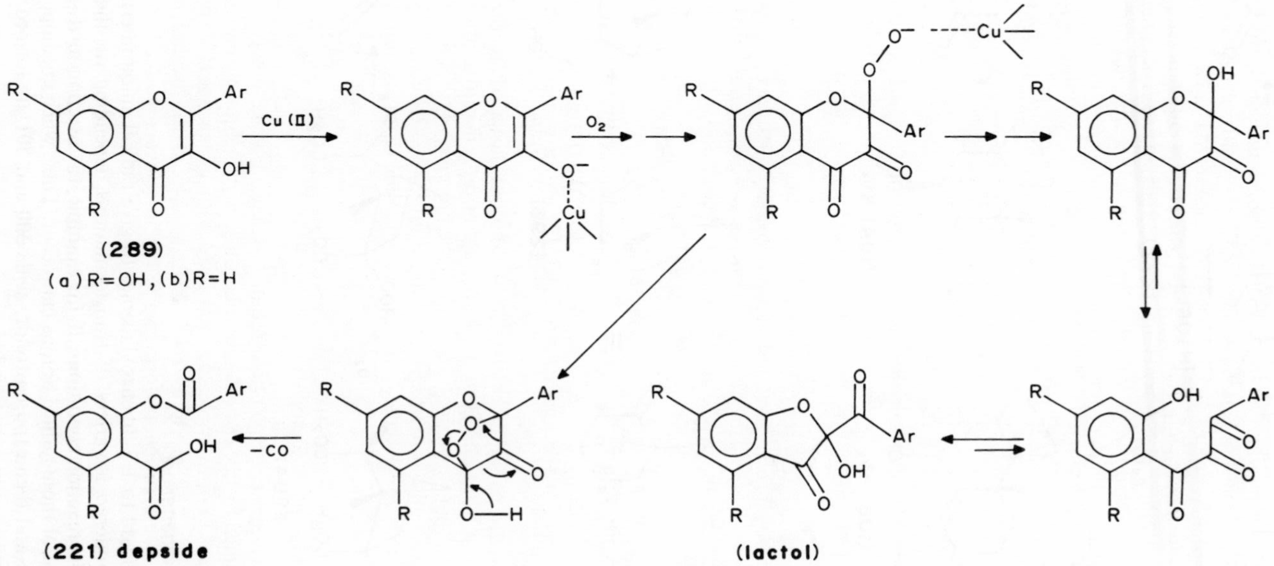
Cu (II)
O2
−CO
(289)
(a) R=OH, (b) R=H
(221) depside
(lactol)

(111)

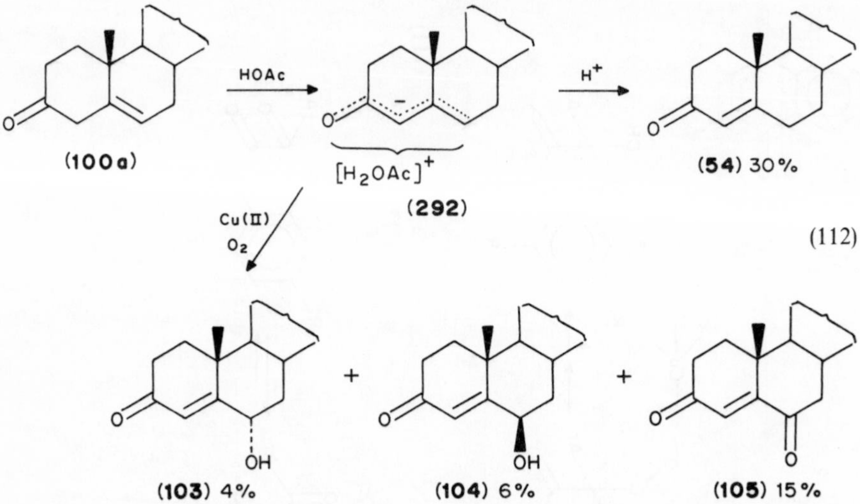

(112)

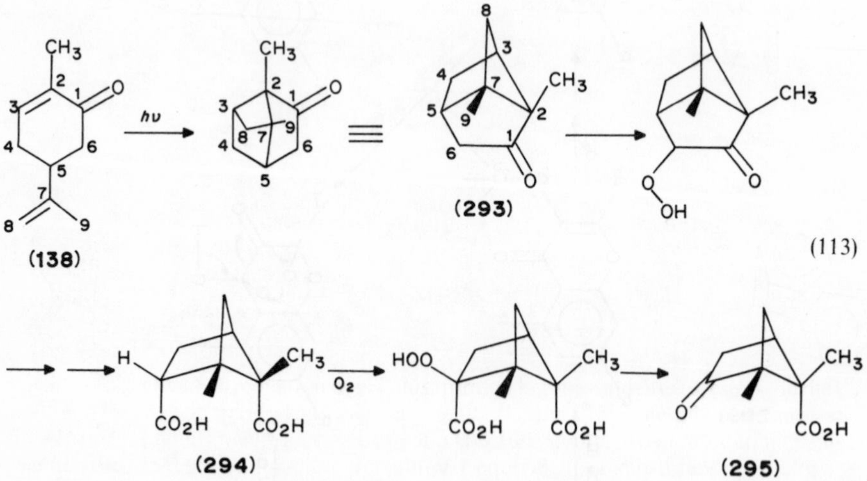

(113)

6. Reductive oxygenation

Alkali metals add to 1,2-dibenzoylalkenes to give radical anion intermediates, which subsequently undergo a variety of transformations depending on the nature of the substrate and the reaction conditions. If, for example, the reaction mixture is exposed to air, oxygenation of these radical anions can occur. Thus, when enedione **299** is reduced with potassium and then exposed to air, acids **300** and **301** are formed[258]. A plausible mechanism is outlined in equation 115.

(114)

(115)

IV. SINGLET MOLECULAR OXYGEN

A. Modes of Reaction[259-262]

Unlike $^3O_2$, which displays a biradical character, all the electrons in $^1O_2$ are paired. Hence, the type of reaction it undergoes is expected to involves electron pairs. What's more, it is convenient to think of $^1O_2$ as the oxygen analogues of ethylene. Indeed, each of the three modes in which $^1O_2$ reacts with unsaturated compounds finds a precedent in one of the reaction pathways of ethylene.

The first of these modes is a [2 + 2] cycloaddition to a double bond to form a 1, 2-dioxacyclobutane or dioxetane (equation 116). These cyclic peroxides are sometimes of moderate stability but readily cleave thermally or photochemically into two carbonyl-containing fragments. The cleavage is quite often accompanied by chemiluminescence.

$$\text{(116)}$$

The second mode bears a striking resemblence to the Alder 'ene' reaction[263,264]. In the $^1O_2$ ene reaction, olefins containing an allylic hydrogen are oxidized to the corresponding allylic hydroperoxides in which the double bond has shifted to a position adjacent to the original double bond (equation 117).

$$\text{(117)}$$

The third and final mode involves a [4 + 2] Diels–Alder-type addition of singlet oxygen to a diene producing endoperoxides (equation 118).

$$\text{(118)}$$

The question of the mechanism in these three reaction types has been the subject of much heated debate over the past decade. The highlights of this long-standing controversy have been reviewed by this author[28b,28d] and a detailed discussion is beyond the scope of this review.

A variety of factors has been shown to control all singlet oxygen reactions[262]. The rate of reaction within a homologous series of compounds is generally inversely proportional to their ionization potential. This suggests that singlet oxygen is mildly electrophilic and sensitive to the nucleophilicity of the olefinic bond[265]. Thus as a rule, alkyl substitution increases the reactivity of olefins 10–100-fold per group. Solvent has only a minimal effect on the rate of reaction; changes in rate are commonly due to solvent effects on the lifetime of singlet oxygen. Because of the low activation energy for singlet oxygen processes (1–5 kcal)[266,267] little if any temperature effect on the rate of reaction is observed. Regarding the mode of reaction, electron-rich olefins (such as vinyl sulfides, enol ethers and enamines) as well as sterically hindered alkenes (such as 2, 2-biadamantylidene[268] and 7,7-binorbornylidene[269]) tend to prefer dioxetane formation, though two modes often compete. Finally, the direction of singlet oxygen attack is predominantly, if not exclusively, from the less hindered side of the molecule.

B. Singlet Oxygen Sources

1. General

An impressive variety of physical and chemical sources of $^1O_2$ is now available for laboratory-scale purposes. These include photosensitization, oxidation of H_2O_2, decomposition of phosphite ozonides and endoperoxides, and microwave discharge. These various sources have been extensively reviewed[28a,270,271]. Of all the techniques available for generating $^1O_2$, photosensitization is clearly the most convenient and, by far, the most commonly used, since it is applicable to a large spectrum of reaction temperatures, solvents and sensitizers. Most importantly for unreactive substrates, this physical method, unlike the chemical methods mentioned above, requires no additional reagents, merely longer photolysis times. It is for this reason that we focus briefly on this method in particular.

2. Photosensitization

By the beginning of the twentieth century there were several reports describing the oxidation of organic and biological substrates in the presence of oxygen, light and a photosensitizer. It has become apparent during the last two decades that there are in fact two general classes of photooxidations. In the first, called Type I, the sensitizer serves as a photochemically activated free-radical initiator. In its excited state, the sensitizer reacts with a molecule of a substrate, resulting in either hydrogen atom abstraction or electron transfer. The radicals thus formed react further with $^3O_2$ or other molecules. In the second class of reactions, dubbed Type II, the sensitizer triplet (sens^3), formed via intersystem crossing (ISC) of the excited singlet state sensitizer (sens^1*), interacts with oxygen, most commonly by transferring excitation, to produce $^1O_2$ (equations 119 and 120). The direct absorption of light by $^3O_2$ to produce $^1O_2$ is a spin-forbidden process. Type II generally predominates with colored sensitizers (dyes), such as methylene blue (MB), tetraphenylporphyrin (TPP) and rose Bengal (RB), which absorb visible light and have triplet energies (E_T) ranging from 30 to 46 kcal mol^{-1}. Type I processes are favored by high-energy, UV-absorbing sensitizers.

$$\text{sens}^1 \xrightarrow{h\nu} \text{sens}^1* \xrightarrow{\text{ISC}} \text{sens}^3 \tag{119}$$

$$\text{sens}^3 + {}^3O_2 \rightarrow \text{sens}^1 + {}^1O_2 \tag{120}$$

A variety of photochemical apparatus and procedures has been described[272,273]. In a typical reaction, the substrate and the sensitizer ($10^{-3}–10^{-5}$ M) are dissolved in an appropriate solvent and photolyzed (250–1000 W) while oxygen is bubbled through the reaction mixture. Alternatively, the solution is rapidly stirred under an oxygen atmosphere with the uptake of oxygen followed by means of a gas buret. A UV cutoff filter is often placed between the light source and the reaction vessel to prevent the initiation of free-radical reactions.

Recently, the use of polymer-based or adsorbant-bound sensitizers[274-279] has become quite popular and several products are commmercially available. Problems such as solubility, removal, recovery and bleaching, often confronted with unbound sensitizers, are eliminated by using this heterogeneous photooxygenation method. The polymer-based sensitizer need simply be suspended in any (mostly organic) solvent which will 'wet' the polymer. Upon conclusion of the photolysis, the sensitizer may be filtered off, washed and reused if so desired.

C. Reaction of Singlet Oxygen with α, β-Unsaturated Carbonyl Compounds

1. Simple systems

   a. s-trans conformation. Despite the intense investigation of $^1O_2$ reaction over the past two decades, there were, until recently, only relatively few examples of the successful oxidation of alkenes that are substituted with electron-withdrawing groups[262]. This is consistent with the observation that singlet oxygen is weakly electrophilic[265]. Numerous examples of the attempted photosensitized oxidation of 3-keto-Δ^4-steroids (155), their 4-methyl analogs (302) and their 7-keto-Δ^5 analog (303) have shown that these enone systems are unreactive towards singlet oxygen[262,280] (equation 121).

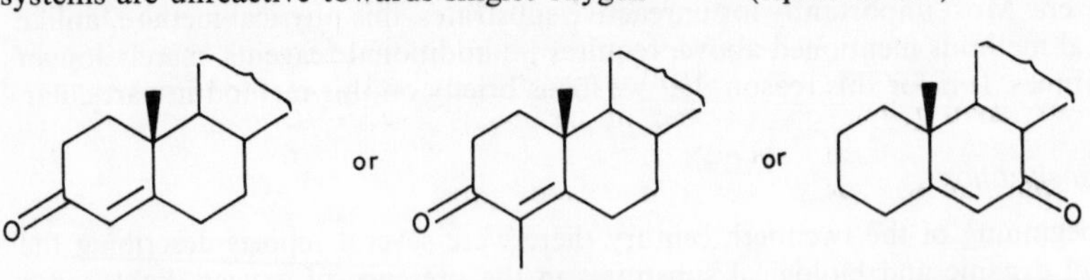

(155)   or   (302)   or   (303)

$$\xrightarrow{^1O_2} \text{no reaction} \qquad (121)$$

   Further research by Ensley's group[280,281] has revealed that the reactivity of α, β-unsaturated carbonyls towards $^1O_2$ is strongly dependent on the conformation of the saturated system. Thus, those enones which prefer or are constrained to an *s-trans* conformation react slowly, if at all. For example, in addition to steroids 155, 302 and 303, cyclohexenones 304a–c[280,281], 3-methoxyflavones 82[104], cyclopentenones 305a, b[280,281], cyclopropenone 306[282] and cyclobutenones[283] are unreactive.

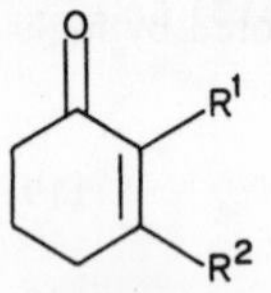

(304)

(a) R^1=Et, R^2=CH$_3$
(b) R^1=H, R^2=OCH$_3$
(c) R^1=CH$_3$, R^2=OCH$_3$

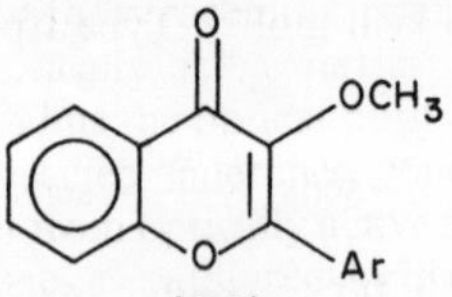

(82)

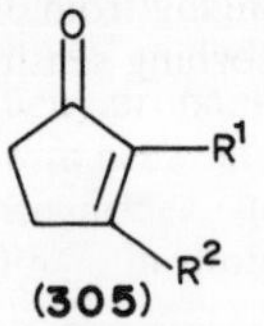

(305)

(a) R^1= n-C$_5$H$_{11}$, R^2=CH$_3$
(b) R^1=CH$_3$, R^2=OCH$_3$

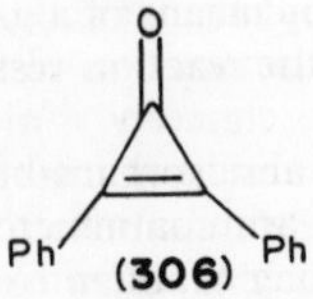

(306)

The only known class of exceptions are α- and β-hydroxyenones whose base-catalyzed singlet oxygenation will be discussed in Section IV.C.2. It should be noted, however, that Wamhoff and coworkers[284] find that the singlet oxygenation of dihaloketones **307** to vicinal triketones **310** proceeds via the corresponding α- and/or β-haloenones **308** and **309**. A dioxetane mechanism is invoked (equation 122).

$$(122)$$

b. s-cis conformation. Ensley's group[280,281,285-287] has further demonstrated that those α, β-unsaturated carbonyl systems which prefer or are constrained to the *s-cis* conformation are rapidly oxidized by singlet oxygen to yield ene reaction products. Thus pulegone[281,285-287] **167** yields allylic hydroperoxides **311**, **312** and **313** (equation 123). β-Hydroperoxy ketone **311** cyclizes spontaneously to peroxide **314**.

$$(123)$$

What is interesting is that in this and the related cases[281,288-293] of α, β-unsaturated ketones, aldehydes, acids, esters and lactones, as well as β-alkoxy enones, the reaction product formed preferentially, if not exclusively, is always the *conjugated* carbonyl. Put somewhat differently, allylic hydrogen abstraction in the ene reaction is preferred from the group geminal to the carbonyl. This 'geminal effect' is surprising since singlet oxygen reactions do not normally show a strong Markownikoff directing effect[259]. Nor can the reactivity of the *s-cis* conformations be explained on the basis of ionization potential[280]. Finally, although singlet oxygen normally abstracts allylic hydrogens from the most crowded side of the olefin ('*cis* effect'), in enone systems the geminal effect takes precedence[291].

Ensley[280,281] has proposed (equation 124) that the initial step involves a $[4 + 2]$ cycloaddition of $^1O_2$ to the enone system generating a 1,2,3-trioxine **315**. (A related intermediate has been invoked in the singlet oxygenation of pyrazolium-4-olate and dithiolium-4-olate[294].) Thermolysis of **315** yields diradical **316** which rearranges directly to the major product, conjugated carbonyl **317**, or via perepoxide **318** to the minor product, unconjugated isomer **319** (equation 124).

$$(124)$$

Chan and colleagues[292,293] have presented evidence for an alternative mechanism for the singlet oxygen reaction of dihydropyrancarboxylic acid **320**. In benzene, hydroperoxides **321** and **322** are formed in a 9:1 ratio as expected by the 'geminal effect' (equation 125). However, there is a profound solvent effects. In proceeding from benzene ($\varepsilon = 2.3$) to CH_2Cl_2 ($\varepsilon = 9.1$) to CH_3CN ($\varepsilon = 37.5$), product ratio flips from 90:10 to 35:65.

$$(125)$$

	(321)	(322)
Benzene/TPP	90%	10%
CH_2Cl_2/TPP	50%	50%
CH_3CN/MB	35%	65%

This role of solvent is not explained by the trioxine mechanism of Ensley. Chan's group suggests, along the lines of Frimer and Bartlett and coworkers[295], that singlet oxygen adds to the double bond to form either an extended or collapsed perepoxide. It is the former which is preferentially stabilized by the polar solvent, and leads to conjugated enones **322** (equation 126).

$$(126)$$

It should be noted, however, that this profound solvent effect, so typical of enol ethers[295-298], is not observed in the case of pulegone[296] and, hence, **320** may prove to be the exception rather than the rule.

In a gas-phase low-temperature ($-190\,^{\circ}$C to $-150\,^{\circ}$C) study of the singlet oxygenation of acrolein and crotonaldehyde (**323**), Carmier and Deglise[299] present IR spectral evidence suggesting that the reaction proceeds via a dioxetane (**324**), which rearranges to an epoxy enol formate (**325**) (equation 127). Such a transformation is completely unprecedented and this reaction deserves further investigation.

$$(127)$$

R = H or CH$_3$

We close this section by pointing out a few anomalies in steroidal systems that have yet to be explained. For example, although **326** takes up one equivalent of oxygen under photooxidative conditions, **327** is not reactive—though it too is locked into a *cis* configuration[262]. Similarly, **328** is inert to $^1O_2$ although an s-*cis* conformation should be readily attainable[262]. Perhaps subtle steric or conformational effects are at play.

2. Keto enols

We have described above the $^1O_2$ ene reaction of olefins containing at least one allylic hydrogen. In this process, allylic hydroperoxides are generated in which the double bond has shifted to a position adjacent to the original double bond. In its most general form, the normal $^1O_2$ ene reaction (equation 129) can be written as shown in equation 128, where 'A' is CH_2 and 'B' is H. Silyloxyolefins (in equation 128, A = O, B = $SiMe_3$) also undergo an ene reaction with $^1O_2$ producing silylperoxy ketones (equation 130)[300-305]. In this transformation, the trimethylsilyl group takes the place of an allylic hydrogen while oxygen replaces the allylic carbon.

$$(128)$$

$$(129)$$

$$(130)$$

$$(131)$$

$$(132)$$

In the same fashion, enols (A = O, B = H) and enolates (A = O, B = $^-$) have been shown to undergo ene-type reactions (equations 131 and 132 respectively)[306]. For example, Matsuura reported that the photosensitized oxygenation of the stable keto enols, 3-hydroxyflavones **263**, like the enzymatic oxidation (Section III.F.2; see Scheme 19) and the corresponding BCA (Section III.E.3.b), yielded depsides **266**. However, in this case both carbon monoxide and carbon dioxide were formed[61b,104,307]. From the fact that CO is stable under the reaction conditions and that the photosensitized oxygenation of *p*-methoxyphenylglyoxylic acid gives anisic acid and carbon dioxide in good yield, it was concluded that two mechanistic pathways are operative (path a and path b in Scheme 19). The initially formed 3-hydroperoxy-1, 2-diketone **264** can decompose (see Section III.C.4) to depside **266** via either cyclic peroxide **265** (path a) or dioxetane **269** (path b). It should be noted that the corresponding 3-methoxyflavones (**82**, Section III.D.2.a) are inert to singlet oxygen[61b,104,307].

Simultaneously with Matsuura's study of the enol of α-diketones, Young and Hart[308,309] observed that β-hydroxy enones (enols of β-diketones) and δ-hydro-xydienones (enols of α,β-unsaturated-δ-diketones) can be oxygenated in basic methanol. Thus, the enolate of diacetylfilicinic acid **329a** reacts with $^1O_2$ giving hydroxy ester **334**. The latter is presumably formed via α-hydroperoxy ketone **331**, which undergoes oxidative cleavage (see Section III.C.4) to α-diketone **332**. Benzilic acid rearrangement of the latter generates **334** (equation 133). A similar reaction is observed with monoacetyl-filicinic acid **329b** with a Kornblum–DeLaMare reaction (Section III.C.2) converting **331** to **332**.

(133)

Similarly, dimedone **335** is oxidized[72,310] under these conditions to a mixture of products containing enol **336** and esters of 3, 3-dimethylglutaric acid **337** (equation 134).

Wasserman and Pickett[69,72] have recently reinvestigated the photooxidation of enols and have discovered that fluoride ion catalyzes this process giving higher yields of the oxidations products and cleaner reaction mixtures. Enols stemming from β-diketones, β-keto esters and α-diketones have been photooxidized under these conditions and the yields of the resulting hydroperoxy diketones are generally around 70% after only 2 h of reaction when carried out in aprotic media (e.g. CHCl$_3$). The uptake of oxygen is sluggish at most in the absence of fluoride. The latter presumably hydrogen bonds with the enol hydrogen, thereby increasing the electron density on the oxygen and the nucleophilicity of the double bond. In the case of β-diketones **338**, the resulting hydroperoxide **340** dehydrates to the corresponding vicinal triketone **341**. The latter undergoes enolization to **342** or solvent addition to **343** (equation 135).

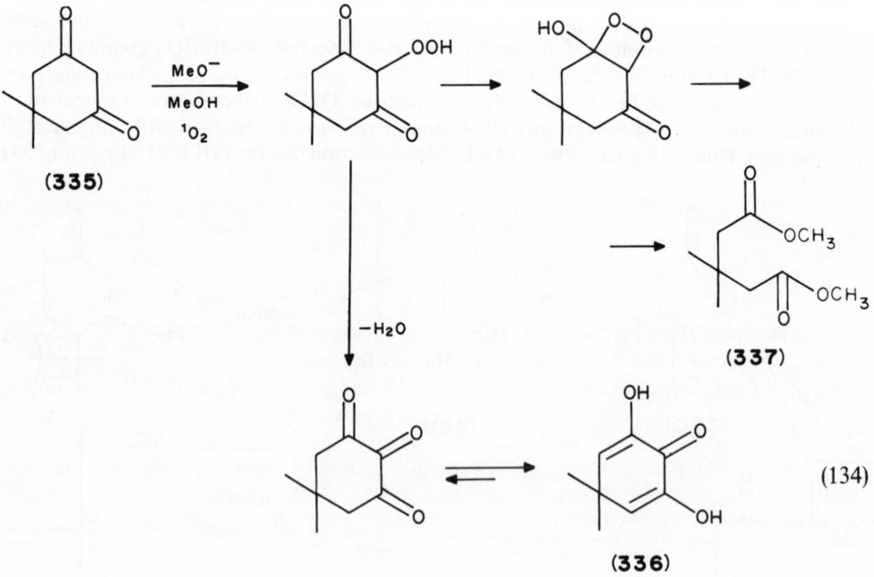

(134)

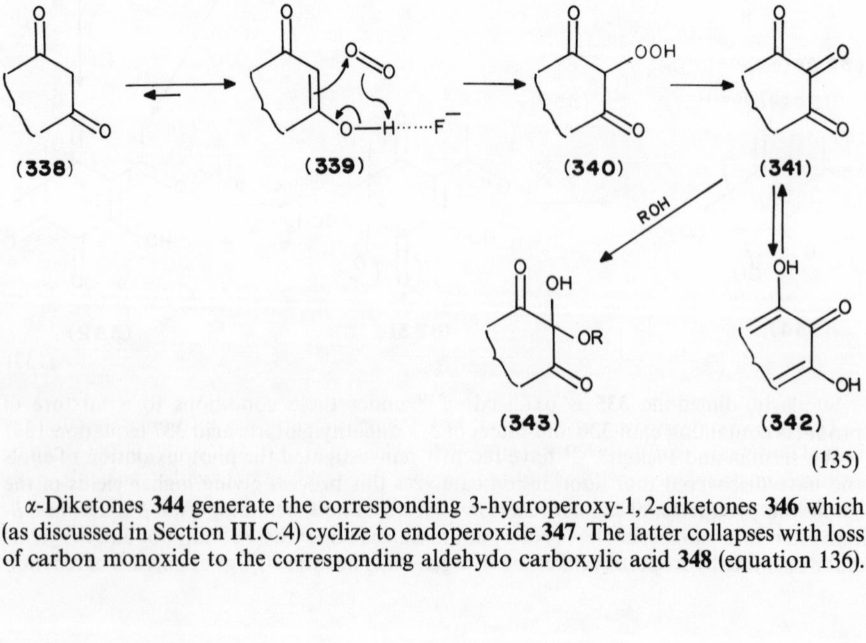

(135)

α-Diketones **344** generate the corresponding 3-hydroperoxy-1,2-diketones **346** which (as discussed in Section III.C.4) cyclize to endoperoxide **347**. The latter collapses with loss of carbon monoxide to the corresponding aldehydo carboxylic acid **348** (equation 136).

In contradistinction to the 3-hydroxyflavone oxygenation, no CO_2 was detected, which rules out the intermediacy of a dioxetane (see Scheme 19, path b).

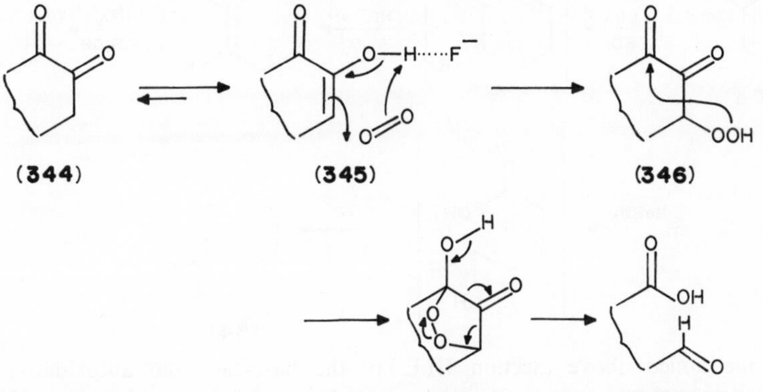

(344) (345) (346)

(347) (348)

(136)

Takeda and coworkers have found that 1, 2-cyclohexanediones **349a**[70,73,75,311] and 1, 2-cyclopentanediones **349b**[76] undergo this singlet oxygenation in methanol in the absence of fluoride to yield oxoalkanoic acids **352** and hydroxy acids **353**. Interestingly, though the exact product distribution is highly dependent upon the reaction temperature, **352** is the major product in the case of the cyclohexanediones **349a**, while **353** predominates in the case of the cyclopentanediones **349b**. The latter product results from the solvent trapping of cyclic peroxide **351** (equation 137). It has yet to be explained why the five-membered ring diones several times slower than their higher homologs.

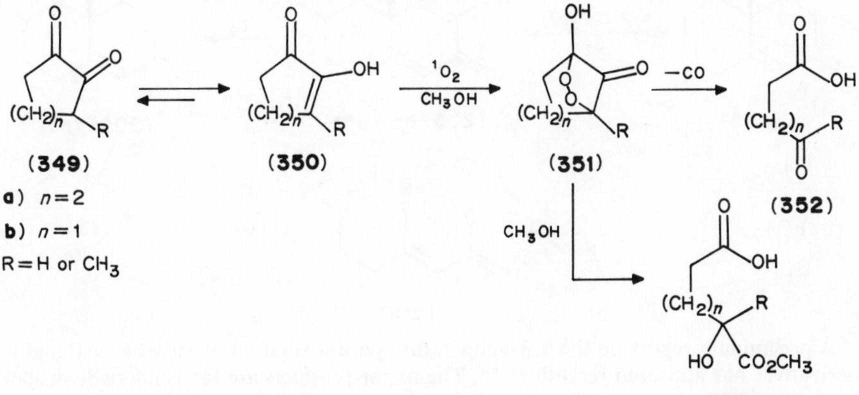

(349) (350) (351) (352) (353)

(a) $n=2$

(b) $n=1$

R=H or CH_3

(137)

The rate of oxygenation of 3-alkyl-1, 2-cyclohexanediones is approximately equal to that of the tetrasubstituted olefin tetramethylethylene (TME)[73]. As a result, the enol can be oxidized in preference to disubstituted olefinic linkages present in the molecule. This observation has enabled Takeda's group[73,75,311] to carry out a new synthetic approach to jasmine lactone and related δ-lactones (**354**) from 1, 2-cyclohexanedione, as outlined in equation 138.

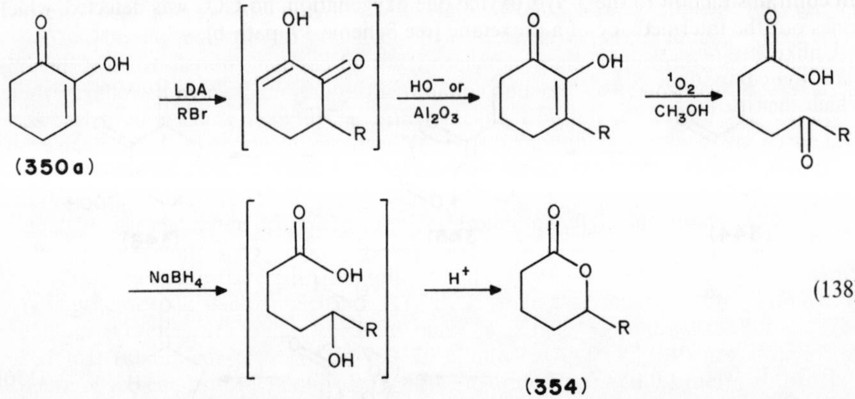

(350a)

(354)

(138)

We mentioned above (Section III.E.3.b) the base-catalyzed autoxidative (BCA) approach Frimer and coworkers[201] have used to convert 3-oxo-Δ^4 steroids **155** to the pharmacologically important 2-oxa analogs **218** (equation 88). At the center of this reaction sequence is the BCA conversion of enol **208** to lactol **209**. This step requires strongly basic conditions and several days of reaction. Using the much milder Wasserman and Pickett procedure[72], this conversion has been carried out on the enols of cholestenone, testosterone, 17α-methyltestosterone, 17α-hydroxyprogesterone, progesterone cortisolone–BMD and cortisone–BMD. Yields are generally 75% and the oxygenation requires only a few hours (equation 139)[312a]. Photooxidation as suggested by Takeda's group proved ineffective.

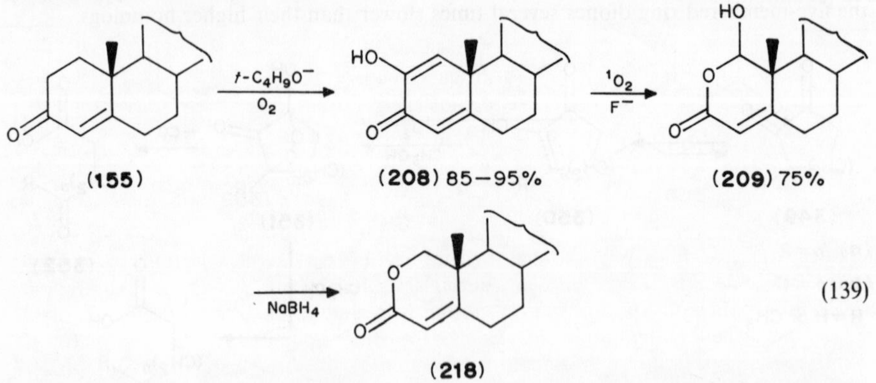

(155) (208) 85–95% (209) 75%

(218)

(139)

A preliminary report on the low-temperature photooxidation of ascorbic acid and its derivatives has appeared recently[312b,c]. The major products are the isomeric hydroperoxyketones and, as expected, oxygenation occurs on the less hindered face of the ring, i.e. opposite to the 'R' group.

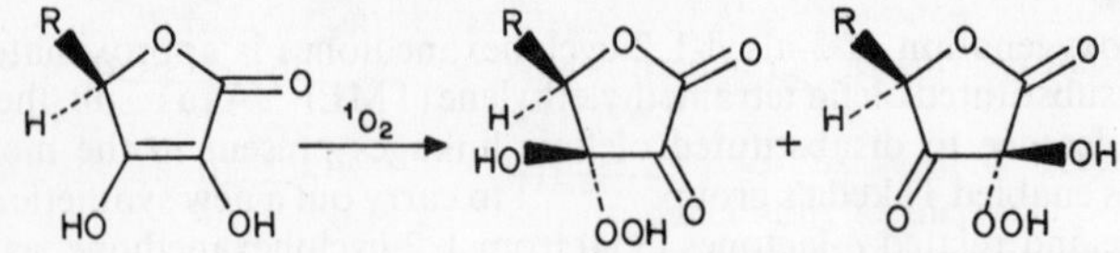

The photooxidation of tropolones is discussed in Section IV.C.5.c.

3. Enamino carbonyl systems

Unlike carbonyl systems which react through an ene mechanism, simple enamino carbonyl compounds 355 with an electron-rich double bond react via dioxetanes 356, which then cleave to the corresponding α-dicarbonyl compounds 357 (equation 140).

$$(140)$$

Wasserman and Ives[313-317] have used this technique to prepare various α-keto derivatives of lactones, esters, amides, lactams and ketones. The sequence is particularly simple since the enamine 355 need not be isolated, and can be converted to the corresponding diketone 357 using polymer-bound rose bengal which facilitates its isolation. The utility of this method is illustrated by the conversion of methone 358 to the corresponding enol 359 in 81% yield (equation 141). Ziegler and coworkers[318] utilized this procedure in the synthesis of a transient α-diketone 361, which subsequently rearranged to 363 (equation 142).

$$(141)$$

$$(142)$$

Wasserman and Han[317] have further extended this technique to the preparation of vicinal tricarbonyl systems (364). The β-dicarbonyl precursors are reacted with DMF acetal to form enamines, which are cleaved by photooxidation (equation 143). This method has been successfully applied to the formation of carbacepham 365 (equation 144)[317].

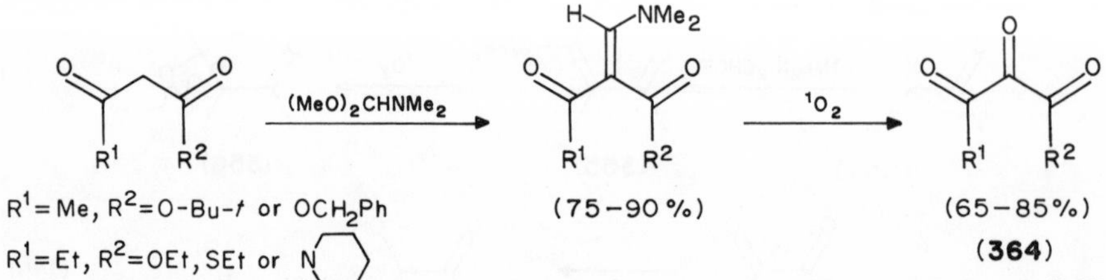

$$R^1 = Me, R^2 = O-Bu-t \text{ or } OCH_2Ph$$
$$R^1 = Et, R^2 = OEt, SEt \text{ or } N\langle\rangle$$

(143)

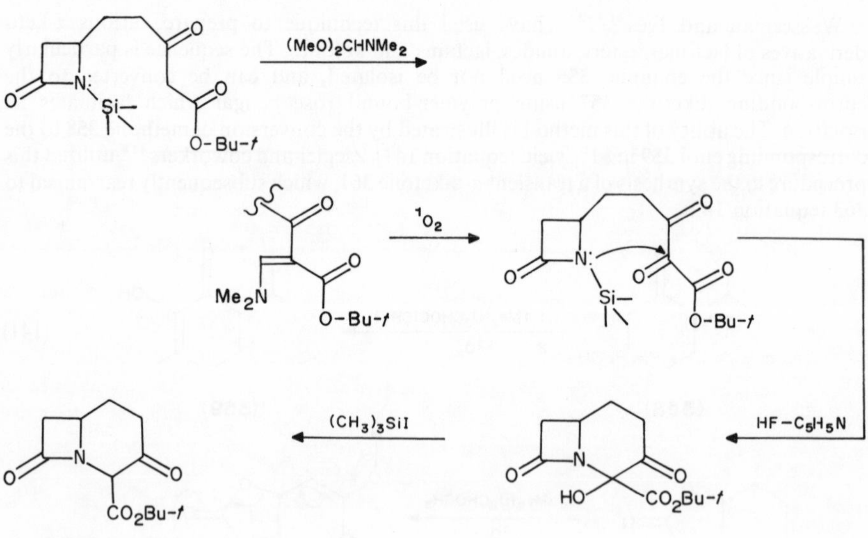

(144)

More complicated or conjugated enamino systems tend to react by an ene mode generating zwitterionic intermediates, which may cyclize in turn to the related dioxetanes. For example, the pyrimidine, 5,6-diphenyluracil 366, on sensitized photooxidation in liquid ammonia at $-70\,°C$, gives an unstable peroxide 367 which, on warming, breaks down to the cleavage product 369, presumably via dioxetane 368 (equation 145)[319].

The pyrimidinium-4-olates of type 370 or 374 react with $^1O_2$ generated chemically or photochemically to yield a 1,4-dipolar cycloaddition product, namely stable endoperoxides 373 and 375 in high yields[320] (equations 146 and 147). Although the authors hesitate to suggest a mechanism, we believe that the intermediate here as well is the zwitterion 371, which cyclizes in this case to endoperoxide 373.

Orito and colleagues[321] report that enamino ketone 376 is oxidatively rearranged to ketolactone 379. The mechanism of this process has been the subject of some disagreement and speculation[38b,316,321,322]. Presumably the initially formed hydroperoxide or peroxyanion **377** rearranges to hydroxydioxetane **378**, which collapses to ketolactone **379** (equation 148).

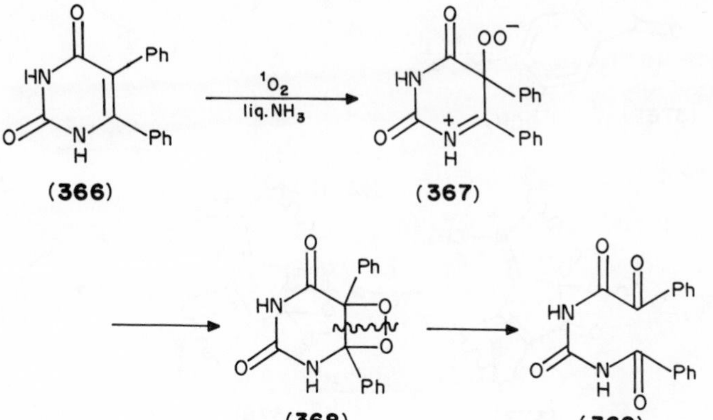

(366) (367) (368) (369)

(145)

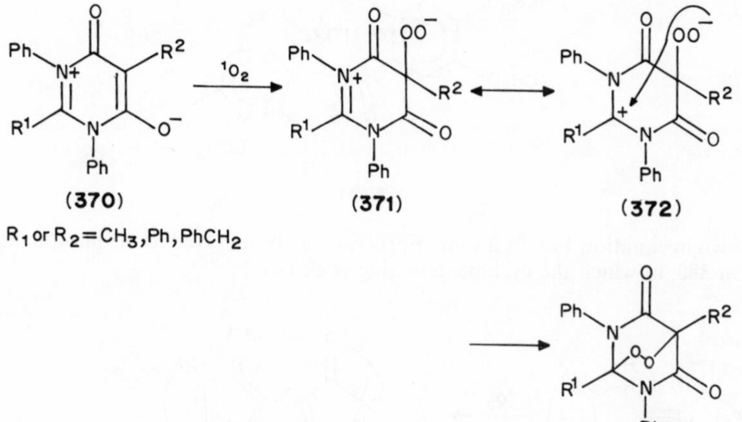

(370) (371) (372) (373)

R_1 or $R_2 = CH_3, Ph, PhCH_2$

(146)

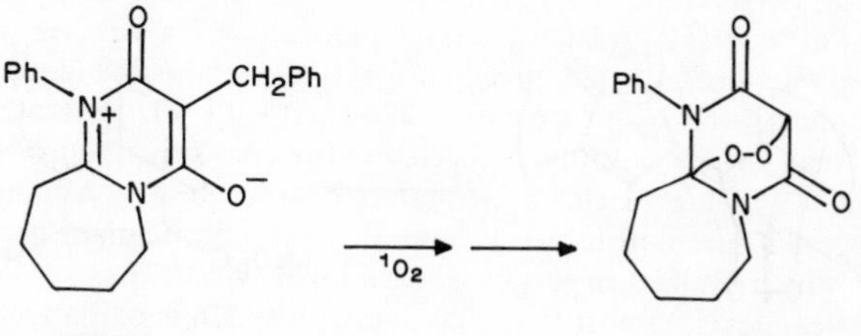

(374) (375)

(147)

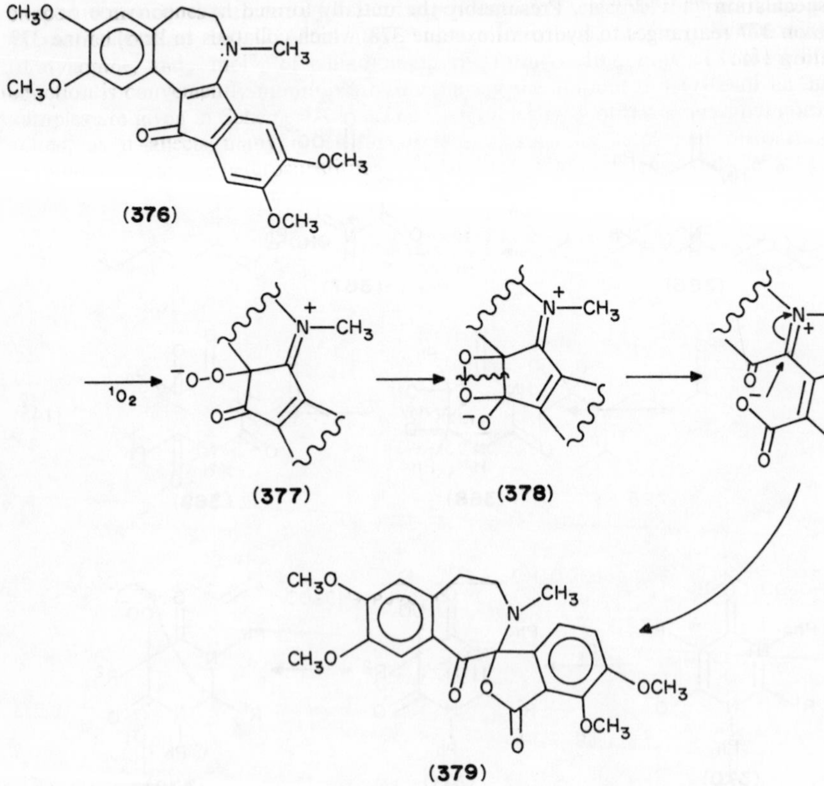

(376)

(377)   **(378)**

(148)

(379)

As shown in equation 149[316], a similar process may be involved in the photooxidation of chlorin **380**, in which the cyclopentene ring is cleaved[323,324].

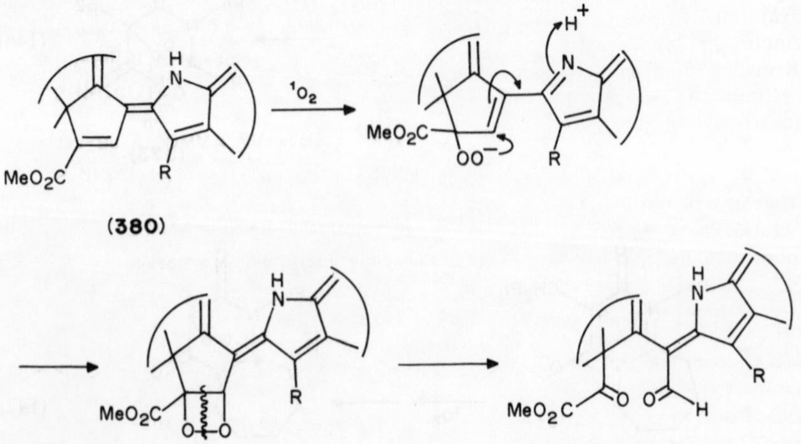

(380)

(149)

Related to the photooxidation of enamino β-diketones[317] is the singlet oxygenation of 2-(2-quinolyl)indan-1,3-dione **381**, which yields phthalic acid **382**, quinoline-2-carbaldehyde **383** and quinoline-2-carboxylic acid **384**. Again, in this conjugated enamino system, an initial ene mode is observed (equation 128, A = N, B = H), not dioxetane formation (equation 150)[324,325].

(381)   **(385)**   **(382)**   **(383)**

$$(150)$$

(384)

All the enamino ketones described above had the amine group at the β olefinic carbon with respect to the carbonyl. Machin and Sammes[326] have reported on the photosensitized oxidations of 3-benzylidenepiperazine-2,5-diones **386a–c** to the corresponding piperazinetriones **389** (equation 151).

NMR evidence was presented for the intermediacy of dioxetane **388**. Interestingly, however, both the E and Z isomers (**386a** and **386b** respectively) yield the *same* dioxetane. This suggests that oxidation of at least one of the arylmethylene isomers proceeds with inversion of configuration, and hence that non-concerted dioxetane formation via zwitterion **387** is involved. To further verify the involvement of acyliminium derivative **387**, nitrogen participation was inhibited by preparing imidate ether **390**. The latter indeed proved inert to $^1O_2$ (equation 152).

The photooxidation of purines and other nitrogen heterocycles has been extensively reviewed[327,328].

4. Chalcones

Chalcones (1,3-diarylpropenones, **391**) which lack alkyl groups on the double bond are precluded from undergoing a $^1O_2$ ene reaction. Nevertheless Chawla and coworkers[329,330] report that prolonged photosensitized irradiation (70–115 h) of these compounds under air lead to oxidative cleavage products, such as **394–397**, which these

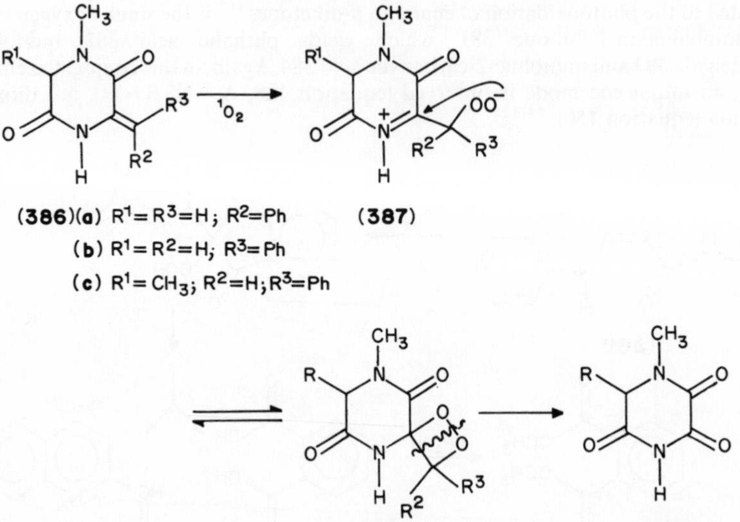

(386)(a) $R^1 = R^3 = H$; $R^2 = Ph$ (387)

(b) $R^1 = R^2 = H$; $R^3 = Ph$

(c) $R^1 = CH_3$; $R^2 = H$; $R^3 = Ph$

(151)

(388)   (389)

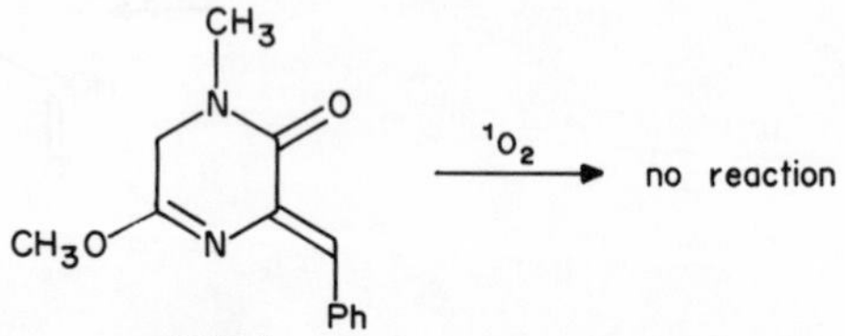

(152)

(390)

authors presume result from dioxetane **392** formation (see Scheme 22). The involvement of $^1O_2$ in this reaction was verified by the anticipated quenching of the reaction with DABCO. When, however, *both* the 2' and 6' positions bear hydroxyl groups, 5-hydroxyflavonols **400** (in a 25–50% yield) are formed as well. The mechanism proposed by Chawla and colleagues[329,330] (Scheme 22, path b) invokes a nucleophilic attack of the 2'-hydroxy group on the enone system. Nucleophilic attack is facilitated by the hydrogen bonding of the carbonyl with the 6'-hydroxy group, which on the one hand increases the electrophilicity of the enone moiety and secondly 'locks' the attacking 2'-hydroxy group in close proximity to the enone system.

Wong[331] has recently reexamined the photosensitized oxidation of 2',4',4-trihydroxychalcones and finds not only the 5-hydroxyflavonol **400** but also products **403** and **404** (Scheme 23). Interestingly, these same products are observed in the enzymatic[240–245] oxidation of chalcones (see Section III.F.3) which is a free radical process. These authors conclude that although singlet oxygen is formed in this reaction, the oxygenation of the enones is initiated by the formation of a phenoxy radical which results[327] in turn from the phenol–singlet oxygen interaction (Scheme 23; cf. Scheme 20). Nucleophilic attack by the phenol group on the dioxetane at either the α or β carbons leads to product **403** and **404** (Scheme 23, path c) or **400** (path b).

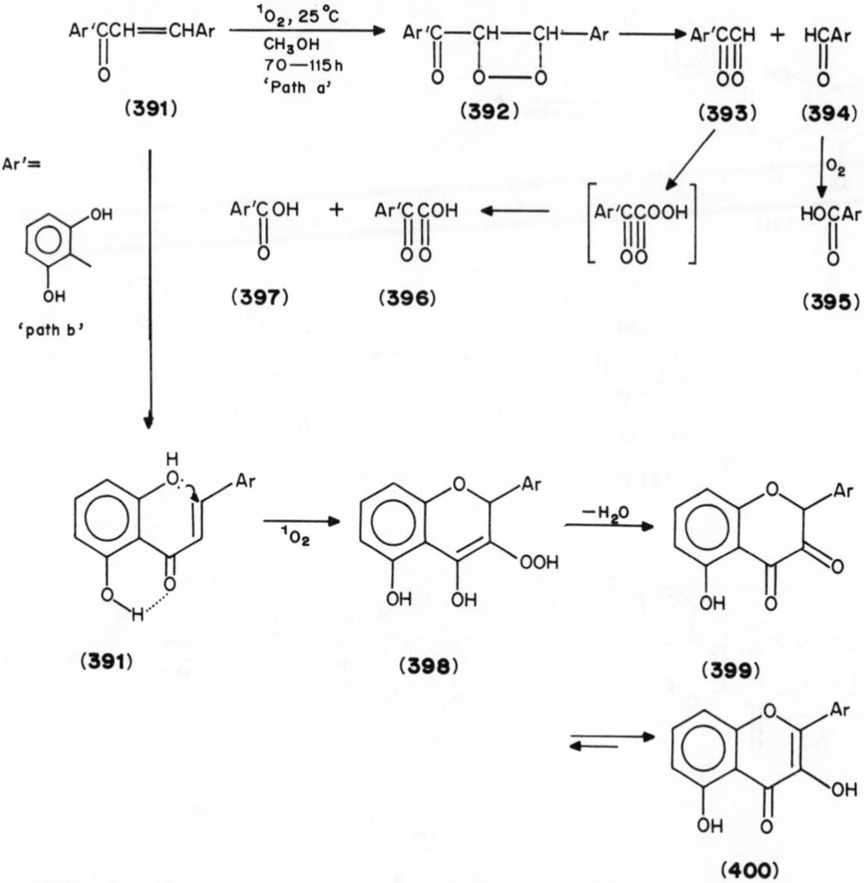

SCHEME 22. Chawla[329,330] mechanism for the singlet oxygenation of chalcones

5. Retinoids and acyclic polyenoates

Polyene carbonyls related to the retinoid family were extensively studied by Mousseron-Canet[332-341] and others[262,342-346] in the late 1960s. Because allylic hydrogens and diene moieties are present, both ene and Diels–Alder modes are observed. For example, ester 405 yields endoperoxide 406 and two ene products, diallylic hydroperoxide 407 and allene 408[338] (equation 153). The formation of allene 408 is quite surprising, since it requires the abstraction of a vinyl hydrogen.

More recently, acyclic conjugated polyenoates have been investigated by Matsumoto and Kuroda[347]. These researchers find that monoolefinic esters such as ethyl crotonate 409 and even dienoates such as ethyl sorbate 410 are essentially inert to $^1O_2$ (equation 154). However, the introduction of an additional methyl group to 410 yields 411 which reacts to generate an endoperoxide product 412 (equation 155). The next-higher vinylog of 411, trienoate 413 is reactive to $^1O_2$ as well, yielding endoperoxide 414 (equation 156). The introduction of a methyl group to 413 yields trienoate 415, which reacts by all three singlet oxygen modes (equation 157).

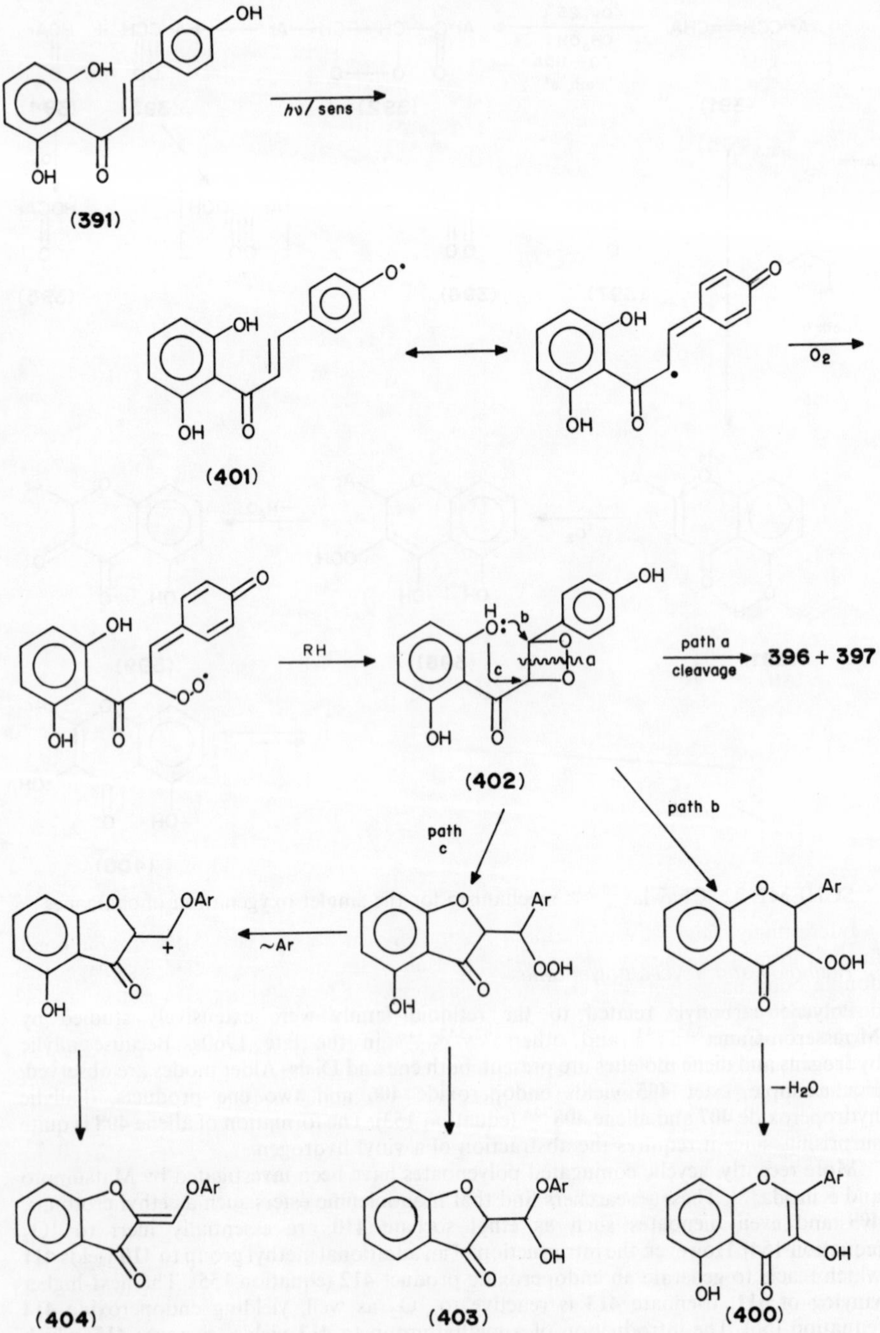

SCHEME 23. The Wong[331] mechanism for the singlet oxygenation of chalcones

$$(153)$$

(409)   (410)   no reaction   (154)

$$(155)$$

$$(156)$$

Interestingly, singlet-oxygen attacks occur in all the above cases at the double bond furthest from the carbonyl. This may be a result of the increased nucleophilicity of the double bond as we get further from the carbonyl. These authors also note that the distribution of the products in the photooxidation of 415a and 415b among the three singlet-oxygen reaction modes (i.e. ene, dioxetane and Diels–Alder) is remarkably affected by the solvent used and the reaction temperature. In particular, polar solvents favor ene-product formation while low temperatures favor dioxetane product.

6. Polyenic steroids

We have noted above (Section IV.C.1.a) that the enone moiety in Δ^4- and $\Delta^{1,4}$-2-oxosteroids is inert and other olefinic linkages present in the molecule are oxidized in preference to them[262]. Thus, the α, β-double bond in the enone moiety of steroid 420 (and in the corresponding $\Delta^{1,4}$-analog) remains unaffected by photooxidation[262] (equation 158).

Interestingly 19-nor-$\Delta^{4,9(10)}$-3-oxo steroids do react via a $^1O_2$-ene mode at the terminal double bond. Thus, steroid 422 is oxidized to phenol 424[348] (equation 159).

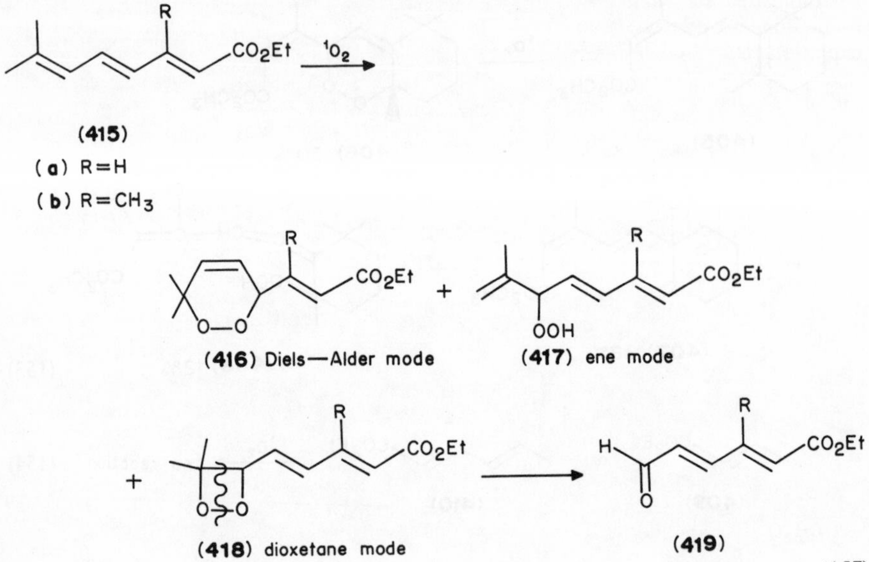

(415)

(a) R=H

(b) R=CH₃

(416) Diels—Alder mode + (417) ene mode

+ (418) dioxetane mode → (419)

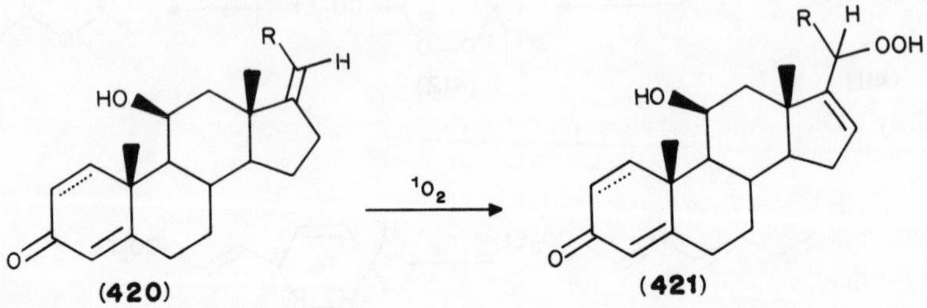

(157)

(420) $\xrightarrow{^1O_2}$ (421)

(158)

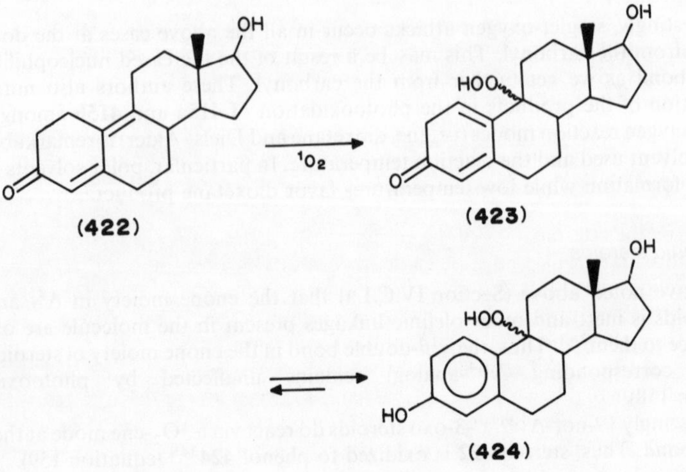

(422) $\xrightarrow{^1O_2}$ (423)

(159)

(424)

Conjugated steroidal polyenones undergo a $2 + 4$ cycloaddition with $^1O_2$ at the homoannalar diene. Thus 7-oxo-$\Delta^{1,3,5}$ steroid **425** yields endoperoxide **426** (equation 160)[349] while tetraenone **427** yields **428** (equation 161)[350,351].

(160)

(425)   **(426)**

(427)   **(428)**

(161)

Clearly, a Diels–Alder-type addition at the homoannular diene moiety is preferred over other modes possible, particularly ene reaction.

7. Homoannular polyenones

a. Cyclones. Homoannular dienones as a class are extremely susceptible to Diels–Alder $^1O_2$ reactions. The smallest ring in this class is cyclopentadienone and, indeed, the photooxidation of arylated cyclopentadienones (cyclones, **429**) has been known for nearly half a century[352-359]. These compounds are generally colored and hence their photo-chemical singlet oxygenation is often a self-sensitized process. Singlet oxygen for these reactions has also been generated by chemical means including triphenylphosphite ozonide and H_2O_2/NaOCl. As shown in equation 162, the primary product is the corresponding endoperoxide **430** which is generally unstable and loses carbon monoxide yielding the *cis*-ene dione **431**. If structurally feasible, the latter will be converted photochemically (UV) or chemically (traces of acid or base) to the more stable *trans* isomer **(432)**[357,359-362].

To verify the intermediacy of the endoperoxide **430**, Chaney and Brown[359] carried out the photooxidation using a molecular oxygen mixture containing $^{16}O_2$ and $^{18}O_2$. The results indicated that indeed the photooxidation of tetracyclone proceeded by a one-molecule mechanism, whereby both oxygen atoms in the resulting dibenzoylstilbene are derived from the same molecule of molecular oxygen.

Bikales and Becker[356] have studied the photooxidation of tetracyclone **429a** and report that, in addition to dibenzoylstilbenes **431** and **432**, they succeeded in isolating pyrone **434**. (A similar product was obtained by Dilthey and coworkers[352] in their study of **429b**.) Bikales and Becker suggest that these pyrones result from a side-reaction with ozone

(429) → (430)

(a) $R^1 = R^2 = R^3 = R^4 = Ph$

(b) $R^1 = R^4 = Ph; R^2 + R^3 =$

(c) $R^1 = R^4 = Ph; R^2 + R^3 =$

(d) $R^1 = R^4 = i\text{-}C_3H_7; R^2 + R^3 =$

(e) $R^1 = R^4 = -(CH_2)_{12}-; R^2 + R^3 =$

(431) → (432)   (162)

formed during the prolonged UV irradiation. We suggest, however, that it results from a Baeyer–Villiger reaction with peroxides formed during the 7–14 days of irradiation required for the completion of this reaction (equation 163). The conversion of cyclones to pyrones with peracids is known[352,358].

(429) → (433) → (434)   (163)

b. *Cyclohexadienones.* Simple cyclohexadienones exist as the corresponding phenols. The singlet-oxygen chemistry of the latter has been discussed recently[306] and is beyond the scope of this review. Surprisingly, however, the photosensitized oxygenation of 6,6-disubstituted cyclohexa-2,4-dien-1-ones does not seem to have been explored extensively. Koch[363], who investigated the thermodynamics of singlet-oxygen reactions, reports that the energy of activation of the reaction of 6,6-dimethylcyclohexa-2,4-dien-1-one is $3.6\,\text{kcal mol}^{-1}$, assuming that endoperoxide product is formed in both cases. Nevertheless, no product study seems to have been carried out.

The research groups of Adam[364,365] and Schuster[366] have shown that α-pyrone endoperoxides **435** are conveniently accessible through singlet oxygenation of α-pyrones **434**. These endoperoxides are hyperenergetic and chemiluminescence accompanies their thermal decomposition (equation 164).

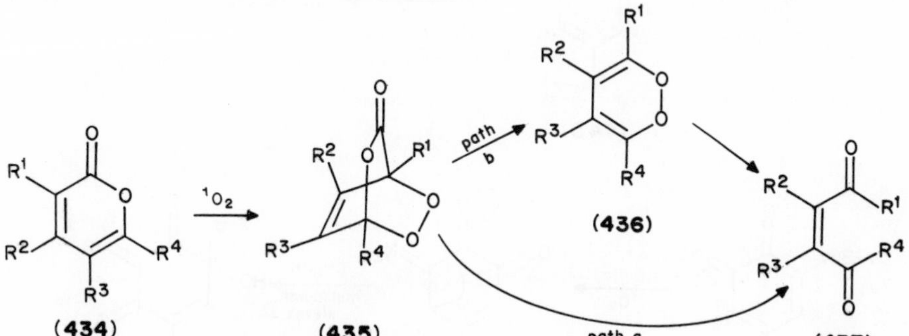

(164)

Work by Schuster and Smith[366] on the benzopyrone system **438** suggests that the decomposition of endoperoxide **439** proceeds via the interesting *o*-oxylylene peroxide **440**, which can be trapped by maleic anhydride (equation 165).

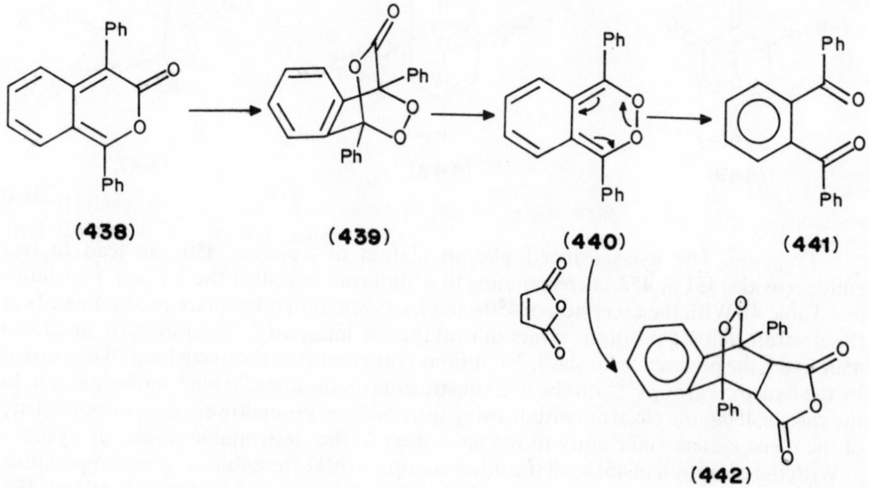

(165)

In the non-benzo analogs, however, Adam and Erden[364] were unsuccessful in trapping the corresponding dioxin **436** even with such reactive dienophiles as 4-phenyl-1,2,4-triazoline-3,5-dione. This suggests that either **436** is not formed (equation 164, path a) or that it suffers valence isomerization to diacylethylene **437** (equation 164, path b) before bimolecular trapping can occur.

3,5-di-*t*-Butyl-*o*-benzoquinone, **443**, is unreactive to $^1O_2$ (equation 166)[367]. However, the corresponding diazo compound **444** yielded endoperoxide **445** as the primary product (equation 167)[367-369]. Peroxide **445** is quite labile and can be transformed to various products **446–449**.

$$ (166) $$

(443)

$$ (167) $$

c. Tropones. The dye-sensitized photooxidation of tropones **450** can lead to two endoperoxides **451** or **452**, corresponding to addition across either the 2,5 or 4,7 positions (see Table 4). With the exception of **450e** and **f**, oxygenation takes place predominantly at the electron-rich 2,5 positions rather than at the less-hindered 4,7 positions. In the case of **450e** and **f**, the preference for the 4,7 additions is attributed to the quenching effect exerted by the hydroxyl groups[374] on the C-2 substituents on the approaching singlet oxygen. In the case of **450g**, the electron-withdrawing nitrobenzoyl group lowers the nucleophilicity of the diene system sufficiently to render it inert to the electrophilic singlet dioxygen.

With the exception of **451b**, all the other endoperoxides are stable at room temperature. *In situ* reduction of the labile peroxy linkage in **451b** yields 5-hydroxytropolone **453**. Peroxide **451b** rearranges in CS_2 to lactone **454** which, upon workup, isomerizes to **455**. If

TABLE 4. Product distribution in the photooxidation of tropones **450**

(450)	(451)	(452)	Total yield	Reference
(a) $R^1 = R^2 = H$	100%	—	90%	370
(b) $R^1 = OCH_3$; $R^2 = H$	100%[a]	—	88%	371, 372
(c) $R^1 = Ar(Ph, p\text{-Tol}, p\text{-An or } p\text{-ClC}_6H_4)$; $R^2 = H$	70%	30%	100%	373
(d) $R^1 = CH_2Ph$; $R^2 = H$	50%	50%	63%	374
(e) $R^1 = CH_2C_6H_4NO_2\text{-}p$; $R^2 = H$	40%	60%	80%	374
(f) $R^1 = CH_2COC_6H_4NO_2\text{-}p$; $R^2 = H$	36%	64%	84%	374
(g) $R^1 = COC_6H_4NO_2\text{-}p$; $R^2 = H$	—	—	no reaction	374
(h) $R^1 = Cl$; $R^2 = OEt$	100%	—	95%	375
(i) $R^1 = R^2 = OCH_3$	100%	—	85%	375
(j) $R^1 = R^2 = Ph$	b	b	c	376

[a]Ustable at room temperature. Similar results are observed with 4- and 6-isopropyl-2-methoxy tropone[371].
[b]Mixture of positional isomers.
[c]No yield reported.

5% methanol is added to the CS$_2$, **451b** is converted to diester **456**. The simplest mechanism for these transformations is outlined in equation 168 (though Forbes and Griffiths have presented data suggesting the presence and perhaps intermediacy of a ketene[372]).

Benztropones **457** undergo photosensitized oxidation to give high yields of lactones **458** (equation 169)[372] via a process analogous to path b of equation 168.

Finally, 2,3-homotropone **459** yields the corresponding endoperoxide **460**. As with other endoperoxides, thiourea reduction yields diol **461** (equation 170)[377].

d. Tropolones. Tropolones are 2-hydroxytropones, and as with other tropones endoperoxides are expected to be the primary products of singlet oxygenation. Although a variety of tropolones have been reacted, no stable endoperoxides have been isolated. For example, 5-hydroxytropolone **462** yields not endoperoxide **463**, but the tautomeric hydroperoxide **464** (equation 171)[375].

Tropolone (**465**) itself yielded cyclohepta-3,6-diene-1,2,5-trione (**467**, equation 172)[373]. In this case, a Kornblum–DeLaMare reaction (see Section III.C.2) of hydroperoxide **466** may be involved, though other mechanisms have been suggested[375].

Takeshita and coworkers[371] have obtained utakin **472** in 21% yield from the photooxygenation of 4-*i*-propyltropolone **463**. The proposed mechanism is outlined in equation 173.

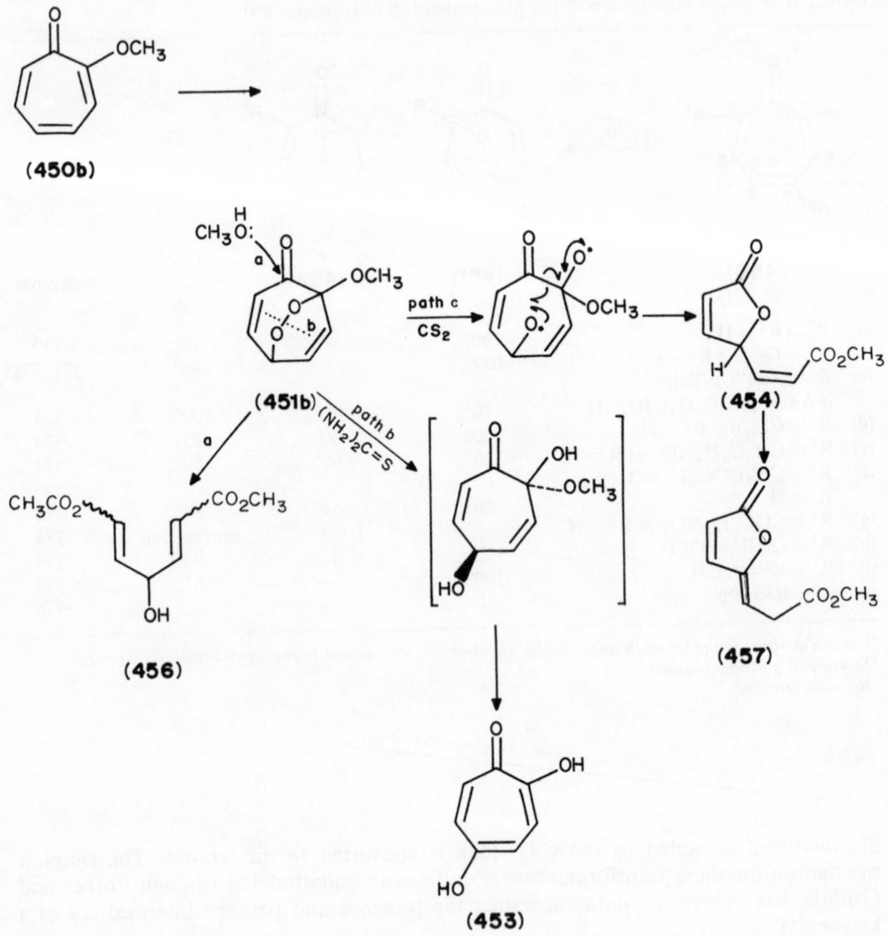

(168)

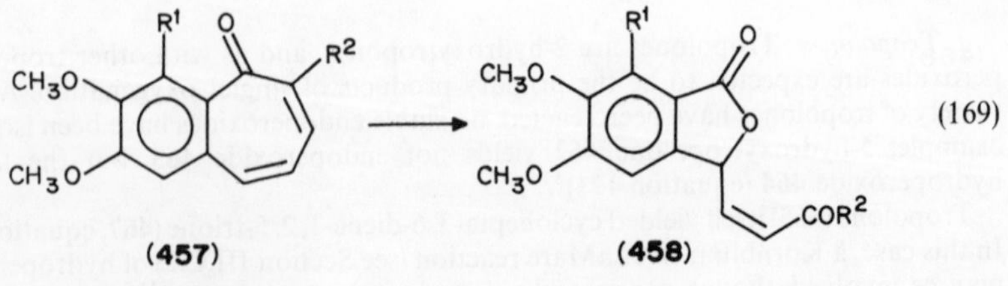

(169)

(a) $R^1=H$, $R^2=OMe$
(b) $R^1=OH$, $R^2=OMe$
(c) $R^1=OH$, $R^2=H$

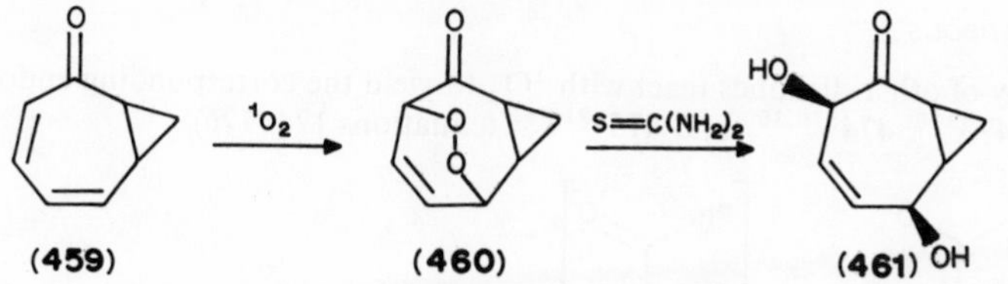

(170)

(459) (460) (461)

(171)

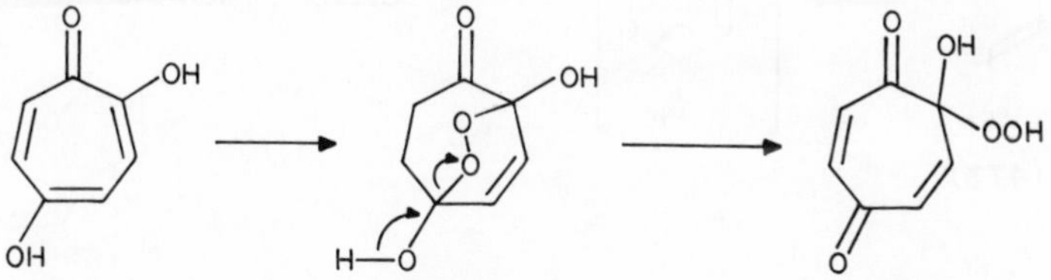

(462) (463) (464)

(172)

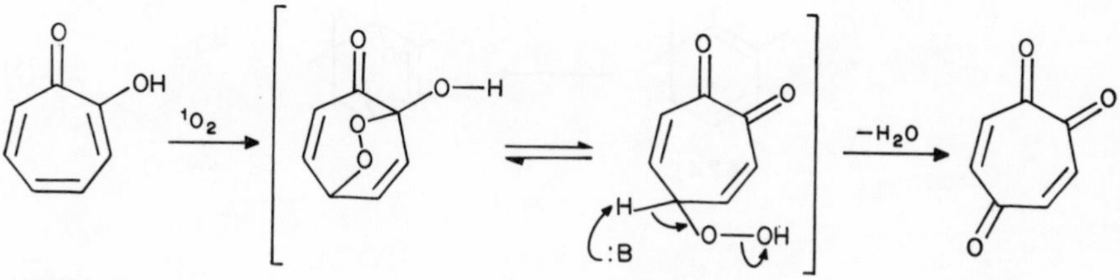

(465) (466) (467)

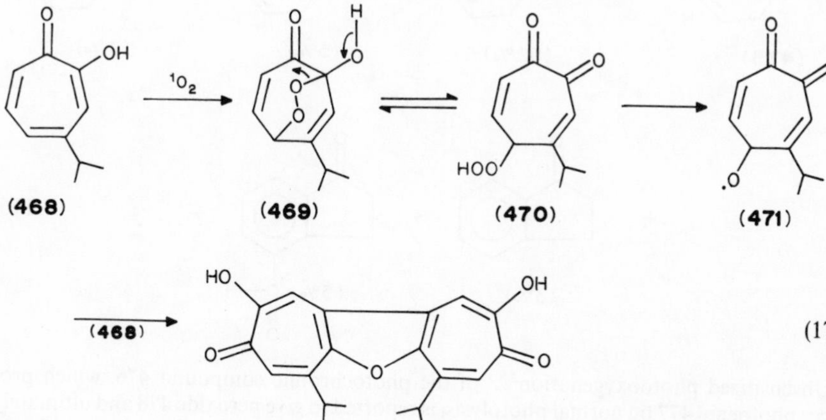

(468) (469) (470) (471)

(173)

(472)

8. Miscellaneous

A variety of other dienones react with $^1O_2$ to yield the corresponding endoperoxides, including **473**[378], **474**[379,380] and **475**[381,382] (equations 174–176).

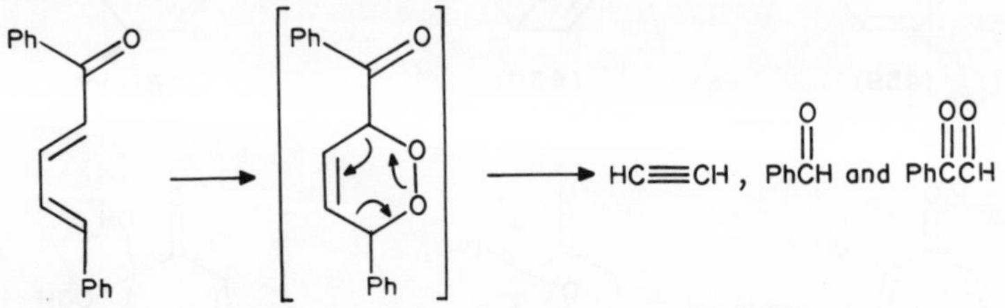

$$HC{\equiv}CH, \quad PhCH \quad and \quad PhCCH \qquad (174)$$

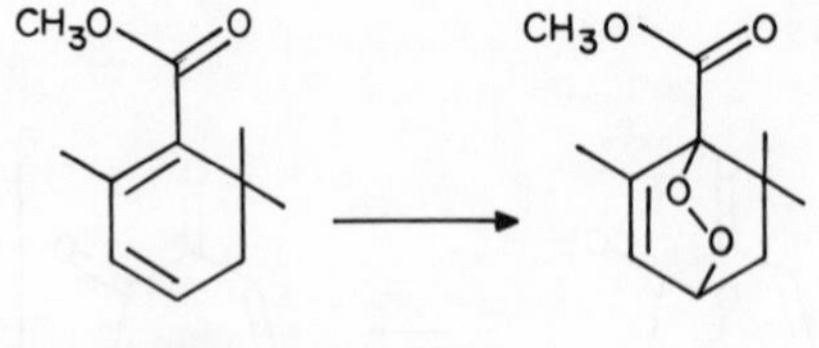

(473)

$$(175)$$

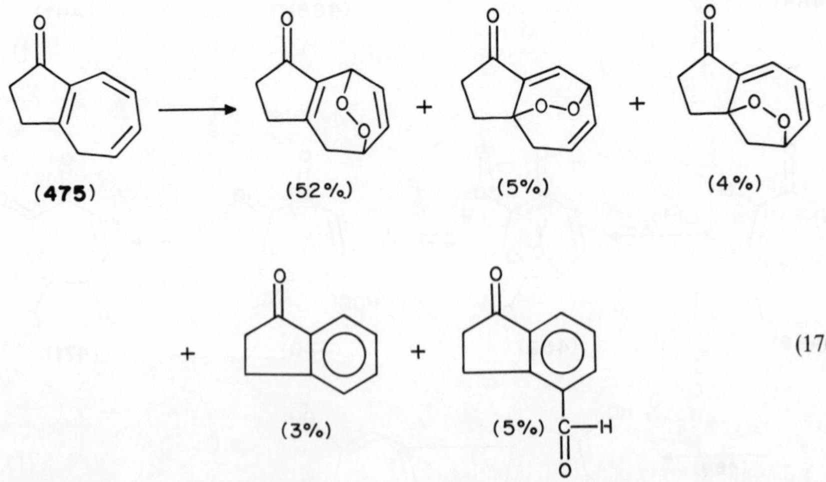

(474)

(475)   (52%)   (5%)   (4%)

$$(176)$$

(3%)   (5%)

Unsensitized photooxygenation[383] of the photochromic compound **476**, which produces photoenol **477** on normal photolysis, is reported to give peroxide **478** and ultimately dione **479**. The mechanism suggested[306] is outlined below and involves attack of $^1O_2$ on the photoenol **477**.

(476) R = H or Ph   **(477)**   **(478)**

(479)

Ketones **480**[384] and **481**[385] are reported to have low-lying triplets, are efficient physical quenchers of $^1O_2$ and are presumably unreactive.

(480)   **(481)**

Finally, polyene carboxylate diester **482**[386] as well as esters **483–485**[387] are all reportedly inert to $^1O_2$.

(482)   **(483)**   **(484)**   **(485)**

D. Reaction of $^1O_2$ with β, γ-Unsaturated Carbonyl Compounds

1. Simple systems

We have noted above the general sluggishness or inertness of simple α, β-unsaturated enones towards singlet oxygen, with the notable exception of those in an *s-cisoid* conformation. This is consistent with the weak electrophilicity of singlet oxygen. Not surprisingly, therefore, the related β,γ-unsaturated enones serve as good substrates for $^1O_2$, generally generating the corresponding γ-hydroperoxy-α,β-unsaturated enone as the primary, if not sole product. This is true for enones 486[388,389], 488[390], 491[389,391,392], 493[393] and 495[394] (equations 177–181).

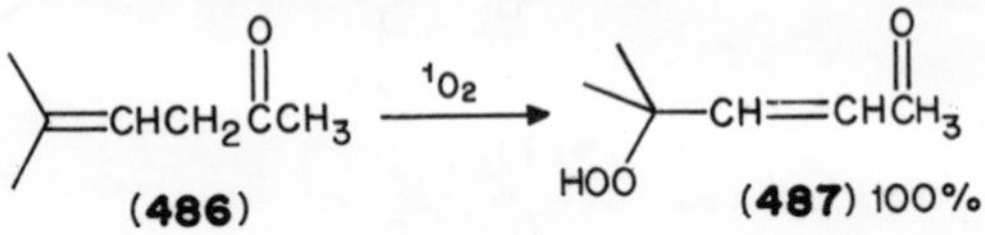

(177)

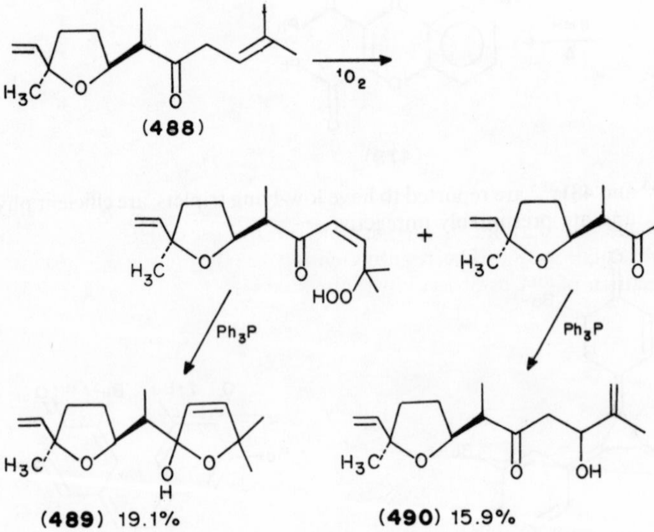

(178)

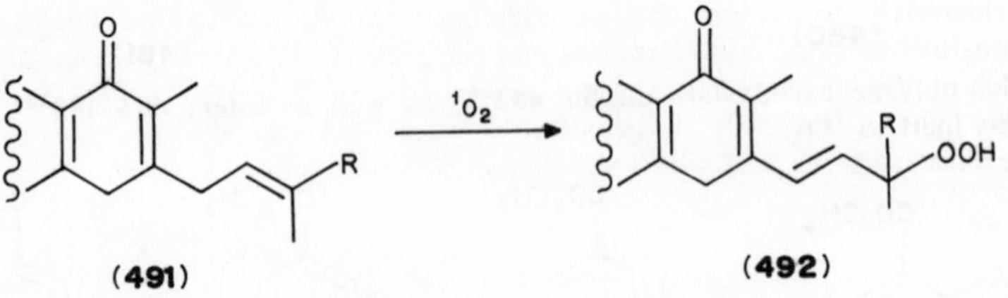

(179)

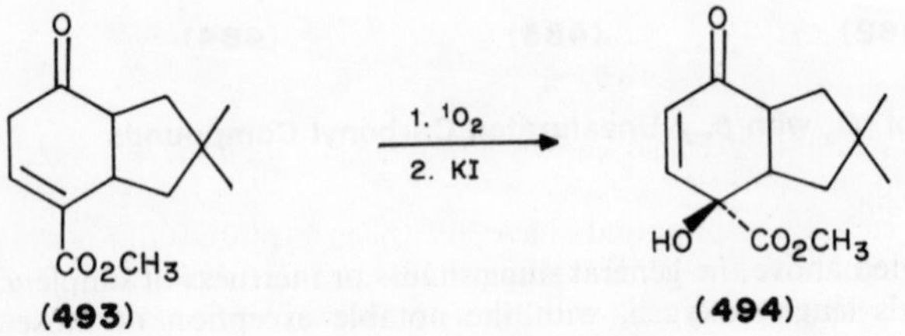

(180)

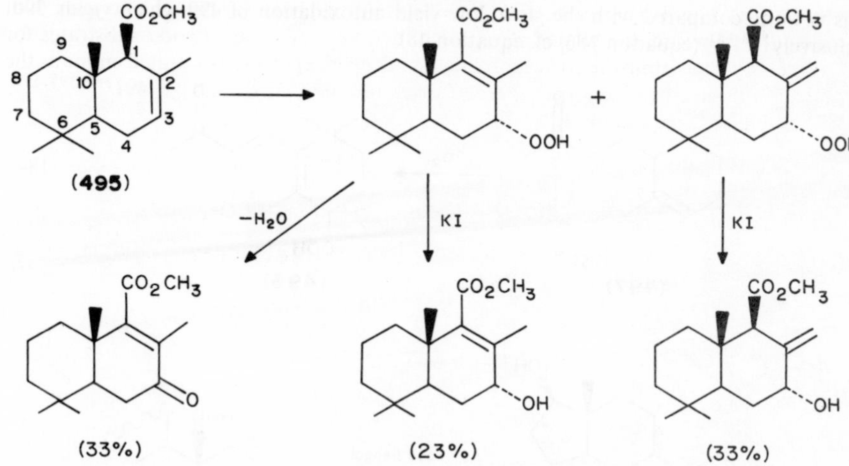

(181)

In the case of **493**, the authors[392] suggest that the species undergoing oxidation is actually a dienol (equation 182). The resulting endoperoxide presumably opens to the hydroperoxy precursor of **494**, as observed with tropolones (cf. equations 171–173).

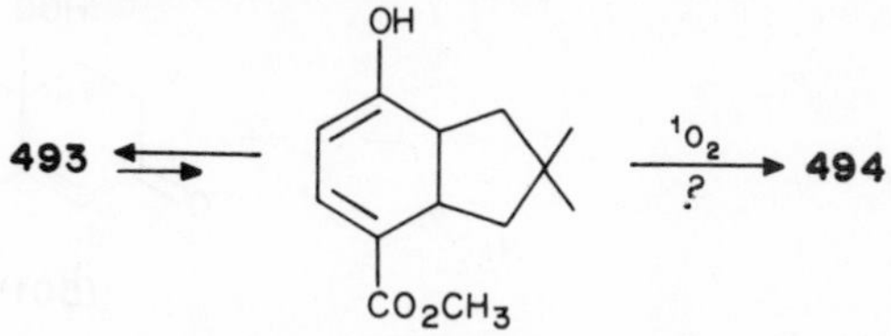

(182)

From compounds **488** and **495** both conjugated and non-conjugated products are formed, though the former predominate. In the case of **495**, the oxygen approaches the ring exclusively *trans* to the C-10 angular methyl group. As noted above, singlet oxygen is quite sensitive to steric considerations and the methyl and carboxylate groups inhibit approach of $^1O_2$ to the top side of the compound.

In the light of the facile oxygenation of esters **493** and **495**, it is surprising that cyclohexenecarboxylic acid **496** is unreactive to $^1O_2$[262] (equation 183).

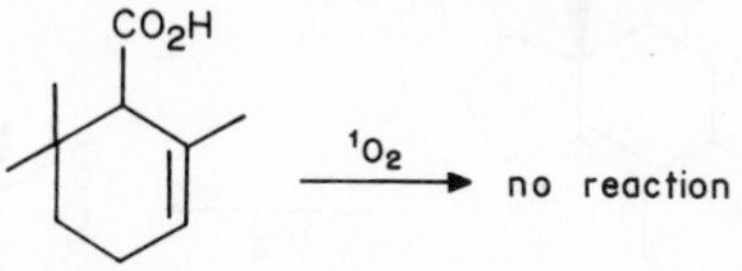

(183)

Vinylogous reactions are also known. Thus $\alpha, \beta, \delta, \varepsilon$-dienone **497** yields the conjugated **498**[336] (equation 184).

A series of steroidal compounds have also been explored and again the conjugated product predominates. 17β-Hydroxyester-5(10)-en-3-one **499** reacts rapidly with singlet oxygen to give a high yield of 10β- and 10α-hydroperoxides **500** and **501** (equation 185)[395].

This is to be compared with the slow, low-yield autoxidation of **499** which yields **500** exclusively[130,131] (equation 186; cf. equation 48).

$$\text{(497)} \xrightarrow{\ ^1O_2\ } \text{(498)} \tag{184}$$

$$\text{(499)} \xrightarrow[\ ^1O_2\ ,1.5h]{h\upsilon\ /\ rose\ bengal} \text{(500) 70\%} \qquad + \qquad \text{(501) 20\%} \tag{185}$$

$$\mathbf{499} \xrightarrow[\ ^3O_2,90\,h\ ]{h\upsilon} \mathbf{500} \quad (31\%) \tag{186}$$

Similarly 6-oxo-$\Delta^{8(14)}$ steroids **502** yield the 14α-hydroxy-7-en-6-ones **(503)** (equation 187)[388,396].

$$\text{(502)} \xrightarrow{\ ^1O_2\ } \text{(503)} \tag{187}$$

Finally, Furutachi and colleagues[388] report that while the autoxidation of Δ^5-cholesten-3-one **100a** produces a mixture of the conjugated 6α- and 6β-hydroperoxides **101** and **102**, singlet oxygenation generates **101** exclusively[388]. The stereoselectivity of this reaction was confirmed by deuterium labelling (equation 188). It should be noted that there are earlier reports[397] suggesting that both **101** and **102** are obtained in hematoporphyrin-sensitized reactions. These, however, probably involve free-radical processes.

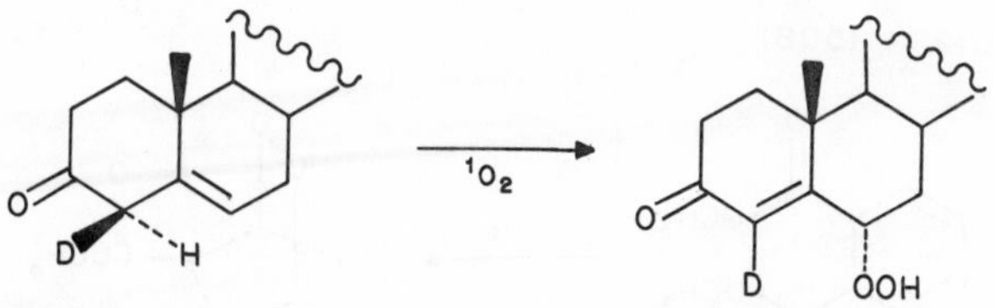

$$\tag{188}$$

(100d) 90% D   **(101d)** 85% D

2. Non-conjugated polyene carbonyls

We have already mentioned above the oxygenation of dienone retinoid **497** in which one of the double bonds is conjugated with the carbonyl group, while the remote double bond is not. In this case, reaction occurs so as to bring the second double bond into conjugation as well (equation 184)[336]. In steroid **112**, on the other hand, both double bonds are out of conjugation with the carbonyl but are conjugated to one another. Because of the *s-trans* conformation, Diels–Alder reaction is precluded. Singlet oxygen reacts with this system via an ene mode to give as the initial product diene **504**, in which the olefinic linkages remain conjugated. Since they are now cisoid, a rapid $^1O_2$ Diels–Alder addition ensues producing a mixture of α- and β-endoperoxides **505** (equation 189)[398].

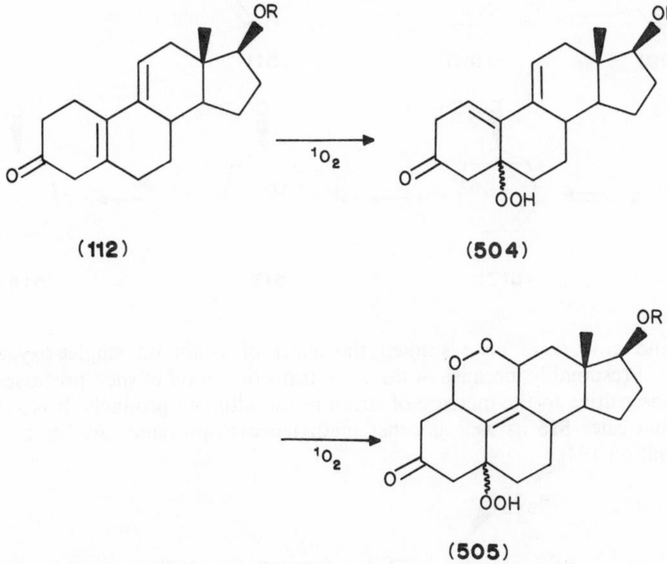

(112)   **(504)**

$$\tag{189}$$

(505)

As expected, where an *s-cis* conformation is feasible 2 + 4 cycloaddition is observed. Thus 3,5-cycloheptadienone **506**[399a] and cyclohexadiene **507**[399b] yield the related endoperoxides (equations 190 and 191).

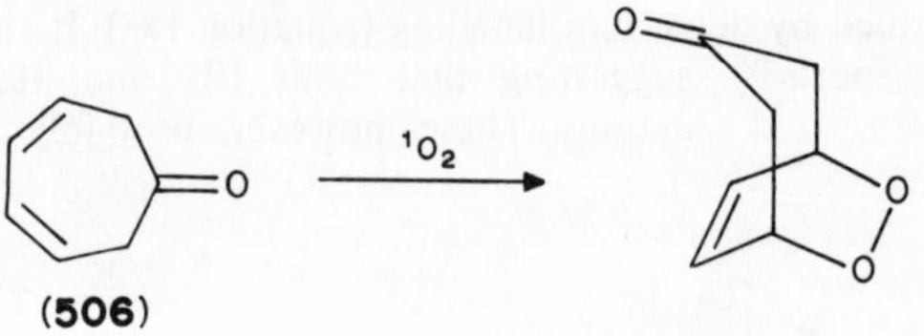

$$(190)$$

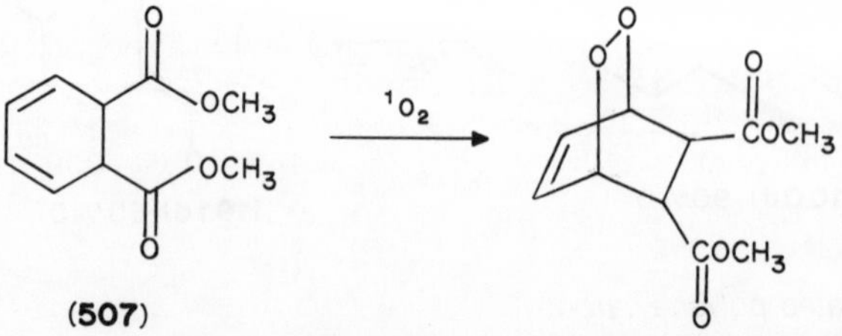

(506)

$$(191)$$

(507)

Of the four isomeric cycloheptatrienecarboxylic acid esters (**508–511**), only the 2, 4, 6-isomer **508** proved reactive to singlet oxygen generated via photosensitization or microwave discharge (equation 192 and 193)[387,400]. In both cases, the norcaradiene endoperoxide **513** was the major product; however, under photosensitization the corresponding diepoxide **514** is formed.

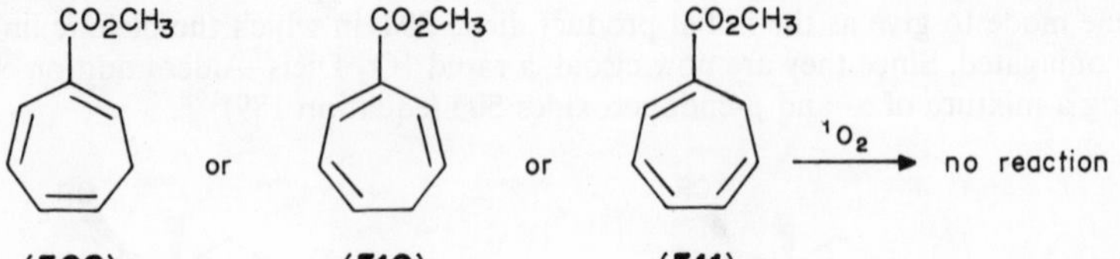

$$\text{no reaction} \quad (192)$$

(509)   **(510)**   **(511)**

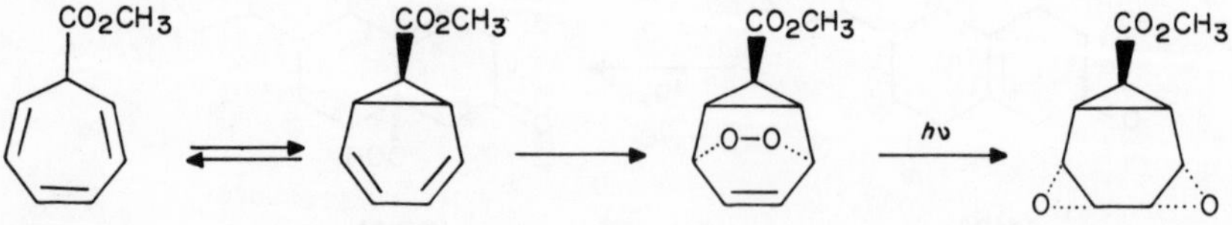

(508)   **(512)**   **(513)**   **(514)**

$$(193)$$

Frimer and coworkers have studied the effect of strain on singlet-oxygen reactions[401–405]. Presumably, because of the early transition state of such processes, $^1O_2$ is essentially insensitive to the increase of strain in the ultimate products. It is surprising, therefore, that ester **515** as well as other methylenecyclopropanes are inert to singlet oxygen (equation 194).

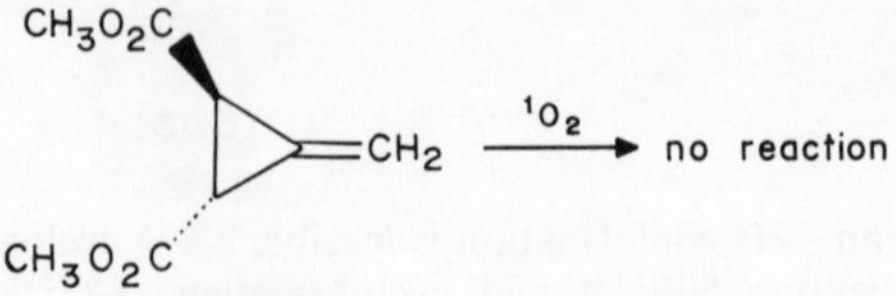

$$\text{no reaction} \quad (194)$$

(515)

Frimer[401,404] has attributed this inactivity to an excessively large interatomic distance between the α-olefinic carbon and the γ-allylic hydrogen, a distance which must be spanned by the molecular oxygen irrespective of mechanism. For methylencyclopropane, the C_α-$H_{allylic}$ distance is 3.27 Å compared to only 3.02 Å in isobutylene. Frimer suggests that as a result of this increment of 0.25 Å, the ring allylic hydrogens are essentially 'out of reach' for the abstracting oxygen atom.

Frimer and Antebi[401,405] also studied the photosensitized oxidation of two other ring-strained β,γ-unsaturated carbonyl compounds, ester **516a** and ketone **516b**. Although the uptake of oxygen proved quite rapid, the reaction was slowed by radical inhibitors but not by $^1O_2$ quenchers. This and other evidence clearly indicates that the primary products, epoxide **518** and allylic hydroperoxide **519**, result not from a singlet-oxygen (Type II) reaction but from free-radical (Type I) processes (equation 195).

The question remains, however, as to why no $^1O_2$ reaction is observed. The answer would seem to be related to the relatively high ionization potential of cyclopropene[406,407] (see Section IV.A). With the rate of $^1O_2$ reaction slowed substantially, the competing photochemically initiated free-radical autoxidation predominates.

E. Ketenes

The low-temperature ($-25\,°C$) reaction of a variety of ketenes **520** with singlet oxygen (chemically generated from triphenylphosphite ozonide) yields the corresponding diox-etanes, α-peroxy lactones **523** (equation 196)[408,409]. The latter cleaves to ketone and CO_2 with fluorescence. The formation of these α-peroxy lactones by photooxygenation at $-78\,°C$ could be established by spectroscopy (characteristic IR absorption $\sim 1880\,cm^{-1}$). However, the yields were generally much lower than those listed in equation 196, presumably because of competing autoxidation yielding peresters (see Section III.D.4) and possibly because of the instability of α-peroxy lactones to these photooxidative reaction conditions. In methanol, α-peroxy lactone formation is completely suppressed and α-methoxyperacetic acids **524** are produced instead. Since the peroxy lactones are stable to methanol, peracids **524** could well result from an intercepted perepoxide **521** or zwitterion **522** intermediate.

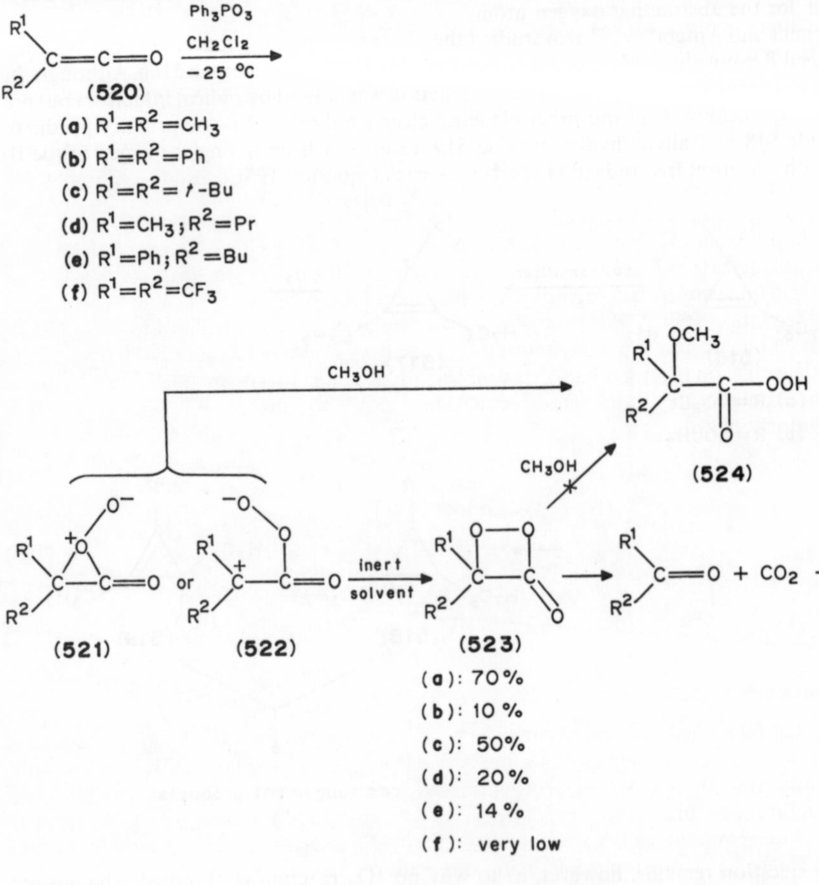
Ph3PO3
CH2Cl2
−25 °C
(520)
(a) R1=R2=CH3
(b) R1=R2=Ph
(c) R1=R2=t-Bu
(d) R1=CH3; R2=Pr
(e) R1=Ph; R2=Bu
(f) R1=R2=CF3
CH3OH
(524)
CH3OH
inert solvent
(521)
or
(522)
(523)
C=O + CO2 + hν
(a): 70%
(b): 10 %
(c): 50%
(d): 20%
(e): 14 %
(f): very low

(196)

V. SUPEROXIDE ANION RADICAL

A. Generation

Despite the omnipresence of one-electron processes in nature, free-radical damage presents a serious and constant threat to living organisms[410–412]. One available source of radicals in the body is the surperoxide radical $O_2^{-\cdot}$, which is formed in a large number of reactions of biological importance in both enzymic and non-enzymic processes[2,3]. It follows then that it is of great value to understand the organic chemistry of $O_2^{-\cdot}$ for, as Fridovich[413] has poignantly noted: 'If we are going to know how it does its dirty work, we have to know what it is capable of doing'. Nevertheless, had convenient methods not been found for generating $O_2^{-\cdot}$ in aprotic organic solvents, progress in this direction would have undoubtedly been slow and tedious.

Two basic approaches have been developed and are presently in use. The first involves *in situ* generation of $O_2^{-\bullet}$ by the electrolytic reduction of molecular oxygen[414-416]. This method permits the controlled generation of low concentrations ($< 10^{-2}$ M) of pure $O_2^{-\bullet}$ and is well suited for mechanistic studies. This is particularly true for cyclic voltammetry, which allows the researcher to follow the course of the reaction and detect unstable intermediates. Efficient product studies, however, require greater $O_2^{-\bullet}$ levels.

An alternate approach utilizes superoxide salts which are well-defined sources of $O_2^{-\bullet}$. The inorganic salts, such as the commercially available potassium superoxide (KO_2), are generally insoluble in aprotic organic solvents, though they are slightly soluble in those of high polarity like DMSO. Nevertheless, solutions of KO_2 have been conveniently prepared in benzene, toluene, acetonitrile, DMSO, pyridine, triethylamine, THF, etc. through the agency of phase-transfer catalysts such as crown ethers[417]. Tetramethyl-ammonium superoxide has also been synthesized and, in contrast to its alkali-metal analogues, is quite soluble in a number of aprotic solvents[418,419].

It should be noted that the halogenated solvents (Freons, CCl_4, $HCCl_3$, CH_2Cl_2) are unsuitable since they react with superoxide[420-424]. Protic media induce the acid-catalyzed disproportionation of superoxide to triplet molecular oxygen ($^3O_2$) and hydroperoxy anion (equation 197)[27]. This process involves primarily two steps (equations 198 and 199) for which kinetic and thermodynamic data have been evaluated by pulse radiolysis[425]. In aprotic media, on the other hand, this solvent-induced disproportionation is absent, while highly unfavorable energetics (equation 200)[426] rule out a spontaneous disproportionation.

$$H^+ + 2O_2^{-\bullet} \rightleftharpoons HO_2^- + O_2 \tag{197}$$

$$H^+ + O_2^{-\bullet} \rightleftharpoons HO_2^\bullet \qquad pK_a(HO_2^\bullet) = 4.69 \tag{198}$$

$$HO_2^\bullet + O_2^{-\bullet} \rightarrow HO_2^- + O_2 \qquad k = 1 \times 10^8 \, M^{-1} s^{-1} \tag{199}$$

$$2O_2^{-\bullet} \rightarrow HO_2^- + O_2^{-2} \qquad \Delta G > 28 \, kcal \, mol^{-1} \tag{200}$$

B. Modes of Reaction

The data obtained over the past decade have led scientists to suggest four basic modes of action for $O_2^{-\bullet}$ in aprotic media, namely electron transfer (equation 201), nucleophilic substitution (equation 202), deprotonation (equation 203) and perhaps hydrogen atom abstraction (equation 204)[27]. It is important to remember, however, that subsequent to each of these primary modes, secondary oxidative processes can and generally do take over. Hence, one must proceed with due caution in any attempt to determine the mechanism of reaction simply based on product analysis. Let us now examine each of these modes in a bit more detail.

$$R + O_2^{-\bullet} \rightarrow R^{-\bullet} + O_2 \tag{201}$$

$$RX + O_2^{-\bullet} \rightarrow RO_2^\bullet + X^- \tag{202}$$

$$RH + O_2^{-\bullet} \rightarrow R^- + HOO^\bullet \tag{203}$$

$$RH + O_2^{-\bullet} \rightarrow R^\bullet + HOO^- \tag{204}$$

1. Electron transfer

Electron transfer is one of the most common modes of $O_2^{-\bullet}$ action in biological systems and is the essence of the disproportionation process (equation 199). This is of course not surprising, considering that the redox potential of the $O_2/O_2^{-\cdot}$ couple (vs. NHE) is $-0.33\,V$ in water and $-0.6\,V$ in organic solvents[427,428]. This quarter of a volt gap between aqueous and aprotic media cannot be attributed to differences in the dielectric constants of the media, since the electrochemical potential of oxygen is relatively insensitive to the differing dielectric constants of a variety of aprotic solvents[427–430]. To understand this and many other phenomena related to $O_2^{-\cdot}$ activity (e.g. nucleophilicity and basicity; *vide infra*), we must recall that $O_2^{-\cdot}$ is a small, hard, non-polarizable anion. In aqueous/protic media it will be highly and tightly solvated and, hence, *thermodynamically more stable* than in aprotic media in which such solvation mechanisms are generally absent.

Superoxide does not interact with simple olefins or aromatic compounds. It will, however, transfer electrons to good electron acceptors such as quinones, which are converted to the corresponding semiquinones (equation 205)[431–435].

$$O=\!\!\left\langle\!\!\left\langle\ \right\rangle\!\!\right\rangle\!\!=\!\!O \xrightarrow{\ O_2^{-\cdot}\ } \ \cdot O\!\!-\!\!\left\langle\!\!\left\langle\ \right\rangle\!\!\right\rangle\!\!-\!\!O^- + O_2 \qquad (205)$$

2. Nucleophilic attack

In aprotic media, $O_2^{-\cdot}$ is an extremely vigorous nucleophile[27]. This 'supernucleophilicity' has been rationalized in terms of the α-effect and is related to the destabilizing effect of the vicinal pairs of non-bonding electrons[27]. Superoxide reacts with halides and sulfonates by an S_N2 process to produce hydroperoxides, peroxides, alcohols or carbonyl compounds depending on the substrate, reaction conditions and work-up procedures, as outlined in equation 206.

$$RX \xrightarrow{\ O_2^{-\cdot}\ } ROO^{\cdot} \xrightarrow[O_2]{\ O_2^{-\cdot}\ } ROO^- \qquad (206)$$

Similarly esters, including linoleates and acyl halides, undergo nucleophilic attack yielding the corresponding diacyl peroxides or carboxylic acids ('saponification'). Evidence has recently been presented, however, which suggests that, in some systems at least, 'saponification' proceeds via an electron transfer process[436,437].

In protic media by comparison, $O_2^{-\cdot}$ reacts by this mode sluggishly, if at all, as a result of the inhibitory effect of the tight hydration sphere surrounding this small, charged anion.

3. Deprotonation

The pK_a of $HOO^{\cdot}$, determined by aqueous radiolytic studies, is 4.69 (equation 198)[425]. Its conjugate base $O_2^{-\cdot}$ should therefore be as weakly basic as acetate. Nevertheless, in aprotic media superoxide has proven to be a powerful mediator of base-catalyzed process. This greater basicity has been rationalized from three perspectives. Firstly, $O_2^-\cdot$ in aprotic media lacks the 'stifling' solvation sphere commonplace for small hard anions in aqueous and protic media. Furthermore, in the poorly solvating aprotic media the equilibrium described by equation 198 should be shifted away from the charged species $O_2^-\cdot$ towards its electroneutral counterpart HOO$\cdot$. Indeed Sawyer[438] has reported a higher pK_a value (~ 12) for HOO$\cdot$ in DMF, and this value should be even higher in nonpolar solvents. More importantly, however, unlike the low fluxes of superoxide obtained when it is generated enzymatically or radiolytically, $O_2^-\cdot$ is longer lived and generally present in much higher concentration in aprotic media. In the presence of excess $O_2^-\cdot$, the facile disproportionation reaction (equation 199) between $O_2^-\cdot$ and its conjugate acid (HOO$\cdot$) drives the unfavorable deprotonation (equation 198) to the right. This in turn raises the effective basicity of $O_2^-\cdot$, i.e. the efficiency with which $O_2^-\cdot$ can effect proton transfer. Indeed, calculations[27] show that $O_2^-\cdot$ can promote proton transfer from substrates to an extent equivalent to that of a conjugate base of an acid with a pK_a of ~ 25.

In light of superoxide's apparent basicity, it is not surprising that it induces elimination reactions as well as base-catalyzed autoxidation (BCA). Qualitatively, Frimer and coworkers[27] have consistently found that the rate of a BCA process depends on the exact nature of the base utilized and decreases in the order: t-butoxide > superoxide > hydroxide.

4. Hydrogen abstraction

Radical reactions are the final modes of action anticipated for superoxide anion radical. Surprisingly, $O_2^-\cdot$ *in aprotic media* turns out to be a rather unreactive radical. Regarding superoxide's ability to abstract hydrogen atoms, thermodynamic calculations[439] suggest that $O_2^-\cdot$ is only likely to do so from those very rare substrates bearing R—H bonds with bond energies as low as 66 kcal. Indeed, the reinvestigation of systems originally assumed to be initiated by hydrogen-atom abstraction mediated by $O_2^-\cdot$ (equation 207, path a) have nearly all turned out to be proton abstraction, followed by oxidation of the resulting anion either by the concomitantly formed hydroperoxy radical or molecular oxygen (equation 207, path b)[27].

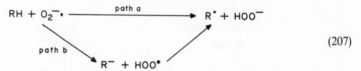

$$(207)$$

We have noted above that superoxide is generally unreactive with olefins, and here again thermodynamic calculations confirm that the radical addition of superoxide to the olefinic double bonds is more than 18 kcal endothermic[27d]. $O_2^-\cdot$ does not initiate free-radical chain polymerization and actually inhibits styrene polymerization[440,441]. Finally, radical–radical coupling, so common and rapid with normal radicals, is quite rare for $O_2^-\cdot$, with the noted exception of superoxide anion-radical cation couplings[27]. By contrast, *in aqueous media* superoxide can serve as an oxidant and thereby initiate radical processes[442,443].

5. Work-up conditions

Frimer and coworkers have described[444] two methods for working-up superoxide reactions. In the first, the reaction is quenched with aqueous acid (e.g. 10% HCl) and the acidic products are extracted by saturated sodium bicarbonate washings. In the second, a tenfold excess of methyl iodide is added to the reaction mixture prior to aqueous work-up. This has the effect of converting the excess superoxide to methoxide or dimethyl ether, in addition to methylating the oxyanions present. As we shall see shortly, this latter method has been found to be particularly useful for detecting and trapping 'saponification' products which, under aqueous workup, simply recyclize back to starting material (equation 208).

$$(208)$$

C. Reaction of Superoxide with Enones Lacking Labile Hydrogens

1. Simple enones

The simple enone moiety *per se* is generally unreactive to O_2^-. Thus, 4, 4, 6, 6-tetrasubstituted cyclohex-2-en-1-ones **188** have proved totally inert to superoxide, even after being in contact for several days[27b,68,164,165] (see equation 81). Some reversible electron transfer does seem to occur, however. Gibian and Russo[445,446] have demonstrated that O_2^- induces the extremely rapid *cis* to *trans* isomerization of *cis*-2, 2, 6, 6-tetramethylhept-4-en-3-one **525** (equation 209). The isomerization to the *trans* isomer **528** presumably proceeds via ketyls **526** and **527**, although radical species could not be observed by CIDNP. It should be noted that *trans* **528** is slowly converted to unidentified products. Neither pivalic acid nor 3, 3-dimethylpyruvic acid were observed, products analogous to those ultimately obtained from chalcones (see next section).

$$(209)$$

2. Aryl enones

When the π system is extended, the electron transfer from O_2^- to substrate becomes a much more facile process. Frimer and Rosenthal[447,448] have studied the oxidative cleavage of chalcones mediated by O_2^-. Carboxylic acids were obtained as the final products and no intermediate epoxide formation could be detected. A Michael-type addition to the enone system was further excluded on the basis of $K^{18}O_2$ experiments which showed very little label incorporation. The mechanism suggested involves electron transfer (equation 210).

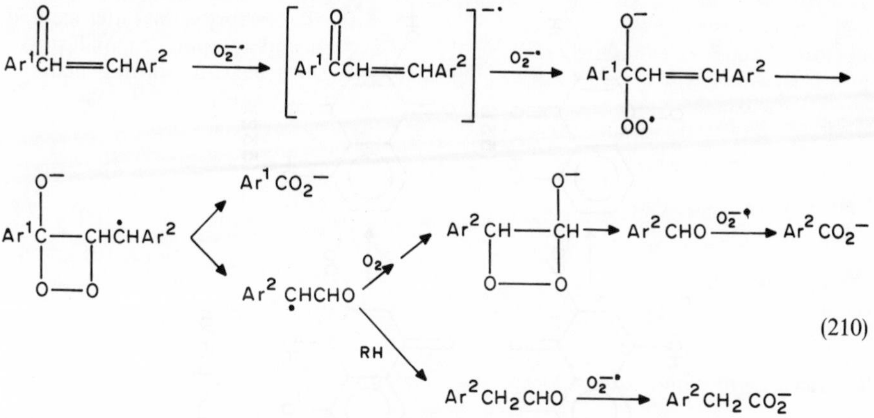

$$
Ar^1CH{=}CHAr^2 \xrightarrow{O_2^{-\cdot}} \left[Ar^1\overset{O}{\overset{\|}{C}}CH{=}CHAr^2 \right]^{-\cdot} \xrightarrow{O_2^{-\cdot}} Ar^1\overset{O^-}{\underset{OO^\cdot}{\overset{|}{C}}}CH{=}CHAr^2 \longrightarrow
$$

(210)

In a related study, dibenzylideneacetone **529** reacted with $O_2^{-\cdot}$ to yield cinnamic acid and benzaldehyde (equation 211)[449].

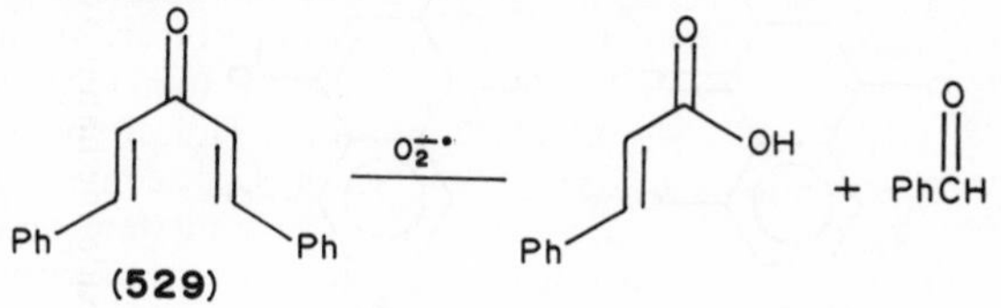

(211)

3. Quinones

Several research groups report that anion radicals can be detected by ESR in the reaction of $O_2^{-\cdot}$ with various benzoquinones[431-435]. However, other than the reversible formation of radical anions, no oxidation products have been isolated in these cases.

De Min and coworkers[435] have recently studied the reaction of superoxide (generated from KO_2 in toluene in the *absence* of 18-crown-6) with juglone (5-hydroxy-1, 4-naphthaquinone; **530**). They report that isomeric enols **531** and **532** are the ultimate oxidation products formed following initial electron transfer of $O_2^{-\cdot}$ to the quinone substrate. The suggested mechanism is outlined in Scheme 24.

Saito and colleagues[450] report that 2,3-dimethyl-1,4-napthaquinone **533** and other vitamin-K-related compounds react with KO_2/18-crown-6 to give the corresponding oxirane **534** and its secondary oxidation product phthalic acid in a 25–35% yield. The remaining products are unidentified and the mechanistic details are unclear. Based on the reactions of other benzoquinones[431-435] initial electron transfer is likely here as well (equation 212), although other mechanisms involving deprotonation have been suggested[27d].

1,2-Naphthaquinone (**536**) is oxidatively cleaved[451], like other α-diketones[27], to the corresponding diacid **537**. Under the basic reaction conditions, the latter undergoes intramolecular Michael addition to furanone **538** (equation 213).

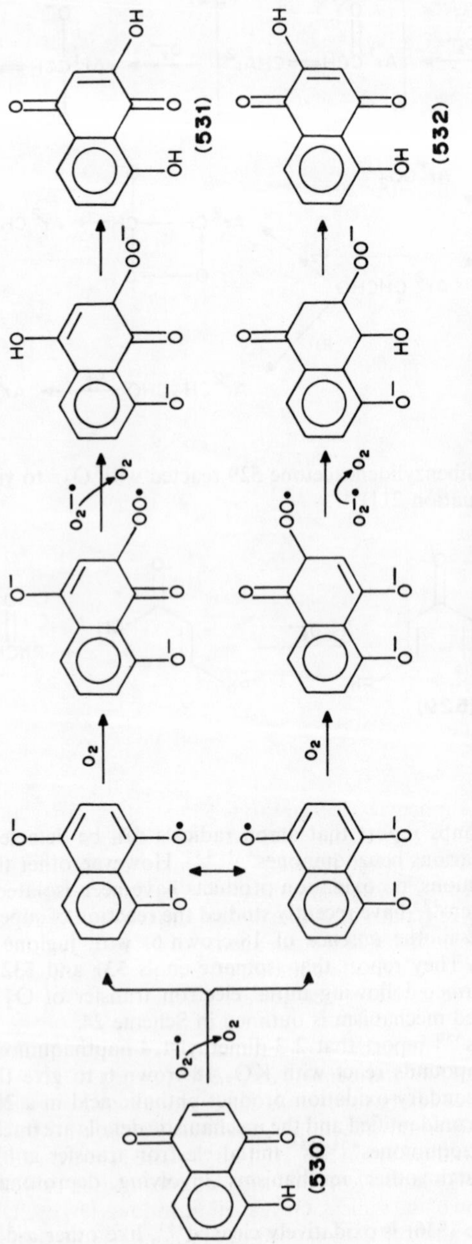

SCHEME 24. Superoxide mediated oxidation of Juglone (530)

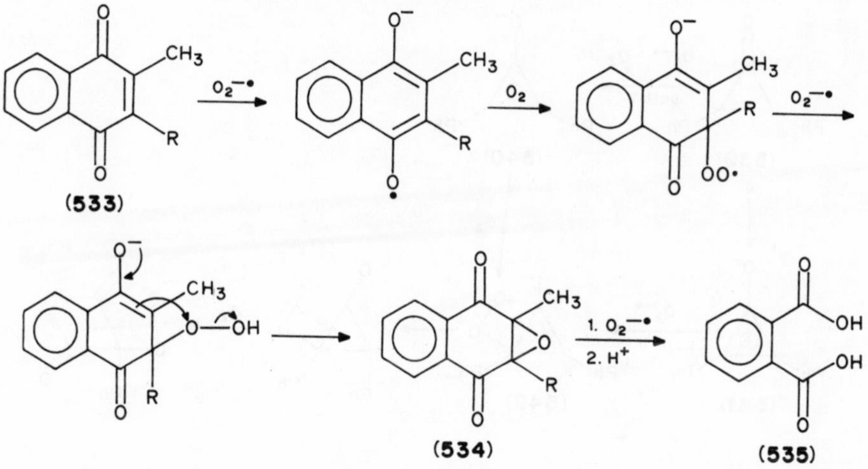

(212)

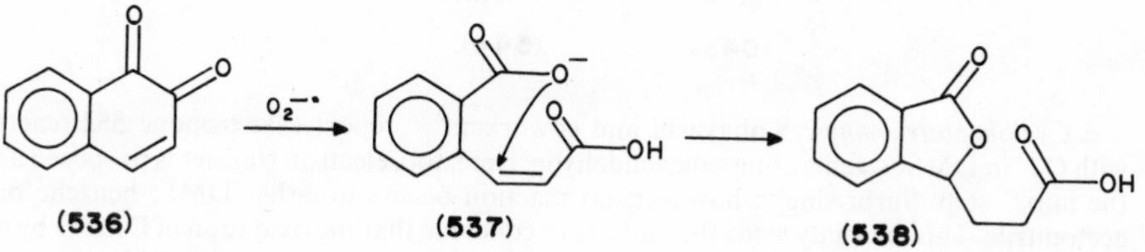

(213)

4. Annelones

a. Cyclopropenones. Diphenylcyclopropenone **539** reacts slowly with superoxide (85% conversion after 7 days)[452] to give benzil **543** (18%) and its oxidative cleavage product[27], benzoic acid **544** (27%). Neckers and Hauck[452] presume that an electron transfer reaction is involved, since the aromatic radical anion **540** is reported to form benzil in the presence of oxygen[453] (equation 214, path a). Nevertheless, diphenylcyclopropenone **539** is also known to react with nucleophiles, as a consequence of the large contribution of aromatic mesomeric structure **541**. Nucleophilic attack of $O_2^{-\cdot}$ on **541** (equation 214, path b) is expected to give the same initial oxygen adduct **542**.

b. Cyclopentadienones. The reaction of tetracyclones **545** with $O_2^{-\cdot}$ leads to 2-pyrone **548**, 2-hydroxyfuranones **549** and benzoic acids **550** and **551**[447,448,452,454] (see Scheme 25). Although the pyrones can be formed via the corresponding epoxides, these were not observed as intermediates in this reaction. As before, electron transfer is presumed to be the first step generating radical anion **546**. Highest unpaired electron density is expected α to the carbonyl (as in **547**) which allows for extended conjugation as well as double allylic and benzylic stabilization of the radical. Oxygenation, followed by cyclization, along path a or path b ultimately leads to pyranone **548** or furanone **549** respectively. The benzoic acids presumably result from oxidative cleavage of various intermediates along the reaction route (see Scheme 25).

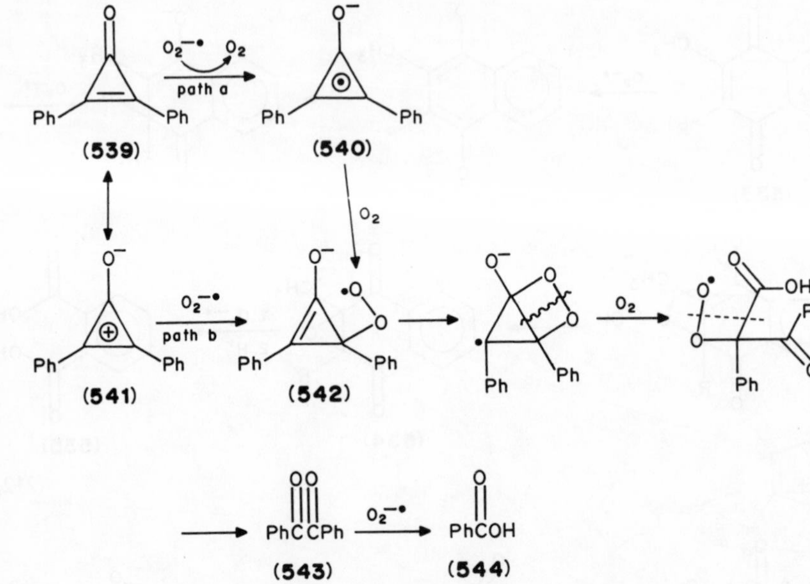

(214)

c. *Cycloheptatrienones*. Kobayashi and coworkers[455] report that tropone **552** reacts with $O_2^{-\cdot}$ in DMSO, generating salicylaldehyde. Here, too, electron transfer is proposed as the initial step. Surprisingly, however, no reaction occurs in either DMF, benzene or acetonitrile. This anomaly leads the authors to conclude that the oxidation of DMSO by a reversibly formed intermediate is a crucial step in this reaction (equation 215).

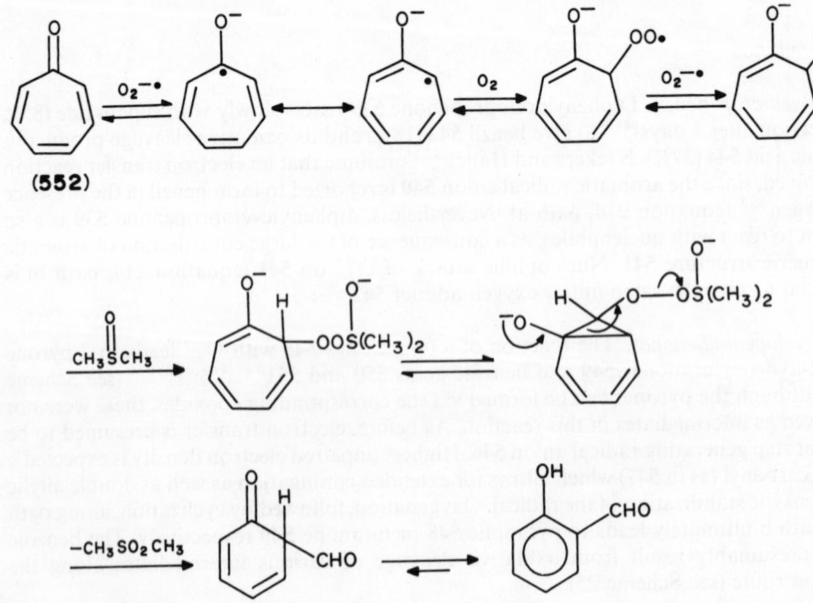

(215)

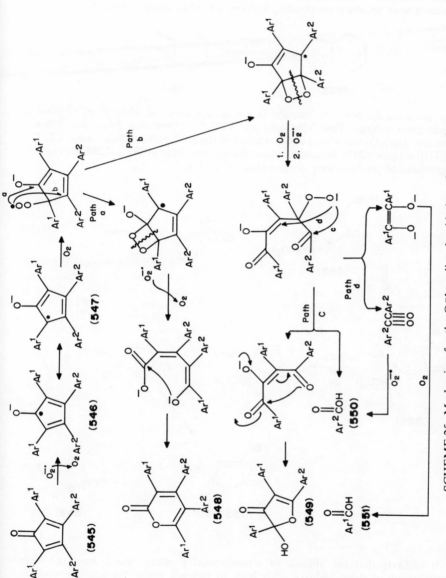

SCHEME 25. Mechanism for the $O_2^{-\cdot}$ mediated oxidation of tetracyclones

5. Lactones

a. 2-Furanones. Lactones, like esters, are expected to undergo $O_2^-\cdot$ mediated saponific-ation[27]. Indeed, dihydrofuranone **553** reacts with KO_2/crown ether to generate, upon aqueous work-up, the corresponding hydroxy acid **554** (equation 216)[456].

$$(216)$$

$$(553) \qquad\qquad (554)$$

Interestingly, however, reaction is not always observed with the corresponding unsaturated analogs. Thus Moro-oka and Foote[457] report that 2-furanone **555** was unreactive to KO_2 introduced to the reaction mixture as a powder dispersion in CH_3CN or THF (equation 217). Related furanone **556** was converted to **557** and **558** but again no saponification products were observed (equation 218).

$$(217)$$

$$(555)$$

$$(557)\ 90\% \qquad\qquad (558)\ 10\%$$

$$(218)$$

It is likely that the absence of saponification products results from the facile recyclization of these hydroxy acids back to lactones under the aqueous work-up conditions (equation 219). In contradistinction to the saturated analogs, free rotation is not allowed and the alcohol and acid fragments are held in proximity to one another. This in turn facilitates lactonization, even under mild conditions. The validity of this hypothesis has been demonstrated in the coumarin system described in the next section.

$$\text{(559)} \quad \xrightleftharpoons[\text{H}^+]{\text{O}_2^{-\cdot}} \quad \text{(560)} \tag{219}$$

b. Coumarins. When the α,β-unsaturated lactone coumarin **561** was reacted with $O_2^{-\cdot}$ in aprotic media followed by aqueous acid work-up, only starting material was isolated. When, however, methyl iodide was used to quench the reaction and convert oxy-anions to methoxy species, an 80% yield of the methyl esters of *o*-methoxy-*cis*-cinnamic acid **562** was obtained (equation 220)[443]. This clearly indicates that the primary process is nucleophilic attack by $O_2^{-\cdot}$ at the lactone carbonyl carbon yielding saponification product **563**.

$$\text{(561)} \quad \xrightleftharpoons[\text{H}^+]{\text{O}_2^{-\cdot}} \quad \text{(563)} \quad \begin{array}{c} \xrightarrow{\times} \text{(564)} \\ \searrow \text{(562) 80\%} \end{array} \tag{220}$$

It is worth noting that no Michael addition of $O_2^{-\cdot}$ has been observed. Furthermore, despite the long reaction time (16 h), there was no evidence for the isomerization of *cis*-**563** (the dianion of *o*-coumarinic acid) to *trans*-**564** (the dianion of *o*-coumaric acid).

By similar processes, methoxycoumarins **565** and **567** are saponified by $O_2^{-\cdot}$, generating esters **566** and **568** (equations 221 and 222)[443].

$$\text{(565)} \quad \xrightarrow{\text{O}_2^{-\cdot}} \quad \text{(566) 94\%} \tag{221}$$

$$\text{(567)} \quad \xrightarrow{\text{O}_2^{-\cdot}} \quad \text{(568) 70\%} \tag{222}$$

D. Reaction with Enones Bearing Labile Hydrogens

1. C—H bonds

In discussing the properties of superoxide (Section V.B.3), we noted that its experimental effective basicity is somewhere in between hydroxide and t-butoxide. It is not surprising, therefore, that the deprotonation of labile hydrogens is probably the most prevalent mode of action for $O_2^{-\cdot}$ in aprotic media. The results obtained in base-catalyzed autoxidation processes mediated by superoxide are generally the same as those obtained with other bases of comparable base strength.

We have already described above the superoxide-mediated oxidation of cyclohexenones[27b,68,164,165] and steroids[199,200,212] (see Sections III.E.3.b and III.E.4). One interesting discrepancy observed in this regard relates to the superoxide-mediated oxidation of the unsubstituted cyclohex-2-en-1-one, **569** (equation 223). Frimer's group[27b,68,164,165], using KO_2/18-crown-6, observed only dimer **572** and trimer formation but no oxidation products. On the other hand, Dietz and coworkers[458], using electrogenerated $O_2^{-\cdot}$, reported a 30% yield of epoxide **573**. *Both* these results have been reconfirmed by Sugawara and Baizer[207] who suggest that, in the case of the electrochemical generation of $O_2^{-\cdot}$, the solvent is saturated with oxygen. In the absence of excess oxygen, the anion of cyclohexenone **570** has too little opportunity to form the hydroperoxide essential for the epoxidation reaction, and reacts by the available alternative pathway, Michael addition.

(223)

To test this hypothesis, the reaction of cyclohexenone with KO_2/18-crown-6 was repeated, but this time the reaction mixture was bubbled vigorously with oxygen[165]. An NMR analysis of the product mixture indicated the presence of a 10% yield of epoxide. A small amount of dimer ($\sim 3\%$) was formed, but no trimer could be detected. The remaining products seem to be the result of multiple oxidation and could not be characterized.

This simple example indicates that oxygenation is not always the preferred reaction course for a carbanion. Gilinsky-Sharon and Frimer[459] report that 5,6-dihydropyrone **574** reacts with superoxide to give dienone **575** in 90% yield, upon methyl iodide work-up of the reaction mixture (equation 224). Such base-induced elimination processes are precedented in the dihydropyrone[460] family.

A similar elimination reaction is observed[459] for ascorbic acid derivative **576**, which is converted by superoxide to alkylidene furanone **577**. This reaction, too, is precedented with other bases[461] (equation 225).

(224)

(225)

We note in passing that in both these cases saponification of the lactone linkage is not observed (cf. Section V.C.5).

2. O—H bonds

a. Enols. In our discussion of base-catalyzed autoxidation we have already described superoxides' ability to deprotonate cyclohexenone and steroidal enols and induce their transformation to the corresponding lactols and/or other oxygenation products (see Section III.E.3.b)[27b,68,164,199,200]. The enols of 1, 3-diones are also rapidly deprotonated by superoxide[205,207], but resist further oxygenation (see equation 72)[171,203−207].

3-Hydroxyflavones **219** undergo O_2^-· mediated autoxidation[180,184,185] to a mixture of depsides **221** and its 'saponification' products **222** and **223** (equation 89). Takahama[180] carried out this reaction in aqueous media using the photooxidation of riboflavin as his O_2^-· source. El-Sukkary and Speier[194] observe the same reaction using KO_2/18-crown-6 in THF. Surprisingly, however, this reaction does not occur when benzene or toluene are the solvents[462]. This raises some serious doubts as to the inertness of THF. The problematic nature of THF was in fact noted two decades ago by two pioneers in the superoxide field, Le Berre and Berguer[463], who discussed the instability of THF superoxide solutions.

In Section V.C.5 we noted that coumarins undergo 'saponification' of the lactone linkage with superoxide. However, in the case of the enols, 4- and 3-hydroxycoumarins, **578** and **581** (equations 226 and 227), deprotonation precedes saponification[68,443]. When **578** is reacted with O_2^-· for one hour, a 30% yield of simple deprotonation product **579** can be isolated subsequent to CH_3I work-up. At longer reaction times (16 h), deprotonation plus saponification affords **580** in 83% yield (equation 226).

Similarly, when **581** is reacted with O_2^-· for one hour, simple deprotonation product **582** and deprotonation plus saponification product **583** are isolated in 20% and 40% yields respectively. The remaining 40% is the oxidative cleavage product *o*-methoxybenzaldehyde **584**, which becomes the major product (93% yield) after 16 h of reaction. The probable mechanism is outlined in equation 227.

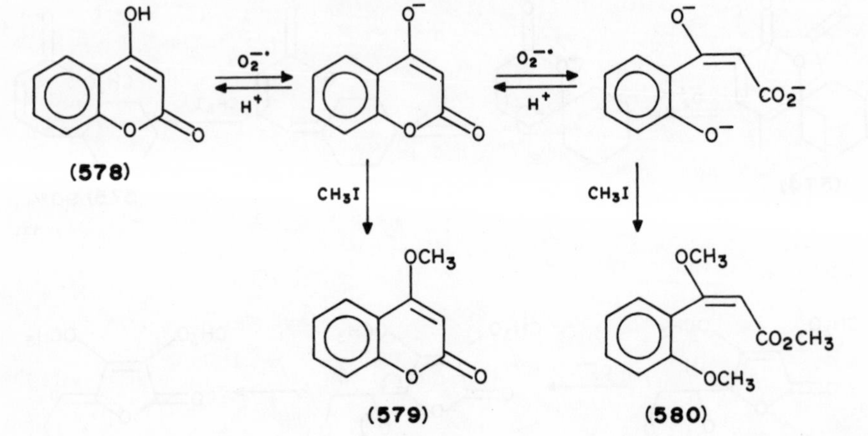

(226)

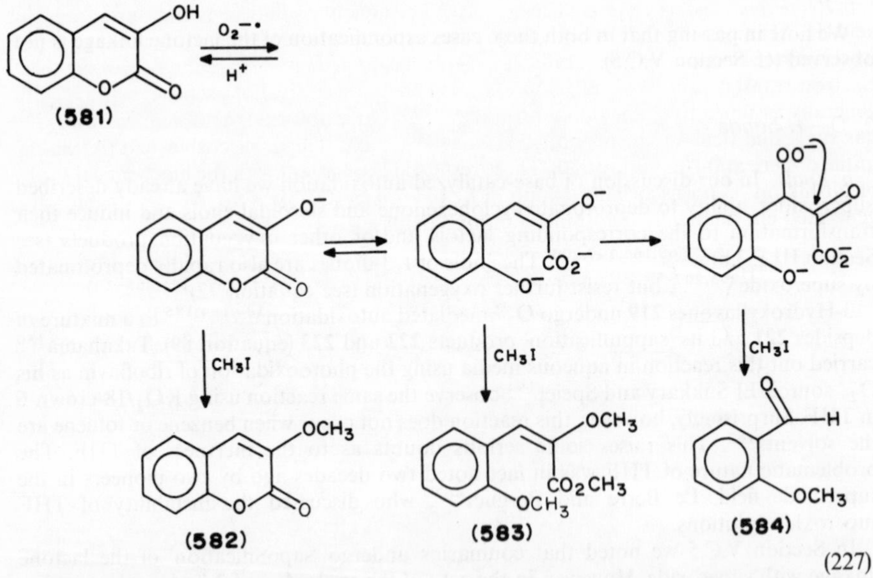

(227)

Gilinsky-Sharon[459] reports that enol **585** reacts with superoxide to give a quantitative yield of acidic products. When the product mixture was diazotized and separated by GLC, three products in a ratio of 2:1:1 were isolated and identified as diesters **590** and **591** and aldehydoester **592**, respectively. These three are presumably formed from the corresponding acids **586** and **587**, and lactone **589** (equation 228).

The formation of lactols from enols is a well-precedented base-catalyzed autoxidation process and hence the formation of lactol **589** is expected. Superoxide is also known to oxidize aldehydes to acids and to effect the oxidative cleavage of diketones to diacids[27]; hence, the oxidation of **588** to **587** and generation of **586** from **585** is not surprising.

$$(228)$$

(590) 49% **(591) 27%** **(592) 24%**

b. Reductones and ascorbic acid derivatives. The aqueous solution oxidation of ascorbic acid (H_2A) and its anion (HA^-) to dehydroascorbic acid (A) by superoxide ion ($O_2^{-\cdot}$) and its conjugate acid, perhydroxyl radical ($HO_2^{-\cdot}$), has been demonstrated to be a direct one-electron transfer process ($k \approx 10^5 \, M^{-1} \, s^{-1}$). The anion radical of ascorbic acid ($A^{-\cdot}$) is generally assumed to be the initial product[464-466] (equation 229). A subsequent study[467] has suggested that $A^{-\cdot}$ disproportionates to HA^- and dehydroascorbic acid (A) via an initial dimerization (equation 230). These results are consistent with the biochemical study of Nishikimi[468] on ascorbate oxidation at pH 7.4 by $O_2^{-\cdot}$ ($k \sim 10^5 \, M^{-1} \, s^{-1}$) that was generated enzymatically (xanthine–xanthine oxidase).

$$(229)$$

$$(230)$$

Comparable studies have been carried out in aprotic media using electrogenerated superoxide to mediate the oxidation of H_2A to A. Sawyer and coworkers[469-472] find that the stoichiometry for this reaction requires three molecules of ascorbic acid and two molecules of superoxide. In addition, superoxide mediates this process without the formation of molecular oxygen. As a result of these observations, Sawyer suggests that the initial rate-determining step is a concerted (equation 231) or rapid sequential (equation 232) transfer of a proton and a hydrogen atom to superoxide generating $A^{-\cdot}$ and H_2O_2 ($k = 2.8 \times 10^4 M^{-1} s^{-1}$). Subsequent reactions involve the proton-induced disproportionation of $A^{-\cdot}$ (equation 233) and oxidation of the resulting HA^- by H_2O_2 to yield A (equation 234). The sum total of these processes (equation 235) has the proper stoichiometry.

$$A\,\begin{array}{c} {}^{H} \\ {}^{H} \end{array} \cdots \cdot O_2 \longrightarrow A^{-\cdot} + H_2O_2 \tag{231}$$

$$A\,\begin{array}{c} {}^{H} \\ {}^{H} \end{array} \cdots^- \cdot O_2 \longrightarrow A^- \begin{array}{c} {}^{H} \\ {}^{O_2} \end{array} \longrightarrow A^{-\cdot} + H_2O_2 \tag{232}$$

$$2A^{-\cdot} + H_2A \rightarrow A + 2HA^- \tag{233}$$

$$HA^- + H_2O_2 \rightarrow A + H_2O + HO^- \tag{234}$$

$$3H_2A + 2O_2^{-\cdot} \rightarrow 3A + 3H_2O + 2HO^- \tag{235}$$

Very recently, Afanas'ev and coworkers[473] have taken issue with Sawyer's mechanism for the superoxide–ascorbic acid system. The Russian group reports that a 50–70% yield of ascorbate anion is formed when electrogenerated superoxide reacts with H_2A in acetonitrile. They posit that this high yield of ascorbate can only be explained by a deprotonation of AH_2 effected by superoxide. They believe, therefore, that deprotonation (equation 236) is the main if not sole pathway of interaction of superoxide with ascorbic acid. Any oxygen generated from the disproportionation of superoxide (equation 237) is presumably converted back to superoxide upon interaction with ascorbate (equation 238). However, the disproportionation is prevented by a series of competing processes (equations 239–242) which eventually convert ascorbate to dehydroascorbic acid.

$$O_2^{-\cdot} + AH_2 \rightarrow HO_2^{\cdot} + AH^- \tag{236}$$

$$HO_2^{\cdot} + O_2^{-\cdot} \rightarrow HO_2^- + O_2 \tag{237}$$

$$AH^- + O_2 \rightarrow AH^{\cdot} + O_2^{-\cdot} \tag{238}$$

$$HO_2^{\cdot} + AH_2 \rightarrow H_2O_2 + AH^{\cdot} \tag{239}$$

$$HO_2^{\cdot} + AH^- \rightarrow H_2O_2 + A^{-\cdot} \tag{240}$$

$$HO_2^{\cdot} + AH^{\cdot} \rightarrow H_2O_2 + A \tag{241}$$

$$HO_2^{\cdot} + A^{-\cdot} \rightarrow HO_2^- + A \tag{242}$$

Sawyer and coworkers[469–472] also report that superoxide reacts further with the dehydroascorbic acid producing oxalate and (by inference) the anion of threonic acid **594**. They suggest that this proceeds via nucleophilic attack of $O_2^{-\cdot}$ at the C_3-carbonyl followed by dioxetane formation and cleavage generating ketoester **593**. Saponification of the latter would yield the observed products (equation 243). It should be noted that products were not isolated in the above studies, and the evidence is based on a combination of spectral and electrochemical data of the reaction mixtures.

Frimer and coworkers[459,474,475] have reacted a variety of reductones with KO_2/crown ether in toluene in the hope that the products isolated would shed some light on the question of mechanism. Gilinsky-Sharon[459,474] reacted ascorbic acid derivatives **595a** and **b** with $O_2^{-\cdot}$ and isolated ketoesters **596** as well as the corresponding threonic acid derivatives **597** (equation 244).

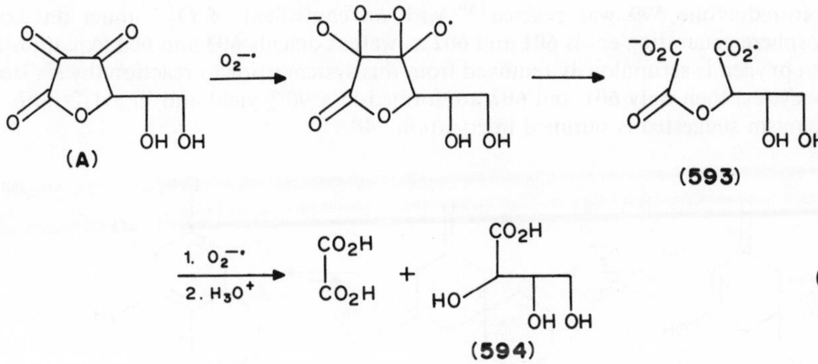
(A)
(593)
1. $O_2^{-\cdot}$
2. H_3O^+
(594)

(243)

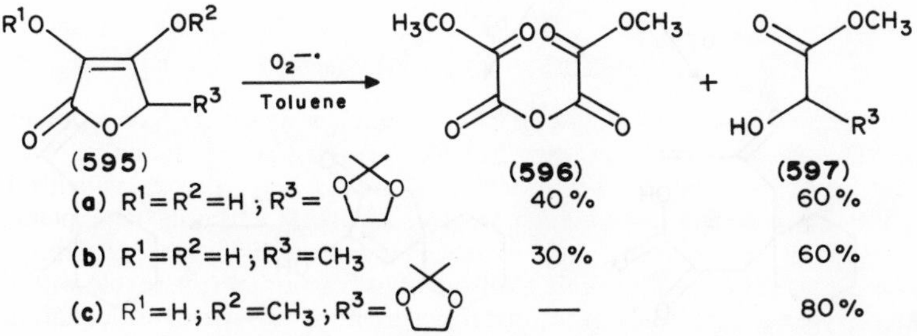
$O_2^{-\cdot}$
Toluene
(595)
(a) $R^1=R^2=H$; $R^3=$
(b) $R^1=R^2=H$; $R^3=CH_3$
(c) $R^1=H$; $R^2=CH_3$; $R^3=$
(596)
40%
30%
—
(597)
60%
60%
80%

(244)

Assuming that reductones **595a** and **595b** are initially oxidized to the corresponding triketones (analogous to dehydroascorbic acid, **A**) and then on to the observed products these results confirm the mechanism for the oxidation of dehydroascorbic acid as outlined in equation 243. Interestingly, however, enol ether **595c** also yields **597c**, though no **596c** was observed. This is somewhat surprising since the 3-methoxy group is expected to prevent the oxidation of **595c** to the corresponding triketone, the precursor required by the mechanism of equation 243. In this case, the course of the reaction can be readily rationalized in terms of an initial deprotonation of the α-hydroxy group (equation 245). Oxygenation then proceeds as described for the base-catalyzed autoxidation of enols to lactols (Section III.E.3.a). In this case, however, loss of carbon monoxide generates carbonate **598**, which further loses CO_2 generating **597c** in high yield and as the sole product.

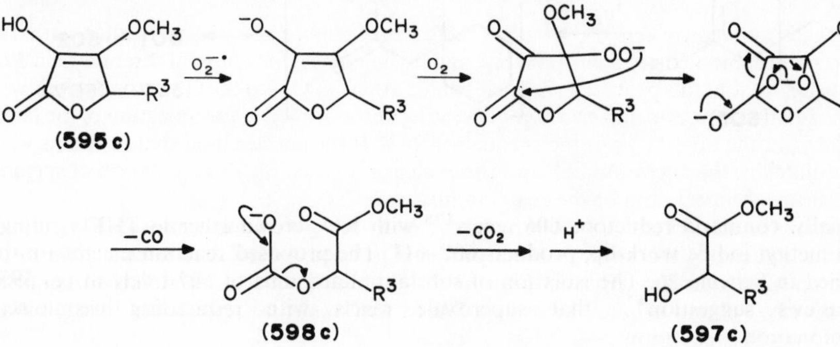
(595c)
$O_2^{-\cdot}$
O_2
−CO
(598c)
$-CO_2$
H^+
(597c)

(245)

Spiroreductone **599** was reacted[459] with an equivalent of $O_2^{-\cdot}$ under an oxygen atmosphere generating enols **601** and **602** as well as diacids **603** and **604** (equation 246). When oxygen is scrupulously removed from this system prior to reaction (by six freeze–thaw cycles) then only **601** and **602** are formed in a 90% yield and in a 1:2 ratio. The mechanism suggested is outlined in equation 246.

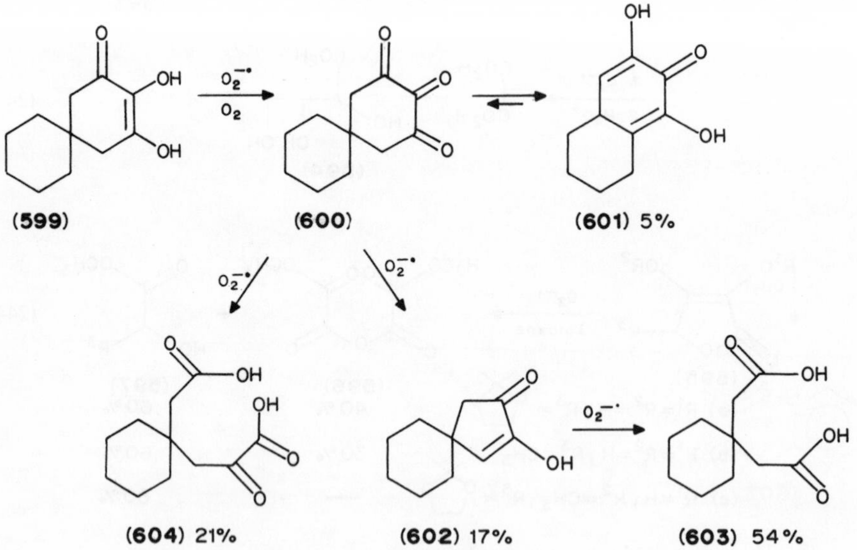

(246)

The initial step involves formation of triketone **600** which tautomerizes to **601**, undergoes benzylic acid rearrangement and decarboxylation to **602**, or cleaves oxidatively (in the fashion of diketones) to **604**. Oxidative cleavage of **602** yields diacid **603**. Gilinsky-Sharon[459] succeeded in synthesizing **601** independently via the fluoride-catalyzed singlet oxygenation[69,72] of enol **605** (equation 247; see Section IV.C.2).

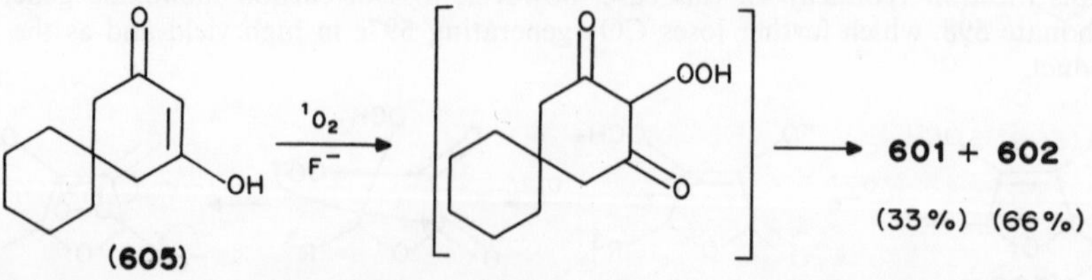

(247)

Finally, coumarin reductone **606** reacts[475] with KO_2/crown ether in THF yielding, upon methyl iodide work-up, products **607–611**. The proposed reaction mechanism is outlined in Scheme 26. The isolation of substantial amounts of **607** tends to confirm Afanas'ev's suggestion[473] that superoxide reacts with reductones via initial deprotonation.

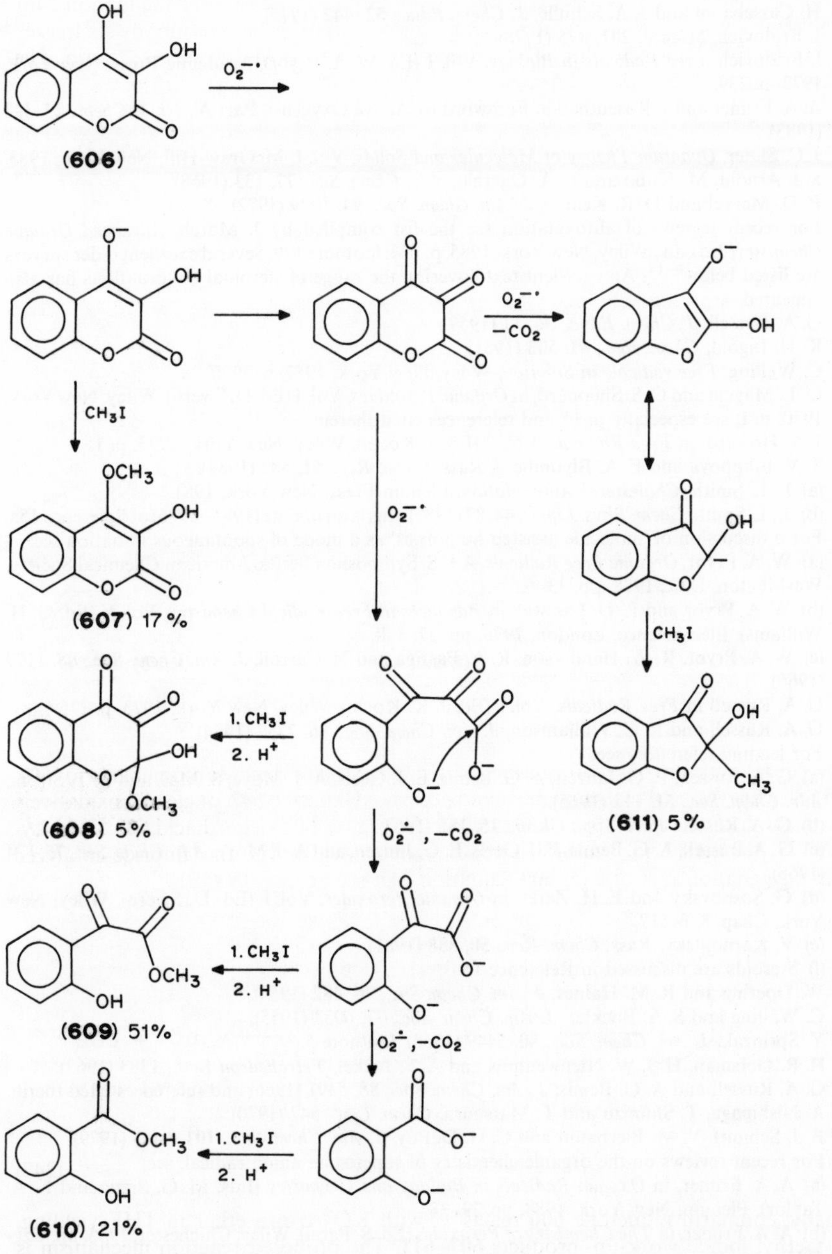

SCHEME 26. Product formation in the $O_2^{-\cdot}$ mediated oxidation of coumarin reductone **606**

VI. REFERENCES

1. H. Cassebaum and J. A. Schufle, *J. Chem. Educ.*, **52**, 442 (1975).
2. I. Fridovich, *Science*, **201**, 875 (1978).
3. I. Fridovich, *Free Radicals in Biology*, Vol. I (Ed. W. A. Pryor), Academic Press, New York, 1972, p. 239.
4. A. A. Frimer and I. Rosenthal, in Foreword to 'Active Oxygen—Part A', *Isr. J. Chem.*, **23**, 398 (1983).
5. J. C. Slater, *Quantum Theory of Molecules and Solids*, Vol. I, McGraw-Hill, New York, 1983.
6. S. J. Arnold, M. Kubo and E. A. Ogryzlo, *Adv. Chem. Ser.*, **77**, 133 (1968).
7. P. D. Merkel and D. R. Kearns, *J. Am. Chem. Soc.*, **94**, 1029 (1972).
8. For recent reviews of autoxidation see the list compiled by J. March, *Advanced Organic Chemistry*, 3rd edn., Wiley, New York, 1985, p. 633, footnote 179. Several excellent older surveys are listed below[9-14]. An excellent text covering the range of steroidal oxygenations has also appeared[15].
9. G. A. Russell, *J. Chem. Educ.*, **36**, 11 (1959).
10. K. U. Ingold, *Chem. Rev.*, **61**, 503 (1961).
11. C. Walling, *Free radicals in Solution*, Wiley, New York, 1957, p. 397ff.
12. O. L. Magelli and C. S. Sheppard, in *Organic Peroxides*, Vol. I (Ed. D. Swern), Wiley, New York, 1970, p. 1; see especially p. 15 and references cited therein.
13. J. A. Howard, in *Free Radicals*, Vol. 2 (Ed. J. Kochi), Wiley, New York, 1973, p. 1.
14. T. V. Filippova and E. A. Blyumberg, *Russ. Chem. Rev.*, **51**, 582 (1982).
15. (a) L. L. Smith, *Cholesterol Autoxidation*, Plenum Press, New York, 1981.
    (b) L. L. Smith, *Chem. Phys. Lipids*, **44**, 87 (1987); this is an update (1981–1986) of Reference 15a.
16. For a discussion of 'molecule assisted homolysis' as a mode of spontaneous initiation see:
    (a) W. A. Pryor, *Organic Free Radicals*, A.C.S. Symposium Series, American Chemical Society, Washington, D.C., 1978, pp. 33–62.
    (b) W. A. Pryor and L. D. Lasswell in *Advances in Free Radical Chemistry*, Vol. V (Ed. G. H. Williams), Elek Science, London, 1975, pp. 27, 37ff.
    (c) W. A. Pryor, R. W. Henderson, R. A. Pastiga and N. Carroll, *J. Am. Chem. Soc.*, **88**, 1199 (1966).
17. G. A. Russell in *Free Radicals*, Vol. I (Ed. J. K. Kochi), Wiley, New York, 1973, p. 275.
18. G. A. Russell and R. C. Williamson, *J. Am. Chem. Soc.*, **86**, 2357 (1964).
19. For leading references see:
    (a) G. A. Russell, E. G. Janzen, A. G. Bemis, E. J. Geels, A. J. Moye, S. Mak and E. T. Strom, *Adv. Chem. Ser.*, **51**, 112 (1965).
    (b) G. A. Russell, *Pure Appl. Chem.*, **15**, 185 (1967).
    (c) G. A. Russell, A. G. Bemis, E. J. Geels, E. G. Janzen, and A. J. Moye, *Adv. Chem. Ser.*, **75**, 174 (1968).
    (d) G. Sosnovsky and E. H. Zaret, in *Organic Peroxides*, Vol. I (Ed. D. Swern), Wiley, New York, Chap. 8, p. 517.
    (e) V. Karnojitsky, *Russ. Chem. Rev.*, **50**, 888 (1981).
    (f) Steroids are discussed in Reference 15.
20. W. Doering and R. M. Haines, *J. Am. Chem. Soc.*, **76**, 482 (1954).
21. C. Walling and S. A. Buckler, *J. Am. Chem. Soc.*, **77**, 6032 (1955).
22. Y. Sprinzak, *J. Am. Chem. Soc.*, **80**, 5449 (1958), footnote 5.
23. H. R. Gersman, H. J. W. Nieuwenhuis and A. F. Bickel, *Tetrahedron Lett.*, 1383 (1963).
24. G. A. Russell and A. G. Bemis, *J. Am. Chem. Soc.*, **88**, 5491 (1966) and references cited therin.
25. A. Nishinaga, T. Shimizu and T. Matsuura, *Chem. Lett.*, 547 (1970).
26. R. J. Schmitt, V. M. Bierbaum and C. H. DePuy, *J. Am. Chem. Soc.*, **101**, 6443 (1979).
27. For recent reviews on the organic chemistry of superoxide anion radical, see:
    (a) A. A. Frimer, in *Oxygen Radicals in Biology and Chemistry* (Eds. M. G. Simic and K. A. Taylor), Plenum, New York, 1988, pp. 29–38.
    (b) A. A. Frimer, in *The Chemistry of Peroxides* (Ed. S. Patai), Wiley, Chichester, 1983, pp. 429–461.
    (c) J. L. Roberts and D. T. Sawyer, *Isr. J. Chem.*, **23**, 430 (1983).
    (d) A. A. Frimer, in *Superoxide Dismutase* Vol. II (Ed. L. W. Oberley), Chemical Rubber Co., Boca Raton, Florida, 1982, pp. 83–125.

28. For recent reviews of the reactions of hydroperoxides see:
   (a) A. A. Frimer, in *The Chemistry of Peroxides* (Ed. S. Patai), Wiley, Chichester, 1983, pp. 201–234.
   (b) A. A. Frimer, *Chem. Rev.*, **79**, 359 (1979).
   (c) R. Hiatt, in *Organic Peroxides*, Vol. 2 (Ed. D. Swern), Wiley, New York, 1971, p. 67.
   (d) A. A. Frimer and L. M. Stephenson, in *Singlet O_2—Volume II: Reaction Modes and Products. Part I* (Ed. A. A. Frimer), CRC Press, Boca Raton, Florida, 1985, Chap. 3.
29. See Reference 28c, top of page 80 and the footnotes to pp. 87 and 96.
30. W. F. Brill, *Adv. Chem. Ser.*, **75**, 93 (1968).
31. A. D. Walsh, *Trans. Faraday Soc.*, **42**, 99, 269 (1946).
32. C. E. Frank, *Chem. Rev.*, **46**, 155, 161 (1950).
33. L. Bateman and H. Hughes, *J. Chem. Soc.*, 4594 (1952).
34. N. Kornblum and H. E. DeLaMare, *J. Am. Chem. Soc.* **73**, 880 (1951).
35. Reference 12c, pp. 51 and 79–80.
36. R. Hiatt, in *Organic Peroxides*, Vol. 3 (Ed. D. Swern), Wiley, New York, 1972, p. 23.
37. R. Criegée, *Ber. Dtsch. Chem. Ges.*, **77**, 722 (1944).
38. R. Criegée and R. Kasper, *Liebigs Ann. Chem.*, **560**, 127 (1948).
39. R. Criegée and H. Dietrich, *Liebigs Ann. Chem.*, **560**, 135 (1948).
40. R. Criegée and A. Schnorrenberg, *Liebigs Ann. Chem.*, **560**, 141 (1948).
41. H. Hock and O. Schrader, *Angew. Chem.*, **49**, 595 (1936).
42. H. Hock and K. Ganicke, *Chem. Ber.*, **71**, 1430 (1938).
43. H. Hock and S. Lang, *Chem. Ber.*, **75**, 300 (1942).
44. H. Hock and S. Lang, *Chem. Ber.*, **77**, 257 (1944).
45. H. Hock and H. Kropf, *Angew. Chem.*, **69**, 313 (1957).
46. H. W. Gardner and R. D. Plattner, *Lipids*, **19**, 294 (1984).
47. S. Muto and T. C. Bruice, *J. Am. Chem. Soc.*, **102**, 7379 (1980).
48. Y. Y. Chan, C. Zhu and H. K. Leung, *J. Am. Chem. Soc.*, **107**, 5274 (1985).
49. Y. Y. Chan, C. Zhu and H. K. Leung, *Tetrahedron Lett.*, **27**, 3737 (1986).
50. E. H. Farmer and A. Sundralingam, *J. Chem. Soc.*, 121 (1942).
51. A. A. Frimer, *J. Org. Chem.*, **42**, 3194 (1977).
52. P. D. Bartlett and A. A. Frimer, *Heterocycles*, **11**, 419 (1978).
53. Y. Sawaki and Y. Ogata, *J. Am. Chem. Soc.*, **100**, 856 (1978).
54. Y. Sawaki and Y. Ogata, *J. Am. Chem. Soc.*, **97**, 6983 (1975).
55. Y. Sawaki and Y. Ogata, *J. Am. Chem. Soc.*, **99**, 5412 (1977).
56. C. W. Jefford, W. Knöpfel and P. A. Cadby, *J. Am. Chem. Soc.*, **100**, 6432 (1978).
57. B. L. Feringa, *Recl. Trav. Chim. Pays-Bas*, **106**, 469 (1987).
58. B. L. Feringa and R. J. Butselaar, *Tetrahedron Lett.*, **24**, 1193(1983).
59. I. R. Barker, in *The Chemistry of the Hydroxyl Group* (Ed. S. Patai), Wiley, New York, 1971, p. 219.
60. H. S. Verter, in *The Chemistry of the Carbonyl Group*, Vol. 2 (Ed. J. Zabicky), Wiley, New York, 1970, pp. 71 and 83–86.
61. (a) R. Hanna and G. Ourisson, *Bull. Soc. Chim. Fr.*, 1945 (1961).
   (b) T. Matsuura, H. Matsushima and H. Sakamoto, *J. Am. Chem. Soc.*, **89**, 6370 (1967).
62. A. Nishinaga and T. Matsuura, *J. Chem. Soc., Chem. Commun.*, 9 (1973).
63. A. Nishinaga, T. Tojo, H. Tomita and T. Matsuura, *J. Chem. Soc., Perkin Trans 1*, 2511 (1979).
64. M. Utaka, S. Matsushita, H. Yamasaki and A. Takeda, *Tetrahedron Lett.*, **21**, 1063 (1980).
65. V. Rajanada and S. B. Brown, *Tetrahedron Lett.*, **22**, 4331 (1981).
66. E. Alvarez, C. Betancor, R. Freire, A. Martin and E. Suarez, *Tetrahedron Lett.*, **22**, 4335 (1981).
67. A. A. Frimer and P. Gilinsky, in *Oxygen and Oxy-Radical in Chemistry and Biology* (Eds. M. A. J. Rodgers and E. L. Powers), Academic Press, New York, 1981, p. 639.
68. A. A. Frimer, P. Gilinsky-Sharon and G. Aljadeff, *Tetrahedron Lett.*, **23**, 1301 (1982).
69. H. H. Wasserman and J. E. Pickett, *J. Am. Chem. Soc.*, **104**, 4695 (1982).
70. M. Utaka, M. Nakatani and A. Takeda, *Tetrahedron Lett.*, **24**, 803 (1983).
71. M. Utaka, M. Hojo, Y. Fujii and A. Takeda, *Chem. Lett.*, 635 (1984).
72. H. H. Wasserman and J. E. Pickett, *Tetrahedron*, **41**, 2155 (1985).
73. M. Utaka, M. Nakatani and A. Takeda, *Tetrahedron*, **41**, 2163 (1985).
74. K. Hayakawa, K. Ueyama and K. Kanematsu, *J. Org. Chem.*, **50**, 1963 (1985).
75. M. Utaka, H. Kuriki, T. Sakai and A. Takeda, *J. Org. Chem.*, **51**, 935 (1986).

76. M. Utaka, M. Nakatami and A. Takeda, *J. Org. Chem.*, **51**, 1140 (1986).
77. D. A. Mayers and J. Kagan, *J. Org. Chem.*, **39**, 3147 (1974); see also Reference 56.
78. H. W. -S. Chan, G. Levett and J. A. Matthew, *Chem. Phys. Lipids*, **24**, 245 (1979).
79. N. A. Porter, L. S. Lehman, B. A. Weber and K. J. Smith, *J. Am. Chem. Soc.*, **103**, 6447 (1981).
80. W. F. Brill, *J. Chem. Soc., Perkin Trans 2*, 621 (1972).
81. N. A. Porter and P. Zuraw, *J. Chem. Soc., Chem. Commun.*, 1472 (1985).
82. N.A. Porter and J. S. Wujek, *J. Org. Chem.*, **52**, 5085 (1987).
83. E. G. E. Hawkins, *J. Chem. Soc.*, 3288 (1955) and references cited therein.
84. A. Hornika and K. Naya, *Bull. Chem. Soc. Jpn.*, **52**, 1964 (1979).
85. C. W. Jefford and C. G. Rimbault, *J. Org. Chem.*, **43**, 1908 (1978); for related examples see Reference 28a (end of Section IV.C.3.c therein).
86. S. A. Moslov and E. A. Blyumberg, *Russ. Chem. Rev.*, **45**, 155 (1976).
87. I. G. Tischenko and L. S. Stanishevskii, *Zh. Obshch. Chim.*, **33**, 3751 (1963); *Chem. Abstr.*, **60**, 7911b (1964).
88. I. G. Tischenko and L. S. Stanishevskii, *Geterogennye Reaktsii i Reakts Sposobnost. Sb.*, 254 (1964); *Chem. Abstr.*, **65**, 5357d (1966).
89. H. C. Volger, W. Brackman and J. W. F. M. Lemmers, *Recl. Trav. Chim. Pays-Bas*, **84**, 1203 (1965); see especially note to page 1216.
90. (a) M. R. Sabol, C. Wigelsworth and D. S. Watt, *Synth. Commun.*, **18**, 1 (1966).
     (b) H. H. Gersman, H. J.W. Nieuwenhuis and A. F. Bickel, *Tetrahedron Lett.*, 1383 (1963).
91. R. Howe and F. J. McQuillan, *J. Chem. Soc.*, 1513 (1958).
92. G. A. S. Ansari and L. L. Smith, *Chem. Phys. Lipids*, **22**, 55 (1978).
93. M. J. Kulig and L. L. Smith, *J. Org. Chem.*, **39**, 3398 (1974).
94. R. F. Majewski, J. M. Berdahl, L. D. Jost, T. A. Martin, J. C. Simms, J. G. Schmidt and J. R. Corrigan, *Steroids*, **16**, 15 (1970).
95. N. L. Allinger and F. Wu, *Tetrahedron*, **27**, 5093 (1971).
96. A. C. Campbell, J. McLean and W. Lawrie, *Tetrahedron Lett.*, 483 (1969).
97. C. M. Siegmann and M. S. Dewinter, *Recl. Trav. Chim. Pays-Bas*, **89**, 442 (1970).
98. R. B. Woodward and R. H. Eastman, *J. Am. Chem. Soc.*, **72**, 399 (1950) regarding N. Sernagiotto, *Gazz. Chim. Ital.*, **47**, 150 (1917).
99. H. O. House, R. J. Outcalt, J. L. Haak and D. Van Derveer, *J. Org. Chem.*, **48**, 1654 (1983).
100. H. O. House, in *Stereochemistry and Reactivity of Systems Containing π Electrons* (Methods in Stereochemical Analysis—Volume 3) (Ed. W. H. Watson), Verlag Chemie, Deerfield Beach, Florida, 1983, pp. 279–317.
101. A. C. Waiss Jr. and J. Corse, *J. Am. Chem. Soc.*, **87**, 2068 (1965).
102. A. C. Waiss Jr., R. E. Ludin, A. Lee and J. Corse, *J. Am. Chem. Soc.*, **89**, 6213 (1967).
103. T. Matsuura and H. Matsushima, *Tetrahedron*, **24**, 6615 (1968).
104. T. Matsuura, *Tetrahedron*, **33**, 2869 (1977).
105. K. Hayakawa, K. Ueyama and K. Kanematsu, *J. Org. Chem.*, **50**, 1963 (1985).
106. G. Bouchoux and Y. Hoppilliard, *Nouv. J. Chem.*, **11**, 225 (1987) and references cited therein.
107. H. Brederiek and G. Bauer, *Liebigs Ann. Chem.*, **739**, 117 (1970).
108. K. Schank, *Synthesis*, 176 (1972).
109. G. Hesse, in *Houben-Weyl: Methoden der Organischen Chemie*, Vol. VI/1d (eds. H. Kropf and G. Hesse), Verlag, Stuttgart, 1978, pp. 217–298.
110. P. P. Barnes and V. J. Tulane, *J. Am. Chem. Soc.*, **62**, 894 (1940).
111. G. Hesse and B. Wehling, *Liebigs Ann. Chem.*, **679**, 100 (1964).
112. W. Mayer, R. Bachmann and F. Kraus, *Chem. Ber.*, **88**, 316 (1955).
113. P. A. Seib and B. M. Tolbert (eds.), '*Ascorbic Acid: Chemistry, Metabolism and Uses*', *Adv. Chem. Ser.*, **200** (1980).
114. *Second Conference on Vitamin C, Ann. N.Y. Acad. Sci.*, **258** (1975).
115. R. S. Harris, in *The Vitamins: Chemistry, Physiology, Pathology and Methods*, Vol. 1 (Eds. W. H. Sebrell, Jr. and R. S. Harris), Academic Press, New York, 1967, p. 305.
116. B. H. J. Bielsky in Reference 113, p. 81.
117. A. Weissberger, J. E. Luvalle and D. S. Thomas, Jr., *J. Am. Chem. Soc.*, **65**, 1934 (1943).
118. M. Ohmri and M. Takagi, *Argic. Biol. Chem.*, **42**, 173 (1978).
119. B. M. Tolbert and J. B. Ward, in Reference 115, pp. 101, 103.
120. H. Dietz, *Liebigs Ann. Chem.*, **738**, 206 (1970).
121. K. Puget and A. M. Michelson, *Biochimie*, **56**, 1255 (1974).

122. A. Rigo, M. Scarpa, E. Argese, P. Ugo and P. Viglino, in *Oxygen Radicals in Chemistry and Biology* (Eds. W. Bors, M. Saran and D. Tait), Walter de Grytes, Berlin, 1984, p. 17.
123. M. M. T. Khan and A. E. Martell, *J. Am. Chem. Soc.*, **89**, 4176 (1967).
124. cf. I. B. Afanas'ev, V. V. Grabovetskii and N. S. Kuprianova, *J. Chem. Soc., Perkin Trans. 2*, 281 (1987).
125. M. Niclause, *Selecta Chimica*, **15**, 57 (1956).
126. L. F. Fieser, T. W. Greene, F. Bischoff, G. Lopez and J. J. Rupp, *J. Am. Chem. Soc.*, **77**, 3929 (1955).
127. L. F. Fieser, F. Alvarez and A. J. Cox, unpublished results reported in L. F. Fieser and M. Fieser, *Steroids*, Reinhold Publishing Co., New York, 1959, p. 235.
128. A. J. Cox, *J. Org. Chem.*, **30**, 2052 (1965).
129. J. T. Teng and L. L. Smith, *J. Steroid Biochem.*, **7**, 577 (1976).
130. E. Shapiro, T. Legatt and E. P. Oliveto, *Tetrahedron Lett.*, 663 (1964).
131. E. Shapiro, L. Finckenor and H. L. Herzog, *J. Org. Chem.*, **33**, 1673 (1968).
132. P. H. Yu and L. Tan, *J. Steroid Biochem.*, **8**, 825 (1977).
133. A. Nickon and W. L. Mendelson, *J. Org. Chem.*, **30**, 2087 (1965).
134. P. B. D. de la Mare and R. D. Wilson, *J. Chem. Soc., Perkin Trans. 2*, 157 (1977)
135. R. Y. Kirdani and D. S. Layne, *Biochemistry*, **4**, 331 (1965).
136. A. Afonso, *Can. J. Chem.*, **47**, 3693 (1969).
137. K. Croshaw, R. C. Newstead and N. A. J. Rogers, *Tetrahedron Lett.*, 2307 (1964).
138. J. J. Brown and S. Bernstein, *Steroids*, **1**, 113 (1963).
139. J. J. Brown and S. Bernstein, *Steroids*, **8**, 87 (1966).
140. R. Joly, J. Warnant, J. Joly and J. Mathieu, *C. R. Acad. Sci. Paris*, **258**, 5669 (1964).
141. M. Debono and R. M. Molloy, *Steroids*, **14**, 219 (1969).
142. N. J. Turro, M.-F. Chow and Y. Ito, *J. Am. Chem. Soc.*, **100**, 5580 (1978).
143. P. D. Bartlett and R. E. McCluney, *J. Org. Chem.*, **48**, 4165 (1983).
144. S. K. Malhotra and H. J. Ringold, *J. Am. Chem. Soc.*, **86**, 1997 (1964).
145. H. O. House, *Modern Synthetic Reactions*, 2nd edn. W. A. Benjamin, Menlo Park, CA, 1972, Chap. 7, pp. 492–509.
146. J. d'Angelo, *Tetrahedron*, **32**, 2979 (1976).
147. L. Nedilec, J. C. Gase and R. Bucourt, *Tetrahedron*, **30**, 3263 (1974).
148. R. A. Lee, C. McAndrews, K. M. Patel and W. Reusch, *Tetrahedron Lett.*, 965 (1973).
149. M. Kawanisi, Y. Itoh, T. Hieda, S. Kozima, T. Hitomi and K. Kobayashi, *Chem. Lett.*, 647 (1985) and references cited therein.
150. P. T. Lansbury, R. W. Erwin and S. A. Jeffrey, *J. Am. Chem. Soc.*, **102**, 1602 (1980).
151. M.-E. Tran Huu Dau, M. Fetizon and N. Trong Anh, *Tetrahedron Lett.*, 851, 855 (1973) and references cited therein.
152. H. E. Zimmerman, *Acc. Chem. Res.*, **20**, 263 (1987) and references cited therein.
153. C. Harries, *Chem. Ber.*, **34**, 2105 (1901).
154. A. Stahler, *Liebigs Ann. Chem.*, **330**, 264 (1904).
155. W. Treibs, *Chem. Ber.*, **63**, 2423 (1930).
156. W. Treibs, *Chem. Ber.*, **64**, 2178 (1931).
157. W. Treibs, *Chem. Ber.*, **64**, 2545 (1931).
158. W. Treibs, *Chem. Ber.*, **65**, 163 (1932).
159. W. Treibs, *Chem. Ber.*, **65**, 1314 (1932).
160. W. Treibs, *Chem. Ber.*, **66**, 610 (1933).
161. W. Treibs, *Chem. Ber.*, **66**, 1483 (1933).
162. H. O. House and W. Gilmore, *J. Am. Chem. Soc.*, **83**, 3972 (1961).
163. R. W. Mouk, K. M. Patel and W. Reusch, *Tetrahedron*, **31**, 13 (1975).
164. A. A. Frimer and P. Gilinsky, *Tetrahedron Lett.*, 4331 (1979).
165. A. A. Frimer and P. Gilinsky-Sharon, J. Hameiri-Buch and Z. Rosental, unpublished results (1988).
166. M. Sugawara and M. M. Baizer, *J. Org. Chem.*, **48**, 4931 (1983).
167. F. Jensen and C. S. Foote, *Photochem. Photobiol.*, **46**, 325 (1987).
168. D. Arigoni, D. H. R. Barton, E. J. Corey and O. Jeger, *Experientia [Base]*, **16**, 41 (1960).
169. D. H. R. Barton, S. K. Pradhan, S. Sternhell and J. F. Templeton, *J. Chem. Soc.*, 255 (1961).
170. E. J. Bailey, D. H. R. Barton, J. Elks ànd J. F. Templeton, *J. Chem. Soc.*, 1578 (1962).
171. W. E. Doering and R. M. Haines, *J. Am. Chem. Soc.*, **76**, 482 (1954).

172. B. Camerino, B. Patelli and R. Sciaky, *Tetrahedron Lett.*, 554 (1961).
173. B. Camerino, B. Patelli and R. Sciaky, *Gazz. Chim. Ital.*, **92**, 693 (1962).
174. J. B. Jones and K. D. Gordon, *Can. J. Chem.*, **50**, 2712 (1972).
175. R. J. Langenbach and H. W. Knoche, *Steroids*, **11**, 123 (1963).
176. H. C. Holland, U. Daum and E. Riemland, *Tetrahedron Lett.*, **22**, 5127 (1981).
177. H. C. Holland, U. Daum and E. Riemland, *Can. J. Chem.*, **60**, 1919 (1982).
178. A. Hornika, E. Yo, O. Mori and K. Naya, *Bull. Chem. Soc. Jpn.*, **52**, 2732 (1979).
179. J. E. Baldwin, D. H. R. Barton and J. K. Sutherland, *J. Chem. Soc.*, 3312 (1964).
180. U. Takahama, *Plant Cell Physiol.*, **28**, 953 (1987).
181. R. Hanna and G. Ourisson, *Bull. Soc. Chim. Fr.*, 3742 (1967).
182. J. Pusset, D. Guenard and R. Beugelmans, *Tetrahedron*, **27**, 2939 (1971).
183. R. Sandmeier and C. Tamm, *Helv. Chim. Acta*, **56**, 2238 (1973).
184. M. M. A. El-Sukkary and G. Speier, *J. Chem. Soc., Chem. Commun.*, 745 (1981); in a personal communication from Prof. Speier regarding this paper it was noted that the isolated product is not 2-benzoyloxyphenylglyoxylic acid but the depside hydrolysis products. The former was only detected in the mass spectrum of the reaction mixture.
185. See Reference 71, footnote 9.
186. R. G. Curtis and R. Schoenfeld, *Aust. J. Chem.*, **8**, 258 (1955).
187. R. Hirschman, G. A. Bailey, R. Walker and J. M. Chemedra, *J. Am. Chem. Soc.*, **81**, 2822 (1959).
188. M. Rajic, T. Rull and G. Ourisson, *Bull. Soc. Chim. Fr.*, 1213 (1961).
189. G. R. Chandry, T. G. Halsall and E. R.H. Jones, *J. Chem. Soc.*, 2725 (1961).
190. H. R. Nace and M. Inaba, *J. Org. Chem.*, **27**, 4024 (1962).
191. J. F. Biellmann and M. Rajic, *Bull. Soc. Chim. Fr.*, 441 (1962).
192. J. B. Davis and B. C. L. Weedon, *Proc. Chem. Soc.*, 182 (1960).
193. R. Kuhn, J. Stene and N.A. Sorensen, *Chem. Ber.*, **78B**, 1688 (1939).
194. W. Kreiser and W. Ulrich, *Ann. Chem.*, **761**, 121 (1972).
195. R. L. Clarke, *J. Am. Chem. Soc.*, **82**, 4629 (1960).
196. P. N. Rao and L. R. Axelrod, *J. Am. Chem. Soc.*, **82**, 2830 (1960).
197. R. E. Lack and A. B. Ridley, *J. Chem. Soc. (C)*, 3017 (1968).
198. J. N. Gardner, F. E. Carlon and O. Gnoj, *J. Org. Chem.*, **33**, 3294 (1968); 17-hydroperoxyprogesterone was synthesized via the enol ether[170].
199. A. A. Frimer, P. Gilinsky-Sharon, J. Hameiri and G. Aljadeff, *J. Org. Chem.*, **47**, 2819 (1982).
200. A. A. Frimer and P. Gilinsky, in *Oxygen and Oxy-Radicals in Chemistry and Biology* (Eds. E. L. Powers and M. A. J. Rodgers), Academic Press, New York, 1981, pp. 639–640.
201. A. A. Frimer, J. Hameiri-Buch, S. Ripstos and P. Gilinsky-Sharon, *Tetrahedron*, **42**, 5693 (1986).
202. J. Hameiri, M. S. Thesis, Bar-Ilan University, Ramat Gan, Israel, 1982.
203. G. A. Russell, *J. Am. Chem. Soc.*, **76**, 1595 (1954).
204. G. A. Russell, A. J. Moye and K. L. Nagpal, *J. Am. Chem. Soc.*, **84**, 4154 (1962).
205. M. Lissel and E. V. Dehmlow, *Tetrahedron Lett.*, 3689 (1978).
206. P. M. Allen, V. Hess, C. S. Foote and M. M. Baizer, *Synth. Commun.*, **12**, 123 (1982).
207. M. Sugawara and M. M. Baizer, *J. Org. Chem.*, **48**, 4931 (1983).
208. A. A. Frimer, G. Aljadeff and P. Gilinsky-Sharon, *Isr. J. Chem.*, **27**, 39 (1986).
209. R. Y. Young, *J. Chem. Soc., Chem. Commun.*, 704 (1970).
210. L. Canonica, B. Danieli, G. Lesma, G. Palmisano and A. Mugnoli, *Helv. Chim. Acta*, **70**, 701 (1987).
211. H. H. Holland, E. Riemland and U. Daum, *Can. J. Chem.*, **60**, 1919 (1982).
212. B. Stern, M. S. Thesis, Bar-Ilan University, Ramat Gan, Israel, 1987.
213. A. A. Frimer and B. Stern, unpublished results (1987).
214. Sherico Ltd., Netherland patent appl. 6400153 (1964); *Chem. Abstr.*, **62**, 9201 (1965).
215. R. Joly, J. Warnant, J. Joly and J. Matlhieu, *C.R. Acad. Sci. Paris*, **258**, 5669 (1964).
216. J. J. Brown and S. Bernstein, *Steroids*, **8**, 87 (1966).
217. G. Buchi, W. PickenHagen and H. Wuest, *J. Org. Chem.*, **37**, 4192 (1972).
218. E. E. van Tamelen and G. T. Hildahl, *J. Am. Chem. Soc.*, **78**, 4405 (1956).
219. (a) A. G. Schering, Fr. patent 2190427 (1974); *Chem. Abstr.*, **81**, 37731 (1974).
    (b) K. Crowshaw, R. C. Newstead and N. A. Roger, *Tetrahedron Lett.*, 2307 (1964).
220. B. Muckensturm, *Tetrahedron*, **31**, 1933 (1975).
221. H. C. Volger and W. Brackman, *Recl. Trav. Chim. Phys-Bas*, **84**, 1017 (1965).

222. H. C. Volger and W. Brackman, *Recl. Trav. Chim. Pays-Bas*, **84**, 579 (1965).
223. H. C. Volger, W. Brackman and J. W. F. M. Lemmens, *Recl. Trans. Chim. Pays-Bas*, **84**, 1203 (1965).
224. H. C. Volger and W. Brackman, *Recl. Trav. Chim. Pays-Bas*, **84**, 1233 (1965).
225. A. Lachman, *J. Am. Chem. Soc.*, **45**, 1509 (1923).
226. P. Harter, in *Houben-Weyl: Methoden der Organischen Chemie*, Vol. IV/1a (Ed. H. Kropf), Verlag, Stuttgart, 1981, pp. 963–1146.
227. W. Charney and H. L. Herzog, *Microbial Transformations of Steroids: A Handbook*, Academic Press, New York, 1967.
228. Reference 15a, Chap. 7.
229. R. A. Johnson, in *Oxidation in Organic Chemistry, Part C* (Ed. W. S. Trahanovsky), Academic Press, New York, 1978, Chap. 2.
230. F. Drawert, H. Barton and J. Beier, in Reference 109, pp. 299–452.
231. S. H. Epstein, P. D. Meister, H. Marian Leigh, D. H. Peterson, H. C. Murray, L. M. Reineke and A. Weintraub, *J. Am. Chem. Soc.*, **76**, 3174 (1954).
232. M. Hayano, in *Oxygenases* (Ed. O. Hayaishi), Academic Press, New York, 1962, pp. 182–240.
233. H. L. Holland and B. J. Auret, *Can. J. Chem.*, **53**, 845 (1975).
234. H. L. Holland and B. J. Auret, *Can. J. Chem.*, **53**, 2041 (1975).
235. H. L. Holland and P. R. P. Diakow, *Can. J. Chem.*, **56**, 694 (1978).
236. H. L. Holland and P. R. P. Diakow, *Can. J. Chem.*, **57**, 436 (1979).
237. J. I. Teng and L. L. Smith, *J. Steroid Biochem.*, **7**, 577 (1976).
238. T. Matsuura, H. Matsushima and R. Nakashima, *Tetrahedron*, **26**, 435 (1970) and references cited therein.
239. T. Matsuura, *Tetrahedron*, **33**, 2869 (1977) and references cited therein.
240. W. G. Rathmell and D. S. Bendall, *Phytochemistry*, **11**, 873 (1972).
241. W. G. Rathmell and D. S. Bendall, *Biochem. J.*, **127**, 125 (1972).
242. E. Wong and J. M. Wilson, *Phytochemistry*, **15**, 1325 (1976).
243. J. M. Wilson and E. Wong, *Phytochemistry*, **15**, 1333 (1976).
244. E. Wong, *Tetrahedron Lett.*, **25**, 2631 (1984).
245. M. J. Begley, L. Crombie, M. London, J. Savin and D. A. Whiting, *J. Chem. Soc., Chem. Commun.*, 1319 (1982).
246. J. E. Baldwin, J. C. Swallow and H. W.-S. Chan, *J. Chem. Soc., Chem. Commun.*, 1407 (1971).
247. H. W. -S. Chan, *J. Am. Chem. Soc.*, **93**, 2357 (1971).
248. T. Akihisa, T. Matsumoto, H. Sakamaki, M. Take and Y. Ichinohe, *Bull. Chem. Soc. Jpn.*, **59**, 680 (1986).
249. M. Utaka, H. Watabu and A. Takeda, *Chem. Lett.*, 1475 (1985).
250. M. Utaka and A. Takeda, *J. Chem. Soc., Chem. Commun.*, 1824 (1985).
251. A. Nishinaga, T. Tojo and T. Matsuura, *J. Chem. Soc., Chem. Commun.*, 896 (1974).
252. W. G. Nigh, in *Oxidation in Organic Chemistry—Part B* (Ed. W.S. Trahanovsky), Academic Press, New York, 1973, pp. 1–96.
253. C. Fabre and C. Lapinte, *Nouv. J. Chim.*, **7**, 123 (1983).
254. A. Weissberger and J. E. LuValle, *J. Am. Chem. Soc.*, **66**, 700 (1944).
255. E. Sernagiotto, *Gazz. Chim. Ital.*, **48**, 52 (1918).
256. E. Sernagiotto, *Gazz. Chim. Ital.*, **47**, 153 (1917).
257. D. Creed, *Tetrahedron Lett.*, **22**, 2039 (1981) and references cited therein.
258. B. Pandey, M. P. Mahajan and M. V. George, *Angew. Chem., Int. Ed. Engl.*, **19**, 907 (1980).
259. For recent volumes and reviews on singlet oxygen chemistry see References 28a, 28b and 260–262.
260. A. A. Frimer (ed.), *Singlet O_2*, Vols. 1–4, CRC Press, Boca Raton, Florida, 1984–1985.
261. H. H. Wasserman and R. W. Murray (eds.), *Singlet Oxygen*, Academic Press, New York, 1979.
262. R. W. Denny and A. Nickon, *Org. React.*, **20**, 133 (1973).
263. K. Alder, F. Pascher and A. Schmitz, *Ber. Dtsch. Chem. Ges.*, **76**, 27 (1943).
264. H. M. R. Hoffman, *Angew. Chem., Int. Ed. Engl.*, **8**, 556 (1969).
265. A. Nickon and W. L. Mendelson, *J. Am. Chem. Soc.*, **87**, 3921 (1965).
266. E. Koch, *Tetrahedron*, **24**, 6295 (1968).
267. R. D. Ashford and E. A. Ogryzlo, *J. Am. Chem. Soc.*, **91**, 5358 (1969).
268. J. H. Wieringa, J. Strating and H. Wynberg, *Tetrahedron Lett.*, 169 (1972).
269. P. D. Bartlett and M. S. Ho, *J. Am. Chem. Soc.*, **96**, 627 (1974).

270. I. Rosenthal, in *Singlet O_2—Volume I: Physical Chemical Aspects* (Ed. A. A. Frimer), CRC Press, Boca Baton, Florida, 1984, Chap. 2.
271. R. W. Murray, in Reference 261, Chap. 3.
272. A. A. Frimer, P. D. Bartlett, A. F. Boschung and J. E. Jewett, *J. Am. Chem. Soc.*, **99**, 7977 (1977).
273. W. R. Adams, in *Oxidation*, Vol. 2 (Eds. R. G. Augustine and D. J. Trecker), Marcel Dekker, New York, 1971, Chap. 2.
274. J. R. Williams, G. Orton and L. R. Unger, *Tetrahedron Lett.*, 4603 (1973).
275. E. C. Blossey, D. C. Neckers, A. L. Thayer and A. P. Schaap, *J. Am. Chem. Soc.*, **95**, 5820 (1973).
276. R. Nilsson and D. R. Kearns, *Photochem. Photobiol.*, **19**, 181 (1974).
277. A. P. Schaap, A. L. Thayer, E. C. Blossey and D. C. Neckers, *J. Am. Chem. Soc.*, **97**, 3741 (1975).
278. C. Lewis and W. H. Scouten, *Biochem. Biophys. Acta*, **444**, 326 (1976).
279. A. P. Schaap, A. I. Thayer, K. A. Kaklika and P. C. Valenti, *J. Am. Chem. Soc.*, **101**, 4016 (1979).
280. H. E. Ensley, R. V. C. Can, R. S. Martin and T. E. Pierce, *J. Am. Chem. Soc.*, **102**, 2836 (1980).
281. H. E. Ensley, P. Balakrishnan and B. Ugarkar, *Tetrahedron Lett.*, **24**, 5189 (1983).
282. D. C. Neckers and G. Hauck, *J. Org. Chem.*, **48**, 4691 (1983).
283. J. Weiss and A. A. Frimer, unpublished results (1988).
284. M. Refat Mahran, W. M. Abdov, M. M. Sidky and H. Wamhoff, *Synthesis*, 506 (1987).
285. H. E. Ensley and R. V. C. Can. *Tetrahedron Lett.*, 513 (1977).
286. See also K. H. Schulte-Elte, M. Gadola and B. L. Muller, *Helv. Chim. Acta*, **54**, 1870 (1971).
287. See also W. Skorianetz, H. Giger and G. Ohloff, *Helv. Chim. Acta*, **54**, 1797 (1971).
288. W. Adam and A. Greisbeck, *Angew. Chem., Int. Ed. Engl.*, **24**, 1070 (1985).
289. W. Adam and A. Greisbeck, *Synthesis*, 1050 (1986).
290. W. Adam. A. Greisbeck and D. Kappes, *J. Org. Chem.*, **51**, 4479 (1986).
291. M. Orfanopoulos and C. S. Foote, *Tetrahedron Lett.*, **26**, 5991 (1985).
292. Y. Y. Chan, C. Zhu and H. -K. Leung, *J. Am. Chem. Soc.*, **107**, 5274 (1985).
293. Y. Y. Chan, C. Zhu and H. -K. Leung, *Tetrahedron Lett.*, **27**, 3741 (1986).
294. H. Gotthardt and K. -H. Schenk, *Chem. Ber.*, **119**, 762 (1986).
295. See references 28d, 272 and L. M. Stephenson, M. J. Grdina and M. Orfanopoulous, *Acc. Chem. Res.*, **13**, 419 (1980).
296. P. D. Bartlett and A. A. Frimer, *Heterocycles*, **11**, 419 (1978).
297. P. D. Bartlett, G. D. Mendenball and A. P. Schaap, *Ann. N.Y. Acad. Sci.*, **171**, 79 (1970).
298. P. D. Bartlett and A. P. Schaap, *J. Am. Chem. Soc.*, **92**, 3223 (1970).
299. J. -C. Carmier and X. Deglise, *C.R. Acad. Sci. Paris*, **278**, 215 (1974).
300. W. Adam and H. C. Steinmetzer, *Angew. Chem., Int. Ed. Engl.*, **11**, 540 (1972).
301. G. M. Rubottom and M. I. Lopez Nieves, *Tetrahedron Lett.*, 2423 (1972).
302. W. Adam and J.-C. Liu, *J. Am. Chem. Soc.*, **94**, 2894 (1972).
303. C. W. Jefford and C. G. Rimbault, *Tetrahedron Lett.*, 2375 (1977).
304. C. W. Jefford and C. G. Rimbault, *J. Am. Chem. Soc.*, **100**, 6437 (1978).
305. W. Adam, A. Alzerreca, J. -C. Liu and F. Yang, *J. Am. Chem. Soc.*, **99**, 5768 (1977).
306. See I. Saito and T. Matsuura in Reference 261, Chap. 10.
307. T. Matsuura, H. Matsushima and R. Nakashima, *Tetrahedron*, **26**, 435 (1970).
308. R. H. Young and H. Hart, *J. Chem. Soc., Chem. Commun.*, 827 (1967).
309. R. H. Young and H. Hart, *J. Chem. Soc., Chem. Commun.*, 828 (1967).
310. R. H. Young, *J. Chem. Soc., Chem. Commun.*, 704 (1970).
311. M. Utaka, H. Kuriki, T. Sakai and A. Takeda, *Chem. Lett.*, 911 (1983).
312. (a) S. Ripshtos, M.S. Dissertation, Bar-Ilan University, Ramat Gan, Israel, 1988.
     (b) B. Kwon and C. S. Foote, *Photochem. Photobiol.*, **47** (Suppl.), 475 (1988).
     (c) B.-M. Kwon and C. S. Foote, *J. Am. Chem. Soc.*, **110**, 6582 (1988).
313. H. H. Wasserman and J. L. Ives, *J. Org. Chem.*, **50**, 3573 (1985).
314. H. H. Wasserman and J. L. Ives, *J. Org. Chem.*, **43**, 3238 (1978).
315. H. H. Wasserman and J. L. Ives, *J. Am. Chem. Soc.*, **98**, 7868 (1976).
316. H. H. Wasserman and J. L. Ives, *Tetrahedron*, **37**, 1819 (1981).
317. H. H. Wasserman and W. T. Han, *Tetrahedron Lett.*, **25**, 3743 (1984).
318. F. E. Ziegler, M. A. Cady, R. V. Nelson and J. M. Photis, *Tetrahedron Lett.*, 2741 (1979).
319. R. S. Vickers and C. S. Foote, *Boll. Chim. Farm.*, **109**, 599 (1970).
320. H. Gotthardt and K. -H. Schenk, *Tetrahedron Lett.*, **24**, 4669 (1983).
321. K. Orito, R. H. Manske and R. Rodrigo, *J. Am. Chem. Soc.*, **96**, 1944 (1974).
322. M. J. S. Dewar and W. Thiel, *J. Am. Chem. Soc.*, **97**, 3978 (1975).
323. R. B. Woodward, *Pure Appl. Chem.*, **2**, 383 (1961); *J. Am. Chem. Soc.*, **82**, 3800 (1960).
324. N. Kuramoto and T. Kitao, *J. Chem. Soc., Chem. Commun.*, 379 (1979).

325. N. Kuramoto and T. Kitao, *J. Chem. Soc., Perkin Trans. 2*, 1569 (1980).
326. P. J. Machin and P. G. Sammes, *J. Chem. Soc., Perkin Trans. 1*, 628 (1976).
327. H. H. Wasserman and B. H. Lipshutz, in Reference 261, Chap. 9.
328. M. U. George and V. Bhat, *Chem. Rev.*, **79**, 447 (1979).
329. H. M. Chawla and S. S. Chebber, *Tetrahedron Lett.*, 2171 (1976).
330. H. M. Chawla and K. Chakrabarty, *J. Chem. Soc., Perkin Trans. 1*, 1511 (1984).
331. E. Wong, *Phytochemistry*, **26**, 1544 (1987).
332. M. Mousseron-Canet, J. C. Mani, J. L. Olivé and J. P. Dalle, *C. R. Acad. Sci. Paris, Ser. C*, **262**, 1397 (1966).
333. M. Mousseron-Canet, D. Lerner and J. C. Mani, *Bull. Soc. Chim. Fr.*, 2144 (1966).
334. M. Mousseron-Canet, J. C. Mani, J. P. Dalle and J. L. Olivé, *Bull. Soc. Chim. Fr.*, 3874 (1966).
335. M. Mousseron-Canet, J. C. Mani and J. P. Dalle, *Bull. Soc. Chim. Fr.*, 608 (1967).
336. M. Mousseron-Canet, J. P. Dalle and J. C. Mani, *Tetrahedron Lett.*, 6037 (1968).
337. M. Mousseron-Canet, J. P. Dalle and J. C. Mani, *Bull. Soc. Chim. Fr.*, 1561 (1968).
338. J. P. Dalle, M. Mousseron-Canet and J. C. Mani, *Bull. Soc. Chim. Fr.*, 232 (1969).
339. J. L. Olivé and M. Mousseron-Canet, *Bull. Soc. Chim. Fr.*, 3252 (1969).
340. M. Mousseron-Canet, J. P. Dalle and J. C. Mani, *Photochem. Photobiol.*, **9**, 91 (1969).
341. D. A. Lerner, J. C. Mani and M. Mousseron-Canet, *Bull. Soc. Chim. Fr.*, 1968 (1970).
342. S. Isoe, S. B. Hyeon, H. Ichikawa, S. Katsumura and T. Sakan, *Tetrahedron Lett.*, 5561 (1968).
343. S. Isoe, S. Katsumura, S. B. Hyeon and T. Sakan, *Tetrahedron Lett.*, 1089 (1971).
344. C. S. Foote and M. Brenner, *Tetrahedron Lett.*, 6041 (1968).
345. E. Demole and P. Enggist, *Helv. Chim. Acta*, **51**, 481 (1968).
346. K. Gollnick and H. J. Kuhn, in Reference 261, Chap. 8, pp. 287ff.
347. M. Matsumoto and K. Kuroda, *Tetrahedron Lett.*, **23**, 1285 (1982).
348. M. Maumy and J. Rigaudy, *Bull. Soc. Chim. Fr.*, 1879 (1975).
349. H. B. Herbest and R. A. L. Wilson, *Chem. Ind. (London)*, 86 (1956).
350. P. Bladon and T. Sleigh, *Proc. Chem. Soc.*, 183 (1962).
351. P. Bladon and T. Sleigh, *J. Chem. Soc.*, 6991 (1965).
352. W. Dilthey, S. Hinkels and M. Leonhard, *J. Prakt. Chem.*, **151**, 97 (1938).
353. G. O. Schenck, *Z. Elektrochem.*, **56**, 855 (1952).
354. C. Dufraisse, A. Etienne and J. Aubry, *Bull. Soc. Chim. Fr.*, 1201 (1954).
355. C. Dufraisse, A. Etienne and J. Aubry, *C.R. Acad. Sci. Paris*, **239**, 1170 (1954).
356. D. M. Bikales and E. I. Becker, *J. Org. Chem.*, **21**, 1405 (1956).
357. C. F. Wilcox and M. P. Stevens, *J. Am. Chem. Soc.*, **84**, 1258 (1962).
358. C. F. H. Allen and J. A. Van Allan, *J. Org. Chem.*, **18**, 882 (1953).
359. B. D. Chaney and S. B. Brown, *Photochem. Photobiol.*, **28**, 339 (1978).
360. C. S. Foote, W. Wexler, W. Ando and R. Higgins, *J. Am. Chem. Soc.*, **90**, 975 (1968).
361. R. W. Murray and M. L. Kaplan, *J. Am. Chem. Soc.*, **91**, 5358 (1969).
362. J. E. Baldwin, J. C. Swallow and H. W. S. Chan, *J. Chem. Soc., Chem. Commun.*, 1407 (1971).
363. E. Koch, *Tetrahedron*, **24**, 6295 (1968).
364. W. Adam and I. Erden, *J. Am. Chem. Soc.*, **101**, 5692 (1979).
365. W. Adam and I. Erden, *Angew. Chem., Int. Ed. Engl.*, **90**, 211 (1978).
366. J. P. Smith and G. B. Schuster, *J. Am. Chem. Soc.*, **100**, 2564 (1978).
367. H. -S. Ryang and C. S. Foote, *J. Am. Chem. Soc.*, **103**, 4951 (1981).
368. W. Ando, H. Miyazaki, K. Veno, H. Nakanishi, T. Sakurai and K. Kobayashi, *J. Am. Chem. Soc.*, **103**, 4949 (1981).
369. H. S. Ryang and C. S. Foote, *Tetrahedron Lett.*, **23**, 2551 (1982).
370. M. Oda and Y. Kitahara, *Tetrahedron Lett.*, 3295 (1969).
371. H. Takeshita, T. Kusaba and M. Mori, *Chem. Lett.*, 1371 (1983).
372. E. J. Forbes and J. Griffiths, *J. Chem. Soc. (C)*, 575 (1968).
373. T. Tezuka, R. Miyamoto, T. Mukai, C. Kabuto and Y. Kitahara, *J. Am. Chem. Soc.*, **94**, 9280 (1972).
374. A. Mori, H. Suizu, and H. Takeshita, *Bull. Chem. Soc. Jpn.*, **60**, 3817 (1987).
375. S. Ito, Y. Shoji, H. Takeshita, M. Hirama and K. Takahashi, *Tetrahedron Lett.*, 1075 (1975).
376. T. Tezuka, R. Miyamoto, M. Nagayama and T. Mukai, *Tetrahedron Lett.*, 327 (1975).
377. Y. Ito, M. Oda and Y. Kithara, *Tetrahedron Lett.*, 239 (1975).
378. G. Rio and J. Berthelot, *Bull. Soc. Chim. Fr.*, 2938 (1971).
379. M. Mousseron-Canet, J. C. Mani, J. P. Dalle and J. L. Olivé, *Bull. Soc. Chim. Fr.*, 3874 (1966).
380. M. Mousseron-Canet, J. C. Mani and J. L. Olivé, *C.R. Acad. Sci. Paris, Ser. C*, **262**, 1725 (1966).
381. N. Akbulut, A. Menzck and M. Balci, *Tetrahedron Lett.*, **28**, 1689 (1987).

382. L. T. Scott and C. M. Adams, *J. Am. Chem. Soc.*, **106**, 4857 (1984).
383. W. W. Henderson and E. F. Ullman, *J. Am. Chem. Soc.*, **87**, 5424 (1955).
384. C. S. Foote, in *Free Radicals in Biology* (Ed. W. A. Pryor), Vol. 2, Academic Press, New York, 1976, p. 85.
385. L. Taimr and J. Pospisil, *Angew. Makromol. Chem.*, **52**, 31 (1976).
386. G. O. Schenck, *Angew. Chem.*, **64**, 12 (1952).
387. A. Ritter, P. Bayer, J. Lutich and G. Schomburg, *Liebigs Ann. Chem.*, 835 (1974).
388. N. Furutachi, Y. Nakadaira and K. Nakanishi, *J. Chem. Soc., Chem. Commun.*, 1625 (1968).
389. C. D. Snyder and H. Rapoport, *J. Am. Chem. Soc.*, **91**, 731 (1969).
390. A. F. Thomas and R. Dubini, *Helv. Chim. Acta*, **57**, 2076 (1974).
391. H. Morimoto, I. Imada and G. Goto, *Liebigs Ann. Chem.*, **735**, 65 (1970).
392. M. Ohmae and G. Katsui, *Vitamins*, **35**, 116 (1967).
393. D. Hellinger, P. de Mayo, M. Nye, L. Westfelt and R. B. Yeats, *Tetrahedron Lett.*, 349 (1970).
394. Y. Kithara, T. Kato, T. Suzuki, S. Kanno and M. Tanemura, *J. Chem. Soc., Chem. Commun.*, 342 (1969).
395. M. Maumy and J. Rigaudy, *Bull. Soc. Chim. Fr.*, 2021 (1976); cf. Reference 388.
396. L. Canonica, B. Danieli, G. Lesma, G. Palmisano and A. Mugnoli, *Helv. Chim. Acta*, **70**, 701 (1987).
397. See Reference 133 and discussion therein regarding the related work of G. O. Schenck, K. Gollnick and O. Neumuller, *Liebigs Ann. Chem.*, **603**, 46 (1957).
398. M. Maumy and J. Rigaudy, *Bull. Soc. Chim. Fr.*, 1487 (1974).
399. (a) W. Adam and I. Erden, *Tetrahedron Lett.*, 1975 (1979).
     (b) G. O. Schenck, *Angew. Chem.*, **64**, 12 (1952).
400. For related cases see:
     (a) W. Adam, M. Balci and B. Pietrzak, *J. Am. Chem. Soc.*, **101**, 6285 (1979).
     (b) W. Adam, M. Balci and J. Rivera, *Synthesis*, 807 (1979).
     (c) T. Asao, M. Yagihara and Y. Kithara, *Heterocycles*, **15**, 985 (1985).
401. A. A. Frimer, *Isr. J. Chem.*, **21**, 194 (1981) and references cited therein.
402. A. A. Frimer, D. Rot and M. Sprecher, *Tetrahedron Lett.*, 1927 (1977).
403. A. A. Frimer and D. Rot, *J. Org. Chem.*, **44**, 3882 (1979).
404. A. A. Frimer, T. Farkash and M. Sprecher, *J. Org. Chem.*, **44**, 989 (1979).
405. A. A. Frimer, and A. Antebi, *J. Org. Chem.*, **45**, 2334 (1980).
406. D. W. Turner, *Molecular Photoelectron Microscopy*, Wiley, New York, 1970.
407. D. H. Aue, M. J. Mishishnek and D. F. Shelhamer, *Tetrahedron Lett.*, 4799 (1973).
408. L. J. Bollyky, *J. Am. Chem. Soc.*, **92**, 3230 (1970).
409. N. J. Turro, Y. Ito, M.-F. Chow, W. Adam, O. Rodrigues and F. Yang, *J. Am. Chem. Soc.*, **99**, 5936 (1977).
410. W. A. Pryor, *Photochem. Photobiol.*, **28**, 787 (1978).
411. L. Parker and J. Walton, *Chem. Technol.*, **7**, 278 (1977).
412. J. Bland, *J. Chem. Educ.*, **55**, 151 (1978).
413. I. Fridovich, as cited in J. D. Spikes and H. M. Swartz, *Photochem. Photobiol.*, **28**, 921, 930 (1978).
414. D. T. Sawyer and M. J. Gibian, *Tetrahedron*, **35**, 1471 (1979).
415. J. Wilshire and D. T. Sawyer, *Acc. Chem. Res.*, **12**, 105 (1979).
416. S. Torii, *Synthesis*, 873 (1986).
417. J. S. Valentine and A. B. Curtis, *J. Am. Chem. Soc.*, **97**, 224 (1975).
418. A. D. McElroy and J. S. Hasman, *Inorg. Chem.*, **40**, 1798 (1964).
419. J. W. Peters and C. S. Foote, *J. Am. Chem. Soc.*, **98**, 873 (1976).
420. (a) J. L. Roberts, Jr., T. C. Calderwood and D. T. Sawyer, *J. Am. Chem. Soc.*, **105**, 7691 (1983).
     (b) S. Matsumoto, H. Sugimoto and D. T. Sawyer, *Chem. Res. Toxicol.*, **1**, 19 (1988).
421. B. Kenion, *J. Chem. Soc., Chem. Commun.*, 731 (1982).
422. J. R. Kanofsky, *J. Am. Chem. Soc.*, **108**, 2977 (1986).
423. C. A. Long and B. H. J. Bielsky, *J. Phys. Chem.*, **84**, 555 (1980).
424. D. K. Akutagawa, N. Furukawa and S. Oae, *Bull. Chem. Soc. Jpn.*, **57**, 1104 (1984).
425. B. H. J. Bielsky, *Photochem. Photobiol.*, **28**, 645 (1978).
426. A. D. Goolsby and D. T. Sawyer, *Anal. Chem.*, **40**, 83 (1968).
427. J. A. Free and J. S. Valentine, in *Superoxide and Superoxide Dismutase* (Eds. A. M. Michelson, J. M. McCord and I. Fridovich), Academic Press, New York, 1977, p. 19.
428. D. T. Sawyer and E. J. Nanni, Jr., in *Oxygen and Oxy-Radicals in Chemistry and Biology* (Eds. M. A. J. Rodgers and E. L. Powers), Academic Press, New York, 1981, pp. 15–44.

429. J. Chevalit, F. Rouelle, L. Gierst and J. P. Lambert, *J. Electroanal. Chem.*, **39**, 201 (1972).
430. R. Deitz, M. E. Peover and P. Rothbaum, *Chem.-Ing.-Tech.*, **42**, 185 (1970).
431. R. Poupko and I. Rosenthal, *J. Phys. Chem.*, **77**, 1722 (1973).
432. K. B. Patel and R. L. Willson, *J. Chem. Soc., Faraday Trans. 1*, **69**, 814 (1973).
433. A. Anne and J. Moiroux, *Nouv. J. Chim.*, **8**, 259 (1984).
434. I. Rosenthal and T. Bercovici, *J. Chem. Soc.*, 200 (1973).
435. M. De Min, M. T. Maurette, E. Oliveros, M. Hocquax and B. Jaquet, *Tetrahedron*, **42**, 4953 (1986).
436. A. R. Forrester and V. Purushotham, *J. Chem. Soc., Chem. Commun.*, 1505 (1984).
437. A. R. Forrester and V. Purushotham, *J. Chem. Soc., Perkin Trans. 1*, 945 (1987).
438. D. -H. Chin, G. Chiercato, Jr., E. J. Nanni, Jr. and D. T. Sawyer, *J. Am. Chem. Soc.*, **104**, 1296 (1982).
439. J. F. Liebman and J. S. Valentine, *Isr. J. Chem.*, **23**, 439 (1983).
440. J. P. Stanley, *J. Org. Chem.*, **45**, 1413 (1980).
441. G. Feroci and S. Roffia, *J. Electroanal, Chem.*, **81**, 387 (1977).
442. See articles of I. Fridovich and J. Fee and the subsequent discussion in *Oxygen and Oxy-Radicals in Chemistry and Biology*, (Eds. M. A. J. Rodgers and E. L. Powers), Academic Press, New York, 1981, pp. 197–239.
443. See the exchange of correspondence between I. Fridovich and D. T. Sawyer and J. S. Valentine, *Acc. Chem. Res.*, **15**, 200 (1982).
444. A. A. Frimer, G. Aljadeff and P. Gilinsky-Sharon, *Isr. J. Chem.*, **27**, 39 (1986).
445. M. J. Gibian and S. Russo, *J. Org. Chem.*, **49**, 4304 (1984).
446. See discussion in Z. V. Todres, *Tetrahedron*, **43**, 3839, 3844 (1987).
447. I. Rosenthal and A. A. Frimer, *Tetrahedron Lett.*, 2805 (1976).
448. A. A. Frimer and I. Rosenthal, *Potochem. Photobiol.*, **28**, 711 (1978).
449. E. Lee-Ruff, *Chem. Soc. Rev.*, **6**, 195 (1977).
450. I. Saito, T. Otsuki and T. Matsuura, *Tetrahedron Lett.*, 1693 (1979).
451. A. Le Berre and Y. Berguer, *Bull. Soc. Chim. Fr.*, 2368 (1966).
452. D. C. Neckers and G. Hauck, *J. Org. Chem.*, **48**, 4691 (1983).
453. P. Furderer, F. Berson and A. Krebs, *Helv. Chim. Acta*, **60**, 1226 (1977).
454. I. Rosenthal and A. A. Frimer, *Tetrahedron Lett.*, 3731 (1975).
455. S. Kobayashi, T. Tezuka and W. Ando, *J. Chem. Soc., Chem. Commun.*, 508 (1979).
456. J. San Fillipo, Jr., L. J. Romano, C. -I. Chern and J. S. Valentine, *J. Org. Chem.*, **41**, 586 (1976).
457. Y. Moro-oka and C. S. Foote, *J. Am. Chem. Soc.*, **98**, 1510 (1976).
458. R. Dietz, A. E. Forno, B. E. Larcombe and M. E. Peover, *J. Chem. Soc.*, 816 (1970).
459. P. Gilinsky-Sharon, Ph.D. Thesis, Bar-Ilan University, Ramat Gan Israel, 1984.
460. R. W. Dugger and C. H. Heathcock, *J. Org. Chem.*, **45**, 1189 (1980) and references cited therein.
461. R. H. Hall, K. Bischofberger, S. J. Eitelman and A. Jordaan, *J. Chem. Soc., Perkin Trans. 1*, 2236 (1977).
462. A. A. Frimer and G. Aljadeff, unpublished results (1984); confirmed in personal communication by G. Speier (1984).
463. A. Le Berre and Y. Berguer, *Bull. Soc. Chim. Fr.*, 2363 (1968).
464. B. H. J. Bielski and H. W. Rechter, *J. Am. Chem. Soc.*, **99**, 3019 (1977).
465. D. E. Cabelli and B. H. J. Bielski, *J. Phys. Chem.*, **87**, 1809 (1983).
466. A. D. Nadezhdin and H. B. Dunford, *Can. J. Chem.*, **57**, 3017 (1979).
467. B. H. J. Bielski, A. D. Allen and H. A. Schwartz, *J. Am. Chem. Soc.*, **103**, 3516 (1981).
468. M. Nishikimi, *Biochem. Biophys. Res. Commun.*, **63**, 463 (1975).
469. D. T. Sawyer, D. T. Richens, E. J. Nanni, Jr. and M. D. Stallings, in *Chemical and Biochemical Aspects of Superoxide and Superoxide Dismutase* (Eds. W. H. Bannister and H. A. O. Hill), Elsevier, Holland, 1980, p. 1.
470. E. J. Nanni, Jr., M. D. Stallings and D. T. Sawyer. *J. Am. Chem. Soc.*, **102**, 4481 (1980).
471. D. T. Sawyer, G. Chiercato, Jr. and T. Tsuchiya, *J. Am. Chem. Soc.*, **104**, 6273 (1982).
472. D. T. Sawyer, T. S. Calderwood, C. C. Johlman and C. L. Wilkins, *J. Org. Chem.*, **50**, 1409 (1985).
473. I. B. Afanas'ev, U. V. Grabovetskii and N. S. Kuprianova, *J. Chem. Soc., Perkin Trans. 2*, 281 (1987).
474. A. A. Frimer and P. Gilinsky-Sharon, 50th Annual Conference of the Israel Chemical Society, Jerusalem, 1985, abstract TD10, p. 82.
475. A. A. Frimer and V. Marks, unpublished results (1987).

The Chemistry of Enones
Edited by S. Patai and Z. Rappoport
© 1989 John Wiley & Sons Ltd

CHAPTER **18**

Reduction of α, β-unsaturated carbonyl compounds

EHUD KEINAN and NOAM GREENSPOON

Department of Chemistry, Technion—Israel Institute of Technology, Technion City, Haifa 32000, Israel

I. INTRODUCTION

The two main reduction modes of α, β-unsaturated aldehydes and ketones involve formal hydride attack at either the C-1 or C-3 of the enone system, leading to allylic alcohol

923

or saturated carbonyl compound, respectively. It has been suggested that the relative importance of these paths depends on the relative 'hardness' or 'softness' of the substrate, defined in terms of coefficients of the lowest unoccupied molecular orbital (LUMO) (*vide infra*). While the 1, 2 addition is considered to be a more charge-controlled process, 1, 4 addition is a frontier-orbital controlled process.

In addition to these two reduction modes, which involve formal addition of a single hydrogen molecule to the substrate, it is also possible to add two hydrogen molecules, yielding the corresponding saturated alcohol. Alternatively, formal addition of two molecules of hydrogen may completely deoxygenate the substrate, giving the unsaturated hydrocarbon. Finally, total reduction with three hydrogen molecules would provide the saturated hydrocarbon.

The synthetic application of a given reduction method should be considered primarily in terms of its regioselectivity, stereochemical control and chemoselectivity. Regioselectivity refers mainly to selection between the 1,4- and 1,2-reduction modes. Stereochemical control refers to the relative and absolute configuration of the newly formed sp^3 centers at positions 1,2 or 3 of the enone system. Chemoselectivity refers to the opportunity of selectively reducing the desired functionality in a complex molecule containing other easily reducible functional groups. Other important factors, particularly for reactions to be carried out in large scale, are the availability and cost of the given reducing system as well as convenience and simplicity of the procedures.

Available methods for reduction of carbonyl functionalities and, in particular, α, β-unsaturated ones may be divided conveniently into four classes, based on historic considerations. The earliest procedures, extensively used prior to the discovery of catalytic hydrogenation and metal hydride reductions, employed dissolving metals. In the broader sense, more recent developments, such as reduction with low-valent transition-metal compounds and electrochemical processes, may also be included in this category as they all proceed, in the mechanistic sense, via sequential addition of electrons and protons to the substrate molecule.

Catalytic hydrogenation may be regarded as the second generation of reducing systems. Indeed, both heterogeneous and homogeneous catalytic hydrogenation replaced many of the earlier dissolving metal techniques, although the latter are still used due to selectivity characteristics or convenience.

The discovery of metal hydrides and complex metal hydrides, particularly those of boron and aluminum in the early 1940s, have revolutionized the reduction of organic functional groups. These reagents may be regarded as the third generation of reducing systems. Extensive studies over the past fifty years have led to a broad variety of hydridic reagents whose reducing power and selectivity are controlled by appropriate modification of the ligands in the metal coordination sphere[1]. Hydridic reagents today include other main-group metal hydrides, such as silicon and tin derivatives, as well as a variety of transition-metal hydrides that are employed in stoichiometric quantities, such as the iron, copper, chromium and cobalt compounds.

The advent of organo-transition-metal chemistry within the past thirty years has generated a plethora of novel synthetic methods that provide new opportunities for selective reduction. Composite reducing systems comprised of a transition-metal catalyst and a relatively nonreactive hydride donor represent the fourth generation of reductants. The generally high selectivities provided by such systems arise from two main facts: (a) specific interaction between the transition-metal catalyst and the substrate functionality, and (b) selective, facile hydride transfer from the hydride-donor to the transition metal, and hence to the substrate. Many of the transfer-hydrogenation methods may be included within this fourth category as well. Therefore, although in many respects several transfer-hydrogenation techniques resemble regular catalytic hydrogenations, they are discussed in Section VI that deals with composite reducing systems.

II. ELECTRON-TRANSFER REDUCTIONS

A. Dissolving-metal Reductions

A variety of organic functional groups are reduced by active metal either in the presence of a proton donor or followed by treatment with a proton donor. This approach is one of the earliest reduction procedures in organic chemistry. Although its importance has decreased with the development of catalytic hydrogenation and metal hydride reduction, there remain a substantial number of dissolving metal reductions still in use due to their advantageous selectivity of reduction. Dissolving metal reductions of α, β-unsaturated carbonyl compounds have been discussed in several review articles[2–10].

Metals commonly utilized include the alkali metals, mainly lithium, sodium and potassium, and also calcium, zinc, magnesium, tin and iron. Alkali metals and calcium have been used in liquid ammonia[10], in low-molecular-weight aliphatic amines[11], in hexamethylphosphoramide[12], in ether or in THF containing crown ethers[13c], or in very dilute solutions in polyethers such as 1,2-dimethoxyethane (DME)[11a,13a,b]. Reactions with metal solutions in liquid ammonia often use a cosolvent, such as ether, THF or DME, to increase solubility of the organic substrate in the reaction mixture. These same metals as well as zinc and magnesium have also been used as suspensions in various solvents including ether, toluene, xylene, etc. In all procedures a proton source (frequently ethanol, isopropanol, t-butanol or even water) is provided in the reaction medium, or together with the substrate, or during the workup procedure.

Sodium amalgam, aluminum amalgam, zinc, zinc amalgam, tin and iron have been added directly to solutions of the substrate in hydroxylic solvents such as ethanol, isopropanol, butanol, isoamyl alcohol, acetic acid, water or aqueous mineral acid. With hydroxylic solvents, and especially with relatively acidic ones, metal amalgams are often used rather than free metals to minimize the release of hydrogen gas side-product.

The dissolving-metal reductions are better classified as 'internal' electrolytic reductions in which an electron is transferred from the metal surface (or from the metal in solution) to the substrate. Reduction with low-valent metal ions may also be included in this general class (*vide infra*).

The generally accepted mechanism for dissolving-metal reduction of enones (Scheme 1)[10] involves reversible addition of an electron to a vacant orbital of the substrate (S), yielding a radical anion (S$^{-\cdot}$). The latter can be protonated to give a neutral radical, which may either dimerize or accept another electron and a proton. Alternatively, stepwise or simultaneous reversible addition of two electrons to S can give a dianion capable of accepting two protons. The sequence and timing of these steps should depend upon the substrate, the homogeneity and reduction potential of the medium, and the presence and nature of proton donors in the medium, among other factors.

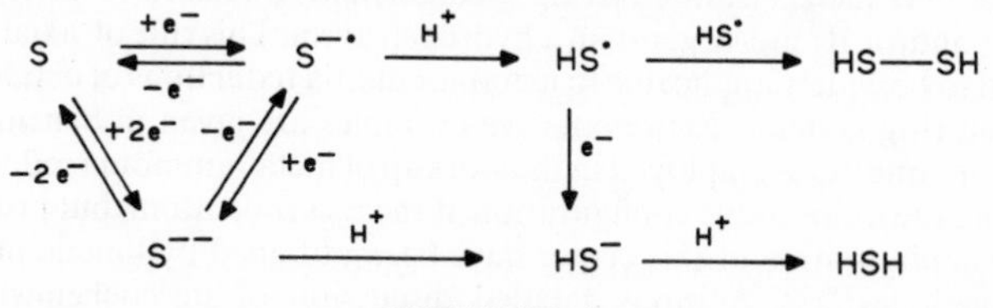

SCHEME 1

The stereochemistry of reduction has been extensively studied. Metal-ammonia reduction of steroid and terpenoid enones with a β carbon at the fusion of two six-membered rings leads, in general, to the thermodynamically more stable isomer at that position[14]. Stork has formulated a more general rule, namely that the product will be the more stable of the two isomers having the newly introduced β-hydrogen axial to the ketone ring[15]. This rule has correctly predicted the stereochemical outcome of many metal-ammonia reductions, with very few exceptions. The rule is rationalized in terms of stereoelectronic effects in the transition state (either the radical anion or the dianion stage). For example, in reduction of octalones of the type shown in Scheme 2, only two (**A** and **B**) of three possible anionic transition states involving a half-chair conformation of the enone-containing ring would be allowed stereochemically[15].

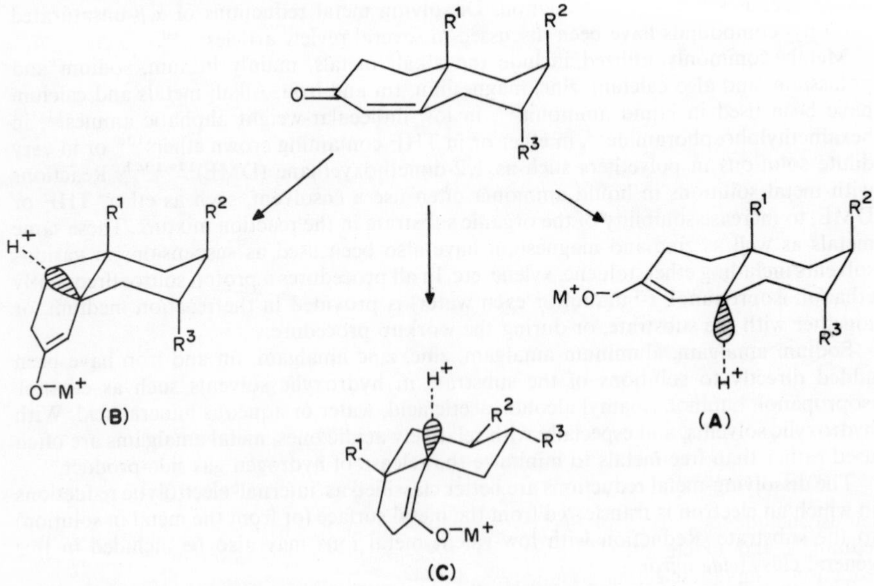

SCHEME 2

In these two conformers the orbital of the developing C—H bond overlaps with the remainder of the π-system of the enolate. The alternative conformer **C** is not allowed because it does not fulfill the overlap requirement. The *trans* transition-state **A** is generally more stable than the *cis* **B**, and the *trans*-2-decalone reduction product would be obtained, despite the fact that the *cis* isomer having a conformation related to **C** should be more stable when R^2 and/or R^3 are larger than a hydrogen atom. This rule of 'axial protonation' has been found to be widely applicable to metal-ammonia reductions of octalones, steroids and other fused-ring systems. Representative examples are given in Scheme 3[15-18].

Generally, the conditions employed in the workup of metal-ammonia reductions lead to products having the more stable configuration at the α-carbon atom, but products having the less stable configuration at this center have been obtained by kinetic protonation of enolate intermediates[19,20]. A more detailed discussion of stereochemistry in metal-ammonia reduction of α, β-unsaturated carbonyl compounds is given in Reference 10.

Scope and limitations. Before the introduction of metal-ammonia solutions for the reduction of α, β-unsaturated carbonyl compounds[10], sodium, sodium amalgam or zinc in protic media were most commonly employed for this purpose. Some early examples of

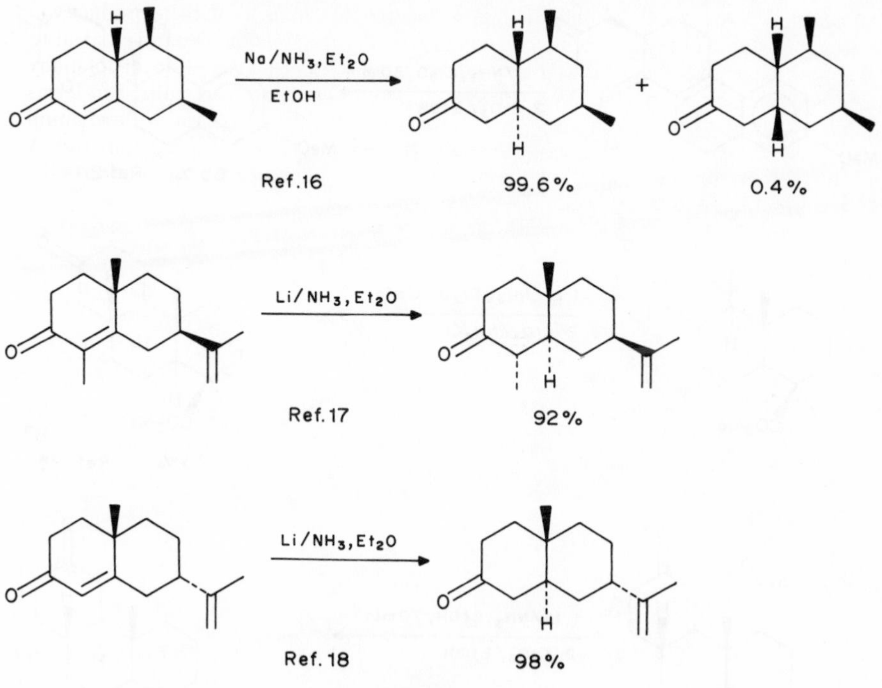

Ref.16   99.6%   0.4%

Ref.17   92%

Ref.18   98%

SCHEME 3

their use include the conversion of carvone to dihydrocarvone with zinc in acid or alkaline medium[21], and of cholest-4-en-3-one to cholestanone with sodium in alcohol[22,23]. Reductions using these earlier methods may be complicated by a variety of side-reactions, such as over-reduction, dimerization, skeletal rearrangements, acid- or base-catalyzed isomerizations and aldol condensations, most of which can be significantly minimized by metal-ammonia reduction.

Ketones ranging from simple acyclic varieties to complex polycyclic ones such as steroids, terpenoids and alkaloids have been reduced to saturated ketones, usually in good yield, by metal solutions, mainly in liquid ammonia. A few examples are given in Scheme 4[10,24–26]. The reduction is applicable to compounds with any degree of substitution on the double bond. Although only two equivalents of these metals are required for the conversion of an enone to a saturated ketone, it is often convenient to employ the metal in excess. Proton donors are often employed to prevent competing side-reactions, such as dimerization. The presence of proton donors in the medium may lead to the conversion of an α,β-unsaturated ketone to the saturated alcohol. Obviously, at least four equivalents of metal must be present for that type of reduction to take place.

Alcohols, such as methanol and ethanol, lead to the sole formation of saturated alcohols from unsaturated ketones when the former are present in excess during the reduction. Mixtures of ketone and alcohol are generally formed when one equivalent of these proton donors is employed[27]. These alcohols have acidity comparable to that of saturated ketones, and when they are present, equilibrium can be established between the initially formed metal enolate and the saturated ketone. The latter is then reduced to the saturated alcohol. Such reductions generally do not occur to a very significant extent when one equivalent of t-butanol[28] or some less acidic proton donor, such as triphenylcarbinol[27], is

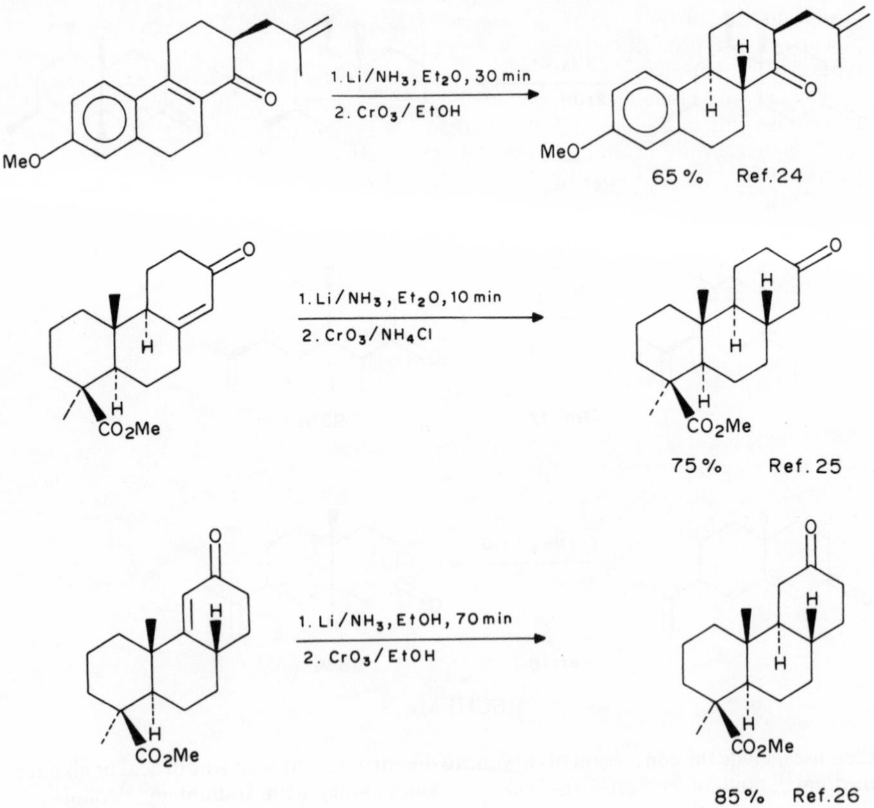

1. Li/NH₃, Et₂O, 30 min
2. CrO₃/EtOH

65 %   Ref. 24

1. Li/NH₃, Et₂O, 10 min
2. CrO₃/NH₄Cl

75 %   Ref. 25

1. Li/NH₃, EtOH, 70 min
2. CrO₃/EtOH

85 %   Ref. 26

SCHEME 4

employed. The acidity of the ketone involved as well as the solubility of the metal enolate in the reaction medium are of importance in determining whether alcohols are formed.

Even though the reaction conditions may lead to formation of the metal enolate in high yield, further reduction may occur during the quenching step of the reaction. Alcohols such as methanol and ethanol convert metal enolates to saturated ketones much faster than they react with metals in ammonia[29,30], and quenching of reduction mixtures with these alcohols will usually lead to partial or complete conversion to alcoholic product rather than the saturated ketone. Rapid addition of excess solid ammonium chloride is the commonly employed quench procedure if ketonic products are desired[31].

To prevent alcohol formation, other reagents that destroy solvated electrons before reaction mixture neutralization may be employed. These include sodium benzoate[32], ferric nitrate[33,34], sodium nitrite[35], bromobenzene[36], sodium bromate[37], 1,2-dibromoethane[4], and acetone[14].

Reduction-alkylation. The versatility of metal-ammonia reduction was considerably advanced by the discovery that the lithium enolates of unsymmetrical ketones generated during reduction can undergo C-alkylation with alkyl halides and carbonation with carbon dioxide[38,39]. These enolate trapping reactions allow regiospecific introduction of groups at the carbon atoms of unsymmetrical ketones via the appropriate enone precursors. This procedure has been widely employed for ketones of a variety of structural types[28,38-44]. The procedure usually involves generation of a specific lithium enolate of an unsymmetrical ketone by reduction of the corresponding α,β-unsaturated ketone with two equivalents of lithium in liquid ammonia that contains no proton donor or just a single equivalent of alcohol. This enolate is then reacted with excess alkylating agent (Scheme 5).

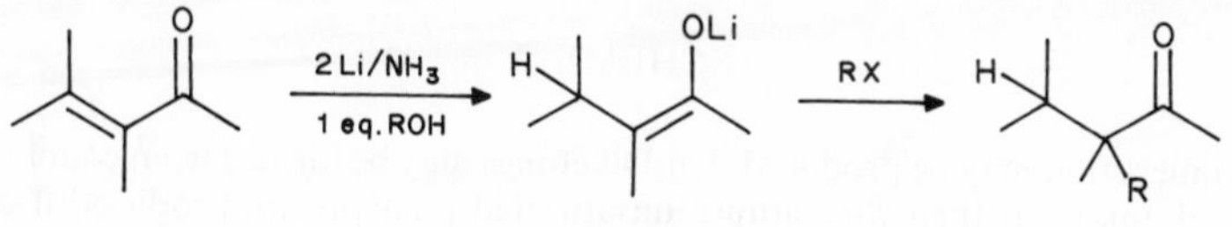

SCHEME 5

This reduction-alkylation sequence has been extensively used in the total synthesis of natural products. The two transformations shown in Scheme 6 represent key steps in the synthesis of d,l-progesterone[45] and lupeol[46].

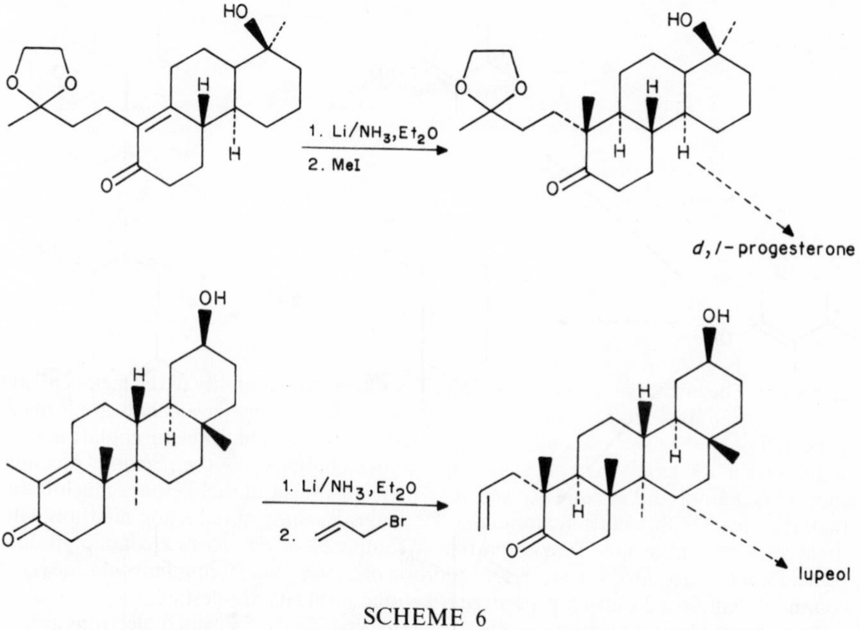

SCHEME 6

If the ammonia is removed and replaced by anhydrous ether, the intermediate lithium enolate can be converted to β-keto ester by carbonation, followed by acidification and treatment with diazomethane, as illustrated in Scheme 7[47].

Dimerization processes. Because of the intermediacy of radical anions and/or hydroxy-allyl free radicals in dissolving-metal reductions of enones, dimerization processes involving these species may compete with simple reduction. Scheme 8 shows the three

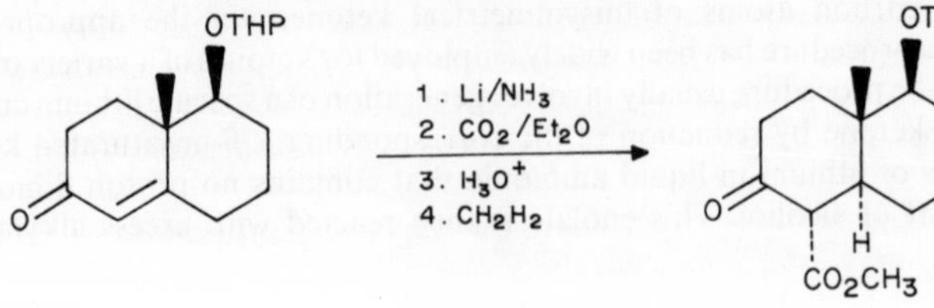

OTHP

1 . Li/NH₃
2 . CO₂/Et₂O
3 . H₃O⁺
4 . CH₂H₂

OTHP

O

H
CO₂CH₃

SCHEME 7

types of dimers that may be produced. 1,6-Diketones may be formed from coupling of the two radical anions at their β-positions; unsaturated pinacols are produced if coupling occurs at the carbonyl carbon atoms; and unsaturated γ-hydroxy ketones are produced by nonsymmetrical coupling of the β-carbon of one radical anion and the carbonyl carbon of a second such intermediate.

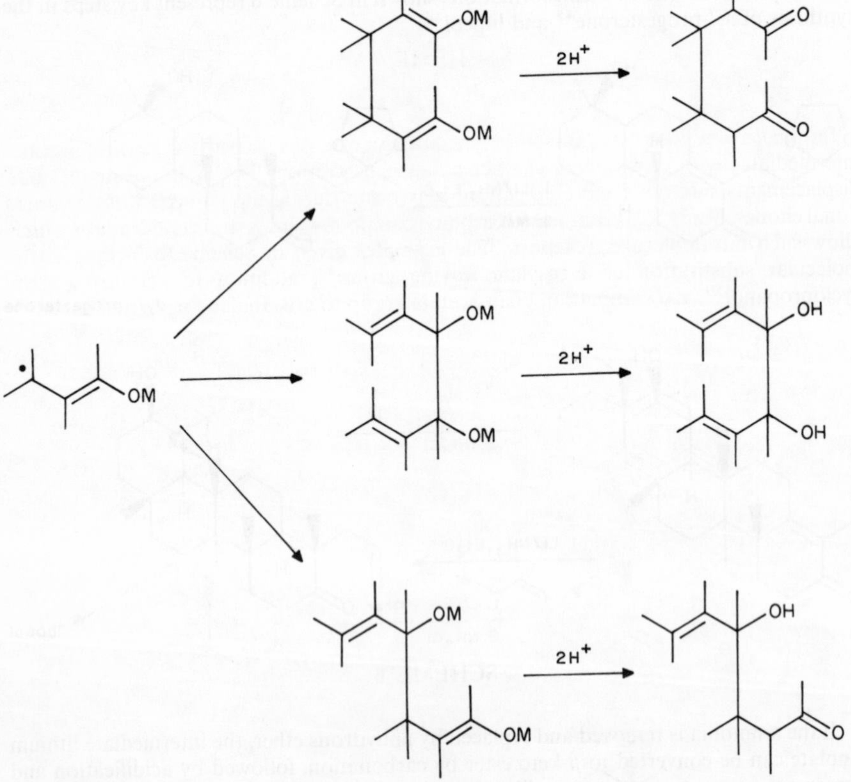

OM

OM

2H⁺

O

O

OM

OM

2H⁺

OH

OH

·

OM

OM

OM

2H⁺

OH

O

SCHEME 8

The dimerization products shown in Scheme 8 are generally the major ones obtained in electrochemical reductions[48-51] (*vide infra*) or reductions at metal surfaces[48,52], in which radical anion intermediates must diffuse to a surface before further electron transfer can occur. In metal-ammonia solutions, however, simple reduction is generally favored over dimerization. These solutions provide high concentrations of available electrons, favoring the probability of the radical ion or hydroxyallyl radical to accept a second electron.

Olefin synthesis. Appropriate quenching of a reductively formed lithium enolate with a carboxylic acid anhydride[53,54], chloride[55], methyl chloroformate[56] or diethyl phosphorochloridate yields the corresponding enol esters, enol carbonates or enol phosphates. These derivatives may be transformed into specific olefins via reductive cleavage of the vinyl oxygen function[57], as illustrated by the example in Scheme 9.

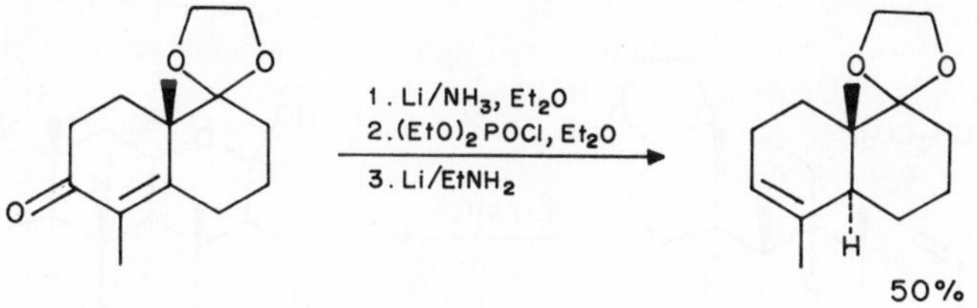

1. Li/NH₃, Et₂O
2. (EtO)₂POCl, Et₂O
3. Li/EtNH₂

50%

SCHEME 9

Intramolecular reactions. Dissolving-metal reduction of unsaturated ketones involve intermediates with carbanionic character at the β-position. Therefore, intramolecular displacements, additions and eliminations may occur during the reduction of polyfunctional enones. Many α, β-unsaturated carbonyl compounds have structural features which allow such intramolecular reactions. The examples given in Scheme 10 include intramolecular substitution of a tosylate leaving group[58], addition to ketone to form cyclopropanol[59], and elimination of an acetate group to give the unconjugated enone[60].

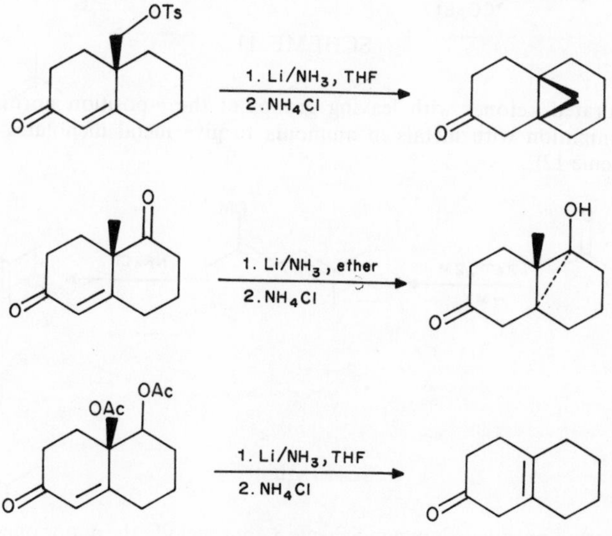

1. Li/NH₃, THF
2. NH₄Cl

1. Li/NH₃, ether
2. NH₄Cl

1. Li/NH₃, THF
2. NH₄Cl

SCHEME 10

The examples given in Scheme 11 include synthesis of a perhydroindanedione skeleton via intramolecular addition to an ester group[61], a related formation of a stable steroidal hemiacetal[62], and lithium-ammonia conversion of a bicyclic unsaturated triester into a tricyclic keto diester[63].

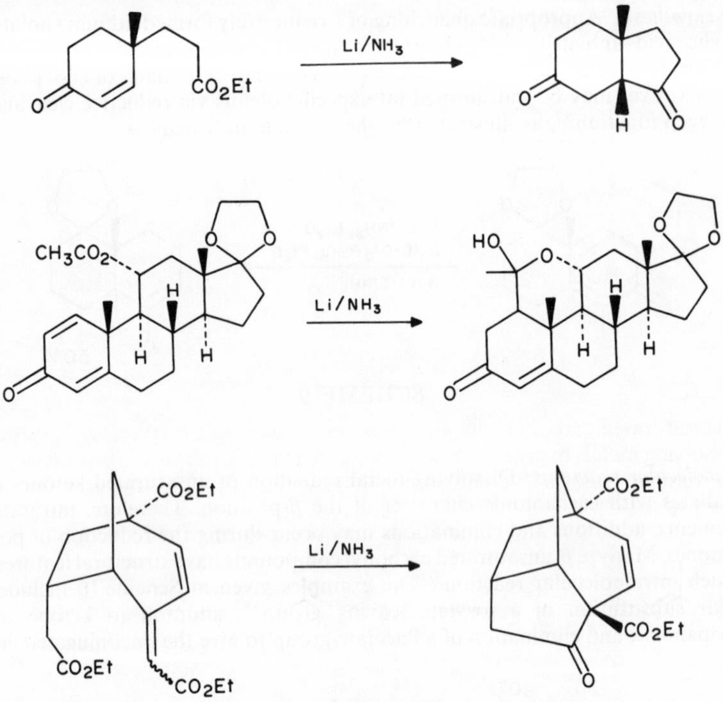

SCHEME 11

α, β-Unsaturated ketones with leaving groups at the γ-position normally undergo reductive elimination with metals in ammonia to give metal dienolates as an initial product (Scheme 12).

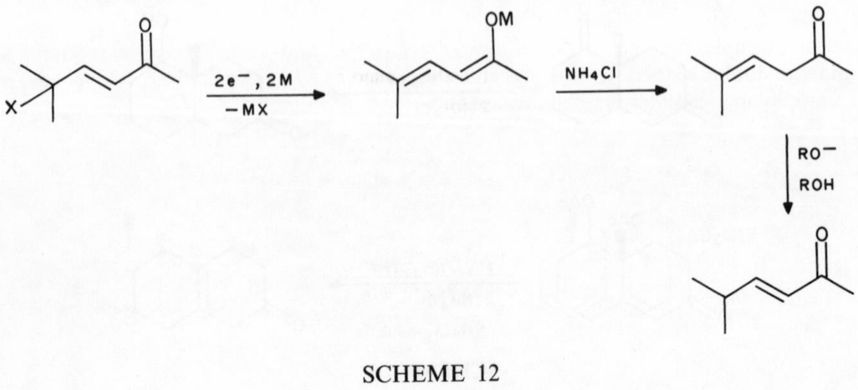

SCHEME 12

Quenching these enolates with ammonium chloride allows the isolation of the β, γ-unsaturated ketone. The latter can isomerize under basic conditions to the conjugated enone. Such processes have been reported with a broad variety of leaving groups, such as hydroxide anion[64,65], alkoxide[66], and acetate[60], as well as during fission of a lactone[67-69] or an epoxide ring[70]. An example involving elimination of hydroxide ion from solidagenone[65] is shown in Scheme 13.

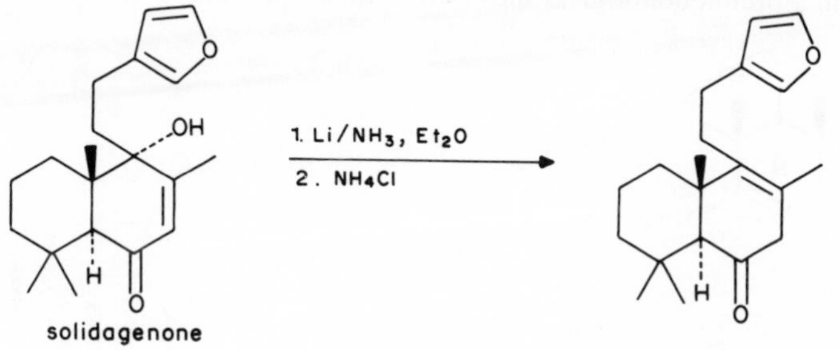

solidagenone

SCHEME 13

α, β-Unsaturated carbonyl compounds having a leaving group at the β position react with dissolving metals to give metal enolates, which may undergo elimination to yield new α, β-unsaturated carbonyl compounds that are susceptible to further reduction (Scheme 14)[43,71-77].

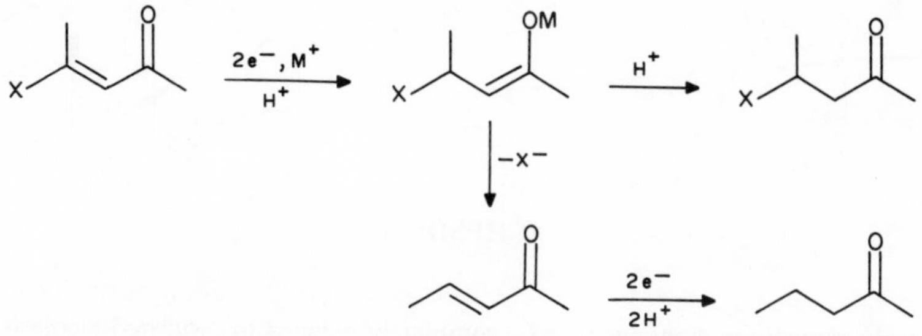

SCHEME 14

For example, β-alkoxy-α, β-unsaturated esters[72,73] and acids[78] have been found to undergo double reduction. This procedure was used as a key step in the total synthesis of eremophilane sesquiterpenes (Scheme 15)[72].

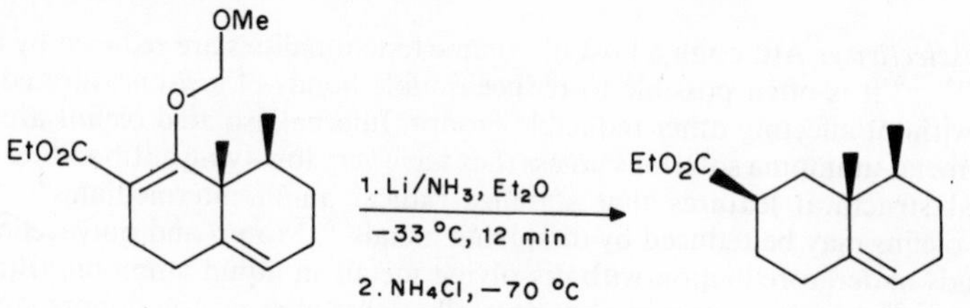

SCHEME 15

Both linear and cross-conjugated dienones are reduced by solutions of metals in liquid ammonia. For example, steroidal 4, 6-dien-3-ones (Scheme 16) and related compounds are reduced initially to 3, 5-dienolates[44,79-86]. While addition of ammonium chloride to the latter leads to formation of the nonconjugated 5-en-3-one system[83], addition of proton donors such as ethanol or water initiates isomerization leading to the more stable, conjugated 4-en-3-one skeleton[80,81]. Treatment of the dienolate with excess methyl iodide rather than a proton donor gives the 4, 4-dimethyl-5-en-3-one[44,87].

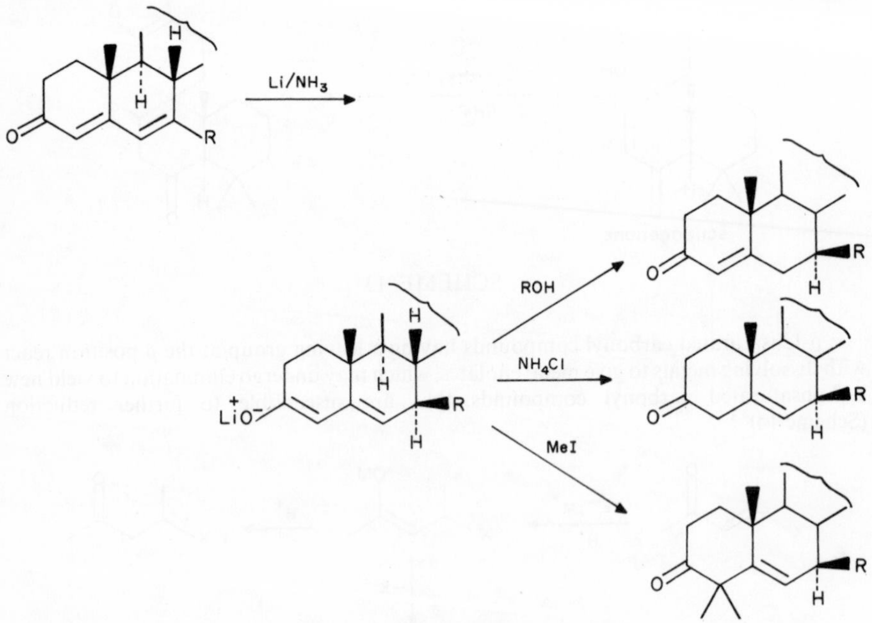

SCHEME 16

Linearly conjugated dienones may be completely reduced to saturated alcohols using excess lithium in liquid ammonia[88]. In variously substituted dienones, the less substituted double bond is often selectively reduced under these conditions. For example, treatment of steroidal 14, 16-dien-20-one with lithium in liquid ammonia (with or without propanol) leads mainly to reduction of the 16, 17 double bond (Scheme 17)[89,90]. Accordingly, the less substituted double bond of cross-conjugated steroidal dienones[4,44,91,92], santonin or related substrates is selectively reduced under these conditions (Scheme 17)[67-69,93].

Chemoselectivity. Although a host of organic functionalities are reduced by dissolving metals[2,3,5-7,9] it is often possible to reduce double bonds of α, β-unsaturated carbonyl systems without affecting other reducible groups. Internal, isolated olefins are normally stable to metal-ammonia solutions unless they have very low-lying antibonding orbitals[94] or special structural features that stabilize radical anion intermediates[95]. However, terminal olefins may be reduced by dissolving metals[96]. Mono- and polycyclic aromatic compounds undergo reduction with dissolving metals in liquid ammonia (Birch reduction)[2,3,5,8,97,98], but these reactions are generally slow unless proton donors are added. It is therefore possible to reduce α, β-unsaturated ketones selectively in the presence of aromatic rings[99–102]. Selective reduction preserving a reducible indole ring is illustrated in Scheme 18[103].

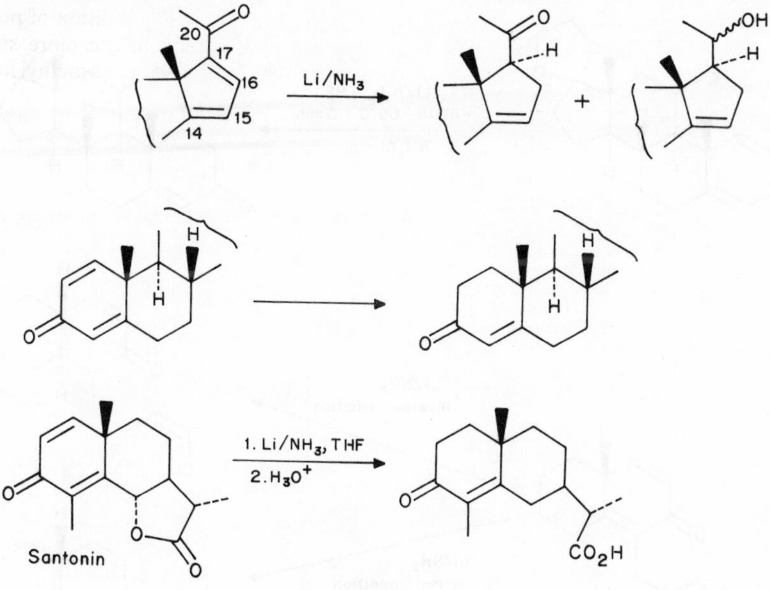

Santonin

SCHEME 17

47%   25%

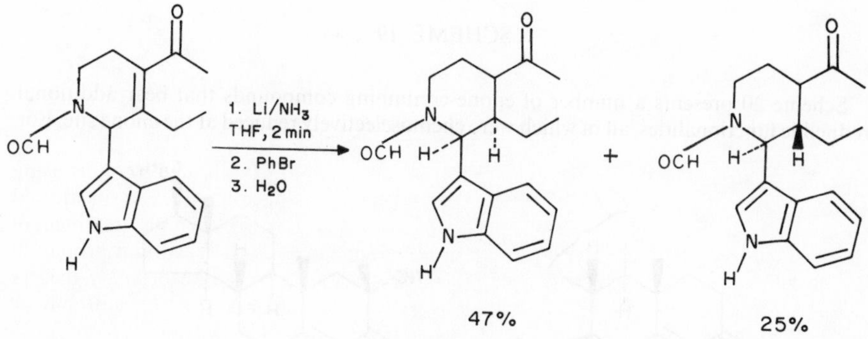

SCHEME 18

Ethynyl carbinols are reduced to allyl alcohols and eventually to olefins with metal-ammonia solutions containing proton donors[104]. However, by excluding proton donors, selective reduction of conjugated enones has been carried out despite the presence of ethynyl carbinol groups[34,105–107]. Similarly, selective reduction of conjugated enones containing allylic alcohols has also been achieved[34,105,107]. Carbon–halogen bonds of alkyl and vinyl halides are readily cleaved by metals in ammonia[5,8,9]. Yet, as shown in Scheme 19, fluoride substituent may be retained by limiting reaction times[92] and a rather sensitive vinyl chloride functionality is preserved by using an inverse addition technique[108].

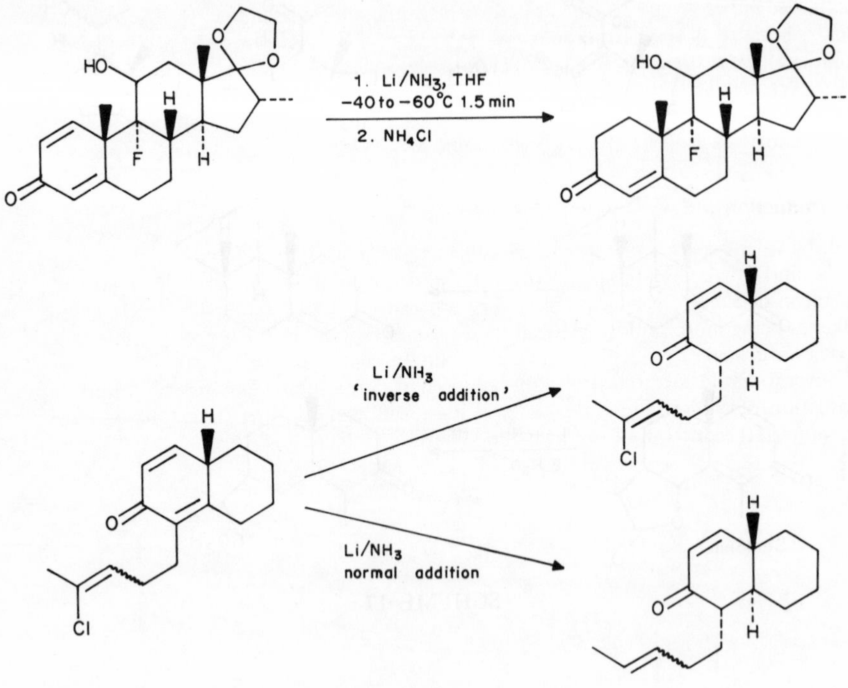
HO
F
H
1. Li/NH3, THF
−40 to −60°C 1.5 min
2. NH4Cl
Li/NH3
'inverse addition'
Li/NH3
normal addition
Cl

SCHEME 19

Scheme 20 presents a number of enone-containing compounds that bear additional reducible functionalities, all of which were chemoselectively reduced at the enone site. For

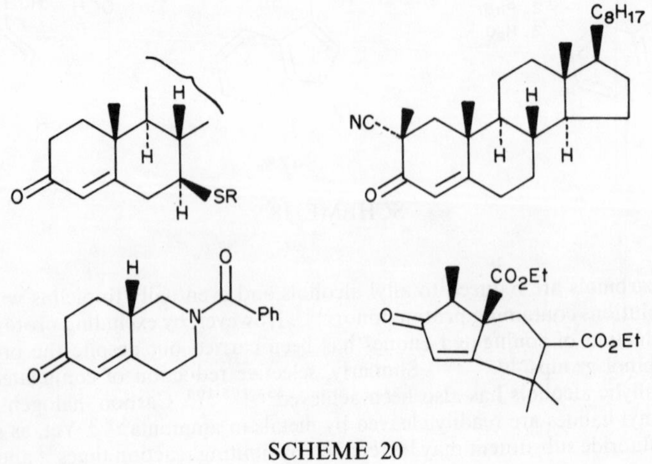
C8H17
SR
NC
Ph
N
CO2Et
CO2Et

SCHEME 20

example, the C—S bond of many thioethers and thioketals are readily cleaved by dissolving metals[5,8,9,109]. Yet, there are examples of conjugate reduction of enones in the presence of a thioalkyl ether group[109,110]. Selective enone reduction in the presence of a reducible nitrile group was illustrated with another steroidal enone[111]. While carboxylic acids, because of salt formation, are not reduced by dissolving metals, esters[112] and amides[2,8] are easily reduced to saturated alcohols and aldehydes or alcohols, respectively. However, metal-ammonia reduction of enones is faster than that of either esters or amides. This allows selective enone reduction in the presence of esters[113] and amides[36,114,115] using short reaction times and limited amounts of lithium in ammonia.

B. Reduction with Low-valent Transition Metals

Low-valent species of early transition metals, such as chromium(II)[116], titanium(II), titanium(III)[117], vanadium, molybdenum and tungsten, are useful reducing agents[118]. Electron-deficient olefins and acetylenes are easily reduced by chromium(II) sulfate, Z-alkenes being more rapidly reduced than the corresponding E-isomers[119]. Titanium(III) species are weaker reducing agents, exhibiting higher chemoselectivity[120].

Several steroid enediones have been reduced by chromium(II) chloride[121]. Interestingly, reduction of cholest-4-ene-3,6-dione yields a different product than that obtained by titanium(III) reduction of the identical substrate (Scheme 21)[120c].

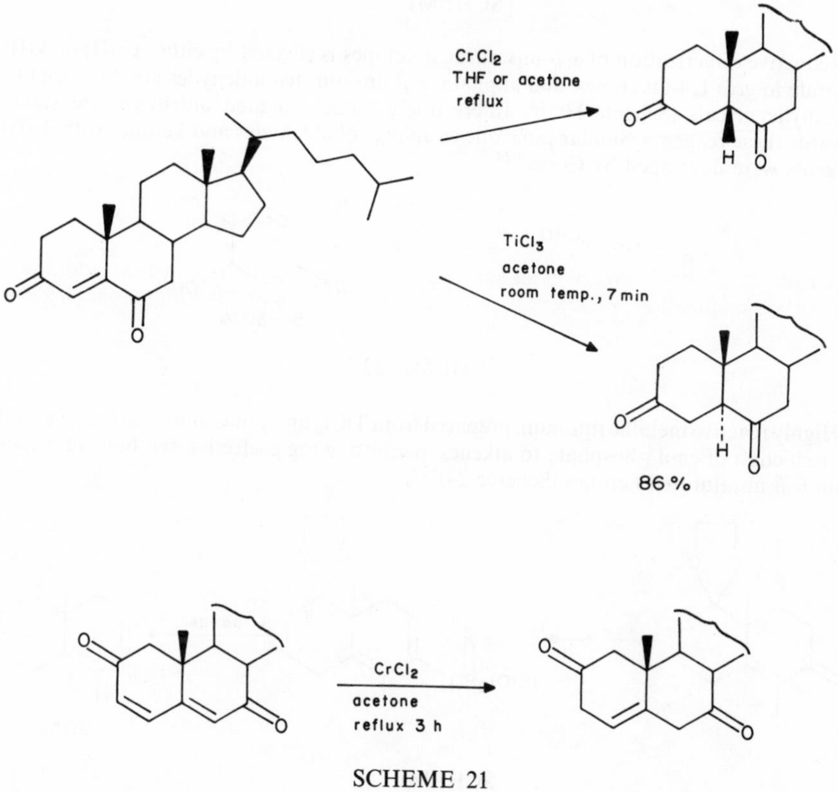

SCHEME 21

Solutions of chromium-bis(ethylenediamine)diacetate complex in methanol are capable of reducing simple α, β-unsaturated ketones to the corresponding saturated ketones. Useful yields are obtained, provided a proton donor (AcOH) and a good hydrogen donor (BuSH) are present in the reaction mixture (Scheme 22)[122].

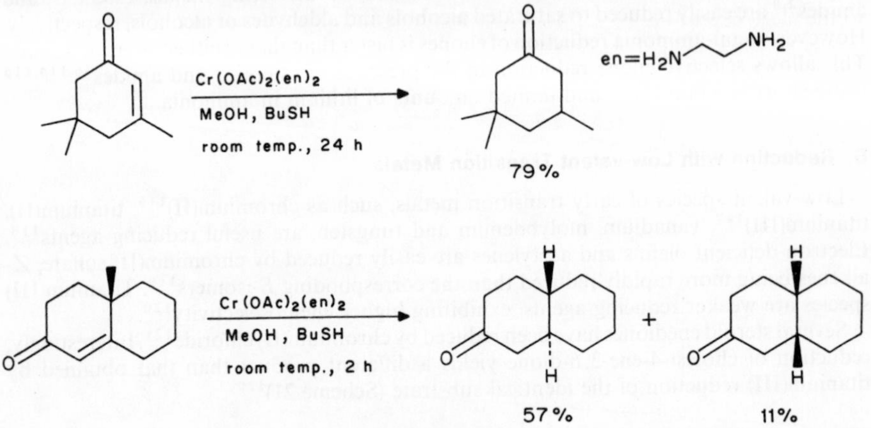

SCHEME 22

Reductive dimerization of α, β-unsaturated ketones is effected by either Cr(II) or V(II) chloride to give 1,4-diketones, and aliphatic α, β-unsaturated aldehydes are dimerized to the allylic glycals (Scheme 23)[123]. Interestingly, nonconjugated aldehydes are stable towards these reagents. Similar pinacolic couplings of aldehydes and ketones with Ti(II) reagents were developed by Corey[124].

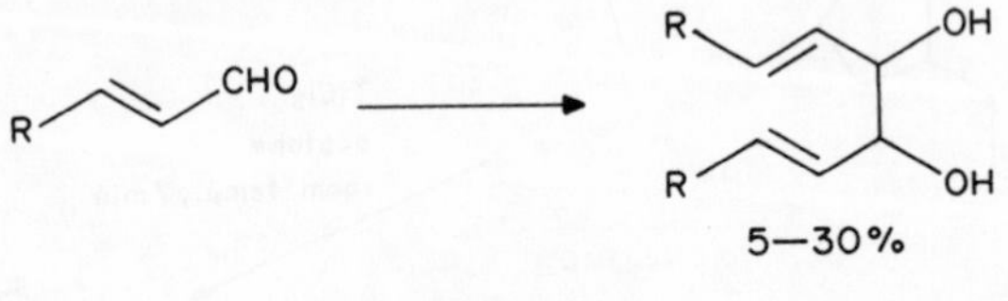

SCHEME 23

Highly reactive metallic titanium, prepared from TiCl$_3$ and potassium, was found useful for reduction of enol phosphate to alkenes, permitting regioselective synthesis of dienes from α, β-unsaturated ketones (Scheme 24)[125].

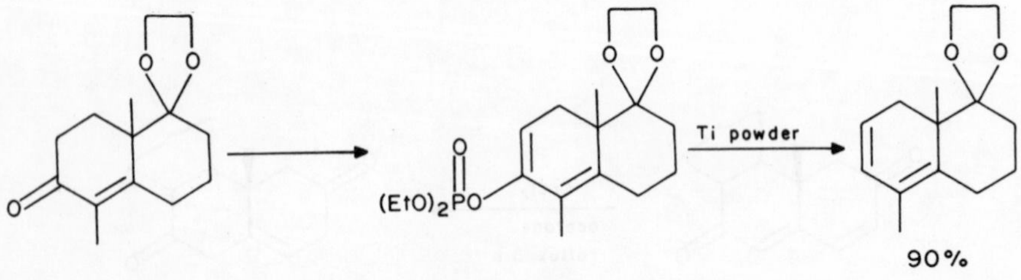

SCHEME 24

C. Electrochemical Reductions

The electrochemical reduction of α, β-unsaturated ketones and related compounds[5] in aprotic media in the absence of metal cations can, in some cases, lead to relatively stable anion radicals[12c,126]. However, in the presence of proton donors the latter are protonated to form hydroxyallyl radicals, which tend to dimerize more rapidly than they diffuse back to the electrode to undergo further reduction (Scheme 25)[12c].

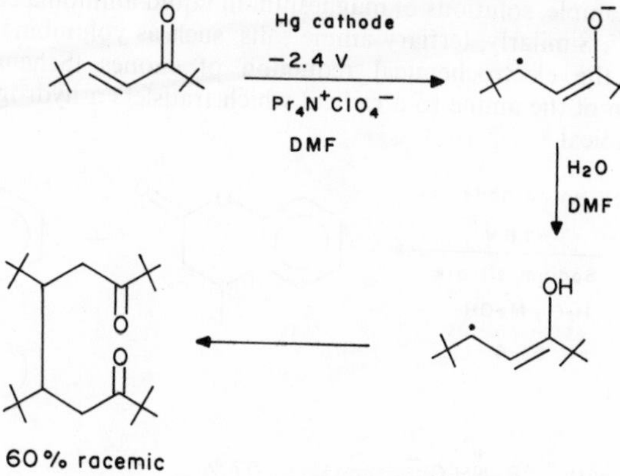

60% racemic

SCHEME 25

Although these allyl radicals prefer to dimerize by coupling at the β-position, if this position is sterically hindered, as in the case of cholest-4-en-3-one, coupling at the carbonyl carbon may be observed yielding a pinacol (Scheme 26)[127].

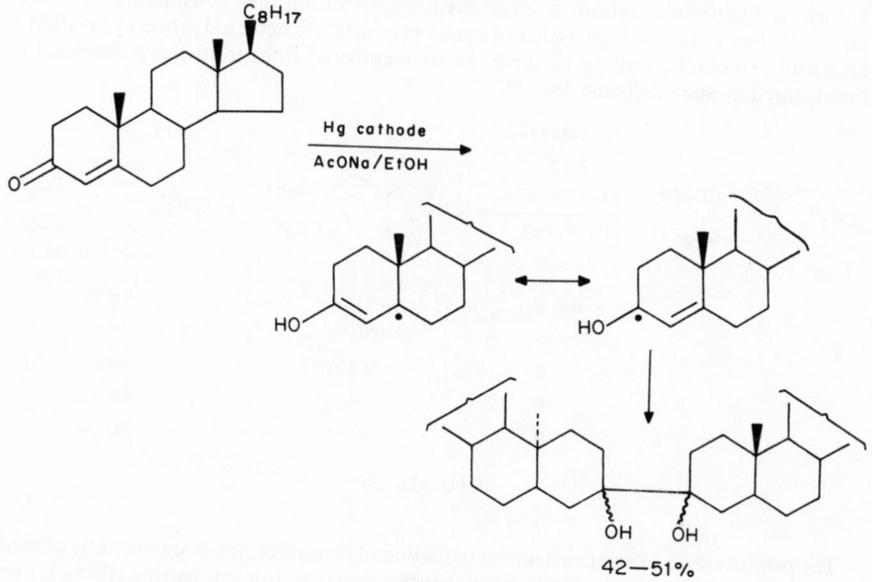

42—51%

SCHEME 26

As noted above, such reductive dimerizations have been recorded when unsaturated carbonyl compounds are reacted with various metals, such as lithium, sodium, sodium amalgam, potassium, aluminum amalgam, zinc or magnesium[128,129]. Formation of monomeric reduction products is impeded in these reactions because the intermediate allylic radical must diffuse back to the electrode surface or metal particle for further reduction. A possible solution to this problem might be concurrent electrochemical generation of a soluble reducing agent that can intercept radical intermediates before their dimerization. For example, solutions of magnesium in liquid ammonia can be generated electrochemically[130c]. Similarly, tertiary amine salts, such as yohimbine hydrochloride, can participate in the electrochemical reduction of enones (Scheme 27)[130a,b], via concurrent reduction of the amine to a radical which transfers a hydrogen atom to the intermediate allyl radical.

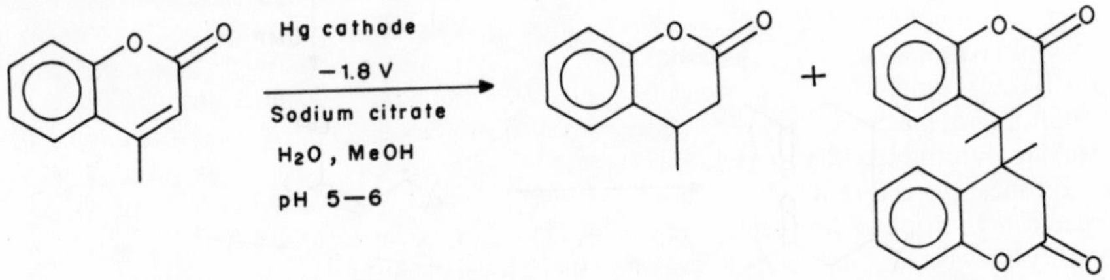

with R₃NH⁺ Cl⁻	57%	42%
without R₃NH⁺ Cl⁻	4%	93%

SCHEME 27

Reductive dimerization of enones to form a new carbon–carbon bond at the β-position, known as hydrodimerization or electrohydrodimerization, has considerable synthetic utility[131]. For example, high yields of cyclic products are achieved when cyclization is kinetically favorable, leading to three- to six-membered rings from the corresponding unsaturated diesters (Scheme 28)[131d].

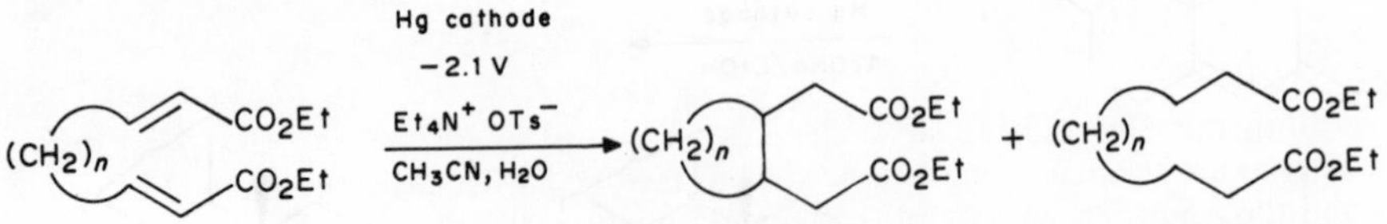

$n = 2$	41%	48%
3	100%	—
4	90%	—
6	—	43%
8	—	50%

SCHEME 28

The product ratio in electrochemical reduction of benzalacetone is significantly altered by surfactants and various cations, which cause micellar and/or ion-pairing effects. Using these additives, it is possible to control the partitioning of the initially formed radical anion between the two main reaction pathways: either dimerization or further reduction to the saturated ketone[132]. Additionally, micellar surfactants allow the use of aqueous media without cosolvents.

III. CATALYTIC HYDROGENATION

Addition of molecular hydrogen to α, β-unsaturated carbonyl compounds has been extensively reviewed[5,133-135]. Enones can be converted to saturated ketones or to unsaturated or saturated alcohols. Usually, double bonds conjugated to the carbonyl moiety are reduced prior to nonconjugated ones. 1,2-Reduction to allylic alcohols via catalytic hydrogenation is quite rare, and this transformation is more conveniently performed with hydridic reducing agents, such as boron- and aluminum-hydrides (*vide infra*). Nevertheless, there are a number of reported cases where 1,2-reduction is preferred over 1,4-selectivity. Citronellal, for example, is reduced preferentially at the carbonyl function using nickel on silica-gel as a catalyst, while hydrogenation catalyzed by Pd/BaSO$_4$ yields the corresponding saturated aldehyde[136]. Reduction to the saturated alcohol is achieved by catalytic hydrogenation over nickel[137], copper chromite[138], or nickel–aluminum alloy in NaOH[139].

Enones are reduced to saturated ketones by catalytic hydrogenation, provided the reaction is stopped following the absorption of 1 mole of hydrogen[140]. A number of catalysts were found useful for this, including platinum[141], platinum oxide[142,143], Pt/C[140], Pd/C[140,144], Rh/C[140], tris(triphenylphosphine)rhodium chloride[145,146], nickel–aluminum alloy in 10% aqueous NaOH[147], and zinc-reduced nickel in an aqueous medium[148]. Mesityl oxide is formed from acetone and reduced in a single pot to methyl isobutyl ketone using a bifunctional catalyst comprised of palladium and zirconium phosphate (Scheme 29)[149].

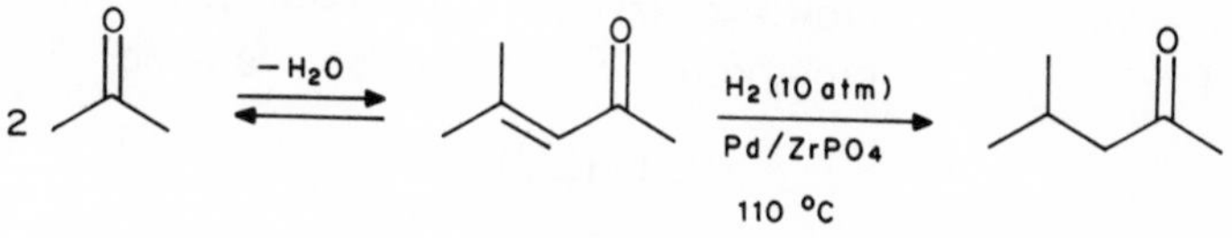

SCHEME 29

Both the ease and the stereochemical course of hydrogenation of α, β-unsaturated ketones are strongly influenced by various factors, particularly the nature of the solvent and the acidity or basicity of the reaction mixture. It is usually difficult to predict the product distribution in a particular reaction under a given set of conditions. Some efforts have been made to rationalize the effect of the various parameters on the relative proportions of 1,2- to 1,4-addition, as well as on the stereochemistry of reduction[150].

For example, the product distribution in β-octalone hydrogenation in neutral media is related to the polarity of the solvent if the solvents are divided into aprotic and protic groups. The relative amount of cis-β-decalone decreases steadily with decreasing dielectric constant in aprotic solvents, and increases with dielectric constant in protic solvents, as exemplified in Scheme 30 (dielectric constants of the solvents are indicated in parentheses)[151]. Similar results were observed in the hydrogenation of cholestenone and on testosterone[152]. In polar aprotic solvents 1,4-addition predominates, whereas in a nonpolar aprotic solvent hydrogenation occurs mainly in the 1,2-addition mode.

Acids and bases have a crucial effect on product stereochemistry in hydrogenation of ring-fused enone systems, as illustrated in Scheme 31[153].

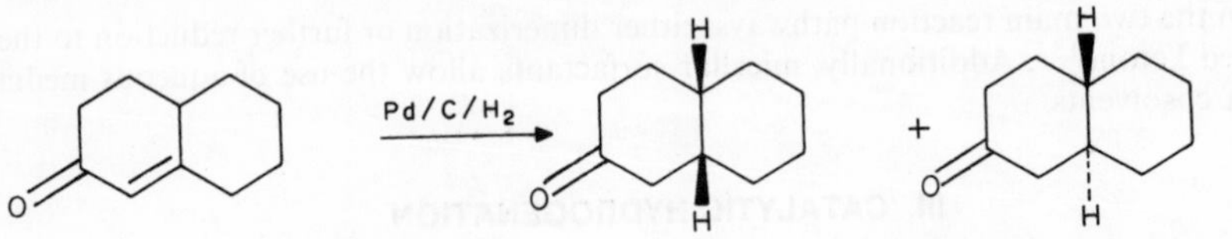

DMF (38)	79 : 21	
AcOEt (6)	57 : 43	
diethyl ether (4.34)	58 : 42	
hexane (1.89)	48 : 52	
methanol (33.6)	41 : 59	
propanol (21.8)	68 : 32	
t-butanol (10.9)	91 : 9	

SCHEME 30

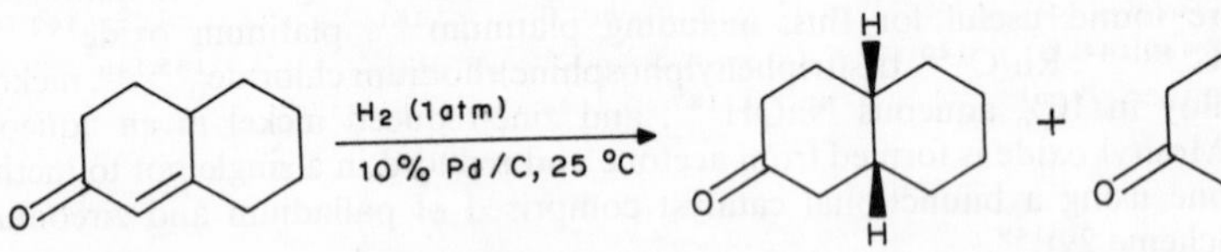

EtOH	53 : 47	
EtOH, H₂O, HCl	93 : 7	
EtOH, KOH	35—50 : 65—50	

SCHEME 31

The increased amounts of *trans*-fused product obtained in basic solutions was suggested to arise from hydrogenation of the relatively flat enolate ion which adsorbs irreversibly onto the catalyst surface. Hydrogenation proceeds by hydride ion-transfer from the metal catalyst, followed by protonation. Conversely, in acidic medium, protonation occurs initially, followed by irreversible adsorption on the catalyst, and then transfer of a hydride ion[150]. Stereochemistry of reduction is also related to catalyst activity, catalyst concentration, pressure and stirring rate, as they all affect hydrogen availability at the catalyst surface. Under conditions of low hydrogen availability a reversible adsorption is favorable, and therefore the product stereochemistry is determined by the relative stability of the *cis*- and *trans*-adsorbed species. However, under conditions of high hydrogen availability, product stereochemistry is determined mainly by the nature of the initial adsorption[150,151]. Platinum catalysts, more than palladium varieties, give products determined by the initial adsorption.

Substrate structure has an important influence on stereoselectivity of hydrogenation. For example, hydrogenation of hydrindanone having a trisubstituted double bond gives mainly the *cis* product (Scheme 32)[154], whereas similar compounds with a tetrasubstituted olefin tend to give the *trans* isomer. This phenomenon has been rationalized in terms of preferred conformation of the adsorbed enone, which minimizes steric interactions[154,155].

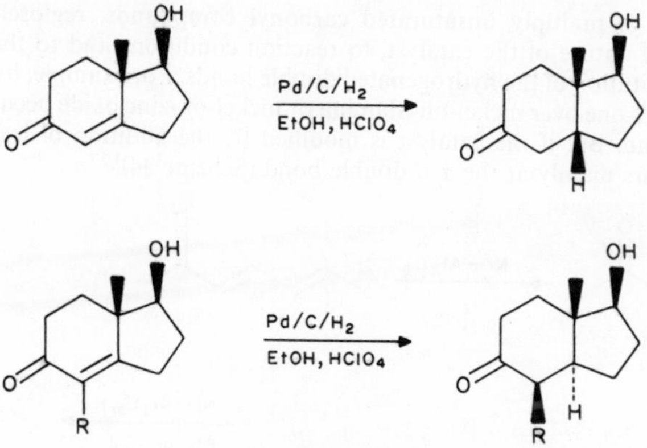

SCHEME 32

The key step in the synthesis of 2-deoxycrustecdysone from the corresponding 20-oxo steroid is the stereoselective catalytic hydrogenation of the α, β-unsaturated lactone shown in Scheme 33 to afford a 2:3 mixture of δ- and γ-lactones, respectively[156]. This crude product was converted into the thermodynamically more stable γ-lactone by treatment with aqueous NaOH.

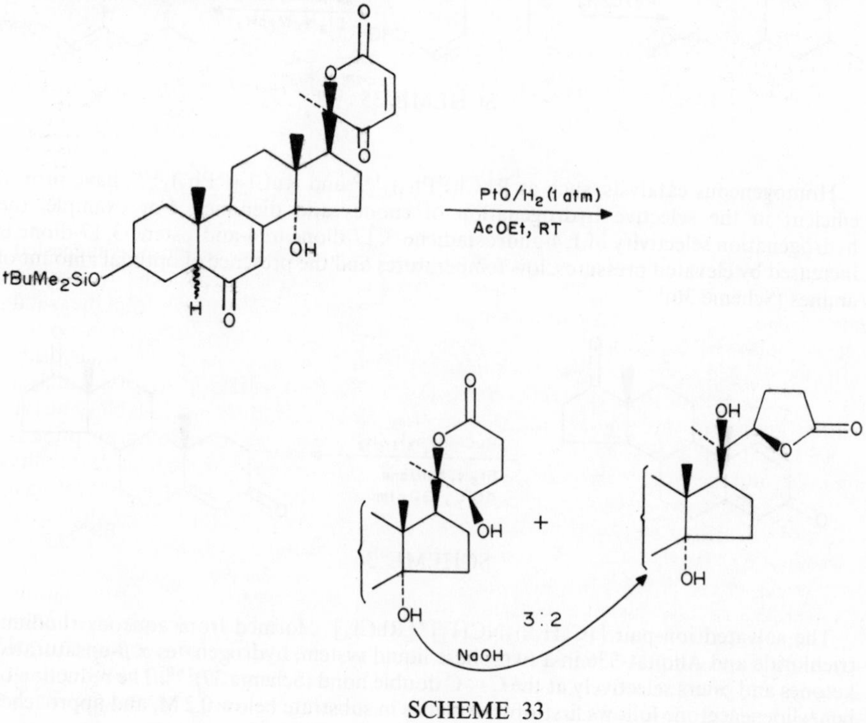

SCHEME 33

In the case of multiply unsaturated carbonyl compounds, regioselectivity is also sensitive to the nature of the catalyst, to reaction conditions and to the structure and degree of substitution of the hydrogenated double bonds. For example, hydrogenation of 3, 5-heptadien-2-one over nickel-on-alumina or nickel-on-zinc oxide occurs mainly at the γ, δ-double bond. But if the catalyst is modified by the addition of lead or cadmium, reduction occurs mainly at the α, β-double bond (Scheme 34)[157].

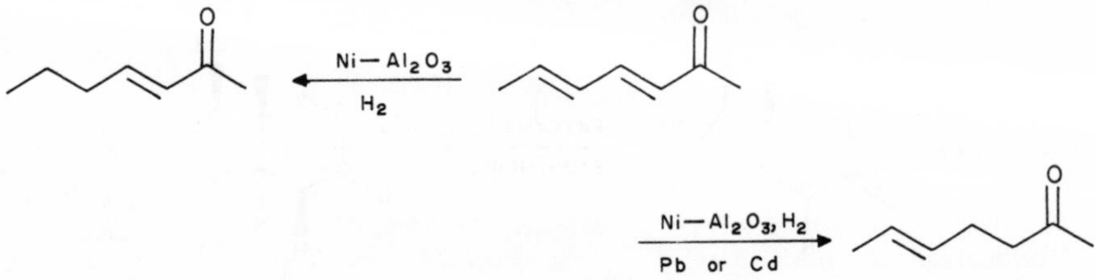

SCHEME 34

Selective reduction the γ, δ-double bond of the dienal shown in Scheme 35 was achieved by hydrogenation over palladium-on-carbon inhibited by quinoline and sulfur. Without inhibition, hydrogenation to the saturated aldehyde was observed[158].

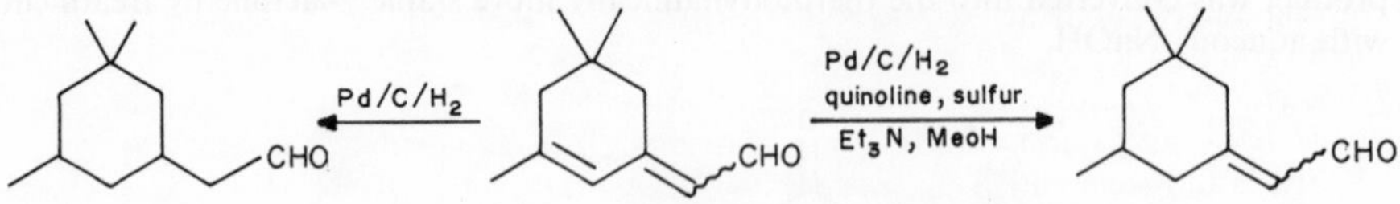

SCHEME 35

Homogeneous catalysts, such as $RhCl(PPh_3)_3$[146] and $RuCl_2(PPh_3)_3$[159], have proved efficient in the selective hydrogenation of enones and dienones. For example, the hydrogenation selectivity of 1, 4-androstadiene-3, 17-dione to 4-androstene-3, 17-dione is increased by elevated pressures, low temperatures and the presence of optimal amount of amines (Scheme 36)[159].

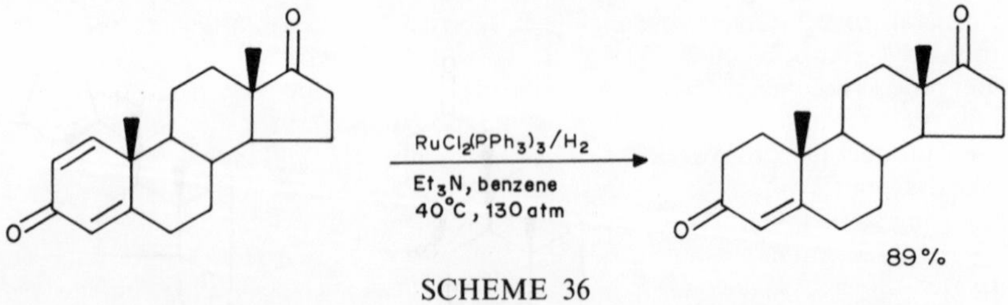

SCHEME 36

The solvated ion-pair $[(C_8H_{17})_3NCH_3]^+[RhCl_4]^-$, formed from aqueous rhodium trichloride and Aliquat-336 in a two-phase liquid system, hydrogenates α, β-unsaturated ketones and esters selectively at the C=C double bond (Scheme 37)[160]. The reduction of benzylideneacetone follows first-order kinetics in substrate below 0.2 M, and approaches second-order in hydrogen at partial pressures below 0.12 atm. The catalysis is also depends on the nature of the solvent, the phase-transfer catalyst and stirring rates.

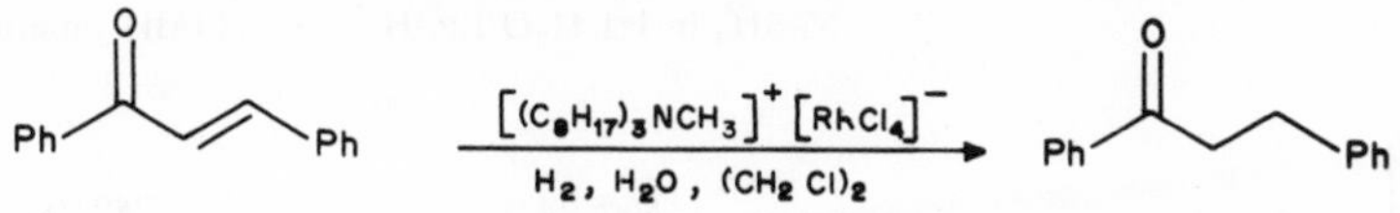

SCHEME 37

The homogeneous water-soluble hydrogenation catalyst $K_3(Co(CN)_5H)$ is very active for hydrogenating conjugated dienes and α, β-unsaturated ketones under phase-transfer reaction conditions[161]. Thus, conjugated dienes are converted into monoenes, generally with overall 1, 4-addition to yield E-olefins, and α, β-unsaturated ketones are reduced to saturated ketones in high yields. These conditions are not useful with α, β-unsaturated aldehydes, as they lead to polymerization of the starting material.

IV. REDUCTIONS WITH MAIN-GROUP METAL HYDRIDES

A. Boron Hydrides

Although $NaBH_4$ does not attack isolated olefins, $C{=}C$ double bonds conjugated to strong anion-stabilizing groups may be reduced by this reagent[162-164].

Rationalization of the regioselectivity of borohydride reduction of α, β-unsaturated aldehydes and ketones has been attempted using the 'hard' and 'soft' acid-base concept[165] (*vide infra*, discussion of aluminum hydrides). It is assumed that the relatively 'soft' hydrides add preferentially to the enone system via a 1, 4-mode while 'hard' reagents attack the carbonyl carbon. Borohydrides are considered softer than the corresponding aluminum hydrides. Replacement of a hydride group on boron by alkoxide makes it a harder reagent. Lithium salts are harder than sodium species. Thus, $LiAlH_4$ gives more 1, 2-attack than $LiBH_4$, which, in turn, gives more than $NaBH_4$. $NaBH(OMe)_3$ yields more 1, 2-reduction product than $NaBH_4$, and when production of alkoxyborates is prevented, 1, 4-reduction predominates. This implies that slow addition of borohydride to a substrate solution should help to build up alkoxyborate species and increase the relative amount of 1, 2-reduction. Generally, aldehydes undergo more 1, 2-reduction than the corresponding ketones.

The reduction of α, β-unsaturated aldehydes and ketones by sodium borohydride leads, in general, to substantial amounts of fully saturated alcohols. In alcoholic solvents, saturated β-alkoxy alcohols are formed via conjugate addition of the solvent[166]. This latter process becomes the main reaction path when reduction is performed in isopropanol in the presence of sodium isopropoxide. In a base, a homoallylic alcohol can become the major product of borohydride reduction of an enone[166].

Analysis of the influence of substrate structure on $NaBH_4$ reduction has shown that increasing steric hindrance on the enone increases 1, 2-attack (Table 1)[166].

$NaBH_4$ reduction of 3-substituted 5, 5-dimethylcyclohex-2-enones in alkaline solution of water–dioxane occurs exclusively at the 1, 2-positions. The rate of reduction is strongly dependent on the 3-substituent. A Hammett-type correlation revealed similar reaction characteristics to those of borohydride reduction of substituted acetophenones[167].

In order to study the factors determining the regioselectivity of sodium borohydride reduction of α, β-unsaturated ketones, reactions with 3-methylcyclohexenone, carvone and cholestenone were carried out in 2-propanol, diglyme, triglyme or pyridine[168]. Mixtures of 1, 2- and 1, 4-reduction products were obtained in the alcoholic and etheric

TABLE 1. The effect of the structure of α, β-unsaturated ketones and aldehydes on their reduction with NaBH$_4$ and LiAlH$_4$[a]

Substrate	NaBH$_4$ in 1:1 H$_2$O/EtOH	LiAlH$_4$ in ether
(methyl vinyl ketone)	86(57:43)	79(92:8)
(pent-3-en-2-one)	90(65:35)	85(99:1)
(4-methylpent-3-en-2-one)	89(92:8)	82(100:0)
(cyclohex-2-enone)	90(59:41)	97(98:2)
(3-methylcyclohex-2-enone)	90(70:30)	88(100:0)
(5-methylcyclohex-2-enone)	100(49:51)	99(91:9)
(5-tert-butylcyclohex-2-enone)	100(42:58)	99(93:7)
CHO (acrolein)	70(85:15)	70(98:2)
CHO (crotonaldehyde)	91(92:8)	94(100:0)
CHO (methacrolein)	100(>99:<1)	98(100:0)
CHO (3-methylbut-2-enal)	95(>99:<1)	82(100:0)

[a]The numbers represent the overall reduction yield (%), the numbers in parentheses represent the ratio of 1,2- to 1,4-attack.

solvents, whereas pure 1,4-reduction was observed in pyridine. Addition of triethyl amine to $NaBH_4$ in diglyme led to formation of triethylamine borine, Et_3NBH_3. Similarly, with pyridine, pyridine-borine could be isolated, leading to exclusive 1,4-reductions.

The results were interpreted in terms of steric requirements of the actual reducing species. It was suggested that attack of BH_4^- proceeds exclusively along the 1,4-reduction mode, whereas alkoxyborohydrides (formed as reaction products) prefer the 1,2-reduction mode. The pyridine-borine itself does not reduce enones under the reaction conditions, but it inhibits formation of alkoxyborohydrides[168]. The same trend was observed with aluminum hydride reductions. When $LiAlH_4$ was first reacted with pyridine to form lithium tetrakis(dihydro-N-pyridyl) aluminate, 1,4-reduction predominated[168].

Low regioselectivity is observed in reduction of enones with a 2:1 mixture of sodium cyanoborohydride and zinc chloride in ether at room temperature[169]. A mixture containing 1,2- and 1,4-reduction products is obtained in a ratio that is greatly dependent upon substrate.

TABLE 2. Reduction of α, β-unsaturated carboxylic acid derivatives with $NaBH_4$

Substrate	Yield (%)
	59
	74
	69
	81
	80
	25
	79

From the reduction in methanol of a series of substituted 2-aryl-(Z)- and (E)-cinnamates by NaBH$_4$ at room temperature, it was concluded that the facile reduction to give dihydrocinnamates proceeds through an early transition state of considerable polarity[162]. A few more examples are given in a related study (Table 2)[170].

Several organoborohydrides were found to effect the selective 1,4-reduction of enones. For example, lithium and potassium tri-*sec*-butylborohydrides (L- and K-Selectride) and lithium triethylborohydride were found useful for conjugate reduction of α, β-unsaturated ketones and esters. In general, β-unsubstituted cyclohexenones undergo exclusive 1,4-reduction to the corresponding ketone enolate, which can be protonated or alkylated in high yields. Ketones such as 5-*t*-butylcyclohex-2-en-1-one are cleanly reduced to the saturated ketone using K-Selectride at − 78 °C in THF (Scheme 38)[171]. This regioselectivity, however, is not general, but is a result of steric hindrance of the olefin, as well as the size of the ring. Thus alkyl substitution at the β-position completely suppresses the 1,4-reduction mode. While enones in 5- and 7-membered rings are reduced preferably in a 1,2-manner, 6-membered ring enones are reduced in a 1,4-mode. Trapping the intermediate enolate by an alkylating agent (e.g. MeI, allyl bromide) results in an efficient reductive alkylation. Accordingly, when the reduction of α, β-unsaturated esters is performed in dry ether solvents, the major reaction product arises from carbonyl condensation. However, addition of a proton source such as *t*-butanol results in 1,4-reduction.

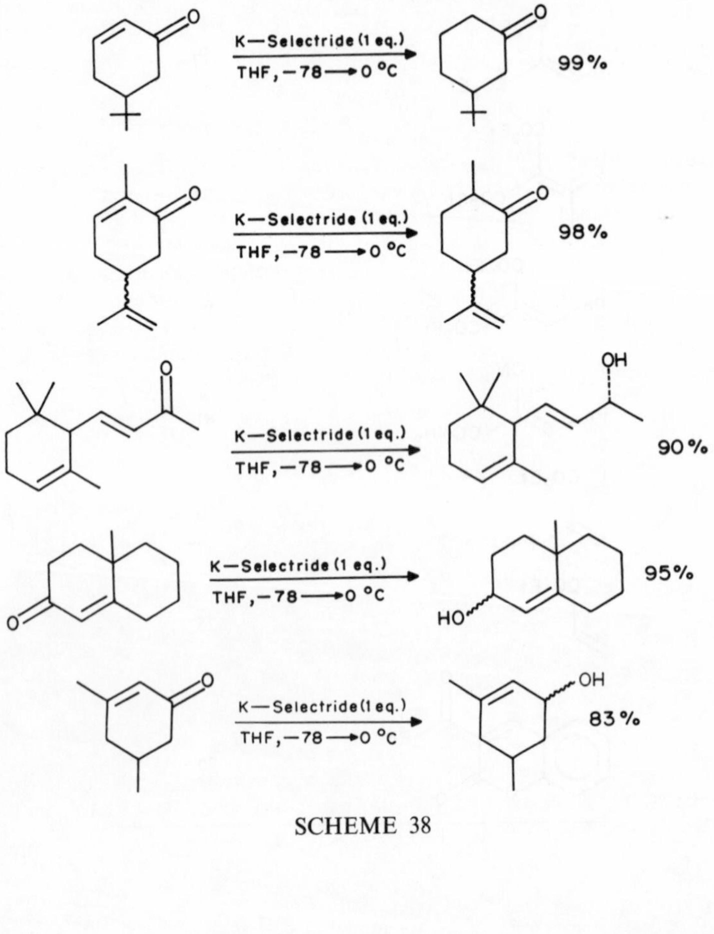

SCHEME 38

Reduction of α, β-unsaturated aldehydes and ketones with 9-borabicyclo[3.3.1]nonane (9-BBN) proceeds selectively and cleanly to form the corresponding allylic alcohols (Scheme 39)[172]. The reaction tolerates a large variety of functionalities, such as nitro, carboxylic acid, amide, nitrile, sulfide, disulfide, epoxide, etc. Hydroboration of the double bond is a much slower reaction, which does not interfere with carbonyl reduction. For example, 1, 2-reduction of cyclohexenone at room temperature with excess of 9-BBN in THF is completed within 10 minutes, while hydroboration of the double bond requires 3 days.

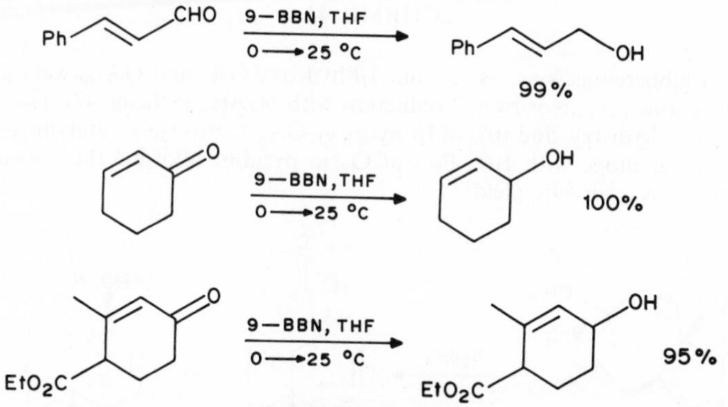

SCHEME 39

Borohydride reduction of α, β-unsaturated carbonyl compounds has been widely applied in natural product chemistry. A number of α, β-unsaturated ketone derivatives of gibberellins are reduced to the corresponding saturated alcohols by NaBH$_4$[173-176].

Sodium borodeuteride reduction of gibberellin A$_3$ 3-ketone affords gibberellin A$_1$ and its 3-epimer (Scheme 40)[173,174]. Attack of hydride proceeds stereospecifically from the β-face at C-1. Protonation at C-2 proceeds with limited selectivity. Thus, reduction of the above-mentioned gibberellin with either NaBH$_4$–CuCl in deuterated methanol or NaBH$_4$–LiBr followed by treatment with D$_2$O gave 2-deuteriogibberellin A$_1$ methyl ester together with some 3-epi-GA$_4$ with approximately 2:1 ratio of the 2β:2α deuterides.

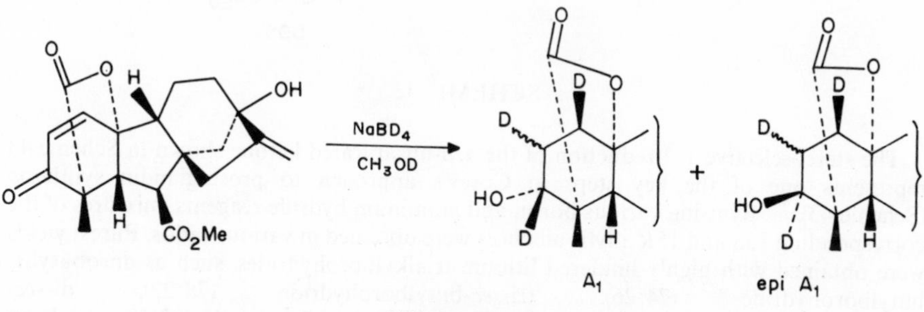

SCHEME 40

Using L-Selectride for the reduction of a similar gibberellin enone derivative resulted mainly in the 1, 2-reduction product, affording the 3α-allylic and saturated alcohols in 47% and 30% yields, respectively (Scheme 41)[175].

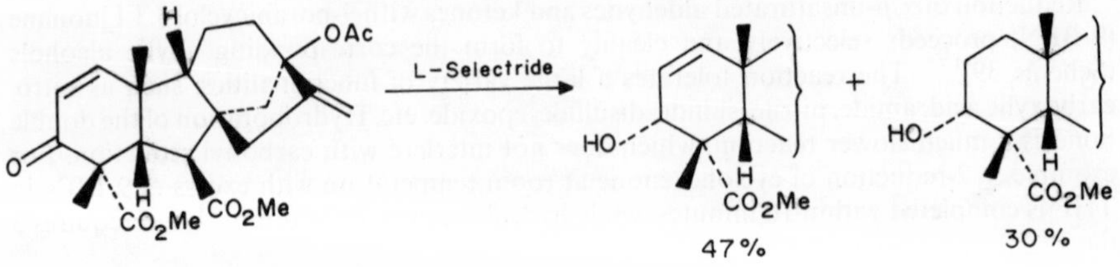

SCHEME 41

Substituted gibberellins, such as 1α- and 1β-hydroxy GA_5 and GA_{20}, were prepared from a single enone precursor by 1, 2-reduction with $NaBH_4$ (Scheme 42). The reaction yielded 33% of 1α-hydroxy- and 10% of 1β-hydroxy-GA_5. Conversely, catalytic hydrogenation of the same enone with 10% $Pd/CaCO_3$ in pyridine afforded the 1, 4-reduction product, 1-oxo-GA_{20}, in 59% yield[176].

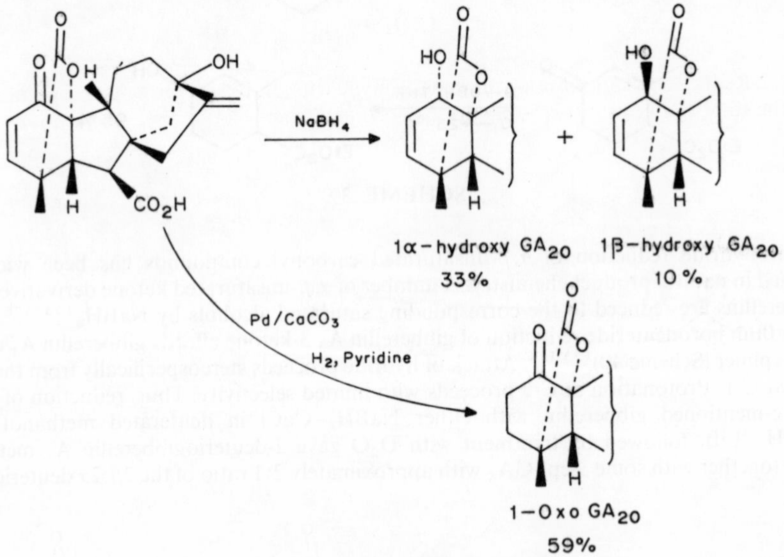

SCHEME 42

The stereoselective 1, 2-reduction of the α, β-unsaturated ketone shown in Scheme 43 represents one of the key steps in Corey's approach to prostaglandin synthesis (Scheme 43)[177]. By using various boron and aluminum hydride reagents, mixtures of the corresponding 15S and 15R allylic alcohols were obtained in various ratios. Purest yields were obtained with highly hindered lithium trialkylborohydrides, such as diisobutyl-*t*-butylborohydride (74:26), tri-*sec*-butylborohydride (78:22), di-*sec*-butylthexylborohydride (80:20), the reagent indicated in Scheme 43 (82:18), etc. Even better stereoselectivity was achieved with *p*-phenylphenylurethane ($R = p$-PhC_6H_4NHCO) as a directing group. This derivative was reduced with thexyl-di-*sec*-butylborohydride and tri-*sec*-butylborohydride with 15S:15R ratios of 88:12 and 89:11, respectively[177].

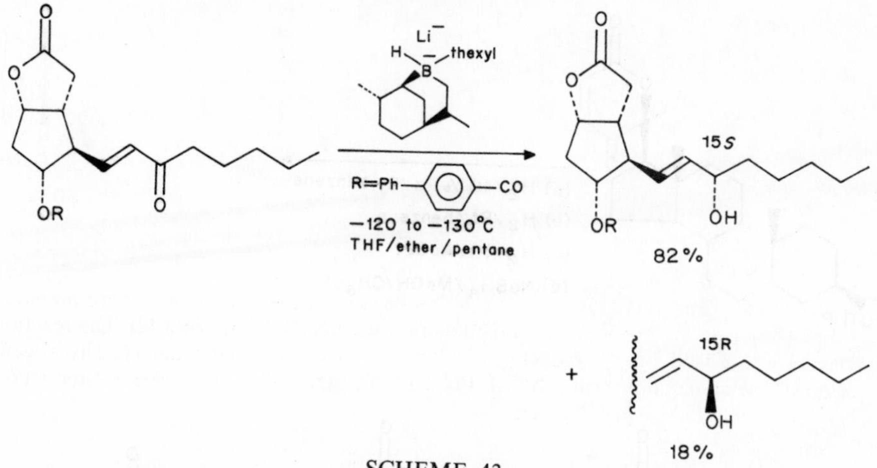

SCHEME 43

1, 2-Reduction of an α, β-unsaturated aldehyde with NaBH₄ represents one of the steps in the total synthesis of 6, 15-dihydroxyperezone (Scheme 44)[178].

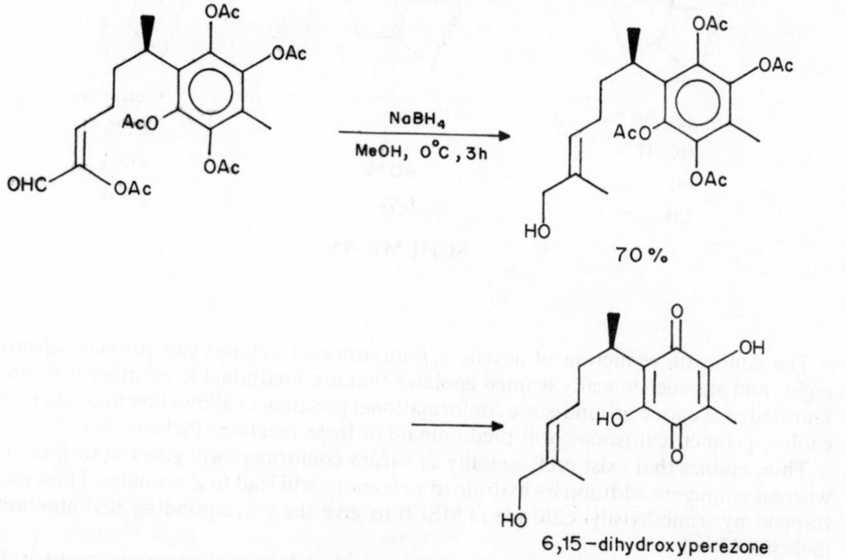

SCHEME 44

Stereoselective reduction of an enono-lactone was a key step in the construction of the 20-hydroxyecdysone side-chain. Totally different mixtures of products were obtained when the reduction was carried out with sodium borohydride or by catalytic hydrogenation (Scheme 45)[156]. In all cases, the 1, 4-reduction mode is preferred. With borohydride, however, this process is followed by a subsequent reduction of the saturated ketone and base-catalyzed rearrangement of the δ-lactone into a γ-lactone.

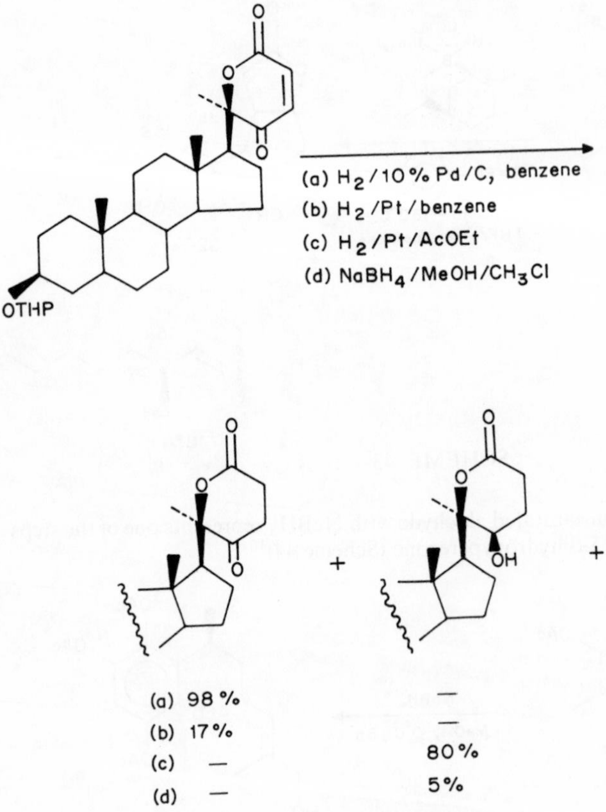

(a) $H_2/10\% \, Pd/C$, benzene
(b) $H_2/Pt/benzene$
(c) $H_2/Pt/AcOEt$
(d) $NaBH_4/MeOH/CH_3Cl$

+   +

(a) 98 %
(b) 17 %
(c) —
(d) —

    —
    —
  80 %
  5 %

81%
20%
78%

SCHEME 45

The conjugate reduction of acyclic α, β-unsaturated ketones can provide selectively regio- and stereochemically defined enolates that are unattainable by other methods. A knowledge of enone ground-state conformational preferences allows one to predict which enolate geometrical isomer will predominate in these reactions (Scheme 46)[179].

Thus, enones that exist preferentially as s-trans conformers will give rise to E-enolates whereas conjugate addition by hydride to s-cis enone will lead to Z-enolates. These can be trapped by trimethylsilyl chloride (TMSCl) to give the corresponding silyl enol ethers (Scheme 47)[179].

Sodium cyanoborohydride (NaBH₃CN) or tetrabutylammonium cyanoborohydride in acidic methanol or acidic HMPT reduces α, β-unsaturated aldehydes and ketones to the corresponding allylic alcohol (Scheme 48)[180]. This system is limited to enones in which the double bond is not further conjugated, in which case the allylic hydrocarbon is formed in substantial amounts. Thus, reduction of chalcone gives mainly 1, 3-diphenylpropene (48%) as well as 26% of the allylic ether. Cyclic enones are also not good substrates, as competing 1,4-addition gives large fractions of saturated alcohols[180].

Lithium butylborohydride is prepared by reacting equimolar amounts of butyl lithium and borane-dimethylsulfide complex[181]. This reagent effectively reduces enones in

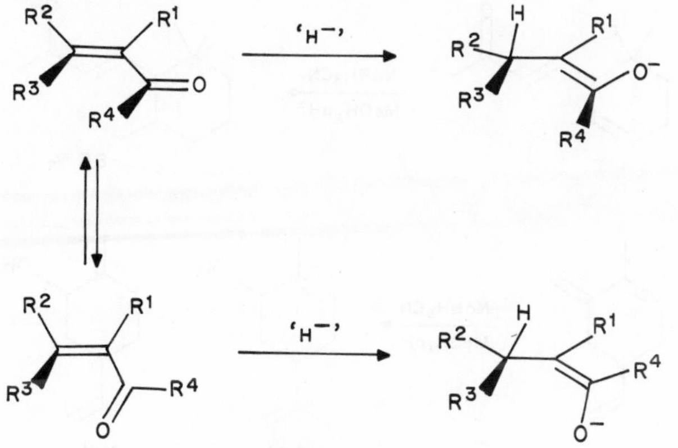

SCHEME 46

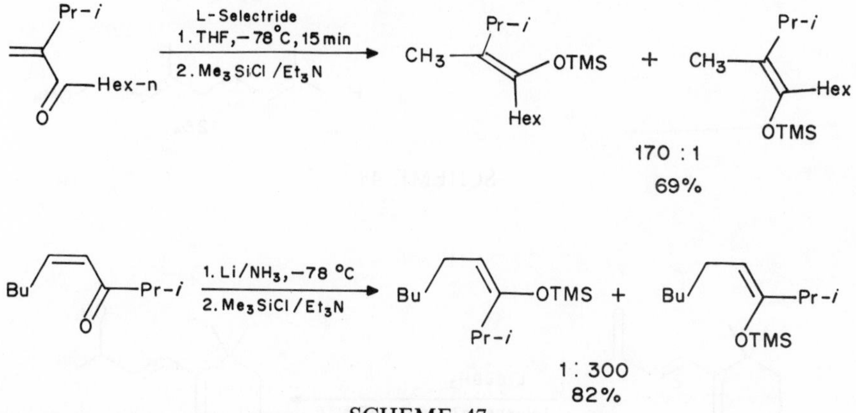

SCHEME 47

toluene–hexane mixtures at $-78\,°C$ to give, in most cases, high yields of the corresponding allylic alcohols (Scheme 49)[181]. Conjugated cyclopentenones, however, give mixtures of 1, 2- and 1, 4-reduction products. Under identical reaction conditions, saturated ketones are reduced to alcohols. The latter process can take place in the presence of simple esters.

Regioselective 1, 2-reduction of enones to the corresponding allylic alcohols is achieved with $NaBH_4$ in the presence of lanthanide ions, such as La^{3+}, Ce^{3+}, Sm^{3+}, Eu^{3+}, Yb^{3+} and Y^{3+}[182]. This procedure is complementary to those giving predominantly 1, 4-selectivity, such as $NaBH_4$ in pyridine[168]. The general utility of $NaBH_4$–$CeCl_3$ selective reduction is illustrated by the conversion of cyclopentenone to cyclopentenol in 97% yield and only 3% of cyclopentanol, although conjugate reduction of cyclopentenone systems by most hydride reagents is usually highly favored (Scheme 50).

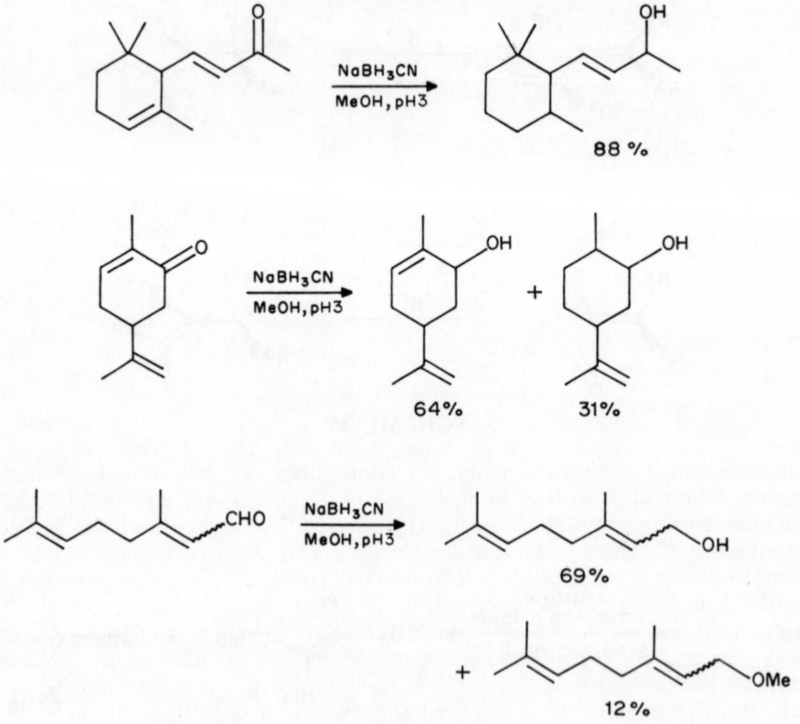

SCHEME 48

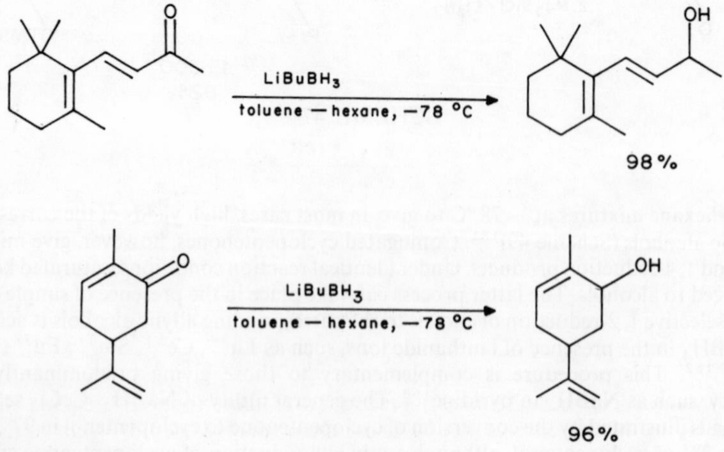

SCHEME 49

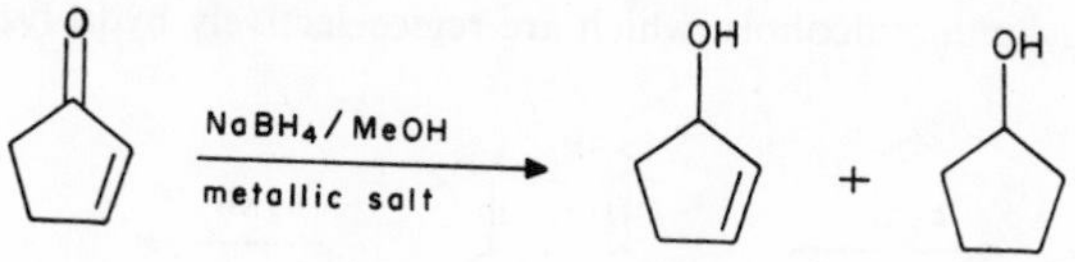

SCHEME 50

Thus, reaction of equimolar amounts of α, β-unsaturated ketones and either samarium or cerium chloride hexahydrate in methanol with sodium borohydride produced high yields of the corresponding allylic alcohols (Scheme 51)[182]. This approach was applied in the synthesis of 7, 7-dimethylnorbornadiene, whereas reduction of 4, 4-dimethylcyclopent-2-enone with sodium borohydride and cerium chloride in methanol afforded dimethylcyclopentenol in 93% yield[183].

A mechanistic study of the role of the lanthanide cations suggests that they catalyze decomposition of borohydride by the hydroxylic solvent to afford alkoxyborohydrides, which may be responsible for the observed regioselectivity. The stereoselectivity of the process is also modified by the presence of Ln^{3+} ions, in that axial attack of cyclohexenone systems is enhanced[182].

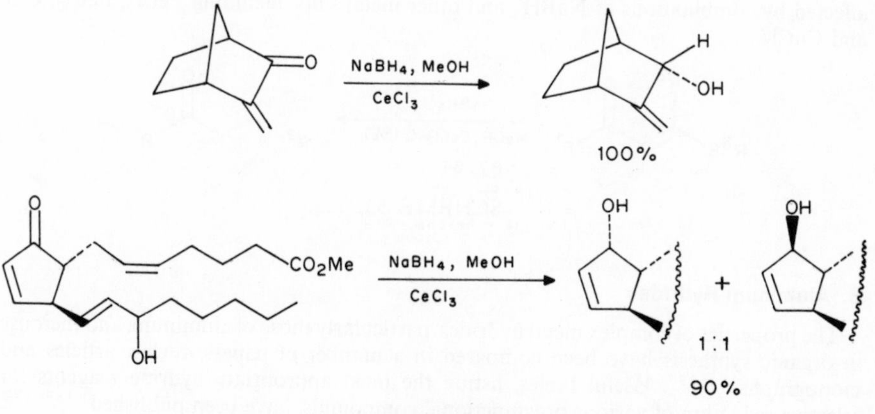

SCHEME 51

β-Dialkylamino conjugated enones are reduced to the corresponding γ-amino alcohols with $NaBH_4$ in the presence of $FeCl_3$. These aminoalcohols could be converted into conjugated enones by chromic acid oxidation and deamination (Scheme 52)[184]. On the other hand, β-acylamino conjugated enones are reduced by $NaBH_4$ to afford β, γ-

unsaturated γ-acylamino alcohols, which are regioselectively hydrolyzed to conjugated enones.

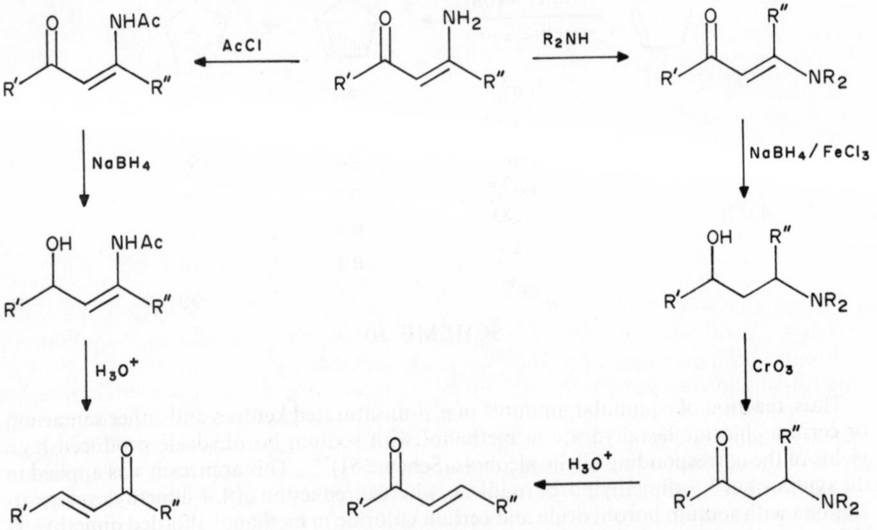

SCHEME 52

Reduction of β-sulfenylated α, β-unsaturated ketones with NaBH₄ in the presence of catalytic amounts of CoCl₂ or NiCl₂ in methanol produces the corresponding desulfenylated, saturated ketones (Scheme 53)[185]. These substrates, however, were not affected by combinations of NaBH₄ and other metal salts, including FeCl₂, FeCl₃, CuI and CuCl₂.

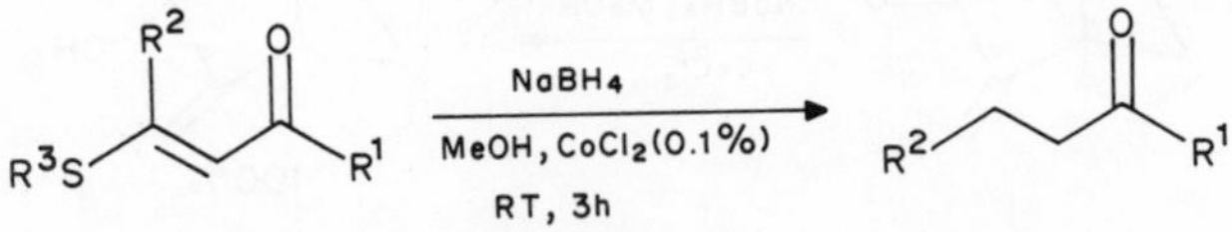

SCHEME 53

B. Aluminum Hydrides

The properties of complex metal hydrides, particularly those of aluminum, and their use in organic synthesis have been compared in a number of papers, review articles and monographs[186-190]. Useful tables, listing the most appropriate hydride reagents for selective reduction of various polyfunctional compounds, have been published[1,189-192]. Use of chiral metal alkoxyaluminum hydride complexes in asymmetric synthesis has also been reviewed[193].

The two modes of reduction of α, β-unsaturated aldehydes and ketones, 1, 2- and 1, 4-addition of metal hydride to the enone system, lead respectively to either an allylic alcohol or a saturated ketone. It has been suggested that the relative importance of these paths depends upon substrate 'hardness' or 'softness', as defined in terms of the coefficients of the lowest unoccupied molecular orbital (LUMO) (vide supra, the discussion of borohydrides).

TABLE 3. Ratio of 1,4- to 1,2-reduction products

	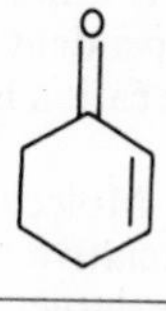	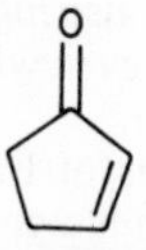	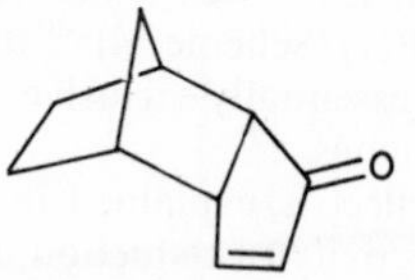
LiAlH(OMe)$_3$	5:95	10:90	24:76
LiAlH$_4$	22:78	86:14	100:0
LiAlH(SMe)$_3$	56:44	95:5	
LiAlH(OBu-t)$_3$	78:22	100:0	100:0
LiAlH(SBu-t)$_3$	95:5	100:0	

While 1,2-addition is considered to be a mainly charge-controlled process, 1,4-addition is a frontier orbital-controlled process[194]. These considerations predict, for example, that the 1,4-addition of a given metal hydride to cyclopentenone should always be faster than a similar addition to cyclohexenone[195]. Moreover, in cases where the enone system is further conjugated to a phenyl ring, as in cinnamaldehyde, increased frontier-orbital control should render the enone more prone to 1,4-addition[196]. Obviously, the course of reduction of conjugated carbonyl compounds is also highly influenced by the nature of the metal hydride. According to Pearson's concept of 'soft' and 'hard' acids and bases[197,198], hard metal hydrides add preferentially to the 2-position and soft metal hydrides to the 4-position of the conjugated enone system[194-196]. As shown in Table 3, these predictions agree well with representative experimental results[195,199].

Because of their electrophilic nature, Li$^+$ cations accelerate the reduction of carbonyl compounds by LiAlH$_4$ or NaBH$_4$, an effect that is significantly inhibited by Li$^+$-complexing agents, such as cryptands, crown ethers or polyamines, which decrease the rate of reduction[200]. In the case of α, β-unsaturated ketones, this slowdown is associated with altered regioselectivity. For example, LiAlH$_4$ reduction of cyclohexenones in the absence

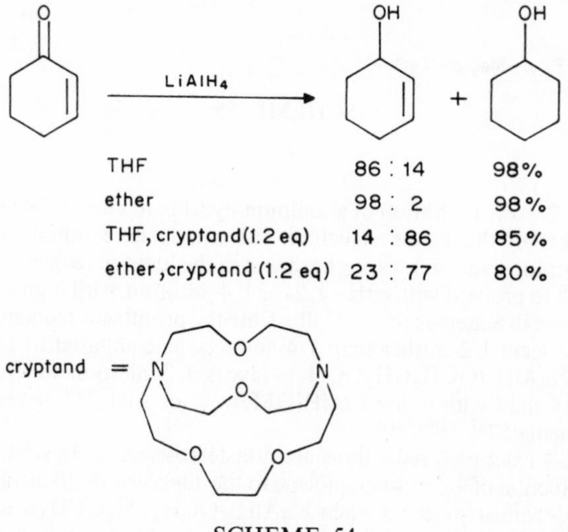

THF	86 : 14	98%
ether	98 : 2	98%
THF, cryptand(1.2 eq)	14 : 86	85%
ether, cryptand (1.2 eq)	23 : 77	80%

SCHEME 54

of the cryptand proceeds predominantly with 1, 2-reduction. In the presence of the cryptand, 1, 4-attack is favored. This selectivity is more pronounced with LiAlH$_4$ than with NaBH$_4$ (Scheme 54)[200] and is also highly dependent on solvent. In diethyl ether, 1, 2-attack is essentially exclusive. However, when the cation is complexed, 1, 4-addition again predominates.

This effect is explained in terms of Frontier Molecular Orbitals treatment[200]. The regioselectivity of reduction depends upon the relative values of the C$_1$ and C$_3$ atomic coefficients in the LUMO. The atom with the larger coefficient corresponds to the predominant site of attack. When Li$^+$ is complexed by the α-enone, the C$_1$ coefficient is larger than that of C$_3$, and C$_1$ attack is favored. In the absence of such complexation, the C$_3$ coefficient is larger, leading to 1, 4-attack. The strength of carbonyl–Li$^+$ interaction is strongly dependent upon the solvent, the nature of the complexing agent and the interaction between the Li$^+$ ion and the reducing agent. Thus, in strongly coordinated solvents such as pyridine[168], 1, 4-reduction predominates.

Steric and electronic factors in the enone substrate may also alter selectivity. For example, the high tendency of LiAlH(OBu-t) to undergo 1, 4-addition with simple enones is modified in the two examples given in Scheme 55[201].

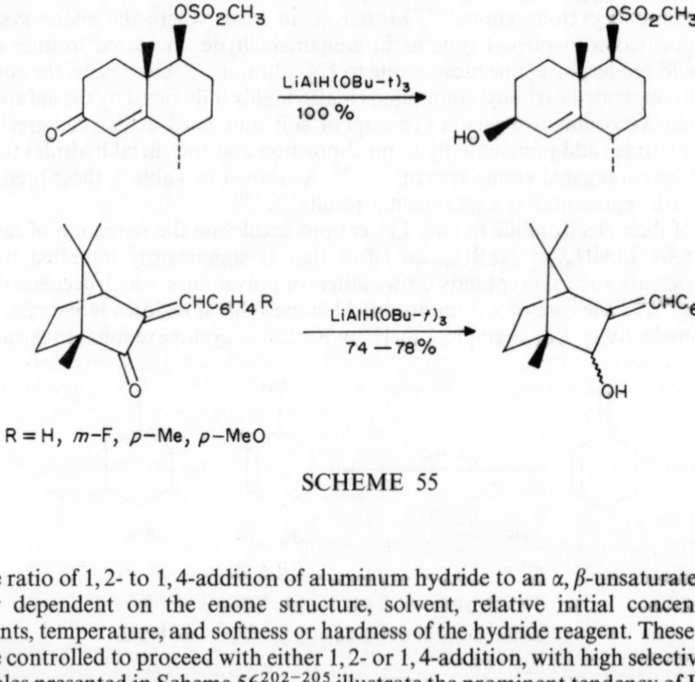

R = H, m–F, p–Me, p–MeO

SCHEME 55

The ratio of 1, 2- to 1, 4-addition of aluminum hydride to an α, β-unsaturated ketone is highly dependent on the enone structure, solvent, relative initial concentrations of reactants, temperature, and softness or hardness of the hydride reagent. These reductions can be controlled to proceed with either 1, 2- or 1, 4-addition, with high selectivity[186]. The examples presented in Scheme 56[202-205] illustrate the prominent tendency of LiAlH$_4$ and LiAlH(OMe)$_3$ to yield 1, 2- rather than 1, 4-adducts, as compared to LiAlH(OBu-t)$_3$.

The reagent NaAlH$_2$(OCH$_2$CH$_2$OCH$_3$)$_2$ favors 1, 2-addition to cyclic enones with greater selectivity than with either LiAlH(OMe)$_3$[195] or AlH$_3$[199]. Several examples are presented in Scheme 57[203,206-210].

In most of these examples, reductions are nonstereoselective. In some cases, however, such as in the reduction of 9-oxoisolongifolene to the allylic 9α- or 9β-alcohols (Scheme 58), reversal of stereochemistry occurs when NaAlH$_2$(OCH$_2$CH$_2$OCH$_3$)$_2$ is used instead of LiAlH$_4$ or NaBH$_4$[211]. While the latter two reagents lead to formation of the thermody-

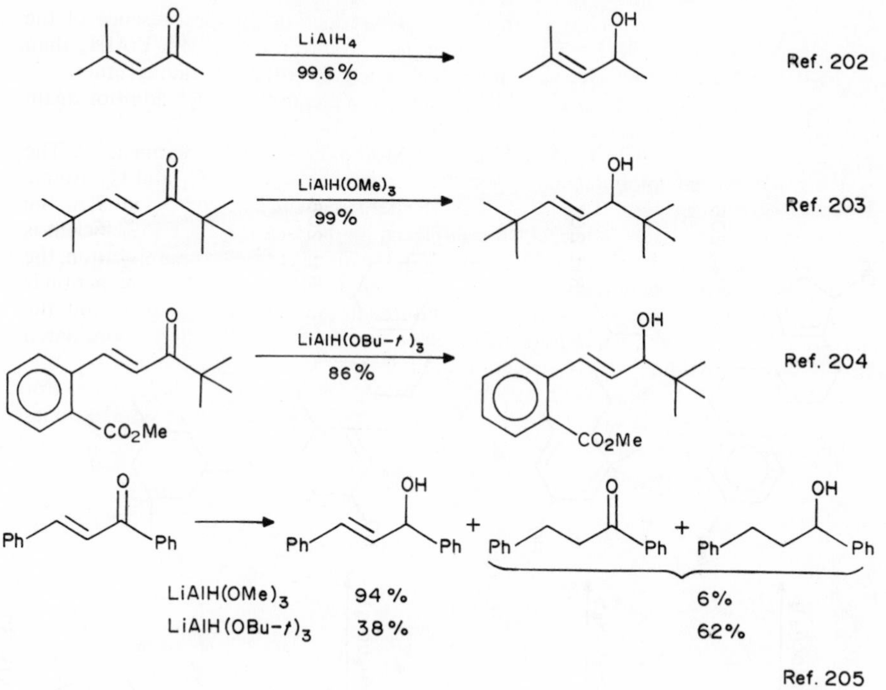

Ref. 202

Ref. 203

Ref. 204

LiAlH(OMe)$_3$	94%	6%
LiAlH(OBu-t)$_3$	38%	62%

Ref. 205

SCHEME 56

namically more stable α-alcohol as the major product, increased steric bulk of the former seems to favor the less stable β-isomer.

Sterically unhindered enones, such as cyclohexenone, are reduced by LiAlH(OBu-t)$_3$ to give predominantly the corresponding saturated ketone[195]. More sterically congested systems are cleanly reduced via the 1,2-mode to give the allylic alcohol, usually with high stereoselectivity (Scheme 59)[212–215].

1,2-Reduction has been reported for other hydride reagents, such as diisobutylaluminum hydride[194,216,217], aluminum hydride[199] and 9-borabicyclononane (9-BBN)[218], as illustrated by the example in Scheme 60.

1,4-Reduction of enones can be effected with high selectivity with AlH(OBu-t)$_2$, AlH(OPr-i)$_2$, AlH(NPr$_2^i$)$_2$ or HBI$_2$, forming saturated ketones in 90–100% yield. AlH(NPr$_2^i$)$_2$ exhibited the lowest selectivity, as no 1,4-reduction of mesityl oxide or isophorone is observed with this reagent. The same reagent with methyl vinyl ketone or cyclohexenone led to mixture of products. Trans-chalcone also undergoes quantitative 1,4-reduction with the above-mentioned hydrides[217]. Similarly, reduction of 9-anthryl styryl ketone or anthracene-9,10-diyl-bis(styryl ketone) with LiAlH(OBu-t)$_3$ affords the saturated ketone as the sole product[219]. Hydrides such as LiAlH(OBu-t)$_3$ and LiAlH(SBu-t)$_3$ favor 1,4-reduction in cyclopentenones[195,196,199,220–223]. An example is given in Scheme 61, where steric factors allow only exo approach of the bulky hydride[224,225].

Scheme 62 illustrates an interesting two-step selective reduction of an enone system, first with sodium hydride and NaAlH$_2$(OCH$_2$CH$_2$OCH$_3$)$_2$ and then with the same reagent in the presence of 1,4-diazabicyclo[2.2.2]octane. Specific reduction, however, is not achieved with NaBH$_4$, LiBH$_4$, LiBH(s-Bu)$_3$ or 9-BBN[226].

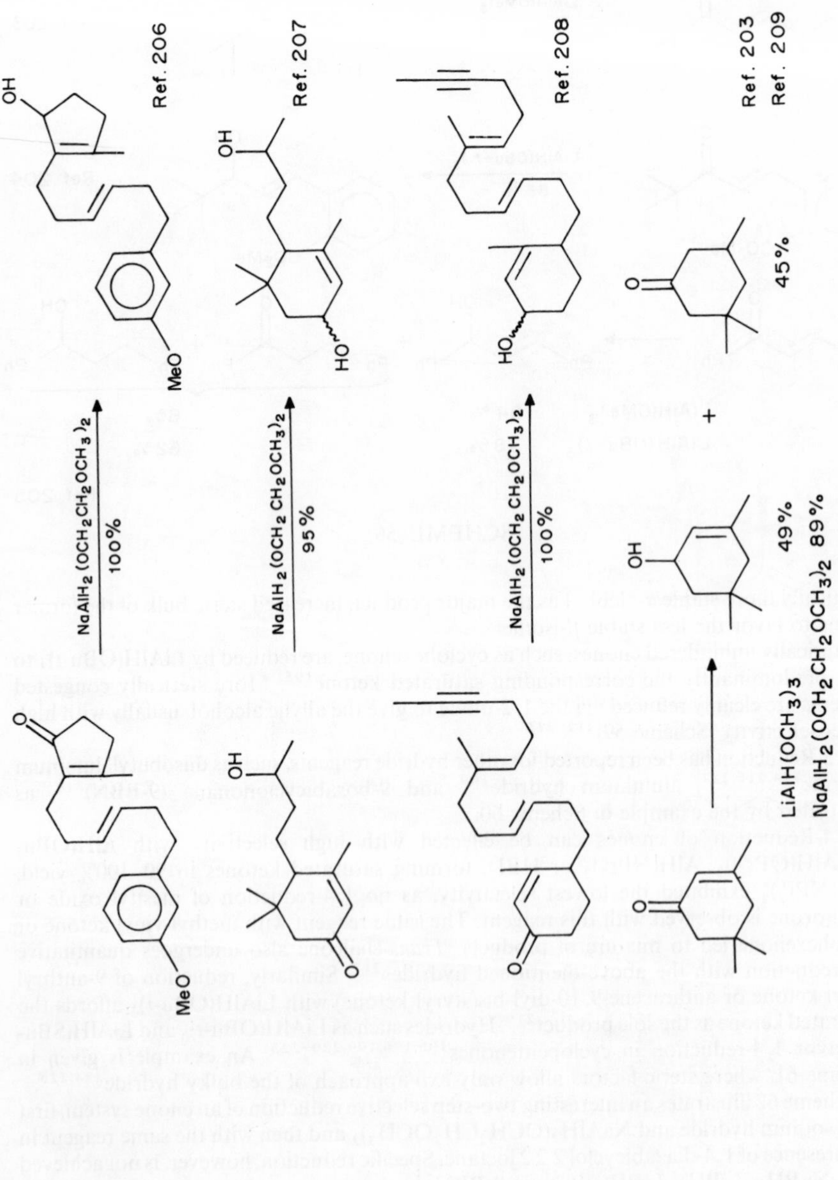

MeO
OH
$NaAlH_2(OCH_2CH_2OCH_3)_2$
100%
Ref. 206
95%
Ref. 207
HO
Ref. 208
O
$LiAlH(OCH_3)_3$
$NaAlH_2(OCH_2CH_2OCH_3)_2$ 89%
49%
+
45%
Ref. 203
Ref. 209

SCHEME 57

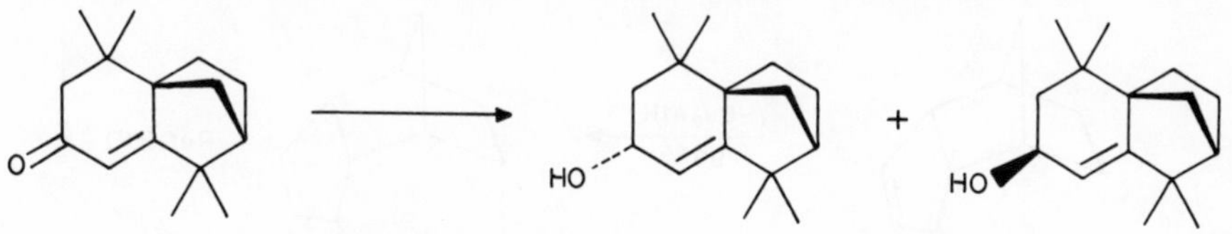

	LiAlH$_4$	76 : 24
	NaBH$_4$	76 : 24
	NaAlH$_2$(OCH$_2$CH$_2$OCH$_3$)$_2$	18 : 82

SCHEME 58

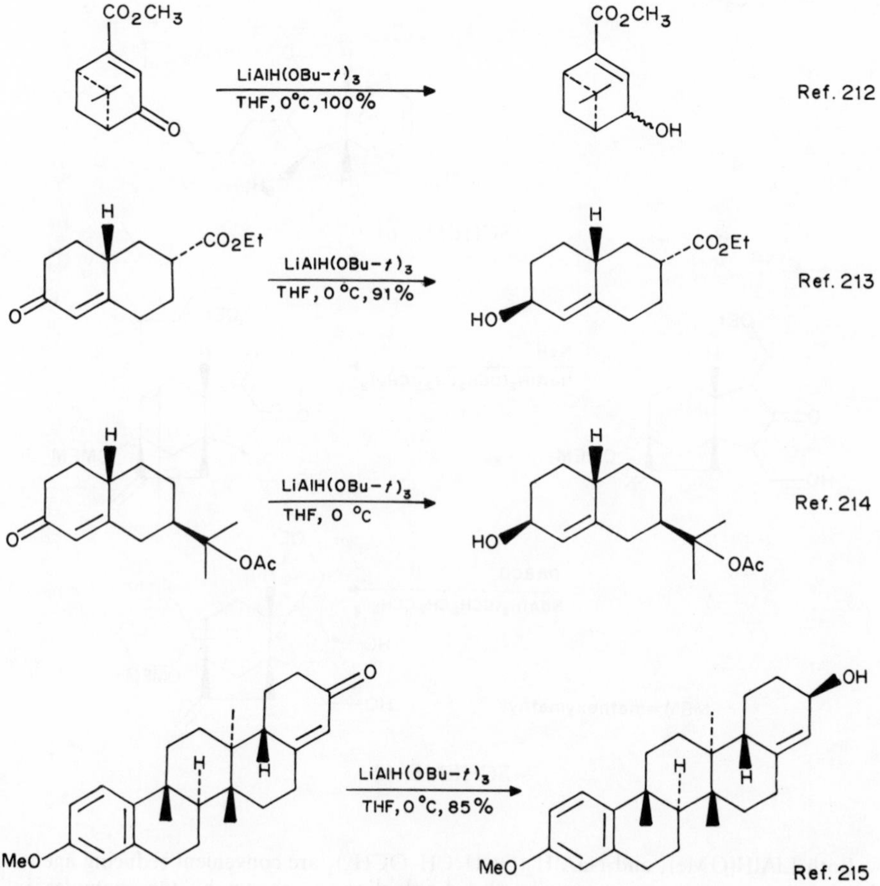

$\xrightarrow[\text{THF, 0°C, 100\%}]{\text{LiAlH(OBu-}t\text{)}_3}$ Ref. 212

$\xrightarrow[\text{THF, 0°C, 91\%}]{\text{LiAlH(OBu-}t\text{)}_3}$ Ref. 213

$\xrightarrow[\text{THF, 0 °C}]{\text{LiAlH(OBu-}t\text{)}_3}$ Ref. 214

$\xrightarrow[\text{THF, 0°C, 85\%}]{\text{LiAlH(OBu-}t\text{)}_3}$ Ref. 215

SCHEME 59

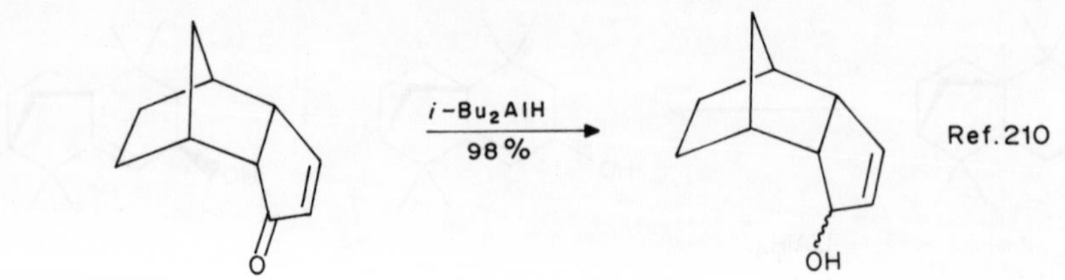

Ehud Keinan and Noam Greenspoon

SCHEME 60

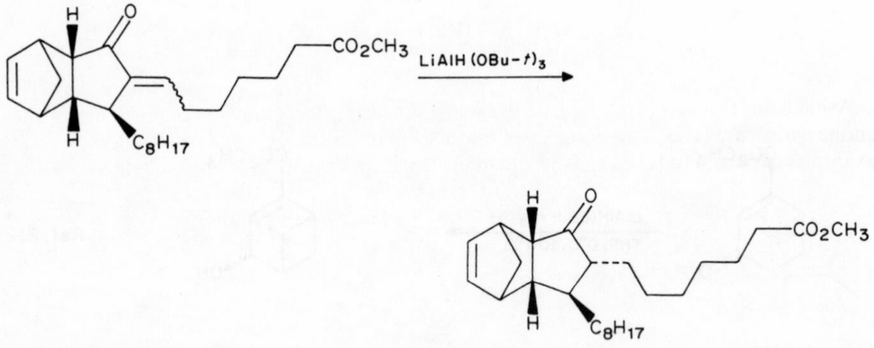

SCHEME 61

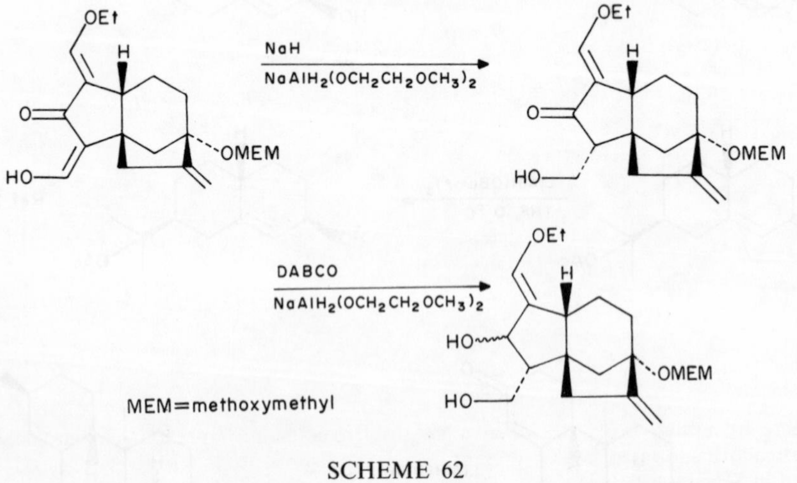

MEM = methoxymethyl

SCHEME 62

Both $LiAlH(OMe)_3$ and $NaAlH_2(OCH_2CH_2OCH_3)_2$ are convenient reducing agents for low-temperature, copper-mediated 1,4-reduction, as shown by the examples in Scheme 63[203,227].

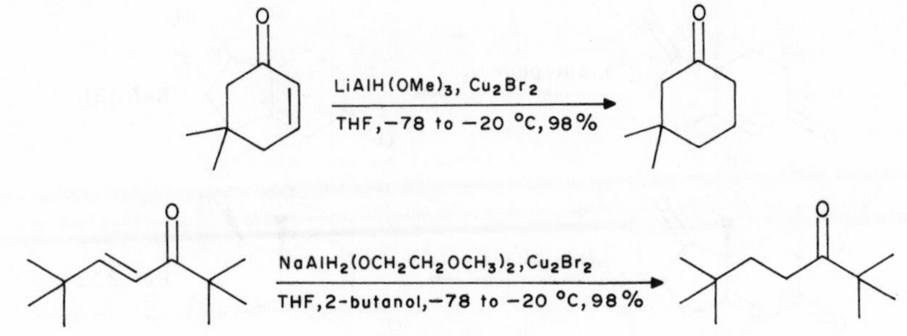

SCHEME 63

Aside from the nature of the hydride reagent, steric effects and lower reactivity of the enone substrate affect the course of reduction in polyfunctional molecules. Several examples of partial reduction of cyclopentenedione systems are given in Scheme 64[228-230].

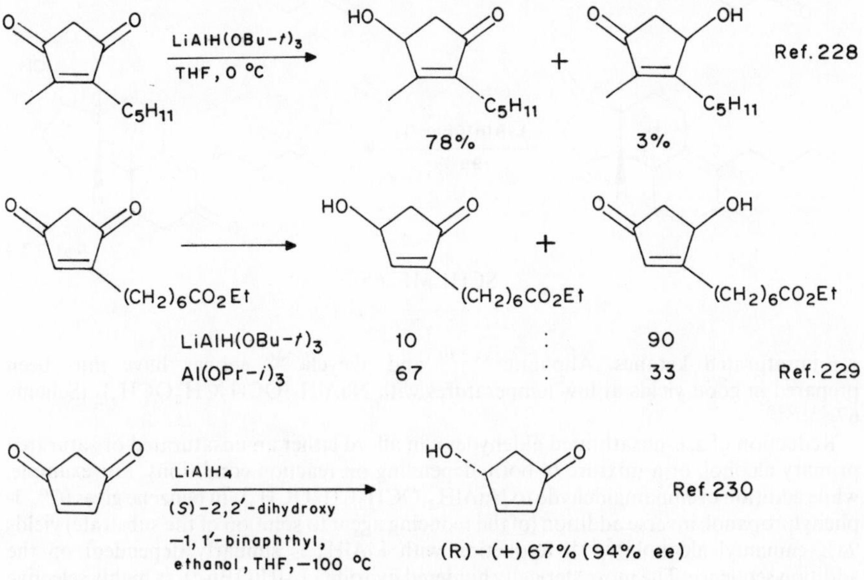

SCHEME 64

There are a number of cases where a less reactive enone group remains intact while a more reactive saturated ketone present in the same substrate is selectively reduced, as shown in Scheme 65[231-234].

Alternatively, there are a number of examples of simultaneous reduction of both saturated and unsaturated ketones or of preferential reduction of the unsaturated one (Scheme 66)[235-237].

Reduction of enol ethers or enol esters of 1,3-diketones followed by acid-catalyzed allylic rearrangement of the reduction product (see p. 85 in Reference 5) is a useful route to

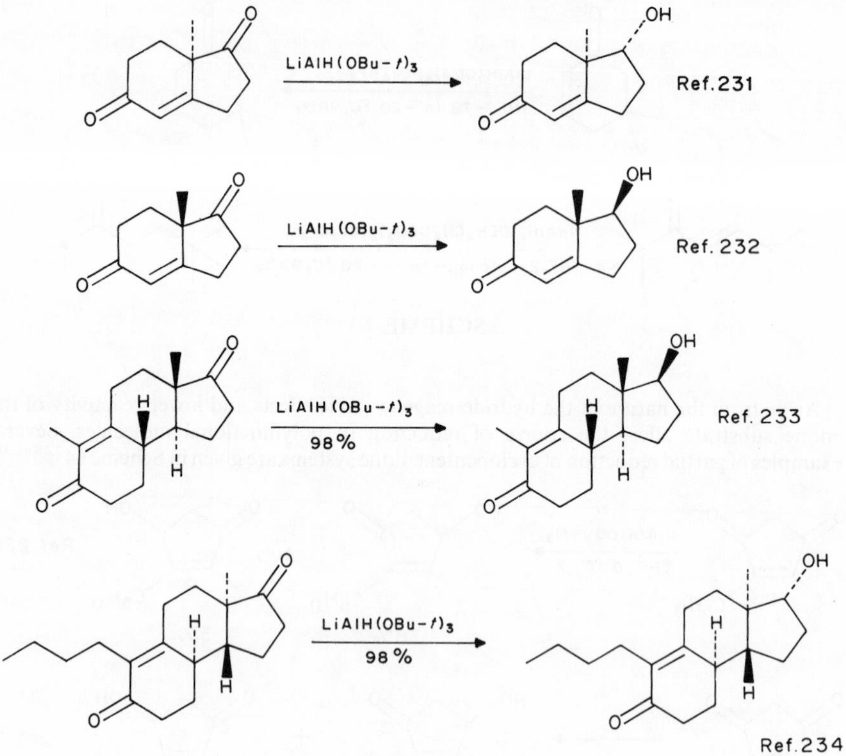

Ehud Keinan and Noam Greenspoon

SCHEME 65

α, β-unsaturated ketones. Aliphatic[238,239] and alicyclic[240] enones have thus been prepared in good yields at low temperatures with $NaAlH_2(OCH_2CH_2OCH_3)_2$ (Scheme 67)[241,242].

Reduction of α, β-unsaturated aldehydes can afford either an unsaturated or saturated primary alcohol, or a mixture of both, depending on reaction conditions. For example, while addition of cinnamaldehyde to $NaAlH_2(OCH_2CH_2OCH_3)_2$ in benzene gives 97% 3-phenylpropanol, inverse addition (of the reducing agent to solution of the substrate) yields 94% cinnamyl alcohol[243,244]. Reduction with $LiAlH_4$ is similarly dependent on the addition sequence. The more sterically hindered hydride $LiAlH(OBu-t)_3$ is highly selective for 1,2-reduction of aldehydes, even under conditions of normal addition. For example, it reduces cinnamaldehyde cleanly to cinnamyl alcohol, without affecting the olefinic bond[245-247]. Similar behavior is exhibited by $NaAlH_2(OCH_2CH_2OCH_3)_2$, which reduces 2-butenal to 2-butenol in 97% yield[244]. On the other hand, hydrides such as $LiAlH(OMe)_3$[187,245,246] and $NaAl_2H_4(OCH_2CH_2NMe_2)_3$[248] usually yield the saturated primary alcohol. Other examples of 1,2-reduction of α, β-unsaturated aldehydes with these reagents are given in Scheme 68[249-251].

Regioselectivity of enone reduction with diisobutylaluminum hydride (DIBAH) is very susceptible to minor structural changes in the substrate. While five-membered exocyclic enones provide the allylic alcohols which are the normal products for this reagent, reduction of chromones possessing exocyclic six-membered enones yield saturated

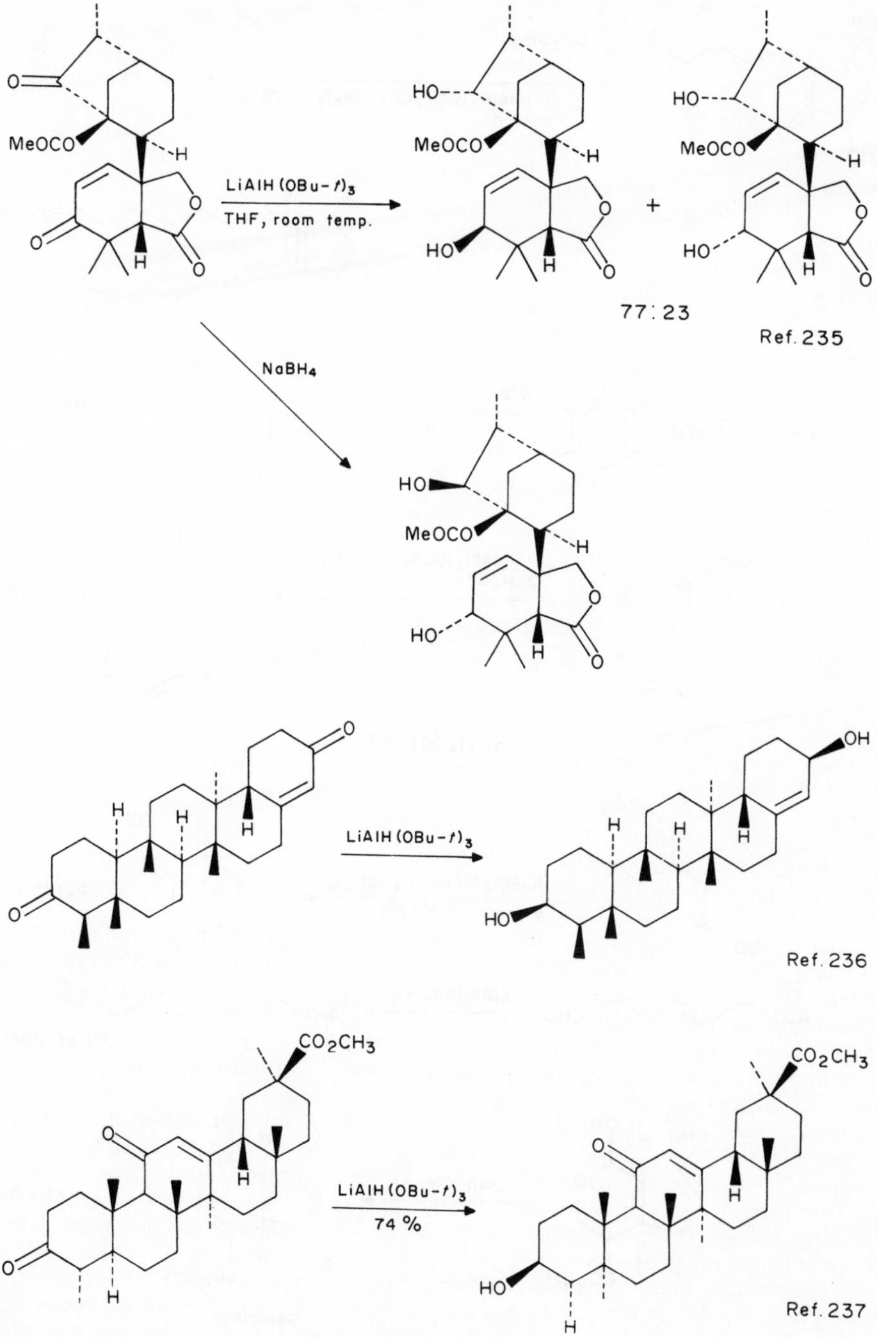

77:23

Ref. 235

LiAlH(OBu-t)₃
THF, room temp.

NaBH₄

LiAlH(OBu-t)₃

Ref. 236

LiAlH(OBu-t)₃
74 %

Ref. 237

SCHEME 66

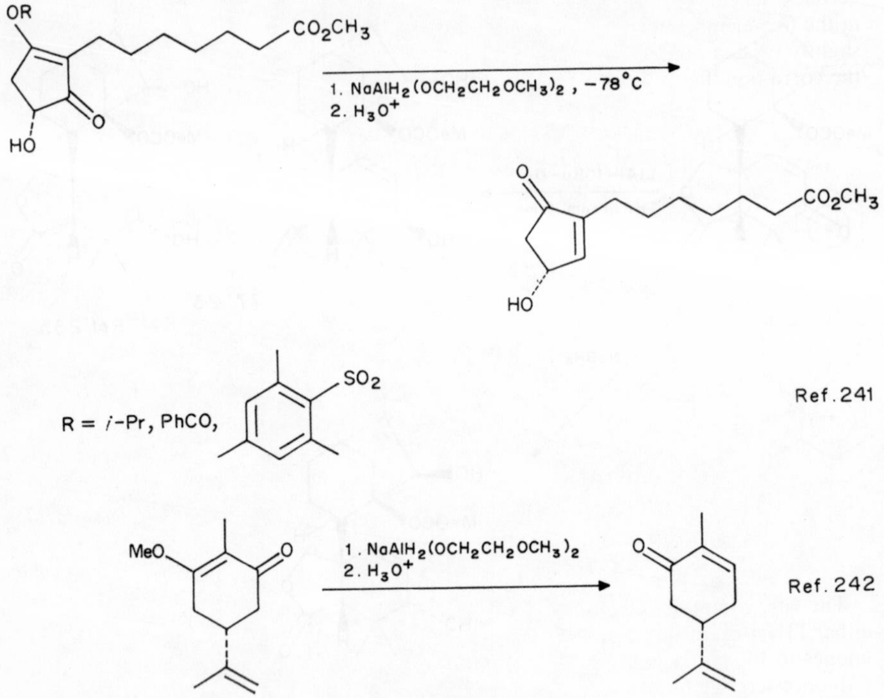
OR
CO_2CH_3
HO
1. $NaAlH_2(OCH_2CH_2OCH_3)_2$, −78°C
2. H_3O^+
O
CO_2CH_3
HO
R = i-Pr, PhCO,
SO_2
Ref. 241
MeO
O
1. $NaAlH_2(OCH_2CH_2OCH_3)_2$
2. H_3O^+
O
Ref. 242

SCHEME 67

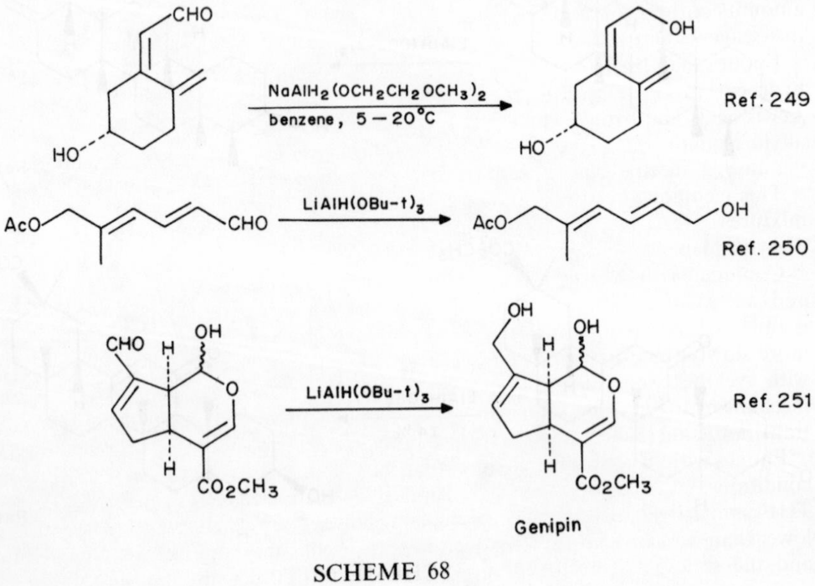
CHO
HO
$NaAlH_2(OCH_2CH_2OCH_3)_2$
benzene, 5 − 20°C
OH
HO
Ref. 249
AcO
CHO
LiAlH(OBu-t)$_3$
AcO
OH
Ref. 250
CHO
H
OH
O
H
CO_2CH_3
LiAlH(OBu-t)$_3$
OH
H
OH
O
H
CO_2CH_3
Genipin
Ref. 251

SCHEME 68

ketones (Scheme 69)[252]. This was explained by the strict coplanarity of the enone function in the five-membered structure, whereas the enones giving rise to saturated ketones are slightly twisted. Reduction of isoflavones with DIBAH under these conditions provides the corresponding isoflavan-4-ones in very high selectivity[252].

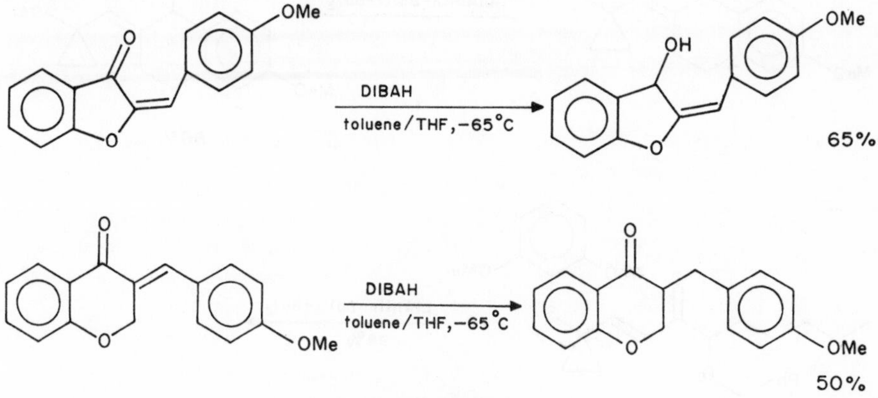

SCHEME 69

The 'ate' complex LiAlH(n-Bu)(i-Bu)$_2$ is prepared from DIBAH and butyllithium in either THF or toluene–hexane. This reagent is more effective for selective 1, 2-reduction of enones to the corresponding allylic alcohol than is DIBAH alone[253]. The reagent also reduces esters, lactones and acid chlorides to the corresponding alcohols, and epoxides to the respective alcohols. α, β-Unsaturated ketones derived from dehydration of aldol products from 1-(arylthio)cyclopropanecarboxaldehydes and ketones were selectively reduced by this 'ate' complex or by DIBAH itself, yielding the allylic alcohols with minor amounts of the 1, 4-reduction product (Scheme 70)[254]. Yields were typically higher with this reagent than with DIBAH.

Enones may be deoxygenated with LiAlH$_4$/AlCl$_3$ to give the corresponding olefinic hydrocarbons. The reactive species seem to be AlHCl$_2$ or AlH$_2$Cl, which act as both Lewis acids and hydride donors. The reaction involves initial 1, 2-reduction to form the allylic alcohol, followed by substitution of the allylic hydroxyl group by hydride (mainly via an S$_N$2' mechanism) to form the corresponding mixture of alkenes (Scheme 71)[255].

This technique has been applied to the deoxygenation of natural products. By using mixtures of LiAlH$_4$ and AlCl$_3$, flavanone and chalcone were transformed into flavan and diarylpropenes, respectively (Scheme 72)[256].

Conjugate reduction is the major pathway of enone reduction with a mixture of LiAlH$_4$ and excess CuI in THF[257]. It has been shown that the active reducing agent in this mixture is an H$_2$AlI species and not the copper hydride. Enones of cis geometry are reduced much more slowly than the corresponding trans compounds, and no reduction was observed with cyclohexenone and 3, 3, 5-trimethylcyclohexenone. These results suggest that the mechanism involves coordination of the metal to the carbonyl, forming a six-center transition state (Scheme 73)[257].

Enones with two alkyl groups at the β-position are reduced very sluggishly under these conditions. Other metal salts, such as HgI$_2$, TiCl$_3$ and HgCl$_2$, premixed with LiAlH$_4$ in THF, similarly give rise to 1, 4-reduction. Yields and selectivities were found to be much lower than with CuI. H$_2$AlI was found to react in the exact same manner as LiAlH$_4$–CuI, and the series H$_2$AlI, HAlI$_2$, H$_2$AlBr, HAlBr$_2$, H$_2$AlCl and HAlCl$_2$ was therefore

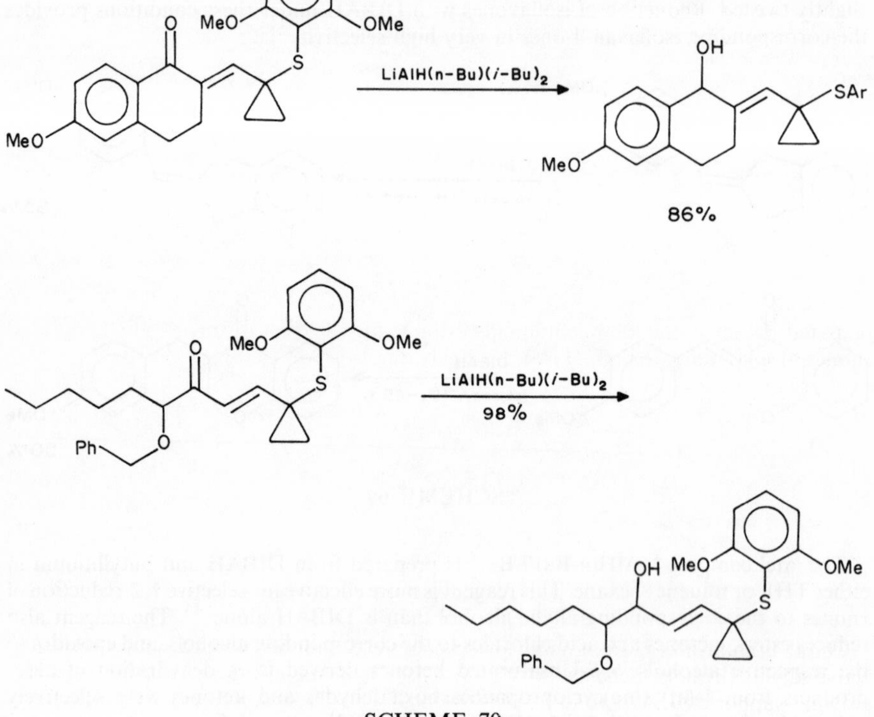

86%

LiAlH(n-Bu)(i-Bu)₂
98%

SCHEME 70

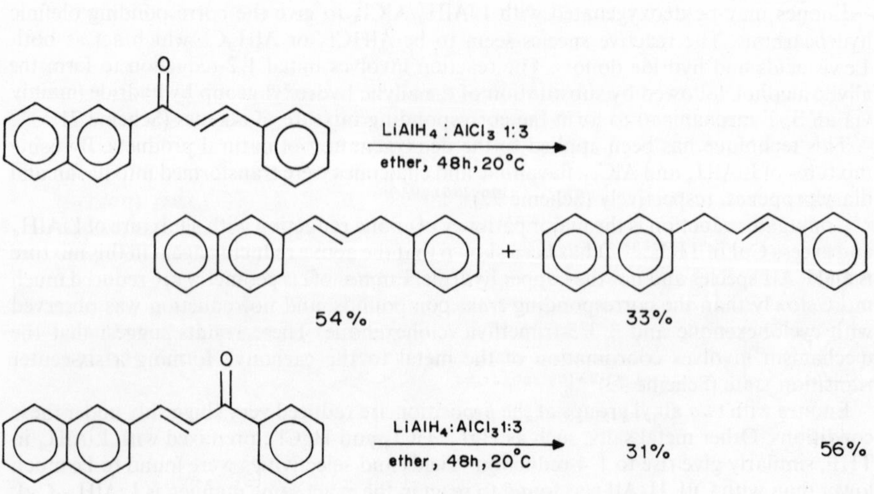

LiAlH₄ : AlCl₃ 1:3
ether, 48h, 20°C

54%   +   33%

LiAlH₄:AlCl₃ 1:3
ether, 48h, 20°C

31%   56%

SCHEME 71

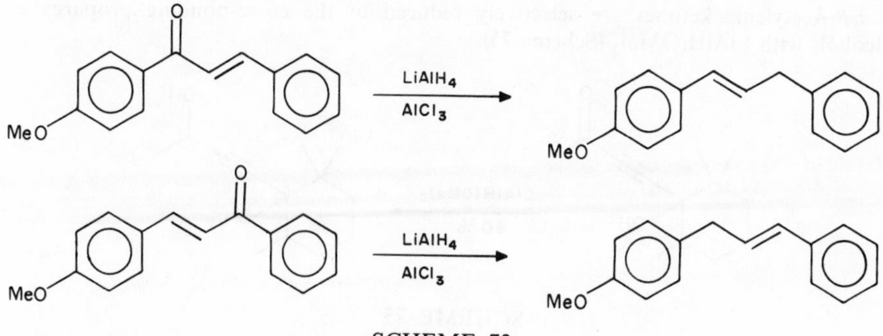

SCHEME 72

prepared. Of these, the iodo compounds exhibited the highest reactivity. HAlI$_2$ reduces enones at a slower rate than H$_2$AlI, probably due to steric factors.

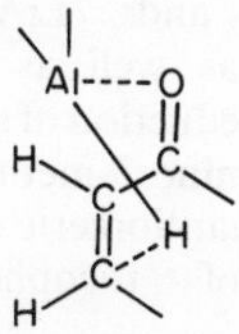

SCHEME 73

Chiral lithium alkoxyaluminumhydride complexes can be used to obtain optically active allylic alcohols (Scheme 74)[258-261]. These reagents are more selective than the polymer-supported LiAlH$_4$ and LiAlH$_4$–monosaccharide complexes[262].

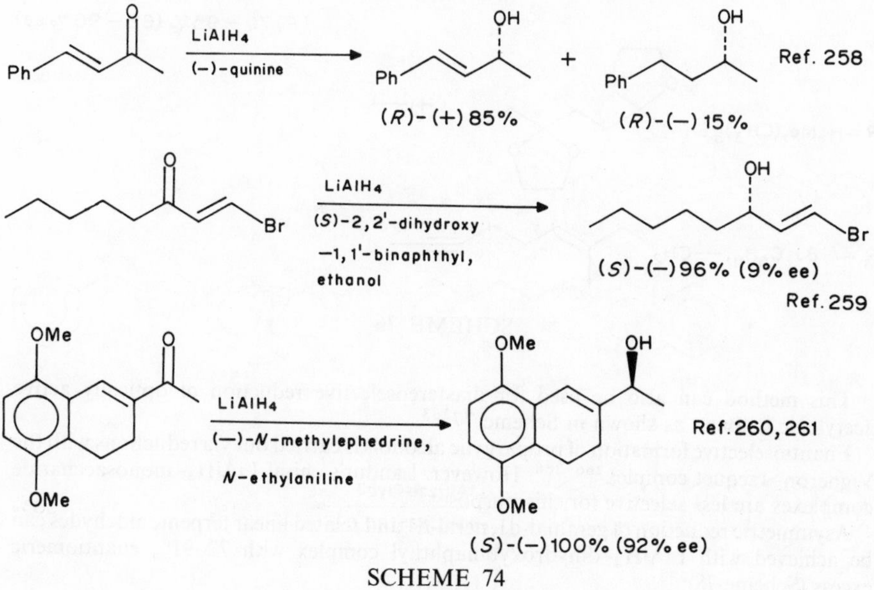

SCHEME 74

α, β-Acetylenic ketones are selectively reduced to the corresponding propargylic alcohols with LiAlH(OMe)$_3$ (Scheme 75).

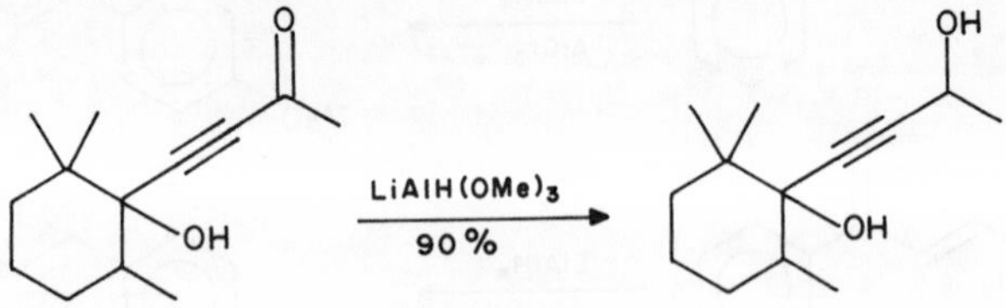

SCHEME 75

Asymmetric 1, 2-reduction of acetylenic ketones is an effective method for preparing optically active propargylic alcohols in high yield and high enantioselectivity. Common chiral reductants for this purpose include the Mosher–Yamaguchi reagent[263–265], the Vigneron–Jacquet complex[266–268] and LiAlH$_4$/2, 2'-dihydroxy-1, 1'-binaphthyl/ methanol (R and S) complexes[269], as well as the LiAlH$_4$-N-methylephedrine/N-ethylaniline complex[260]. For example, reduction of simple acetylenic ketones (Scheme 76) with LiAlH$_4$/(2S, 3R)-(+)-4-dimethylamino-3-methyl-1, 2-diphenyl-2-butanol results in propargylic (R)-alcohols in 62–95% enantiomeric excess. These chiral building blocks were used in the synthesis of tocopherol, prostaglandins and 11α-hydroxyprogesterone[264,265].

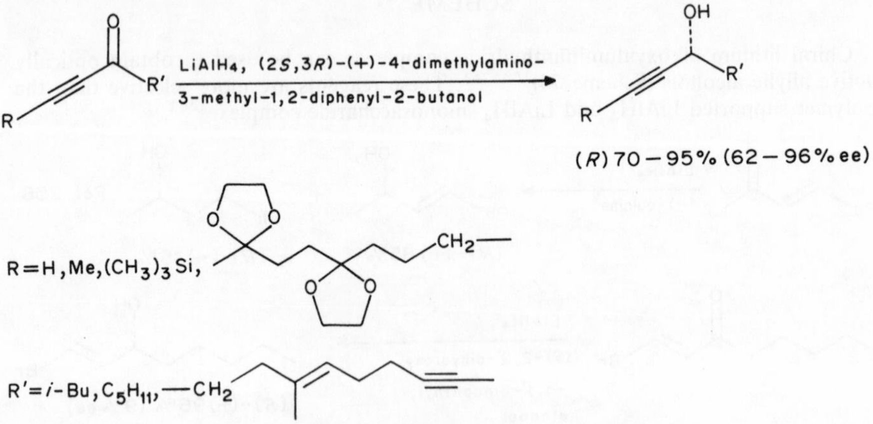

SCHEME 76

This method can also be used for diastereoselective reduction of optically active acetylenic ketones, as shown in Scheme 77[263].

Enantioselective formation of propargylic alcohols is carried out via reductions with the Vigneron–Jacquet complex[266–268]. However, Landor's chiral LiAlH$_4$–monosaccharide complexes are less selective for this purpose[270–272].

Asymmetric reduction of geranial-d1, neral-d1 and related linear terpenic aldehydes can be achieved with LiAlH$_4$–dihydroxybinaphthyl complex with 72–91% enantiomeric excess (Scheme 78)[273].

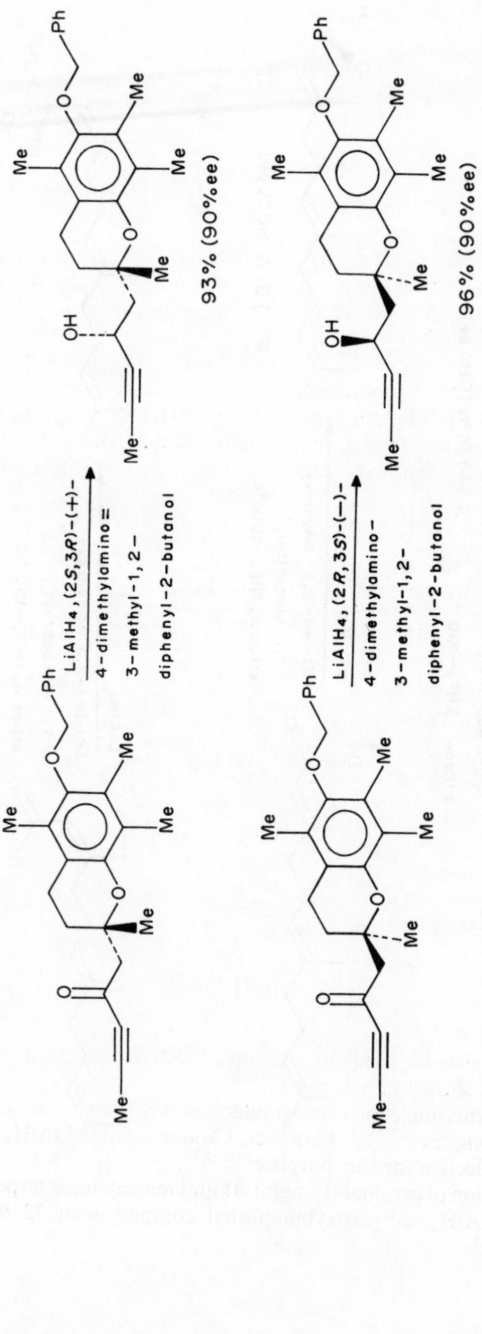

SCHEME 77

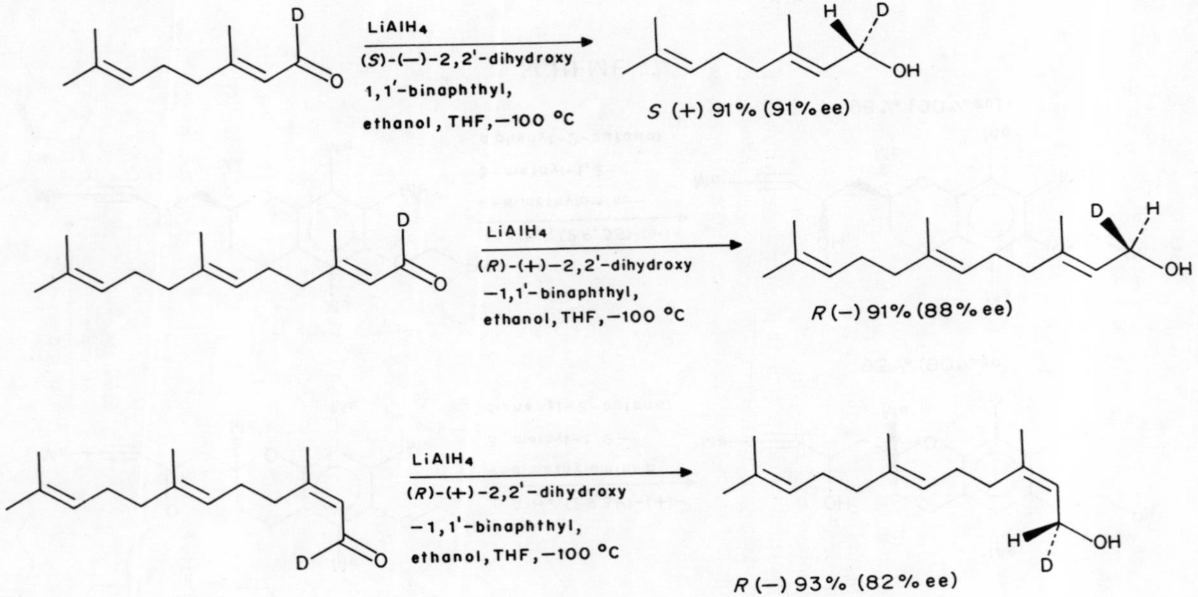
LiAlH4
(S)-(−)-2,2'-dihydroxy
1,1'-binaphthyl,
ethanol, THF, −100 °C
S (+) 91% (91% ee)
LiAlH4
(R)-(+)-2,2'-dihydroxy
−1,1'-binaphthyl,
ethanol, THF, −100 °C
R (−) 91% (88% ee)
LiAlH4
(R)-(+)-2,2'-dihydroxy
−1,1'-binaphthyl,
ethanol, THF, −100 °C
R (−) 93% (82% ee)

SCHEME 78

Asymmetric reduction of prochiral α,β-unsaturated ketones with chiral hydride reagents derived from LiAlH$_4$ and (S)-4-anilino- and (S)-4-(2,6-xylidino)-3-methylamino-1-butanol gives (S)- and (R)- allylic alcohols, respectively, in high chemical and optical yields (Scheme 79)[274].

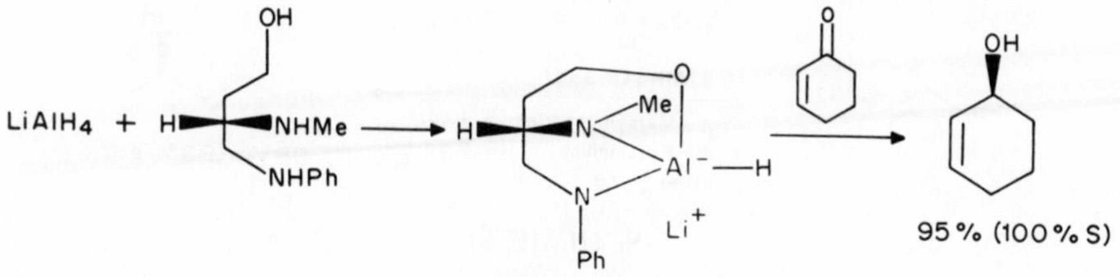

SCHEME 79

A modified aluminum hydride is prepared by treating LiAlH$_4$ in THF with equimolar amounts of ethanol and optically pure S-($-$)-2,2'-dihydroxy-1,1'-binaphthyl. Allylic alcohols of very high optical purity are obtained in high yield by reduction of α,β-unsaturated ketones with this reagent[275]. Of particular interest are the attractive opportunities provided by this reagent in prostaglandin synthesis. For example, some of the chemical transformation shown in Scheme 80[275] are more effective in both terms of chemical and optical yields than standard microbiological reduction[276].

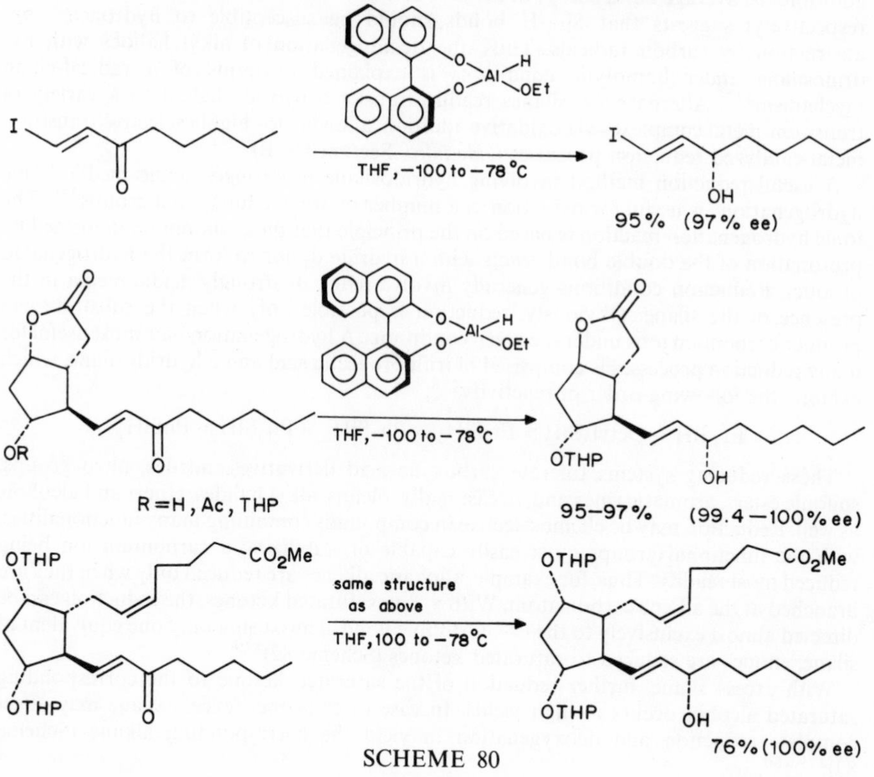

SCHEME 80

Asymmetric reduction of α, β-unsaturated ketones is achieved with LiAlH$_4$, partially decomposed by $(-)$-N-methylephedrine and ethylaniline (Scheme 81)[260]. This reagent converts open chain enones into the corresponding optically active allylic alcohols in high chemical (92–100%) and optical yields (78–98% ee).

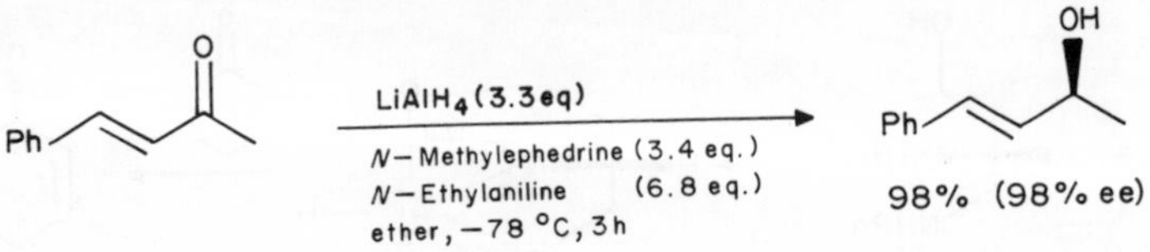

SCHEME 81

C. Silicon Hydrides

The hydrogen in the Si—H bond is slightly hydridic in nature, as would be expected from the relative electronegativities of silicon (1.7) and hydrogen (2.1). Therefore, silanes may function as hydride transfer agents toward highly electrophilic species such as carbonium ions. The hydridic nature of the Si—H bond may be significantly increased upon interaction with strong anionic ligands, such as fluoride and alkoxides (vide infra). In addition, the average bond energy of the Si—H and C—H bonds (70 and 99 kcal mol^{-1}, respectively) suggests that Si—H bonds should be susceptible to hydrogen atom abstraction by carbon radicals. Thus, the dehalogenation of alkyl halides with hydridosilane under homolytic conditions is explained in terms of a radical-chain mechanism[277]. Alternatively, silanes readily transfer a hydride ligand to a variety of transition-metal complexes via oxidative addition, allowing for highly selective transition metal-catalyzed reduction processes (vide infra, Section IV, B).

A useful reduction method involving hydridosilane in strongly acidic media, 'ionic hydrogenation', is useful for reduction of a number of organic functional groups[278]. The ionic hydrogenation reaction is based on the principle that the carbonium ion formed by protonation of the double bond reacts with a hydride donor to form the hydrogenated product. Reduction conditions generally involve reflux in strongly acidic media in the presence of the silane. Obviously, reduction is possible only when the substrate can produce carbonium ions under the given conditions. A hydrogenation pair most useful for many reduction processes is comprised of trifluoroacetic acid and a hydridosilane, which exhibits the following order of reactivity[278]:

$$\text{Et}_3\text{SiH} > \text{Octyl}_3\text{SiH} > \text{Et}_2\text{SiH}_2 > \text{Ph}_2\text{SiH}_2 > \text{Ph}_3\text{SiH} > \text{PhSiH}_3$$

These reducing systems tolerate carboxylic acid derivatives, nitriles, nitro groups, sulfonic esters, aromatic rings and, occasionally, olefins, alkyl halides, ethers and alcohols as well. Reduction may be chemoselective in compounds containing many functionalities, with the functional groups most easily capable of stabilizing a carbonium ion being reduced most readily. Thus, for example, aliphatic alkenes are reduced only when they are branched at the alkene carbon atom. With α, β-unsaturated ketones, the reduction can be directed almost exclusively to the C—C double bond. Thus, using only one equivalent of silane, enones are reduced to saturated ketones (Scheme 82)[279].

With excess silane, further reduction of the saturated ketone to the corresponding saturated alcohol occurs in high yields. In case of chalcones, excess silane may affect complete reduction and deoxygenation to yield the corresponding alkane (Scheme 83)[279,280].

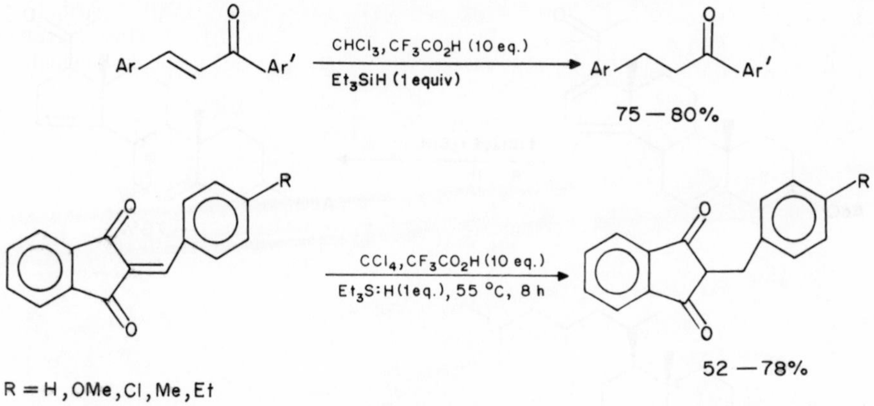

Ar' $\xrightarrow{\begin{array}{c} CHCl_3, CF_3CO_2H \ (10 \ eq.) \\ Et_3SiH \ (1 \ equiv) \end{array}}$ Ar'

75 — 80%

R = H, OMe, Cl, Me, Et

$\xrightarrow{\begin{array}{c} CCl_4, CF_3CO_2H \ (10 \ eq.) \\ Et_3S:H \ (1 \ eq.), \ 55 \ °C, \ 8 \ h \end{array}}$

52 — 78%

SCHEME 82

Ph $\xrightarrow{\begin{array}{c} CHCl_3, CF_3CO_2H \ (6eq.) \\ Et_3SiH \ (3eq.) \\ 5 \ h, \ 50 \ °C \end{array}}$ Ph 85%

$\xrightarrow{\begin{array}{c} CHCl_3, CF_3CO_2H \ (6 \ eq.) \\ Et_3S:H \ (3 \ eq.), \ 4 \ h, \ 55 \ °C \end{array}}$ 90%

Ar' $\xrightarrow{\begin{array}{c} CHCl_3CF_3CO_2H \ (10 \ eq.) \\ Et_3SiH \ (3 \ eq.) \\ 60 \ °C, \ 7 \ h \end{array}}$ Ar'

70—82%

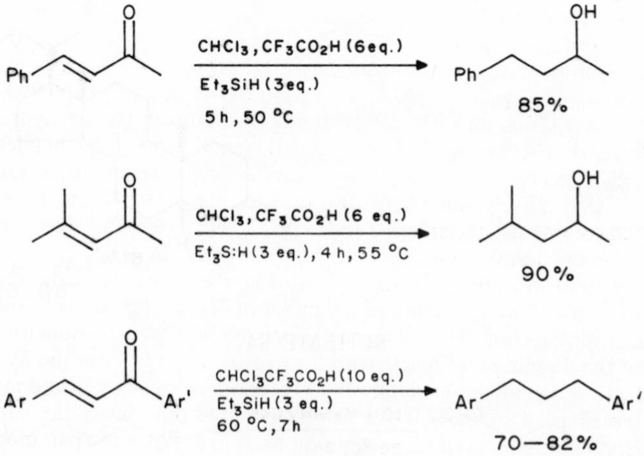

SCHEME 83

The reaction of conjugated enones and dienones with trimethyl- and triethylsilane in the presence of $TiCl_4$ followed by aqueous workup produces the corresponding saturated ketones. This Lewis acid catalysis is particularly useful for conjugated reduction of sterically hindered systems (Scheme 84)[281]. α, β-Unsaturated esters are not reduced under these conditions.

Anionic activation of Si—H bonds[282] by fluorides, such as KF or CsF, or by potassium phthalate, $KHCO_3$, KSCN, etc., yields powerful hydridic reagents that reduce the carbonyl group of aldehydes, ketones and esters[283]. It was postulated that the active species in these reactions is a pentacoordinated or even hexacoordinated hydridosilane. 1, 2-Reductions of α, β-unsaturated aldehydes and ketones occur with very high selectivity to give allylic alcohols (Scheme 85)[283]. The analogous activation of hydridosilanes by fluoride ions is also achieved under acidic conditions with boron trifluoride etherate, in which the latter compound is consumed and fluorosilanes are formed[284].

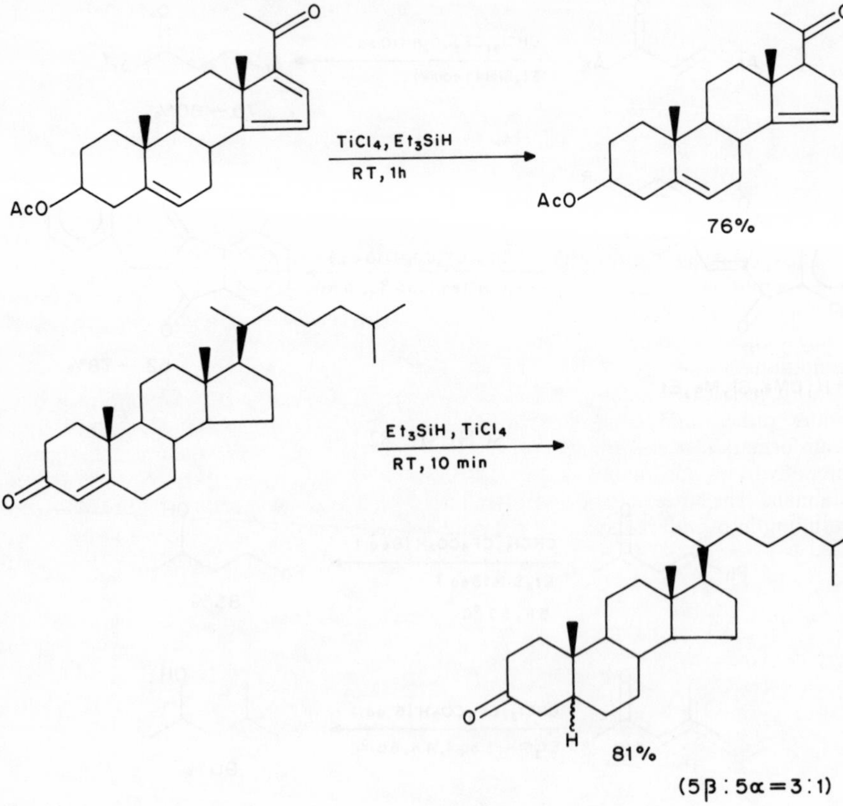

$$\xrightarrow[\text{R T, 1h}]{\text{TiCl}_4, \text{Et}_3\text{SiH}}$$

76%

$$\xrightarrow[\text{R T, 10 min}]{\text{Et}_3\text{SiH, TiCl}_4}$$

81%

$(5\beta : 5\alpha = 3 : 1)$

SCHEME 84

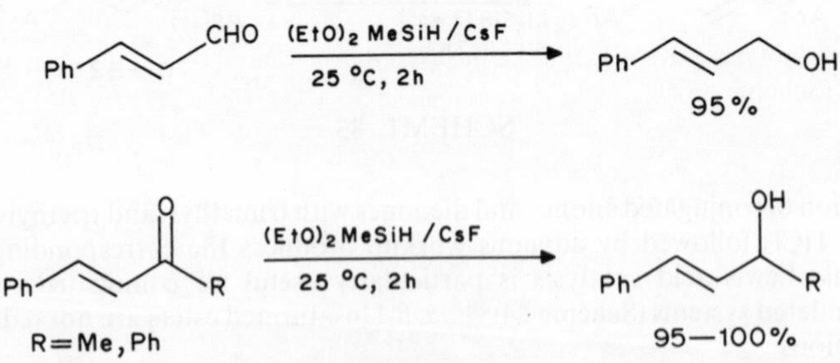

$$\text{Ph}\diagup\diagdown\text{CHO} \xrightarrow[\text{25 °C, 2h}]{\text{(EtO)}_2\text{MeSiH/CsF}} \text{Ph}\diagup\diagdown\text{OH}$$

95%

$$\text{Ph}\diagup\diagdown\overset{\text{O}}{\underset{}{\text{C}}}\text{R} \xrightarrow[\text{25 °C, 2h}]{\text{(EtO)}_2\text{MeSiH/CsF}} \text{Ph}\diagup\diagdown\overset{\text{OH}}{\underset{}{\text{C}}}\text{R}$$

R = Me, Ph   95—100%

SCHEME 85

Effective anionic activation of trichlorosilane can be carried out with either catechol or 2,2'-dihydroxybiphenyl in THF yielding bis(diolato)hydridosilicates (Scheme 86)[285]. Such reagents exhibit reducing power that is reminiscent of the complex aluminum hydrides. Even tertiary amines are useful activators of trichlorosilane, enhancing its hydridic character[286].

SCHEME 86

D. Tin Hydrides

The special characteristics of organotin hydrides as reducing agents are rationalized by the fact that the tin–hydrogen bond is both weaker and less polar than the B—H or Al—H bonds[287]. These characteristics are manifested in reactions that proceed by either a free radical chain or polar mechanism, depending on the substrate, catalyst and reaction conditions.

α, β-Unsaturated aldehydes and ketones are readily reduced by organotin hydrides under rather mild conditions, but the reaction is often obscured by subsequent transformation of the adducts[288]. On heating or under UV irradiation, the organotin monohydrides add mainly at the 1,4-positions of the enone system to form the enol stannane. The latter may be hydrolyzed or cleaved by a second equivalent of tin hydride, resulting in overall reduction of the double bond (Scheme 87)[287,288].

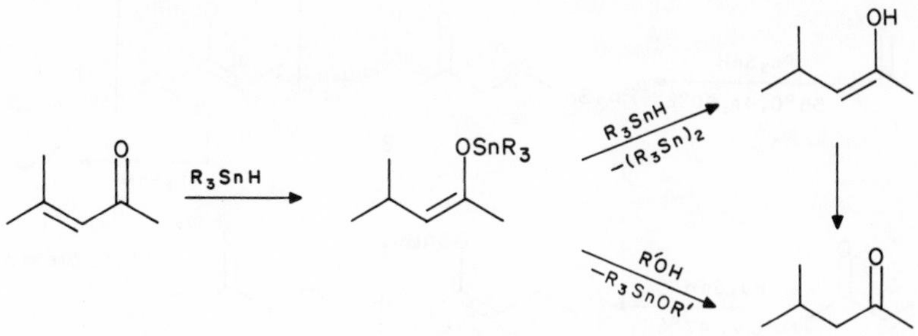

SCHEME 87

The protonolysis pathway was demonstrated in reactions carried out in deuteriated methanol (Scheme 88)[289].

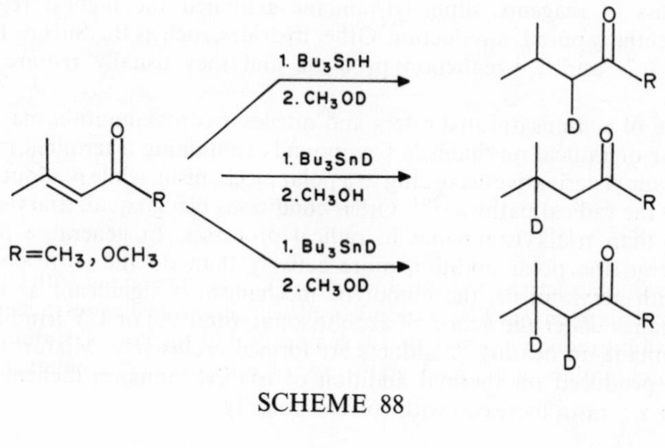

SCHEME 88

Enolate cleavage by a second equivalent of tin hydride is illustrated in Scheme 89[2881]. With Bu_3SnH the reaction proceeds no further, whereas the more electrophilic Ph_3SnH leads to hydrostannolysis of the tin enolate.

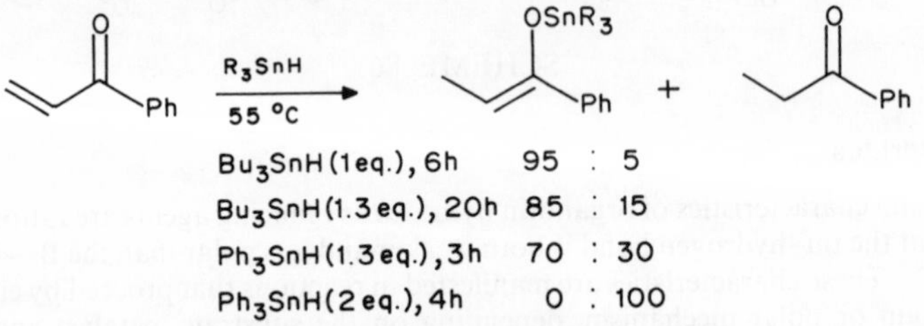

Bu_3SnH (1eq.), 6h	95	: 5
Bu_3SnH (1.3 eq.), 20h	85	: 15
Ph_3SnH (1.3 eq.), 3h	70	: 30
Ph_3SnH (2 eq.), 4h	0	: 100

SCHEME 89

Sterically nonhindered enones may produce mixtures of products, including carbon-stannylated species. For example, methyl vinyl ketone gives rise to significant quantities of the inverted 1,4-adduct, where tin binds at the 4-position, leading to β-stannyl ketone. In the case of methyl propenyl ketone, addition occurs at position 3 and 4, producing α-stannyl ketone (Scheme 90)[288j].

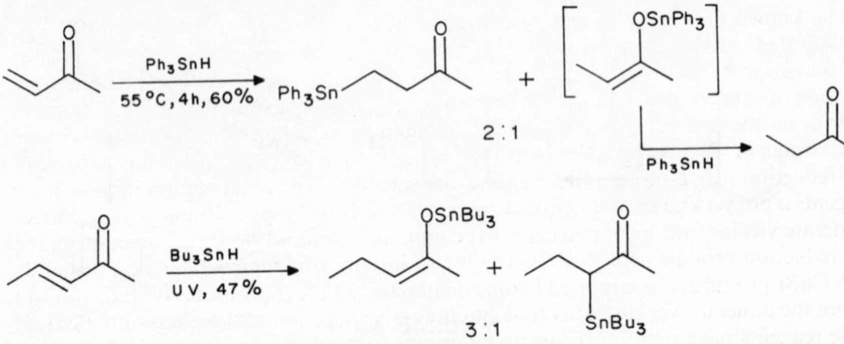

SCHEME 90

In this class of reagents, diphenylstannane exhibited the highest regioselectivity, affording essentially pure 1,2-reduction. Other hydrides, such as Bu_3SnH or Ph_3SnH, give mixtures of 1,2- and 1,4-reduction products and they usually require free radical initiation[290].

In the case of α,β-unsaturated esters and nitriles, hydrostannation may proceed via either a polar or radical mechanism. Compounds containing a terminal multiple bond form the α-stannyl derivative according to a polar mechanism, while β-adducts are formed according to the radical pathway[291]. Other conditions being equal, triarylstannanes are more active than trialkylstannanes in radical processes. In general, α,β-unsaturated nitriles undergo the polar addition more actively than do the corresponding esters. However, with acrylonitrile, the homolytic mechanism is significant as well[292]. With trialkylstannanes under the action of azobis(isobutyronitrile) or UV irradiation or with triphenylstannane on heating, β-adducts are formed exclusively. Mixtures of α- and β-adducts are produced on thermal addition of trialkylstannanes (Scheme 91)[292]. Expectedly, the α/β ratio increases with solvent polarity.

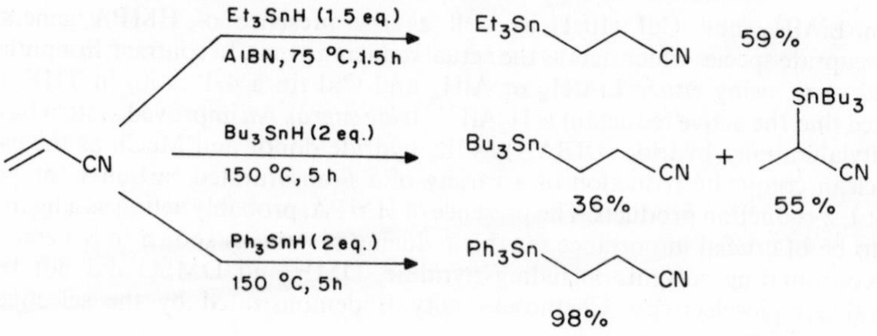

SCHEME 91

Hydrostannation of α-acetylenic esters generally produces a mixture of products. For more details, see Reference 287.

V. REDUCTIONS WITH STOICHIOMETRIC AMOUNTS OF TRANSITION-METAL HYDRIDES

A. Copper Hydrides

The known preference of organo-copper reagents to engage in 1,4-addition to α, β-unsaturated carbonyl compounds[293] prompted an extensive search for analogous hydrido-copper reagents that would undergo conjugate addition to enones. Indeed, reaction of cuprous bromide with either two equivalents of lithium trimethoxyaluminum hydride or one equivalent of sodium bis(2-methoxyethoxy)aluminum dihydride ('Vitride' by Eastman or 'Red-Al' by Aldrich) in THF produces a heterogeneous mixture capable of 1,4-reduction of α, β-unsaturated ketones and esters[294]. The exact composition of these reagents is not yet known. Reductions usually take place between -20 and $-78\,°C$ to give moderate yields of the saturated carbonyl compound along with varying amounts of the 1,2-reduction product (Scheme 92). The use of lithium trimethoxyaluminium deuteride with CuBr produces the saturated ketone deuteriated at the β-position. Addition of D_2O before the aqueous workup leads to deuterium incorporation at the α-position. Because these reagents react with other functional groups (saturated ketones and aldehydes and alkyl bromides being reduced almost as rapidly as enones), their chemoselectivity is limited. The reagent has also been used for the conjugate reduction of α, β-unsaturated nitriles[295].

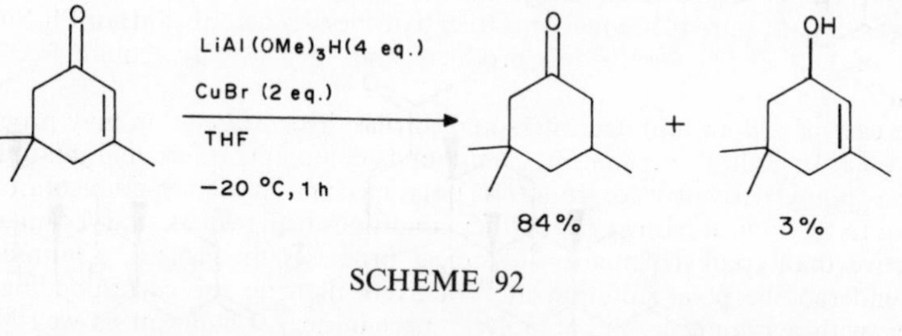

SCHEME 92

Combination of $LiAlH_4$ and catalytic amounts of CuI in HMPA/THF (1:4) is useful for 1,4-reduction of α, β-unsaturated ketones, aldehydes and esters[296]. Reactions carried out at $-78\,°C$ for 1 hour resulted predominantly in the 1,4-reduction product, but traces of the saturated and allylic alcohols were also formed[296]. It was claimed that the ratio

Ehud Keinan and Noam Greenspoon

between LiAlH$_4$ and CuI (10:1) as well as the presence of HMPA generates a
hydridocuprate species which acts as the actual reducing agent. In contrast, in a previously
reported work using either LiAlH$_4$ or AlH$_4$ and CuI (in a 4:1 ratio) in THF, it was
suggested that the active reductant is H$_2$AlI[257] (vide supra). An improved system based on
diisobutylaluminum hydride (DIBAH) as the hydride donor and MeCu as the catalyst
effects clean conjugate reduction of a variety of α, β-unsaturated carbonyl compounds
without 1, 2-reduction products. The presence of HMPA, probably acting as a ligand, was
found to be of crucial importance for this reducing system, as shown in Scheme 93[297].
Other coordinating solvents including pyridine, DMF and DMSO did not lead to
comparable regioselectivity. Chemoselectivity is demonstrated by the selective 1, 6-

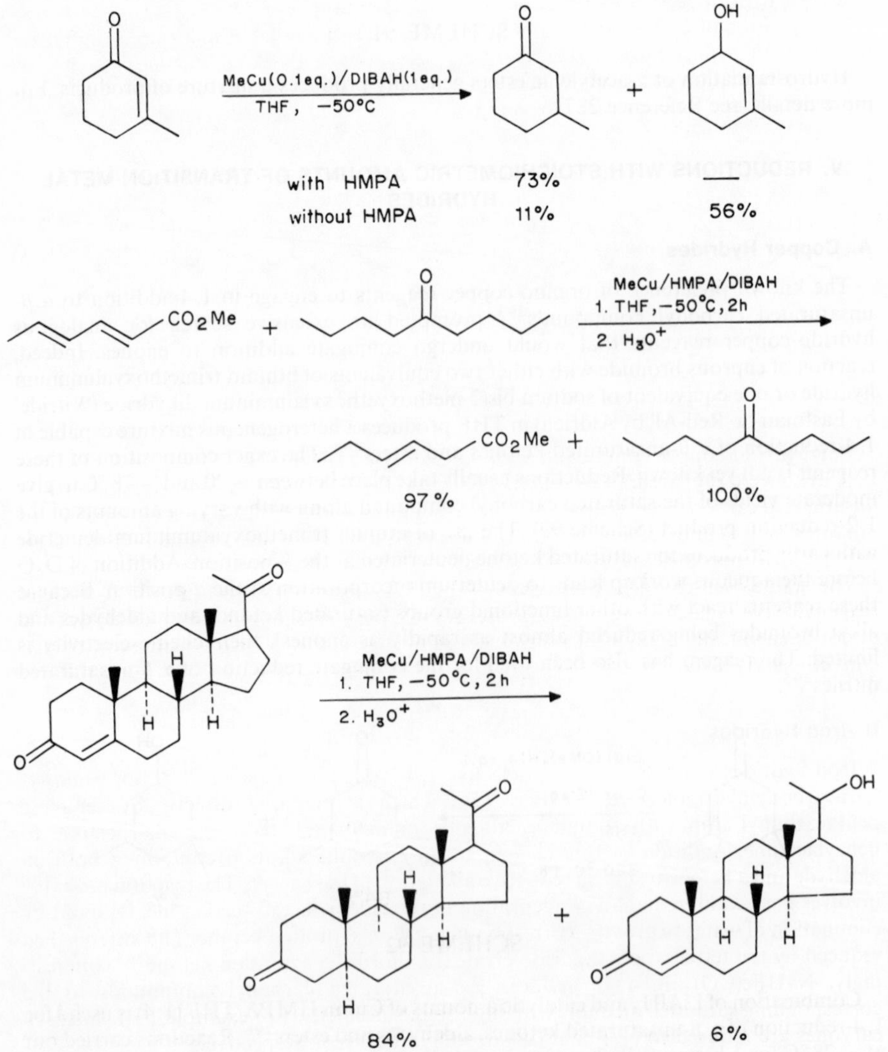

with HMPA	73%	—
without HMPA	11%	56%

SCHEME 93

reduction of methyl sorbate in the presence of a saturated ketone, and the conjugate reduction of the enone of progesterone with only minor reduction of the saturated ketone in this molecule.

A series of heterocuprate complexes Li^+HRCu^-, with R representing a nontransferable ligand such as 1-pentynyl, t-BuO^- or PhS^-, was generated in toluene from DIBAH and CuI by addition of RLi. These reagents were used for clean 1,4-reduction of α, β-unsaturated ketones and esters[298]. Yields, however, were quite low in several cases due to the strong basicity of these reagents. Although HMPA was found to facilitate 1,4-reduction in substrates where the β-carbon is highly substituted, enone reduction in multifunctional compounds resulted in low yields (Scheme 94). In a related, independent study, the hydridocuprate complex was prepared by addition of RLi (R = alkyl or alkynyl) to a suspension of CuH in ether or in THF. These reagents were used for clean conjugate reduction of α, β-unsaturated carbonyls[299], however with poor chemoselectivity, as saturated aldehydes and ketones were reduced under these conditions to the corresponding alcohols, and various tosylates and bromides were reductively cleaved.

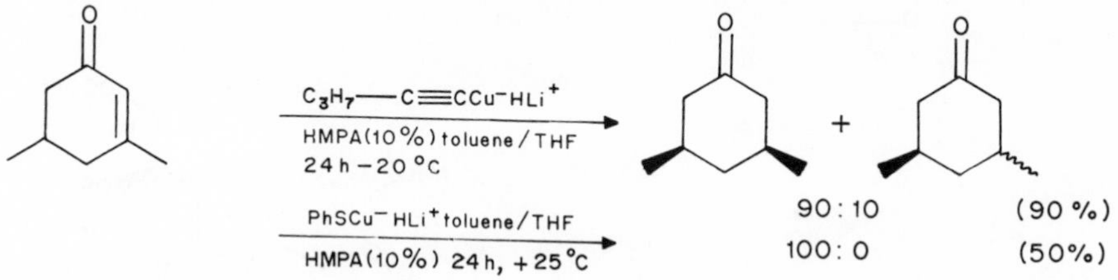

SCHEME 94

Polyhydrido-copper complexes, such as $LiCuH_2$, Li_2CuH_3, Li_3CuH_4, Li_4CuH_5 and Li_5CuH_6, were prepared[300] by $LiAlH_4$ reduction of $Li_nCu(CH_3)_{n-1}$. Reduction of α, β-unsaturated carbonyl compounds with any of these hydrides in ether or in THF produced mixtures of 1,4- and 1,2-reduction products. These reagents also reduce ketones, alkyl halides, alkyl tosylates and aryl halides.

The stable, well-characterized copper(I) hydride cluster $((PPh_3)CuH)_6$[301] is a useful reagent for conjugate reduction of α, β-unsaturated carbonyl compounds[302]. This hydride donor is chemically compatible with chlorotrimethylsilane, allowing formation of silyl enol ethers via a reductive silation process (Scheme 95).

B. Iron Hydrides

Iron hydrides were also used for selective 1,4-reduction of enones[287b]. For example, tetracarbonylhydridoferrate, $NaHFe(CO)_4$, which is prepared directly by refluxing pentacarbonyl iron with sodium methoxide in methanol, reduces benzalacetone to benzylacetone. Addition of this reagent to an ethanolic solution containing both an aldehyde and a ketone results in reductive alkylation of the ketone. The reaction probably involves base-catalyzed aldol condensation of the aldehyde and the ketone, followed by elimination of water to give the corresponding α, β-unsaturated ketone. The latter is then reduced by the tetracarbonylhydridoferrate, to afford the saturated ketone[303]. Interestingly, $NaHFe(CO)_4$ in THF reduces α, β-unsaturated carbonyl compounds to the corresponding saturated alcohols with high stereospecificity. For example, (+)- and (−)-carvones are reduced to (−)- and (+)-neodihydrocarveol, respectively[304].

The binuclear hydride $NaHFe_2(CO)_8$[305,306], which is prepared by addition of AcOH to a slurry of $Na_2Fe_2(CO)_8$ in THF, is also useful for clean conjugate reductions. This reagent

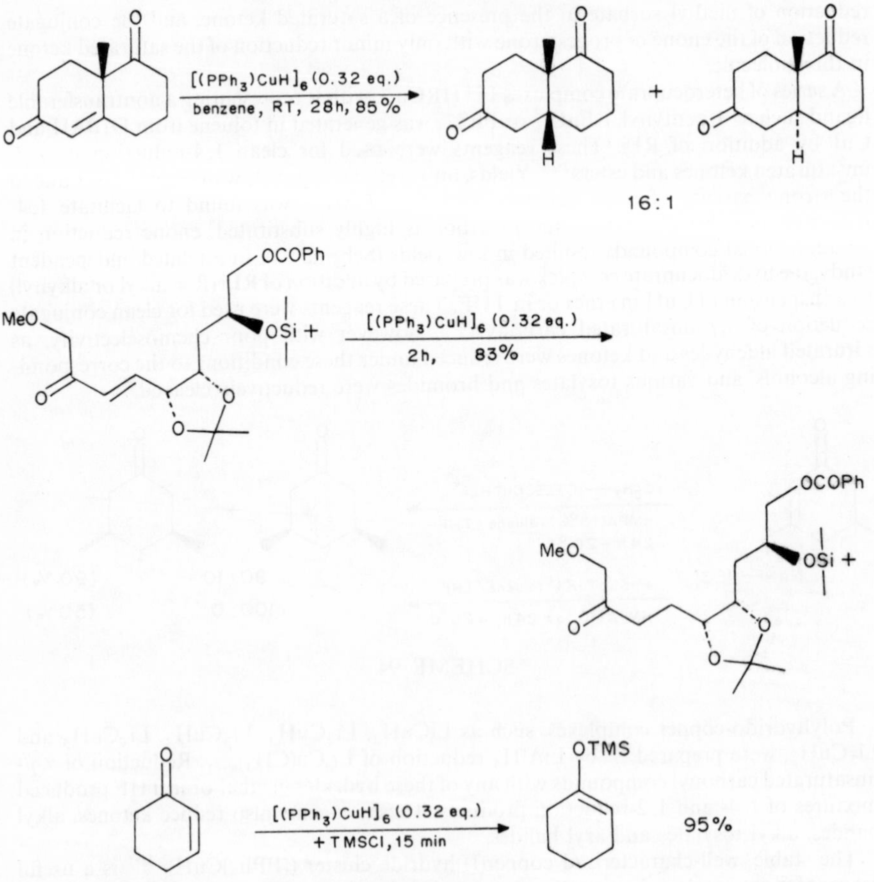

16 : 1

SCHEME 95

is capable of selective 1, 4-reduction of α, β-unsaturated ketones, aldehydes, esters, nitriles, amides and lactones in good yields (Scheme 96). Reductions are generally performed at − 50 °C in a THF solution of $NaHFe_2(CO)_8$ and HOAc. Usually, two or more equivalents of the reagent are required for the reduction of 1 equivalent of substrate.

According to a detailed mechanistic study[306], the reaction involves concerted, reversible, regiospecific addition of $NaHFe_2(CO)_8$ to the C=C double bond of the enone, affording the corresponding binuclear iron enolate. Cleavage of the latter to the mononuclear iron enolate represents the rate determining step. Finally, protonolysis of this iron enolate by acetic acid provides the saturated ketone (Scheme 97).

C. Other Transition-metal Hydrides

The intermetallic hydride $LaNi_5H_6$ was found to be an effective reagent for conjugate reduction of enones. Reduction of the resulting saturated carbonyl compound occurs very slowly with this reagent, giving high yields of the 1, 4-reduction product[307].

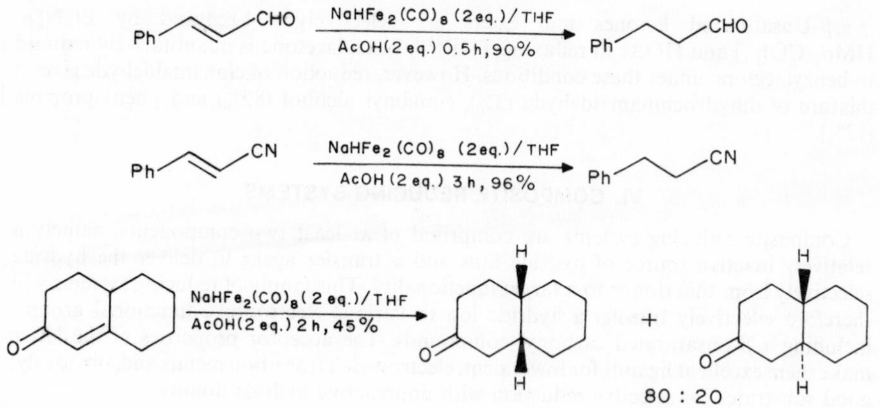

Ph⌒⌒CHO $\xrightarrow[\text{AcOH(2 eq.) 0.5 h, 90\%}]{\text{NaHFe}_2\text{(CO)}_8 \text{ (2eq.)/THF}}$ Ph⌒⌒⌒CHO

Ph⌒⌒CN $\xrightarrow[\text{AcOH (2 eq.) 3 h, 96\%}]{\text{NaHFe}_2\text{(CO)}_8 \text{ (2eq.)/THF}}$ Ph⌒⌒⌒CN

$\xrightarrow[\text{AcOH(2 eq.) 2 h, 45\%}]{\text{NaHFe}_2\text{(CO)}_8 (\text{2 eq.)/THF}}$ +   80 : 20

SCHEME 96

$\text{NaHFe}_2\text{(CO)}_8$ + R⌒⌒COR′ ⇌ ... $\text{HFe}_2\text{(CO)}_8$ Na

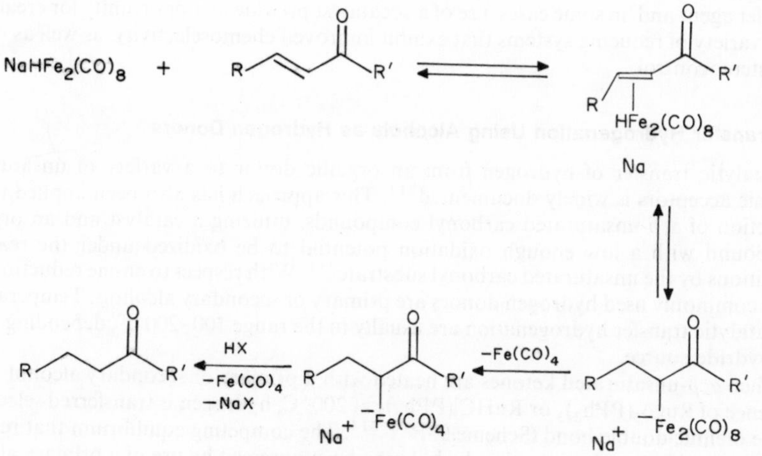

SCHEME 97

α, β-Unsaturated carbonyl compounds are reduced selectively and in good yields (55–80%) to the corresponding saturated derivatives by the hydridochromium complex $\text{NaHCr}_2\text{(CO)}_{10}$ in THF at 66 °C. This latter complex is prepared by stirring chromium-hexacarbonyl with potassium graphite (C_8K) in dry THF with subsequent addition of water[308].

Excess hydridocobaltcarbonyl reduces α, β-unsaturated ketones and aldehydes in moderate yield and good regioselectivity. The reaction involves complexation of the double bond to cobalt, followed by migratory insertion of hydride into the enone, forming an oxa-allyl cobalt complex[309]. Poor chemoselectivity is one of the major drawbacks of this reaction, as simple olefins are rapidly hydroformylated to the corresponding aldehyde under the reaction conditions (25 °C, 1 atm of CO).

α, β-Unsaturated ketones and esters are selectively 1,4-reduced by $Et_4N[\mu-HMo_2(CO)_{10}]$ and HOAc in refluxing THF[310]. Benzalacetone is quantitatively reduced to benzylacetone under these conditions. However, reduction of cinnamaldehyde gives a mixture of dihydrocinnamaldehyde (3%), cinnamyl alcohol (85%) and phenylpropane (12%).

VI. COMPOSITE REDUCING SYSTEMS

Composite reducing systems are comprised of at least two components, namely a relatively inactive source of hydride ions and a transfer agent to deliver the hydride selectively from that donor to a target functionality. This family of reducing systems will therefore selectively transfer a hydride ion to various electrophilic functional groups, including α, β-unsaturated carbonyl compounds. The acceptor properties of the latter make them excellent ligands for low-valent, electron-rich transition metals and, obviously, good substrates for selective reduction with nonreactive hydride donors.

Such multiple-component reducing systems offer high flexibility because they involve a large number of independent variables that can be tailored to various synthetic tasks, especially in comparison to metal hydride reduction which utilizes a single reagent. Thus, appropriate modification of the hydride donor, judicious selection of a transition metal transfer agent and, in some cases, use of a cocatalyst provide an opportunity for creating a wide variety of reducing systems that exhibit improved chemoselectivity, as well as regio- and stereocontrol.

A. Transfer Hydrogenation Using Alcohols as Hydrogen Donors

Catalytic transfer of hydrogen from an organic donor to a variety of unsaturated organic acceptors is widely documented[311]. This approach has also been applied to the reduction of α, β-unsaturated carbonyl compounds, utilizing a catalyst and an organic compound with a low enough oxidation potential to be oxidized under the reaction conditions by the unsaturated carbonyl substrate[311]. With respect to enone reduction, the most commonly used hydrogen donors are primary or secondary alcohols. Temperatures for catalytic transfer hydrogenation are usually in the range 100–200 °C, depending upon the hydride source.

When α, β-unsaturated ketones are heated with a primary or secondary alcohol in the presence of $RuCl_2(PPh_3)_3$ or $RuHCl(PPh_3)_3$ at 200 °C, hydrogen is transferred selectivity to the olefinic double bond (Scheme 98)[312–314]. The competing equilibrium that reduces the saturated ketone back to the alcohol may be suppressed by use of a primary alcohol such as benzyl alcohol or, more conveniently, by the use of boiling ethylene glycol, since saturated ketones are readily separated from insoluble glyoxal polymers[315]. Polyvinyl alcohol can also be used as convenient hydrogen donor[316]. α, β-Unsaturated ketones give higher yields than the corresponding aldehydes, which undergo self-condensation. α, β-Unsaturated esters undergo transesterification side-reactions with the donor alcohol.

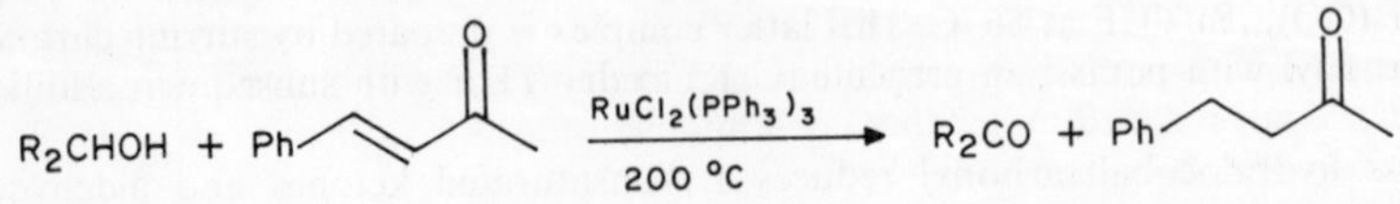

$$R_2CHOH + Ph \diagup\!\!\!\diagdown\!\!\!\diagup\!\!\!\diagdown\!\!\!\overset{O}{\overset{\|}{C}} \xrightarrow[\text{200 °C}]{RuCl_2(PPh_3)_3} R_2CO + Ph\diagup\!\!\!\diagdown\!\!\!\diagup\!\!\!\diagdown\!\!\!\overset{O}{\overset{\|}{C}}$$

SCHEME 98

Studies on the role of a Ru(II) catalyst as well as the mechanism of hydrogen transfer in enone reduction with benzyl alcohol at 170–190 °C revealed that $RuCl_2(PPh_3)_3$ is converted by the primary alcohol into $RuH_2(CO)(PPh_3)_3$, which then hydrogenates benzylideneacetone[317]. The kinetic data are compatible with the expression:

$$\text{reaction rate} = k_{obs}[Ru][enone][alcohol]$$

The rate-determining step of this reaction is generally assumed to be hydrogen transfer from the alcohol to a ruthenium species[317].

Transfer hydrogenation catalyzed by $RuCl_2(PPh_3)_3$ has been applied to the synthesis of cyclododecane-1, 2-dione in 53% yield from the corresponding 1, 2-diol using benzyl-ideneacetone as the hydrogen acceptor[318]. 5, 5-Dimethylcyclohexa-1, 3-dione reacts via its enol tautomer on heating with ethylene glycol in the presence of $RuCl_2(PPh_3)_3$ to give 3, 3-dimethylcyclohexanol, 3, 3-dimethylcyclohexanone and its corresponding ketal (Scheme 99)[319].

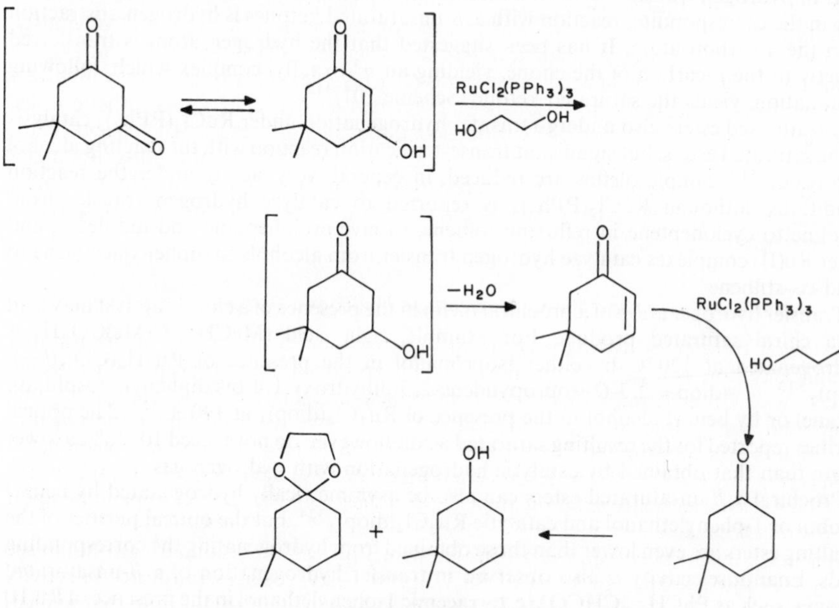

SCHEME 99

Vinyl ketones, such as methylvinyl ketone, are not reduced in the presence of $RuCl_2(PPh_3)_3$ on heating with common primary or secondary alcohols, but they are reduced on heating with allylic alcohols, such as hex-1-en-3-ol, using hydrated $RuCl_3$, $RuCl_2(PPh_3)_3$, $RuHCl(PPh_3)_3$, $RuH(OAc)(PPh_3)_3$ or, most efficiently, $Ru_3O(OAc)_7$ (Scheme 100)[320]. Surprisingly, other ketones, including acetophenone or benzylidene-acetone, are not reduced under these conditions.

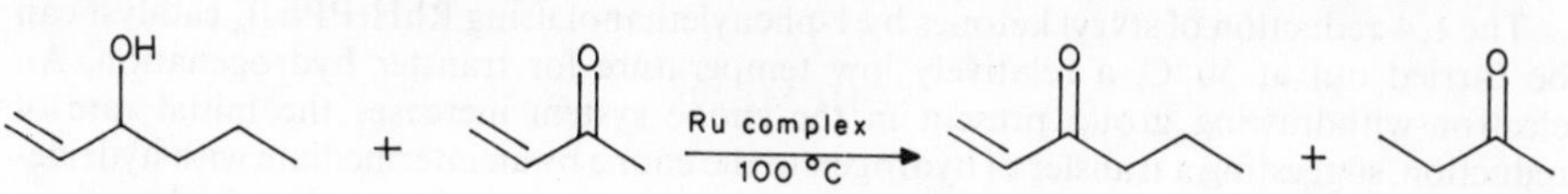

SCHEME 100

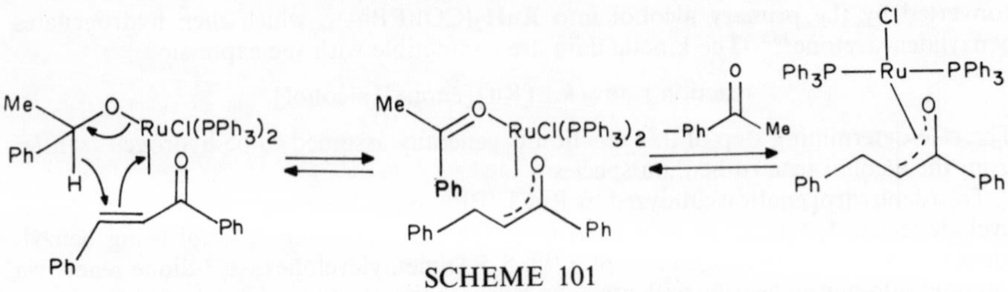

SCHEME 101

As in hydrogen transfer between alcohols and saturated ketones, the rate-determining step in the corresponding reaction with α, β-unsaturated ketones is hydrogen abstraction from the α-carbon atom. It has been suggested that the hydrogen atom is transferred directly to the β-carbon of the enone, yielding an η^3-oxaallyl complex which, following protonation, yields the saturated ketone (Scheme 101)[312].

Unsaturated esters also undergo trasnfer hydrogenation under $RuCl_2(PPh_3)_3$ catalysis to the saturated esters, but significant transesterification reaction with the reacting alcohol also occurs[313]. Simple olefins are reduced, in general, very slowly under the reaction conditions, although $RuCl_2(PPh_3)_3$ is reported to catalyze hydrogen transfer from indoline to cycloheptene in refluxing toluene, to give cycloheptane and indole[321], and other Ru(II) complexes catalyze hydrogen transfer from alcohols to diphenylacetylene to yield cis-stilbene[322].

Transfer hydrogenation of a prochiral olefin in the presence of a chiral catalyst may lead to a chiral saturated product. For example, tiglic acid ($MeCH=C(Me)CO_2H$) is hydrogenated at 120 °C by either isoprôpanol in the presence of $Ru_4H_4(CO)_8((-)$-diop)$_2$[323] (diop = 2, 3-O-isopropylidene-2, 3-dihydroxy-1, 4-bis(diphenylphosphino)-butane) or by benzyl alcohol in the presence of $Ru_2Cl_4(diop)_3$ at 190 °C[324]. The optical purities reported for the resulting saturated acids, however, do not exceed 10–15%, a lower figure than that obtained by catalytic hydrogenation with hydrogen gas.

Prochiral α, β-unsaturated esters can also be asymmetrically hydrogenated by benzyl alcohol or 1-phenylethanol and catalytic $Ru_2Cl_4(diop)_3$[324], but the optical purities of the resulting esters are even lower than those obtained from hydrogenating the corresponding acids. Enantioselectivity is also observed in transfer hydrogenation of α, β-unsaturated ketones, such as $PhCH=CHCOMe$, by racemic 1-phenylethanol in the presence of Ru(II) chloro complexes containing optically active tertiary phosphines, including diop and neomenthyldiphenylphosphine. Thus the optical purity of 1-phenylpropan-1-ol enriched in the S-$(-)$-isomer is 11% when reacted under these conditions with benzylideneacetone[325].

Asymmetric hydrogen transfer from optically active monosaccharides, such as 1, 2-α-D-glucofuranose, to prochiral enones is catalyzed by $RuCl_2(PPh_3)_3$ in diphenyl ether at 180 °C or by $RuH_2(PPh_3)_4$ in toluene at 100 °C (Scheme 102)[326].

Catalytic hydrogen transfer from sugars with free anomeric hydroxyl groups was studied with 2, 3; 5, 6-di-O-isopropylidene-D-mannofuranose and $RuH_2(PPh_3)_4$. In an excess of enone acceptor, these sugars were converted in high yields into the corresponding lactones (Scheme 103)[327].

The 1, 4-reduction of styryl ketones by 1-phenylethanol using $RhH(PPh_3)_4$ catalyst can be carried out at 50 °C, a relatively low temperature for transfer hydrogenation. An electron-withdrawing group present in the enone system increases the initial rate of reduction, suggesting a transfer of hydrogen to the enone by an intermediate with hydride-ion character[328]. Isotope labeling of the alcohol donors shows that hydrogen is regioselectively transferred from the carbinol carbon to the β-carbon of the enone, with the

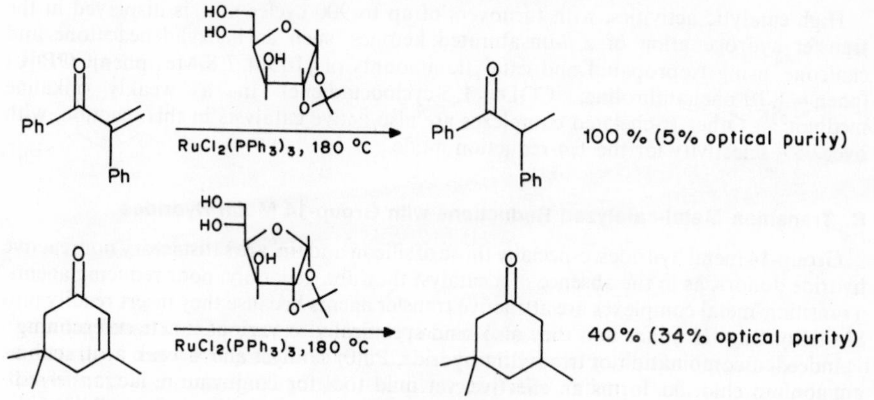

SCHEME 102

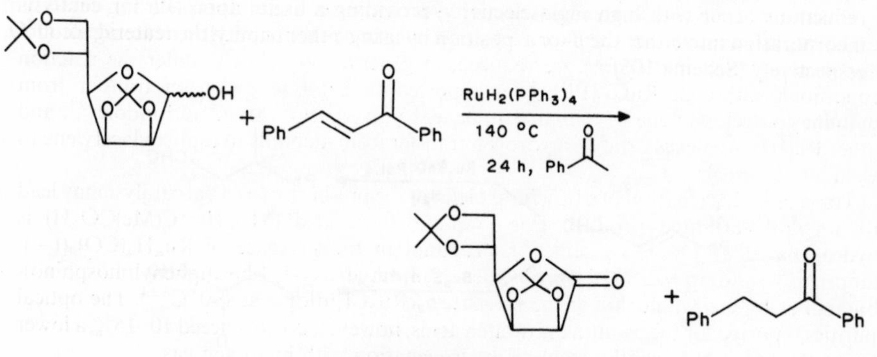

SCHEME 103

hydroxylic proton being transferred to the α-position (Scheme 104). Cleavage of an O—H bond is the rate-determining step in this reaction[329].

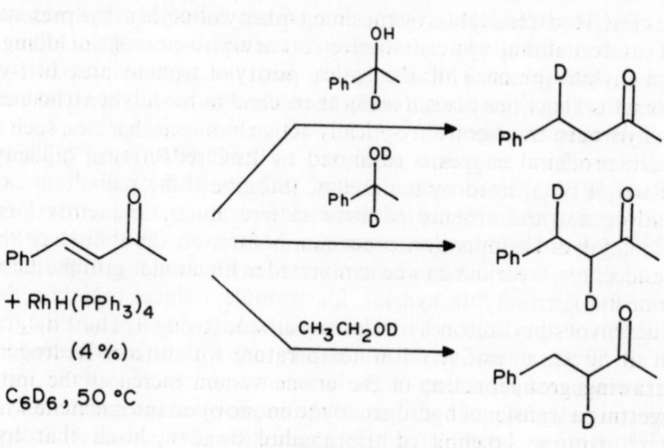

SCHEME 104

High catalytic activities, with turnovers of up to 900 cycles/min, is displayed in the transfer hydrogenation of α, β-unsaturated ketones, such as benzylideneacetone and chalcone, using isopropanol and catalytic amounts of [Ir(3, 4, 7, 8-Me$_4$-phen)COD]Cl (phen = 1, 10-phenanthroline; COD = 1, 5-cyclooctadiene) in a weakly alkaline medium[330]. Other Ir-chelated complexes are also active catalysts in this reaction, with over 95% selectivity for the 1, 4-reduction mode.

B. Transition Metal-catalyzed Reductions with Group-14 Metal Hydrides

Group-14 metal hydrides, especially those of silicon and tin, are satisfactory nonreactive hydride donors, as in the absence of a catalyst they are, generally, poor reducing agents. Transition-metal complexes are attractive transfer agents because they insert readily into Si—H or Sn—H bonds and they also bind specifically to various functional groups.

Indeed, a combination of tributyltin hydride, Pd(0) catalyst and a weak acid, such as ammonium chloride, forms an effective, yet mild tool for conjugate reduction of α, β-unsaturated aldehydes and ketones[331]. Similar results are obtained with other acidic cocatalysts, such as zinc chloride, acetic acid and tributyltin triflate[332]. With this system, reductions occur with high regioselectivity, providing a useful approach for deuterium incorporation into either the β- or α-position by using either tributyltin deuteride or D$_2$O, respectively (Scheme 105)[331].

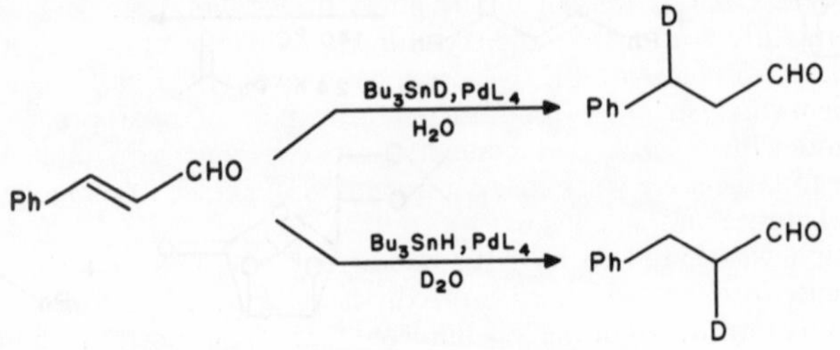

SCHEME 105

The above-described reducing system comprising tributyltin hydride and a soluble palladium(0) catalyst also allows chemoselective reductive cleavage of allylic heterosubstituents, even in the presence of aldehydes, benzylic acetate and benzylic chloride groups. These latter functions are normally as reactive as the allylic structure when using standard hydride reducing agents[333].

Silicon hydrides offer even greater selectivity in these reductions[334]. Their superiority over tin hydrides is manifested by the greater stability of the palladium catalyst in the reaction solution, and the absence of diene side-products, frequently formed via the competing Pd-catalyzed elimination processes. Moreover, the difference in reactivities between tin and silicon hydrides can be exploited for functional-group differentiation. In the presence of Pd(0), tributyltin hydride, for example, reduces rapidly α, β-unsaturated ketones and aldehydes but silicon hydrides are unable to do so. Thus, the treatment of a mixture of an allylic acetate and an unsaturated ketone with tin hydride and Pd(0) catalyst results in total conjugate reduction of the latter and nonreacted allylic acetate (Scheme 106)[334]. In contrast, employment of silicon hydride provided complementary chemoselectivity: allylic reduction was completed before reduction of the Michael acceptor could be detected.

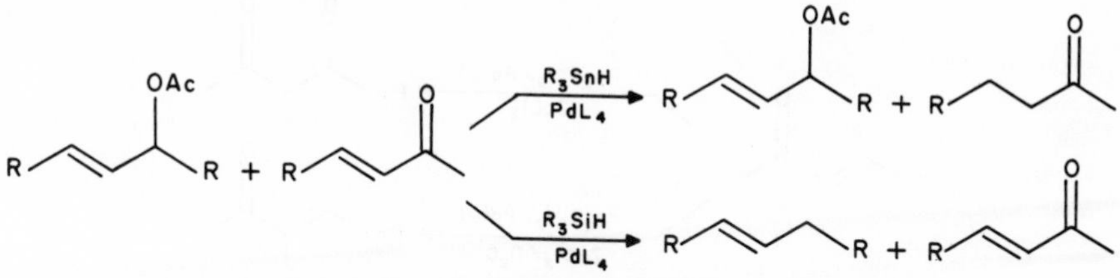

SCHEME 106

When using either tin or silicon hydrides, allylic substitution occurs with absolute inversion of configuration at the carbon, implying that hydride is initially transferred to palladium and from there to the allylic ligand via migratory insertion[333,334c]. This behavior is reminiscent of the proposed mechanism of the palladium-catalyzed conjugate reduction of enones (vide infra).

The useful flexibility characteristic of these multicomponent reducing systems is well illustrated by the silicon hydride/Pd(0) mixture. As mentioned above, this combination is essentially useless for reduction of electron-deficient olefins. However, addition of catalytic amounts of zinc chloride fundamentally alters the situation and creates a new three-component mixture that enables rapid conjugate reduction of α,β-unsaturated ketones and aldehydes[335]. In fact, soluble palladium complexes of various oxidation states were equally efficient catalysts, an obvious practical advantage of this approach. The generality of the method with respect to the substrate, its experimental simplicity and its easy applicability to large-scale work make it a method of choice for conjugate reduction of unsaturated ketones and aldehydes.

The reaction was found to be both regio- and stereoselective. In all cases where diphenyldideuteriosilane was used to reduce unsaturated ketones, deuterium was stereoselectively introduced at the less-hindered face of the substrate and regioselectively at the β-position (Scheme 107). Conversely, when reductions were carried out in the presence of traces of D_2O, deuterium incorporation occurred at the α-position[335].

Interestingly, this method is highly selective for unsaturated ketones and aldehydes, as reduction of corresponding α,β-unsaturated carboxylic acid derivatives, such as esters, amides and nitriles, is very sluggish under the conditions used. Thus, benzylideneacetone was selectively and cleanly reduced in the presence of methyl cinnamate, cinnamonitrile or cinnamamide[335].

Based on deuterium-incorporation experiments and ^{1}H NMR studies, a multistep catalytic cycle was postulated (Scheme 108) in which the first step is rapid, reversible coordination of the Pd(0)-phosphine complex to the electron-deficient olefin, resulting in complex I. Oxidative addition of silicon hydride to palladium in that complex forms hydrido-palladium olefin complex II. Migratory insertion of hydride into the electrophilic β-carbon of the coordinated olefin produces intermediate palladium enolate III which, via reductive elimination of the silicon moiety and enolate ligand, completes the catalytic hydrosilation cycle, resulting in silyl enol ether IV. The latter is prone to acid-catalyzed hydrolysis, yielding the saturated ketone[335].

The role of the Lewis acid cocatalyst is not yet fully understood. One may envision a number of points at which intervention of a Lewis acid could promote the reaction. It seems that in addition to its obvious role in catalyzing hydrolysis of the silyl enol ether, $ZnCl_2$ polarizes the substrate, thereby facilitating migratory insertion of hydride into the olefin (II to III in Scheme 108).

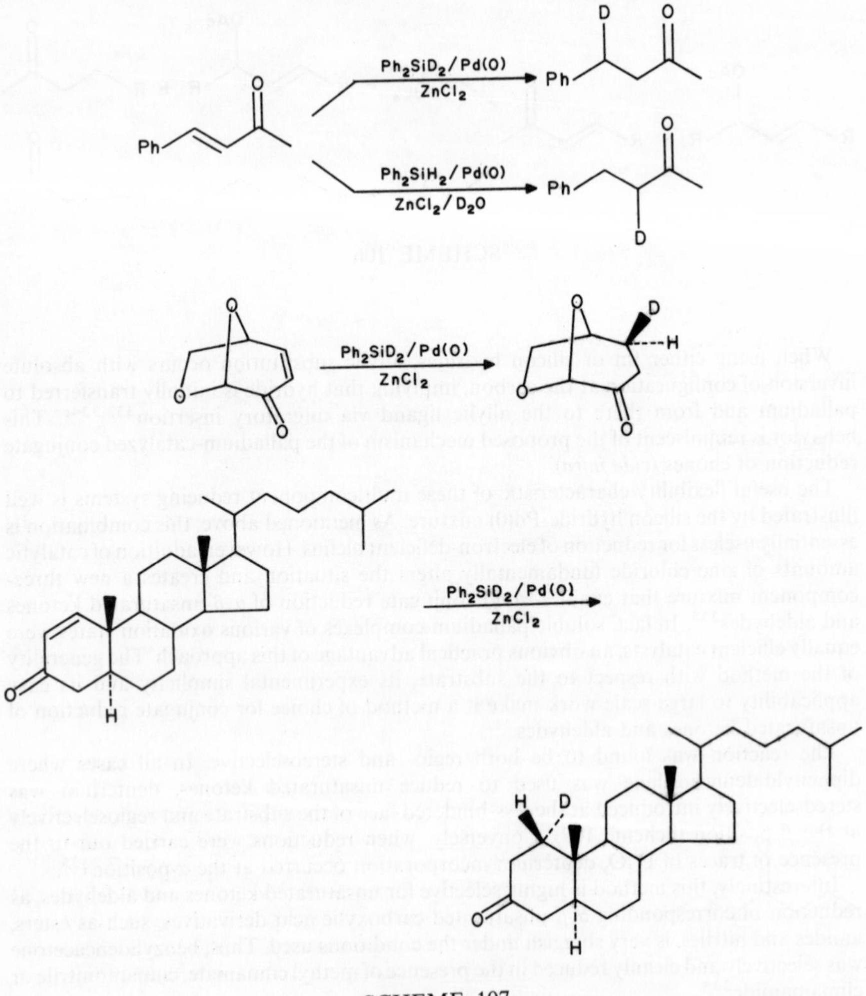

SCHEME 107

Combination of silicon hydrides with catalytic amounts of a ruthenium(II) complex in tetrahydrofuran, chloroform or benzene has afforded a new reducing system capable of efficient reduction of α, β-unsaturated carboxylic acids, esters, amides, etc[336]. Addition of a weak proton source, such as a sterically-hindered phenol, significantly increases reaction rates. The ruthenium mixture was found to exhibit the same regioselectivity observed with the above-described palladium systems.

The order of reactivity of this Ru/silane combination to various functional groups differs greatly from that of its Pd/silane/ZnCl$_2$ analog. While the latter is very useful for allylic reductions and essentially useless for unsaturated esters, the Ru-based system exhibits exactly opposite reactivity. A convincing demonstration of this complementary chemoselectivity is illustrated by the reduction of cinnamyl cinnamate (Scheme 109), a substrate containing both an allylic carboxylic and an α, β-unsaturated ester[336]. Each of

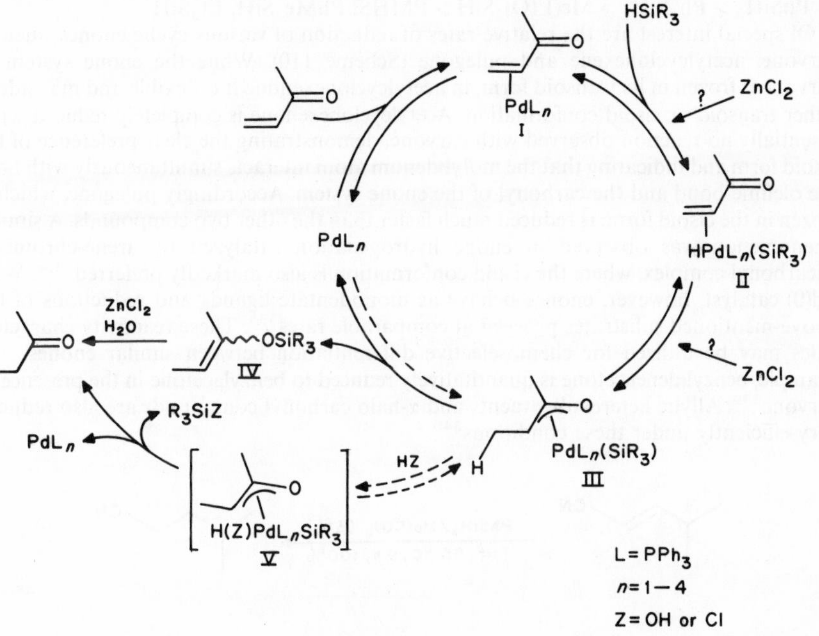

SCHEME 108

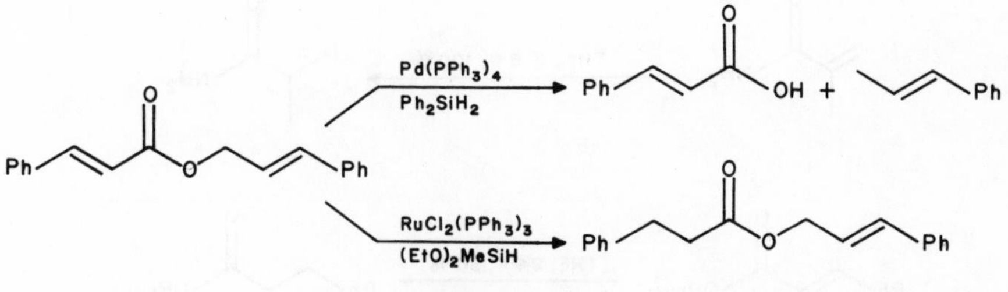

SCHEME 109

these can be reduced separately by silicon hydride and the appropriate transition-metal catalyst.

Early transition-metal complexes, including those of group 6, have been rarely used to catalyze transfer hydrogenation[337] and hydrogenation with hydrogen gas[338] and, in particular, little is known about hydrosilation with these catalysts. Under mild thermal conditions, catalytic amounts of $Mo(CO)_6$ and phenylsilane engender a powerful reducing system, suitable for conjugate reduction of α, β-unsaturated ketones, carboxylic acids, esters, amides, etc. The mixture is especially useful for conjugate reduction of unsaturated nitriles, usually difficult to reduce with other media (Scheme 110)[339]. Although the reaction also works with mono- and dihydridosilanes, the general order of silane reactivity is: $PhSiH_3 > Ph_2SiH_2 > Me(EtO)_2SiH > PMHS, PhMe_2SiH, Et_3SiH$.

Of special interest are the relative rates of reduction of various cyclic enones, such as carvone, acetylcyclohexene and pulegone (Scheme 110). While the enone system in carvone is frozen in its transoid form, in acetylcyclohexenone it is flexible and may adopt either transoid or cisoid conformation. Acetylcyclohexenone is completely reduced while essentially no reaction observed with carvone, demonstrating the clear preference of the cisoid form and indicating that the molybdenum atom interacts simultaneously with both the olefinic bond and the carbonyl of the enone system. Accordingly pulegone, which is frozen in the cisoid form, is reduced much faster than the other two compounds. A similar phenomenon was observed in enone hydrogenation catalyzed by arene-chromium tricarbonyl complex, where the cisoid conformation is also markedly preferred[338c]. With Pd(0) catalyst, however, enones behave as monodentate ligands and reductions of the above-mentioned substrates proceed at comparable rates[335]. These reactivity character-istics may be utilized for chemoselective differentiation between similar enones. For example, benzylideneacetone is quantitatively reduced to benzylacetone in the presence of carvone[339]. Allylic heterosubstituents and α-halo carbonyl compounds are also reduced very efficiently under these conditions[340].

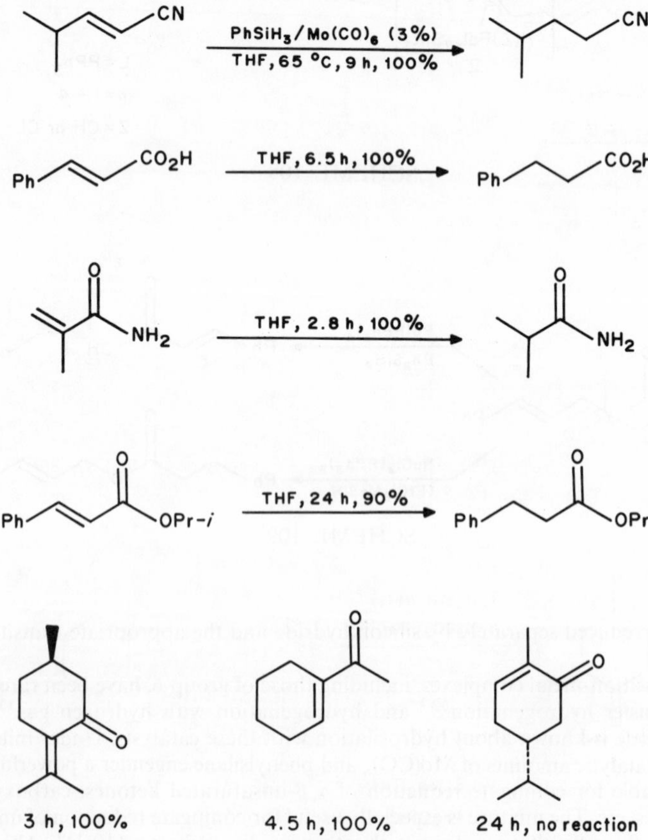

SCHEME 110

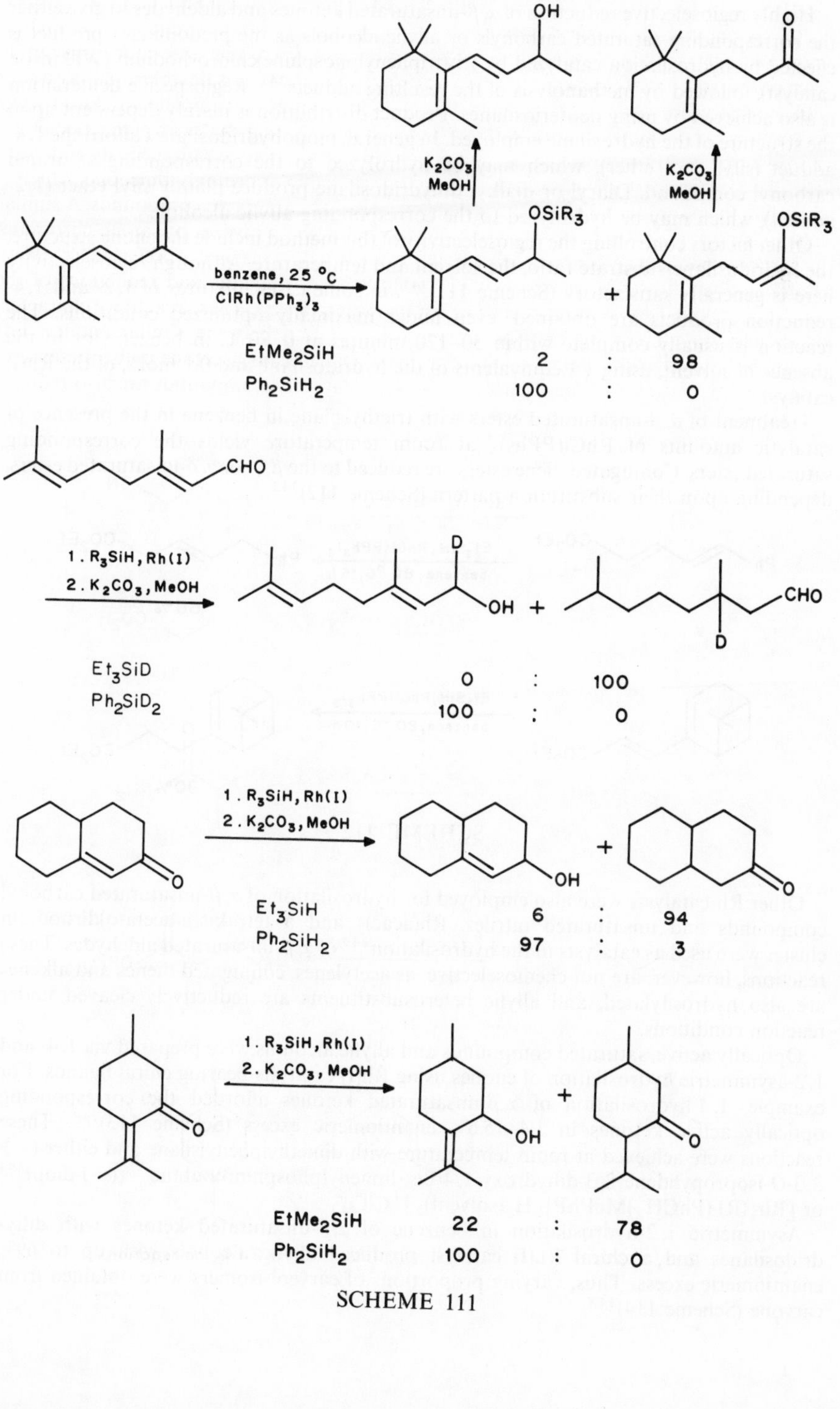
OH
O
K_2CO_3
MeOH
K_2CO_3
MeOH
O
$OSiR_3$
$OSiR_3$
benzene, 25 °C
$ClRh(PPh_3)_3$
+
$EtMe_2SiH$ 2 : 98
Ph_2SiH_2 100 : 0
CHO
1. R_3SiH, Rh(I)
2. K_2CO_3, MeOH
D
OH
+
CHO
D
Et_3SiD 0 : 100
Ph_2SiD_2 100 : 0
1. R_3SiH, Rh(I)
2. K_2CO_3, MeOH
O
OH
+
O
Et_3SiH 6 : 94
Ph_2SiH_2 97 : 3
1. R_3SiH, Rh(I)
2. K_2CO_3, MeOH
O
OH
+
O
$EtMe_2SiH$ 22 : 78
Ph_2SiH_2 100 : 0

SCHEME 111

Highly regioselective reduction of α, β-unsaturated ketones and aldehydes to give either the corresponding saturated carbonyls or allylic alcohols as the predominant product is effected by hydrosilation catalyzed by tris(triphenylphosphine)chlororhodium (Wilkinson catalyst), followed by methanolysis of the resulting adducts[341]. Regiospecific deuteriation is also achieved by using deuteriosilanes. Product distribution is mainly dependent upon the structure of the hydrosilane employed. In general, monohydridosilanes afford the 1, 4-adduct (silyl enol ether), which may be hydrolyzed to the corresponding saturated carbonyl compound. Diaryl or dialkyl dihydridosilane produce mainly silyl ether (1, 2-adduct), which may be hydrolyzed to the corresponding allylic alcohol.

Other factors controlling the regioselectivity of this method include the enone structure, the hydridosilane/substrate ratio, the solvent and temperature. Although regioselectivity here is generally satisfactory (Scheme 111)[341], in some cases mixtures of 1, 2- and 1, 4-reduction products are obtained, even under maximally optimized conditions. The reaction is usually complete within 30–120 minutes at 0–80 °C in benzene, or in the absence of solvent, using 1.1 equivalents of the hydridosilane and 0.1 mol% of the Rh(I) catalyst.

Treatment of α, β-unsaturated esters with triethylsilane in benzene in the presence of catalytic amounts of $RhCl(PPh_3)_3$ at room temperature yields the corresponding saturated esters. Conjugated diene esters are reduced to the β, γ- or γ, δ-unsaturated esters, depending upon their substitution pattern (Scheme 112)[342].

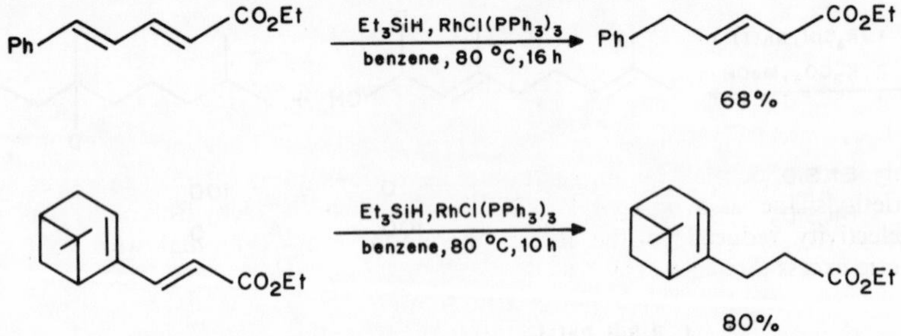

SCHEME 112

Other Rh catalysts were also employed for hydrosilation of α, β-unsaturated carbonyl compounds and unsaturated nitriles. $Rh(acac)_2$ and a tetrakis(μ-acetato)dirhodium cluster were used as catalysts in the hydrosilation[343] of α, β-unsaturated aldehydes. These reactions, however, are not chemoselective, as acetylenes, conjugated dienes and alkenes are also hydrosilylated, and allylic heterosubstituents are reductively cleaved under reaction conditions.

Optically active, saturated compounds and allylic alcohols were prepared via 1, 4- and 1, 2-asymmetric hydrosilation of enones using Rh(I) catalysts bearing chiral ligands. For example, 1, 4-hydrosilation of α, β-unsaturated ketones afforded the corresponding optically active ketones in 1.4–15.6% enantiomeric excess (Scheme 113)[344]. These reactions were achieved at room temperature with dimethylphenylsilane and either (−)-2, 3-O-isopropylidene-2, 3-dihydroxy-1, 4-bis(diphenylphosphino)butane ((−)-diop)[344] or $[Rh\{(R)-(PhCH_2)MePhP\}_2 H_2(solvent)_2]^+ ClO_4^-$.

Asymmetric 1, 2-hydrosilation in benzene of α, β-unsaturated ketones with dihydridosilanes and a chiral Rh(I) catalyst produced allylic alcohols with up to 69% enantiomeric excess. Thus, varying proportions of carveol isomers were obtained from carvone (Scheme 114)[345].

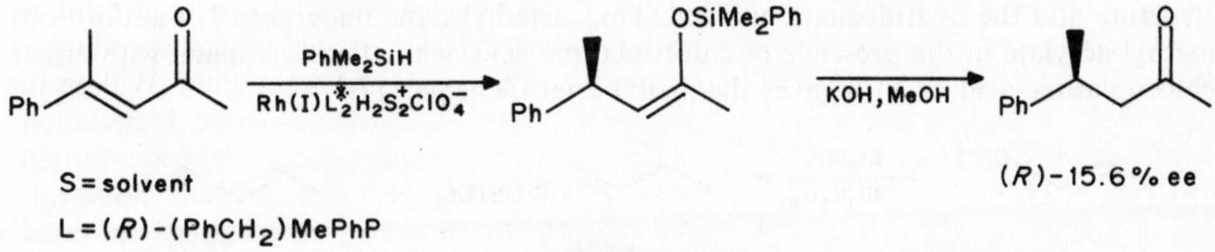
PhMe2SiH
Rh(I)L*2H2S2+ClO4−
OSiMe2Ph
KOH, MeOH
(R)-15.6% ee
S = solvent
L = (R)-(PhCH2)MePhP

SCHEME 113

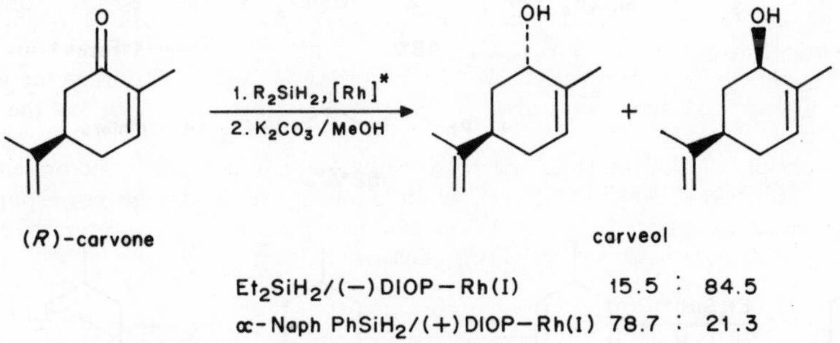
1. R2SiH2, [Rh]*
2. K2CO3/MeOH
(R)-carvone
carveol
Et2SiH2/(−)DIOP−Rh(I) 15.5 : 84.5
α-Naph PhSiH2/(+)DIOP−Rh(I) 78.7 : 21.3

SCHEME 114

Highly stereoselective 1,2-hydrosilation of an α, β-unsaturated aldehyde was achieved with triethylsilane and nonchiral Wilkinson catalyst[346]. Dehydrofaranal was thus stereoselectivity reduced to the insect pheromone (3*S*, 4*R*)-faranal with 85% diastereomeric excess (Scheme 115).

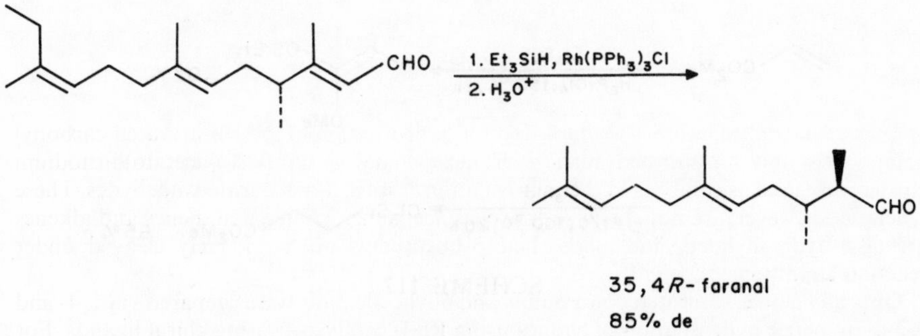
CHO
1. Et3SiH, Rh(PPh3)3Cl
2. H3O+
CHO
3S,4R-faranal
85% de

SCHEME 115

The main product in hydrosilation of α, β-unsaturated ketones and aldehydes catalyzed by chloroplatinic acid, platinum on alumina, or metallic nickel is the corresponding silyl enol ether[347]. With nickel catalyst, product distribution is highly dependent on the enone structure, as exemplified in Scheme 116[348].

Hydridosilanes add to α, β-unsaturated esters, producing the corresponding silyl enolate as well as carbon silylated products. The course of addition depends on substrate structure and the hydridosilane utilized. Thus, triethylsilane undergoes 1,4-addition to methyl acrylate in the presence of chloroplatinic acid, while trichlorosilane with either chloroplatinic acid or Pt/C gives the β-silyl ester (Scheme 117)[349].

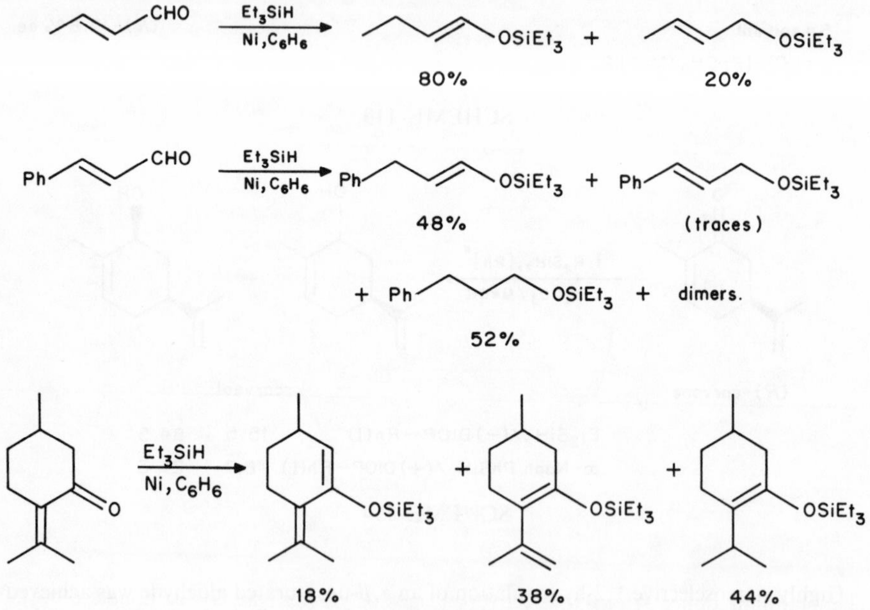

80%   20%

48%   (traces)

+ Ph⌃⌃⌃OSiEt₃ + dimers.

52%

18%   38%   44%

SCHEME 116

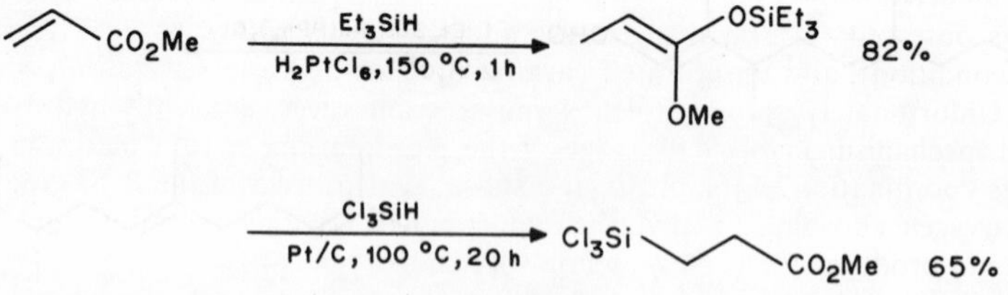

SCHEME 117

This approach was successfully applied to the total synthesis of d,l-muscone[350]. Treatment of the α,β- and β,γ-enone mixture (Scheme 118) with triethylsilane in refluxing glyme containing catalytic amounts of chloroplatinic acid afforded 1-triethyl-silyloxycyclotetradecene. The two isomeric enones rapidly equilibrate under these conditions.

Selective reduction of pregna-14,16-dien-20-ones to pregn-14-en-20-ones is achieved via hydrosilation with tetramethyldisiloxane and catalytic amounts of chloroplatinic acid (Scheme 119)[351]. α,β-Unsaturated esters are also reduced to the corresponding saturated esters under these conditions[352].

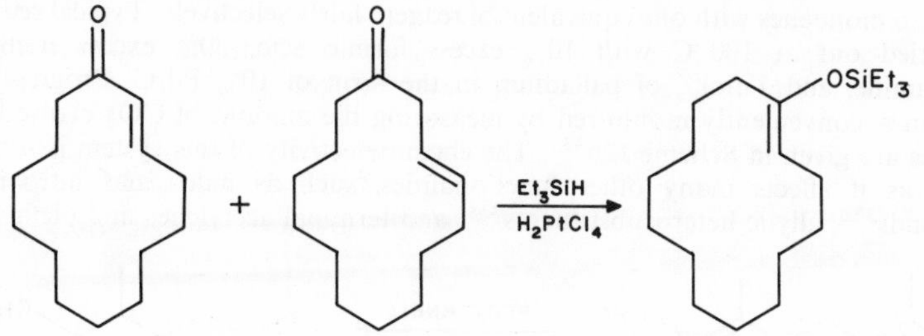

SCHEME 118

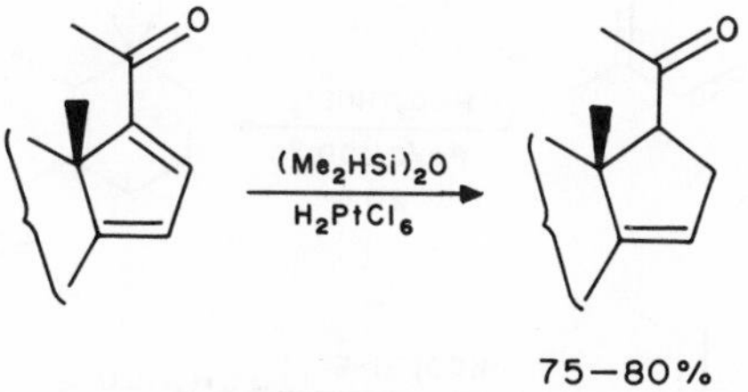

75—80%

SCHEME 119

The platinum dimer $(Pt(\mu\text{-}H)(SiR_3)(PR'_3))_2$ also catalyzes the hydrosilation of α, β-unsaturated aldehydes and ketones. Several aldehydes and ketones were hydrosilated in high yield in the presence of this dimer[353] at 60–100 °C and trialkylsilanes, including $MePh_2SiH$, $EtMe_2SiH$ and Et_3SiH. Triethoxysilane, was inert under these reaction conditions. Excellent regioselectivity was generally observed except in cases of highly sterically hindered enones such as tetraphenylcyclopentadienone, where the 1,2-reduction mode was observed. Saturated aldehydes and ketones were not reduced under these reaction conditions, and unsaturated carboxylic acids and esters were only sluggishly reduced. Unfortunately, terminal olefins and acetylenes were efficiently hydrosilated. A suggested mechanism involves cleavage of the platinum dimer to a platinum hydride species, its coordination to the olefin, and subsequent transfer of the R_3Si group to the carbonyl oxygen, affording a π-allyl platinum complex. Hydride migration from Pt to the allylic ligand produces the corresponding silyl enol ether.

C. Transition Metal-catalyzed Reductions with Other Hydrogen Donors

Aldehydes such as α-naphthaldehyde, p-tolualdehyde or p-chlorobenzaldehyde and DMF can serve as hydrogen donors and transfer their formyl hydrogen to α, β-unsaturated ketones in the presence of $RuCl_2(PPh_3)_3$. However, in some cases, decarbonylation of the aldehyde is so severe that no transfer hydrogenation is observed[354].

A particularly convenient hydrogen donor is formic acid, which not only hydrogenates α, β-unsaturated ketones[355], but also terminal olefins in the presence of a variety of ruthenium complexes under mild conditions[356].

Trialkylammonium formate and catalytic amounts of palladium on carbon form a convenient reducing system for reduction of a number of organic functional groups, including α, β-unsaturated aldehydes, ketones and esters[357]. Conjugated dienes are reduced to monoenes with one equivalent of reagent fairly selectively. Typical reductions are carried out at 100 °C with 10% excess formic acid, 30% excess triethyl- or tributylamine, and 1 mol% of palladium in the form of 10% Pd/C. Progress of the reduction is conveniently monitored by measuring the amount of CO_2 evolved. Some examples are given in Scheme 120[357]. The chemoselectivity of this system is somewhat limited, as it affects many other functionalities, such as halo- and nitroaromatic compounds[358], allylic heterosubstituents[359], and terminal acetylenes and olefins[357].

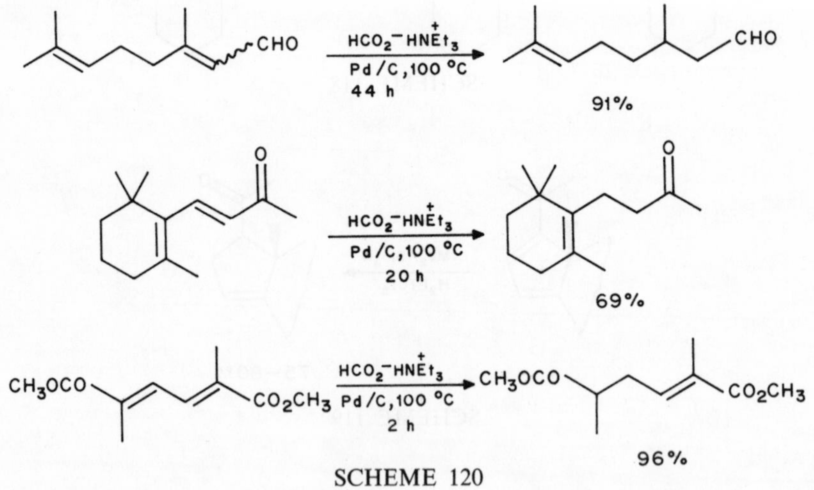
CHO
HCO2⁻ HNEt3
Pd/C, 100 °C
44 h
CHO
91%
HCO2⁻ HNEt3
Pd/C, 100 °C
20 h
69%
CH3OCO
CO2CH3
HCO2⁻ HNEt3
Pd/C, 100 °C
2 h
CH3OCO
CO2CH3
96%

SCHEME 120

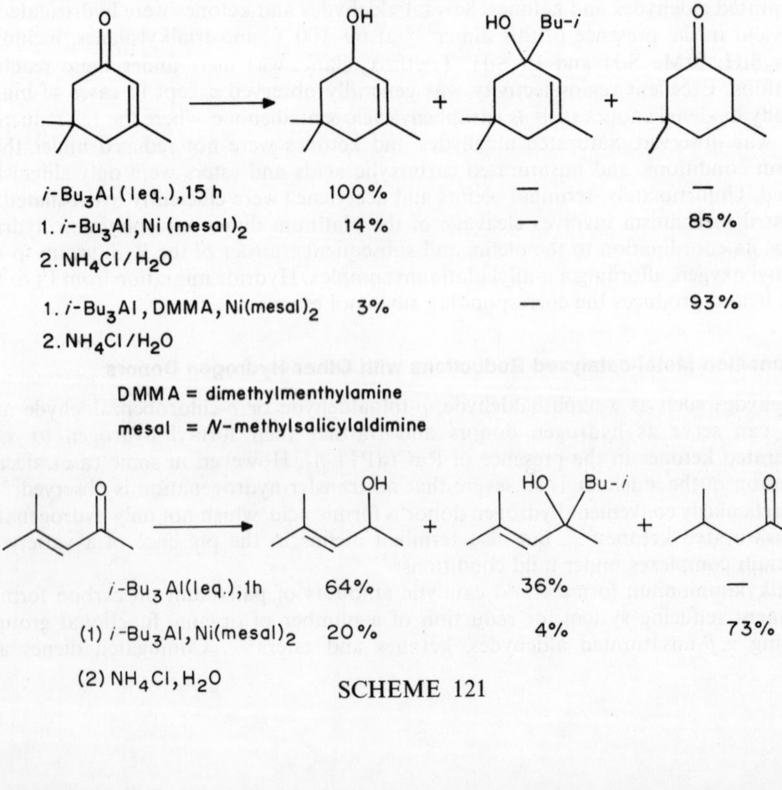
O
OH
HO Bu-i
O
i-Bu3Al (1eq.), 15 h 100% — —
1. i-Bu3Al, Ni(mesal)2 14% — 85%
2. NH4Cl/H2O
1. i-Bu3Al, DMMA, Ni(mesal)2 3% — 93%
2. NH4Cl/H2O
DMMA = dimethylmenthylamine
mesal = N-methylsalicylaldimine
O
OH
HO Bu-i
O
i-Bu3Al(1eq.), 1h 64% 36% —
(1) i-Bu3Al, Ni(mesal)2 20% 4% 73%
(2) NH4Cl, H2O

SCHEME 121

The reaction between triisobutylaluminum and α, β-unsaturated ketones, in pentane at room temperature, leads to products which correspond to a 1, 2-addition processes. The extent of such reactions depends both on the structure of the enone and of the concentration ratio between reagent and substrate. Under these experimental conditions, bis(N-methylsalicylaldimine)nickel catalyzes conjugate reduction of α, β-unsaturated ketones by triisobutylaluminum[360]. The cyclic and acyclic saturated ketones are obtained in 40–90% yield, the lower figure corresponding to enones substituted at the α-position (Scheme 121). In all cases, 1, 2-reduction products were also obtained (probably via noncatalyzed reduction) and, in some cases, side-products containing an isobutyl group were also formed. The reaction is interpreted in terms of a catalytic cycle involving a hydridonickel intermediate formed by reaction of i-Bu₃Al with the nickel complex. Addition of the hydridonickel to the olefin affords a nickel enolate that undergoes transmetallation, to aluminum enolate. The latter is finally hydrolyzed to the saturated ketone.

A number of composite reducing systems comprised of heterogeneous mixtures of transition metal salts, sodium alkoxides and sodium hydride were developed, which are

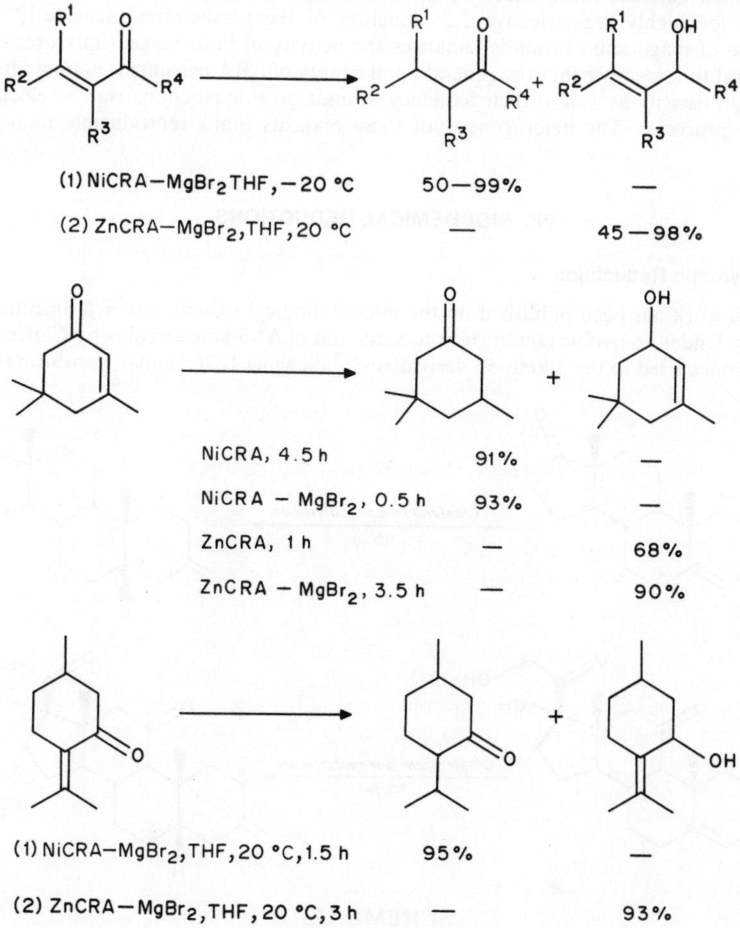

(1) NiCRA—MgBr₂THF, −20 °C	50—99%	—
(2) ZnCRA—MgBr₂, THF, 20 °C	—	45—98%

NiCRA, 4.5 h	91%	—
NiCRA − MgBr₂, 0.5 h	93%	—
ZnCRA, 1 h	—	68%
ZnCRA − MgBr₂, 3.5 h	—	90%

(1) NiCRA—MgBr₂, THF, 20 °C, 1.5 h	95%	—
(2) ZnCRA—MgBr₂, THF, 20 °C, 3 h	—	93%

SCHEME 122

useful for reduction of various organic functional groups[361]. In organic chemistry, sodium hydride is generally used as a base for proton abstraction. Although some substrates can be reduced by NaH, it is by itself a poor reducing agent.

Typical reducing systems (known as complex reducing agents, CRA)[361] are prepared from a transition-metal chloride or acetate, sodium *tert*-amyloxide and sodium hydride (in 1:1:4 ratio) in either THF or DME. Obviously, neither the exact structure of the actual reducing entity nor their reduction mechanism is fully understood.

The CRA reagents involving nickel salts exhibit reducing properties that are significantly different from those of the corresponding CRA prepared from zinc or magnesium salts. It was demonstrated that the three-component mixture, $NaH/RONa/Ni(OAc)_2$ (NiCRA), reduces carbon–carbon double bonds[362]. Conversely, the mixture $NaH/RONa/ZnCl_2$ (ZnCRA) reduces olefins poorly but effectively reduces saturated carbonyl functionalities, particularly when mixed with alkaline- or alkaline earth-metal salts[363]. These observations led to the expected complementary regioselectivity when reducing α, β-unsaturated carbonyl compounds with these reagents.

Indeed, NiCRA exhibits very high regioselectivity for 1,4-reduction of a number of α, β-unsaturated ketones, while under the same conditions ZnCRA is an effective reagent mixture for highly regioselective 1,2-reduction of these substrates (Scheme 122)[364]. Addition of magnesium bromide enhances the activity of both reagent mixtures. It is important to remember that the general applicability of CRA reagents is limited, due to their high basicity as well as their tendency to undergo side-reactions via one electron-transfer processes. The heterogeneity of these reagents limits reproducible reduction yields.

VII. BIOCHEMICAL REDUCTIONS

A. Enzymatic Reductions

Much work has been published on the microbiological reduction of α, β-unsaturated ketones. Under anaerobic conditions the reduction of Δ^4-3-keto steroids by *Clostridium paraputrificum* led to the 3-keto-5β derivatives[365] (Scheme 123). Similar transformations

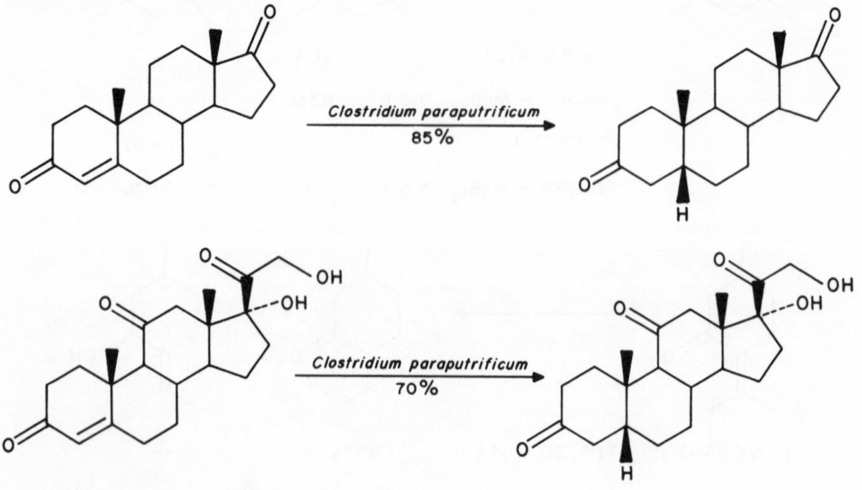

Clostridium paraputrificum
85%

Clostridium paraputrificum
70%

SCHEME 123

were observed previously with *Bacillus putrificus*[366], *Penicillium decumbens*[367], *Rhizopus nigricans*[368] or *Aspergillus niger*[369]. In most cases further reduction led to the corresponding 3α-hydroxy-5β derivatives.

Highly enantioselective conjugate reductions of substituted cyclopentenones and cyclohexenones were reported by Kergomard using *Beauveria sulfurescens* (ATCC 7159) under anaerobic conditions[370]. The reaction takes place only with substrates containing a small substituent in the α-position and hydrogen in the β-position. The saturated ketones obtained were, in some cases, accompanied by saturated alcohols. A number of useful transformations, including enantioselective reductions of acyclic substrates, are illustrated in Scheme 124.

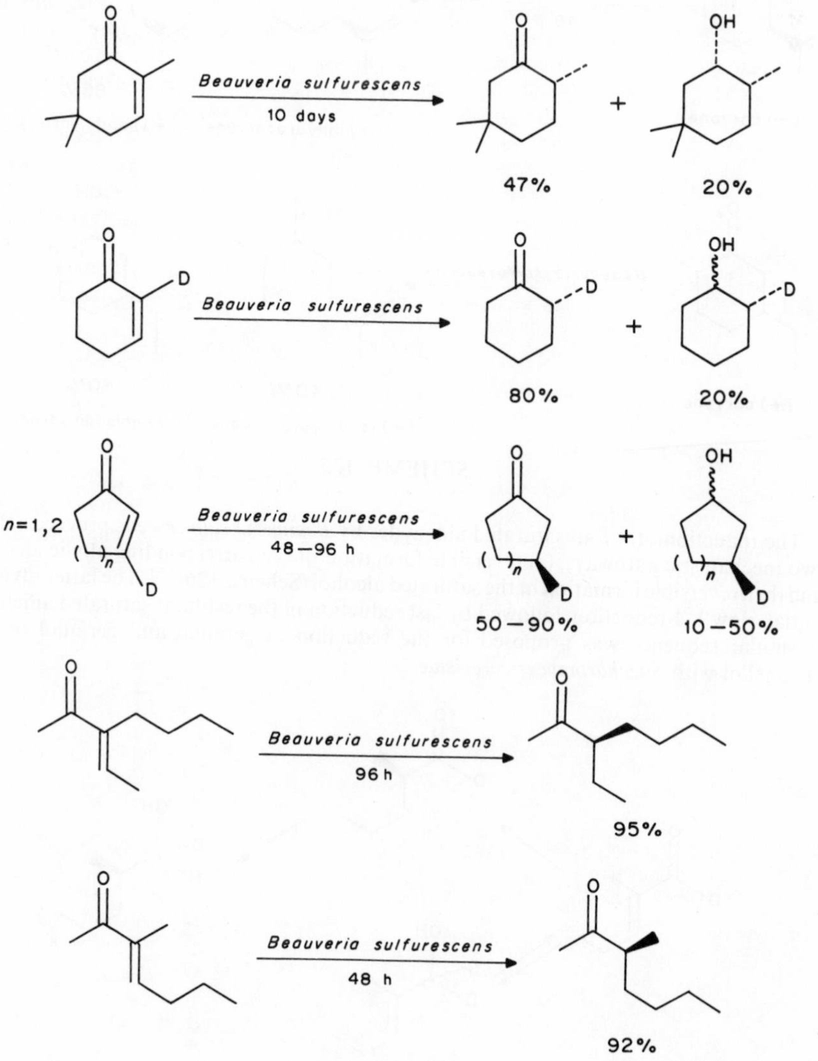

SCHEME 124

Both naturally occurring enantiomers of carvone were selectively reduced by *B. sulfurescens* (Scheme 125). (−)-Carvone was reduced to (+)-dihydrocarvone (*trans*) and further to (−)-neodihydrocarveol, whereas (+)-carvone was reduced to (−)-isodihydrocarvone (*cis*), which was then converted to (−)-neoisodihydrocarveol[371]. Similar reductions with identical stereoselectivities were observed earlier with *Pseudomonas ovalis* (strain 6–1) and with a strain of *Aspergillus niger*[371].

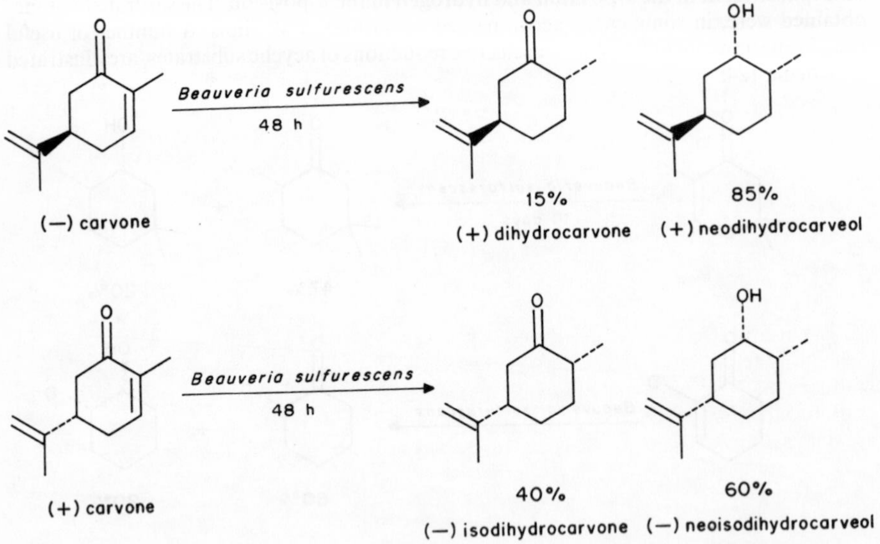

(−) carvone

(+) dihydrocarvone   (+) neodihydrocarveol

15%   85%

(+) carvone

(−) isodihydrocarvone   (−) neoisodihydrocarveol

40%   60%

SCHEME 125

The reduction of α, β-unsaturated aldehydes by *Beauveria sulfurescens* proceeds along two mechanistic pathways: (a) reversible formation of the corresponding allylic alcohols and (b) irreversible formation of the saturated alcohol (Scheme 126)[372]. The latter involves initial, slow 1,4-reduction, followed by fast reduction of the resultant saturated aldehyde. A similar sequence was proposed for the reduction of geranial and geraniol to (R)-citronellol with *Saccharomyces cerevisiae*.

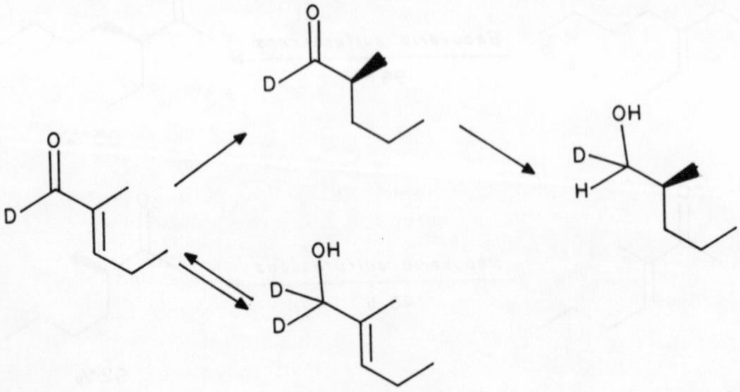

SCHEME 126

The above-described reducing characteristics of *B. sulfurescens* were found to be a general phenomenon exhibited by many types of eukaryotic organisms (six fungi) and prokaryotes (more than 20 Actinomycetes and Clostridium species)[373]. For example, in conjugate reduction of cyclohexenone derivatives the addition of two hydrogen atoms across the olefin occurs with *trans* stereochemistry, as shown in Scheme 127 where X represents a small alkyl group and Y a hydrogen atom. In all cases, the 1,4-reduction mode was completed within 48 hours. As these characteristics are shared by many organisms, it was suggested that they all contain very similar reducing enzymes[373].

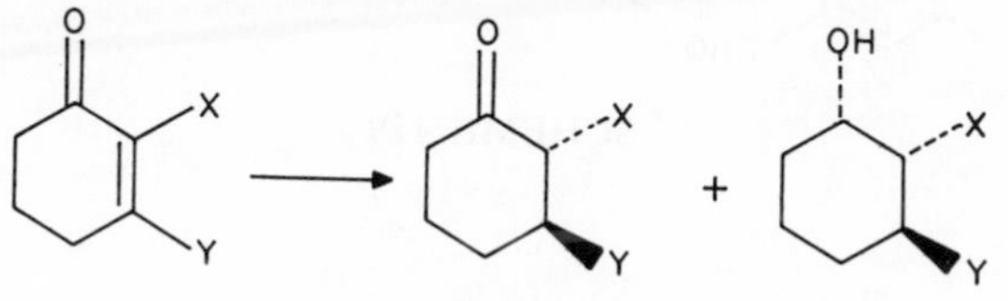

SCHEME 127

α,β-Unsaturated ketones bearing perfluoroalkyl groups are reduced by baker's yeast (Scheme 128)[374]. Perfluoroalkyl alkenyl ketones give mainly the saturated ketone, along with a small amount of optically active saturated alcohol. Substrates having a perfluoroalkyl group attached to the alkene moiety give mixtures of optically active allylic as well as saturated alcohols, whose relative concentration is time-dependent.

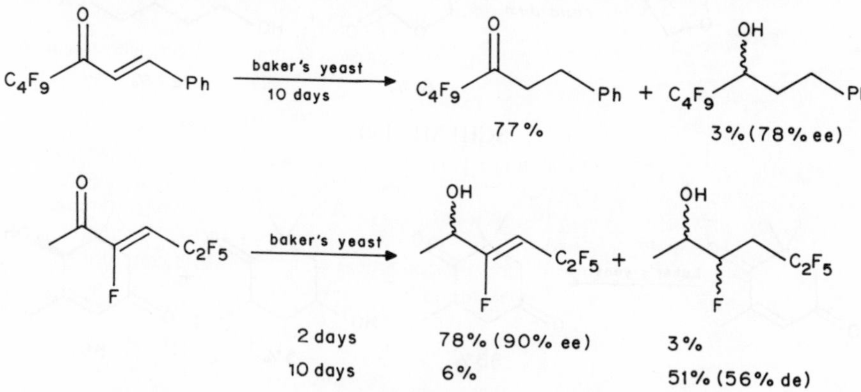

SCHEME 128

Unsaturated aldehydes derived from citronellol and geraniol are also reduced by baker's yeast to the corresponding saturated primary alcohols with very high enantioselectivity (Scheme 129)[375].

Two key chiral building blocks used in the total synthesis of α-tocopherol were prepared via microbial reduction of unsaturated carbonyl compounds with baker's yeast and with *Geotrichum candidum*, as illustrated in Scheme 130[376].

Similarly, a key intermediate in the total synthesis of optically active natural carotenoids was prepared by microbial reduction of oxo-isophorone with baker's yeast (Scheme 131)[377].

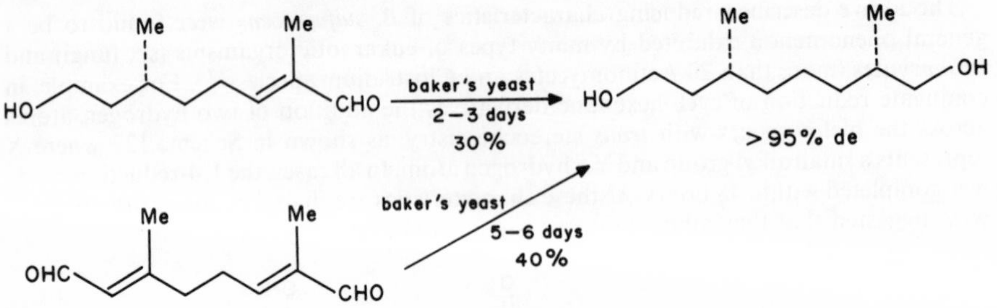

SCHEME 129

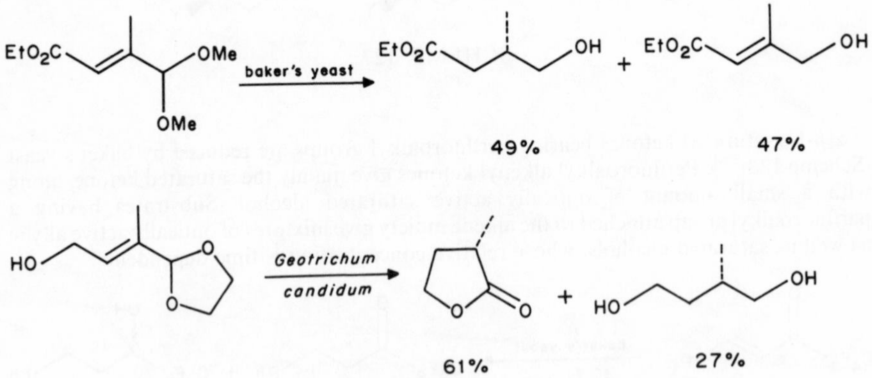

SCHEME 130

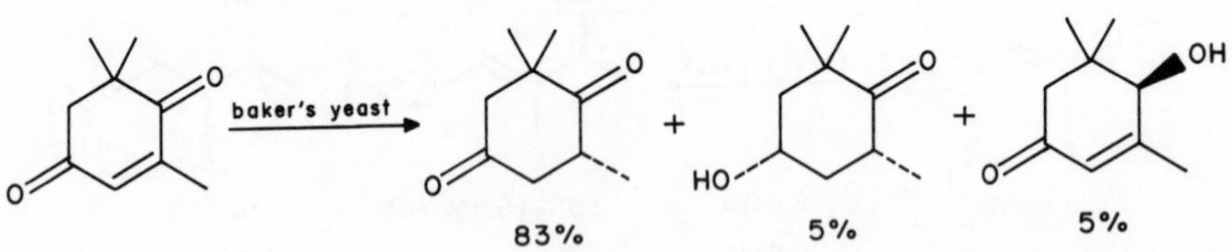

SCHEME 131

An alternative approach to the synthesis of α-tocopherol employs a chiral building block that was obtained by baker's yeast reduction of 2-methyl-5-phenylpentadienal (Scheme 132)[378].

SCHEME 132

Microbial reduction of enones has been applied to prostaglandin synthesis. For example, enantioselective reduction of the enone system in $\Delta^{8(12)}$-15-dehydro-PGE$_1$ with *Flavobacterium* sp. (NRRL B-3874) provided optically pure $(-)$-15-*epi*-$\Delta^{8(12)}$-PGE$_1$ (Scheme 133)[379].

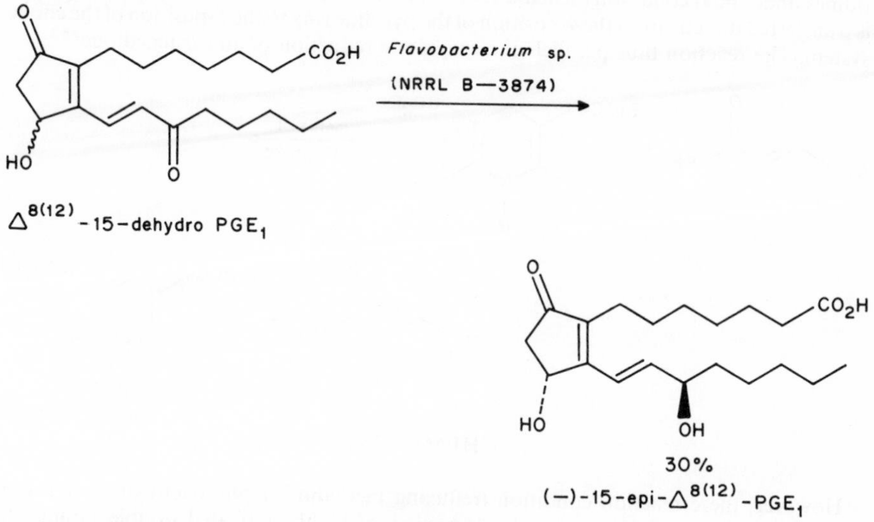

$\Delta^{8(12)}$ -15-dehydro PGE$_1$

30%

$(-)$-15-epi-$\Delta^{8(12)}$-PGE$_1$

SCHEME 133

As a general rule of enzymatic reductions, the 1,4-reduction of enones is preferred over the 1,2-reduction mode. However, when an electronegative substituent, such as halogen, is introduced that stabilizes the double bond, enzymatic reduction to allylic alcohols may be achieved[276]. A 1,2-reduction of a β-iodo enone is illustrated in Scheme 134.

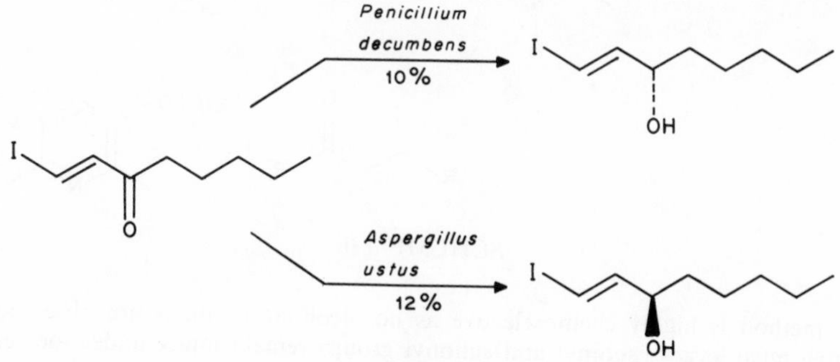

Penicillium decumbens
10%

Aspergillus ustus
12%

SCHEME 134

B. Biomimetic Reductions with NAD(P)H Models

A number of pyridine nucleotide-linked dehydrogenases catalyze the reversible hydrogenation–dehydrogenation of the double bond in α, β-unsaturated ketones[380]. Similar biomimetic conjugate reduction of α, β-unsaturated aldehydes and ketones occurs

Ehud Keinan and Noam Greenspoon

with NAD(P)H models, such as 3,5-dicarboethoxy-2,6-dimethyl-1,4-dihydropyridine (Hantzsch ester). With highly electron-deficient olefins, such as maleic acid, maleic anhydride, diethyl maleate, diethyl fumarate, etc., reductions proceed well[381]. Similarly, the olefinic bond of 1-phenyl-4,4,4-trifluoro-2-buten-1-one is reduced by dihydropyridines under mild condition (Scheme 135)[382]. Tracer experiments showed that hydrogen is transferred directly from the 4-position of the pyridine ring to the β-position of the enone system. The reaction thus parallels the enzymatic reduction of androstenedione[383].

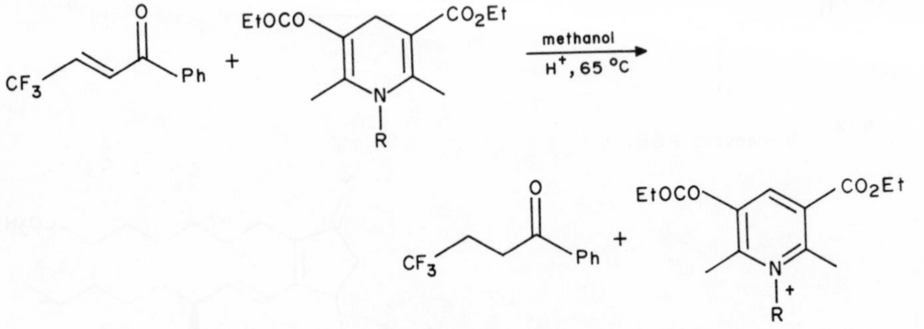

SCHEME 135

However, these reaction condition (refluxing methanol or photoactivation at room temperature) are useful only for the reduction of highly activated double bonds[384]. Nevertheless, it was found that the reaction is promoted by silica gel[385], broadening the scope of reducible enone substrates (Scheme 136).

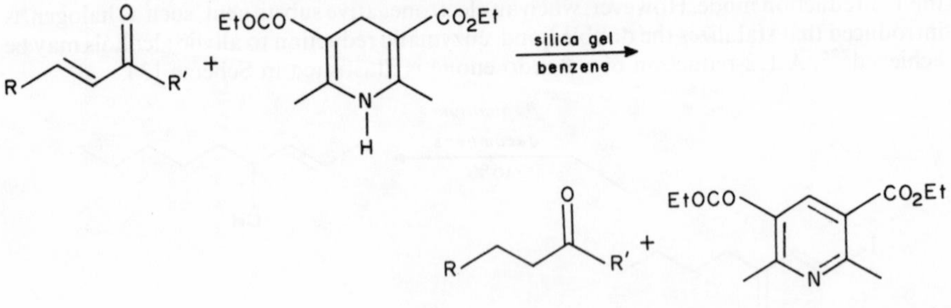

SCHEME 136

The method is highly chemoselective as no alcoholic products are observed, and carbonyl, nitro, cyano, sulfinyl and sulfonyl groups remain intact under the reaction conditions (Scheme 137).

Pandit has provided evidence for the Lewis-acid catalysis postulated to operate in these reductions[386]. The reduction of various cinnamoylpyridines by 1,4-dihydropyridine derivatives to the corresponding saturated ketones is catalyzed by zinc or magnesium cations. The reduction rate was fastest in the case of 2-cinnamoylpyridine, in which the metal ion can complex simultaneously to both the nitrogen and oxygen sites (Scheme 138). This example is regarded as a model of Lewis-acid catalysis of the NADH-dependent enzymatic reduction of δ^4-3-ketosteroids.

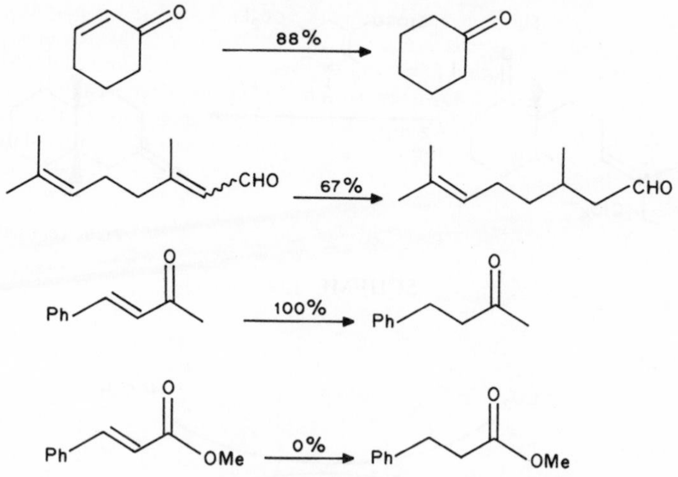

SCHEME 137

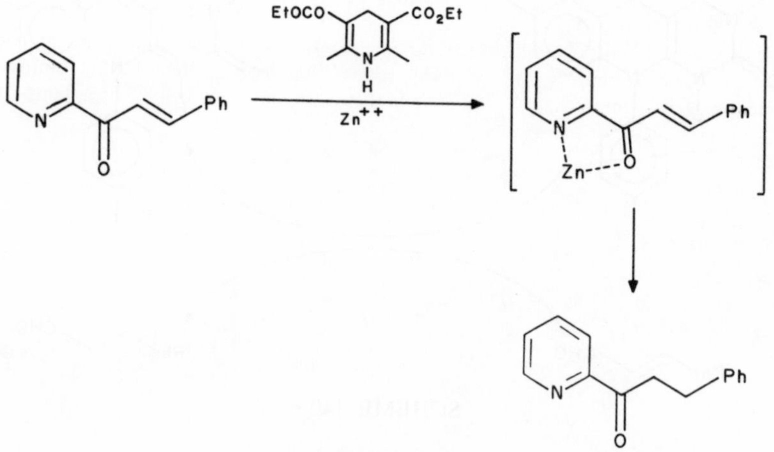

SCHEME 138

In a similar manner, iminium salts derived from α, β-unsaturated aldehydes and ketones are reduced by Hantzsch ester (Scheme 139)[387]. The ratio between the 1, 4- and 1, 2-reduction products depends upon the pK_a of the amine component.

An autorecycling system for the specific 1, 4-reduction of α, β-unsaturated ketones and aldehydes was based on 1, 5-dihydro-5-deazaflavin, which can be regarded as an NADH model[388]. The reaction occurs on heating the substrate with catalytic amounts of 5-deazaflavin in 98% formic acid, typically at 120 °C for 24 h (Scheme 140).

The iminium salts of 3, 3, 5-trimethylcyclohex-2-en-1-one were reduced with 1, 4-dihydronicotinamide sugar pyranosides to give the corresponding optically active

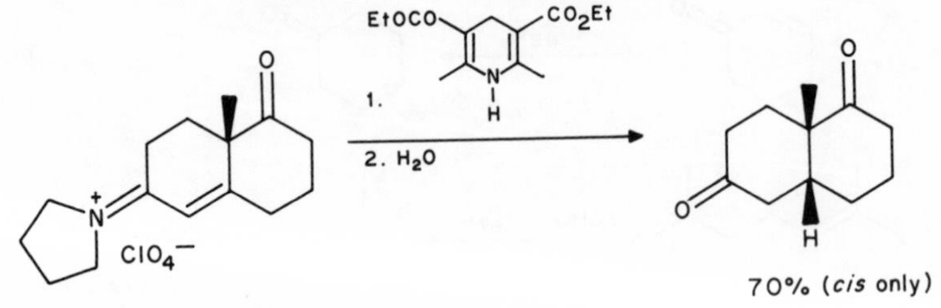

SCHEME 139

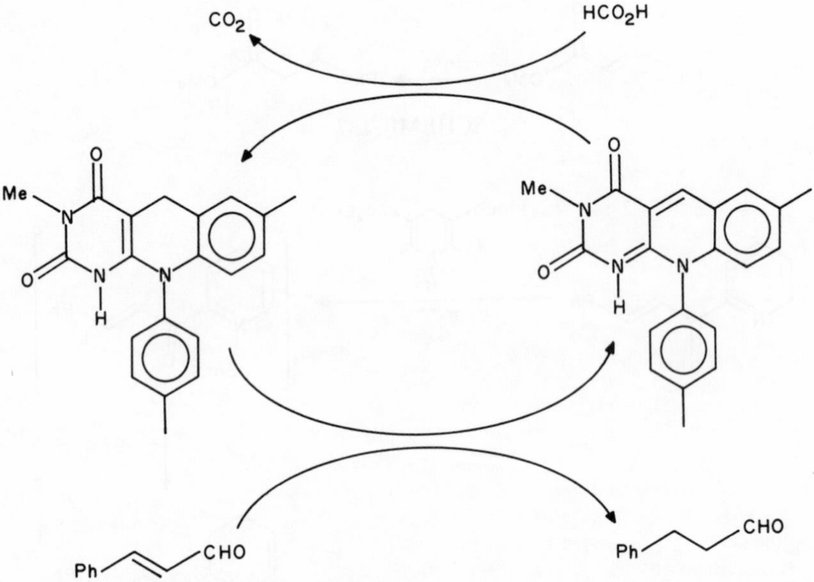

SCHEME 140

saturated ketone in enantiomeric excess ranging over 3–31%. The product stereochemistry changed sensitively with structural variations in the sugar residues (Scheme 141)[389].

The cob(I)alamin catalyzed reduction of α-methyl-α,β-unsaturated carbonyl compounds produces the corresponding saturated derivatives having an S configuration at the α-carbon (Scheme 142)[390]. The highest enantiomeric excess (33%) is exhibited by the Z-configurated methyl ketone. The E-configurated enone is reduced by this system to the corresponding R-product with poor enantiomeric excess.

VIII. MISCELLANEOUS REDUCING AGENTS

Several techniques utilizing miscellaneous reagents, that were not mentioned in the preceding sections, have been reported to effect the 1,4-reduction of α,β-unsaturated aldehydes and ketones.

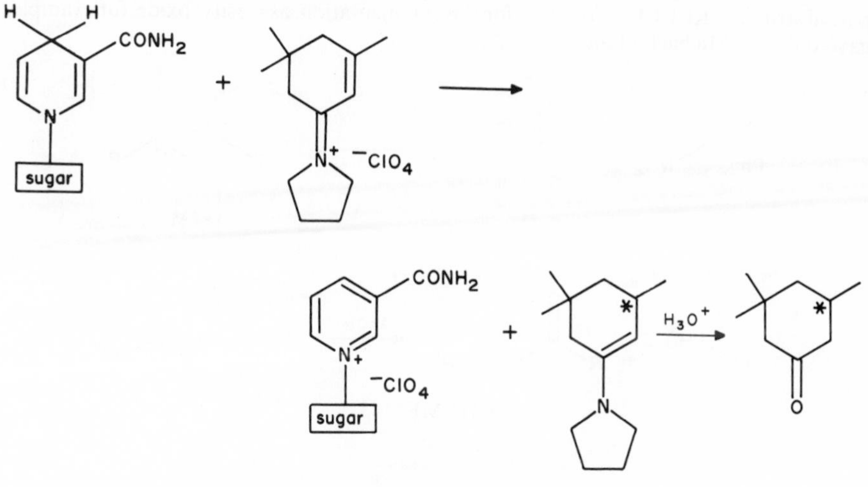

SCHEME 141

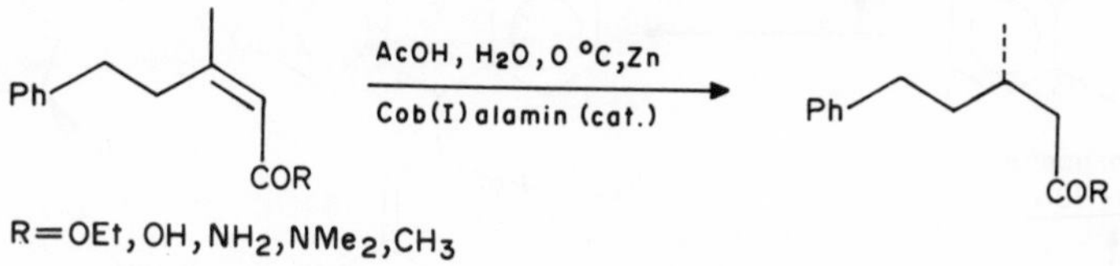

SCHEME 142

Sodium dithionite under nitrogen atmosphere at 80 °C in a water–benzene mixture and in the presence of a phase-transfer catalyst was shown to be a useful reducing agent. Dienoic carboxylic acids and esters were reduced in a 1,6-mode using this approach[391].

2-Phenylbenzothiazoline reduced α,β-unsaturated carbonyl compounds in a 1,4-fashion in the presence of stoichiometric amounts of aluminum chloride[392]. No 1,2-reduction products or saturated alcohols were detected. The reagent reduces unsaturated esters and aldehydes much less effectively.

Condensation of an α,β-unsaturated ketone with benzylamine gives the corresponding Schiff base. Treatment with a base, such as potassium t-butoxide, affects rearrangement to a benzaldehyde derivative, as shown in Scheme 143[393]. Hydrolysis of the latter with dilute acetic acid furnishes the corresponding saturated ketone with concomitant formation of benzaldehyde.

A reagent prepared from tellurium powder and sodium borohydride in ethanol engenders 1,4-reduction of α,β-unsaturated aldehydes, ketones and esters in high yield and with good regio- and chemoselectivity (no 1,2-reduction and no reduction of isolated double bonds)[394].

Anthracene hydride (the anion derived from 9,10-dihydroanthracene) reacts rapidly with chalcone to form an anionic Michael adduct along with a chalcone dimerization product (Scheme 144)[395]. Prolonged reaction in the presence of anthracene hydride cleaves the Michael adduct into anthracene and the enolate of the saturated ketone. The partial structure RCCCO is essential for this fragmentation, as mesityl oxide, for example, gave only the Michael adduct.

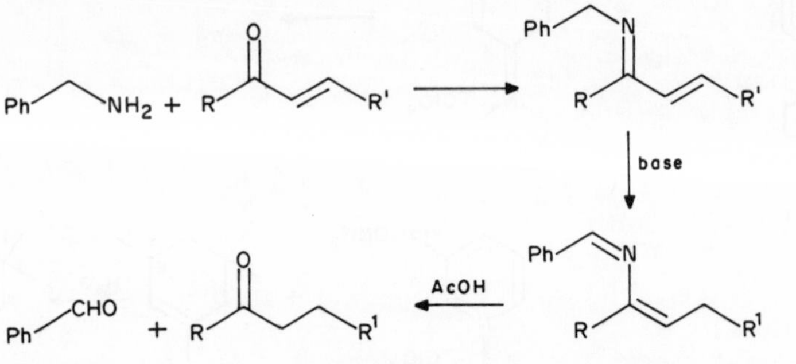

SCHEME 143

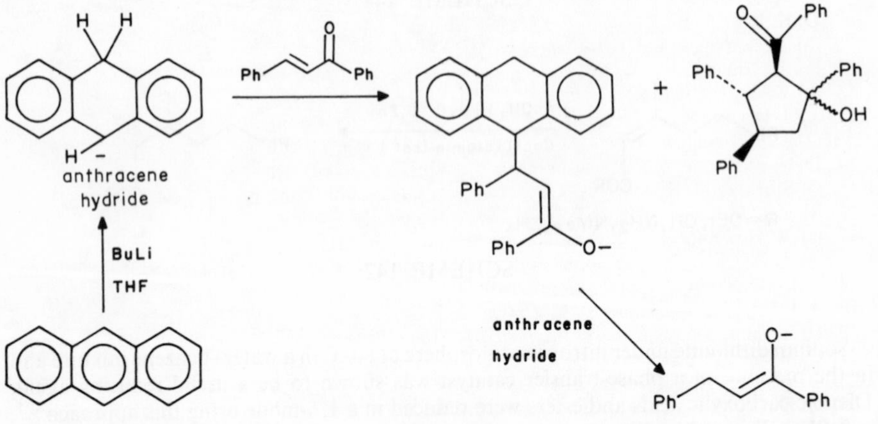

SCHEME 144

Photolysis of 4a-methyl-4, 4a, 9, 10-tetrahydro-2-(3H)-phenanthrone in isopropanol gave rearranged and 1, 4-reduction products, along with traces of 1, 2-reduction and small amounts of coupling products[396].

2-Propanol doped on dehydrated alumina reduces at room temperature various aldehydes and ketones to the corresponding alcohols[397]. α, β-Unsaturated aldehydes are selectively reduced under these conditions to the corresponding allylic alcohols. For example, citral is converted to geraniol in 88% yield.

α, β-Unsaturated nitriles are reduced to saturated nitriles with triethylamineformic acid azeotrope in DMF[398].

α, β-Unsaturated ketones are reduced to allylic alcohols with β-branched trialkyl-aluminum compounds, such as (i-Bu)$_3$Al and tris-((S)-2-methylbutyl)aluminum. The latter reagent reduces prochiral enones to optically active allylic alcohols with 7–15% enantiomeric excess[399].

IX. REFERENCES

1. H. C. Brown and S. Krishnamurthy, *Tetrahedron*, **35**, 567 (1979).
2. (a) A. J. Birch and H. Smith, *Quart. Rev.*, **12**, 17 (1958).
   (b) A. J. Birch and G. Subba-Rao, in *Advances in Organic Chemistry* (Ed. E. C. Taylor), Vol. 8, Wiley, New York, 1972, p. 1.
3. C. Djerassi, *Steroid Reactions*, Holden-Day, Inc., San Francisco, 1963, pp. 299–325.
4. H. L. Dryden, Jr., in *Organic Reactions in Steroid Chemistry* (Eds. J. Fried and J. A. Edwards), Vol. I, Van Nostrand Reinhold Co., New York, 1972, p. 1.
5. H. O. House, *Modern Synthetic Reactions*, 2nd ed., Benjamin, Menlo Park, California, 1972.
6. F. Johnson, *Chem. Rev.*, **68**, 375 (1968).
7. F. J. McQuillin, in *Techniques of Organic Chemistry* (Ed. A. Weissberger), Vol. XI, Part I, Interscience, New York, 1963, Chap. 9.
8. H. Smith, *Organic Reactions in Liquid Ammonia*, Wiley, New York, 1963.
9. M. Smith, in *Reduction* (Ed. R. L. Augustine), Marcel Dekker, New York, 1968, Chap. 2.
10. D. Caine, *Org. React.*, **23**, 1 (1976).
11. (a) M. C. R. Symons. *Quart. Rev.*, **13**, 99 (1959).
    (b) U. Schindewolf, *Angew. Chem., Int. Ed. Engl.*, **7**, 190 (1968).
    (c) J. L. Dye, *Acc. Chem. Res.*, **1**, 306 (1968).
12. (a) H. Normant, *Angew. Chem., Int. Ed. Engl.*, **6**, 1046 (1967).
    (b) H. Normant, *Bull. Soc. Chim. Fr.*, 791 (1968).
    (c) K. W. Bowers, R. W. Giese, J. Grimshaw, H. O. House, N. H. Kolodny, K. Kronberger and D. K. Roe, *J. Am. Chem. Soc.*, **92**, 2783 (1970).
    (d) M. Larcheveque, *Ann. Chim. (Paris)*, **5**, 129 (1970).
13. (a) J. L. Down, J. Lewis, B. Moore and G. Wilkinson, *J. Chem. Soc.*, 3767 (1959).
    (b) C. Agami, *Bull. Soc. Chim. Fr.*, 1205 (1968).
    (c) J. L. Dye, M. G. DeBacker and V. A. Nicely, *J. Am. Chem. Soc.*, **92**, 5226 (1970).
    (d) C. D. Pedersen, *J. Am. Chem. Soc.*, **89**, 7017 (1967); **92**, 386, 391 (1970).
14. D. H. R. Barton and C. H. Robinson, *J. Chem. Soc.*, 3054 (1954).
15. G. Stork and S. D. Darling, *J. Am. Chem. Soc.*, **82**, 1512 (1960); **86**, 1761 (1964).
16. M. J. T. Robinson, *Tetrahedron*, **21**, 2475 (1965).
17. (a) R. Howe and F. J. McQuillin, *J. Chem. Soc.*, 2670 (1956).
    (b) G. L. Chetty, G. S. Krishna Rao, S. Dev and D. K. Banerjee, *Tetrahedron*, **22**, 2311 (1966).
18. F. J. McQuillin, *J. Chem. Soc.*, 528 (1955).
19. A. J. Birch, H. Smith and R. E. Thornton, *J. Chem. Soc.*, 1339 (1957).
20. H. E. Zimmerman, *J. Am. Chem. Soc.*, **78**, 1168 (1956).
21. O. Wallach, *Ann. Chem.*, **279**, 377 (1894).
22. O. Wallach, *Ann. Chem.*, **275**, 111 (1893).
23. O. Diels and E. Abderhalden, *Chem. Ber.*, **39**, 884 (1906).
24. L. H. Knox, E. Blossy, H. Carpio, L. Cervantes, P. Crabbe, E. Velarde and J. A. Edwards, *J. Org. Chem.*, **30**, 2198 (1965).
25. J. A. Barltrop and A. C. Day, *Tetrahedron*, **22**, 3181 (1966).
26. T. A. Spencer, R. A. J. Smith, D. L. Storm and R. M. Villarica, *J. Am. Chem. Soc.*, **93**, 4856 (1971).
27. L. E. Hightower, L. R. Glasgow, K. M. Stone, D. A. Albertson and H. A. Smith, *J. Org. Chem.*, **35**, 1881 (1970).
28. H. A. Smith, B. J. L. Huff, W. J. Powers and D. Caine, *J. Org. Chem.*, **32**, 2851 (1967).
29. J. F. Eastham and D. R. Larkin, *J. Am. Chem. Soc.*, **81**, 3652 (1959).
30. H. O. House, *Rec. Chem. Prog.*, **28**, 98 (1967).
31. W. L. Jolly and L. Prizant, *Chem. Commun.*, 1345 (1968).
32. A. P. Krapcho and A. A. Bothner-By, *J. Am. Chem. Soc.*, **81**, 3658 (1959).
33. D. C. Burke, J. H. Turnbull and W. Wilson, *J. Chem. Soc.*, 3237 (1953).
34. I. N. Nazarov and I. A. Gurvich, *J. Gen. Chem. USSR*, **25**, 921 (1955).
35. A. J. Birch, E. Pride and H. Smith, *J. Chem. Soc.*, 4688 (1958).
36. G. Buchi, S. J. Gould and F. Naf, *J. Am. Chem. Soc.*, **93**, 2492 (1971).
37. M. E. Kuehne, *J. Am. Chem. Soc.*, **83**, 1492 (1961).
38. G. Stork, P. Rosen and N. L. Goldman, *J. Am. Chem. Soc.*, **83**, 2965 (1961).
39. G. Stork, P. Rosen, N. L. Goldman, R. V. Coombs and J. Tsuji, *J. Am. Chem. Soc.*, **87**, 275 (1965).

40. M. J. Weiss, R. E. Schaub, G. R. Allen, Jr., J. F. Poletta, C. Pidacks, R. B. Conrow and C. J. Coscia, *Tetrahedron*, **20**, 367 (1964); *Chem. Ind.* (*London*), 118 (1963).
41. A. Coulombeau, *Bull. Soc. Chim. Fr.*, 4407 (1970).
42. R. Deghenghi, C. Revesz and R. Gaudry, *J. Med. Chem.*, **6**, 301 (1963).
43. R. M. Coates and R. L. Sowerby, *J. Am. Chem. Soc.*, **93**, 1027 (1971).
44. (a) R. Deghenghi and R. Gaudry, *Tetrahedron Lett.*, 489 (1962).
    (b) R. E. Schaub and M. J. Weiss, *Chem. Ind.* (*London*), 2003 (1961).
45. G. Stork and J. E. McMurry, *J. Am. Chem. Soc.*, **89**, 5464 (1967).
46. G. Stork, S. Uyeo, T. Wakamatsu, P. Grieco and J. Labovitz, *J. Am. Chem. Soc.*, **93**, 4945 (1971).
47. T. A. Spencer, T. D. Weaver, R. M. Villarica, R. J. Friary, J. Posler and M. A. Schwartz, *J. Org. Chem.*, **33**, 712 (1968).
48. K. W. Bowers, R. W. Giese, J. Grimshaw, H. O. House, N. H. Kolodny, K. Kronberger and D. K. Roe, *J. Am. Chem. Soc.*, **92**, 2783 (1970).
49. C. L. Perrin, *Prog. Phys. Org. Chem.*, **3**, 165 (1965).
50. M. M. Baizer and J. P. Petrovich, *Adv. Phys. Org. Chem.*, **7**, 189 (1970).
51. D. Miller, L. Mandell and R. A. Day, Jr., *J. Org. Chem.*, **36**, 1683 (1971).
52. J. Weimann, S. Risse and P. -F. Casals, *Bull. Soc. Chim. Fr.*, 381 (1966).
53. B. J. L. Huff, Ph.D. Dissertation, Georgia Institute of Technology, 1969, *Diss. Abstr. B*, **29** (12), 4589 (1969).
54. G. Stork, M. Nussim and B. August, *Tetrahedron, Suppl.*, **8**, 105 (1966).
55. P. Angibeaund, H. Riviere and B. Tchoubar, *Bull. Soc. Chim. Fr.*, 2937 (1968).
56. T. A. Spencer, R. J. Friary, W. W. Schmiegel, J. F. Simeone and D. S. Watt, *J. Org. Chem.*, **33**, 719 (1968).
57. R. E. Ireland and G. Pfister, *Tetrahedron Lett.*, 2145 (1969).
58. G. Stork and J. Tsuji, *J. Am. Chem. Soc.*, **83**, 2783 (1961).
59. P. S. Venkataramani, J. E. Karoglan and W. Reusch, *J. Am. Chem. Soc.*, **93**, 269 (1971).
60. T. A. Spencer, K. K. Schmiegel and W. W. Schmiegel, *J. Org. Chem.*, **30**, 1626 (1965).
61. R. G. Carlson and R. G. Blecke, *J. Chem. Soc., Chem. Commun*, 93 (1969).
62. M. Tanabe, J. W. Chamberlin and P. Y. Nishiura, *Tetrahedron Lett.*, 601 (1961).
63. B. M. Trost, Abstracts of Papers, Joint Conference CIC–ACS, Toronto, Canada, May 24–29, 1970, Organic Section, Paper No. 42.
64. C. Amendolla, G. Rosenkranz and F. Sondheimer, *J. Chem. Soc.*, 1226 (1954).
65. T. Anthonsen, P. H. McCabe, R. McGrindle and R. D. H. Murray, *Tetrahedron*, **25**, 2233 (1969).
66. T. Masamune, A. Murai, K. Orito, H. Ono, S. Numata and H. Suginome, *Tetrahedron*, **25**, 4853 (1969).
67. H. Bruderer, D. Arigoni and O. Jeger, *Helv. Chim. Acta*, **39**, 858 (1956).
68. R. Howe, F. J. McQuillin and R. W. Temple, *J. Chem. Soc.*, 363 (1959).
69. K. S. Kulkarni and A. S. Rao, *Tetrahedron*, **21**, 1167 (1965).
70. K. Irmscher, W. Beerstecher, H. Metz, R. Watzel and K. -H. Bork, *Chem. Ber.*, **97**, 3363 (1964).
71. A. Spassky-Pasteur, *Bull. Soc. Chim. Fr.*, 2900 (1969).
72. R. M. Coates and J. E. Shaw, *Tetrahedron Lett.*, 5405 (1968); *J. Org. Chem.*, **35**, 2597 (1970).
73. R. M. Coates and J. E. Shaw, *J. Org. Chem.*, **35**, 2601 (1970).
74. R. E. Ireland and J. A. Marshall, *J. Org. Chem.*, **27**, 1615 (1962).
75. M. Vandewalle and F. Compernolle, *Bull. Soc., Chim. Belg.*, **75**, 349 (1966).
76. M. Vandewalle and F. Compernolle, *Bull. Soc. Chim. Belg.*, **76**, 43 (1967).
77. D. S. Watt, J. M. McKenna and T. A. Spencer, *J. Org. Chem.*, **32**, 2674 (1967).
78. J. E. Shaw and K. K. Knutson, *J. Org. Chem.*, **36**, 1151 (1971).
79. J. A. Campbell and J. C. Babcock, *J. Am. Chem. Soc.*, **81**, 4069 (1959).
80. A. F. Daglish, J. Green and V. D. Poole, *J. Chem. Soc.*, 2627 (1954).
81. F. Johnson, G. T. Newbold and F. S. Spring, *J. Chem. Soc.*, 1302 (1954).
82. J. A. Marshall and H. Roebke, *J. Org. Chem.*, **33**, 840 (1968).
83. M. Nussim, Y. Mazur and F. Sondheimer, *J. Org. Chem.*, **29**, 1120 (1964).
84. H. Van Kamp, P. Westerhof and H. Niewind, *Rec. Trav. Chim. Pays-Bas*, **83**, 509 (1964).
85. E. Wenkert, A. Afonso, J. B. Bredenberg, C. Kaneko and A. Tahara, *J. Am. Chem. Soc.*, **86**, 2038 (1964).
86. P. Westerhof and E. H. Reerink, *Rec. Trav. Chim. Pays-Bas*, **79**, 771 (1960).
87. K. P. Dastur, *Tetrahedron Lett.*, 4333 (1973).
88. A. Zurcher, H. Heusser, O. Jeger and P. Geistlich, *Helv. Chim. Acta*, **37**, 1562 (1954).

89. G. Bach. J. Capitaine and Ch. R. Engel, *Can. J. Chem.*, **46**, 733 (1968).
90. H. Heusser, M. Roth, O. Rohr and R. Anliker, *Helv. Chim. Acta*, **38**, 1178 (1955).
91. W. F. Johns, *J. Org. Chem.*, **36**, 711 (1971).
92. E. Shapiro, T. Legatt, L. Weber, M. Steinberg and E. P. Oliveto, *Chem. Ind.(London)*, 300 (1962).
93. W. Cocker, B. Donnelly, H. Gobinsingh, T. B. H. McMurry and N. A. Nisbet, *J. Chem. Soc.*, 1262 (1963).
94. B. R. Ortiz de Montellano, B. A. Loving, T. C. Shields and P. D. Gardner, *J. Am. Chem. Soc.*, **89**, 3365 (1967).
95. D. J. Marshall and R. Deghenghi, *Can. J. Chem.*, **47**, 3127 (1969).
96. T. G. Halsall, D. W. Theobald and K. B. Walshaw, *J. Chem. Soc.*, 1029 (1964).
97. A. J. Birch, *Quart. Rev.*, **4**, 69 (1950).
98. R. G. Harvey, *Synthesis*, 161 (1970).
99. W. Nagata, T. Terasawa, S. Hirai and K. Takeda, *Tetrahedron Lett.*, 27 (1960); *Chem. Pharm. Bull. (Tokyo)*, **9**, 769 (1961).
100. W. S. Johnson, E. R. Rogier, J. Szmuszkovicz, H. I. Hadler, J. Ackerman, B. K. Bhattacharyya, B. M. Bloom, L. Stalmann, R. A. Clement, B. Bannister and H. Wynberg, *J. Am. Chem. Soc.*, **78**, 6289 (1956).
101. W. F. Johns, *J. Org. Chem.*, **28**, 1856 (1963).
102. W. S. Johnson, J. M. Cox, D. W. Graham and H. W. Whitlock, Jr., *J. Am. Chem. Soc.*, **89**, 4524 (1967).
103. M. V. R. Koteswara Rao, G. S. Krishna Rao and S. Dev., *Tetrahedron*, **22**, 1977 (1966).
104. F. B. Colton, L. N. Nysted, B. Riegel and A. L. Raymond, *J. Am. Chem. Soc.*, **79**, 1123 (1957).
105. A. Bowers, H. J. Ringold and E. Denot, *J. Am. Chem. Soc.*, **80**, 6115 (1958).
106. I. A. Gurvich, V. F. Kucherov and T. V. Ilyakhina, *J. Gen. Chem. USSR*, **31**, 738 (1961).
107. P. S. Venkataramani, J. P. John, V. T. Ramakrishnan and S. Swaminathan, *Tetrahedron*, **22**, 2021 (1966).
108. P. T. Lansbury, P. C. Briggs, T. R. Demmin and G. E. DuBois, *J. Am. Chem. Soc.*, **93**, 1311 (1971).
109. H. Kaneko, K. Nakamura, Y. Yamoto and M. Kurokawa, *Chem. Pharm. Bull. (Tokyo)*, **17**, 11 (1969).
110. R. E. Schaub and M. J. Weiss, *J. Org. Chem.*, **26**, 3915 (1961).
111. P. Beak and T. L. Chaffin, *J. Org. Chem.*, **35**, 2275 (1970).
112. E. Wenkert and B. G. Jackson, *J. Am. Chem. Soc.*, **80**, 217 (1958).
113. G. Stork and F. H. Clarke, *J. Am. Chem. Soc.*, **77**, 1072 (1955); **83**, 3114 (1961).
114. S. Dube and P. Deslongchamps, *Tetrahedron Lett.*, 101 (1970).
115. W. G. Dauben, W. W. Epstein, M. Tanabe and B. Weinstein, *J. Org. Chem.*, **28**, 293 (1963).
116. J. R. Hanson, *Synthesis*, 1 (1974).
117. J. E. McMurry, *Acc. Chem. Res.*, **7**, 281 (1974).
118. T.-L. Ho, *Synthesis*, 1 (1979).
119. (a) C. E. Castro, R. D. Stephens and S. Moje, *J. Am. Chem. Soc.*, **88**, 4964 (1966).
    (b) A. Zurqiyah and C. E. Castro, *Org. Synth. Coll. Vol.*, **5**, 993 (1973).
120. (a) E. Knecht, *Ber. Dtsch. Chem. Ges.*, **36**, 166 (1903).
    (b) P. Karrer, Y. Yen and I. Reichstein, *Helv. Chim. Acta*, **13**, 1308 (1930).
    (c) L. C. Blaszczak and J. E. McMurry, *J. Org. Chem.*, **39**, 258 (1974).
121. (a) J. R. Hanson and E. Premuzic, *J. Chem. Soc. (C)*, 1201 (1969).
    (b) J. R. Hanson and E. Premuzic, *Angew. Chem., Int. Ed. Engl.*, **7**, 247 (1968).
122. H. O. House and E. F. Kinloch, *J. Org. Chem.*, **39**, 1173 (1974).
123. J. B. Conant and H. B. Cutter, *J. Am. Chem. Soc.*, **48**, 1016 (1926).
124. E. J. Corey, R. L. Danheiser and S. Chandrasekaran, *J. Org. Chem.*, **41**, 260 (1976).
125. S. C. Welch and M. E. Walters, *J. Org. Chem.* **43**, 2715 (1978).
126. (a) A. Berndt, *Angew. Chem., Int. Ed. Engl.*, **6**, 251 (1967).
    (b) A. Berndt, *Tetrahedron Lett.*, 177 (1970).
127. (a) P. Bladon, J. W. Cornforth and R. H. Jaeger, *J. Chem. Soc.*, 863 (1958).
    (b) H. Lund, *Acta Chim. Scand.*, **11**, 283 (1957).
128. R. C. Fuson, *Rec. Chem. Prog.*, **12**, 1 (1951).
129. (a) C. G. Overberger and A. M. Schiller, *J. Org. Chem.*, **26**, 4230 (1961).
    (b) E. L. Totton, N. C. Camp III, G. M. Cooper, B. D. Haywood and D. P. Lewis, *J. Org. Chem.*, **32**, 2033 (1967).

(c) H. Rosen, Y. Arad, M. Levy and D. Vofsi, *J. Am. Chem. Soc.*, **91**, 1425 (1969).

(d) P. Matsuda, *Tetrahedron Lett.*, 6193 (1966).

(e) A. Zysman, G. Dana and J. Wiemann, *Bull. Soc. Chim. Fr.*, 1019 (1967).

(f) J. Wiemann, M. R. Monot, G. Dana and J. Chuche, *Bull. Soc. Chim. Fr.*, 3293 (1967).

(g) E. Touboul, F. Weisbuch and J. Wiemann, *Bull. Soc. Chim. Fr.*, 4291 (1967).

(h) C. Glacet, *Compt. Rend.*, **227**, 480 (1948).

(i) J. Wiemann and R. Nahum, *Compt. Rend.*, **238**, 2091 (1954).

130. (a) R. N. Gourley, J. Grimshaw and P. G. Miller, *J. Chem. Soc.* (*C*), 2318 (1970).

(b) L. Horner and D. H. Skaletz, *Tetrahedron Lett.*, 3679 (1970).

(c) A. Spassky-Pasteur, *Bull. Soc. Chim. Fr.*, 2900 (1969).

131. (a) M. M. Baizer, *J. Org. Chem.*, **29**, 1670 (1964); **31**, 3847 (1966).

(b) M. M. Baizer and J. D. Anderson, *J. Org. Chem.*, **30**, 1348, 1351, 1357, 3138 (1965).

(c) J. D. Anderson, M. M. Baizer and E. J. Prill, *J. Org. Chem.*, **30**, 1645 (1965).

(d) J. D. Anderson, M. M. Baizer and J. P. Petrovich, *J. Org. Chem.*, **31**, 3890, 3897 (1966).

(e) J. H. Wagenknecht and M. M Baizer, *J. Org. Chem.*, **31**, 3885 (1966).

(f) M. R. Ort and M. M Baizer, *J. Org. Chem.*, **31**, 1646 (1966).

(g) M. M. Baizer and J. D. Anderson, *J. Electrochem. Soc.*, **111**, 223, 226 (1964); M. M. Baizer, *J. Electrochem.*, **111**, 215 (1964).

(h) For reviews, see: M. M. Baizer, J. D. Anderson, J. H. Wagenknecht, M. R. Ort and J. P. Petrovich, *Prog. Electrochem. Acta*, **12**, 1377 (1967); J. D. Anderson, J. P. Petrovich and M. M. Baizer, *Adv. Org. Chem.*, **6**, 257 (1969); M. M. Baizer and J. P. Petrovich, *Prog. Phys. Org. Chem.*, **7**, 189 (1970).

132. D. A. Jaeger, D. Bolikal and B. Nath, *J. Org. Chem.*, **52**, 276 (1987).

133. B. R. James, *Homogeneous Hydrogenation*, Wiley–Interscience, New York, 1973.

134. (a) P. S. Rylander, *Hydrogenation Methods*, Academic Press, London, 1985.

(b) P. S. Rylander, *Catalytic Hydrogenation in Organic Syntheses*, Academic Press, London, 1979.

135. M. Freifelder, *Catalytic Hydrogenation in Organic Synthesis*, Willey–Interscience, New York, 1978.

136. G. R. Ames and W. Davey, *J. Chem. Soc.*, 3001 (1956).

137. J. J. Brunet, P. Gallois and P. Caubere, *J. Org. Chem.*, **45**, 1937, 1946 (1980).

138. H. Adkins and R. Connor, *J. Am. Chem. Soc.*, **53**, 1091 (1931).

139. N. F. Hayes, *Synthesis*, 702 (1975).

140. E. Breitner, E. Roginski and P. N. Rylander, *J. Org. Chem.*, **24**, 1855 (1959).

141. A. Skita, *Chem. Ber.*, **48**, 1486 (1915).

142. C. Weygand and W. Meusel, *Chem. Ber.*, **76**, 498 (1943).

143. R. Adams, J. W. Kern and R. L. Shriner, *Org. Synth. Coll. Vol.*, **1**, 101 (1932).

144. R. L. Augustine, *J. Org. Chem.*, **23**, 1853 (1958).

145. R. E. Harmon, J. L. Parsons, D. W. Cooke, S. K. Gupta and J. Schoolenberg, *J. Org. Chem.*, **34** 3684 (1969).

146. (a) C. Djerassi and J. Gutzwiller, *J. Am. Chem. Soc.*, **88**, 4537 (1966).

(b) A. J. Birch and K. A. M. Walker, *J. Chem. Soc.* (*C*) 1894 (1966).

147. P. L. Cook, *J. Org. Chem.*, **27**, 3873 (1962).

148. K. Sakai and K. Watanabe, *Bull. Chem. Soc. Jpn.*, **40**, 1548 (1967).

149. Y. Watanabe, Y. Matsumura, Y. Izumi and Y. Mizutani, *Bull. Chem. Soc. Jpn.*, **47**, 2922 (1974).

150. R. L. Augustine, *Adv. Catal.*, **25**, 63 (1976) and references cited therein.

151. R. L. Augustine, *Ann. N.Y. Acad. Sci.*, **145**, 19 (1967).

152. F. J. McQuillin, W. O. Ord and P. L. Simpson, *J. Chem. Soc.*, 5996 (1963).

153. (a) R. L. Augustine, *J. Org. Chem.*, **23**, 1853 (1958).

(b) R. L. Augustine and A. D. Broom, *J. Org. Chem.*, **25**, 802 (1960).

(c) R. L. Augustine, D. C. Migliorini, R. E. Foscante, C. S. Sodano and M. J. Sisbarro, *J. Org. Chem.*, **34**, 1075 (1969).

(d) S. Nishimura, M. Shimahara and M. Shiota, *J. Org. Chem.*, **31**, 2394 (1966).

(e) M. G. Combe, H. B. Henbest and W. R. Jackson, *J. Chem. Soc.* (*C*), 2467 (1967).

(f) H. B. Henbest, W. R. Jackson and I. Malunowicz, *J. Chem. Soc.* (*C*), 2469 (1967).

(g) I. Gardine, R. W. Howsam and F. J. McQuillin, *J. Chem. Soc.* (*C*), 260 (1969).

(h) H. J. E. Loewenthal, *Tetrahedron*, **6**, 269 (1959).

(i) L. Velluz, J. Valls and G. Nomine, *Angew. Chem., Int. Ed. Engl.*, **4**, 181 (1965).

154. T. C. McKenzie, *J. Org. Chem.*, **39**, 629 (1974).
155. Z. J. Hajos and D. R. Parrish *J. Org. Chem.*, **38**, 3239 (1973).
156. (a) T. Kametani, M. Tsubuki, H. Furuyama and T. Honda, *J. Chem. Soc., Perkin Trans. 1*, 557 (1985).
     (b) T. Kametani, M. Tsubuki, K. Higurashi and T. Honda, *J. Org. Chem.*, **51**, 2932 (1986).
157. (a) N. V. Borunova, L. K. Friedlin, L. I. Gvinter, T. Atabekov, V. A. Zamureenko and I. M. Kustanovich, *Izv. Akad. Nauk SSSR, Ser. Khim.*, **6**, 1299 (1972); *Chem. Abstr.*, **77**, 87461 (1972).
     (b) L. K. Friedlin, L. I. Gvinter, N. V. Borunova, S. F. Dymova and I. M. Kustanovich, *Katal Reakts. Zhidk. Faze*, 309 (1972); *Chem. Abstr.*, **79**, 115066z (1973).
158. P. C. Traas, H. Boelens and H. J. Takken, *Synth. Commun.*, **6**, 489 (1976).
159. (a) S. Nishimura and K. Tsuneda, *Bull. Chem. Soc. Jpn.*, **42**, 852 (1969).
     (b) S. Nishimura, T. Ichino, A. Akimoto and K. Tsuneda, *Bull. Chem. Soc. Jpn.*, **46**, 279 (1973).
     (c) S. Nishimura, T. Ichino, A. Akimoto, K. Tsuneda and H. Mori, *Bull. Chem. Soc. Jpn.*, **48**, 2852 (1975).
160. J. Azran, O. Buchman, I. Amer and J. Blum, *J. Mol. Catal.*, **34**, 229 (1986).
161. D. L. Reger, M. M. Habib and D. J. Fauth, *J. Org. Chem.*, **45**, 3860 (1980).
162. J. H. Schauble, G. J. Walter and J. G. Morin, *J. Org. Chem.*, **39**, 755 (1974).
163. A. Hassner and C. Heathcock, *J. Org. Chem.*, **29**, 1350 (1964).
164. E. Schenker, in *Newer Methods of Preparative Organic Chemistry* (Ed. W. Forest), Vol. IV, Academic Press, New York, 1968, p. 196.
165. J. Bottin, O. Eisenstein, C. Minot and N. T. Anh, *Tetrahedron Lett.*, 3015 (1972).
166. M. K. Johnson and B. Rickborn, *J. Org. Chem.*, **35**, 1041 (1970).
167. S. Geribaldi, M. Decouzon, B. Boyer and C. Moreau, *J. Chem. Soc., Perkin Trans. 2*, 1327 (1986).
168. W. R. Jackson and Z. Zurquiyah, *J. Chem. Soc., Chem. Commun.*, 5280 (1965).
169. S. Kim, C. H. Oh, J. S. Ko, K. H. Ahn and Y. J. Kim, *J. Org. Chem.*, **50**, 1927 (1985) and references cited therein.
170. S. B. Kadin, *J. Org. Chem.*, **31**, 620 (1966).
171. B. Ganem and J. M. Fortunato, *J. Org. Chem.*, **41**, 2194 (1976).
172. S. Krishnamurthy and H. C. Brown, *J. Org. Chem.*, **40**, 1864 (1975).
173. Z. J. Duri and J. R. Hanson, *J. Chem. Soc., Perkin Trans. 2*, 363 (1984).
174. J. MacMillan and C. L. Willis, *J. Chem. Soc., Perkin Trans. 2*, 357 (1984).
175. (a) M. H. Beale, *J. Chem. Soc., Perkin Trans. 1*, 1151 (1985).
     (b) M. H. Beale, J. MacMillan, C. R. Spray, D. A. Taylor and B. O. Phinney, *J. Chem. Soc., Perkin Trans. 1*, 541 (1984).
176. B. Voigt and G. Adam, *Tetrahedron*, **39**, 449 (1983).
177. E. J. Corey, K. B. Becker and R. K. Varma, *J. Am. Chem. Soc.*, **94**, 8616 (1972).
178. P. Joseph-Nathan, M. E. Garibay and R. L. Santillan, *J. Org. Chem.*, **52**, 759 (1987).
179. A. R. Chamberlin and S. H. Reich, *J. Am. Chem. Soc.*, **107**, 1440 (1985).
180. R. O. Hutchins and D. Kandasamy, *J. Org. Chem.*, **40**, 2530 (1975).
181. S. Kim, Y. C. Moon and K. H. Ahn, *J. Org. Chem.*, **47**, 3311 (1982).
182. (a) J. -L. Luche, *J. Am. Chem. Soc.*, **100**, 2226 (1978).
     (b) J. -L. Luche and A. L. Gemal, *J. Am. Chem. Soc.*, **101**, 5848 (1979).
     (c) A. L. Gemal and J. -L. Luche, *J. Am. Chem. Soc.*, **103**, 5454 (1981).
183. C. W. Jefford, T. W. Wallace, N. T. H. Can and C. G. Rimbault, *J. Org. Chem.*, **44**, 689 (1979).
184. (a) C. Kashima and Y. Yamamoto, *Chem. Lett.*, 1285 (1978).
     (b) C. Kashima, Y. Yamamoto and Y. Tsuda, *J. Org. Chem.*, **40**, 526 (1975).
185. (a) T. Nishio and Y. Omote, *Chem. Lett.*, 1223 (1979).
     (b) T. Nishio and Y. Omote, *J. Chem. Soc., Perkin Trans. 1*, 934 (1981).
186. J. Malek, *Org. React.*, **34**, 1 (1985).
187. H. C. Brown and P. M. Weissman, *J. Am. Chem. Soc.*, **87**, 5614 (1965).
188. (a) H. C. Brown, S. C. Kim and S. Krishnamurthy, *J. Org. Chem.*, **45**, 1 (1980).
     (b) H. C. Brown, P. K. Jadhav and A. K. Mandal, *Tetrahedron*, **37**, 3547 (1981).
     (c) M. Fieser and L. F. Fieser, *Reagents for Organic Synthesis*, Vols. I–XIII, Wiley–Interscience, New York, 1967–1988.
     (d) S. I. Yamada and K. Koga, in *Selective Organic Transformations*, Vol. I (Ed. B. S. Thyagarajan), Wiley–Interscience, New York, 1970.
     (e) B. D. James, *Rec. Chem. Prog.*, **31**, 199 (1970).
     (f) J. Vit, *Eastman Org. Chem. Bull.*, **42**, 1 (1970); *Chem. Abstr.*, **74**, 99073p (1971).

(g) D. M. S. Wheeler and M. M. Wheeler, in *Organic Reactions in Steroid Chemistry* (Eds. J. Fried and J. A. Edwards), Vol. I, Van Nostrand Reinhold, New York, 1972, Chap. 2.
(h) J. Malek and M. Cerny, *Synthesis*, 217 (1972).
(i) A. S. Kushner and T. Vaccariello, *J. Chem. Educ.*, **50**, 154, 157 (1973).
(j) H. Mishima, *Yuki Gosei Kagaku Kyokai Shi*, **32**, 1014 (1974); *Chem. Abstr.*, **82**, 138613b (1975).
(k) C. F. Lane, *Chem. Rev.*, **76**, 773 (1976).
(l) C. F. Lane, in *Aspects of Mechanistic Organometallic Chemistry* (Proceedings of Symposium) (Ed. J. H. Brewster), Plenum Press, New York, 1978, pp. 181–198.
(m) J. R. Boone and E. C. Ashby, *Top. Stereochem.*, **11**, 53 (1979).
(n) P. A. Bartlett, *Tetrahedron*, **36**, 2 (1980).
(o) S. O. Kim, *Hwakhak Kwa Kongop Ui Chinbo*, **20**, 293 (1980); *Chem. Abstr.*, **94**, 102222g (1981).
189. Reference 5, Chap. 2.
190. E. R. H. Walker, *Chem. Soc. Rev.*, **5**, 23 (1976).
191. (a) A. Hajos, *Komplexe Hydride*, VEB Deutscher Verlag der Wissenschaften, East Berlin, 1966.
(b) A. Hajos, *Complex Hydrides and Related Reducing Agents in Organic Synthesis*, Elsevier Scientific Publ. Co., Amsterdam, 1979.
192. H. C. Brown, *Boranes in Organic Chemistry*, Cornell University Press, Ithaca, New York, 1972, Chaps. 12–13.
193. (a) D. R. Boyd and M. A. McKervey, *Quart. Rev.*, **22**, 95 (1968).
(b) J. Mathieu and J. Weill-Raynal, *Bull. Soc. Chim. Fr.*, 1211 (1968).
(c) T. D. Inch, *Synthesis*, 466 (1970).
(d) J. D. Morrison and H. S. Mosher, *Asymmetric Organic Reactions*, Prentice Hall, Englewood Cliffs, N. J., 1971, pp. 116–132, 202–215, 386–389; Reprint ed., American Chemical Society, Washington, D.C., 1976.
(e) H. J. Schneider and R. Haller, *Pharmazie*, **28**, 417 (1973).
(f) J. W. Scott and D. Valentine, Jr., *Science*, **184**, 943 (1974).
(g) D. Valentine, Jr. and J. W. Scott, *Synthesis*, 329 (1978).
(h) J. W. ApSimon and R. P. Seguin, *Tetrahedron*, **35**, 2797 (1979).
194. O. Eisenstein, J. M. Lefour, C. Minot, N. T. Anh and G. Soussan, *C.R. Acad. Sci. Paris, Ser. C*, **274**, 1310 (1972).
195. J. Durand, N. T. Anh and J. Huet, *Tetrahedron Lett.*, 2397 (1974).
196. J. Bottin, O. Eisenstein, C. Minot and N. T. Anh, *Tetrahedron Lett.*, 3015 (1972).
197. R. G. Pearson, *J. Chem. Educ.*, **45**, 581 (1968).
198. J. Seyden-Penne, *Bull. Soc. Chim. Fr.*, 3871 (1968).
199. H. C. Brown and H. M. Hess, *J. Org. Chem.*, **34**, 2206 (1969).
200. A. Loupy and J. Seyden-Penne, *Tetrahedron*, **36**, 1937 (1980).
201. (a) J. C. Richer and A. Rossi, *Can. J. Chem.*, **50**, 438 (1972).
(b) J. A. Marshall and J. A. Ruth, *J. Org. Chem.*, **39**, 1971 (1974).
202. M. E. Cain, *J. Chem. Soc.*, 3532 (1964).
203. M. F. Semmelhack, R. D. Stauffer and A. Yamashita, *J. Org. Chem.*, **42**, 3180 (1977).
204. J. E. Baldwin, R. C. Thomas, L. I. Kruse and L. Silberman, *J. Org. Chem.*, **42**, 3846 (1977).
205. (a) P. L. Southwick, N. Latif, B. M. Fitzgerald and N. M. Zaczek, *J. Org. Chem.*, **31**, 1 (1966).
(b) J. Durand and J. Huet, *Bull. Soc. Chim. Fr.*, Pt. 2, 428 (1978).
206. P. A. Bartlett and W. S. Johnson, *J. Am. Chem. Soc.*, **95**, 7501 (1973).
207. G. D. Prestwich, F. B. Whitfield and G. Stanley, *Tetrahedron*, **32**, 2945 (1976).
208. (a) R. L. Markezich, W. E. Willy, B. E. McCarry and W. S. Johnson, *J. Am. Chem. Soc.*, **95**, 4414 (1973).
(b) W. S. Johnson, B. E. McCarry, R. L. Markezich and S. G. Boots, *J. Am. Chem. Soc.*, **102**, 352 (1980).
209. P. C. Traas, H. Boellens and H. J. Takken, *Recl. Trav. Chim. Pays-Bas*, **95**, 57 (1976).
210. K. E. Wilson, R. T. Seidner and S. Masamune, *J. Chem. Soc., Chem. Commun.*, 213 (1970).
211. D. V. Banthorpe, A. J. Curtis and W. D. Fordham, *Tetrahedron Lett.*, 3865 (1972).
212. N. Lander and R. Mechoulam, *J. Chem. Soc., Perkin Trans. 1*, 484 (1976).
213. R. A. Finnegan and P. L. Bachman, *J. Org. Chem.*, **30**, 4145 (1965).
214. D. Caine, P. C. Chen, A. S. Frobese and J. T. Gupton, *J. Org. Chem.*, **44**, 4981 (1979).
215. R. E. Ireland, M. I. Dawson, S. C. Welch, A. Hagenbach, J. Bordner and B. Trus, *J. Am. Chem. Soc.*, **95**, 7829 (1973).

216. E. Winterfeldt, *Synthesis*, 617 (1975).
217. E. C. Ashby and J. J. Lin, *Tetrahedron Lett.*, 3865 (1976).
218. H. C. Brown, U. S. NTIS, AD Rep. AD-A026132 (1976); *Chem. Abstr.* **85**, 176353m (1976).
219. H. J. Williams, *J. Chem. Soc., Perkin Trans. 1*, 1852 (1973).
220. W. L. Dilling and R. A. Plepys, *J. Chem. Soc., Chem. Commun.*, 417 (1969).
221. W. L. Dilling and R. A. Plepys, *J. Org. Chem.*, **35**, 1971 (1970).
222. J. P. Bugel, P. Ducos, O. Gringore and F. Rouessac, *Bull. Soc. Chim. Fr.*, 4371 (1972).
223. P. R. Story and S. R. Fahrenholtz, *J. Am. Chem. Soc.*, **87**, 1623 (1965).
224. J. B. Wiel and F. Rouessac, *J. Chem. Soc., Chem. Commun.*, 446 (1976).
225. J. B. Wiel and F. Rouessac, *Bull. Soc. Chim. Fr.*, Pt. 2, 273 (1979).
226. E. J. Corey and J. Gorzynski Smith, *J. Am. Chem. Soc.*, **101**, 1038 (1979).
227. M. F. Semmelhack and R. D. Stauffer, *J. Org. Chem.*, **40**, 3619 (1975).
228. M. Vandewalle and E. Madeleyn, *Tetrahedron*, **26**, 3551 (1970).
229. C. J. Sih, R. G. Salomon, P. Price, R. Sood and G. Peruzzotti, *J. Am. Chem. Soc.*, **97**, 857 (1975).
230. M. Suzuki, T. Kawagishi, T. Suzuki and R. Noyori, *Tetrahedron Lett.*, **23**, 4057 (1982).
231. P. A. Grieco, N. Fukamiya and M. Miyashita, *J. Chem. Soc., Chem. Commun.*, 573 (1976).
232. (a) Z. G. Hajos, D. R. Parrish and E. P. Oliveto, *Tetrahedron Lett.*, 6495 (1966).
    (b) Z. G. Hajos, D. R. Parrish and E. P. Oliveto, *Tetrahedron*, **24**, 2039 (1968).
233. G. Saucy, R. Borer and A. Furst, *Helv. Chim. Acta*, **54**, 2034 (1971).
234. G. Saucy and R. Borer, *Helv. Chim. Acta*, **54**, 2121 (1971).
235. E. Fujita, T. Fujita and Y. Nagao, *Tetrahedron*, **25**, 3717 (1969).
236. R. E. Ireland and D. M. Walba, *Tetrahedron Lett.*, 1071 (1976).
237. K. F. Cohen, R. Kazlauskas and J. T. Pinhey, *J. Chem. Soc., Perkin Trans. 1*, 2076 (1973).
238. G. Stork, G. A. Kraus and G. A. Garcia, *J. Org. Chem.*, **39**, 3459 (1974).
239. G. Stork and G. A. Kraus, *J. Am. Chem. Soc.*, **98**, 2351 (1976).
240. (a) R. Pappo and P. W. Collins, *Tetrahedron Lett.*, 2627 (1972).
    (b) R. Pappo and C. J. Jung, Ger. Offen. 2,321,984 (1973); *Chem. Abstr.* **80**, 26827b (1974).
    (c) M. M. S. Bruhn and R. Pappo, Ger. Offen 2,415,765 (1974); *Chem. Abstr.*, **82**, 86119y (1975).
    (d) R. Pappo and C. J. Jung, U. S. Pat. 3,969,391 (1976); *Chem. Abstr.*, **86**, 55057e (1977).
    (e) C. J. Sih, J. B. Heather, G. P. Peruzzotti, P. Price, R. Sood and L. F. Hsu Lee, *J. Am. Chem. Soc.*, **95**, 1676 (1973).
    (f) C. J. Sih, J. B. Heather, R. Sood, P. Price, G. P. Peruzzotti, L. F. Hsu Lee and S. S. Lee, *J. Am. Chem. Soc.*, **97**, 865 (1975).
    (g) C. J. Sih and J. B. Heather, U. S. Pat. 3, 968, 141 (1976); *Chem. Abstr.*, **86**, 29416b (1977).
241. (a) C. J. Sih, J. B. Heather, G. P. Peruzzotti, P. Price, R. Sood and L. F. Hsu Lee, *J. Am. Chem. Soc.*, **95**, 1676 (1973).
    (b) C. J. Sih, J. B. Heather, R. Sood, P. Price, G. P. Peruzzotti, L. F. Hsu Lee and S. S. Lee, *J. Am. Chem. Soc.*, **97**, 867 (1975).
242. Y. Asaka, T. Kamikawa and T. Kubota, *Tetrahedron Lett.*, 1597 (1972).
243. V. Bazant, M. Capka, M. Cerny, V. Chvalovsky, K. Kochloefl, M. Kraus and J. Malek, *Tetrahedron Lett.*, 3303 (1968).
244. M. Capka, V. Chvalovsky, K. Kochloefl and M. Kraus, *Collect. Czech. Chem. Commun.*, **34**, 118 (1969).
245. H. C. Brown and N. M. Yoon, *J. Am. Chem. Soc.*, **88**, 1464, (1966).
246. H. C. Brown, U. S. Clearinghouse, *Fed. Sci. Tech. Inform.*, AD 645581 (1966); *Chem. Abstr.*, **67**, 99306x (1967).
247. H. C. Brown and P. M. Weissman, *Isr. J. Chem.*, **1**, 430 (1963).
248. O. Kriz, J. Machacek and O. Strouf, *Collect. Czech. Chem. Commun.*, **38**, 2072 (1973).
249. J. V. Forsch, I. T. Harrison, B. Lythgoe and A. K. Saksena, *J. Chem. Soc., Perkin Trans. 1*, 2005 (1974).
250. W. Sucrow, *Tetrahedron Lett.*, 4725 (1970).
251. G. Buchi, B. Gubler, R. S. Schneider and J. Wild, *J. Am. Chem. Soc.*, **89**, 2776 (1967).
252. S. Antus, A. Gottsegen and M. Nogradi, *Synthesis*, 574 (1981).
253. S. Kim and K. H. Ahn, *J. Org. Chem.*, **49**, 1749 (1984).
254. B. M. Trost and L. N. Jungheim, *J. Am. Chem. Soc.*, **102**, 7910 (1980).
255. H. J. Williams, *Tetrahedron Lett.*, 1271 (1975).
256. (a) M. M. Bokadia, B. R. Brown, D. Cobern, A. Roberts and G. A. Somerfield, *J. Chem. Soc.*, 1658 (1962).

(b) J. Broome, B. R. Brown, A. Roberts and A. M. S. White, *J. Chem. Soc.*, 1406 (1960).
257. (a) E. C. Ashby and J. J. Lin, *Tetrahedron Lett.*, 4453 (1975).
(b) E. C. Ashby J. J. Lin and R. Kovar, *J. Org. Chem.*, **41**, 1941 (1976).
258. (a) O. Cervinka, O. Kriz and J. Cervenka, *Z. Chem.*, **11**, 109 (1971).
(b) O. Cervinka and O. Kriz, *Collect. Czech. Chem. Commun.*, **38**, 294 (1973).
259. R. Noyori, I. Tomino and M. Nishizawa, *J. Am. Chem. Soc.*, **101**, 5843 (1979).
260. S. Terashima, N. Tanno and K. Koga, *J. Chem. Soc., Chem. Commun.*, 1026 (1980).
261. (a) S.Terashima, N. Tanno and K. Koga, *Tetrahedron Lett.*, **21**, 2753 (1980).
(b) S. Terashima, N. Tanno and K. Koga, *Chem. Lett.*, 981 (1980).
262. J. Huton, M. Senior and N. C. A. Wright, *Synth. Commun.*, **9**, 799 (1979).
263. N. Cohen, R. J. Lopresti, C. Neukom and G. Saucy, *J. Org. Chem.*, **45**, 582 (1980).
264. R. S. Brinkmeyer and V. M. Kapoor, *J. Am. Chem. Soc.*, **99**, 8339 (1977).
265. W. S. Johnson, R. S. Brinkmeyer, V. M. Kapoor and T. M. Yarnell, *J. Am. Chem. Soc.*, **99**, 8341 (1977).
266. J. P. Vigneron and V. Bloy, *Tetrahedron Lett.*, 2683 (1979).
267. J. P. Vigneron and V. Bloy, *Tetrahedron Lett.*, **21**, 1735 (1980).
268. J. P. Vigneron and J. M. Blanchard, *Tetrahedron Lett.*, **21**, 1739 (1980).
269. M. Nishizawa, M. Yamada and R. Noyori, *Tetrahedron Lett.*, **22**, 247 (1981).
270. S. R. Landor, B. J. Miller and A. R. Tatchell, *J. Chem. Soc.* (C), 1822 (1966).
271. S. R. Landor, B. J. Miller and A. R. Tatchell, *Proc. Chem. Soc.*, 227 (1964).
272. S. R. Landor, B. J. Miller and A. R. Tatchell, *J. Chem. Soc.* (C), 2339 (1971).
273. M. Nishizawa and R. Noyori, *Tetrahedron Lett.*, **21**, 2821 (1980).
274. T. Sato, Y. Gotoh, Y. Wakabayashi and T. Fujisawa, *Tetrahedron Lett.*, **24**, 4123 (1983).
275. R. Noyori, I. Tomino and M. Nishizawa, *J. Am. Chem. Soc.*, **101**, 3843 (1979).
276. C. J. Sih, J. B. Heather, R. Sood, P. Price, G. Peruzzotti, L. F. Hsu Lee and S. S. Lee, *J. Am. Chem. Soc.*, **97**, 865 (1975).
277. Y. Nagai, *Intra-Sci. Chem. Rep.*, **4**, 115 (1970).
278. D. N. Kursanov, Z. N. Parnes and N. M. Loim, *Synthesis*, 633 (1974).
279. (a) Z. N. Parnes, N. M. Loim, V. A. Baranova and D. N. Kursanov, *Zh. Org. Khim.*, **7**, 2066 (1977); *Chem. Abstr.*, **76**, 13495 (1972).
(b) D. N. Kursanov *et al.*, *Izv. Akad. Nauk SSSR, Ser. Khim.*, 843 (1974).
280. D. N. Kursanov, N. M. Loim, V. A. Baranova, L. V. Moiseeva, L. P. Zalukaev and Z. N. Parnes, *Synthesis*, 420 (1973).
281. E. Yoshii, T. Koizumi, I. Hayashi and Y. Hiroi, *Chem. Pharm. Bull.*, **25**, 1468 (1977).
282. G. G. Furin, O. A. Vyazankina, B. A. Gostevsky and N. S. Vyazankin, *Tetrahedron*, **44**, 2675 (1988).
283. R. J. P. Corriu, R. Perz and C. Reye, *Tetrahedron*, **39**, 999 (1983).
284. M. P. Doyle, C. T. West, S. J. Donnelly and C. C. McOsker, *J. Organomet. Chem.*, **117**, 129 (1976).
285. M. Kira, K. Sato and H. Sakurai, *J. Org. Chem.*, **52**, 948 (1987).
286. R. A. Benkeser, *Acc. Chem. Res.*, **4**, 94 (1971).
287. (a) H. G. Kuivila, *Synthesis*, 499 (1970).
(b) A. Hajos, *Complex Hydrides and Related Reducing Agents in Organic Synthesis*, Amsterdam, Elsevier, 1979.
(c) Y. I. Baukov and I. F. Lutsenko, *Organomet. Chem. Rev., A*, **6**, 355 (1970).
288. (a) H. G. Kuivila and O. F. Beumel, *J. Am. Chem. Soc.*, **83**, 1246 (1961).
(b) H. G. Kuivila and O. F. Beumel, *J. Am. Chem. Soc.*, **80**, 3798 (1958).
(c) G. J. M. Van Der Kerk, J. G. A. Luijten and J. G. Noltes, *Chem. Ind.*, 352 (1956).
(d) G. J. M. Van Der Kerk, J. G. Noltes and J. G. A. Luijten, *J. Appl. Chem.*, **7**, 356 (1957).
(e) G. J. M. Van Der Kerk and J. G. Noltes, *J. Appl. Chem.*, **9**, 106 (1959).
(f) J. G. Noltes and G. J. M. Van Der Kerk, *Chem. Ind.*, 294 (1959).
(g) I. F. Lutsenko, S. V. Ponomarev and O. P. Petri, *Obshch. Khim.*, **32**, 896 (1962).
(h) M. Pereyre and J. Valade, *C.R. Acad. Sci. Paris*, **258**, 4785 (1964).
(i) M. Pereyre and J. Valade, *C.R. Acad. Sci. Paris*, **260**, 581 (1965).
(j) M. Pereyre and J. Valade, *Bull. Soc. Chim. Fr.*, 1928 (1967).
(k) M. Pereyre, G. Colin and J. Valade, *Tetrahedron Lett.*, 4805 (1967).
(l) A. J. Leusink and J. G. Noltes, *Tetrahedron Lett.*, 2221 (1966).

289. M. Pereyre and J. Valade, *Tetrahedron Lett.*, 489 (1969).
290. H. Laurent, P. Esperling and G. Baude, *Ann. Chem.*, 1996 (1983).
291. (a) B. R. Laliberte, W. Davidson and M. C. Henry, *J. Organomet. Chem.*, **5**, 526 (1966).
     (b) A. J. Leusink and J. G. Noltes, *J. Organomet. Chem.*, **16**, 91 (1969).
     (c) W. P. Neumann, H. Niermann and R. Sommer, *Ann. Chim.*, **659**, 27 (1962).
     (d) M. Pereyre, G. Colin and J. Valade, *Bull. Soc. Chim. Fr.*, 3358 (1968).
     (e) S. Matsuda, Sh. Kikkava and I. Omae, *J. Organomet. Chem.*, **18**, 95 (1969).
292. (a) A. J. Leusink and J. G. Noltes, *Tetrahedron Lett.*, 335 (1966).
     (b) W. P. Neumann and R. Sommer, *Ann. Chim.*, **675**, 10 (1964).
293. G. A. Posner, *Org. React.*, **19**, 1 (1972).
294. (a) M. F. Semmelhack and R. D. Stauffer, *J. Org. Chem.*, **40**, 3619 (1975).
     (b) M. F. Semmelhack, R. D. Stauffer and A. Yamashita, *J. Org. Chem.*, **42**, 3180 (1977).
295. M. E. Osborn, J. F. Pegues and L. A. Paquette, *J. Org. Chem.*, **45**, 167 (1980).
296. T. Saegusa, K. Kawasaki, T. Fujii and T. Tsuda, *J. Chem. Soc., Chem. Commun.*, 1013 (1980).
297. T. Tsuda, T. Hayashi, H. Suton, T. Kamamoto and T. Saegusa, *J. Org. Chem.*, **51**, 537 (1986).
298. R. K. Boeckman, Jr. and R. Michalak, *J. Am. Chem. Soc.*, **96**, 1623 (1974).
299. S. Masamune, G. S. Bates and P. E. Georghiou, *J. Am. Chem. Soc.*, **96**, 3686 (1974).
300. E. C. Ashby, J. J. Lin and A. B. Goel, *J. Org. Chem.*, **43**, 183 (1978).
301. T. H. Lemmen, K. Folting, J. C. Huffman and K. G. Caulton, *J. Am. Chem. Soc.*, **107**, 7774 (1985).
302. W. S. Mahoney, D. M. Brestensky and J. M. Stryker, *J. Am. Chem. Soc.*, **110**, 291 (1988).
303. (a) G. F. Cainelli, M. Panunzio and A. Umani-Ronchi, *J. Chem. Soc., Perkin Trans 1*, 1273 (1975).
     (b) G. F. Cainelli, M. Panunzio and A. Umani-Ronchi, *Tetrahedron Lett.*, 2491 (1973).
304. M. Yamashita, K. Miyoshi, Y. Okada and R. Suemitsu, *Bull. Chem. Soc. Jpn.*, **55**, 1329 (1982).
305. G. P. Boldrini and A. Umani-Ronchi, *J. Organomet. Chem.*, **171**, 85 (1979).
306. (a) J. P. Collman, R. G. Finke, P. L. Matlock, R. Wahren and J. I. Brauman, *J. Am. Chem. Soc.*, **98**, 4685 (1976).
     (b) J. P. Collman, R. G. Finke, P. L. Matlock, R. Wahren, R. G. Komoto and J. I. Brauman, *J. Am. Chem. Soc.*, **100**, 1119 (1978).
307. T. Imamoto, T. Mita and M. Yokomoto, *J. Chem. Soc., Chem. Commun.*, 163 (1984).
308. G. P. Boldrini and A. Umani-Ronchi, *Synthesis*, 596 (1976).
309. R. W. Goetz and M. Orchin, *J. Am. Chem. Soc.*, **85**, 2782 (1963).
310. P. H. Gibson and Y. S. El-Omrani, *Organometallics*, **4**, 1473 (1985).
311. (a) A. W. Johnstone, A. H. Wilby and I. D. Entwistle, *Chem. Rev.*, **85**, 129 (1985).
     (b) G. Brieger and T. J. Nestrick, *Chem. Rev.*, **74**, 567 (1974).
     (c) G. W. Parshall, *Catal. Rev.*, **23**, 107 (1981).
312. Y. Sasson and J. Blum, *J. Org. Chem.*, **40**, 1887 (1975).
313. Y. Sasson and J. Blum, *Tetrahedron Lett.*, 2167 (1971).
314. V. Z. Sharf, L. K. Freidlin, I. S. Shekoyan and V. N. Krutii, *Izv. Akad. Nauk SSSR, Ser. Khim.*, 575 (1976); 834 (1977) [*Bull. Acad. Sci. USSR, Div. Chem. Sci.*, **25**, 557 (1976); **26**, 758 (1977)].
315. Y. Sasson, M. Cohen and J. Blum, *Synthesis*, 359 (1973).
316. G. Descotes and J. Sabadie, *Bull. Soc. Chim. Fr.*, Pt-2, 158 (1978).
317. G. Speier and L. Marko, *J. Organomet. Chem.*, **210**, 253 (1981).
318. S. L. Regen and G. M. Whitesides, *J. Org. Chem.*, **37**, 1832 (1972).
319. Y. Sasson, J. Blum and E. Dunkelblum, *Tetrahedron Lett.*, 3199 (1973).
320. Y. Sasson and G. L. Rempel, *Can. J. Chem.*, **52**, 3825 (1974).
321. T. Nishiguchi, H. Imai, Y. Hirose and K. Fukuzumi, *J. Catal.*, **41**, 249 (1976).
322. A. Dobson, D. S. Moore and S. D. Robinson, *J. Organomet. Chem.*, **177**, C8 (1979).
323. M. Bianchi, U. Matteoli, G. Menchi, P. Frediani, F. Piacenti and C. Botteghi, *J. Organomet. Chem.*, **195**, 337 (1980).
324. K. Ohkubo, I. Terada and K. Yoshinaga, *Inorg. Nucl. Chem. Lett.*, **15**, 421 (1979).
325. (a) K. Ohkubo, K. Hirata, K. Yoshinaga and M. Okada, *Chem. Lett.*, 183 (1976).
     (b) K. Ohkubo, K. Hirata and K. Yoshinaga, *Chem. Lett.*, 577 (1976).
     (c) K. Ohkubo, T. Shoji, I. Terada and K. Yoshinaga, *Inorg. Nucl. Chem. Lett.*, **13**, 443 (1977).
326. G. Descotes and D. Sinou, *Tetrahedron Lett.*, 4083 (1976).
327. G. Descotes, J. P. Praly and D. Sinou, *J. Mol. Catal.*, **6**, 421 (1979).
328. (a) D. Beaupere, P. Bauer and R. Uzan, *Can. J. Chem.*, **57**, 218 (1979).

    (b) D. Beaupere, L. Nadjo, R. Uzan and P. Bauer, *J. Mol. Catal.*, **14**, 129 (1982).
    (c) D. Beaupere, P. Bauer, L. Nadjo and R. Uzan, *J. Organomet. Chem.*, **231**, C49 (1982).
    (d) D. Beaupere, P. Bauer, L. Nadjo and R. Uzan, *J. Mol. Catal.*, **18**, 73 (1983).
329. D. Beaupere, L. Nadjo, R. Uzan and P. Bauer, *J. Mol. Catal.*, **20**, 185, 195 (1983).
330. A Camus, G. Mestroni and G. Zassinovich, *J. Organomet. Chem.*, **184**, C10 (1980).
331. E. Keinan and P. A. Gleize, *Tetrahedron Lett.*, **23**, 477 (1982).
332. (a) P. Four and F. Guibe, *Tetrahedron Lett.*, **23**, 1825 (1982).
    (b) Y. T. Xian, P. Four, F. Guibe and G. Balavoine, *Nouv. J. Chim.*, **8**, 611 (1984).
333. E. Keinan and N. Greenspoon, *Tetrahedron Lett.*, **23**, 241 (1982).
334. (a) E. Keinan and N. Greenspoon, *J. Org. Chem.*, **48**, 3545 (1983).
    (b) E. Keinan and N. Greenspoon, *Isr. J. Chem.*, **24**, 82 (1984).
    (c) N. Greenspoon and E. Keinan, *J. Org. Chem.*, **53**, 3723 (1988).
335. (a) E. Keinan and N. Greenspoon, *J. Am. Chem. Soc.*, **108**, 7314 (1986).
    (b) E. Keinan and N. Greenspoon, *Tetrahedron Lett.*, **26**, 1353 (1985).
336. E. Keinan, N. Godinger and N. Greenspoon, unpublished results.
337. (a) T. Tatsumi, M. Shibagaki and H. Tominaga, *J. Mol. Catal.*, **13**, 331 (1981).
    (b) T. Tatsumi, K. Hashimoto, H. Tominaga, Y. Mizuta, K. Hata, M. Hidai and Y. Uchida, *J. Organomet. Chem.*, **252**, 105 (1983).
    (c) Y. Lin and X. Lu, *J. Organomet. Chem.*, **251**, 321 (1983).
338. (a) L. Marko and Z. Nagy-Magos, *J. Organomet. Chem.*, **285**, 193 (1985).
    (b) E. N. Frankel, *J. Org. Chem.*, **37**, 1549 (1972).
    (c) M. Sodeoka and M. J. Shibasaki, *J. Org. Chem.*, **50**, 1147 (1985).
339. E. Keinan and D. Perez, *J. Org. Chem.*, **52**, 2576 (1987).
340. D. Perez, N. Greenspoon and E. Keinan, *J. Org. Chem.*, **52**, 5570 (1987).
341. (a) I. Ojima, T. Kogure and Y. Nagai, *Tetrahedron Lett.*, 5035 (1972).
    (b) I. Ojima and T. Kogure, *Organometallics*, **1**, 1390 (1982).
    (c) I. Ojima, M. Nihonyanagi, T. Kogure, M. Kumagai, S. Horiuchi and K. Nakatsugawa, *J. Organomet. Chem.*, **94**, 449 (1975).
342. H. J. Liu and B. Ramani, *Synth. Commun.*, **15**, 965 (1985).
343. A. J. Cornish, M. F. Lappert, G. L. Filatvos and T. A. Nile, *J. Organomet. Chem.*, **172**, 153 (1979).
344. T. Hayashi, K. Yamamoto and M. Kumada, *Tetrahedron Lett.*, 3 (1975).
345. (a) T. Kogure and I. Ojima, *J. Organomet. Chem.*, **234**, 249 (1982).
    (b) I. Ojima and T. Kogure, *Chem. Lett.*, 985 (1975).
346. M. Kobayashi, T. Koyama, K. Ogura, S. Seto, F. J. Ritter and I. E. M. Bruggemann-Rotgans, *J. Am. Chem. Soc.*, **102**, 6602 (1980).
347. (a) D. L. Bailey, U. S. Patent 2,917,530 (1959), *Chem. Abstr.*, **54**, 6549 (1960); U. S. Patent 2,970,150 (1961), *Chem. Abstr.*, **55**, 16423 (1961).
    (b) E. Y. Lukevits, *Izv. Akad. Nauk Latv. SSSR*, 111 (1963).
    (c) A. D. Petrov and S. I. Sadykh-Zade, *Dokl. Akad. Nauk SSSR*, **121**, 119 (1959).
    (d) A. D. Petrov, V. F. Mironov, V. A. Ponomarenko, S. I. Sadykh-Zade and E. A. Chernyshov, *Izv. Akad. Nauk SSSR*, 954 (1968).
    (e) S. I. Sadykh-Zade and A. D. Petrov, *Zh. Obshch. Khim.*, **29**, 3194 (1959).
348. (a) E. Frainnet, *Pure Appl. Chem.*, **19**, 489 (1969).
    (b) E. Frainnet and R. Bourhis, *Bull. Soc. Chim. Fr.*, 2134 (1966).
    (c) R. Bourhis, E. Frainnet and F. Moulines, *J. Organomet. Chem.*, **141**, 157 (1977).
349. (a) A. D. Petrov and S. I. Sadykh-Zade, *Bull. Soc. Chim. Fr.*, 1932 (1959).
    (b) A. D. Petrov, S. I. Sadykh-Zade and E. I. Filatova, *Zh. Obshch. Khim.*, **29**, 2936 (1959).
350. G. Stork and T. L. Macdonald, *J. Am. Chem. Soc.*, **97**, 1264 (1975).
351. E. Yoshii, H. Ikeshima and K. Ozaki, *Chem. Pharm. Bull.*, **20**, 1827 (1972).
352. E. Yoshii, Y. Kobayashi, T. Koizumi and T. Oribe, *Chem. Pharm. Bull.*, **22**, 2767 (1974).
353. A. P. Barlow, N. M. Boag and F. G. A. Stone, *J. Organomet. Chem.*, **191**, 39 (1980).
354. J. Blum, Y. Sasson and S. Iflah, *Tetrahedron Lett.*, 1015 (1972).
355. H. Imai, T. Nishiguchu and K. Fukuzumi, *Chem. Lett.*, 655 (1976).
356. (a) M. E. Vol'pin, V. P. Kukolev, V. O. Chernyshev and I. S. Kolomnikov, *Tetrahedron Lett.*, 4435 (1971).
    (b) I. S. Kolomnikov, Y. D. Koreshov, V. P. Kukolev, V. A. Mosin and M. E. Vol'pin, *Izv. Akad. Nauk SSSR, Ser. Khim.*, 175 (1973) [*Bull. Acad. Sci. USSR, Div. Chem. Sci.*, **22**, 180 (1973)].
357. N. A. Cortese and R. F. Heck, *J. Org. Chem.*, **43**, 3985 (1978).

358. N. A. Cortese and R. F. Heck, *J. Org. Chem.*, **42**, 3491 (1977).
359. J. Tsuji and T. Yamakawa, *Tetrahedron Lett.*, 613 (1979).
360. A. M. Caporusso, G. Giacomelli and L. Lardicci, *J. Org. Chem.*, **47**, 4640 (1982).
361. P. Caubere, *Angew. Chem., Int. Ed. Engl.*, **22**, 599 (1983).
362. J. J. Brunet, L. Mordenti, B. Loubinoux and P. Caubere, *Tetrahedron Lett.*, 1069 (1978).
363. J. J. Brunet, L. Mordenti and P. Caubere, *J. Org. Chem.*, **43**, 4804 (1978).
364. L. Mordenti, J. J. Brunet and P. Caubere, *J. Org. Chem.*, **44**, 2203 (1979).
365. A. Fauve and A. Kergomard, *Tetrahedron*, **37**, 899 (1981).
366. L. Mamoli, R. Roch and H. Teschen, *Z. Physiol. Chem.*, **261**, 287 (1939).
367. T. L. Miller and E. J. Hessler, *Biochem. Biophys. Acta*, **202**, 354 (1970).
368. H. C. Murray and D. H. Peterson, US Patent 2659743 (1953); *Chem. Abstr.*, **48**, 13737c (1954).
369. H. C. Murray and D. H. Peterson, US Patent 2649402 (1953).
370. (a) A. Kergomard, M. F. Renard and H. Veschambre, *J. Org. Chem.*, **47**, 792 (1982).
    (b) A. Kergomard, M. F. Renard and H. Veschambre, *Tetrahedron Lett.*, 5197 (1978).
    (c) G. Dauphin, J. C. Gramain, A. Kergomard, M. F. Renard and H. Veschambre, *Tetrahedron Lett.*, **21**, 4275 (1980).
    (d) G. Dauphin, J. C. Gramain, A. Kergomard, M. F. Renard and H. Veschambre, *J. Chem. Soc. Chem. Commun.*, 318 (1980).
371. (a) Y. Noma, S. Nonomura, H. Ueda and C. Tatsumi, *Agric. Biol. Chem.*, **38**, 735 (1974).
    (b) Y. Noma and S. Nonomura, *Agric. Biol. Chem.*, **38**, 741 (1974).
372. (a) ►M. Bostmembrun-Desrutt, G. Dauphin A. Kergomard, M. F. Renard and H. Veschambre, *Tetrahedron*, **41**, 3679 (1985).
    (b) M. Desrut, A. Kergomard, M. F. Renard and H. Veschambre, *Tetrahedron*, **37**, 3825 (1981).
373. (a) A. Kergomard, M. F. Renard and H. Veschambre, *Agric. Biol. Chem.*, **49**, 1497 (1985).
    (b) M. Desrut, A. Kergomard, M. F. Renard and H. Veschambre, *Biochem. Biophys. Res. Commun.*, **110**, 908 (1983).
    (c) A. Kergomard, M. F., Renard and H. Veschambre, *Agric. Biol. Chem.*, **46**, 97 (1982).
    (d) A. Kergomard, M. F., Renard, H. Veschambre, C. A. Groliere and J. Dupy-Blanc, *Agric. Biol. Chem.*, **50**, 487 (1986).
374. T. Kitazume and N. Ishikawa, *Chem. Lett.*, 587 (1984).
375. (a) P. Gramatica, P. Manitto and L. Poli, *J. Org. Chem.*, **50**, 4625 (1985).
    (b) P. Gramatica, P. Manitto, B. M. Ranzi, A. Delbianco and M. Francavilla, *Experientia*, **38**, 775 (1982).
376. H. G. W. Leuenberger, W. Boguth, R. Barner, M. Schmid and R. Zell, *Helv. Chim. Acta*, **62**, 455 (1979).
377. H. G. W. Leuenberger, W. Boguth, E. Widmer and R. Zell, *Helv. Chim. Acta*, **59**, 1832 (1976).
378. C. Fuganti and P. Grasselli, *J. Chem. Soc., Chem. Commun.*, 995 (1979).
379. M. Miyano, C. R. Dorn, F. B. Colton and W. J. Marsheck, *J. Chem. Soc., Chem. Commun.*, 425 (1971).
380. (a) B. Eckstein and A. Nimrod, *Biochim. Biophys. Acta*, **1**, 499 (1977).
    (b) I. A. Watkinson, D. C. Wilton, A. D. Rahimtula and M. M. Akhtar, *Eur. J. Biochem.*, **1**, 23 (1971).
381. E. A. Braude, J. Hannah and R. Linstead, *J. Chem. Soc.*, 3257 (1960).
382. B. E. Norcross, P. E. Klinedinst, Jr. and F. H. Westheimer, *J. Am. Chem. Soc.*, **84**, 797 (1962).
383. J. S. McGuire and G. M. Tompkins, *Fed. Proc.*, **19**, A29 (1960).
384. (a) Y. Ohnishi, M. Kagami and A. Ohno, *Chem. Lett.*, 125 (1975).
    (b) Y. Ohnishi, M. Kagami, T. Numakunai and A. Ohno, *Chem. Lett.*, 915 (1976).
385. K. Nakamura, M. Fujii, A. Ohno and S. Oka, *Tetrahedron Lett.*, **25**, 3983 (1984).
386. R. A. Gase and U. K. Pandit, *J. Am. Chem. Soc.*, **101**, 7059 (1979).
387. (a) M. J. de Nie-Sarink and U. K. Pandit, *Tetrahedron Lett.*, 2449 (1979).
    (b) U. K. Pandit, F. R. Mas Cabre, R. A. Gase and M. J. de Nie-Sarink, *J. Chem. Soc., Chem. Commun.*, 627 (1974).
388. F. Yoneda, K. Kuroda and K. Tanaka, *J. Chem. Soc., Chem. Commun.*, 1194 (1984).
389. N. Baba, T. Makino, J. Oda and Y. Inouye, *Can. J. Chem.*, **58**, 387 (1980).
390. A. Fischli and D. Suss, *Helv. Chim. Acta*, **62**, 2361 (1979).
391. (a) F. Camps, J. Coli, A. Guerrero, J. Guitart and M. Riba, *Chem. Lett.*, 715 (1982).
    (b) O. Louis-Andre and G. Gelbard, *Tetrahedron Lett.*, **26**, 831 (1985).
392. H. Chikashita, M. Miyazaki and K. Itoh, *Synthesis*, 308 (1984).

393. S. K. Malhotra, D. F. Moakley and F. Johnson, *J. Am. Chem. Soc.*, **89**, 2794 (1967).
394. M. Yamashita, Y. Kato and R. Suemitsu, *Chem. Lett.*, 847 (1980).
395. H. Stamm, A. Sommer, A. Onistschenko and A. Woderer, *J. Org. Chem.*, **51**, 4979 (1986).
396. A. C. Chan and D. I. Schuster, *J. Am. Chem. Soc.*, **108**, 4561 (1986).
397. G. H. Posner and A. W. Runquist, *Tetrahedron Lett.*, 3601 (1975).
398. K. Nanjo, K. Suzuki and M. Sekiya, *Chem. Pharm. Bull.*, **25**, 2396 (1977).
399. G. Giacomelli, A. M. Caporusso and L. Lardicci, *Tetrahedron Lett.*, **22**, 3663 (1981).

The Chemistry of Enones
Edited by S. Patai and Z. Rappoport
© 1989 John Wiley & Sons Ltd

CHAPTER **19**

Organometallic derivatives of α, β-unsaturated enones

JAMES A. S. HOWELL

Chemistry Department, University of Keele, Keele, Staffordshire, ST5 5BG, UK

I. INTRODUCTION

The purpose of this chapter is to review the chemistry of α, β-unsaturated enones bound as ligands to low valent mono- and polymetallic transition metal centres. The enone ligand in such complexes may most usefully be classified in terms of the formal number of electrons donated to the metal centre; thus, structures **1** to **7**, for which examples all exist in the literature, represent donation of one, two, three or the maximum of four electrons. For the complexes described here, the set of auxiliary ligands L_n completes the 16- or 18-electron configuration at the metal centre. In general, the normal organic reactivity of the enone is substantially retained in the η^1-structures **2** and **3**, while that of the η^1-acyl structure **1** differs substantially. For low valent metals, η^2-coordination to the C=C bond in **4** is almost invariably preferred relative to coordination to a ketonic lone pair. Three-electron coordination in **5** and **6** is completed by chelation of the C=C bond and a ketonic lone pair respectively, while η^4-complexes contain the enone bound via its 4π-electron system. One may note the potentially facile interconversion of structural types [$1 \rightleftarrows 5, 3 \rightleftarrows 6, 4 \rightleftarrows 7$] through loss or gain of a two-electron auxiliary ligand.

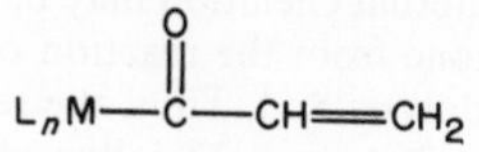

$$L_n M - \overset{\overset{\displaystyle O}{\|}}{C} - CH = CH_2$$

(1)

$$L_n M - \overset{\overset{\displaystyle COR}{|}}{C} = CH_2$$

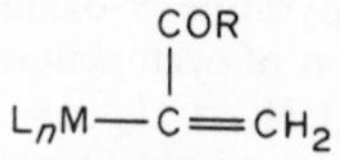

(2)

$$L_nMCH\!\!=\!\!CHCOR$$

(3)

$$L_nM \longleftarrow \overset{CH_2}{\underset{CHCOR}{\|}}$$

(4)

(5)

(6)

(7)

II. COMPLEXES CONTAINING ONE- AND THREE-ELECTRON DONOR LIGANDS

η^1-complexes may be prepared by reaction of metal anion with β-haloenones. Thus, treatment of NaCpFe(CO)$_2$ with MeCOCH=CHCl yields **8**[1,2]; the normal ketonic reactivity of **8** is demonstrated in formation of the hydrazone **9b**[3] and in reaction with Et$_3$OBF$_4$ followed by PhNH$_2$, to give **9a**[4]. Photolysis in the presence of PPh$_3$ yields **11** rather than the product of insertion or internal chelation[5]. Most interesting is the reaction with MeLi, followed by protonation, to yield the carbene complex **10** which shows potential as a cyclopropanation reagent[1].

The isomeric acyl complexes **12** and **13** may be prepared through a similar reaction of acid chloride with metal anion; only the *trans* isomer of **12** is isolated from either *cis* or *trans* acid chloride[1,6,7]. Complex **13** may be photochemically decarbonylated to the vinyl complex **14** (or **15** in the presence of PPh$_3$), and protonation yields the carbene derivatives **16** or **17**[1]. Internal chelation under mild conditions has been observed in the transformation of **18** to **19** on heating in hexane[8], and is also observed in complexes amenable to M—H or M—R insertion. Thus, whereas reaction of Fe(CO)$_4^{2-}$ with CH$_2$=CHCOCl yields the stable acyl anion **20**[9], reaction with *cis*-BrCH=CHCO$_2$Me yields the chelated complex **22a**, presumably via initial formation of the vinyl complex **21** followed by rapid insertion of CO[10]. Complexes of structure **22** are generally more accessible through reaction of alkynes with HFe(CO)$_4^-$, in which the σ-vinyl intermediate is generated by insertion of alkyne into the Fe—H bond[10]. The reactivity of **20** and **22** differs substantially; whereas acidolysis or reaction of **20** with alkyl halide yields aldehyde and ketone respectively[11,12], **22d** is protonated at the carbon β to the metal to give the alkene complex **23**, and is alkylated at oxygen to give the carbene complex **24** which may be oxidized with pyridine-N-oxide to **25**[10]. Transient internal chelation may be responsible for the isolation of cyclopentanone and cyclohexanone from the reaction of Fe(CO)$_4^{2-}$ with Br(CH$_2$)$_n$CH=CH$_2$ ($n = 2, 3$), followed by acidolysis[13,14]. Thus, the reaction may proceed by insertion of CO into the initial σ complex **26** to give **27**, followed by internal cyclization to give **28** and acidolysis to release the cyclic ketone. The reaction is sensitive to

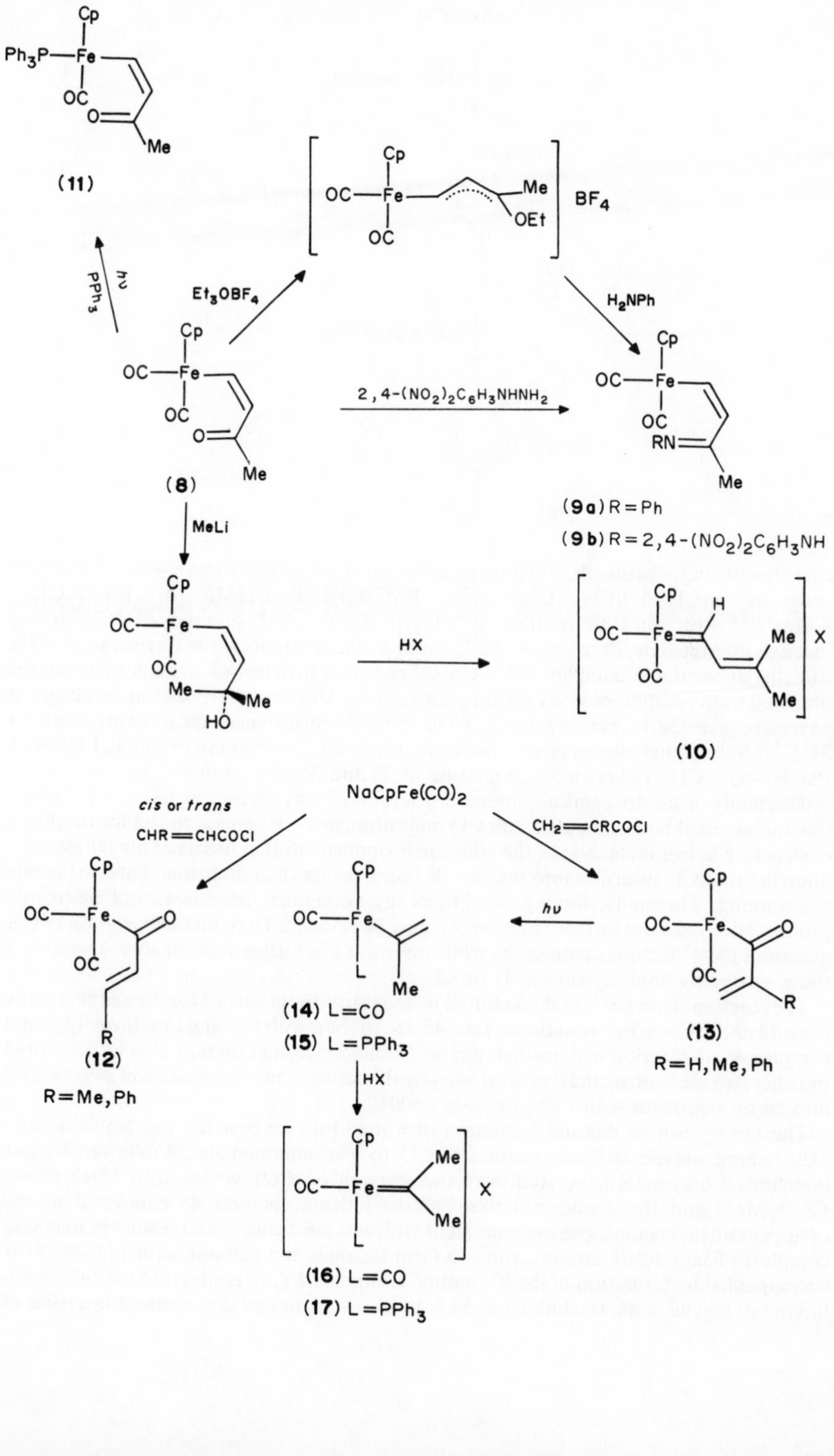

$$NaRe(CO)_5$$

$$\downarrow \;\; PhCOCH{=}CHCl$$

$(CO)_5Re$ ... (+ *trans* isomer)

O= ... Ph

(18)

$$\downarrow \;\; \Delta \; (cis \; only)$$

$(CO)_4Re$ ...

O= ... Ph

(19)

chain length and substituent, and no cyclized product is isolated where $n = 4$, or for halides such as $BrCH(Me)(CH_2)_2CH{=}CH_2$, $Br(CH_2)_3CH{=}CHMe$, or $BrCH_2CH{=}CMe_2$[11,14]. Reaction with the allene $Br(CH_2)_2CH{=}C{=}CH_2$ proceeds in a similar way through intermediate **29** to release **30**[14], though alkylation occurs at oxygen to give the trimethylenemethane complex **31**[15]. Uncyclized intermediates of structure **32** may be obtained from reaction of $[Fe(CO)_4R]^-$ with allene. Alkylation or protonation occurs at oxygen to give **33a, b**; rearrangement of **33b** on mild heating yields the η^4-enone complex **34**[15-19]. Substituted allenes give isomeric mixtures; thus, reaction of $[Fe(CO)_4Et]^-$ with $PhCH{=}C{=}CH_2$ yields an 80:20 mixture of **35** and **36**.

Thermally or photochemically induced insertion of alkynes into metal–acyl bonds, or into metal–alkyl bonds coupled with CO migration, provide general routes to complexes of structure **6**. For metal alkyls, the addition is opposite to that observed for $HFe(CO)_4^-$, implying that CO insertion into the M—R bond, rather than insertion of alkyne, is rate determining. Thermally, forcing conditions are sometimes necessary, and frequently products derived from further reaction of **6** may be isolated. Thus, further insertion of CO generates the η^3-lactone complex **40**, while insertion of a further mole of alkyne generates the η^3- or η^5-pyranyl derivatives **41** or **42**.

The reaction sequence is best illustrated by the transformation of **44**, obtained thermally from **43** and $HC{\equiv}CBu^t$, into the lactone **45** on reaction with CO and into the η^3-pyranyl complex **46** on reaction with further alkyne[20]. Lactone formation may also be promoted by other two-electron ligands, as illustrated by the conversion of complexes of structure **49** into **50** on treatment with PPh_3 or isocyanide[21].

The mechanism of thermal formation of **6** may thus be best represented as a rate-determining, alkyne-assisted insertion of CO to give intermediate **39** followed by fast insertion of alkyne. Kinetic studies of the reaction of $(CO)_5MnMe$ with $MeO_2CC{\equiv}CCO_2Me$[22] and the greater reactivity of the indenyl complex **43** compared to the cyclopentadienyl complex **48** are consistent with this mechanism. The related manganese complexes **53a–c** resist carbonylation to form lactones, but reaction with $PhC{\equiv}CH$ is accompanied by formation of the η^5-pyranyl complex **54**[23]. In contrast to the tail-to-tail linking of alkyne in **46**, the linking in **54** is head-to-tail, implying a reversed insertion of

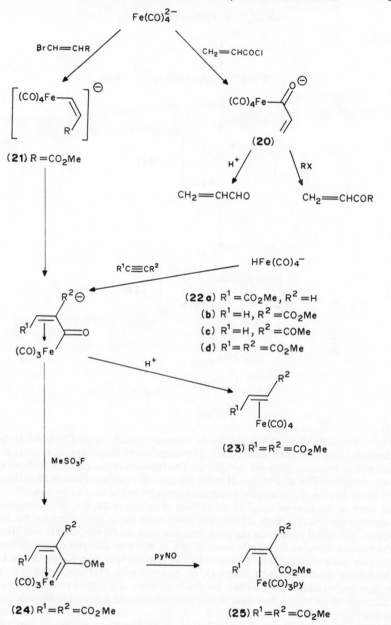

alkyne in the conversion of **39** to **6**. This may be ascribed to the minimized steric hindrance of the Bu^t group in **44**, and to the enhanced electronic stability conferred on the M—C σ bond by the α-phenyl substituent of **53b, c**. Indeed, the indenyl complex **43** reacts photochemically with both MeC≡CH and PhC≡CH to yield the α-substituted complex

$$Fe(CO)_4^{2-} \ + \ Br(CH_2)_nCH{=\!\!=}CH_2$$

$$\Big[(CO)_4Fe(CH_2)_nCH{=\!\!=}CH_2\Big]^{\ominus} \quad (26)$$

$$\left[\begin{array}{c}(CO)_3Fe \quad\quad O \\ \diagdown\quad\diagup\!\!\diagup \\ \diagup\!\!\diagup \diagdown (CH_2)_n\end{array}\right]^{\ominus} \quad (27)$$

$$\left[(CO)_3Fe{-}\!\!\diagdown\!\!\diagup\!\!\diagdown\!\!\diagup^{O}_{(CH_2)_n}\right] \quad (28)$$

H⁺

$$\diagdown\!\!\diagup\!\!\diagdown\!\!\diagup^{O}_{(CH_2)_n} \qquad n=2,3$$

55[24–26]. Similarly, CpFe(CO)$_2$Me reacts photochemically with CF$_3$C≡CH to give exclusively the η^5-pyranyl isomer **56**[27]. The direction of initial insertion is sensitive to metal size in sterically crowded complexes. The reaction of the pentamethylcyclopentadienyl complexes **57** with PhC≡CH yields the sterically preferred isomer **58** in the case of chromium, but the electronically preferred isomer **59** in the case of tungsten[28].

At least in the case of tungsten complexes of structure **48**, the initial stages in the photochemical reaction with alkyne may differ from those postulated for the thermal reaction. The initial product of the photoreaction between CpW(CO)$_3$Me and HC≡CH is the *mono*carbonyl complex (**60**) containing a formal four-electron donor alkyne. This undergoes facile reaction with PMe$_3$ or CO to give the insertion products **61a, b** while more forcing reaction of **61b** with CO, or reaction with P(OMe)$_3$, results in alkyne insertion to give **62a, b**. Use of PMe$_3$ results in addition of a second mole of PMe$_3$ at the α-carbon to give **63**[31–33]. Insertion is also promoted by the reaction of **60** with NOCl to give **61c**[34].

Direct conversion of metal acyl **37** to **6** is accompanied by ligand loss (usually carbon monoxide), and therefore becomes increasingly facile towards the right-hand side of the transition metal series. Thus, whereas **47** or **52** requires elevated temperature and/or long reaction times, reaction of cobalt acyls such as **64** with alkynes occurs more easily to yield directly the lactone complex **65**. Hydrogenation yields the free, saturated lactone **66**, but the reaction may be made catalytic in cobalt if the acyl group R^1 contains an activated hydrogen. Thus, deprotonation of **67** yields the free unsaturated lactone **68** with release of $Co(CO)_4^-$ which may be recycled to **67** as shown[35].

The conversion of acyl chlorides to lactones using $Ni(CO)_4$ may similarly be viewed as proceeding through the intermediates **69** to **71** with final hydrolysis liberating the unsaturated lactone **72**[36]. Part of this reaction has recently been modelled in the conversion of the acyl complex 73 to 74 on reaction with PhC≡CH; heating results in phosphine migration to the α-carbon to give 75[37].

Protonation of 76a in non-coordinating acid (HBF$_4$) yields the η^4-enone complex 77[38], whereas in coordinating acid (CF$_3$COOH), addition of two moles of acid occurs to 76b, c to yield either 79 and the liberated ketone or the η^1-complex 78 in which effective hydrogenation of the C=C bond has occurred[39]. Complexes 76c react with PMe$_3$ to give both the product of carbonyl substitution 80 and the product of addition at the α-carbon 81[39]; molybdenum yields only the addition product 81, whereas further substitution of the tungsten complex occurs to give 82[40,41]. Protonation of the cyclic derivatives 84 also yields stable η^4-enone complexes 85 and 86 from which the free ketone 87 can be released by treatment of the molybdenum complex with CO[42].

A similar rich chemistry is evident in the reactions of metal thiolates with alkynes; the products isolated depend significantly on the metal, the thiolate and the alkyne. Thus, reaction of tungsten thiolates of structure 88 in which R^1 is electron withdrawing yields stable four-electron donor alkyne complexes 89[43,44]; where R^1 is alkyl, thermal reaction occurs under mild conditions to yield complexes such as 90a-c. Isomerisation via a formal 1,3-sulphur shift gives the rearranged products 92a, b and 94. The mechanism is strongly dependent on the alkyne substituent; where R = CF$_3$, the η^2-vinyl complexes 91a, b may be isolated, whereas where R = CO$_2$Me, isolable σ-vinyl complexes such as 93 are formed as intermediates. The 1,3-sulphur shift via the η^2-vinyl structure is promoted by electron-donating groups; thus, reaction of CpW(CO)$_3$SPr^i with CF$_3$C≡CCF$_3$ proceeds directly to 91b. Isomerization of 94 to the more thermodynamically stable isomer 95 occurs on heating, while lactone formation may be induced by reaction of 91b with two-electron ligands to give 96a[45-49]. Under more forcing photochemical activation, CpW(CO)$_3$SMe

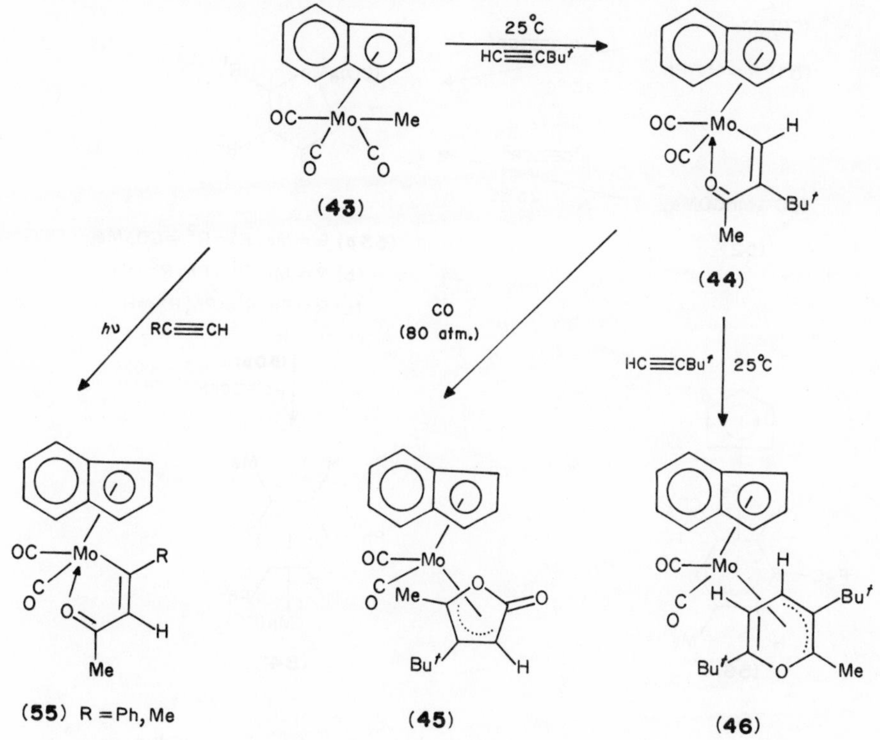
25°C
HC≡CBu^t
(43)
(44)
hν RC≡CH
CO (80 atm.)
HC≡CBu^t 25°C
(55) R = Ph, Me
(45)
(46)

a: M = Mo, 60°C; M = W, hν

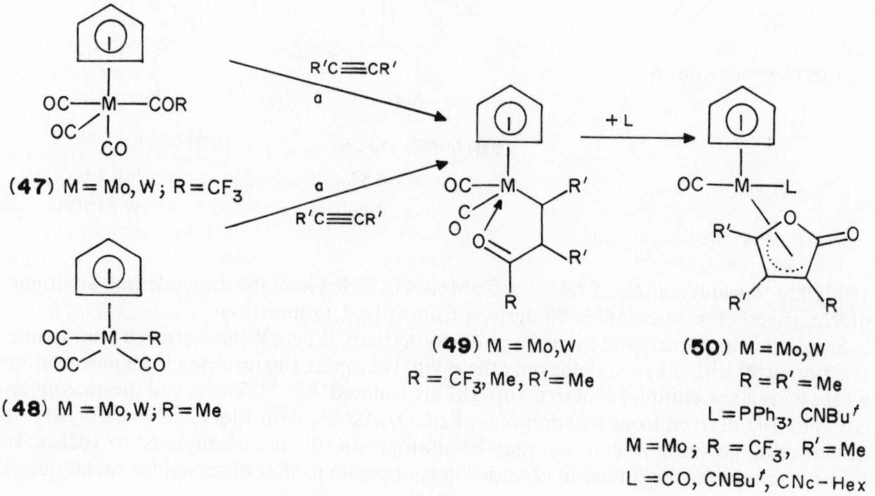
R′C≡CR′
a
+L
(47) M = Mo, W; R = CF3
(48) M = Mo, W; R = Me
(49) M = Mo, W
R = CF3, Me, R′ = Me
(50) M = Mo, W
R = R′ = Me
L = PPh3, CNBu^t
M = Mo; R = CF3, R′ = Me
L = CO, CNBu^t, CNc-Hex

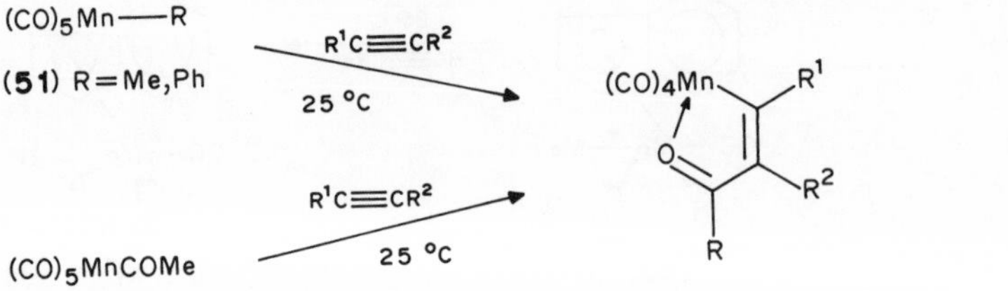

(CO)₅Mn—R

(51) R=Me,Ph

$R^1C\equiv CR^2$ 25 °C

(CO)₅MnCOMe

(52)

$R^1C\equiv CR^2$ 25 °C

(CO)₄Mn ... R¹ / R² / R

(53a) R=Me, $R^1=R^2=CO_2Me$
(b) R=Me, $R^1=Ph$, $R^2=H$
(c) R=Ph, $R^1=Ph$, $R^2=H$

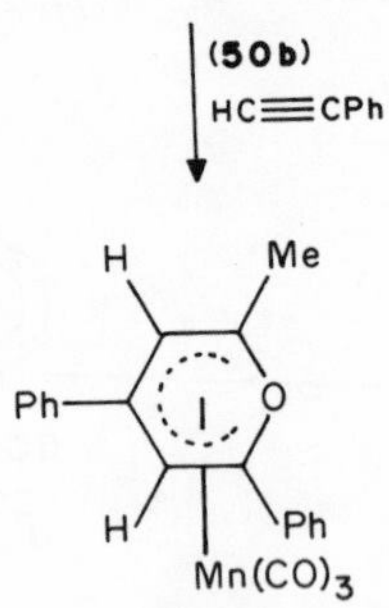

(50b)
$HC\equiv CPh$

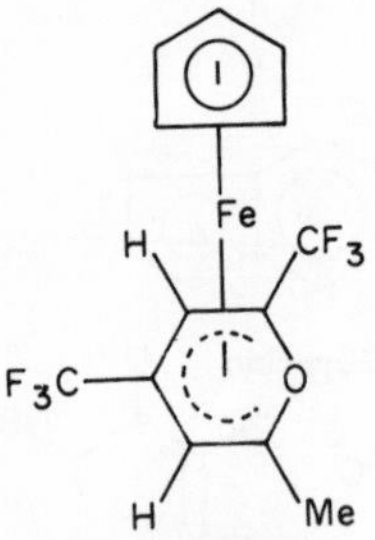

(56)

(54)

PhC≡CH

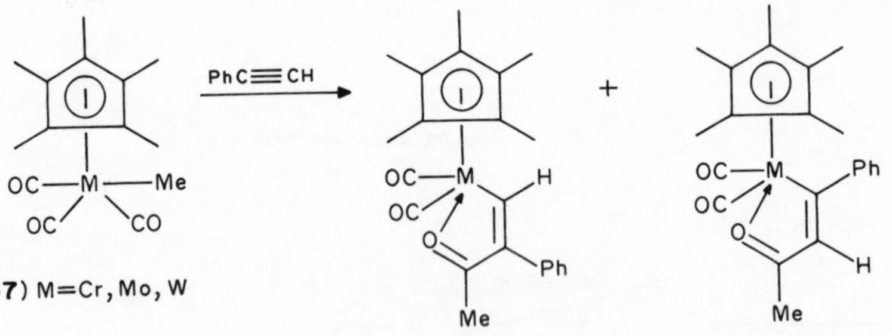

(57) M=Cr, Mo, W

(58) M=Cr 100%
Mo 33%
W 0%

+

(59) M=Cr 0%
Mo 66%
W 100%

yields **92c**, **96b** and complex **97**. Use of CpMo(CO)₃SMe yields the molybdenum analogue of **92c**, together with complex **98** derived from it by CO insertion[50,51].

Similar reactions occur in the analogous iron system. Where R^3 is electron withdrawing, reaction of **99** with alkynes yields only the σ-vinyl complex **100** resulting from insertion[52]; where R^3 is alkyl, complexes of structure **101** are isolated[46,51,53]. These, and the analogous complex **103** derived from reaction of CpFe(CO)₂AsMe₂ with $MeO_2CC\equiv CCO_2Me$[45], do not undergo sulphur shift, but may be photochemically decarbonylated to **102b, c**. It may be noted that the direction of addition is opposite to that observed for metal–alkyl

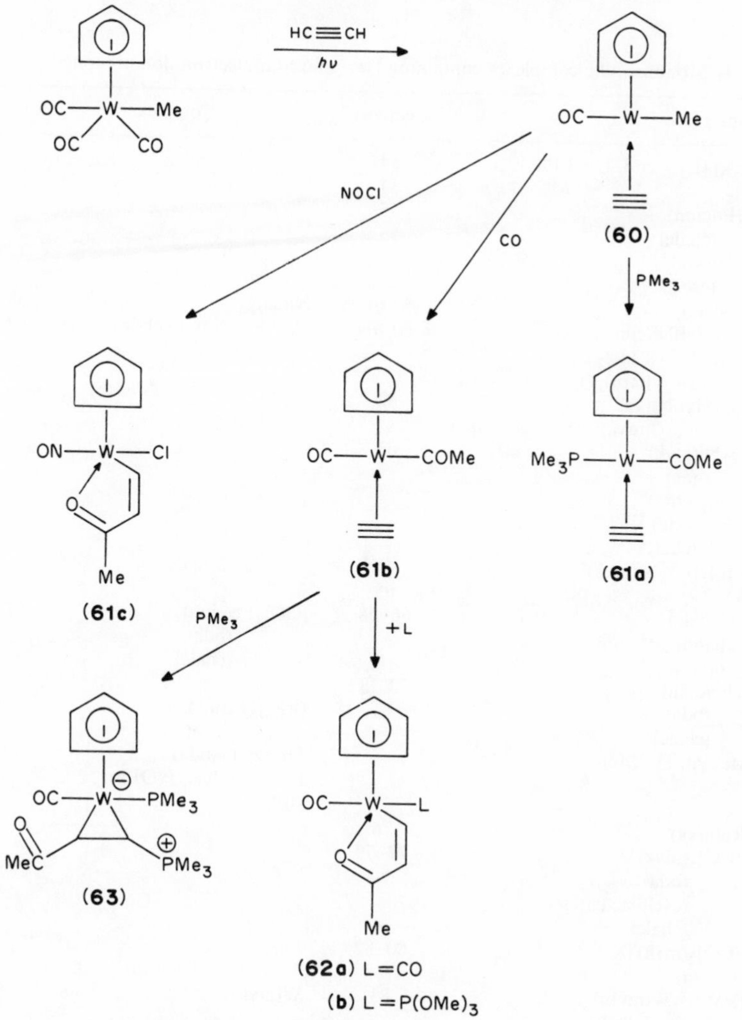

HC≡CH
$h\nu$

NOCl

CO

(60)

PMe₃

OC—W—Me
OC   CO

OC—W—Me

ON—W—Cl
O
Me
(61c)

OC—W—COMe
(61b)

Me₃P—W—COMe
(61a)

PMe₃

+L

OC—W—PMe₃
MeC ⊖ ⊕PMe₃
O
(63)

OC—W—L
O
Me
(62a) L=CO
(b) L=P(OMe)₃

bonds; mechanistically, it has been suggested that this is a result of rate-determining attack of sulphur lone pair at an alkyne carbon to give 104 which may collapse to yield either the metal acyl 105 or the σ-vinyl derivative 106.

III. COMPLEXES CONTAINING TWO- AND FOUR-ELECTRON DONOR LIGANDS

A representative, but not comprehensive, list of monometallic complexes containing two- and four-electron donor enones is given in Table 1, which also shows a list of abbreviations used in this section. With few exceptions, complexes are prepared by interaction of the free enone with an appropriate metal substrate. Exceptions are represented by preparations of complexes 34 and 77 already noted, by the preparation of CpMn(CO)₂(mvk) from the

TABLE 1. Monometallic complexes containing two- and four-electron donor enones[a]

η^2-Complex	Reference	η^4-Complex	Reference
$L_2Pt(mvk)$ ($L_2 = cod$, $L = PPh_3$)	54		
$Pt(mvk)_3$	54		
$(PPh_3)_2Pt(cinn)$	55		
(chalc)	55		
(bda)	55		
(crot)	56		
$(PPh_3)_2Ni(ac)_2$	57, 61	$Ni(ac)_2$	58, 59
$(2,2'$-bipyridyl)Ni(cinn)$_2$	60, 61	$(2,2'$-bipyridyl)Ni(bda)	60
(ac)$_2$			
(crot)$_2$			
$(2,2'$-bipyridyl)Ni(ac)[b]	62		
(mvk)			
$(Bu^tNC)_2Ni(mvk)$[b]	63		
(ac)			
(cinn)			
(bda)			
(chalc)			
$[P(O\text{-}o\text{-tolyl})_3]_2Ni(mvk)$[b]	64		
$Ag(mvk)^+$ [b]	65		
$(CO)_4Fe(chalc)$	66–68	$(CO)_3Fe(cinn)$	67–70
(cinn)		(bda)	
(ac)		(chalc)	
$L(CO)_3Fe(cinn)$	72, 73		
(bda)		$(PF_3)_3Fe(mvk)$	71
(chalc)		(crot)	
$L = PMe_2Ph$, $P(OMe)_3$		$L(CO)_2Fe(bda)$	74, 75
		$L = P(OMe)_3$, $P(OPh)_3$	
		PPh_3	
$(CO)_4Ru(mvk)$	76		
$CpMn(CO)_2(mvk)$	77–79		
(bda)			
(cyclohexenone)			
(chalc)			
$[CpFe(CO)_2(mvk)]X$	80–82		
(ac)			
$(CO)_3(PMe_3)_2W(mvk)$	83	$W(mvk)_3$	84
(ac)			
(cinn)		$[(C_5Me_5)W(CO)_2(bda)]BF_4$	38
(crot)			
$(diphos)_2(CO)Mo(mvk)$	85	$(CO)_2Mo(ac)_2$	86
$Cp_2V(mvk)$	87		
(ac)			
(crot)			

[a] *Abbreviations*: cod = 1,5-cyclooctadiene; mvk = methyl vinyl ketone (CH_2=CHCOMe); ac = acrolein (CH_2=CHCHO); cinn = cinnamaldehyde (*trans*-PhCH=CHCHO); bda = benzylideneacetone (*trans*-PhCH=CHCOMe); chalc = chalcone (*trans*-PhCH=CHCOPh); crot = crotonaldehyde (*trans*-MeCH=CHCHO); diphosethylenebis(diphenylphosphine) (Ph$_2$PCH$_2$CH$_2$PPh$_2$).
[b] Not isolated.

reaction of CpMn(CO)$_2$(THF) with the diazo compound N$_2$C(Me)C(O)Me[79], and the

preparation of [CpFe(CO)$_2$(mvk)]BF$_4$ from NaCpFe(CO)$_2$ and H$_2$C—CHCOMe[80].

These η^2-complexes range from the strongly bound d^2 vanadium derivatives, which may essentially be regarded as metallacyclopropanes[88,89], to weakly bound d^{10} nickel(0) and silver(I) complexes which are stable only in solution in the presence of excess enone. Within the much broader general class of metal–alkene complexes, the conjugative, electron-withdrawing COR substituent lowers particularly the energy of the π^* orbital, thus increasing the π-acceptor capacity of the alkene. Relative to ethene and its alkyl substituted derivatives, or to electron-rich alkenes such as CH$_2$=CHOR, enones form stronger metal–alkene bonds. The difference, however, between the substituents CHO, COR and CO$_2$R in this respect is sufficiently small that the order of stability can depend on the metal or auxiliary ligands. Thus, whereas stability constants for (2,2′-dipyridyl)Ni(alkene) and [P(O—o-tolyl)$_3$]$_2$Ni(alkene) decrease in the order CHO ≫ COR ≈ CO$_2$R, infrared data for (CNBu^t)$_2$Ni(alkene) and (CO)$_3$(PMe$_3$)$_2$W(alkene) are more consistent with the order COR > CHO ≈ CO$_2$R. A much wider variety of metal complexes of mono-ester and di-ester substituted alkenes

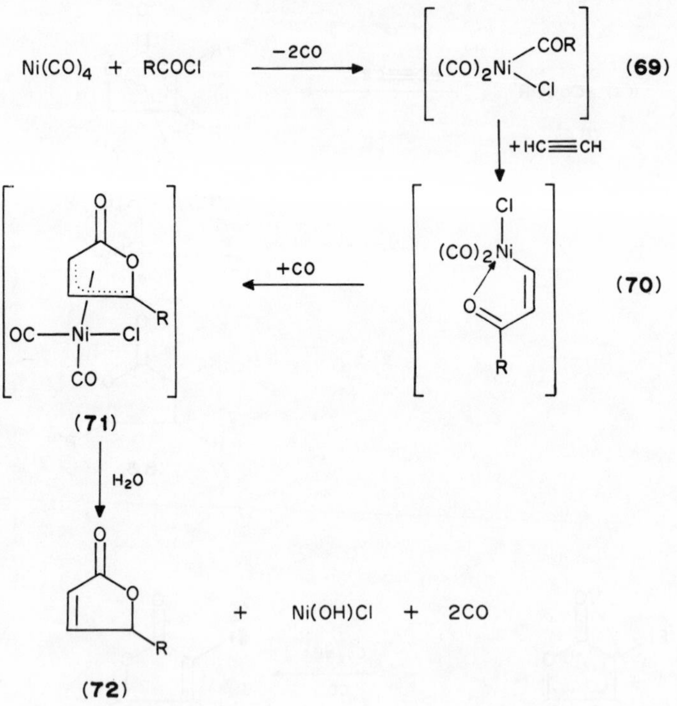

$$Ni(CO)_4 + RCOCl \xrightarrow{-2CO} \left[(CO)_2Ni \begin{smallmatrix} COR \\ Cl \end{smallmatrix} \right] \quad (69)$$

(69) → (70) +HC≡CH

$$(70)$$

(70) → (71) +CO

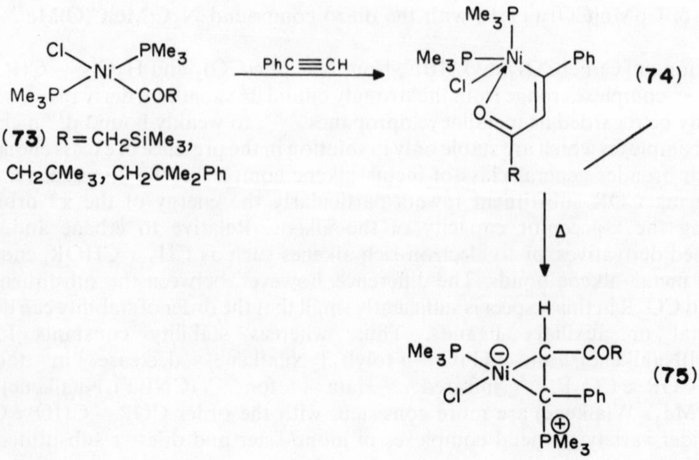

(71)

(71) → (72) H_2O

(72) + $Ni(OH)Cl$ + $2CO$

(73) R = CH₂SiMe₃,
CH₂CMe₃, CH₂CMe₂Ph

(73) → (74) PhC≡CH

(74)

(74) → (75) Δ

(75)

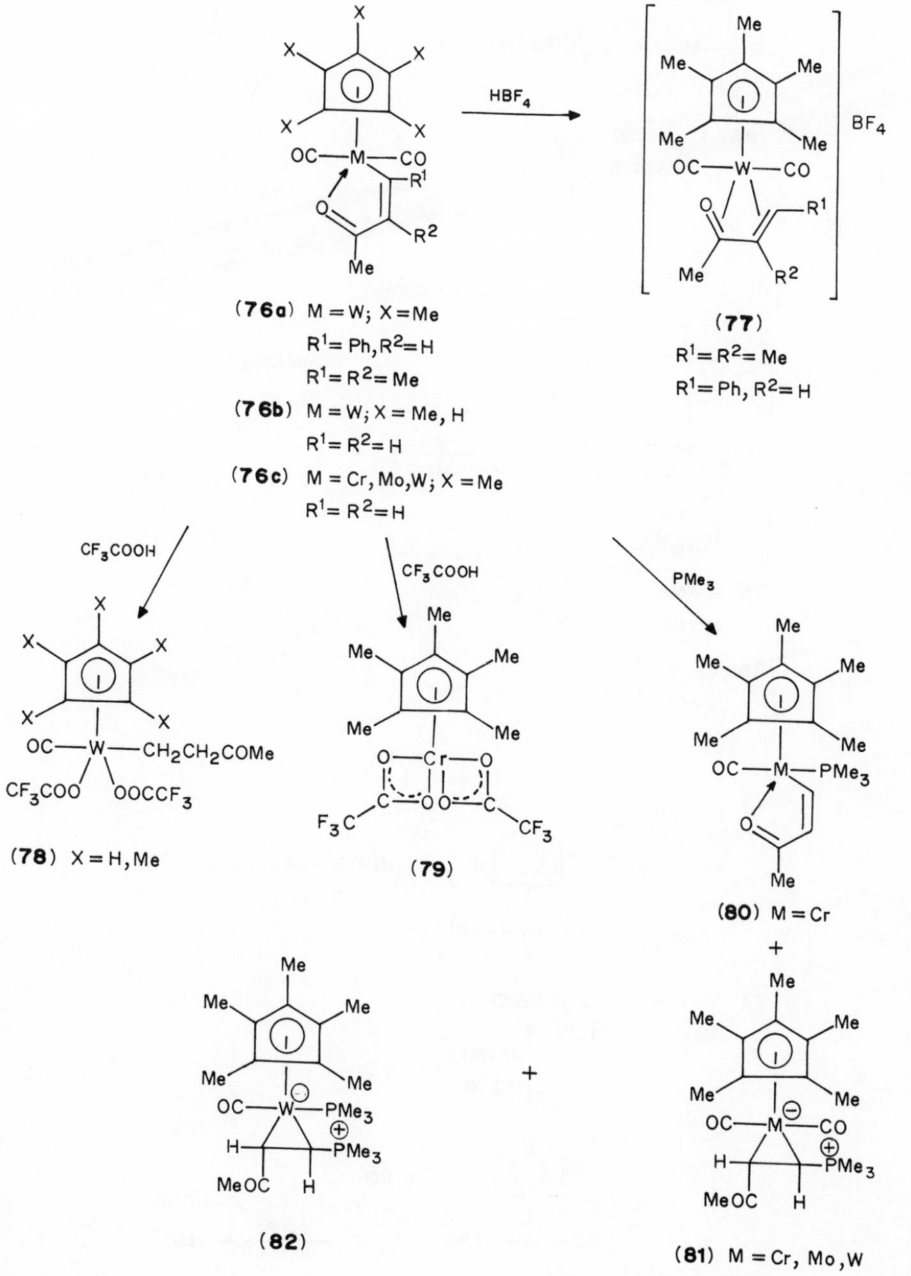

(76a) M = W; X = Me
R¹ = Ph, R² = H

(76a) $M = W$; $X = Me$
$R^1 = Ph, R^2 = H$
$R^1 = R^2 = Me$

(76b) $M = W$; $X = Me, H$
$R^1 = R^2 = H$

(76c) $M = Cr, Mo, W$; $X = Me$
$R^1 = R^2 = H$

(77)
$R^1 = R^2 = Me$
$R^1 = Ph, R^2 = H$

CF_3COOH

CF_3COOH

PMe_3

(78) $X = H, Me$

(79)

(80) $M = Cr$

+

(81) $M = Cr, Mo, W$

(82)

+

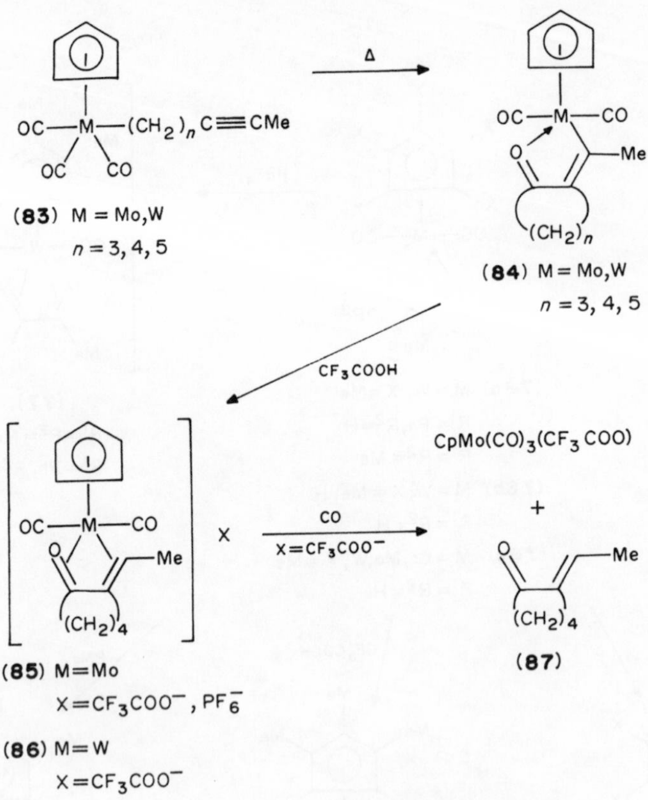

(83) M = Mo, W
n = 3, 4, 5

(84) M = Mo, W
n = 3, 4, 5

CF₃COOH

(85) M = Mo
X = CF₃COO⁻, PF₆⁻

(86) M = W
X = CF₃COO⁻

$CpMo(CO)_3(CF_3COO)$

+

(87)

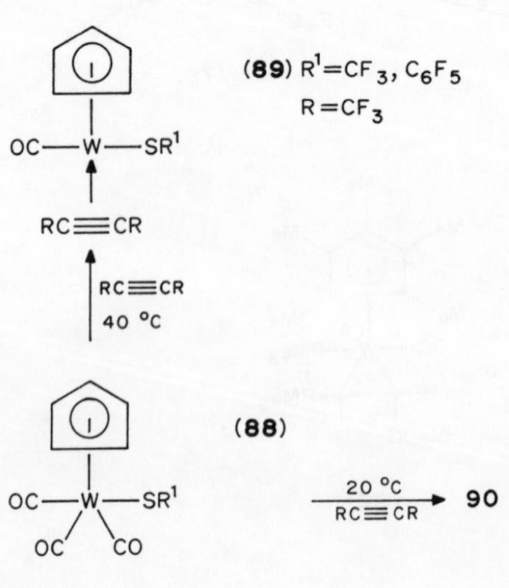

(89) $R^1 = CF_3, C_6F_5$
R = CF₃

(88)

$\xrightarrow[RC\equiv CR]{20\ ^\circ C}$ 90

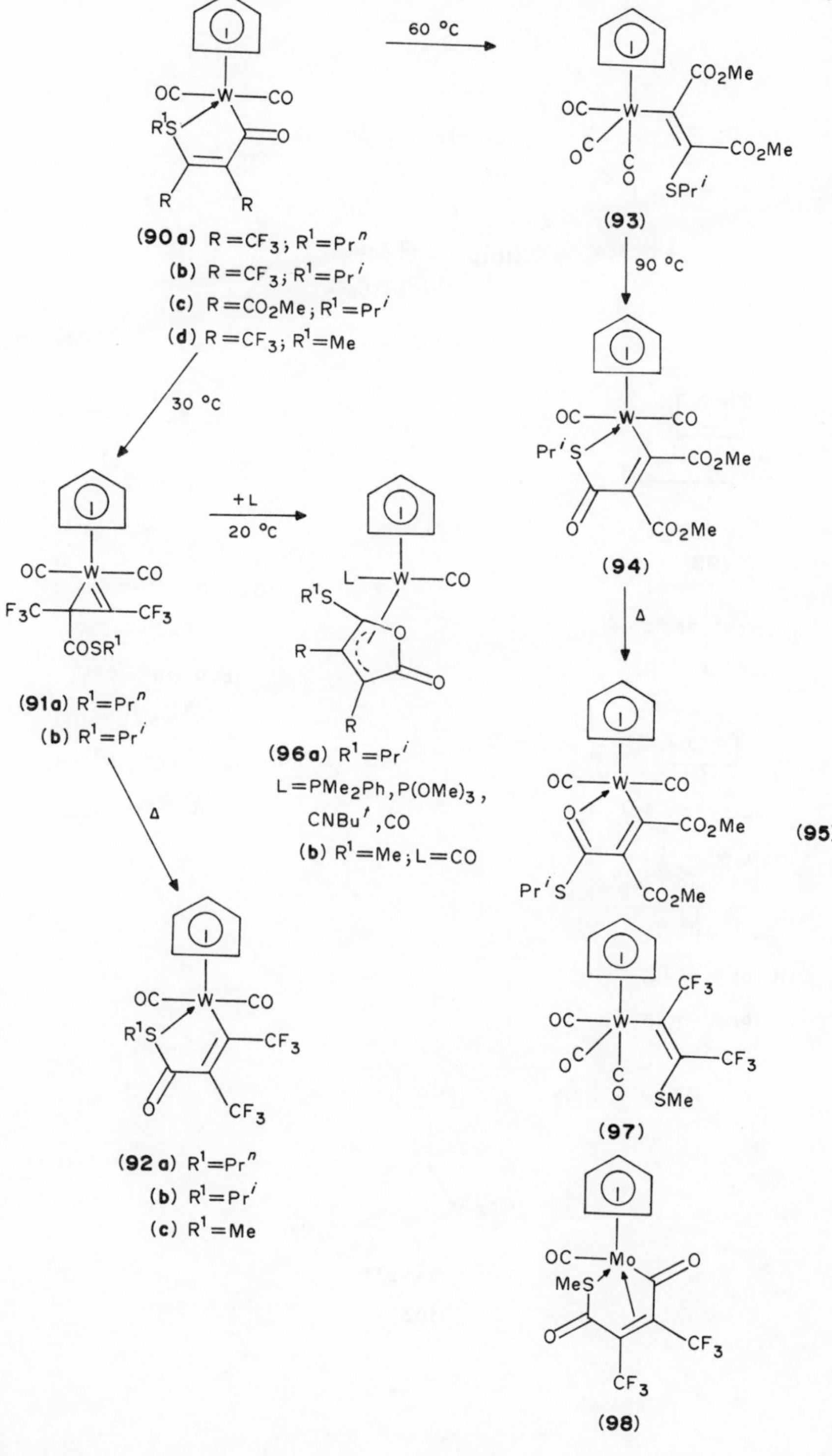

(90a) R = CF₃; R¹ = Prⁿ

(b) R = CF₃; R¹ = Prⁱ

(c) R = CO₂Me; R¹ = Prⁱ

(d) R = CF₃; R¹ = Me

(91a) R¹ = Prⁿ

(b) R¹ = Prⁱ

(92a) R¹ = Prⁿ

(b) R¹ = Prⁱ

(c) R¹ = Me

(96a) R¹ = Prⁱ

L = PMe₂Ph, P(OMe)₃,

CNBuᵗ, CO

(b) R¹ = Me; L = CO

(93)

(94)

(95)

(97)

(98)

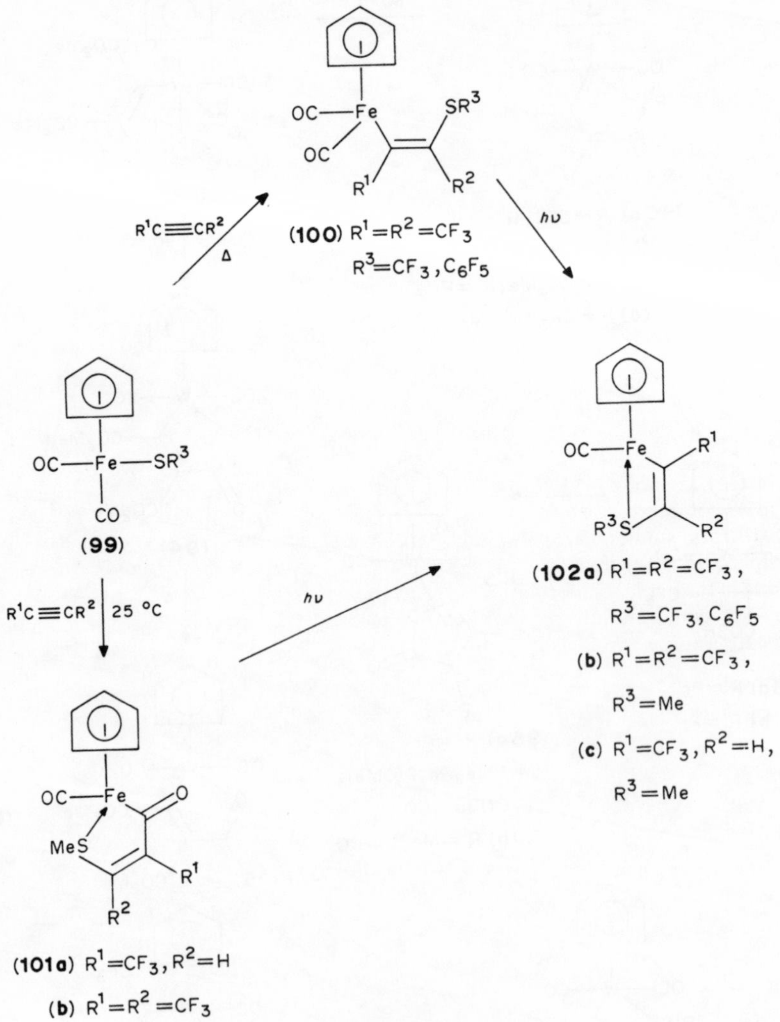

(100) $R^1=R^2=CF_3$
$R^3=CF_3, C_6F_5$

$R^1C\equiv CR^2$
Δ

$h\upsilon$

(99)

$R^1C\equiv CR^2$ | 25 °C

$h\upsilon$

(102a) $R^1=R^2=CF_3$,
$R^3=CF_3, C_6F_5$

(b) $R^1=R^2=CF_3$,
$R^3=Me$

(c) $R^1=CF_3, R^2=H$,
$R^3=Me$

(101a) $R^1=CF_3, R^2=H$

(b) $R^1=R^2=CF_3$

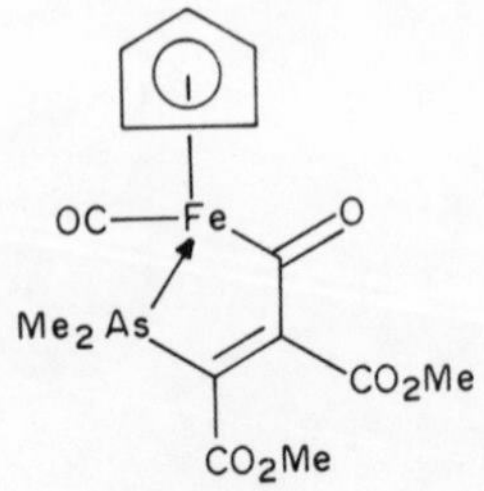

(103)

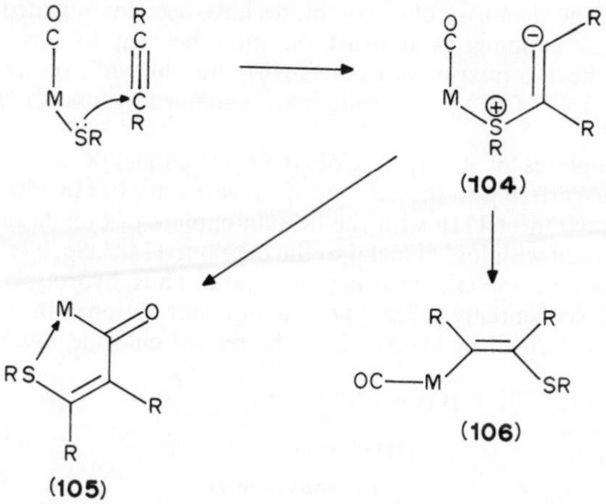

(104)

(105)

(106)

exist; they are not covered here, and are distinguished from enones by their inability to form η^4-complexes.

Structural studies of η^2-complexes indicate a stabilization of the s-*cis* conformation on complexation. Whereas free methyl vinyl ketone and cinnamaldehyde exist predominantly in the s-*trans* conformation[90,91], crystal structures of η^2-mvk complexes reveal only the s-*cis* conformation[92,93], while solution dipole moment studies on $(cinn)Fe(CO)_4$ indicate an s-*cis* $\rightleftharpoons$ s-*trans* equilibrium[94].

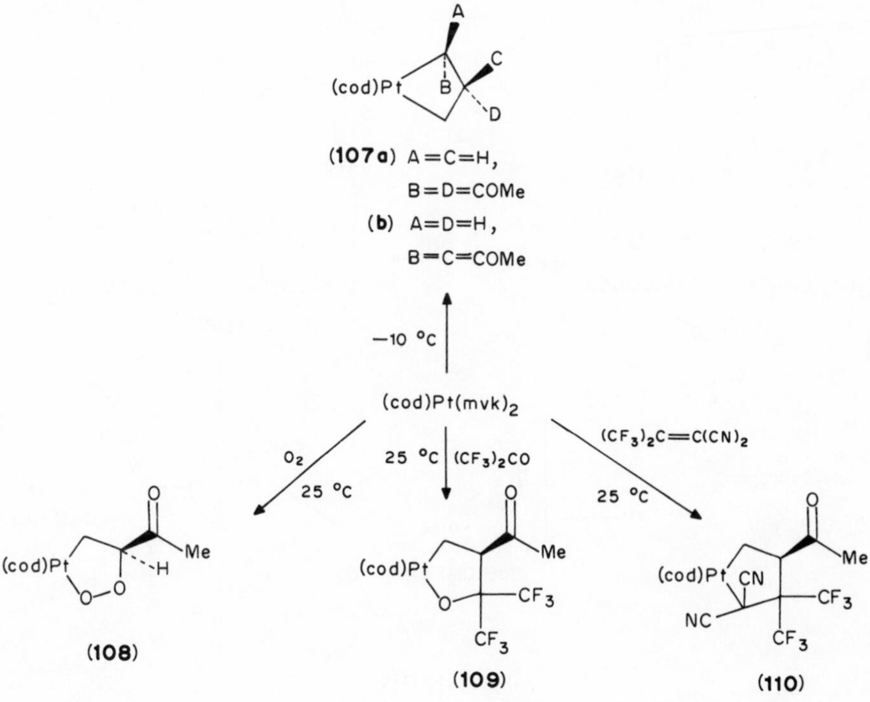

(107a) A = C = H,
       B = D = COMe

(b) A = D = H,
    B = C = COMe

−10 °C

(cod)Pt(mvk)$_2$

O_2
25 °C

25 °C (CF$_3$)$_2$CO

(CF$_3$)$_2$C=C(CN)$_2$
25 °C

(108)

(109)

(110)

Some aspects of the chemistry of η^2-complexes have been investigated. In the electron-rich $(cod)Pt(mvk)_2$, coupling is induced on mild heating to give the head-to-tail metallocyclopentane as a mixture of isomers (**107a** and **b**), while on treatment with O_2, $(CF_3)_2CO$ or $(CF_3)_2C=C(CN)_2$, insertion is accompanied by loss of one mole of mvk to give **108–110**[54].

Like other complexes of its type, $[CpFe(CO)_2(\eta^2\text{-enone})]X$ salts react easily with nucleophiles at the carbon β to the keto group. Treatment of **111a** with $LiCuMe_2$ yields **112**[95], whereas reaction of **111b** with the lithium enolate of cyclohexanone yields **113**, which may be cyclized with loss of metal to the octalone **114**[80]. Such reactions have also been used to generate complexes of structural type **2**. Thus, hydrolysis of the cumulene complex **115** yields sequentially **117** and **118** via initial formation of the unstable enol **116**; complex **118** is also formed by hydrolysis of the related chloride **119**[96,97].

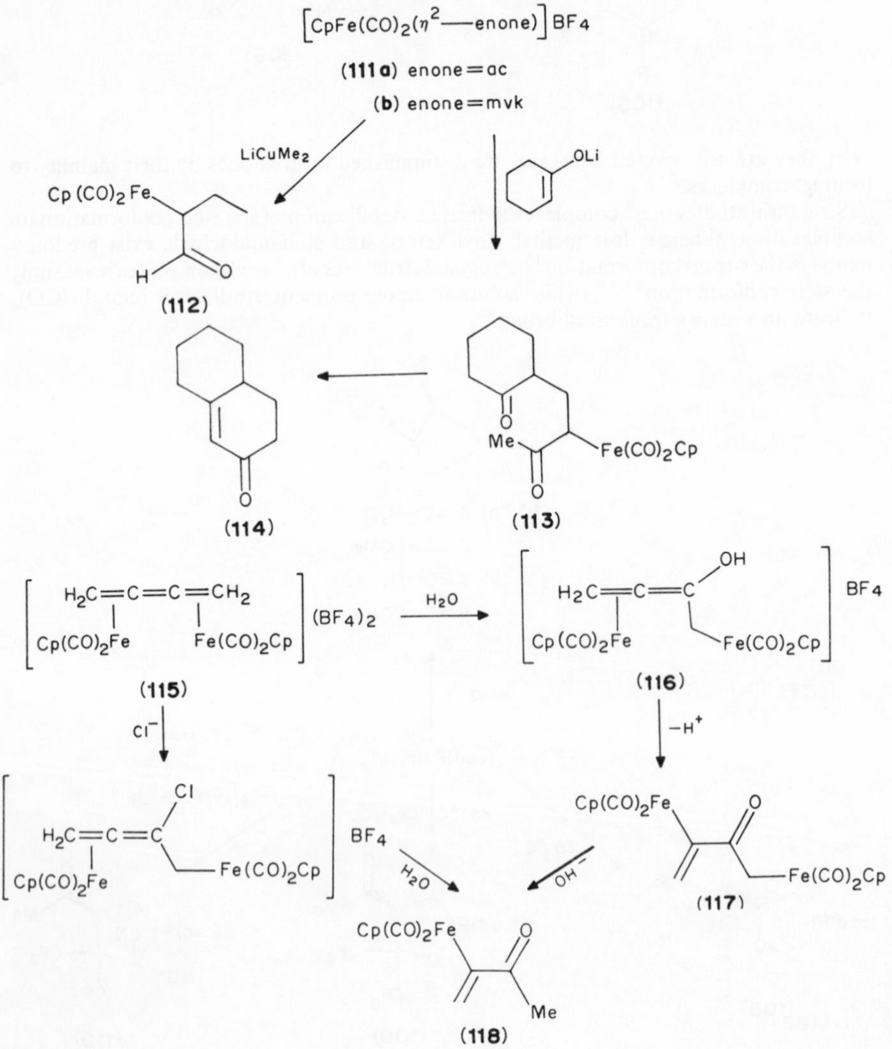

$$\left[CpFe(CO)_2(\eta^2\text{----enone}) \right] BF_4$$

(111a) enone = ac

(b) enone = mvk

(112)

(114)

(113)

(115)

(116)

(117)

(118)

The electrophilic character of the ketonic group is much reduced on complexation of an enone to $Fe(CO)_4$, consistent with the electron-releasing character of this metal fragment; no reaction with amines is observed under conditions where the iron–alkene bond is retained. Adduct formation is, however, observed with BF_3, and further reaction with primary amine generates the carbamoyl chelate complexes **121a, b** which, where $R^2 = Me$, exist in equilibrium with the η^2-structure **122**; in one case, final conversion to the N-bonded derivative **123** is found[98,99]. Acetylation of **120a, c** proceeds to yield a complex best formulated as **124** which on treatment with nucleophiles generates the η^4-derivative **125**; acetylation of **125** reversibly generates **124**[100].

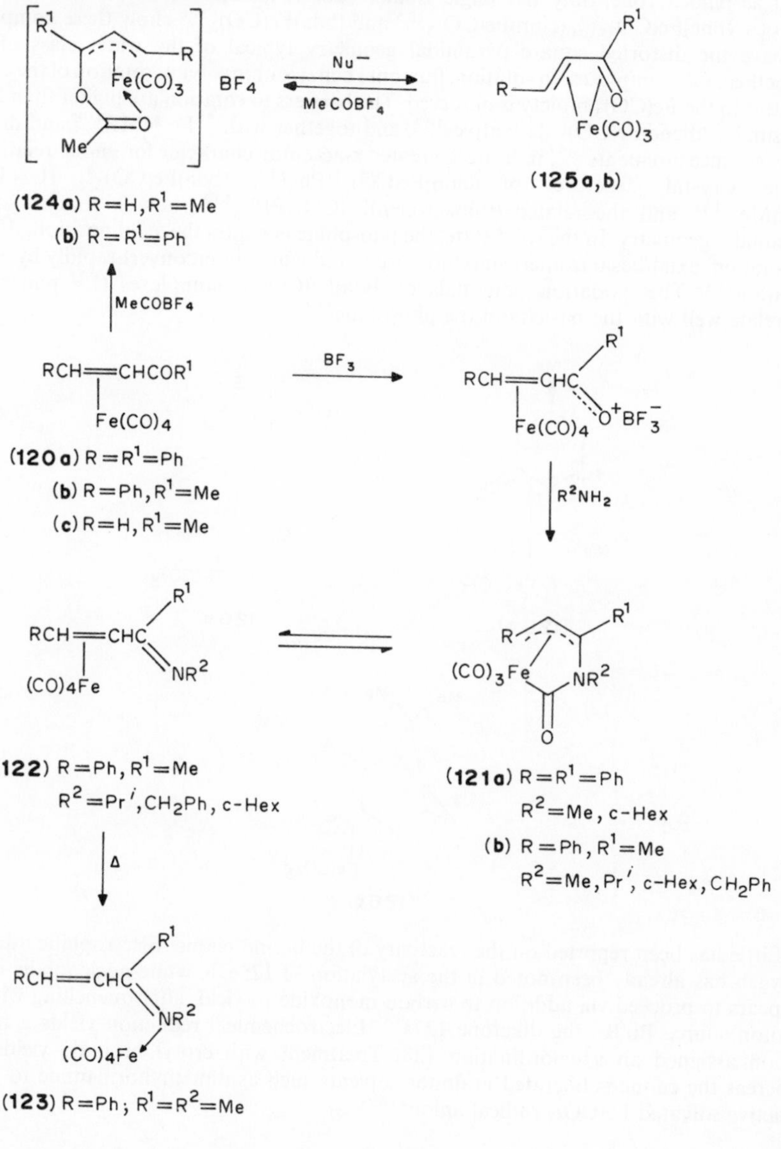

(124a) R=H, R^1=Me
   (b) R=R^1=Ph

(125a,b)

MeCOBF$_4$

RCH=CHCOR^1
$|$
Fe(CO)$_4$

(120a) R=R^1=Ph
   (b) R=Ph, R^1=Me
   (c) R=H, R^1=Me

$\xrightarrow{BF_3}$

RCH=CHC
$\diagdown$R^1
$|$
Fe(CO)$_4$ O$^+$BF$_3^-$

$\downarrow$ R^{2}NH$_2$

RCH=CHC
$\diagdown$R^1
$|$
(CO)$_4$Fe   NR^2

(122) R=Ph, R^1=Me
   R^2=Pr^i, CH$_2$Ph, c-Hex

(CO)$_3$Fe   NR^2
   $\parallel$
   O

(121a) R=R^1=Ph
   R^2=Me, c-Hex
   (b) R=Ph, R^1=Me
   R^2=Me, Pr^i, c-Hex, CH$_2$Ph

$\downarrow \Delta$

RCH=CHC
$\diagdown$R^1
$|$
(CO)$_4$Fe   NR^2

(123) R=Ph; R^1=R^2=Me

The chemistry of η^4-complexes is primarily concerned with those of tricarbonyliron. Such complexes may be prepared from reaction of the free enone either photochemically with $Fe(CO)_5$[101] or thermally with $Fe_2(CO)_9$[67-70,102,103]. In both cases, $(\eta^2$-enone)$Fe(CO)_4$ complexes of varying stability are formed initially; thermally, these undergo transformation on mild heating to the η^4-complex. A kinetic study shows, however, that this proceeds via rate-determining dissociation to enone and $Fe(CO)_4$ followed by rapid CO loss and recoordination of the enone in the η^4-mode[104]. For enones not possessing a symmetry plane, such as pulegone, isomers 126a and 126b may be isolated which differ in the orientation of the $Fe(CO)_3$ moiety, though for sterically crowded enones such as pinocarvone, only the single isomer 126c is isolated[70]. Crystal structures of (pinocarvone)$Fe(CO)_3$[70], (cinn)$Fe(CO)_3$[105] and (bda)$Fe(CO)_3$[106] show these complexes to have the distorted square pyramidal geometry typical of the wider class of $(\eta^4$-diene)$Fe(CO)_3$ complexes. In solution, fluxional behaviour involving rotation of the enone relative to the $Fe(CO)_3$ moiety is observed. The barriers to rotation are higher than those for similar (diene)$Fe(CO)_3$ derivatives[107], and together with ^{57}Fe NMR[108] and dipole-moment measurements[109], indicate a greater π-acceptor character for enone relative to diene. Crystal structures of (cinn)$Fe(CO)_2PPh_3$[110], (bda)$Fe(CO)_2L$ ($L = PEt_3$, $PPhMe_2$)[111], and the related (thioacrolein)$Fe(CO)_2PPh_3$[112] show a similar square pyramidal geometry. In the solid state, the phosphine occupies the axial position, though in solution, axial/basal isomeric mixtures are found which interconvert rapidly by enone rotation[113]. The oxidation potentials of (bda)$Fe(CO)_2L$ complexes (L = phosphine) correlate well with the basicity of the phosphine[114].

(126 a)   (126 b)

(126 c)

Little has been reported on the reactivity of the bound enone. Electrophilic attack at oxygen has already been noted in the acetylation of 125a, b, while nucleophilic attack appears to proceed via addition to carbon monoxide to yield, after quenching with the proton source Bu^tBr, the diketone 127[115]. Electrochemical reduction yields a radical anion assigned an η^2-coordination 128. Treatment with crotyl bromide yields 129, whereas the enone is liberated in donor solvents such as dimethylformamide to give a reactive solvated $Fe(CO)_3$ radical anion[116].

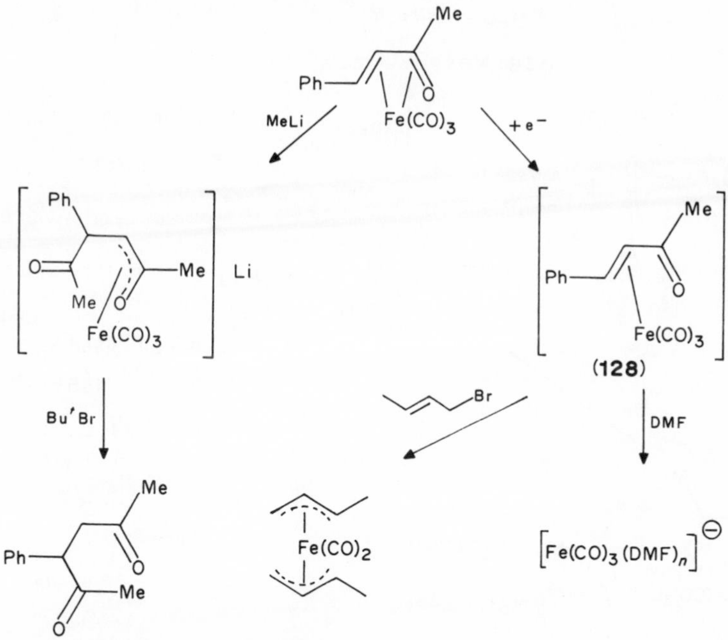

(128)

(127)   (129)

The great utility of these complexes lies in their substitutional lability towards group V donors (phosphines and phosphites) and conjugated dienes. Reaction of (bda)Fe(CO)$_3$ with ligand(L) proceeds to yield isolable $(\eta^2\text{-bda})\text{Fe(CO)}_3$L complexes 130a–c[72,73,117] while 130d–f may be observed *in situ*; all reactions proceed to completion except that with SbPh$_3$, in which a (bda)Fe(CO)$_3$/(130f) equilibrium is established[118]. For the phosphite derivatives, re-chelation of the enone occurs smoothly to yield 132a, b[74], whereas with PPh$_3$, the reaction is accompanied by concomitant formation of (PPh$_3$)$_2$Fe(CO)$_3$ 131[118]. Indeed, in the presence of excess ligand, this reaction may be used to advantage to produce L$_2$Fe(CO)$_3$ complexes such as 133 and 134[119,120]. Complexes such as 132c–e are best prepared by photolysis of Fe(CO)$_4$L in the presence of enone[75,111,121]. Reaction of (bda)Fe(CO)$_3$ with *p*-nitrosulphinylaniline or CS$_2$ in the presence of PPh$_3$ yields the novel η^2-complexes 135a, b respectively[122,123].

Exchange with cyclic and acyclic conjugated dienes proceeds smoothly to yield the $(\eta^4\text{-diene})\text{Fe(CO)}_3$ complex[124,125]. The mild conditions required (50–80 °C, toluene) make this the reaction of choice for dienes sensitive to heat or light, and several examples are shown below. Dienes not containing a plane of symmetry can give isomeric mixtures; thus, where R = H, both 142 and 143 are isolated, whereas when R = Me, only the less sterically hindered 142 is obtained. The isolation of exclusively 144 may perhaps be ascribed to initial interaction with the ester group, followed by transfer of Fe(CO)$_3$ to the same face. (Bda)Fe(CO)$_3$ also functions as a reactive source of Fe(CO)$_3$ in the ring opening of methylenecyclopropenes, alkyne coupling to give 149, and CO elimination to give the *o*-xylylene complex 150.

The high selectivity towards cyclic dienes, and cyclohexadiene in particular, may be used in the extraction of unstable tautomers from C$_6$-diene/C$_8$-triene equilibria. Thus, although the concentration of diene in the 151 ⇌ 152 and 154 ⇌ 155 equilibria is small, only 153 and 156 are isolated from reaction with (bda)Fe(CO)$_3$[101,138,139]. The unstable

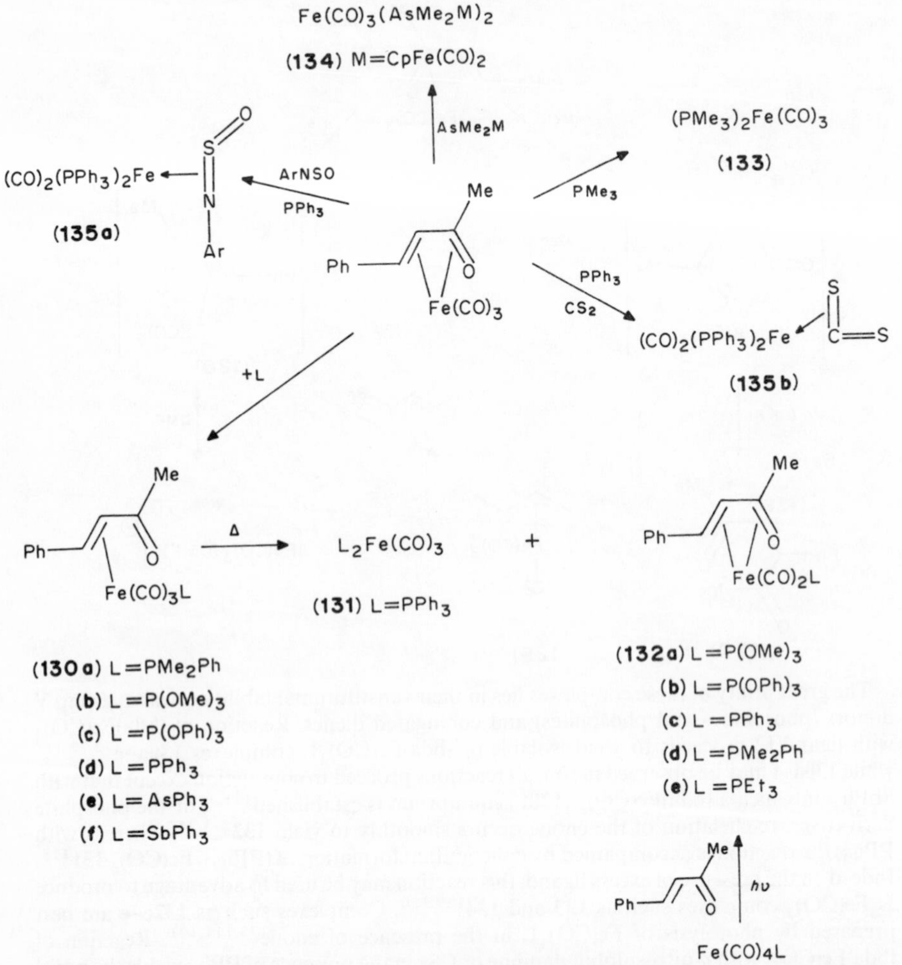

Fe(CO)$_3$(AsMe$_2$M)$_2$

(134) M=CpFe(CO)$_2$

(CO)$_2$(PPh$_3$)$_2$Fe ← ArNSO / PPh$_3$

(135a)

AsMe$_2$M

(PMe$_3$)$_2$Fe(CO)$_3$

(133)

PMe$_3$

PPh$_3$ / CS$_2$

(CO)$_2$(PPh$_3$)$_2$Fe—C=S

(135b)

+L

L$_2$Fe(CO)$_3$

(131) L=PPh$_3$

(130a) L=PMe$_2$Ph
(b) L=P(OMe)$_3$
(c) L=P(OPh)$_3$
(d) L=PPh$_3$
(e) L=AsPh$_3$
(f) L=SbPh$_3$

(132a) L=P(OMe)$_3$
(b) L=P(OPh)$_3$
(c) L=PPh$_3$
(d) L=PMe$_2$Ph
(e) L=PEt$_3$

Fe(CO)$_4$L

tautomers may be released by low-temperature oxidation. Diene exchange is also observed with (enone)Fe(CO)$_2$L complexes such as 132c–e, though at much reduced rates[140]. Lateral coordination of the Fe(CO)$_3$ moiety confers chirality on complexes of unsymmetrically substituted dienes such as 157, and diene exchange using chiral enones such as (−)-cholest-4-ene-3, 6-dione or (−)-3β-(acetyloxy)pregna-5, 16-dien-20-one proceeds with significant asymmetric induction to give enantiomeric excesses of up to 43%[141].

Structural and chemical data on other η^4-complexes is sparse. W(mvk)$_3$, which has a trigonal prismatic geometry in common with W(butadiene)$_3$[142,143], reacts with Ph$_3$PCH$_2$ to yield a complex of formula W(CH$_2$PPh$_3$)$_3$ with loss of mvk[144]. Mo(CO)$_2$(ac)$_2$ is initially isolated as a soluble monomer, but deposits a polymer on standing in which acrolein is thought to act as a bridging ligand between metal atoms. Such a four-electron donation, bridging two metals in the s-*trans* configuration, has been structurally characterized in the two copper complexes Cu$_4$Cl$_4$(mvk)$_4$ (158)[93] and [CuCl(ac)]$_n$ (159)[145]. The enone is weakly bound, and the coordination observed may be relevant to copper-catalysed conjugate addition reactions of enones.

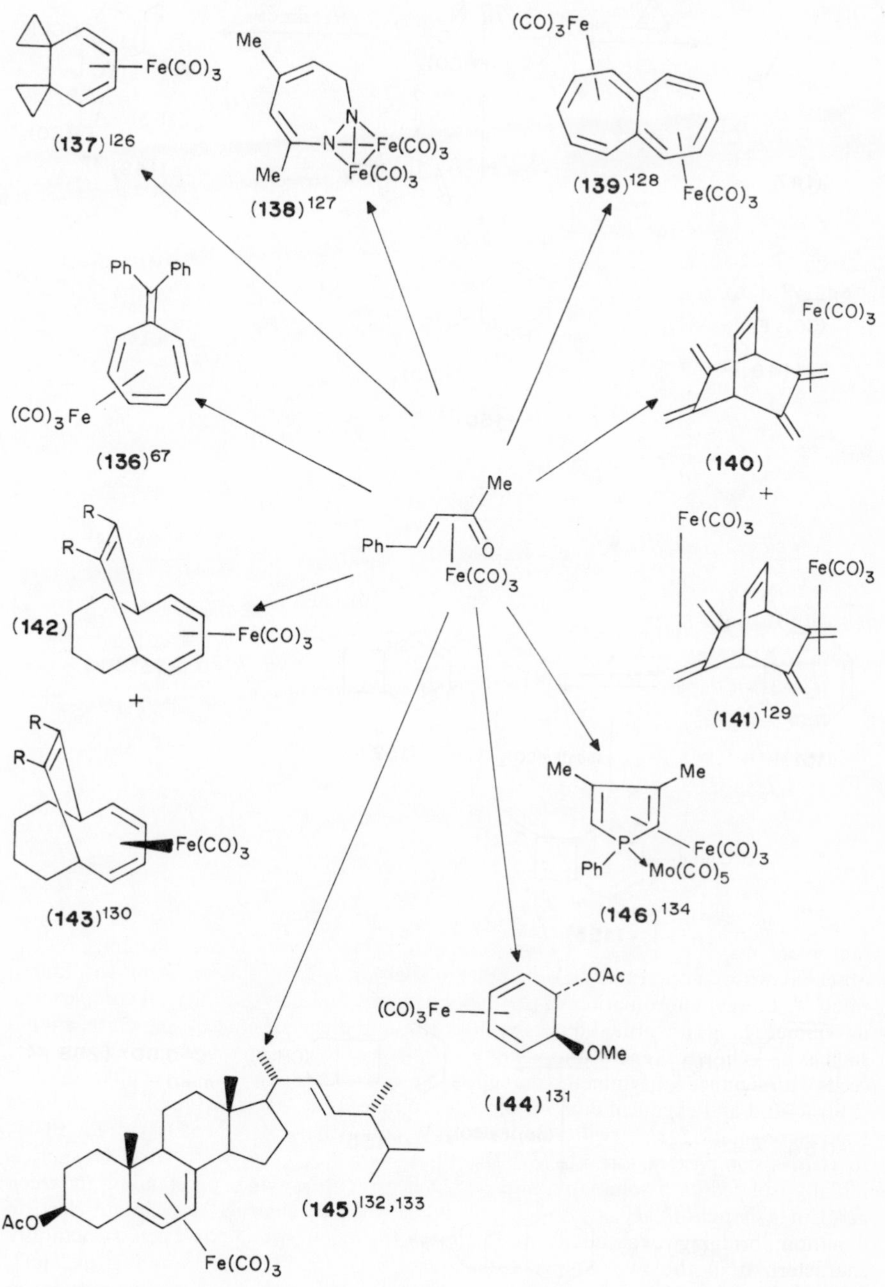

(137)[126]

(138)[127]

(139)[128]

(136)[67]

(140)

+

(141)[129]

(142)

+

(143)[130]

(146)[134]

(144)[131]

(145)[132,133]

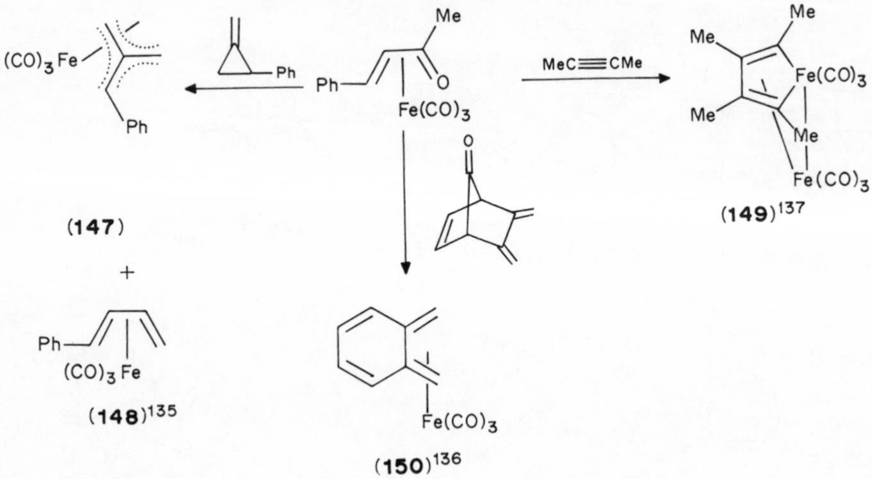

(147)

+

(148)[135]

(150)[136]

(149)[137]

(151) K (152) $K = 0.18\ (373\ K)$

$\xrightarrow{\text{(bda)Fe(CO)}_3}$

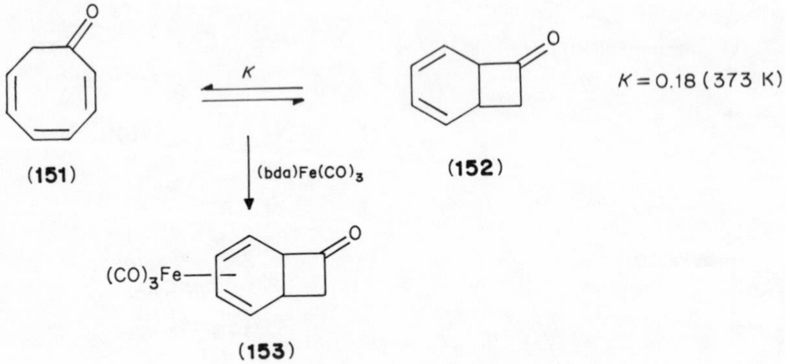

(153)

(154) $n = 0, 2$ K (155) $K = 0.007\ (298\ K)$
$n = 0$

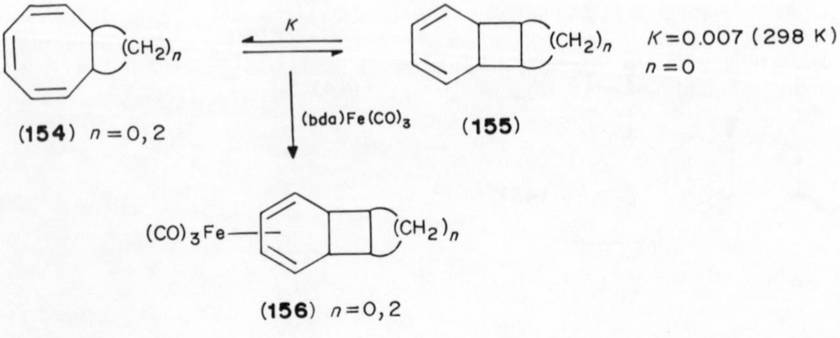

$\xrightarrow{\text{(bda)Fe(CO)}_3}$

(156) $n = 0, 2$

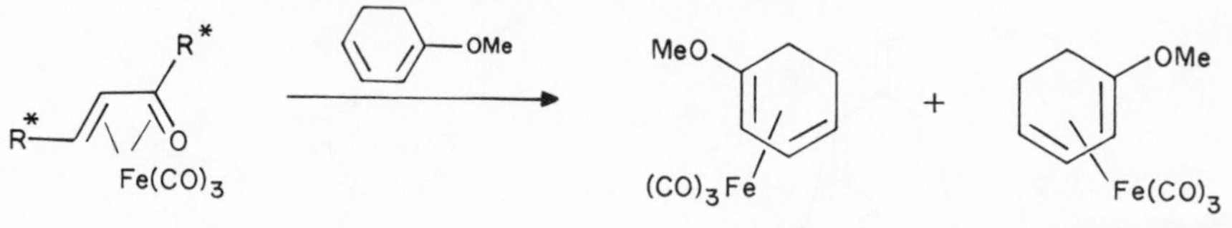

$R^* =$ chiral centre(s)

(157a)   (157b)

$(1R)-(-)$ $(1S)-(+)$

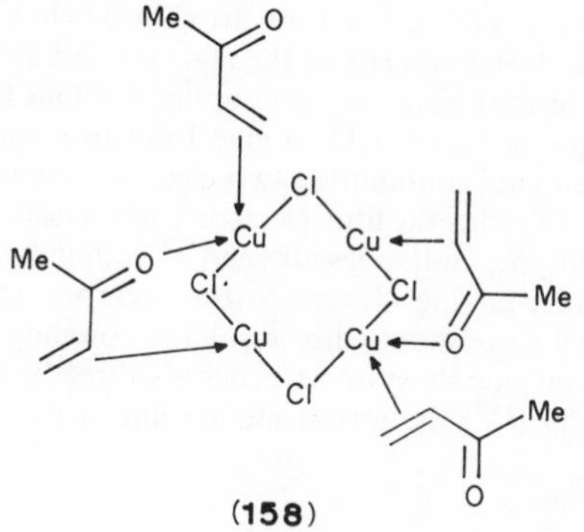

(158)

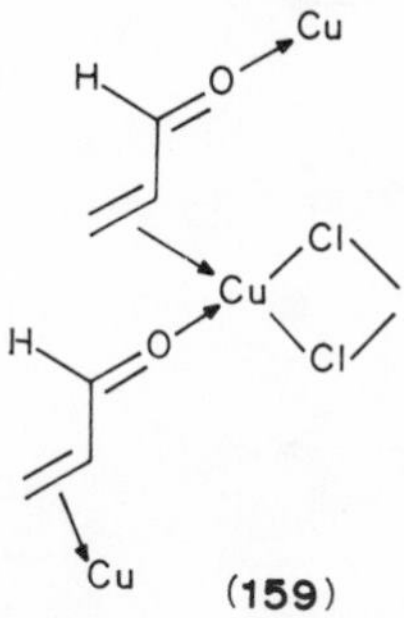

(159)

IV. ENONES IN POLYMETALLIC COMPLEXES

Polymetallic complexes containing bound enones may also be derived from either CO/alkyne coupling or by reaction of an enone with a metal substrate. The former most commonly yields coordination geometries represented in their simplest form by the mono- and dimetallacyclic structures 160 and 161, though structural types 162 and 163 derived formally from addition of a further mole of alkyne have also been observed.

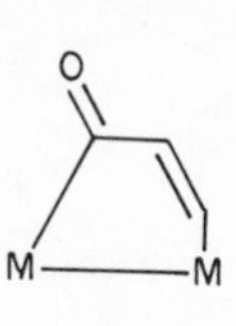

(160) 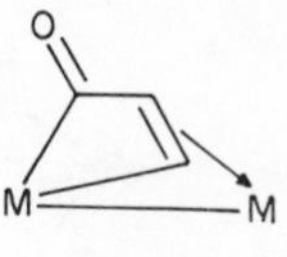   (161)

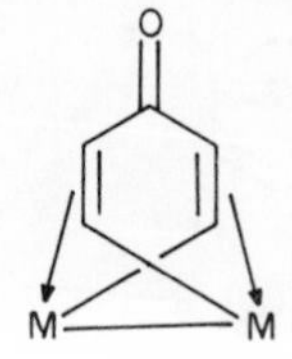

(162) 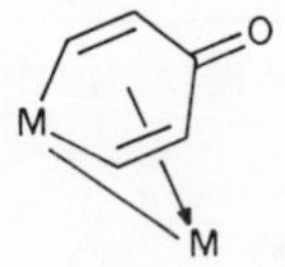   (163)

Reaction sequences which illustrate these structures are shown below. In most cases, the equations greatly underestimate the complexity of the reaction and, in particular, do not show the plethora of products derived from alkyne coupling without CO incorporation. Thus, complex **164** undergoes facile loss of CO to give **165** [also available from direct reaction of $Cp_2Rh_2(CO)_3$ with alkyne] containing a two-electron alkyne bound parallel to the M–M axis. Further loss of CO yields the four-electron transversely bounded complex **166** which, on reaction with 2-butyne, undergoes alkyne/CO coupling to give the 'flyover' complex **167**. Thermolysis results in ring closure to the cyclopentadienone derivative **168**[146,147]. Treatment of **166** with isocyanate also results in coupling to give the amide **169**[148]. Reaction with analogous cobalt systems proceeds directly to the complex of structure **168**[149,150]. $Cp_2W_2(CO)_6$[151], the mixed metal dimer $Cp_2NiMo(CO)_4$[152] and

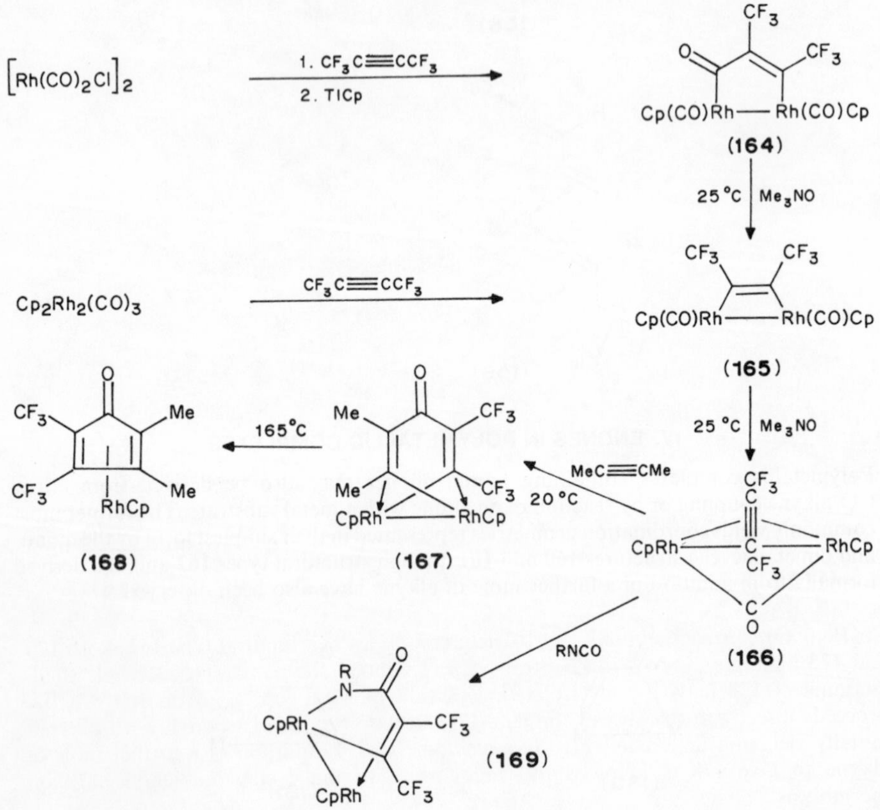

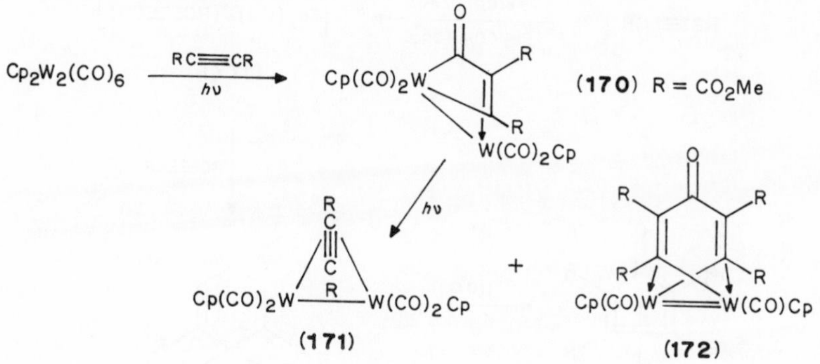
Cp2W2(CO)6
RC≡CR
hν
Cp(CO)2W
W(CO)2Cp
(170) R = CO2Me
hν
Cp(CO)2W
W(CO)2 Cp
(171)
+
Cp(CO)W
W(CO)Cp
(172)

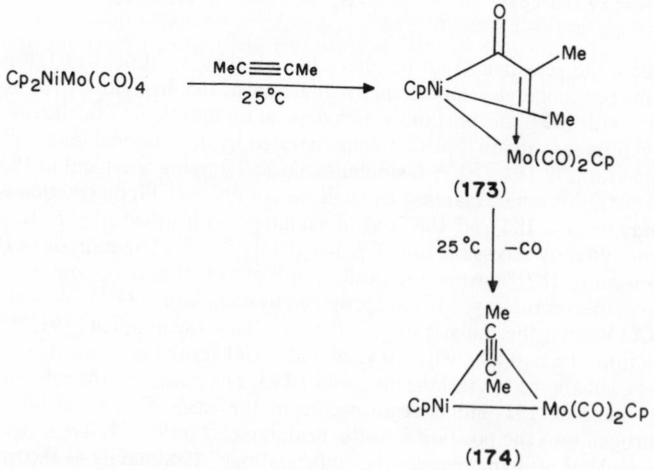
Cp2NiMo(CO)4
MeC≡CMe
25°C
CpNi
Me
Mo(CO)2Cp
(173)
25 °C −CO
CpNi
Mo(CO)2Cp
(174)

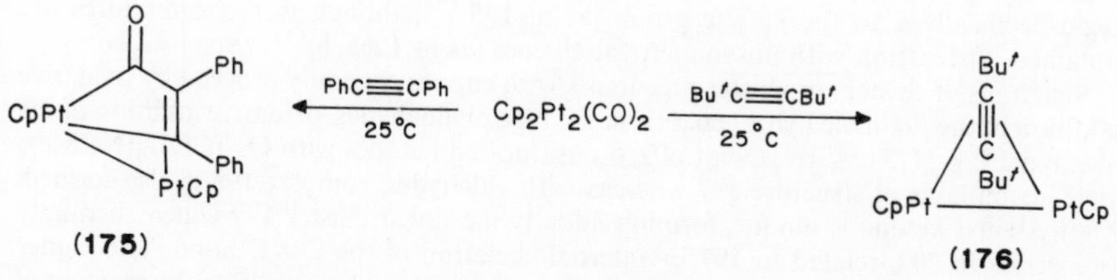
CpPt
Ph
PtCp
(175)
PhC≡CPh
25°C
Cp2Pt2(CO)2
Bu'C≡CBu'
25°C
CpPt
PtCp
(176)

$Cp_2Pt_2(CO)_2$[153] react with alkynes to yield complexes of structural type **162**. Both **170** and **173** undergo CO loss to give transversely bonded alkyne complexes, while small amounts of **172** may also be isolated. Reaction of $Cp_2Pt_2(CO)_2$ with $Bu^tC{\equiv}CBu^t$ proceeds directly to the alkyne complex **176**. Reaction of iron carbonyls with alkynes initially yields unstable complexes of stoichiometry **177**; addition of a further mole of alkyne to give **178** is followed by ring closure to the cyclopentadienone **179** on thermolysis[154–156].

$$RC{\equiv}CR \xrightarrow[\text{Fe}_2(\text{CO})_9/35^\circ\text{C}]{\text{Fe(CO)}_5/h\upsilon} \left[\text{Fe}_2(\text{CO})_7(\text{RC}{\equiv}\text{CR}) \right]$$

(177)

RC≡CR

(179) R = cyclopropyl   (178) R = Me, Ph, cyclopropyl

Reactions can be quite sensitive to an auxiliary ligand; thus, treatment of **180** with alkyne yields **181**, the pentamethylcyclopentadienyl analogue of **165**. In contrast, **181** exists in rapid equilibrium with **182** in proportions which depend on the alkyne substituent; the rapid M—CO/acyl interconversion is further demonstrated by the fluxional interconversion of the rhodium atoms in **182**[157,158]. A similar facile acyl flipping is evident in **183**; in the ruthenium complex, this is also manifest by facile loss of PhC≡CPh on reaction with two-electron ligands to give **184** and **185**, and in exchange with other alkynes to give **186a, b** which are not directly accessible from Cp₂Ru₂(CO)₄[159-161]. Thermolysis of **186b** results in isomerization to **187**[162], while the protonation of **183**[163] may be compared to that of **22d**. Examples of structural type **164** are provided by complexes **189**[164] and **190**; the latter undergoes CO loss on thermolysis to give the metallacyclopentadiene **191**[165,166].

In contrast, reaction of Co₂(CO)₈ with alkynes under CO results in coupling of two moles of CO with one of alkyne to yield the butenolide **193**. The reaction proceeds via the Co₂(CO)₆(alkyne) complex **192** and is regiospecific in the case of terminal alkynes, incorporating hydrogen into the position α to the bridging carbon[167,168]. Treatment of **193** with a further mole of propyne releases the "bifurandione" **194**, mainly as the *trans*-dimethyl isomer[169]. Reaction of the alkyne complex **192** with alkenes provides an efficient synthesis of cyclopentenones; with asymmetric alkynes, the substituent is incorporated regiospecifically α to the ketone group (as in **195**)[170], though isomeric mixtures are obtained on reaction with unsymmetrical alkenes (as in **196a, b**)[171].

Reactions of clusters or cluster precursors with enones generally proceed by oxidative addition of the β-vinylic hydrogen and, in the case of aldehydes, oxidative addition of the aldehydic C—H. Thus, treatment of α, β-unsaturated ketones with Os₃(CO)₁₀(NCMe)₂ yields complexes of structure **197** whereas with aldehydes, complex **198** is also formed. Methyl vinyl ketone is unique, forming initially the linear cluster **199** which thermally isomerizes to **202**, related to **197** by internal chelation of the C=C bond. The higher energy isomeric forms **200** and **201a** may be formed under milder conditions by reaction of HC≡CCOMe with H₂Os₃(CO)₁₀, but isomerize thermally to **199**. Complex **201b** is formed in the analogous reaction of H₂Os₃(CO)₁₀ with PhC≡CCHO[172,173]. The ruthenium dimer **203** with an enone coordination similar to **202** has been prepared[174], while complexes such as **205**, containing an enone coordination similar to **198**, have been isolated from carbon monoxide insertion into bridging vinyl derivatives such as **204**[175]. The bridged phosphido derivative **206** undergoes CO/alkyne coupling on reaction with PhC≡CPh to give **207**, similar in coordination to **169**[176].

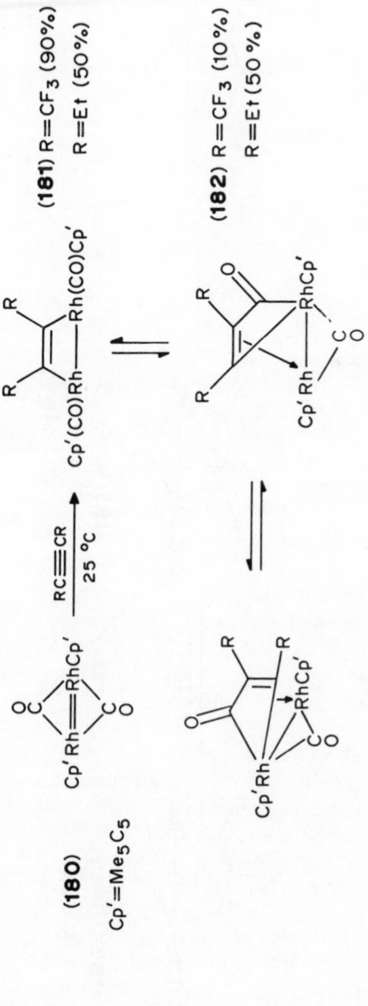
(180)
Cp′=Me$_5$C$_5$
RC≡CR
25 °C
(181) R=CF$_3$ (90%)
R=Et (50%)
(182) R=CF$_3$ (10%)
R=Et (50%)

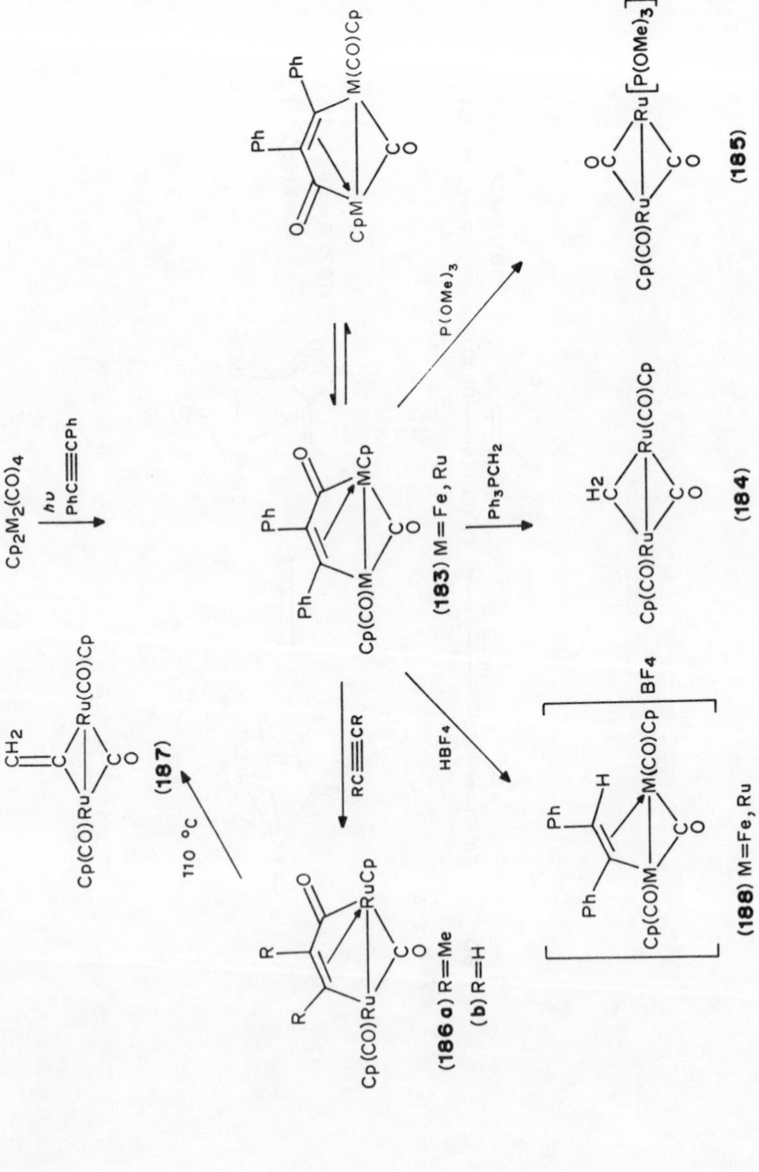

$Cp_2M_2(CO)_4$
hν
PhC≡CPh
(183) M=Fe, Ru
Ph₃PCH₂
P(OMe)₃
RC≡CR
HBF₄
110 °C
(184)
(185)
(186 a) R=Me
(b) R=H
(187)
(188) M=Fe, Ru

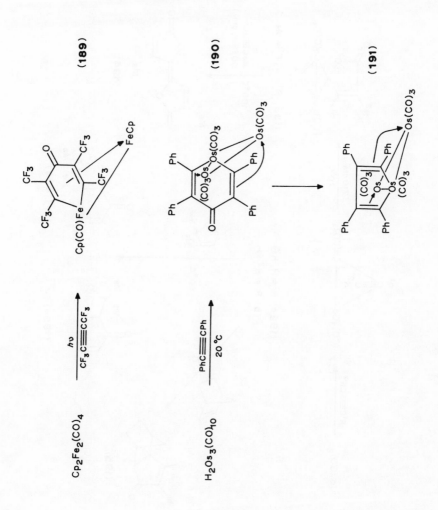

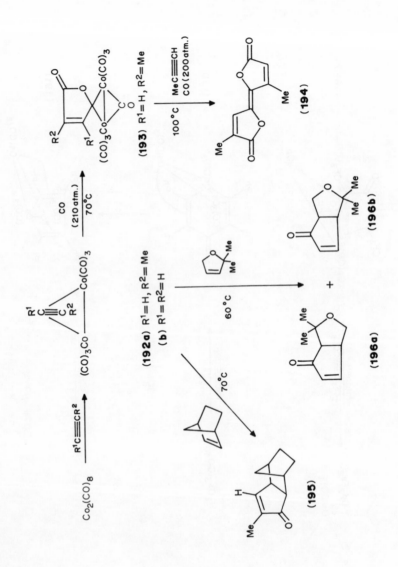

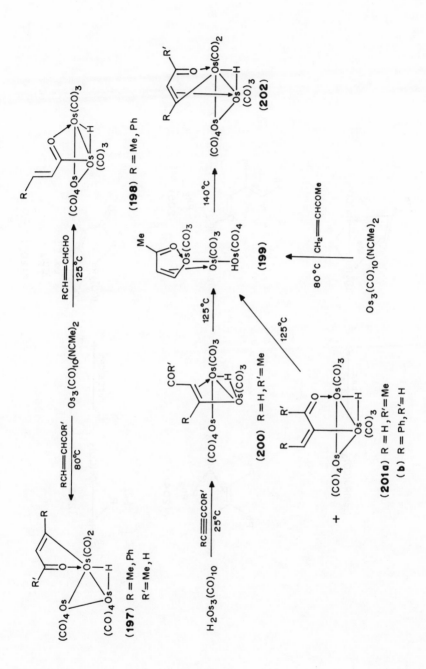

(cod)Ru(CO)$_3$ $\xrightarrow{\text{bda} \atop 70\ ^\circ\text{C}}$ (203)

(204) $\xrightarrow{\text{Bu}^t\text{NC}}$ (205)

(206) $\xrightarrow{\text{PhC}\equiv\text{CPh}}$ (207)

V. REFERENCES

1. C. P. Casey, W. H. Miles and H. Tukuda, *J. Am. Chem. Soc.*, **107**, 2924 (1985).
2. M. I. Rybinskaya, A. N. Nesmeyanov, L. V. Rybin and Y. A. Ustynyk, *J. Gen. Chem. USSR (Engl. Transl.)*, **37**, 1505 (1967).
3. L. V. Rybin, V. S. Kaganovich and M. I. Rybinskaya, *Izv. Akad. Nauk SSSR, Ser. Khim.*, 360 (1969).
4. A. N. Nesmeyanov, E. A. Petrovskaya, L. V. Rybin and M. I. Rybinskaya, *Izv. Akad. Nauk SSSR, Ser. Khim.*, 2045 (1979).
5. A. N. Nesmeyanov, M. I. Rybinskaya, V. S. Kaganovich, Y. A. Ustynyk and I. S. Leshcheva, *Izv. Akad. Nauk SSSR, Ser. Khim.*, 1100 (1969).
6. K. M. Kremer, G. H. Kuo, E. J. O'Connor, P. Helquist and R. C. Kerber, *J. Am. Chem. Soc.*, **104**, 6119 (1982).
7. S. Quinn and A. Shaver, *Inorg. Chim. Acta*, **39**, 243 (1980).
8. A. N. Nesmeyanov, V. S. Kaganovich, L. V. Rybin, P. V. Petrovskii and M. I. Rybinskaya, *Izv. Akad. Nauk SSSR, Ser. Khim.*, 1576 (1971).
9. W. O. Siegl and J. P. Collman, *J. Am. Chem. Soc.*, **94**, 2518 (1972).
10. T. Mitsudo, H. Nakanishi, T. Inubishi, I. Marishima, Y. Watanabe and Y. Takegami, *J. Chem. Soc., Chem. Commun.*, 416 (1976).
11. M. P. Cooke and R. M. Parlman, *J. Am. Chem. Soc.*, **97**, 6863 (1975).
12. J. P. Collman, *Acc. Chem. Res.*, **8**, 342 (1975).
13. J. Y. Merour, J. L. Roustan, C. Charrier, J. Collin and J. Benaim, *J. Organomet. Chem.*, **51**, C24 (1973).
14. J. Y. Merour, J. L. Roustan, C. Charrier, J. Benaim, J. Collin and P. Cadiot, *J. Organomet. Chem.*, **168**, 337 (1979).
15. J. L. Roustan, A. Guinot and P. Cadiot, *J. Organomet. Chem.*, **194**, 191 (1980).
16. A. Guinot, P. Cadiot and J. L. Roustan, *J. Organomet. Chem.*, **128**, C35 (1977).
17. J. L. Roustan, A. Guinot, P. Cadiot and A. Forgnes, *J. Organomet. Chem.*, **194**, 179 (1980).
18. J. L. Roustan, A. Guinot and P. Cadiot, *J. Organomet. Chem.*, **194**, 357 (1980).
19. J. L. Roustan, A. Guinot and P. Cadiot, *J. Organomet. Chem.*, **194**, 367 (1980).
20. M. Bottrill and M. Green, *J. Chem. Soc., Dalton Trans.*, 820 (1979).
21. M. Green, J. Z. Nyathi, C. Scott, F. G. A. Stone, A. J. Welch and P. Woodward, *J. Chem. Soc., Dalton Trans.*, 1067 (1978).
22. B. L. Booth and E. J. R. Lewis, *J. Chem. Soc., Dalton Trans.*, 417 (1982).
23. B. L. Booth and R. G. Hargreaves, *J. Chem. Soc. (A)*, 308 (1970).
24. H. G. Alt, *Z. Naturforsch.*, **32B**, 1139 (1977).
25. H. G. Alt, *Chem. Ber.*, **110**, 2862 (1977).
26. H. G. Alt, *Angew. Chem., Int. Ed. Engl.*, **15**, 759 (1976).
27. M. Bottrill, M. Green, E. O'Brien, L. E. Smart and P. Woodward, *J. Chem. Soc., Dalton Trans.*, 292 (1980).
28. H. G. Alt, G. S. Hermann, H. E. Engelhardt and R. D. Rogers, *J. Organomet. Chem.*, **331**, 329 (1987). See also References 29 and 30.
29. H. G. Alt, H. E. Engelhardt, U. Thewalt and J. Riede, *J. Organomet. Chem.*, **288**, 165 (1985).
30. H. G. Alt and H. I. Hayen, *J. Organomet. Chem.*, **315**, 337 (1986).
31. H. G. Alt, *J. Organomet. Chem.*, **127**, 349 (1977).
32. H. G. Alt and J. Schwarzle, *J. Organomet. Chem.*, **155**, C65 (1978).
33. H. G. Alt, M. E. Eichner and B. M. Jansen, *Angew. Chem., Int. Ed. Engl.*, **21**, 861 (1982).
34. H. G. Alt, H. I. Hayen, H. P. Klein and U. Thewalt, *Angew. Chem.*, **96**, 811 (1984).
35. R. F. Heck, *J. Am. Chem. Soc.*, **86**, 2819 (1964).
36. G. P. Chiusoli and L. Cassar, *Angew. Chem., Int. Ed. Engl.*, **6**, 124 (1967).
37. E. Carmona, E. Guttierrez-Puebla, A. Morge, J. M. Marin, M. Paneque and M. L. Poveda, *Organometallics*, **3**, 1438 (1984).
38. H. G. Alt, G. S. Hermann and U. Thewalt, *J. Organomet. Chem.*, **327**, 237 (1987).
39. H. G. Alt and H. I. Hayen, *J. Organomet. Chem.*, **316**, 105 (1986).
40. H. G. Alt and U. Thewalt, *J. Organomet. Chem.*, **268**, 235 (1984).
41. H. G. Alt, J. A. Schwarzle and F. R. Kreissl, *J. Organomet. Chem.*, **152**, C57 (1978).
42. P. L. Watson and R. G. Bergman, *J. Am. Chem. Soc.*, **101**, 2055 (1979).
43. J. L. Davidson, *J. Chem. Soc., Dalton Trans.*, 2423 (1986).

44. P. S. Braterman, J. L. Davidson and D. W. A. Sharp, *J. Chem. Soc., Dalton Trans.*, 241 (1976).
45. L. Carlton, J. L. Davidson and M. Shiralian, *J. Chem. Soc., Dalton Trans.*, 1577 (1986).
46. J. L. Davidson, M. Shiralian, L. Manojlovic-Muir and K. W. Muir, *J. Chem. Soc., Dalton Trans.*, 2167 (1984).
47. J. L. Davidson and L. Carlton, *J. Chem. Soc., Chem. Commun.*, 964 (1984).
48. L. Manojlovic-Muir and K. W. Muir, *J. Organomet. Chem.*, **168**, 403 (1979).
49. J. L. Davidson, *J. Chem. Soc., Chem. Commun.*, 597 (1979).
50. F. Y. Petillon, F. Le Floch-Perennou, J. E. Guerchais, D. W. A. Sharp, L. Manojlovic-Muir and K. W. Muir, *J. Organomet. Chem.*, **202**, 23 (1980).
51. J. E. Guerchais, F. Le Floch-Perennou, F. Y. Petillon, A. N. Keith, L. Manojlovic-Muir, K. W. Muir and D. W. A. Sharp, *J. Chem. Soc., Chem. Commun.*, 411 (1979).
52. J. L. Davidson and D. W. A. Sharp, *J. Chem. Soc., Dalton Trans.*, 2283 (1975).
53. F. Y. Petillon, F. Le Floch-Perennou, J. E. Guerchais and D. W. A. Sharp, *J. Organomet. Chem.*, **173**, 89 (1979).
54. M. Green, J. A. K. Howard, P. Mitrprachon, M. Pfeffer, J. L. Spencer, F. G. A. Stone and P. Woodward, *J. Chem. Soc., Dalton Trans.*, 306 (1979).
55. W. J. Cherwinski, B. F. G. Johnson and J. Lewis, *J. Chem. Soc., Dalton Trans.*, 1405 (1974).
56. S. Cenini, R. Ugo and G. La Monica, *J. Chem. Soc. (A)*, 409 (1971).
57. G. N. Schrauzer, *Chem. Ber.*, **94**, 642 (1961).
58. G. N. Schrauzer, *J. Am. Chem. Soc.*, **81**, 5311 (1959).
59. H. P. Fritz and G. N. Schrauzer, *Chem. Ber.*, **94**, 651 (1961).
60. E. Dinjus, H. Langbein and D. Walther, *J. Organomet. Chem.*, **152**, 229 (1978).
61. E. Dinjus, I. Gorski, H. Matschiner, E. Uhlig and D. Walther, *Z. Anorg. Allg. Chem.*, **436**, 39 (1977).
62. T. Yamamoto, A. Yamamoto and S. Ikeda, *J. Am. Chem. Soc.*, **93**, 3360 (1971).
63. S. D. Ittel, *Inorg. Chem.*, **16**, 2589 (1977).
64. C. A. Tolman, *J. Am. Chem. Soc.*, **96**, 2780 (1974).
65. T. Fueno, O. Kajimoto and J. Furukawa, *Bull. Chem. Soc. Jpn.*, **41**, 782 (1968).
66. E. Weiss, K. Stark, J. E. Lancaster and H. D. Murdoch, *Helv. Chim. Acta*, **46**, 288 (1963).
67. J. A. S. Howell, B. F. G. Johnson, P. L. Josty and J. Lewis, *J. Organomet. Chem.*, **39**, 329 (1972).
68. A. M. Brodie, B. F. G. Johnson, P. L. Josty and J. Lewis, *J. Chem. Soc., Dalton Trans.*, 2031 (1972).
69. K. Stark, J. E. Lancaster, H. D. Murdoch and E. Weiss, *Z. Naturforsch.*, **19B**, 284 (1964).
70. E. A. K. von Gustorf, F. W. Grevels, C. Kruger, G. Olbrich, F. Mark, D. Schulz and R. Wagner, *Z. Naturforsch.*, **27B**, 392 (1972).
71. T. Kruck and L. Knoll, *Chem. Ber.*, **106**, 3578 (1973).
72. A. Vessieres, D. Touchard and P. Dixneuf, *J. Organomet. Chem.*, **118**, 93 (1976).
73. A. Vessieres and P. Dixneuf, *Tetrahedron Lett.*, 1499 (1974).
74. A. Vessieres and P. Dixneuf, *J. Organomet. Chem.*, **108**, C5 (1976).
75. B. F. G. Johnson, J. Lewis, G. R. Stephenson and E. J. S. Vichi, *J. Chem. Soc., Dalton Trans.*, 369 (1978).
76. F. W. Grevels, J. G. A. Reuvers and J. Takats, *J. Am. Chem. Soc.*, **103**, 4069 (1981).
77. M. Giffard and P. Dixneuf, *J. Organomet. Chem.*, **85**, C26 (1975).
78. M. Giffard, E. Gentric, D. Touchard and P. Dixneuf, *J. Organomet. Chem.*, **129**, 371 (1977).
79. W. A. Hermann, *Chem. Ber.*, **108**, 486 (1975).
80. A. Rosan and M. Rosenblum, *J. Org. Chem.*, **40**, 3621 (1975).
81. A. Cutler, D. Ehntholt, W. P. Giering, P. Lennon, S. Raghu, A. Rosan, M. Rosenblum, J. Tancrede and D. Wells, *J. Am. Chem. Soc.*, **98**, 3495 (1976).
82. E. K. G. Schmidt and C. H. Thiel, *J. Organomet. Chem.*, **209**, 373 (1981).
83. U. Koemm and C. G. Kreiter, *J. Organomet. Chem.*, **240**, 27 (1982).
84. R. B. King and A. Fronzaglia, *Inorg. Chem.*, **5**, 1837 (1966).
85. T. Tatsumi, H. Tominaga, M. Hidai and Y. Uchida, *J. Organomet. Chem.*, **199**, 63 (1980).
86. D. P. Tate, A. A. Buss, J. M. Augl, B. L. Boss, J. G. Graselli, W. M. Ritchie and F. J. Knoll, *Inorg. Chem.*, **4**, 1323 (1965).
87. M. Moran, J. J. Santos-Garcia, J. R. Masaguer and V. Fernandez, *J. Organomet. Chem.*, **295**, 327 (1985).
88. G. Fachinetti, S. del Nero and C. Floriani, *J. Chem. Soc., Dalton Trans.*, 1046 (1976).
89. G. Fachinetti, C. Floriani, A. Chiesa-Villa and C. Guastini, *Inorg. Chem.*, **18**, 2282 (1979).

90. J. R. Durig and T. S. Little, *J. Chem. Phys.*, **75**, 3661 (1980).
91. J. R. Bentley, K. B. Everard, R. J. B. Marsden and L. E. Sutton, *J. Chem. Soc.*, 2957 (1949).
92. G. Le Borgne, E. Gentric and D. Grandjean, *Acta Crystallogr.*, **31B**, 2824 (1975).
93. S. Andersson, M. Hakansson, S. Jagner, M. Nillson and F. Urso, *Acta Chem. Scand.*, **40A**, 195 (1986).
94. S. Sorriso and G. Cardaci, *J. Chem. Soc., Dalton Trans.*, 1041 (1975).
95. P. Lennon, A. M. Rosan and M. Rosenblum, *J. Am. Chem. Soc.*, **99**, 8426 (1977).
96. T. E. Bausch and W. P. Giering, *J. Organomet. Chem.*, **144**, 335 (1978).
97. T. E. Bausch, M. Konowitz and W. P. Giering, *J. Organomet. Chem.*, **114**, C15 (1976).
98. A. N. Nesmeyanov, L. V. Rybin, N. A. Stelzer and M. I. Rybinskaya, *J. Organomet. Chem.*, **182**, 393 (1979).
99. A. N. Nesmeyanov, M. I. Rybinskaya, L. V. Rybin, N. T. Gubenko, N. G. Bokii, A. S. Batsanov and Y. T. Struchkov, *J. Organomet. Chem.*, **149**, 177 (1978).
100. A. N. Nesmeyanov, L. V. Rybin, N. T. Gubenko, M. I. Rybinskaya and P. V. Petrovski, *J. Organomet. Chem.*, **71**, 271 (1974).
101. M. S. Brookhart, G. W. Koszalka, G. O. Nelson, G. Scholes and R. A. Watson, *J. Am. Chem. Soc.*, **98**, 8155 (1976).
102. A. J. P. Domingos, J. A. S. Howell, B. F. G. Johnson and J. Lewis, *Inorg. Synth.*, **16**, 103 (1976).
103. K. Stark, J. E. Lancaster, H. D. Murdoch and E. Weiss, *Z. Naturforsch.*, **19B**, 284 (1964).
104. G. Cardaci, *J. Am. Chem. Soc.*, **97**, 1412 (1975).
105. A. de Cian and R. Weiss, *Acta Crystallogr.*, **28B**, 3273 (1972).
106. P. Huebner, H. Kuehr and E. Weiss, *Cryst. Struct. Commun.*, **10**, 1451 (1981).
107. D. Leibfritz and H. tom Dieck, *J. Organomet. Chem.*, **105**, 255 (1976).
108. T. Kenny, W. von Phillipsborn, J. Kronenbitter and A. Schwenk, *J. Organomet. Chem.*, **205**, 211 (1981).
109. S. Sorriso and G. Cardaci, *J. Organomet. Chem.*, **101**, 107 (1975).
110. M. Sacerdoti, V. Bertolasi and G. Gill, *Acta Crystallogr.*, **36B**, 1061 (1980).
111. E. J. S. Vichi, P. R. Raithby and M. McPartlin, *J. Organomet. Chem.*, **256**, 111 (1983).
112. R. L. Harlow and C. E. Pfluger, *Acta Crystallogr.*, **29B**, 2633 (1973).
113. J. A. S. Howell, D. T. Dixon and J. C. Kola, *J. Organomet. Chem.*, **266**, 69 (1984).
114. A. M. Benedetti, V. M. Noguiera and E. J. S. Vichi, 4th An. Simp. Bras. Eletroquim. Electroanal., 471 (1984); *Chem. Abstr.*, **101**, 139565k (1984).
115. S. E. Thomas, *J. Chem. Soc., Chem. Commun.*, 226 (1987).
116. N. El Murr, M. Riveccie and P. Dixneuf, *J. Chem. Soc., Chem. Commun.*, 552 (1978).
117. G. Cardaci and G. Concetti, *J. Organomet. Chem.*, **90**, 49 (1974).
118. G. Cardaci and G. Bellachioma, *Inorg. Chem.*, **16**, 3099 (1977).
119. S. C. Wright and M. S. Baird, *J. Am. Chem. Soc.*, **107**, 6899 (1985).
120. M. Borner and H. Vahrenkamp, *Chem. Ber.*, **114**, 1382 (1981).
121. E. J. S. Vichi, F. Y. Fujiwara and E. Stein, *Inorg. Chem.*, **24**, 286 (1985).
122. H. C. Ashton and A. R. Manning, *Inorg. Chem.*, **22**, 1440 (1983).
123. H. Le Bozec, P. H. Dixneuf, A. J. Carty and N. J. Taylor, *Inorg. Chem.*, **17**, 2568 (1978).
124. M. Brookhart and G. O. Nelson, *J. Organomet. Chem.*, **164**, 193 (1979).
125. P. M. Burkinshaw, D. T. Dixon and J. A. S. Howell, *J. Chem. Soc., Dalton Trans.*, 999 (1980).
126. D. Wormsbacher, F. Edelmann, D. Kaufmann, U. Behrens and A. de Meijere, *Angew. Chem., Int. Ed. Engl.*, **20**, 696 (1981).
127. C. B. Argo and J. T. Sharp, *Tetrahedron Lett.*, **22**, 353 (1981).
128. E. Vogel, D. Kerimis, N. T. Allinson, R. Zellerhof and J. Wassen, *Angew. Chem., Int. Ed. Engl.*, **18**, 545 (1979).
129. P. Narbel, T. Boschi, R. Roulet, P. Vogel, A. A. Pinkerton and D. Schwarzenbach, *Inorg. Chim. Acta*, **36**, 161 (1979).
130. L. A. Paquette, J. M. Photis and R. P. Micheli, *J. Am. Chem. Soc.*, **99**, 7899 (1977).
131. R. W. Ashworth and G. A. Berchtold, *J. Am. Chem. Soc.*, **99**, 5200 (1977).
132. D. H. R. Barton, A. A. L. Gunatilaka, T. Nakanishi, H. Patin, D. A. Widdowson and B. R. Worth, *J. Chem. Soc., Perkin Trans. 1*, 821 (1976).
133. G. Evans, B. F. G. Johnson and J. Lewis, *J. Organomet. Chem.*, **102**, 507 (1975).
134. C. C. Santini, J. Fischer, F. Matthey and A. Mitschler, *Inorg. Chem.*, **20**, 2848 (1981).
135. A. R. Pinhas, A. G. Samuelson, R. Risemberg, E. V. Arnold, J. Clardy and B. K. Carpenter, *J. Am. Chem. Soc.*, **103**, 1668 (1981).

136. R. B. King and M. N. Ackermann, *J. Organomet. Chem.*, **60**, C57 (1973).
137. J. Ioset and R. Roulet, *Helv. Chim. Acta*, **68**, 236 (1985).
138. B. F. G. Johnson, J. Lewis and D. Wege, *J. Chem. Soc., Dalton Trans.*, 1874 (1976).
139. C. R. Graham, G. Scholes and M. Brookhart, *J. Am. Chem. Soc.*, **99**, 1180 (1977).
140. J. A. S. Howell, J. C. Kola, D. T. Dixon, P. M. Burkinshaw and M. J. Thomas, *J. Organomet. Chem.*, **266**, 83 (1984).
141. A. J. Birch, W. D. Raverty and G. R. Stephenson, *Organometallics*, **3**, 1075 (1984).
142. R. E. Moriarty, R. D. Ernst and R. Bau, *J. Chem. Soc., Chem. Commun.*, 1242 (1972).
143. J. C. Green, M. R. Kelly, P. D. Grebenik, C. E. Briant, N. A. McEvoy and D. M. P. Mingos, *J. Organomet. Chem.*, **228**, 239 (1982).
144. W. C. Kaska, R. F. Reichelderfer and L. Prizant, *J. Organomet. Chem.*, **129**, 97 (1977).
145. S. Andersson, M. Hakansson, S. Jagner, M. Nilsson, C. Ullenius and F. Urso, *Acta Chem. Scand.*, **40A**, 58 (1986).
146. P. A. Corrigan, R. S. Dickson, S. H. Johnson, G. N. Pain and M. Yeoh, *Aust. J. Chem.*, **35**, 2203 (1982).
147. R. S. Dickson, M. C. Nesbit, B. M. Gatehouse and G. N. Pain, *J. Organomet. Chem.*, **215**, 97 (1981).
148. R. S. Dickson, G. D. Fallon, R. G. Nesbit and G. N. Pain, *Organometallics*, **4**, 355 (1985).
149. R. S. Dickson and H. P. Kirsch, *Aust. J. Chem.*, **27**, 61 (1974).
150. R. S. Dickson and S. H. Johnson, *Aust. J. Chem.*, **29**, 2189 (1976).
151. S. R. Finnimore, S. A. R. Knox and G. E. Taylor, *J. Chem. Soc., Dalton Trans.*, 1783 (1982).
152. M. C. Azar, M. J. Chetcuti, C. Eigenbrot and K. A. Green, *J. Am. Chem. Soc.*, **107**, 7209 (1985).
153. N. M. Boag, R. J. Goodfellow, M. Green, B. Hessner, J. A. K. Howard and F. G. A. Stone, *J. Chem. Soc., Dalton Trans.*, 2585 (1983).
154. R. Victor, S. Sarel and V. Usieli, *J. Organomet. Chem.*, **129**, 387 (1977).
155. W. Hubel, in *Organic Syntheses via Metal Carbonyls* (Eds. I. Wender and P. Pino), Interscience, New York, 1968, pp. 273–342.
156. F. A. Cotton, D. L. Hunter and J. M. Troup, *Inorg. Chem.*, **15**, 63 (1976).
157. R. S. Dickson, G. S. Evans and G. D. Fallon, *Aust. J. Chem.*, **38**, 273 (1985).
158. W. A. Herrmann, C. Bauer and J. Weichmann, *J. Organomet. Chem.*, **243**, C21 (1983).
159. D. L. Davies, S. A. R. Knox, K. A. Mead, M. J. Morris and P. Woodward, *J. Chem. Soc., Dalton Trans.*, 2293 (1984).
160. D. L. Davies, A. F. Dyke, S. A. R. Knox and M. J. Morris, *J. Organomet. Chem.*, **215**, C30 (1981).
161. A. F. Dyke, S. A. R. Knox, P. J. Naish and G. E. Taylor, *J. Chem. Soc., Dalton Trans.*, 1297 (1982).
162. R. E. Colborn, D. L. Davies, A. F. Dyke, A. Endesfelder, S. A. R. Knox, A. G. Orpen and D. Plaas, *J. Chem. Soc., Dalton Trans.*, 2661 (1983).
163. A. F. Dyke, S. A. R. Knox, M. J. Morris and P. J. Naish, *J. Chem. Soc., Dalton Trans.*, 1417 (1983).
164. J. L. Davidson, M. Green, F. G. A. Stone and A. J. Welch, *J. Chem. Soc., Chem. Commun.*, 286 (1975).
165. W. G. Jackson, B. F. G. Johnson, J. W. Kelland, J. Lewis and K. T. Schorpp, *J. Organomet. Chem.*, **88**, C17 (1975).
166. R. P. Ferrari, G. A. Vaglio, O. Cambino, M. Valle and G. Cetini, *J. Chem. Soc., Dalton Trans.*, 1998 (1972).
167. G. Varadi, I. Vecsei, I. Otvos, G. Palyi and L. Marko, *J. Organomet. Chem.*, **182**, 415 (1979).
168. G. Palyi, G. Caradi, A. Vizi-Orosz and L. Marko, *J. Organomet. Chem.*, **90**, 85 (1975).
169. D. J. S. Guthrie, I. U. Khand, G. R. Knox, J. Kollmeier, P. L. Pauson and W. E. Watts, *J. Organomet. Chem.*, **90**, 93 (1975).
170. I. U. Khand and P. L. Pauson, *J. Chem. Soc., Perkin Trans. 1*, 30 (1976).
171. D. C. Billington, W. J. Kerr and P. L. Pauson, *J. Organomet. Chem.*, **328**, 223 (1987).
172. A. J. Arce, Y. de Sanctis and A. J. Deeming, *J. Organomet. Chem.*, **295**, 365 (1985).
173. A. J. Deeming, P. J. Manning, I. P. Rothwell, M. B. Hursthouse and N. P. C. Walker, *J. Chem. Soc., Dalton Trans.*, 2039 (1984).
174. A. J. P. Domingos, B. F. G. Johnson, J. Lewis and G. M. Sheldrick, *J. Chem. Soc., Chem. Commun.*, 912 (1973).
175. K. Henrick, J. A. Iggo, M. J. Mays and P. R. Raithby, *J. Chem. Soc., Chem. Commun.*, 209 (1984).
176. R. Regragui, P. H. Dixneuf, N. J. Taylor and A. J. Carty, *Organometallics*, **3**, 814 (1984).

The Chemistry of Enones
Edited by S. Patai and Z. Rappoport
© 1989 John Wiley & Sons Ltd

CHAPTER **20**

Dienols (enolization of enones)

BRIAN CAPON

Chemistry Department, Hong Kong University, Pokfulam Road, Hong Kong

I. INTRODUCTION

This chapter is mainly concerned with the chemistry of 1,3-dien-1-ols (**2**) which may formally be generated by the enolization of α, β- (**1**) or β, γ- (**3**) unsaturated carbonyl compounds. This may be achieved either thermally with the aid of catalysts, or photochemically. As photoenolization was reviewed by Sammes in 1976[1] and by Wagner in 1980[2] only the properties of the dienols obtained by photoenolization will be considered in this chapter, not the details of the photoenolization process.

α, β-Unsaturated carbonyl compounds with a proton attached to the α-carbon (**4**) may formally also undergo enolization to yield 1,2-dienols (**5**), and α, β-unsaturated ketones with a hydrogen at the α'-position (**6**) may formally enolize to 1,3-dien-2-ols (**7**).

For many years dienols and their anions could only be studied indirectly, usually by studying the reactions of enones and inferring the properties of the presumed dienol and dienolate intermediates. Many of the results reported in this chapter will be of this type. However, more recently methods have become available for generating dienols in the gas phase and in solution, so that they can be detected spectroscopically and sometimes isolated; hence direct measurements of their properties are now possible.

(1) (2) (3)

(4) (5)

$R^4 \quad R^3 \quad H \quad R^1$
$\beta \quad \alpha \quad \alpha'$
$R^5 \quad R^2$
(6)

$R^4 \quad R^3 \quad R^1$
$R^5 \quad OH \quad R^2$
(7)

II. THE 1,3-BUTADIEN-1-OLS

A. Conformations and Relative Stabilities

The simplest 1,3-dienols are (Z)- and (E)-1,3-butadien-1-ol (2, R^1, R^2, R^3, R^4, R^5 = H). Four planar conformations for each of these are possible, depending on the orientation around the $O-C_1$ and C_2-C_3 bonds (see Scheme 1). In addition there are an infinite number of non-planar *gauche* conformations. As discussed below (Section II.C) the ^{1}H-NMR spectra indicate that, in solution, the stable conformation of the (E)-isomer is *s-cis, s-trans* and of the (Z)-isomer, *s-trans, s-trans*.

Calculations of the relative energies of (Z)- and (E)-1,3-butadien-1-ols in their less stable (C_2-C_3)*s-cis* conformations have been reported in two papers whose main purpose was to elucidate the mechanism of photoenolization[3,4]. The *ab initio* methods (STO-3G, 4–31G basis sets)[4] give the Z-dienol as more stable by 4–5 kcal mol^{-1}. This is much larger than would be expected on the basis of the experimental heats of formation[5,6], presumably because the enols exist mainly in the (C_2-C_3)*s-trans* conformation. MNDO calculations were reported for these[5,6] and they agree quite well with the experimental heats of formation, but neither these calculations nor the experimental results are sufficiently accurate to determine where (Z)- or (E)-1,3-butadien-1-ol is more stable. However, if the corresponding ethyl ethers can be taken as a model, the (E)-dienol would be expected to be slightly more stable as (E)-1,3-butadienyl ethyl ether is more stable than its (Z)-isomer with $\Delta H° = 0.92$ kcal mol^{-1} at 25 °C in hexane[7].

B. Generation in the Gas Phase

A 1,3-butadien-1-ol was postulated as an intermediate in the gas-phase photo-isomerization of crotonaldehyde into 3-butenal[8] and tentatively assigned as one of the products of the gas-phase photolysis of crotonaldehyde on the basis of its IR spectrum ($v = 3630, 1100$ cm^{-1})[9]. Tureček and coworkers[5] generated (Z)- and (E)-butadien-1-ol by high vacuum flash pyrolysis of the Diels–Alder adducts 8 and 9. The deuterium-labelled dienols with deuterium on oxygen and on C_1 were also generated from labelled precursors.

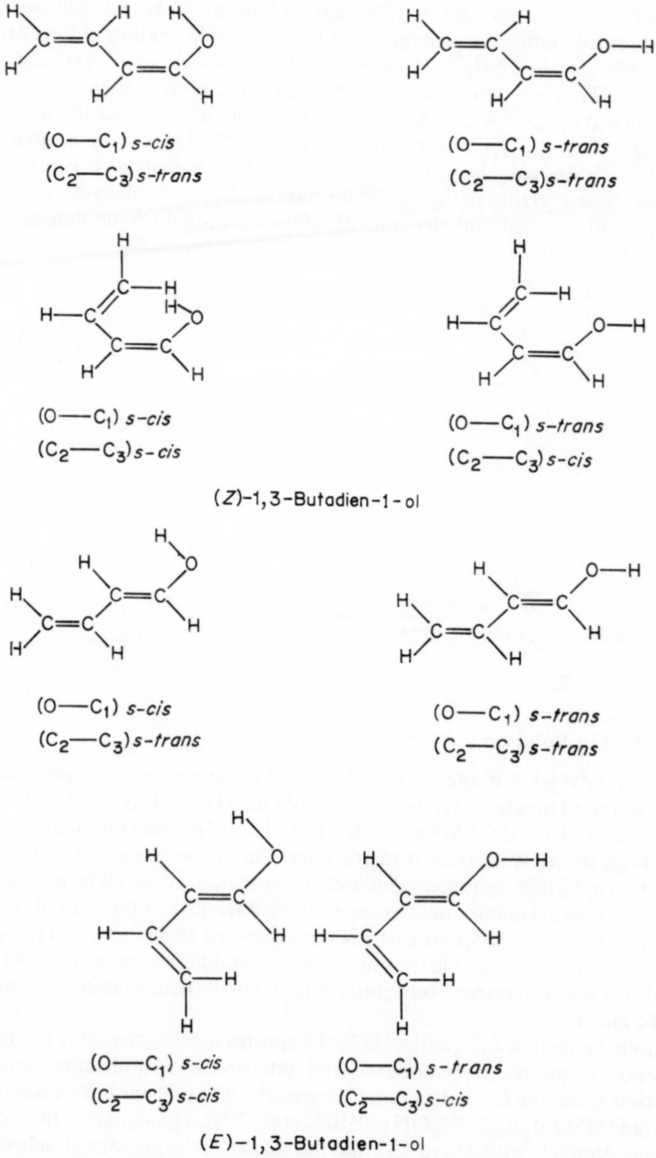

$(O-C_1)$ s-cis
(C_2-C_3) s-trans

$(O-C_1)$ s-trans
(C_2-C_3) s-trans

$(O-C_1)$ s-cis
(C_2-C_3) s-cis

$(O-C_1)$ s-trans
(C_2-C_3) s-cis

(Z)-1,3-Butadien-1-ol

$(O-C_1)$ s-cis
(C_2-C_3) s-trans

$(O-C_1)$ s-trans
(C_2-C_3) s-trans

$(O-C_1)$ s-cis
(C_2-C_3) s-cis

$(O-C_1)$ s-trans
(C_2-C_3) s-cis

(E)-1,3-Butadien-1-ol

SCHEME 1. Planar conformations of (Z)- and (E)-1,3-butadien-1-ols

The dienols, mixed with a maximum of 10–15% crotonaldehyde, were characterized by their 75 eV electron-impact mass spectra and by the collision-induced decomposition spectra of their molecular ions[10]. The EI mass spectra of the (Z)- and (E)-isomers showed only small differences and the main basis for the assignment of stereochemistry was the known stereospecificity of the retro-Diels–Alder reaction and the expected high energy of activation for the E–Z isomerization. The heats of formation of the dienols were estimated from their threshold ionization energies and the heats of formation of the corresponding cation radicals to be $\Delta H_{f,298}^\circ(E) = -21 \pm 2\,\text{kcal mol}^{-1}$ and $\Delta H_{f,298}^\circ(Z) = -21.5 \pm 2\,\text{kcal mol}^{-1}$. They are therefore less stable than the conjugated aldehyde, crotonaldehyde, for which $\Delta H_{f,298}^\circ = -25.6\,\text{kcal mol}^{-1}$ but more stable than the non-conjugated 3-butenal for which $\Delta H_{f,298}^\circ$ was estimated to be $-80\,\text{kJ mol}^{-1}$ by applying Benson's additivity rules. It was calculated that the barrier of intramolecular ketonization was too high for this to be a viable pathway for formation of crotonaldehyde (or its Z-isomer) under the conditions used and that the 10–15% of crotonaldehyde detected must have been formed by a surface catalysed process.

(8)

(9)

C. Generation in Solution

(E)- and (Z)-$[O$-$^2H]$-1, 3-butadien-1-ol (**12** and **13**) have been generated in solution by hydrolysis of their trimethylsilyl derivatives (**10** and **11**) in $CD_3CN:D_2O$ (9:1 v/v) which contained DCl (3.16×10^{-3} M) at 32 °C. The ^1H NMR spectral changes indicated in Scheme 2 took place. With both isomers, after 1 hour the signals of their trimethylsilyl groups at $\delta = ca\,0.2$ had disappeared completely and been replaced by a signal at $\delta = 0.03$ ascribed to trimethylsilanol or hexamethyl disiloxane[11]. Only small changes were observed in the rest of the spectra which were ascribed to the (E)- and (Z)-$[O$-$^2H]$-1, 3-butadien-1-ol. These were stable in solution for several hours but were slowly converted, by γ- and δ-deuteriation respectively, into a mixture of the deuterated (E)-2-butenal and 3-butenal (**14** and **15**).

The coupling constant J_{2-3} in the ^1H NMR spectra of both dienols is 10–11 Hz, similar to that reported for alkoxybutadienes and interpreted as indicating that the stable conformation about the C-2—C-3 bond is s-$trans$[12]. The OH enols were also generated in a mixture of DMSO-d_6 and CH_3OH (90:10 v/v) at 32 °C. The signals of the oxygen-bound protons were doublets with $\delta = ca\,8.91$ and the signals of the protons attached to C-1 were four line signals. The HO—C$_1$—H coupling constants for the (E)- and (Z)-isomers were respectively 8.8 and 5.8 Hz. This is similar to what is found for (E)- and (Z)-1-propenols[13] and suggested that the (E)-isomer is predominantly in the s-cis conformation and the (Z)-isomer is predominantly in the s-$trans$ conformation around the C—O bond. It is therefore concluded that the most stable conformations are s-cis, s-$trans$ (E-isomer) and s-$trans$, s-$trans$ (Z-isomer) (see Scheme 1).

The pH–rate profiles for the ketonization of both dienols in water are U-shaped curves, consistent with there being H_3O^+, HO^- and water catalysed processes. The overall rate

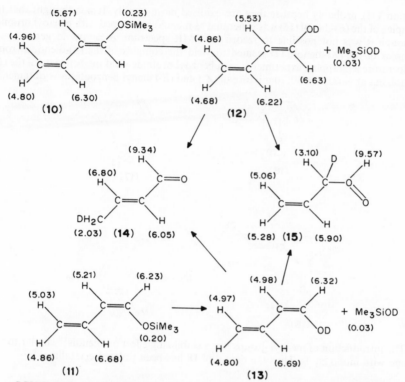

SCHEME 2. The ^{1}H NMR spectral changes that take place on hydrolysis of (E)- and (Z)-1-trimethylsilyloxy-1, 3-butadiene in CD$_3$CN: D$_2$O(9:1 v/v) which contains DCl (3.16 × 10^{-3} M) at 32 °C. The numbers in brackets are δ values

constants for protonation at both the α- and γ-position at the minima of the pH–rate profile (pH 3–5) are 3 × 10^{-3} to 4 × 10^{-3} s^{-1} at 25 °C which correspond to a half life of about 3 to 4 minutes. The (E)-isomer is slightly longer lived than the (Z)-isomer.

The ketonization of the dienols to yield 3-butenal involves α-protonation and is analogous to the ketonization of vinyl alcohol[14,15] but is much slower. Thus the additional double bond causes a decrease in k_{H^+} by factors of 1544 (Z-isomer) and 918 (E-isomer). The value of k_{H^+} for ketonization with γ-protonation is 7.3 times faster with the (E)-isomer than with the (Z)-isomer. This is a little less than can be calculated for the relative rates of protonation of the corresponding (E)- and (Z)-butadienyl ethers in 80% aqueous dioxan[16]. It seems that the transmission of positive charge to the oxygen in the transition state for protonation at the γ-position is more efficient with the (E)- than with the (Z)-isomers of both dienols and dienyl ethers, possibly because it is easier for the former to attain a planar conformation. The greater reactivity at the 4-position of the (E)-isomer is also shown in the water and hydroxide-ion-catalysed reaction of the dienols. These last reactions presumably involve the dienolate ions and relative rates of protonation of these and of the dienols at the α and γ positions are discussed in Sections V and IV.

Attempts have been made to generate the iron–carbonyl complexes of these dienols **16** and **17** by treatment of the corresponding acetates with methyl lithium in diethyl ether, but not with as much success as in the generation of the complex of 1, 3-butadien-2-ol (see

Section VII), probably because they are oxidised more rapidly. It was thought that the complex of the (E)-dienol (16) was generated as it could be trapped with benzoyl bromide although it was not possible to record its NMR spectrum. Attempts to generate the complex of the (Z)-dienol (17) seemed to yield the complex of the (E)-dienol at room temperature as trapping experiments with benzoyl bromide yield the benzoate of the (E)-dienol, but at $-60\,^{\circ}$C a 30:70 mixture of (Z-) and (E-) dienyl benzoates was obtained[17].

(16)

(17)

(18)

The introduction of mesityl groups has a stabilizing effect on dienols[18] similar to that found with mono enols. Thus the dien-diol 18 has been isolated crystalline[18].

D. Generation of 1,3-Butadien-1-olate Anions

The trans-buta-1,3-dien-1-olate ion has been generated as its potassium salt by treatment of crotonaldehyde with potassium amide in liquid ammonia and characterized by its ^{1}H NMR spectrum (see Scheme 3)[19] and as its lithium salt by cleavage of 2-substituted-4,7-dihydro-1,3-dioxepines with butyl lithium[20]. The corresponding cis ion has been obtained by cleavage of 2,5-dihydrofuran either with potassium amide in liquid ammonia to yield the potassium salt[21] or with n-butyl lithium in n-hexane to yield the lithium salt[22,23]. It should be noted that the ^{1}H NMR spectra of the potassium and lithium salts show substantial chemical shift differences (see Scheme 3). The ^{13}C NMR spectrum of the lithium salt of the cis anion has also been reported and π-electron densities calculated by CNDO/2[23]. It has also been generated and its presence inferred on the basis of trapping experiments[24]. In addition the 1,3,5-hexatrien-1-olate anion has been generated and characterized by ^{13}C NMR spectroscopy[23].

III. PHOTOCHEMICALLY GENERATED 1,3-DIEN-1-OLS

A. From o-Substituted Aromatic Carbonyl Compounds

Photoenolization[1] was discovered by Yang and Rivas[25]. In the initial experiments the evidence for a process like that shown in equation 1 was indirect, such as incorporation of deuterium when the solvent was CH_3OD, and trapping with dimethyl acetylene-dicarboxylate. However, later, by the use of flash photolysis, two intermediates were detected on photoenolization of o-benzylbenzophenone in cyclohexane, one of which was

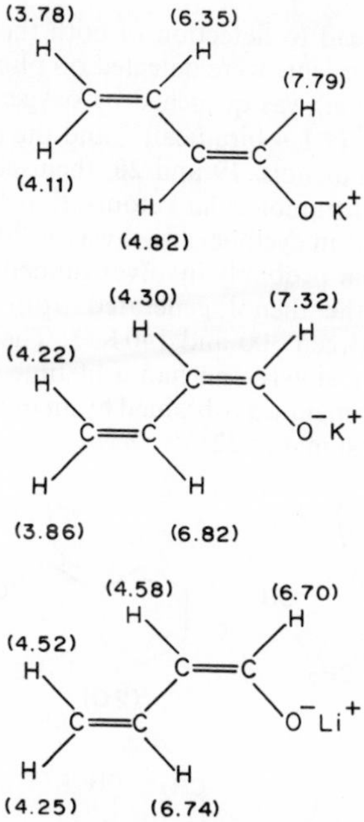

$J_{1,2} = 10.5\text{Hz}, J_{2,3} = 10.6\text{Hz}, J_{3,4t} = 16.5\text{Hz}, J_{3,4c} = 10.7\text{Hz}, J_{4c,4t} = 3.2\text{ Hz}$ (Ref. 19)

$J_{1,2} = 4.5\text{Hz}, J_{2,3} = 11.0\text{Hz}, J_{3,4t} = 17.2\text{Hz}, J_{3,4c} = 10.2\text{Hz}, J_{4c,4t} = 3.3\text{ Hz}$ (Ref. 21)

$J_{1,2} = 6\text{Hz}, J_{2,3} = 10.5\text{Hz}, J_{3,4t} = 17.0\text{Hz}, J_{3,4c} = 10.5\text{Hz}, J_{4c,4t} = 2\text{ Hz}$ (Ref. 22)

SCHEME 3. ^{1}H NMR spectral data of buta-1, 3-dienolate anions. The numbers in brackets are δ values

thought to be an excited state, and the other a dienol.* This latter species reformed the starting material with a rate constant, $9.4 \times 10^{-3}\ \text{s}^{-1}$ (temperature *ca* 20 °C) (equation 2)[26].

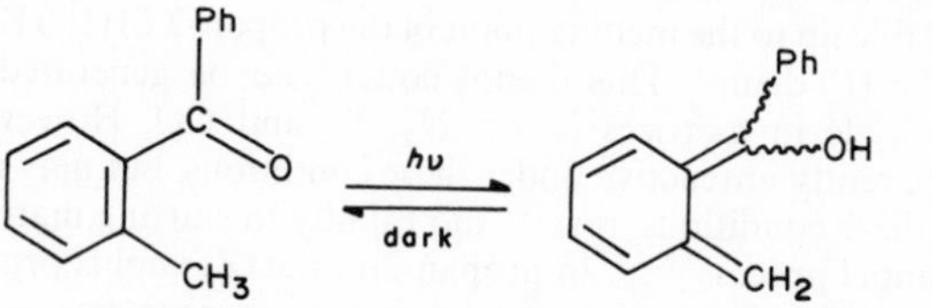

(1)

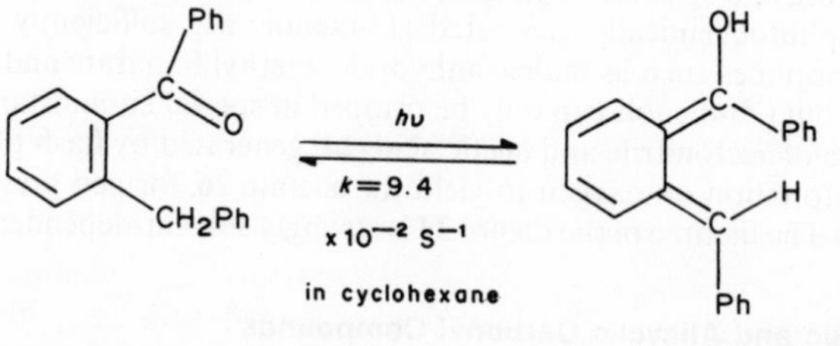

(2)

*These species have four double bonds and thus are really tetraenols, but their formation and ketonization appears to be qualitatively similar to that of analogous species which lack the two endocyclic double bonds (see Section III.B).

Subsequent work using laser flash photolysis lead to detection of both the (Z)- and (E)-dienols in this type of reaction. Thus, three transients were detected on photolysis of 2-methylacetophenone[26], one of which ($\lambda_{max} = 330$ nm) was quenched by oxygen and was ascribed to an excited triplet state of the dienols[27] (a 1,4-biradical)[28] and the other two ($\lambda_{max} = 390$ nm) were ascribed to the (Z)- and (E)-dienols, **19** and **20**, themselves. That ascribed to the (Z)-isomer (**19**), which can undergo intramolecular ketonization, had a very short lifetime which depended on the solvent. Thus in cyclohexane it was < 20 ns but in HMPA, 160 μs[27]. At low temperature this reaction probably involves tunnelling as the isotope effect, k_H/k_D, for the ketonization of the dienol generated from 2-[^{2}H$_3$]-methylacetophenone increased from 3 to 180 between 300 and 140 K[29]. The transient ascribed to the (E)-isomer **20** decayed much more slowly and had a lifetime of several seconds in cyclohexane[30,31]. Support for this assignment was obtained by an investigation of **21** which only yielded a short-lived dienol, presumably **22**[27].

(19)   (20)

(21)   (22)

A dienol thought to be the (Z)-form, **19**, was also detected (by IR spectroscopy) on irradiation of 2-methylacetophenone in propan-2-ol at 77 K[32]. This reverts to the ketone on warming to 100 K. In contrast, similar treatment of 2-methylbenzaldehyde yielded a dienol which is stable up to the melting point of the propan-2-ol (180 K) and was therefore thought to be the (E)-dienol. This dienol could also be generated by irradiation of 2-methylbenzaldehyde in matrices in Ar, N$_2$, Xe and CO. However, 2-methylaceto-phenone was apparently unreactive under these conditions, but may form the (Z)-dienol **19** which, under these conditions, reverts too rapidly to starting material to be detected, probably via a tunnel process[32,29]. In propan-2-ol the (Z)-enol is probably stabilized by hydrogen bonding and so can be detected, but in the other matrices it is not so stabilized.

Frequently, photochemically generated (E)-dienols are sufficiently long-lived to be trapped by dienophiles such as maleic anhydride, methyl fumarate and phenyl fumarate (equation 3)[33], but (Z)-dienols can only be trapped in special circumstances, such as with **24**. In a mixture of acetonitrile and acetic acid, **24**, generated by flash photolysis of **23**, is trapped by protonation on oxygen to yield the acetate **26**, formed presumably from the carbocation **25**. The lifetime of the dienol **24** is strongly solvent-dependent (see Table 1)[34].

B. From Acyclic and Alicyclic Carbonyl Compounds

Photoenolization has also been postulated to occur on irradiation of aliphatic α, β-unsaturated ketones (e.g. equation 4 and 5)[35,36], α, β-unsaturated esters (e.g. equation 6)[37,38] and of alicyclic α, β-unsaturated ketones (e.g. equation 7)[39]. With all of

$$\text{e. g. } B = CO_2R^1 \tag{3}$$

TABLE 1. Lifetime of **24** in different solvents at room temperature[34]

Solvent	Lifetime	Solvent	Lifetime
Cyclohexane	8 μs	Ethanol	0.8 ms
Benzene	40 μs	Acetonitrile + 1% water	0.8 ms
Diethyl ether	2 ms	Acetonitrile + 30% water	1.3 ms
Acetonitrile	3 ms	Acetic acid	26 μs
Tetrahydrofuran	7 ms		
Dimethyl sulphoxide	11 ms		
HMPA	90 ms		

B. Capon

these the first step is the conversion of the (E)- to the (Z)-form. In further steps the (Z)-form is converted into the β,γ-unsaturated isomer through presumed dienol intermediates (27–30). None of these were detected, however. The success of these 'deconjugation' reactions depends on the dienols undergoing some ketonization with protonation at the α-position and on the resulting β,γ-unsaturated carbonyl compounds not undergoing photochemical reversion to their α,β-unsaturated isomers. However, the relative amounts of ketonization with protonation at the α- and γ-positions are not known, although it should in principle be possible to determine them by using a medium such as CH_3OD and measuring the amount of deuterium incorporation at the α- and γ-positions of the products.

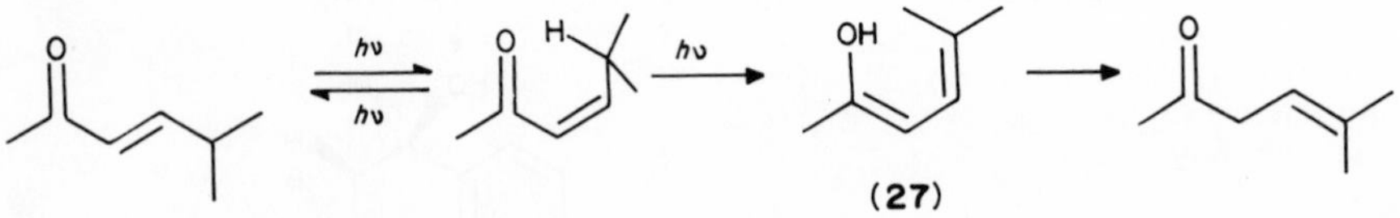

(4)

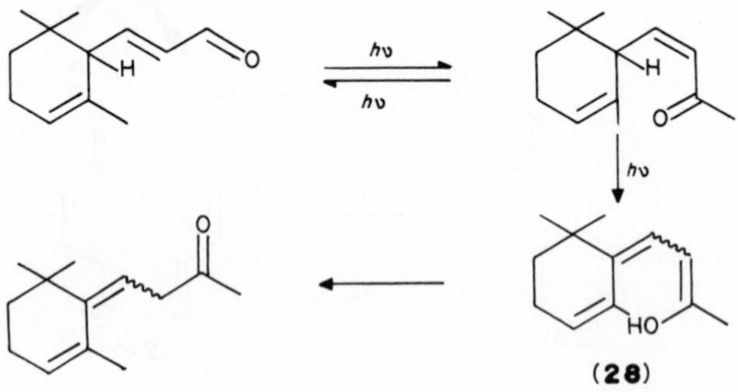

(5)

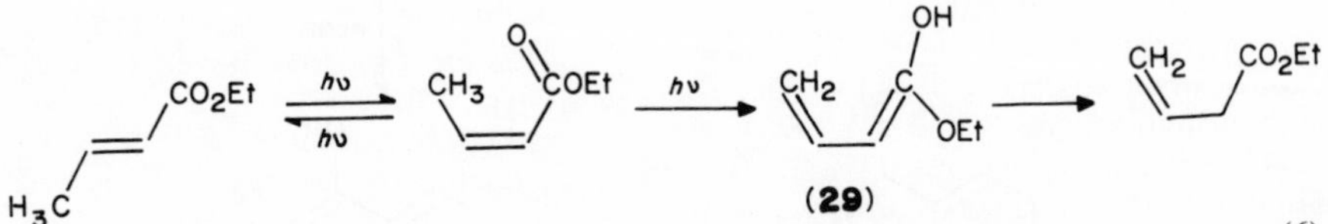

(6)

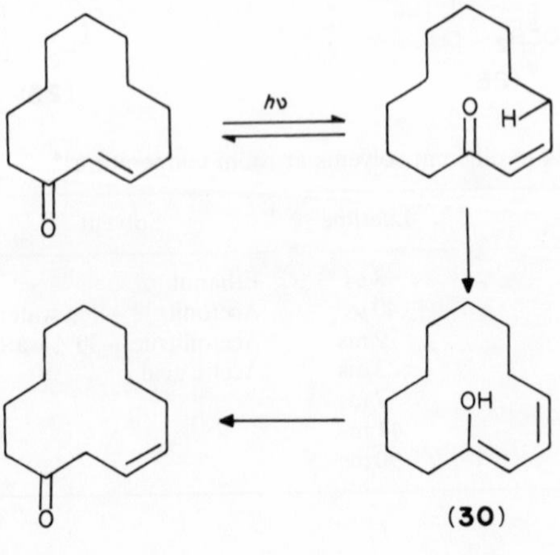

(7)

Photoenolization of α, β-unsaturated ketones does not always lead to formation of their β, γ-unsaturated isomers. Thus ketones **31, 32, 33** and **34** appear to undergo photoenolization in deuteriomethanol, as deuterium is incorporated into the γ-methyl groups, although there was no isomerization to β, γ-unsaturated ketones[40]. Rapid *cis–trans* isomerization about the C—C double bonds also occurs, as with **31, 33** and **34** the rates of incorporation of deuterium into the methyl group *cis* and *trans* to the carbonyl group are almost the same, and with **32** the isomerization was demonstrated by the observation of the signal of the methyl group of its (Z)-isomer in the ^{1}H NMR spectrum. The formation of the dienols was also demonstrated by IR spectroscopy on irradiation of matrices of **31** and **34** in a mixture of methylcyclohexane and 2-methyltetrahydrofuran (2:1) at liquid nitrogen temperatures. It therefore seems that the dienols generated from these α, β-unsaturated ketones have a strong tendency to ketonize by re-forming the starting materials, possibly by an intramolecular 1,5-migration. In the presence of hydrochloric acid, however, **32** and **34** isomerize to their β, γ-unsaturated isomers[41]. Intermolecular protonation at the α-position is now competing with intramolecular protonation at the γ-position.

(31)   (32)   (33)   (34)

Similar behaviour is found in the presence of a base. Thus in the presence of imidazole[42,43] or pyridine[43,44] in DMF, mesityl oxide (**31**) yields the deconjugated ketone **35** upon irradiation. Now the dienols **36a** and **37a** can be trapped by trimethylsilyl chloride to yield the trimethylsilyl ethers **36b** and **37b**. The major product was the (Z)-isomer **36b** which suggests that the reaction proceeds through a singlet excited state which yields the Z-dienol **36a** in a concerted process. The (E)-isomer **37b** which was only detected after *ca* 60% conversion was thought to be formed from the (Z)-isomer **36b** by triplet energy transfer. In the absence of base the dienol **36a** reverts to starting mesityl oxide, possibly by a 1,5-sigmatropic rearrangement, but in the presence of base ketonization involves the dienolate ion (**38**) which undergoes preferential α-protonation. A similar effect of base is found in the 'deconjugation' of α, β-unsaturated esters[45].

(35)   (36)   (37)   (38)
              (a)R=H   (a) R=H
              (b)R=SiMe$_3$ (b) R=SiMe$_3$

The pH dependence of the photochemical deconjugation of mesityl oxide has been reported. The quantum yield increases from 0.007 to 0.1 when the pH is changed from 2.5 to 13. This presumably reflects the higher proportion of α/γ protonation of the dienolate ion (**38**) compared to the dienol (**36a**)[46]. As can be seen from the discussion in Sections IV and V, dienolate ions normally undergo a higher proportion of protonation at the α-position than the corresponding dienols.

The dienolate ion derived from mesityl oxide, **38**, and the corresponding ion (**39**) with a t-butyl substituent have been detected as transients (λ_{max} 290 nm) in flash photolysis experiments[47a]. These transients were only detected at pH > 9.5 and their concentration increase with increasing pH, so they were assigned to the dienolate ions rather than to the dienols themselves. Analysis of the variation of the rate of decay of these transients with pH yields reasonable values for the K_a of the enols, of $3.8 \pm 0.17 \times 10^{-11}$ M (**36a**) and $1.07 \pm 0.5 \times 10^{-11}$ M (**40**), and for the rate constants for the ketonization of the enolate ions at 23 °C: $539 \pm 17\,s^{-1}$ (**38**) and $1184 \pm 21\,s^{-1}$ (**39**). From these results values of k_{HO^-} for the ketonization of the dienols are $2 \times 10^6\,M^{-1}\,s^{-1}$ (**36a**) and $1.2 \times 10^6\,M^{-1}\,s^{-1}$ (**40**) at 23 °C. These should be compared with values for the (E)- and (Z)-1,3-butadien-1-ol of 1.28×10^5 and $1.14 \times 10^5\,M^{-1}\,s^{-1}$ at 25 °C[11]. It was possible to evaluate the rate constant for the uncatalysed ketonization of dienol (**40**) to be $40 \pm 13\,s^{-1}$ at 23 °C. This is much higher than the rate constant for the uncatalysed ketonization of vinyl alcohol[14] or for the uncatalysed ketonization of cyclohexan-1,3-dienol[48] and suggests that there is a special mechanism for the ketonization of this dienol (**40**) which involves a 1,5-hydrogen shift[47a].

(**39**)   (**40**)

Recently the dienols **36a** and **40** themselves have been detected by ^{1}H NMR spectroscopy on irradiation of their keto forms in methanol solution at -76 °C. There was about 50% conversion to the dienols and the ketonization of the latter was followed by NMR spectroscopy at temperatures up to -23 °C[47b].

Detection of long-lived dienol intermediates in photoenolization at room temperature by NMR spectroscopy has also proved possible in certain instances. Thus, when 1-acetylcyclooctene (**41**) in acetonitrile solution is irradiated under nitrogen it is converted to a 5:1 mixture of the dienols **42** and **43** which were characterized by their NMR, IR and UV spectra[49]. Again the first step in this reaction was thought to be an (E) → (Z) isomerization. The dienols **42** and **43** were very sensitive to oxygen, but in the absence of oxygen they were converted slowly (or rapidly in the presence of catalysts) to a mixture of α, β- (**41**) and β, γ- (**44**) unsaturated ketones, the composition of which depended on the conditions. Thus in

(**41**)   (**42**)   +

(**44**)   (**43**)

the presence of a trace of sulphuric acid the $\beta, \gamma/\alpha, \beta$ ratio (α to γ protonation) was 75:25, but in the presence of t-BuOK/t-BuOH it was 100:0. This suggests that protonation of the dienolate anion at the α-position is relatively more favourable than that of the dienol. This is similar to what is found with other dienols and their anions (see Sections IV and V). It is not clear if the stability of these dienols is just kinetic, or whether they are thermodynamically more stable than other dienols which have not been detected, since there are no measurements of the equilibrium constants for enolization.

IV. POSITION OF PROTONATION OF 1,3-DIEN-1-OLS

Most of the evidence on the position of protonation of dienols has been obtained indirectly from studies on the isomerization of β, γ-enones into α, β-dienones for which the dienols are the presumed intermediates. However, the recent preparation of solutions of the simplest conjugated dienols, i.e. 1,3-butadien-1-ols, has enabled their protonation to be studied directly.

The conversion of β, γ-dienones into α, β-dienones falls into two extreme types (equations 8 and 9) depending on whether the enolization of the β, γ-dienone is rapid and reversible (equation 8) or the slow rate-determining step (equation 9). These two situations of course correspond to preferential α-protonation of the dienol ($k_\alpha/k_\gamma \gg 1$) and preferential γ-protonation ($k_\gamma/k_\alpha \gg 1$), respectively. There are two ways of distinguishing between these mechanisms: (i) the solvent isotope effect and (ii) deuterium incorporation. For the mechanism of equation 8 ($k_\alpha/k_\gamma \gg 1$) deuterium incorporation into the β, γ-enone should be detected when a deuteriated medium is used and the deuterium isotope effect for isomerization k_{H^+}/k_{D^+} should be greater than 1 (4 to 6 is common). For the mechanism of equation 9, however, there should be no deuterium incorporation when a deuteriated medium is used and k_{H^+}/k_{D^+} should be less than 1 (0.6–0.9 is common). Examination of the results in Table 2 suggests that the following two structural features favour γ-protonation of dienols: (i) planarity of the dienol system and (ii) the presence of a substituent on the β-carbon. On the other hand, non-planarity of the system favours α-protonation.

$$(8)$$

$$(9)$$

As discussed by Whalen and coworkers[50], the C_γ/C_α protonation ratio must depend on the relative abilities of the two transition states to delocalize the developing positive charge onto the oxygen. This will be a maximum for γ-protonation when the angle ϕ between the double bonds is $0°$, which is probably the situation with the acyclic dienols (Table 2, entries 1, 2, 4) and the cyclopentadienol (Table 2, entry 3). In contrast, this angle has been estimated to be respectively $18°$ and $64°$ for the cyclohexadienol (Table 2, entry 7) and the cyclooctadienol (Table 2, entry 9), so that protonation at the γ-position is disfavoured and protonation occurs mainly at the α-position.

B. Capon

TABLE 2. Position of protonation of 1,3-dienols to form carbonyl compounds[a]

No.	Enol	% α	% γ	k_{H^+}/k_{D^+} for isomerization of β,γ- into α,β-enone	Ref.
				Dienols which are Protonated Mainly at the γ-Position	
1	*(structure)*[b]	9.9	90.1	—	11
2	*(structure)*[b]	30.2	69.8	—	11
3	*(structure)*	0	100[c]	0.91	50
4	*(structure)*	—	—	0.59–0.83	51a
5.	*(structure)*	—	—	0.77	51b
6	*(structure)*[d] Cholesteryl system	< 10	> 90	0.61	52–55
				Dienols which are Protonated Mainly at the α-Position	
7	*(structure)*[e]	98	2	5	51b

TABLE 2. (continued)

No.	Enol	% α	% γ	k_{H^+}/k_{D^+} for isomerization of β,γ- into α,β-enone	Ref.
8		90	10	1.0	51a
9		ca 100	ca 0	—	56
10		75	25	—	40

[a]Except for the first two and the last entries the positions of protonation were determined by studying deuterium incorporation concurrent with the isomerization of the β,γ-enone into the α,β-enone in a deuteriated medium.
[b]Direct measurement of the products from pregenerated dienol (see Section II.C).
[c]No deuterium incorporation into the α-position of the product.
[d]When cholest-5-en-3-one was allowed to isomerize by treatment with the DCl in diglyme–D_2O, the product cholest-4-en-3-one contained less than 0.1 atom of deuterium at C-4. When the reaction was stopped at 70% reaction no deuterium incorporation into the starting material was detected[52]. Similar treatment of 17β-hydroxyandrost-5-en-3-one showed 0.08 atom of deuterium at C-4 of the starting material when it was recovered after 50% reaction[52]. A kinetic investigation of the isomerization of androst-5-ene-3,17-dione into androst-4-ene-3,17-dione and of 17α-ethynyl-17β-hydroxy-5-estren-3-one into 17α-ethynyl-17β-hydroxy-4-estren-3-one in DCl/D_2O showed non-first-order behaviour from which it was concluded that 'partitioning of the dienols' is kinetically significant[54]. It therefore seems that the tendency to α-protonation is greater in the androstenone and estrenone series than in the cholestenone series.
[e]Rate of enolization of β,γ-enone is reported to be about 50 times the rate of isomerization into the α,β-enone.
[f]Rate of enolization of β,γ-enone is reported to be nearly 10 times the rate of isomerization into the α,β-enone. The isotope effect $k_{H^+}/k_{D^+} = 1$ was attributed to it being a complex function of the rate constants for enolization of the β,γ-enone and for ketonization of the dienol to yield β,γ- and α,β-enones; i.e. the mechanism is not the limiting one.
[g]Extensive deuterium incorporation into the β,γ-enone was reported under conditions where no isomerization into the α,β-enone could be detected.
[h]Direct measurement of the products from the photochemically generated dienol (see Section III).

Alkyl substituents at the β-position also favour γ-protonation, as illustrated by entries 5 and 6 in Table 2. This is easily rationalized as resulting from stabilization of one of the canonical structures of the transition state as shown in **45**.

(**45**)

V. POSITION OF PROTONATION OF 1,3-DIEN-1-OLATE ANIONS

Dienolate ions usually undergo protonation faster at the α-position than at the γ-position $(k_\alpha/k_\gamma > 1)$, an observation which can be correlated with the charge densities at these positions[57-59]. The preference for α-protonation is however relatively slight, unless the dienolate system is non-planar. Thus, for deuteriation by $D_2PO_4^-$ of the cyclopentadienolate ion, which is planar, k_α/k_γ is 3.2^{50} and $k_{HO^-}^\alpha/k_{HO^-}^\gamma$ for the ketonization of the (Z)- and (E)-1,3-butadien-1-ols, which presumably pass through the dienolate ions, is 4.1 and 1.2 respectively[11]. It also seems that the value of k_α/k_γ for ion 47 is similar since isomesityl oxide (46) on treatment with trimethyl amine in CH_3OD undergoes exchange about four times faster than it is isomerized into mesityl oxide[60].

In contrast, the values of k_α/k_γ for deuteriation of the non-planar cyclooctadienolate[56] and cyclohexadienolate[50] ions are respectively > 1700 and 575. This last observation is similar to many results reported for steroidal dienolate[54,59,61-65] of general structure 48a or 48b. As discussed by several workers[57-59,64,66], if it is assumed that the transition state for protonation of the dienolate ion is 'early', then there should be a correlation between k_α/k_γ and the charge densities at these positions. The relative charge densities will however depend on how close the dienolate ion is to planarity and the negative charge ratio q_γ/q_α should be a maximum when the system is planar with the dihedral angle ϕ between the double bonds equal to zero. This is probably the situation with the butadienolate and cyclopentadienolate ions and k_α/k_γ is less than about $4^{11,50}$. However, in the cyclohexadienolate ion ϕ is 10 to 15° and k_α/k_γ is 575^{50}, and in the cyclooctadienolate ion ϕ is 64° and k_α/k_γ is greater than 1700^{56}.

(48a)　　　　　(48b)

This apparently tidy picture is complicated by the report that, in aqueous solution, the dienolate ion derived from androst-5-ene-3,17-dione 49 'may pick up a proton at C-4 and C-6 with comparable ease'[67]. The experimental basis for this claim, which seems to be sound, was that a deuterium isotope effect of $k_H/k_D = 3.2$ was measured for the isomerization of 50 into 51 by carrying out measurements early in the reaction, and that the first-order plot for the deuteriated substrate 50b was curved and that the slope eventually became equal to that for the non-deuteriated substrate 50a, which indicates that complete exchange had taken place. The apparent disagreement between this result and other work on steroidal dienolates of this general structure, which indicates that α-

protonation is much faster than γ-protonation[54,59,61-65], was attributed to a difference in solvent[67]. Clearly this warrants further investigation.

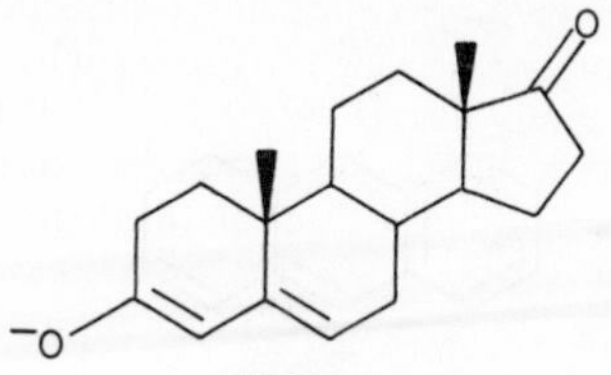

(49)

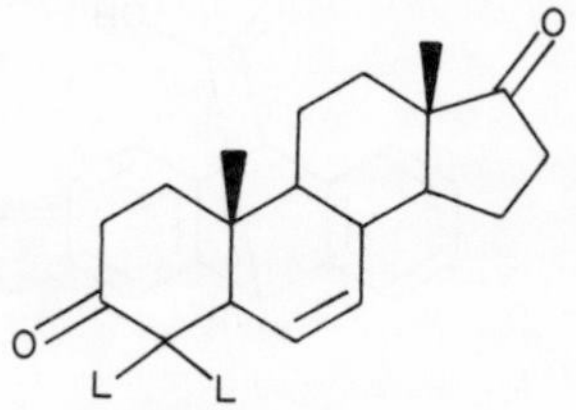

(50)

(a) L=H

(b) L=D

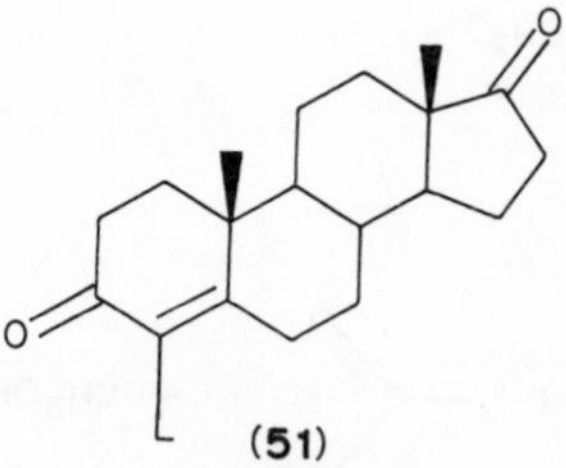

(51)

(a) L=H

(b) L=D

Substituents affect the site of protonation of steroidal dienolate ions. Thus, the percentage of α-protonation of the dienolate ions **52** derived from cholestenone are 95, 60 and $< 5\%$ with the substituents R = H, Me and MeO[66].

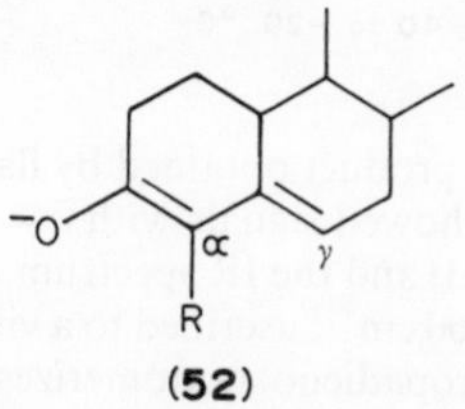

(52)

Dienolate ions derived from α, β-unsaturated esters also undergo predominant protonation at the α-position. These ions are acyclic and, like those derived from acyclic aldehydes and ketones, the preference for α-protonation is not high. Thus k_{α}/k_{γ} for protonation of the dienolate ion derived from ethyl crotonate is 6.7 and that derived from ethyl 3-methyl-2-butenoate is 4.3[68].

VI. 1, 2-DIENOLS

1, 2-Dienols (**5**) are formally derived by enolization of α, β-unsaturated carbonyl compounds which have a proton attached to the α-carbon atom. The simplest 1, 2-dienol, propa-1, 2-dienol, the enol of acraldehyde, has been generated by flash thermolysis of its

Diels–Alder adduct with anthracene (equation 10)[69] and by hydrolysis of the orthoester precursors **53** and **54**[70].

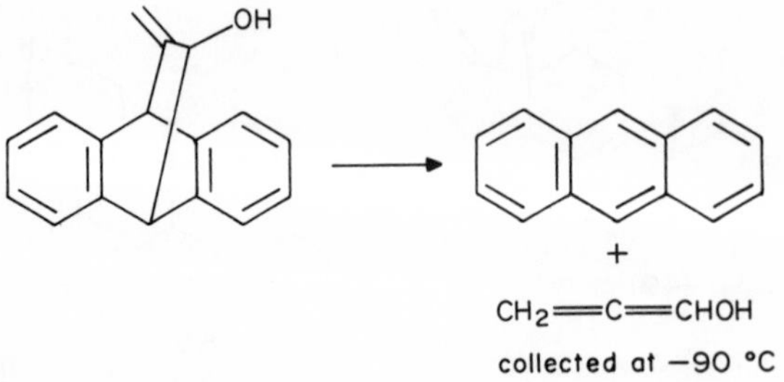

$$\text{(10)}$$

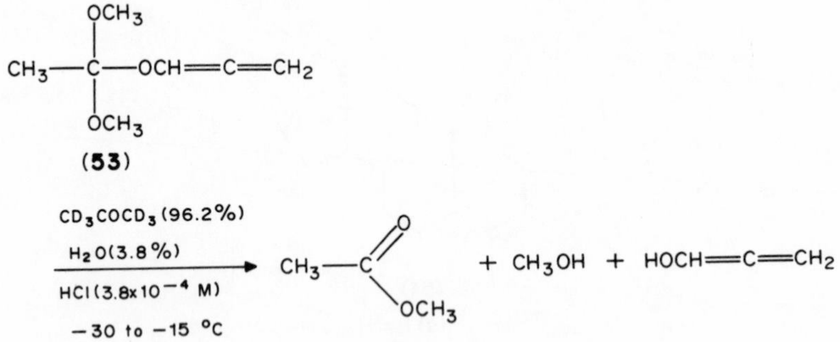

$$+$$

$$CH_2{=}C{=}CHOH$$

collected at $-90\ ^\circ C$

(53)

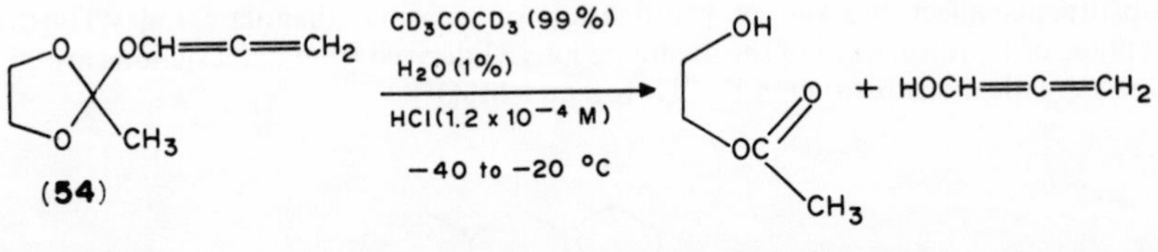

(54)

The ^{1}H NMR spectrum of the product obtained by flash thermolysis was measured in CFCl$_3$ solution at $-90\ ^\circ$C and showed signals with $\delta = 5.3$ (d, 6 Hz, 2H), 6.56 (dt, 6 and 10 Hz, 1H) and 7.00 (d, 10 Hz, OH) and the IR spectrum of the solid material at $-196\ ^\circ$C showed a band with $v = 1980–1960\ \text{cm}^{-1}$ ascribed to a vibration of the $C{=}C{=}C$ system. When prepared in this way the propadienol tautomerizes quantitatively into acraldehyde at $-50\ ^\circ$C.

The same compound was prepared in a mixture of CD$_3$COCD$_3$ and H$_2$O which contained a small amount of HCl at -40 to $-15\ ^\circ$C. When prepared from **54** in CD$_3$COCD$_3$ (99%)–H$_2$O (1%) its ^{1}H NMR spectrum showed the following signals at $-40\ ^\circ$C: $\delta = 5.23$ (d, $J = 5.8$ Hz, 2H), 6.73 (dt, $J = 5.8$ and 9.5 Hz, 1 H) and 7.05 (d, $J = 9.5$ Hz, OH), so the spectra of the compound prepared in the two different ways in two different solvents were very similar. In CD$_3$COCD$_3$–H$_2$O mixture the position of the OH peak depended on the temperature, moving downfield on cooling and upfield on warming with $\delta = 7.5$ at $-100\ ^\circ$C and 6.8 at $-20\ ^\circ$C[70]. The 9.5 Hz coupling between the OH and α-CH protons is similar to that reported for vinyl alcohol[13] and suggests that the *s-cis* conformation, **55**, is the predominant one. The ^{13}C NMR spectrum was also measured and showed $\delta = 87.4$, 116.8 and 203.4 ascribed, respectively, to C-3, C-1 and C-2.

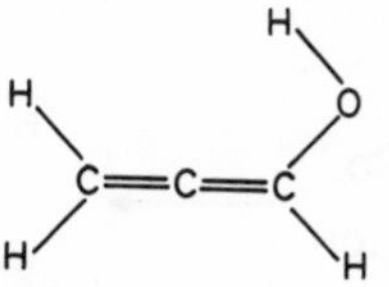

(55)

The kinetics of conversion of propadienol into acraldehyde were measured for an aqueous solution at 15 °C. The pH–rate profile was a bell-shaped curve for which the following rate constants were evaluated: $k_{H^+} = 5.6\ M^{-1}\ s^{-1}$, $k_{HO^-} = 1.11 \times 10^9\ M^{-1}\ s^{-1}$ and $k_{H_2O} = 7.61 \times 10^{-3}\ s^{-1}$. The values of k_{H^+} and k_{H_2O} are slightly smaller than those reported for vinyl alcohol ($k_{H^+} = 20.2\ M^{-1}\ s^{-1}$, $k_{H_2O} = 1.38 \times 10^{-2}\ s^{-1}$) but k_{HO^-} is 74 times greater than that for vinyl alcohol ($k_{HO^-} = 1.5 \times 10^7\ M^{-1}\ s^{-1}$). This was interpreted in terms of a mechanism which involved a rapid and reversible ionization followed by limiting protonation of the dienolate ion (equations 11 and 12). It was thought[70] that the latter and the transition state for its protonation would be more stable than the corresponding structure in the ketonization of vinyl alcohol, since there is a contributing resonance structure **56** which is a vinylic carbanion and these species are normally more stable than alkyl carbanions.

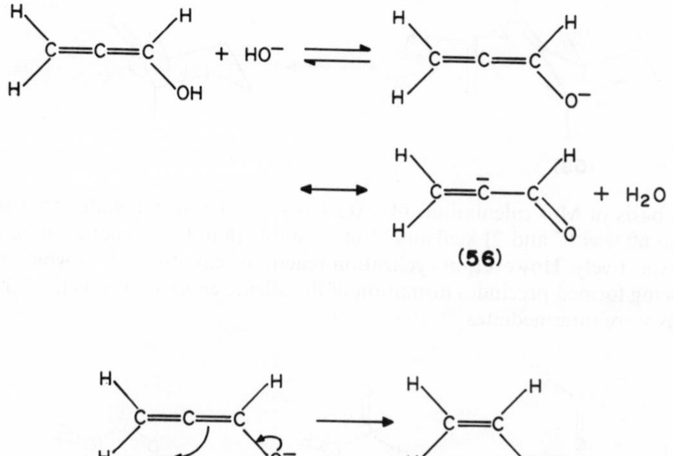

(11)

(56)

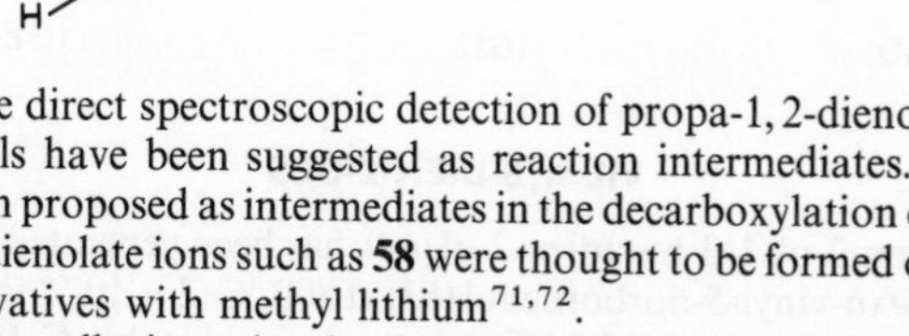

(12)

In addition to the direct spectroscopic detection of propa-1,2-dienol described above, various allenic enols have been suggested as reaction intermediates. Thus gem-allenic dienols **57** have been proposed as intermediates in the decarboxylation of α, β-unsaturated malonic acids and dienolate ions such as **58** were thought to be formed on treatment of the organocopper derivatives with methyl lithium[71,72].

The formation of an allenic enolate ion by the direct removal of the α-proton of an α, β-unsaturated ketone has been proposed to occur in the racemization and deuterium exchange of ketone **59** in methanolic sodium methoxide. The ratio of the rate constants for these processes at 50 °C, $k_e/k_r = 1.43$, indicates about 40% retention of optical activity which was attributed to internal return[73].

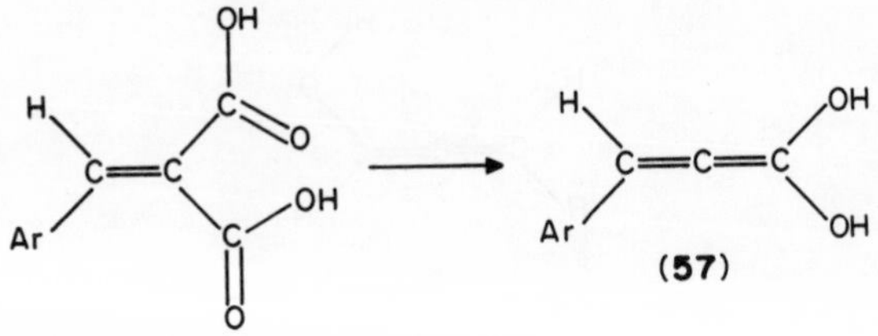

(57)

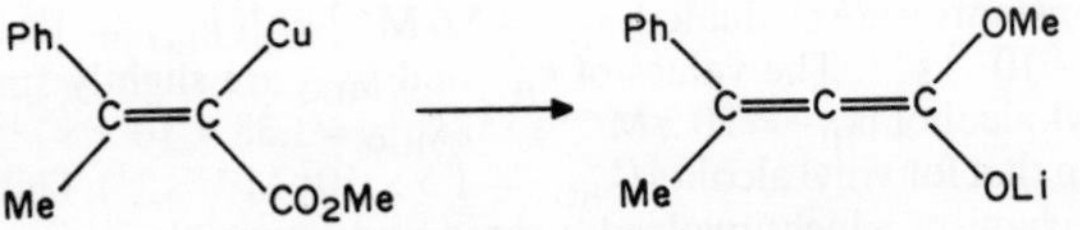

(58)

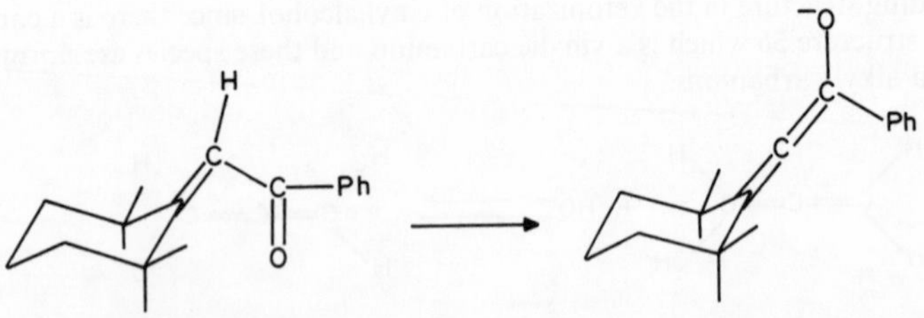

(59)

On the basis of MO calculations (4–31G basis set) it was concluded that the allenic enolate ion **60** was 17 and 21 kcal mol^{-1} more stable than the isomeric enone anions **61** and **62**, respectively. However, in cyclization reactions (equation 13) in which the size of the ring being formed precludes formation of the allenic enolates, it was thought that the enone ions were intermediates[74].

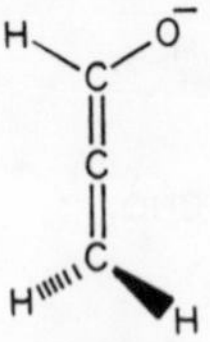

(60) 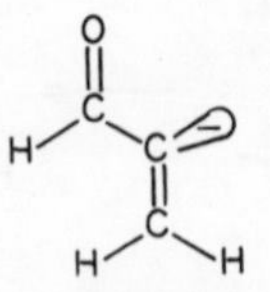   (61) 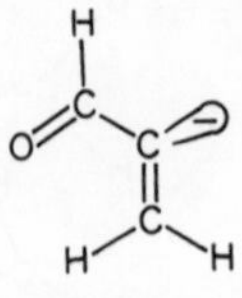   (62)

VII. 1,3-DIEN-2-OLS

The simplest 1,3-dien-2-ol, 1,3-butadien-2-ol (**64**), has been generated in the gas phase by flash pyrolysis of 5-*exo*-vinyl-5-norbornenol (**63**) at 800 °C (2 × 10^{-6} torr). It was reported that the 75 eV mass spectrum of (**64**) 'differs from those of stable C_4H_6O isomers with a C—C—C(O)—C, C—C—C—C—O or cyclic frame'. Methyl vinyl ketone (20–30%, IE = 9.65 eV) was thought to be present as well as **64** (IE = 8.68 eV) on the basis of the deconvoluted ionization efficiency curve. This was thought to be formed by surface-catalysed isomerization. The heat of formation of the dienol was estimated to be − 18.4

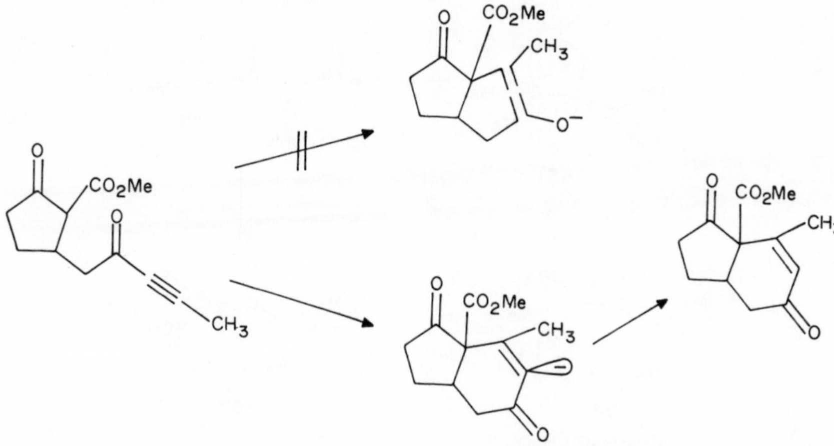

(13)

$\pm 1.2\,\text{kcal mol}^{-1}$ compared to that for the keto form, $-26.8\,\text{kcal mol}^{-1}$, i.e. *ca* $6.2\,\text{kcal mol}^{-1}$ less stable. This compares to a value of $10\,\text{kcal mol}^{-1}$ for the 2-propenol acetone pair[75].

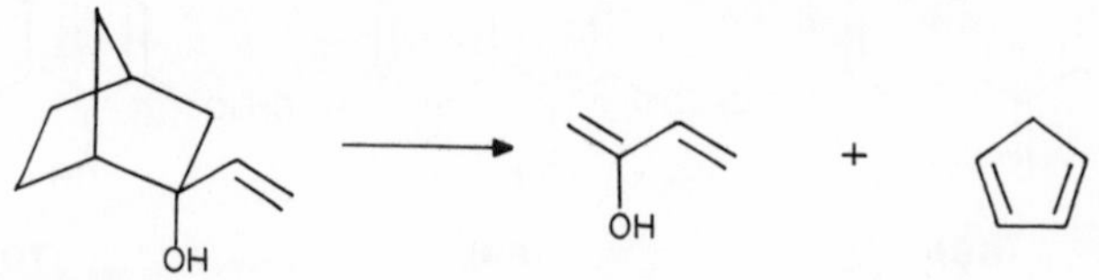

(63)   (64)

The same dienol has been generated as its tricarbonyl iron complex in solution (equation 14). The ^{1}H NMR spectrum had signals with the chemical shifts (δ values) indicated and the pK_a was determined to be 9.24 in 48% aqueous ethanol (estimated 8.5 in water)[17].

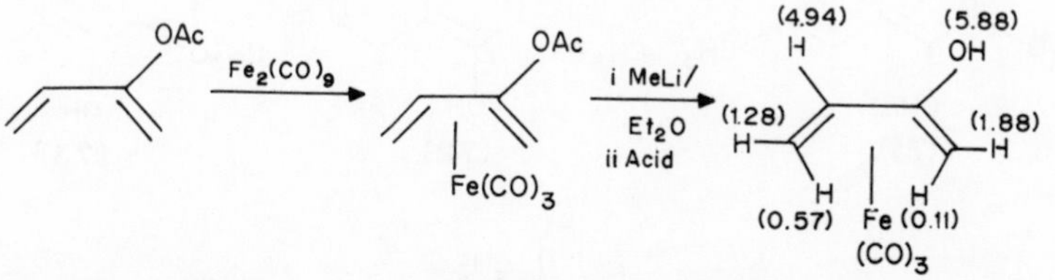

(14)

A more complex 1,3-dien-2-ol, 2,4-dimethyl-1,3-pentadien-3-ol (67), was generated in solution from the amide-acetal precursor (65). The best reagents for the generation of 67 appear to be a slight excess of *t*-butyl alcohol in CCl_4 or dimethyl sulphoxide which contains traces of moisture or acid. If an excess of methanol or water is present and the solvent is CCl_4 monoenols (66) are mainly formed. In dimethyl sulphoxide solution the dienol (67) was stable for several days[76,77].

A series of stable bicyclic 1,3-dien-2-ols (referred to by the authors as enols) has been reported by Reusch[78] and Kanematsu[79] and their coworkers. Thus dienol 69 spontaneously crystallized from a mixture of 68 (its keto form) and an isomer and the dienols 70,

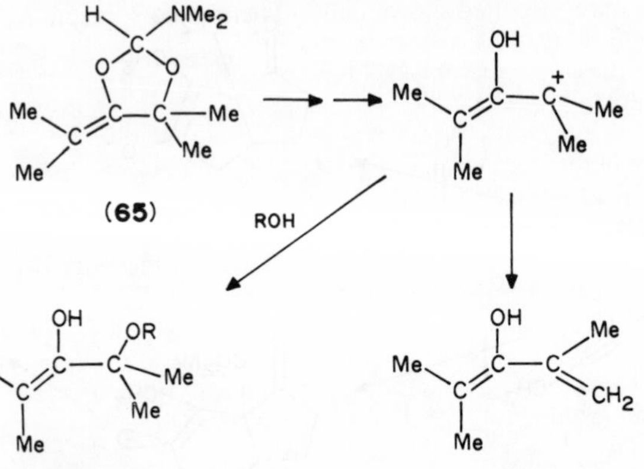

(65)

ROH

(66)

ROH = MeOH, H₂O

(67)

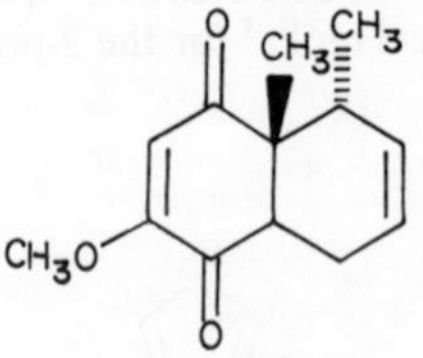

(68)

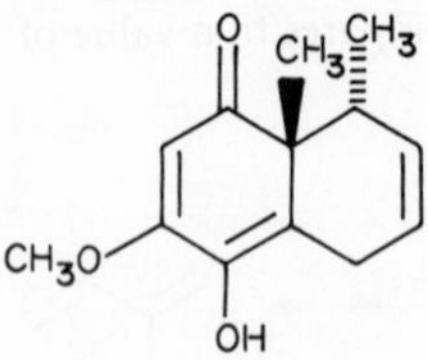

(69)

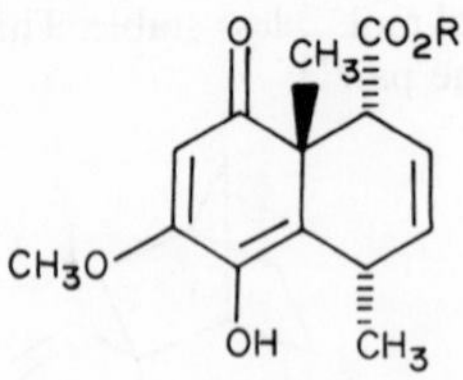

(70)

R = CH₃ or CH₂CH₃

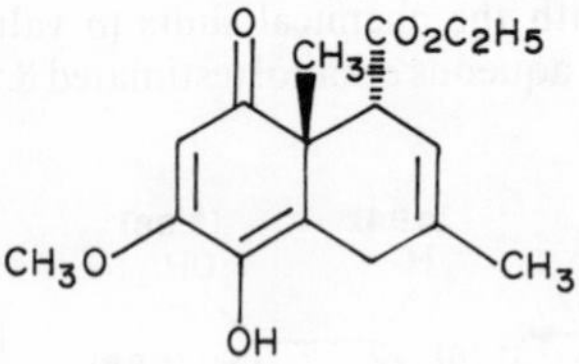

(71)

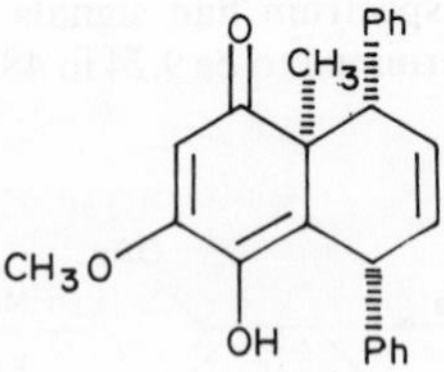

(72)

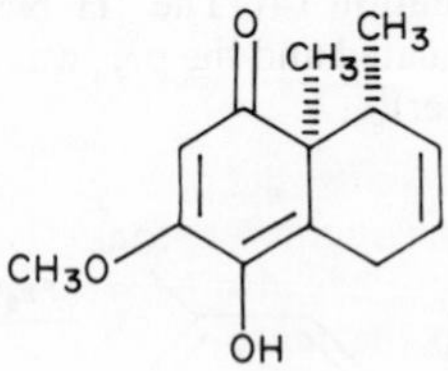

(73)

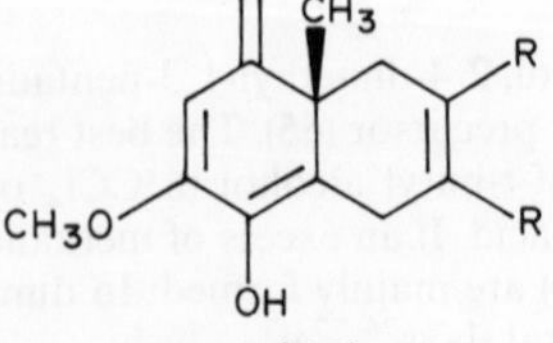

(74)

(a) R = H

(b) R = CH₃

71 and **72** were isolated from the Diels–Alder reaction of 2-methoxy-5-methylbenzoquinone and the corresponding dienes, while dienols **73** and **74** were isolated by treatment of the corresponding *cis*-fused keto forms with base (*t*-BuOK in *t*-BuOH or NaOH in dioxan) followed by rapid acidification. Enols **69, 71** and **72** were also prepared by this method.

This presence of the angular methyl group is essential for dienol formation since, when this is replaced by a hydrogen, base treatment of the keto form leads to aromatization. The presence of the methoxyl group is also necessary for the dienol to be detected or isolated, since base treatment of the *cis*-fused keto form **75** leads to formation of the *trans*-fused keto form **77**, but dienol **76**, the anion of which is presumably an intermediate, could be neither isolated nor detected. When the methoxyl group of **74b** was replaced by a methylthio group, the enol **78** was detected as an intermediate by NMR spectroscopy, but could not be isolated. It was suggested that the heteroatom stabilizied the enol by intramolecular hydrogen bonding and that a methoxyl group was more effective than a methylthio group[79].

(75)　　　　　(76)　　　　　(77)

or its anion

i. NaOH
ii. HCl

CH₃S

OH　(78)
(not isolated pure)

It is possible that the keto group in the γ'-position to the enolic hydroxyl also exerts a stabilizing influence on the enolic form and a simpler 1,3-dien-2-ol with this structural feature has also been isolated. This (**79**) was obtained (along with a ring-closed isomer) from the reaction of 3, 3, 5, 5-tetramethylcyclopentane-1, 2-dione with benzyl methyl ketone (equation 15). However, with **79**, unlike with the bicyclic dienols, there is the possibility of an intramolecular hydrogen bond between the enolic hydroxyl group and the keto group. That there is in fact such a hydrogen bond in the solid state was demonstrated by X-ray crystallography, which indicated a short O–O interatomic distance of 2.55–

$$CH_3COCH_2Ph \longrightarrow \qquad (15)$$

(79)

B. Capon

2.58 Å. The dienolic structure also persists in $CDCl_3$ solution. The ^1H NMR spectrum shows the signal of the enolic proton at $\delta = 11.72$ with a long-range allylic coupling of 1.7 Hz. The intramolecular hydrogen bond also causes a shift in the ^{13}C NMR resonance of the carbonyl group to $\delta = 216.7$ compared with $\delta = 204$ to 208 for other cyclopentenones[80].

VIII. REFERENCES

1. P. G. Sammes, *Tetrahedron*, **32**, 405 (1976).
2. P. J. Wagner, in *Rearrangements in Ground and Excited States* (Ed. P. de Mayo), Academic Press, New York, 1980, p. 427.
3. A. Sevin, B. Bigot and M. Pfau, *Helv. Chim. Acta*, **62**, 699 (1979).
4. J. J. Dannenberg and J. C. Rayez, *J. Org. Chem.*, **48**, 4723 (1983).
5. F. Tureček, Z. Havlas, F. Maquin, N. Hill and T. Gäumann, *J. Org. Chem.*, **51**, 4061 (1986).
6. F. Tureček and Z. Havlas, *J. Org. Chem.*, **51**, 4066 (1986).
7. T. Okuyama, T. Fueno and J. Furakawa, *Tetrahedron*, **25**, 5409 (1969).
8. C. A. McDowell and S. Sifniades, *J. Am. Chem. Soc.*, **84**, 4606 (1962); see footnote 7.
9. J. W. Coomber, J. N. Pitts and R. R. Schrock, *Chem. Commun.*, 190 (1968).
10. The CAD–MIKE mass spectrum of the (E)-isomer has also been reported: S. Arseniyadis, J. Goré, P. Guenot and R. Carrié, *J. Chem. Soc., Perkin Trans 2*, 1413 (1985).
11. B. Capon and B. Z. Guo, *J. Am. Chem. Soc.*, **110**, 5144 (1988).
12. F. Tonnard, S. Odiot, J. R. Dorie and M. L. Martin, *Org. Magn. Reson.*, **5**, 265, 271 (1973).
13. B. Capon and A. K. Siddhanta, *J. Org. Chem.*, **49**, 255 (1984).
14. B. Capon and C. Zucco, *J. Am. Chem. Soc.*, **104**, 7567 (1982).
15. Y. Chiang, M. Hojatti, J. R. Keeffe, A. J. Kresge, N. P. Schepp and J. Wirz, *J. Am. Chem. Soc.*, **109**, 4000 (1987).
16. T. Okuyama, T. Sakagami and T. Fueno, *Tetrahedron*, **29**, 1503 (1973).
17. C. H. DePuy, R. N. Greene and T. E. Schroer, *Chem. Commun.*, 1225 (1968).
18. R. E. Lutz and C. J. Kibler, *J. Am. Chem. Soc.*, **62**, 360 (1940).
19. G. J. Heiszwolf and H. Kloosterziel, *Recl. Trav. Chim. Pays-Bas*, **86**, 807 (1967). See also G. J. Heiszwolf, J. A. A. van Drunen and H. Kloosterziel, *Recl. Trav. Chim. Pays-Bas*, **88**, 1377 (1969).
20. G. Demailly, J. B. Ousset and C. Mioskowski, *Tetrahedron Lett.*, **25**, 4647 (1984).
21. H. Kloosterziel, J. A. A. van Drunen and P. Galama, *Chem. Commun.*, 885 (1969).
22. R. B. Bates, L. M. Kroposki and D. E. Potter, *J. Org. Chem.*, **37**, 560 (1972).
23. F. T. Oakes, F. A. Yang and J. F. Sebastian, *J. Org. Chem.*, **47**, 3094 (1982).
24. V. Rautenstrauch, *Helv. Chim. Acta*, **55**, 594 (1972).
25. N. C. Yang and C. Rivas, *J. Am. Chem. Soc.*, **83**, 2213 (1961).
26. E. F. Zwicker, L. I. Grossweiner and N. C. Yang, *J. Am. Chem. Soc.*, **85**, 2671 (1963). The dienol was formulated as the syn-isomer but on the basis of later work the anti-structure seems more likely.
27. R. Haag, J. Wirz and P. J. Wagner, *Helv. Chim. Acta*, **60**, 2595 (1977).
28. K. Akiyama, Y. Ikegami and S. Tero-Kubota, *J. Am. Chem. Soc.*, **109**, 2538 (1987).
29. K. H. Grellman, H. Weller and E. Tauer, *Chem. Phys. Lett.*, **95**, 195 (1983).
30. In cyclohexane $k = 0.32\,s^{-1}$, in dioxan $k = 1.8\,s^{-1}$, in propan-2-ol $k = 7.9\,s^{-1}$ (temperature *ca* 20 °C?). H. Lutz, E. Brehéret and L. Lindqvist, *J. Chem. Soc., Faraday Trans. 1*, **69**, 2096 (1973).
31. D. M. Findlay and M. F. Chir, *J. Chem. Soc., Faraday Trans. 1*, **72**, 1096 (1976). See also C. V. Kumar, S. K. Chattopadhyay and P. K. Das, *J. Am. Chem. Soc.*, **105**, 5143 (1983). P. K. Das, M. V. Encinas, R. D. Small, Jr. and J. C. Scaiano, *J. Am. Chem. Soc.*, **101**, 6965 (1979).
32. J. Gebicki and A. Krantz, *J. Chem. Soc., Perkin Trans. 2*, 1623 (1984).
33. M. Pfau, S. Combrisson, J. E. Rowe and N. D. Heindel, *Tetrahedron*, **34**, 3459 (1978). M. Pfau, J. E. Rowe and N. D. Heindel, *Tetrahedron*, **34**, 3469 (1978).
34. E. Rommel and J. Wirz, *Helv. Chim. Acta*, **60**, 38 (1977).
35. N. C. Yang and M. J. Jorgenson, *Tetrahedron Lett.*, 1203 (1964).
36. C. P. Visser and H. Cerfontain, *Recl. Trav. Chim. Pays-Bas*, **100**, 153 (1981).
37. D. E. McGreer and N. W. K. Chiu, *Can. J. Chem.*, **46**, 2225 (1968).
38. J. A. Barltrop and J. Wills, *Tetrahedron Lett.*, 4987 (1968). M. J. Jorgenson and L. Gundel, *Tetrahedron Lett.*, 4991 (1968).

39. H. Nozaki, T. Mori and R. Noyori, *Tetrahedron*, **22**, 1207 (1966); see also A. Marchesini, G. Pagani and U. M. Pagnoni, *Tetrahedron Lett.*, 1041 (1973).
40. M. Tada and K. Miura, *Bull. Chem. Soc. Jpn.*, **49**, 713 (1976).
41. T. Sato, personal communication to the authors of Ref. 40. (see footnote 13).
42. C. S. K. Wan and A. C. Weedon, *J. Chem. Soc., Chem. Commun.*, 1235 (1981).
43. R. Ricard, P. Sauvage, C. S. K. Wan, A. C. Weedon and D. F. Wong, *J. Org. Chem.*, **51**, 62 (1986).
44. S. L. Eng, R. Ricard, C. S. K. Wan and A. C. Weedon, *J. Chem. Soc., Chem. Commun.*, 236 (1983).
45. I. A. Skinner and A. C. Weedon, *Tetrahedron Lett.*, **24**, 4299 (1983). A. C. Weedon, *Can. J. Chem.*, **62**, 1933 (1984). R. M. Duhaime, D. A. Lombardo, I. A. Skinner and A. C. Weedon, *J. Org. Chem.*, **50**, 873 (1985). D. A. Lombardo and A. C. Weedon, *Tetrahedron Lett.*, **27**, 5555 (1986).
46. A. Deflandre, A. Lheureux, A. Rioual and J. Lemaire, *Can. J. Chem.*, **54**, 2127 (1976).
47. (a) R. M. Duhaime and A. C. Weedon, *J. Am. Chem. Soc.*, **107**, 6723 (1985).
    (b) R. M. Duhaime and A. C. Weedon, *Can. J. Chem.*, **65**, 1867 (1987).
48. R. M. Pollack, J. P. G. Mack and G. Blotny, *J. Am. Chem. Soc.*, **109**, 3138 (1987).
49. R. Noyori, H. Inoue and M. Katô, *J. Am. Chem. Soc.*, **92**, 6699 (1970). R. Noyori, H. Inoue and M. Katô, *Bull. Chem. Soc. Jpn.*, **49**, 3673 (1976).
50. D. L. Whalen, J. F. Weimaster, A. M. Ross and R. Radhe, *J. Am. Chem. Soc.*, **98**, 7319 (1976).
51. (a) D. S. Noyce and M. Evett, *J. Org. Chem.*, **37**, 397 (1972).
    (b) D. S. Noyce and M. Evett, *J. Org. Chem.*, **37**, 394 (1972).
52. S. K. Malhotra and H. J. Ringold, *J. Am. Chem. Soc.*, **87**, 3228 (1965); **85**, 1538 (1963).
53. P. Talahay and V. S. Wang, *Biochim. Biophys. Acta*, **18**, 300 (1965).
54. S. K. Perera, W. A. Dunn and L. R. Fedor, *J. Org. Chem.*, **45**, 2816 (1980).
55. W. R. Nes, E. Loeser, R. Kirdani and J. Marsh, *Tetrahedron*, **19**, 299 (1963).
56. N. Heap and G. H. Whitham, *J. Chem. Soc. (B)*, 164 (1966).
57. A. J. Birch, *Faraday Discuss. Chem. Soc.*, **2**, 246 (1947).
58. A. J. Birch, *J. Chem. Soc.*, 1551 (1950).
59. H. J. Ringold and S. K. Malhotra, *Tetrahedron Lett.*, 669 (1962).
60. H. C. Volger and W. Brackman, *Recl. Trav. Chim. Pays-Bas*, **84**, 1017 (1965).
61. A. J. Birch, *J. Chem. Soc.*, 2325 (1950).
62. A. J. Birch, P. Hextall and J. A. K. Quartey, *Aust. J. Chem.*, **6**, 445 (1953).
63. W. G. Dauben and J. F. Eastham, *J. Am. Chem. Soc.*, **72**, 2305 (1950); **73**, 4463 (1951). W. G. Dauben, J. F. Eastham and R. A. Micheli, *J. Am. Chem. Soc.*, **73**, 4496 (1951).
64. H. E. Zimmerman, in *Molecular Rearrangements*, Part 1 (Ed. P. de Mayo), Interscience Publishers, New York, pp. 346–347. *Acc. Chem. Res.*, **20**, 263 (1987).
65. B. Belleau and T. F. Gallagher, *J. Am. Chem. Soc.*, **73**, 4458 (1951).
66. G. H. Whitham and J. A. F. Wickramsinghe, *J. Chem. Soc. (C)*, 338 (1968).
67. J. B. Jones and D. C. Wigfield, *Can. J. Chem.*, **47**, 4459 (1969).
68. M. W. Rathke and D. Sullivan, *Tetrahedron Lett.*, 4249 (1972).
69. A. Hakiki, J. L. Ripoll and A. Thuillier, *Tetrahedron Lett.*, **25**, 3459 (1984).
70. B. Capon, A. K. Siddhanta and C. Zucco, *J. Org. Chem.*, **50**, 3580 (1985); C. Zucco, Ph.D. thesis, University of Glasgow, 1982.
71. J. Klein and R. Levene, *J. Chem. Soc., Perkin Trans. 2*, 1971 (1973).
72. J. Klein and A. Y. Meyer, *J. Org. Chem.*, **29**, 1038 (1964).
73. J. F. Arnett and H. M. Walborsky, *J. Org. Chem.*, **37**, 3678 (1972).
74. J. F. Lavallée, G. Berthiaume, P. Deslongchamps and F. Grein, *Tetrahedron Lett.*, **27**, 5455 (1986).
75. F. Tureček, *Tetrahedron Lett.*, **25**, 5133 (1984).
76. H. M. R. Hoffmann and E. A. Schmidt, *J. Am. Chem. Soc.*, **94**, 1373 (1972).
77. E. A. Schmidt and H. M. R. Hoffmann, *J. Am. Chem. Soc.*, **94**, 7832 (1972).
78. J. S. Tou and W. Reusch, *J. Org. Chem.*, **45**, 5012 (1980).
79. K. Hayakawa, K. Ueyama and K. Kanematsu, *J. Chem. Soc., Chem. Commun.*, 71 (1984); *J. Org. Chem.*, **50**, 1963 (1985).
80. T. Simonen and R. Kivekas, *Acta Chem. Scand., Ser. B*, **38**, 679 (1984).

The Chemistry of Enones
Edited by S. Patai and Z. Rappoport
© 1989 John Wiley & Sons Ltd

CHAPTER **21**

Asymmetric synthesis with chiral enones

MICHAEL R. PEEL and CARL R. JOHNSON

Department of Chemistry, Wayne State University, Detroit, Michigan 48202, USA

I. INTRODUCTION

The α,β-unsaturated ketone (enone) functionality enjoys a pivotal position in organic chemistry. The ability to selectively functionalize up to five carbons by conjugate addition, alkylation, Diels–Alder reaction, etc. makes the enone function an attractive subunit for elaboration.

With such diverse chemistry available, the enone function has played a prominent role in the synthesis of many complex molecules in racemic or optically pure form. This chapter will review synthetic applications of enones possessing (non-racemic) chiral centers. Emphasis will be placed on commercially or otherwise readily available homochiral enones.

II. 2-CYCLOPENTENONES

The total synthesis of natural and unnatural prostaglandins has received much attention over the past twenty years[1]. Of the numerous synthetic approaches to the prostaglandins perhaps the most conceptually attractive route involves 'three component coupling'[2]. In such a route conjugate addition of an optically pure lower side-chain synthon to a protected, homochiral 4-hydroxy-2-cyclopenten-1-one is followed by treatment of the resulting enolate with a suitable electrophilic top-chain synthon to afford the basic prostaglandin skeleton in a single operation (Scheme 1).

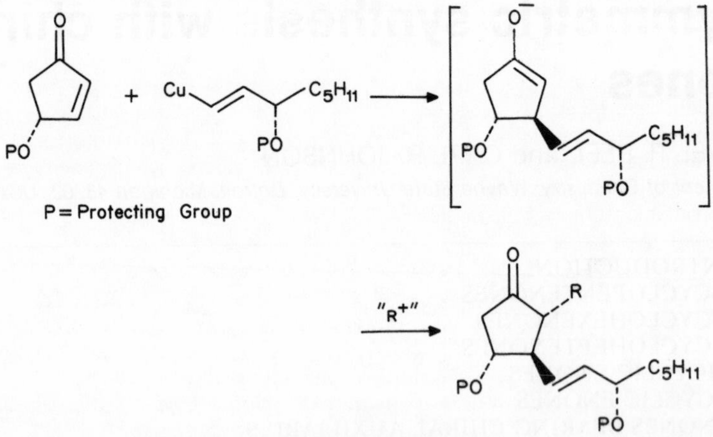

P = Protecting Group

SCHEME 1

This synthetic procedure is dependent on the ready availability of the enone 1 in optically pure form. A number of methods have been developed for preparation of the latter including resolution of the racemic hydroxycyclopentenone, preparation from optically pure natural products, asymmetric synthesis and microbial or enzymatic methods.

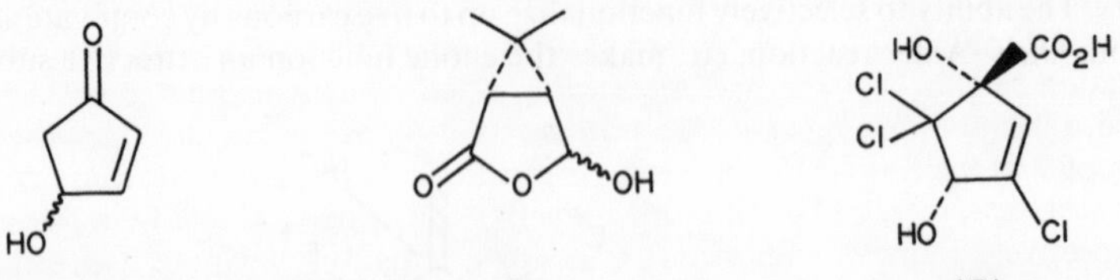

(1)   (2)   (3)

The racemic enone (±)-1 has been prepared in a number of ways[3] from simple starting materials such as cyclopentadiene[4], 2-methylfuran[5], 2-(hydroxymethyl)furan[6], etc. Despite the sensitive nature of the compound, it has been efficiently resolved using the (1S, 3S)-trans-chrysanthmic acid derivative 2[7]. Resolution of the cyclopentenecarboxylic acid 3, derived from phenol, using brucine followed by decarboxylation and removal of the chlorines also led to the optically pure (4R)-1[8].

Optically pure natural products have served as precursors to (R)-1 as demonstrated by Tsuchihashi and coworkers[9] (Scheme 2). The isopropylidene protected diol diiodide 4, prepared in four steps from D-tartaric acid, on condensation with methyl methylthiomethyl sulfoxide afforded the protected cyclopentanone 5. Acid hydrolysis gave (R)-1 in 22% overall yield (from D-tartaric acid) and 85% optical purity.

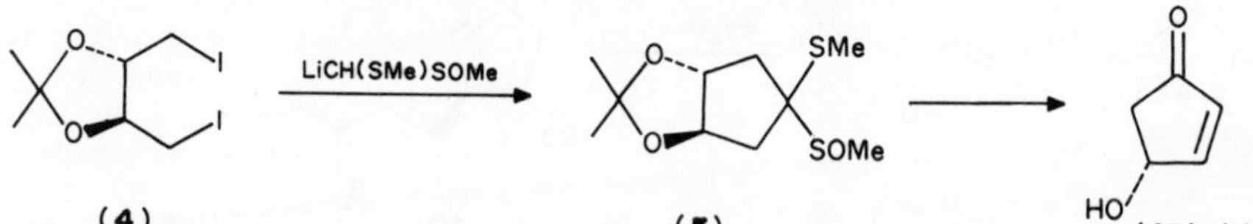

(4) →[LiCH(SMe)SOMe] (5) → (4R)-(1)

SCHEME 2

The enantioselective transformation of prochiral or meso compounds into optically pure products has recently received much attention since, in principle, a prochiral compound can be completely converted into a single enantiomer without the 50% 'waste' inherent in the resolution of a racemate[10]. The chiral 4-hydroxycyclopentenone is a prime candidate for preparation via this 'meso trick' since the prochiral cyclopentenes 6 and 7 are readily available from cyclopentadiene.

(6)   (7)

Noyori and coworkers found that reduction of cyclopentenedione (6) with the chiral reducing agent (S)-Binal-H gave (4R)-hydroxy-2-cyclopenten-1-one (1) in 65% yield and 94% ee[11].

The use of enzymes in organic synthesis is increasing in popularity, especially for the preparation of optically pure compounds. The enantioselective hydrolysis of diacetate 7 has been achieved using a variety of enzymes and microbes to give, ultimately, (4R)-hydroxy-2-cyclopenten-1-one with high purity[12].

The use of organocopper chemistry to effect conjugate addition of nucleophiles to enone systems is well established and has been reviewed extensively[13]. In efforts directed towards the synthesis of prostaglandins, it was found that the ω chain could be introduced very efficiently in homochiral form via the appropriate cuprate; however, extreme difficulty was encountered during the direct alkylation of the intermediate enolate with organic halides. One solution to this problem was recently reported by Noyori and coworkers[14] who employed a lithium (or copper) to tin transmetallation at the enolate stage (Scheme 3). Addition of 9 to 8 followed by transmetallation to tin enolate 10 and reaction with the allylic or acetylenic iodide 11 or 12 afforded the PGE derivatives 13 and 14 in 78% and 82% yield, respectively. Removal of the silyl protecting groups and enzymatic hydrolyses of the esters would complete the shortest prostaglandin synthesis to date.

The problems associated with the direct alkylation of enolate 15 have also been circumvented by the use of more reactive electrophiles to give products which can be readily transformed into the natural prostaglandins. The enolate 15 could be condensed with aldehydes to afford aldol products in good yield[15]. This strategy was effectively employed by Noyori in the synthesis of PGE$_1$ and PGE$_2$ (Scheme 4). Condensation of enolate 15 with methyl 7-oxoheptanoate gave the aldol product 16 which was dehydrated and reduced to give PGE$_1$ (after removal of the protecting groups).

In a similar manner, the enolate 17 was condensed with methyl 7-oxo-5-heptynoate (Scheme 5) to give the aldol product 18. This compound was efficiently deoxygenated to give the protected 5,6-dehydro-PGE$_2$ derivative which served as an intermediate for the

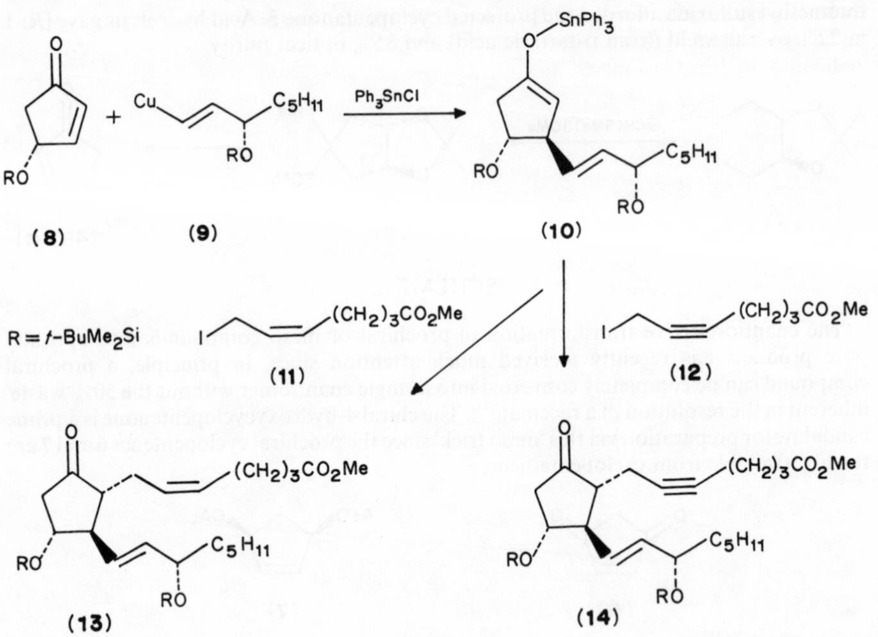

SCHEME 3

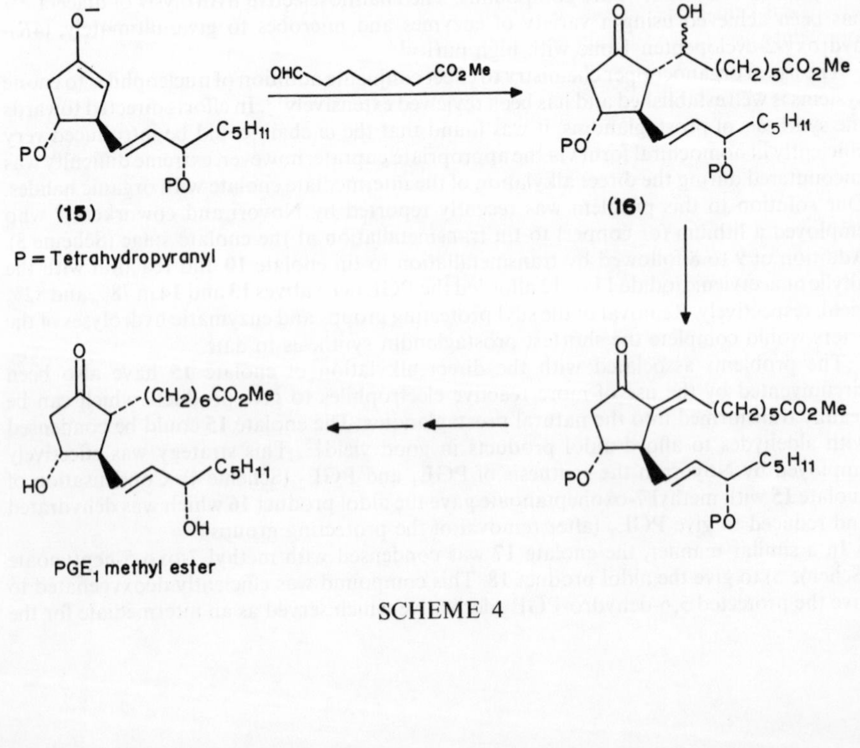

SCHEME 4

synthesis of a variety of primary PGs, e.g. PGE_2 (partial hydrogenation), $PGF_{2\alpha}$ (Bu_2AlH reduction, partial hydrogenation), PGE_1 (saturation of 5, 6-triple bond), $PGF_{1\alpha}$ (Bu_2AlH reduction, hydrogenation), PGD_1 and PGD_2.

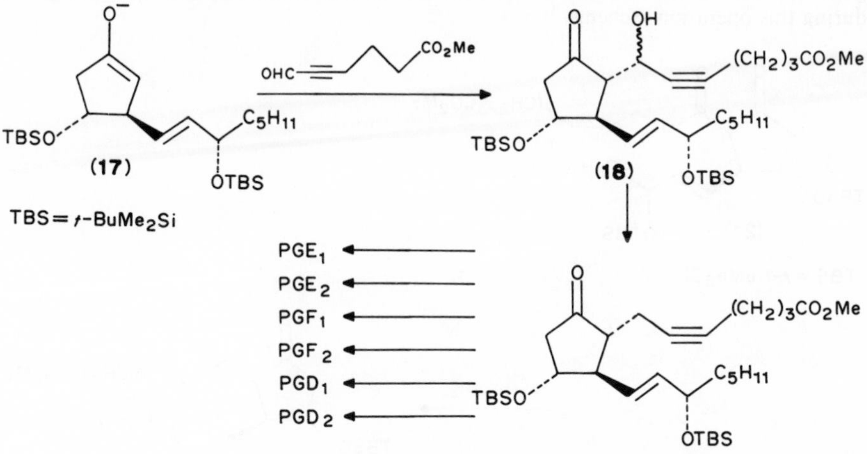

SCHEME 5

The enolate **17** can also be condensed with vinyl nitro compounds to give adducts such as **19** (Scheme 6), which are valuable intermediates in the preparation of both natural PGs and biologically important PG metabolites. For example, the nitro group in **19** is readily removed using tributyltin hydride to give the protected PGE_1 derivative[16]. Alternatively, the nitro group can be transformed into a keto functionality via a modified Nef reaction to give the 6-keto-PGE_1 derivative **20**, a metabolite of PGI_2 (prostacyclin), known to be a powerful vaso-active substance.

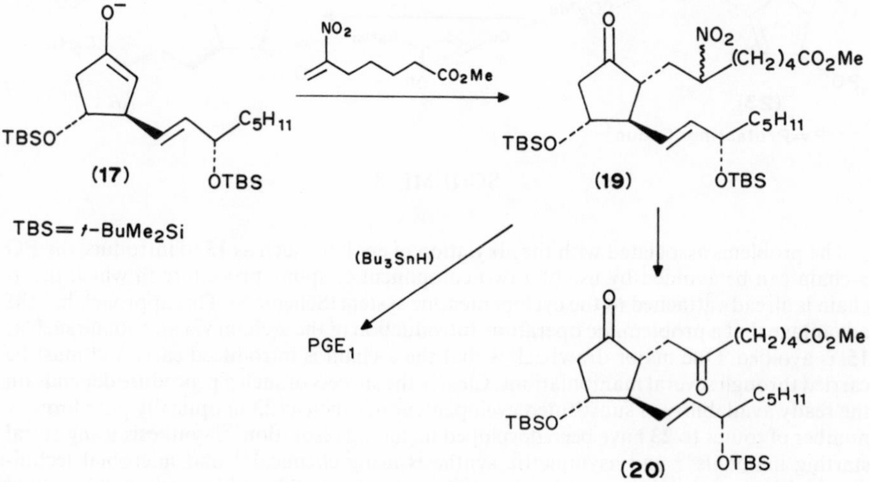

SCHEME 6

An interesting variant on the enolate trapping procedure was presented by Kurozumi and coworkers[17], who utilized a 2-alkenyloxycarbonylimidazole as the electrophile to give the alkenyloxycarbonylated product (21). Palladium-catalyzed decarboxylative allylation gave the PGE_2 derivative 22, however, the 5,6-*cis* stereochemistry was completely lost during this operation (Scheme 7).

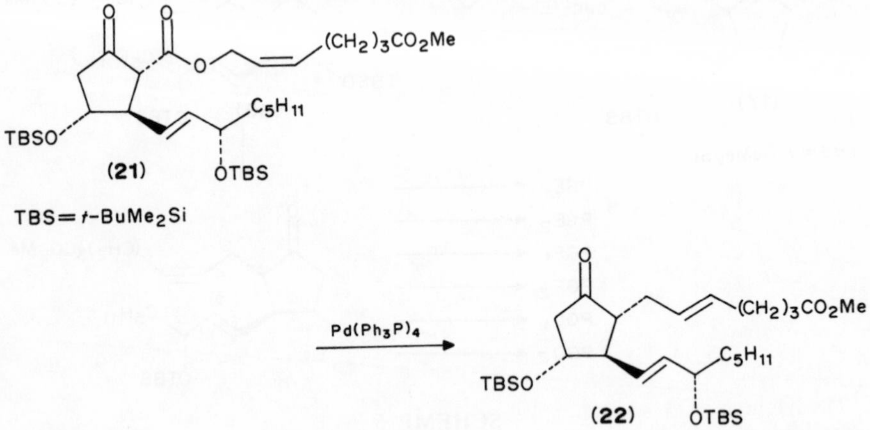

(CH$_2$)$_3$CO$_2$Me

C$_5$H$_{11}$

TBSO

(21)   ÖTBS

TBS = *t*-BuMe$_2$Si

Pd(Ph$_3$P)$_4$

(CH$_2$)$_3$CO$_2$Me

C$_5$H$_{11}$

TBSO

(22)   ÖTBS

SCHEME 7

In addition to the synthesis of natural PGs, the three-component coupling procedure has been exploited for the preparation of a number of physiologically important PG analogues[18].

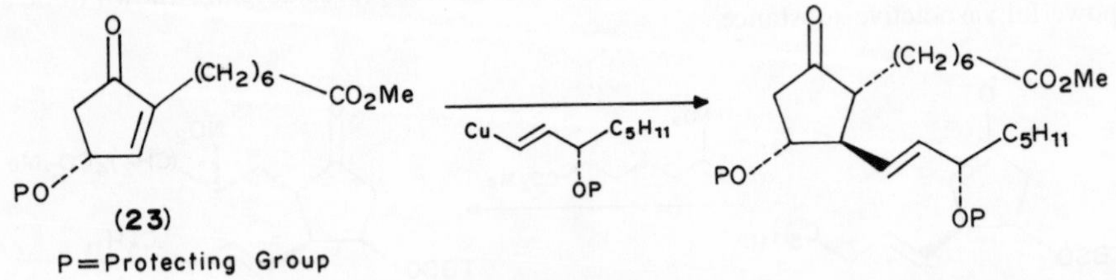

(CH$_2$)$_6$—CO$_2$Me

PO

(23)

Cu   C$_5$H$_{11}$

ÖP

(CH$_2$)$_6$ CO$_2$Me

C$_5$H$_{11}$

PO

ÖP

P = Protecting Group

SCHEME 8

The problems associated with the alkylation of enolates such as 15 to introduce the PG α-chain can be avoided by use of a two-component coupling procedure in which the α-chain is already attached to the cyclopentenone system (Scheme 8). This approach has the advantage that a problematic operation, introduction of the α-chain via an enolate such as 15, is avoided. One major drawback is that the α-chain is introduced early and must be carried through several manipulations. Clearly the success of such a procedure depends on the ready availability of substituted cyclopentenones such as 23 in optically pure form. A number of routes to 23 have been developed including resolution[19], synthesis using chiral starting materials[20] and asymmetric synthesis using chemical[21] and microbial techniques[22]. Much of the pioneering work in this area is due to Sih and coworkers, who found that the cyclopentanetrione 25, available by condensation of ketone 24 with diethyl oxalate, was enantioselectively reduced by the microbe *Dipodascus uninucleatus* to give the hydroxycyclopentanedione **26** (Scheme 9), which was transformed into the cyclopentenone **27**[22b,c,d].

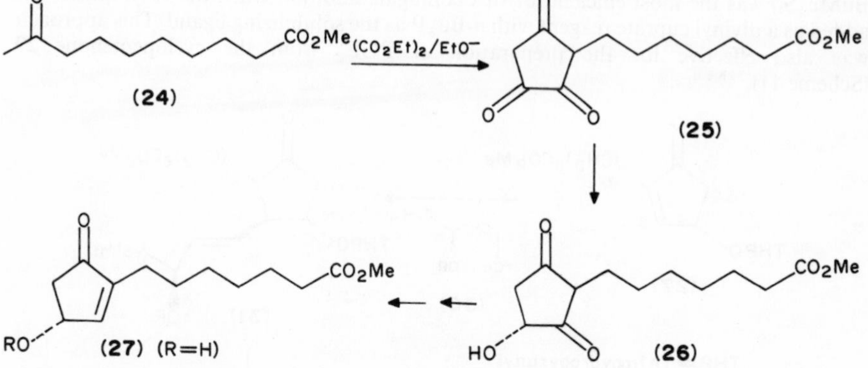

SCHEME 9

The synthesis of **27** has also been achieved via the asymmetric chemical reduction of **25** using lithium aluminum hydride partially decomposed by (−)-*N*-methylephedrine; the optical purity of the product is reported as 54(±6)%[21]. Compound **27** was prepared by Stork and Takahashi in homochiral form from D-glyceraldehyde[20].

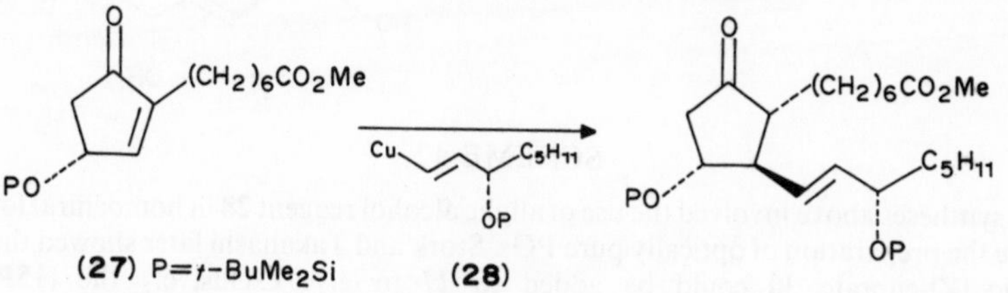

SCHEME 10

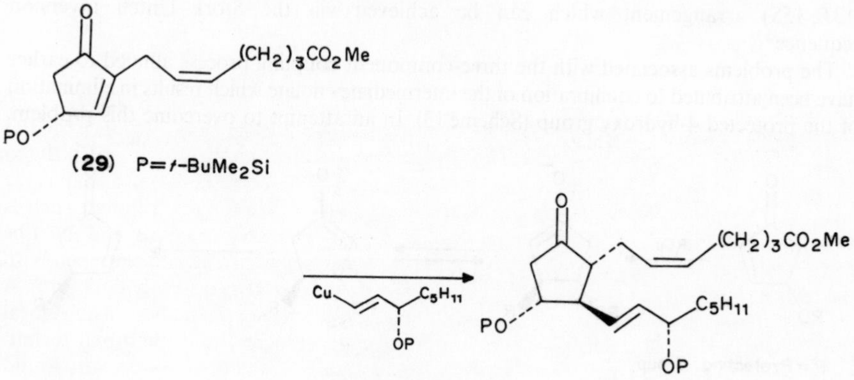

SCHEME 11

Michael R. Peel and Carl R. Johnson

The conjugate addition of optically pure ω-chain **28** to suitably protected (+)-**27** to give the prostaglandin E_1 derivative (Scheme 10) was studied extensively by Sih. He found that the reaction was extremely dependent on the protecting groups on **27** and **28** and also on the type of copper reagent used. A combination of TBS-**27** and TBS-**28** (TBS = t-$BuMe_2Si$) was the most efficient for this conjugate addition when the vinyl anion was added as a divinyl cuprate reagent with n-Bu_3P as the solubilizing ligand. This approach was also effective for the preparation of PGE_2 from the cyclopentenone **29** (Scheme 11).[22b,c,d]

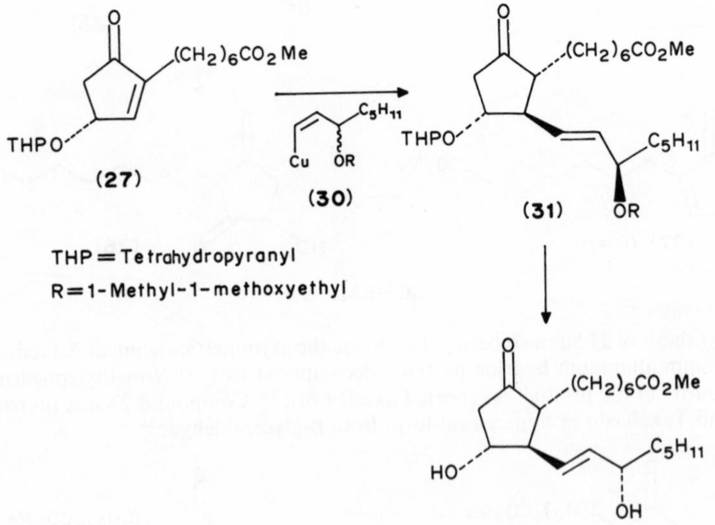

THP = Tetrahydropyranyl

R = 1-Methyl-1-methoxyethyl

SCHEME 12

The syntheses above involved the use of allylic alcohol reagent **28** in homochiral form to achieve the preparation of optically pure PGs. Stork and Takahashi later showed that the racemic (Z)-cuprate **30** could be added to **27** to give exclusively the (15R)-PG derivative **31** through complete kinetic resolution (Scheme 12)[20]. Completion of the synthesis of PGE_1 involved the correction of the (13Z, 15R) side-chain of **31** to the (13E, 15S) arrangement which can be achieved via the Stork–Untch inversion sequence[23].

The problems associated with the three-component coupling process alluded to earlier have been attributed to equilibration of the intermediate enolate which results in elimination of the protected 4-hydroxy group (Scheme 13). In an attempt to overcome this problem,

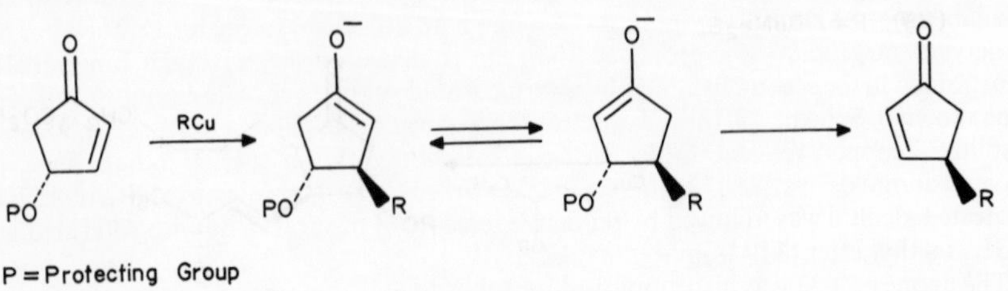

P = Protecting Group

SCHEME 13

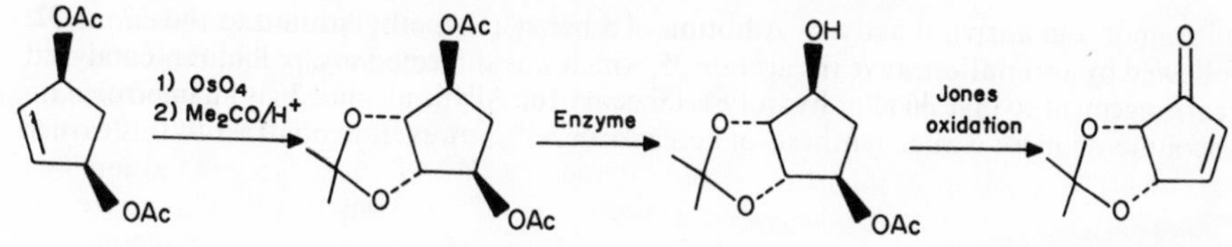

SCHEME 14

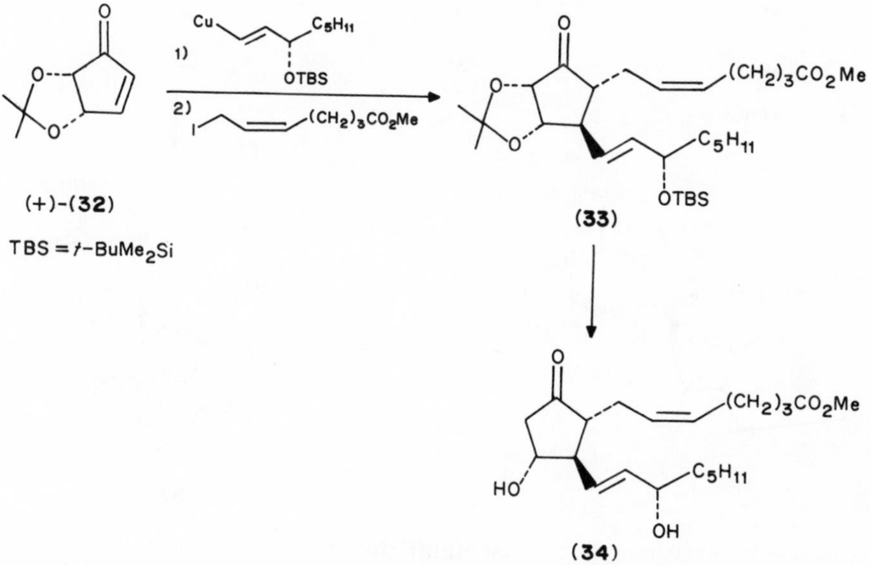

TBS = *t*-BuMe₂Si

SCHEME 15

Johnson and Penning prepared the cyclopentenone **32** as outlined in Scheme 14; the key step in the sequence was the conversion of the *meso*-diacetate to the homochiral monoacetate using electric eel acetylcholinesterase. The overall yield of **32** (98% ee) by the sequence shown in Scheme 14 was 65%. Each carbon of the cyclopentenone framework of ketone **32** is differentially functionalized. The bicyclo[3.3.0] system ensured high or complete diastereoselectivity at the convex face. These factors, coupled with the ready availability of **32** in optically pure form, make it an attractive enone for elaboration to a variety of targets. It was proposed that the presence of the α-oxygen functionality, constrained in the second five-membered ring, would suppress enolate equilibration of the type shown in Scheme 13. Indeed, addition of the lower side-chain as a tributylphosphine stabilized copper reagent followed by alkylation with an allylic iodide gave the prostaglandin derivative (**33**) in 53% yield (Scheme 15). Deprotection of the 15-TBS protected alcohol was followed by reductive removal of the acetonide with Al(Hg) to give PGE₂ methyl ester (**34**)[24].

The enone (+)-**32** was also employed by Johnson and coworkers[25] as an intermediate in an efficient synthesis of neplanocin A, a carbocyclic nucleoside which shows significant antitumor and antiviral activity. Addition of a benzyloxymethyllithium to the enone 32, followed by acetylation, gave the acetate 35, which was subjected to a palladium-catalyzed rearrangement to give 36 after hydrolysis (Scheme 16). Allylic alcohol 36 is an intermediate encountered in an earlier synthesis of neplanocin A[26]; however, in contrast to this earlier

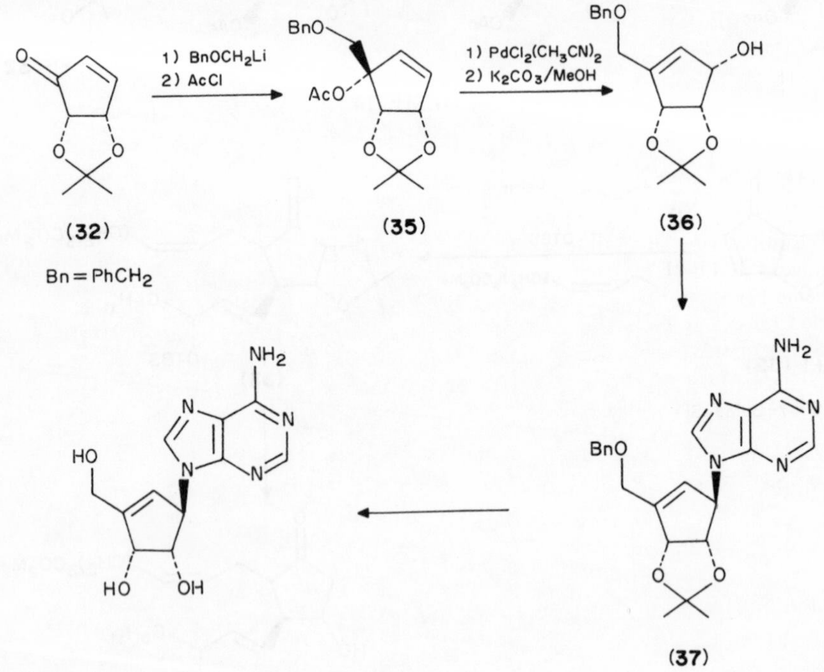

(32)   (35)   (36)

Bn = PhCH₂

Bn $=$ PhCH$_2$

(37)

SCHEME 16

synthesis, Johnson found that the adenine base could be introduced intact by simple displacement of the mesylate derived from 36 to give 37. Deprotection of 37 completed the synthesis of (−)-neplanocin A in 11% yield from cyclopentadiene.

III. 2-CYCLOHEXENONES

The optically pure monoterpene (+)-pulegone (38) has seen frequent and variable use in organic synthesis including: (i) the direct manipulation of the pulegone framework into the derived product; (ii) conversion of pulegone to another, non-enone, compound which is carried on in the synthesis; and (iii) transformation to a compound useful as a temporary chiral auxiliary for a wide variety of asymmetric processes.

The direct incorporation of the pulegone structure into a target molecule is probably the most efficient use of the chiral unit and several syntheses involving this procedure have been reported, most of which involve the exocyclic enone unit as a Michael acceptor.

Two independent approaches to the ionophoric antibiotic aplasmomycin (39) have been reported, both of which depend on (+)-pulegone as the basic starting material and also for the source of optical activity. Both groups recognized the C₂ symmetry present in the aplasmomycin skeleton and made similar initial bond disconnections, however, the subunits were prepared via different routes.

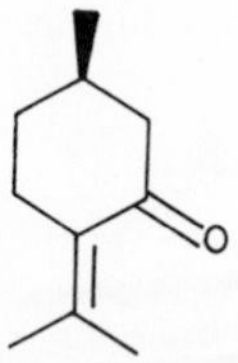

(38)

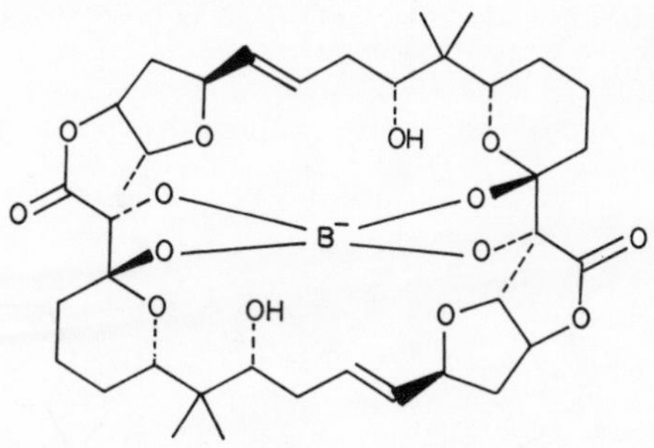

(39)

The initial step in the approach of Corey's group[27] to this molecule involved the conjugate addition of vinyl cuprate to (+)-pulegone to give **40** after equilibration (Scheme 17). An impressive stereoselective hydroxylation (OsO_4) of the vinyl moiety was carried out to give **41**, after suitable manipulation, which was oxidized to a lactone. Cleavage of the lactone using trimethylaluminum and propanedithiol gave keteneth-ioacetal **42** which was transformed to the key intermediate **43**.

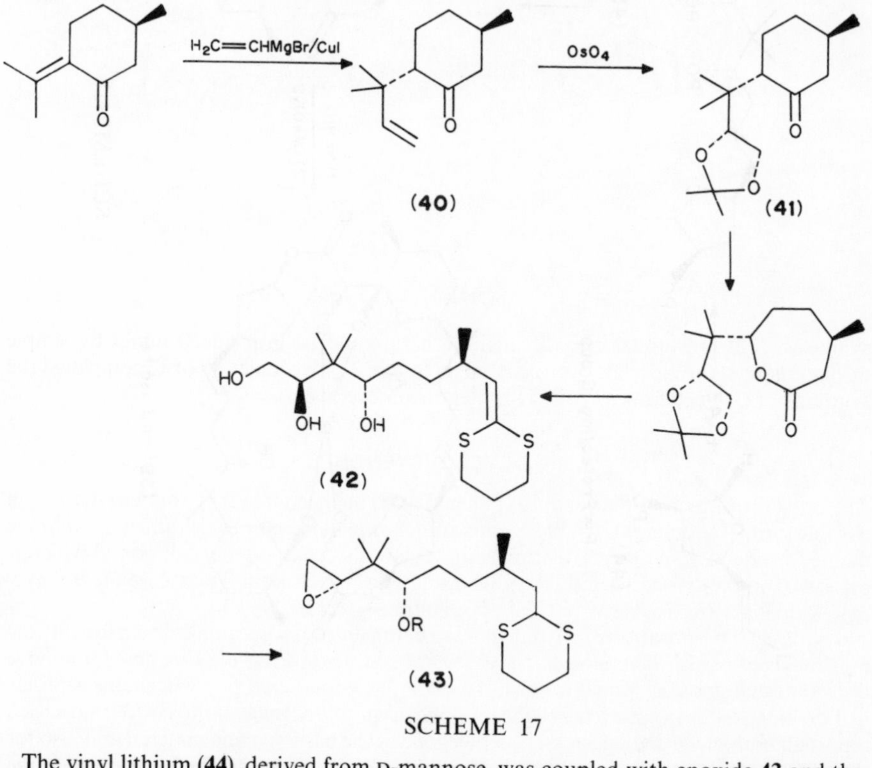

H_2C=CHMgBr/CuI

OsO₄

(40)

(41)

(42)

(43)

SCHEME 17

The vinyl lithium (**44**), derived from D-mannose, was coupled with epoxide **43** and the dithiane moiety was metallated and condensed with dimethyl oxalate to give **45**. The latter represents one half of the aplasmomycin skeleton in suitably protected form (Scheme 18).

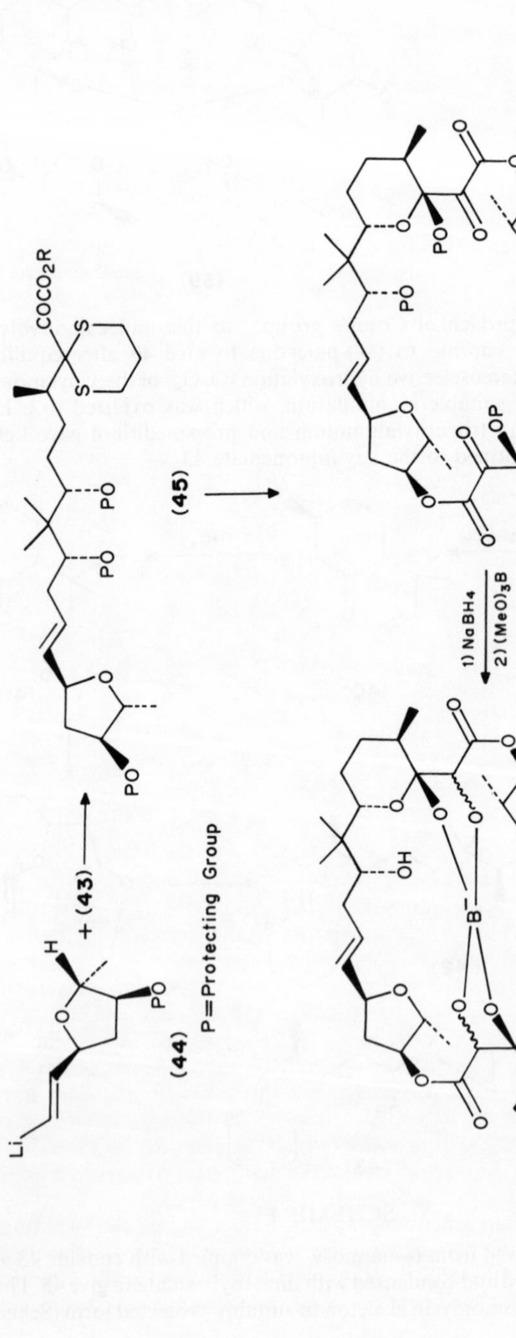

COCO$_2$R

(45) →

(46)

+ **(43)** →

(44)

P=Protecting Group

1) NaBH$_4$
2) (MeO)$_3$B

(39) and isomer

SCHEME 18

Selective deprotection of **45** allowed coupling of two units of **45** to give the cyclic compound **46** with the key macrolactonization being achieved in 71% yield. Completion of the synthesis involved reduction and incorporation of the boron atom, however, no selectivity was achieved during the reduction step.

White and coworkers' approach[28] to aplasmomycin involved the chiral lactone **49**, which served as the C(3)–C(10) segment in each half of the macrocycle. This lactone was prepared either by resolution or, more efficiently, by manipulation of (+)-pulegone as outlined in Scheme 19[28b]. Keto ester **47** was prepared from (+)-pulegone via (i) hydrocyanation followed by hydrolysis or (ii) conjugate addition of vinyl cuprate followed by oxidative cleavage, and was subjected to Baeyer–Villiger oxidation to give lactone **48**. Ring contraction of **48** was achieved using conventional chemistry to give lactone (+)-**49** which was ultimately transformed into (+)-aplasmomycin and also served as an important building block in efforts directed towards a synthesis of boromycin.

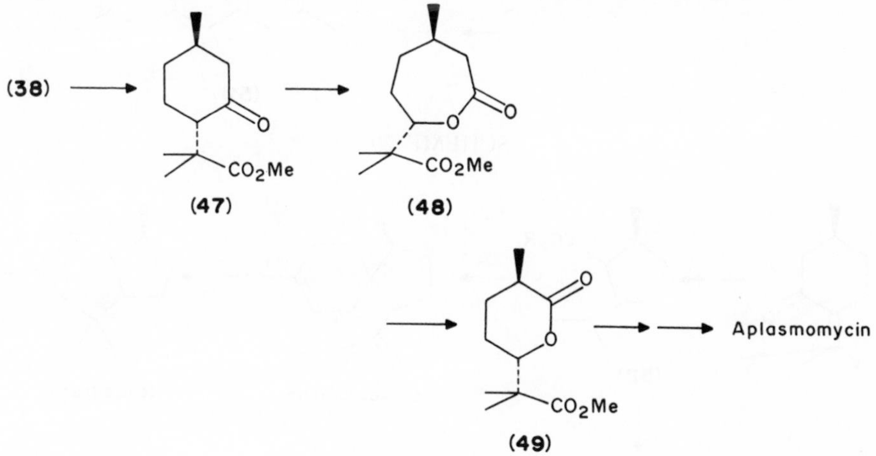

SCHEME 19

(+)-Pulegone has also found use in the synthesis of optically active acyclic compounds such as the vitamin E side-chain **51** as outlined in Scheme 20[29]. The key features of this synthesis involve the selective mono-demethylation of the isopropylidene moiety, via deconjugative ketalization and ozonolysis, and a highly stereoselective Carroll rearrangement of the β-keto ester **50** which serves to establish the stereochemistry at C-(7) of the final product. This example of 1,3-stereocontrol provides a highly efficient and completely stereocontrolled synthesis of the optically pure (3R, 7R) vitamin E side-chain.

Transformation of (+)-pulegone into the optically pure cyclopentane ester **52** can be readily achieved via the Favorskii rearrangement; this cyclopentane ester has found considerable use as a synthetic intermediate[30]. Since the enone functionality of (+)-pulegone plays no significant role in syntheses involving **52**, after the initial Favorskii rearrangement, this chemistry will not be covered in depth here. However, a notation of some syntheses involving cyclopentane **52** is given in Scheme 21.

The use of chiral auxiliaries to achieve asymmetric induction in a chemical transformation is an important process in organic chemistry, and one of the most widely used chiral auxiliaries is (−)-8-phenylmenthol (**53**)[31]. This chiral adjuvant is readily available from (+)-pulegone (Scheme 22) and has proved successful in achieving significant dias-

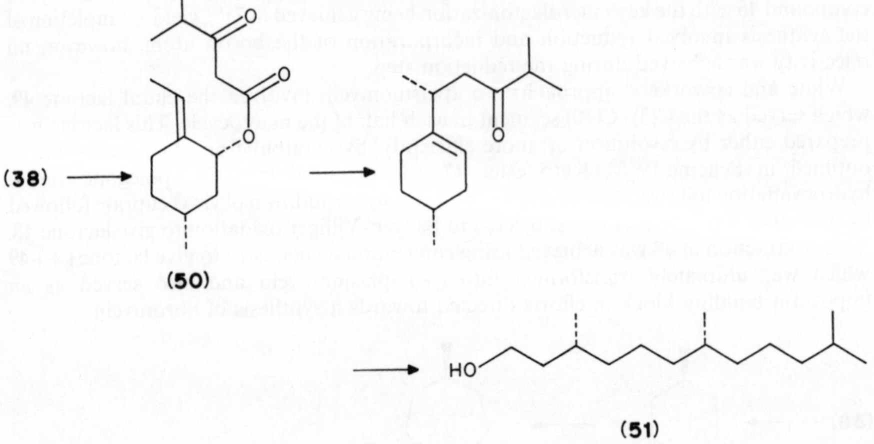

(38) ⟶

(50)

⟶

(51)

SCHEME 20

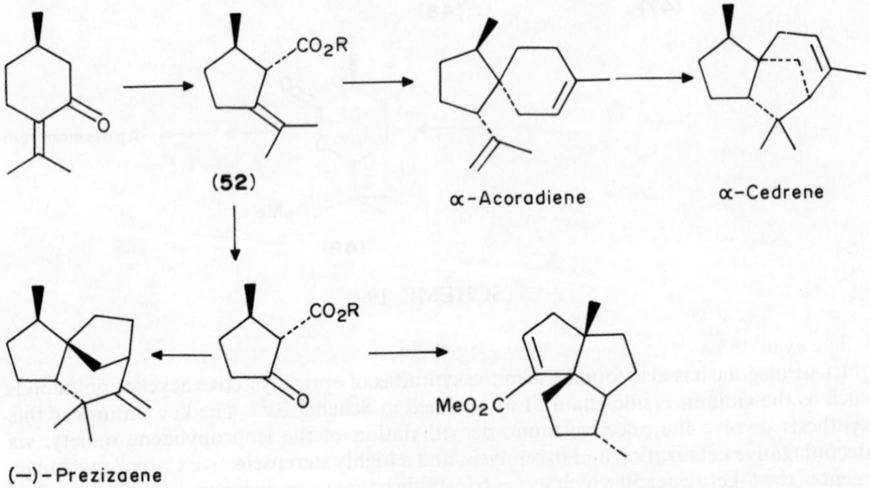

(52)

α-Acoradiene

α-Cedrene

(−)-Prezizaene

MeO₂C

SCHEME 21

tereoselection in a number of reactions, including the Diels–Alder reaction, ene reactions, conjugate addition, alkylations, etc.[32]. The acrylate ester of (+)-8-phenylmenthol was employed by Corey and Ensley[31] as a dienophile for a highly diastereoselective Diels–Alder reaction. The bicyclic product was elaborated to an intermediate which was useful for prostaglandin syntheses.

The Michael acceptor property of (+)-pulegone was employed by Lynch and Eliel[33] to prepare the optically active 1,3-oxathiane 54 which was metallated, condensed with aldehydes and oxidized to give keto oxathianes 55 (Scheme 23). Addition of Grignard reagents to these keto oxathianes occurred with excellent diastereoselectivity and, after cleavage of the oxathiane, led to α-hydroxy aldehydes with a high degree of optical purity.

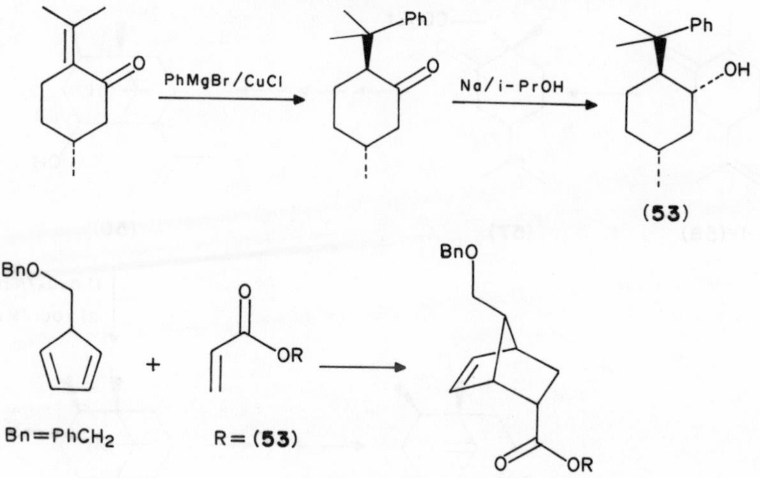

SCHEME 22

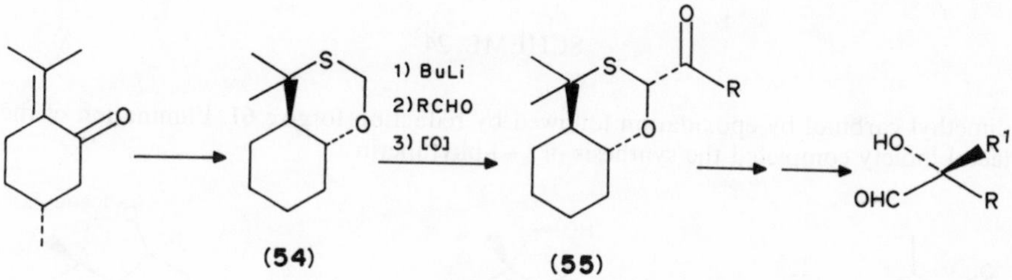

SCHEME 23

The availability of carvone, in either enantiomeric form, along with its diverse array of functionality has rendered this substance an attractive starting material for numerous synthetic endeavors. A classical method for the functionalization of α,β-unsaturated ketones involves a conjugate reduction/alkylation sequence and carvone has proved to be very amenable to this process. Alkylation of the enolate derived by conjugate reduction of (−)-carvone (56) with ethyl bromoacetate (Scheme 24) was the initial step in a recent synthesis of the non-isoprenoid sesquiterpene, upial, which served to establish the absolute configuration of this compound[34]. The keto ester 57 was elaborated to the bicyclic adduct 58, whereupon the isopropenyl group was unmasked to give an ester function which was required for the final transformation.

The above synthesis of upial demonstrates the use of the isopropenyl group of carvone as a latent ester function; an example of the isopropenyl group acting as a dimethylcarbinol equivalent is outlined in Scheme 25, an elegant synthesis of (−)-phytuberin[35]. Condensation of the enolate derived by conjugate reduction of (−)-carvone with formaldehyde gave the hydroxymethylketone 59 as a mixture of diastereomers. Interestingly, the minor isomer could be re-equilibrated by simple thermolysis apparently via a retroaldolization/aldolization sequence. Elaboration of 59 to 60 was achieved via sequential ethynylation and hydration and the isopropenyl group was then converted to a

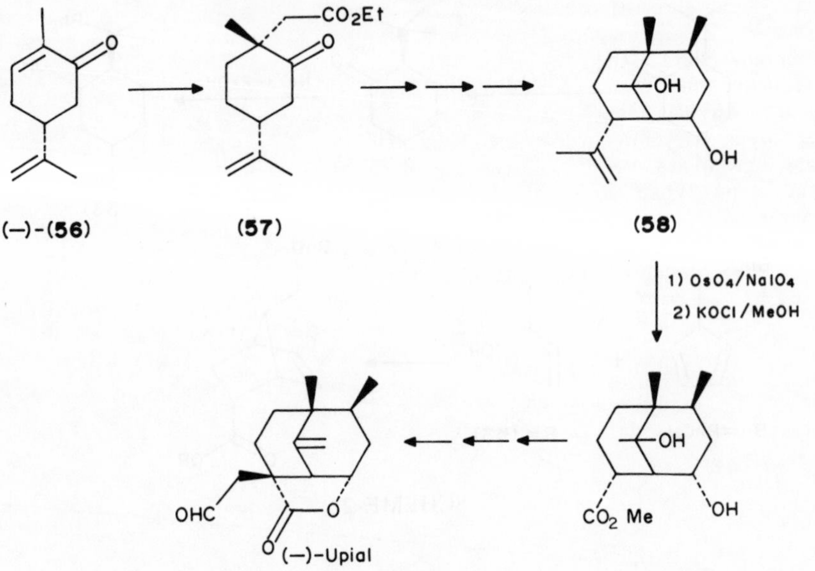
CO2Et
OH
OH
(−)-(56)
(57)
(58)
1) OsO4/NaIO4
2) KOCl/MeOH
OH
OH
CO2 Me
OHC
O
O
(−)-Upial

SCHEME 24

dimethyl carbinol by epoxidation followed by reduction to give **61**. Elimination of the lactol moiety completed the synthesis of (−)-phytuberin.

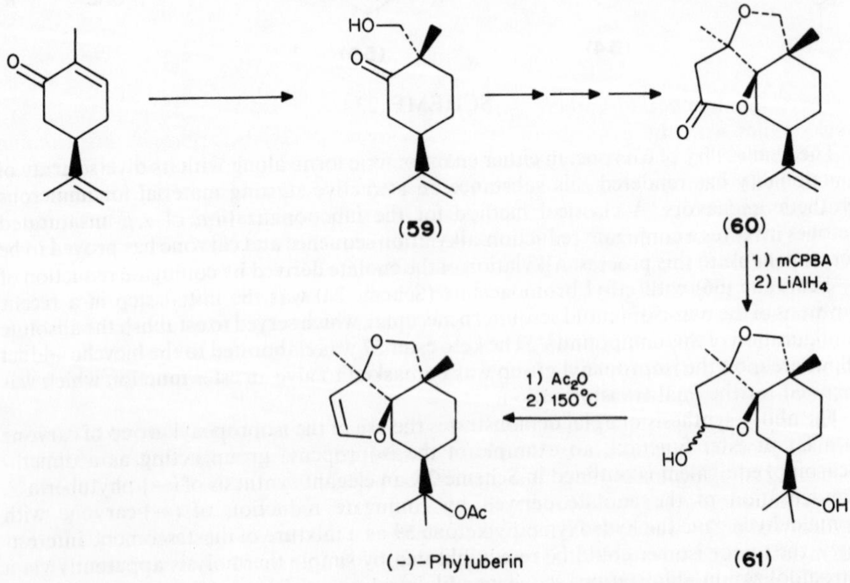
O
HO
O
O
O
O
(59)
(60)
1) mCPBA
2) LiAlH4
O
O
1) Ac2O
2) 150°C
O
O
HO
OAc
OH
(−)-Phytuberin
(61)

SCHEME 25

In studies directed towards a general synthesis of quassinoids, and in particular bruceatin, Ziegler required the keto alcohol 64 in chiral, non-racemic form. This compound could be effectively prepared from (+)-carvone [(+)-56] (2 mol scale) utilizing a reductive annelation sequence to give enone 62 (Scheme 26)[36]. Reduction of the enone 62 to alcohol 63 was followed by transformation of the isopropenyl group to a ketone by an ozonolysis, Baeyer–Villiger oxidation, oxidation sequence. Introduction of a hydroxy-methyl group at C-9 was achieved through multiple transformations to give the desired product 64. While this compound ultimately proved to be unsuitable for further transformation into bruceatin, the chemistry developed by Ziegler demonstrates that carvone can serve as a valuable precursor to highly functionalized decalins via the conjugative reduction/alkylation sequence.

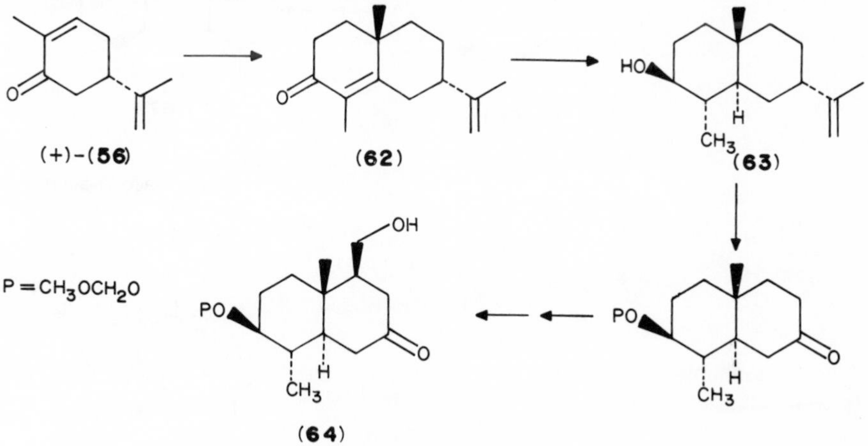

(+)-(56)   (62)   (63)

P = CH₃OCH₂O

(64)

SCHEME 26

The enone functionality in carvone can also serve as a Michael acceptor for carbon nucleophiles with the isopropenyl group serving to direct the approach of incoming nucleophiles. This methodology was employed by Brattesani and Heathcock to prepare the *cis* hydrindanone 68, an intermediate in a proposed synthesis of the sesquiterpene alkaloid dendrobine[37]. Copper-catalyzed addition of 4-butenylmagnesium bromide to (+)-carvotanoacetone (65) occurred with complete diastereoselectivity, *trans* to the isopropyl group, to give adduct 66 (Scheme 27). Ozonolysis and chain extension gave the unsaturated nitrile 67 which, on treatment with base, underwent a stereoselective, intramolecular Michael addition reaction to give *cis* hydrindanone 68. Unfortunately the stereochemistry of the cyanomethyl side-chain is the opposite of that required for elaboration into dendrobine, however, this approach using carvone allows rapid entry into optically pure hydrindanones.

An impressive use of the complete carvone structure in the synthesis of a complex natural product is manifest in the first synthesis of picrotoxin by Corey and Pearce (Scheme 28)[38]. The first step in this synthesis involved α-alkylation of the anion derived by γ-deprotonation of the N,N-dimethylhydrazone derivative 69 of (−)-carvone to give 70. Hydrolysis of 70 was followed by acid-catalyzed intramolecular aldol condensation, ethynylation and intramolecular bromoetherification to give 71. The acetylene 71 was transformed into the corresponding protected aldehyde which was converted to diketone 72 using potassium *tert*-butoxide, dimethyl disulfide and oxygen, methodology developed

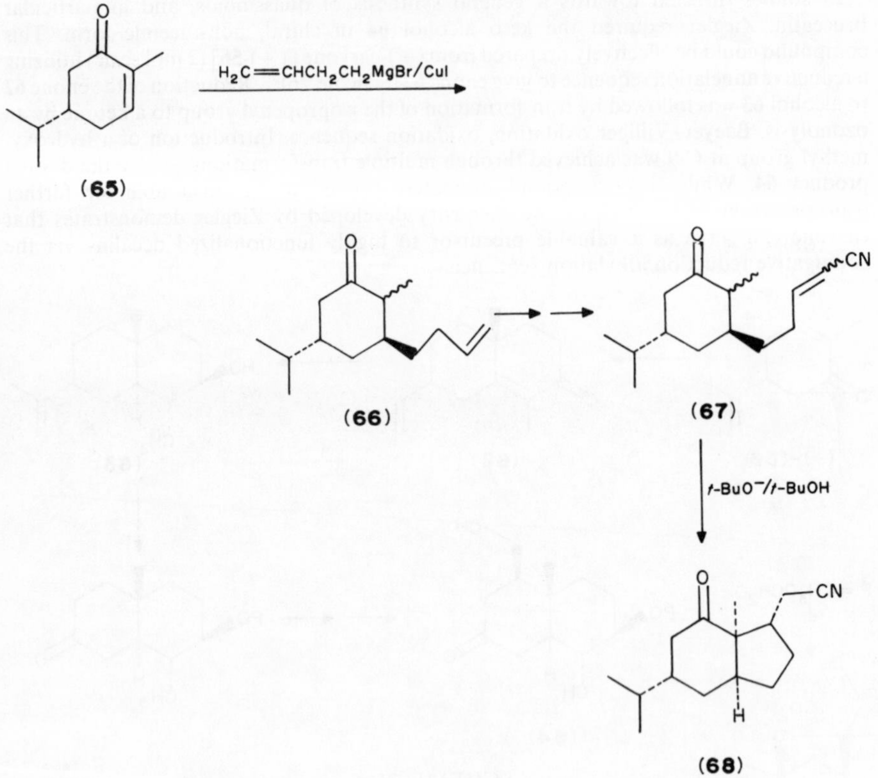

(65)

H₂C=CHCH₂CH₂MgBr/CuI →

(66)   → →   **(67)**

t-BuO⁻/*t*-BuOH ↓

(68)

SCHEME 27

by Barton. Intramolecular aldol condensation of the aldehyde corresponding to **72** gave **73**, which established the hydroindene nucleus of picrotoxin. Oxidative cleavage of diketone **73** to diacid **74** was followed by double lactonization to give **75** which, on elimination of the benzoate, epoxidation and reductive removal of bromine, afforded (−)-picrotoxin. It is perhaps pertinent to note here that an inexpensive starting material with a single chiral center has been stereoselectively transformed into a complex product with eight contiguous chiral centers (three of them quaternary) in homochiral form.

Stereoselective epoxidation of the enone group of carvone, or a carvone derivative, represents a convenient method for the introduction of two new asymmetric centers on carvone. This epoxidation, which occurs with complete selectivity, *trans* to the isopropenyl group, was exploited to establish the stereochemistry required in a total synthesis of a vitamin D metabolite, 1α, 25-dihydroxycholecalciferol [**81** (P = H)] (Scheme 29)[39]. Epoxy ketone **76** was subjected to standard Horner–Emmons conditions and the isopropenyl group was transformed into an alcohol, via oxidative cleavage, Baeyer–Villiger oxidation and hydrolysis, to give **77**. Regioselective cleavage of the epoxide and elimination of the resulting tertiary alcohol gave **78** which was converted into the allylic phosphine oxide **79**. Wittig–Horner reaction between **79** and **80** (for the preparation of **80** see Scheme 46) gave **81**; the stereochemistry of the newly formed double bond was completely that shown.

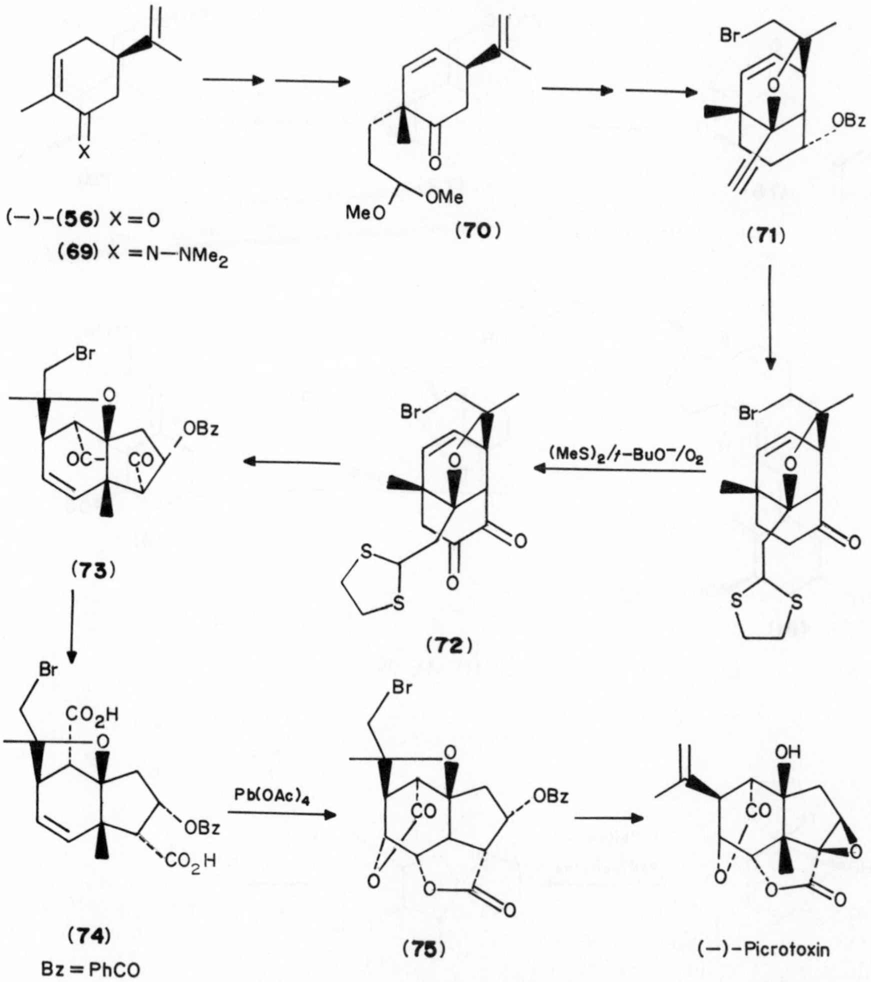

(−)-(56) X = O
(69) X = N—NMe₂

(70)

(71)

(73)

(72)

(MeS)₂/t-BuO⁻/O₂

(74)

Bz = PhCO

Pb(OAc)₄

(75)

(−)-Picrotoxin

SCHEME 28

The enantiomer of **76**, prepared from (−)-carvone was employed by Yoshikoshi and coworkers as the key starting material in a short synthesis of the diterpene taonianone (**84**) (Scheme 30)[40]. The epoxide was transformed into cyclopentene **83** via protection, cleavage of the epoxide, hydrogenation, oxidative cleavage of diol **82** and intramolecular aldol condensation. Elaboration of the aldehyde group of **83** gave (+)-taonianone of known absolute configuration which allowed assignment of the stereochemistry of natural material.

An electrooxidative approach to the important pesticide (1R, 3R)-methyl chrysanthmate (**89**) reported by Torii and coworkers[41] used (+)-carvone as the starting material (Scheme 31). Stereoselective epoxidation of (+)-carvone hydrochloride (**85**) followed by methanolysis gave **86** which was oxidatively degraded using a MeOH–LiClO₄–Pt system

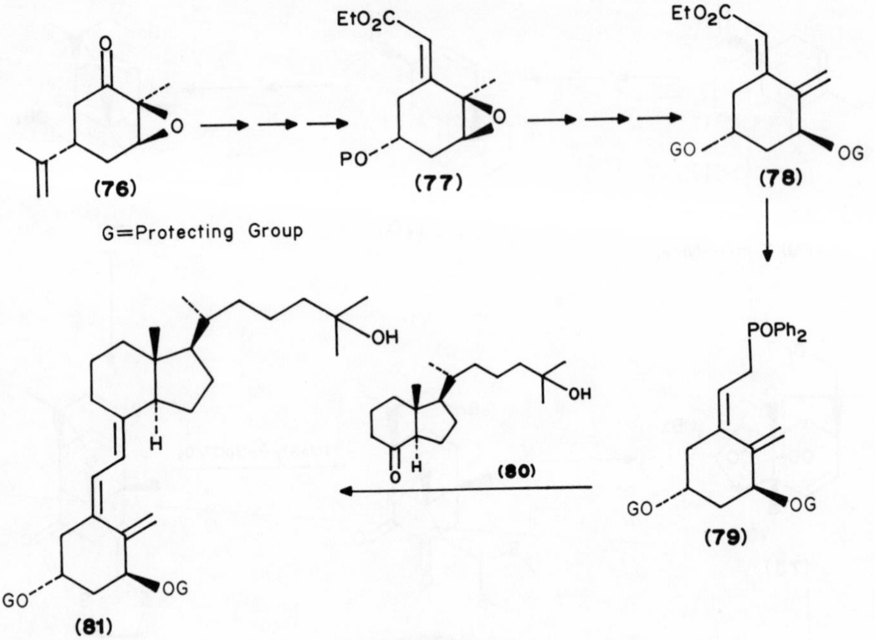

G=Protecting Group

SCHEME 29

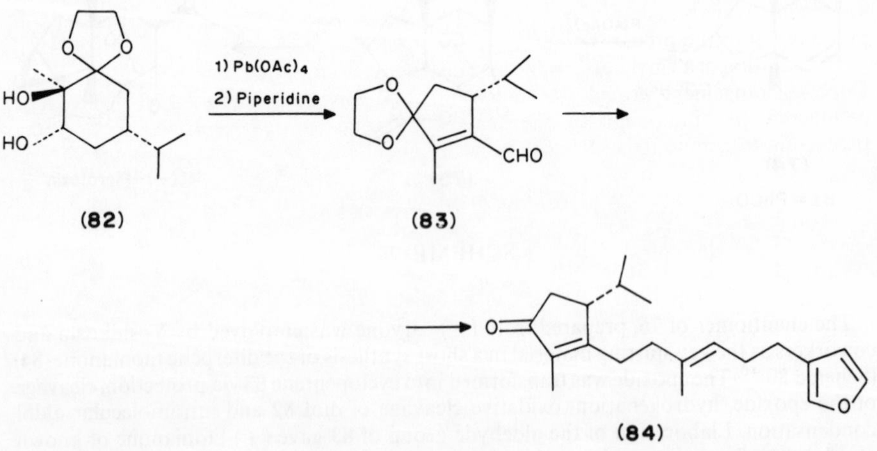

SCHEME 30

to give **87** in high yield. Ester **87** was treated with methyllithium followed by hydrolysis and oxidation to give the key intermediate **88** which is known to be a precursor of methyl chrysanthmate.

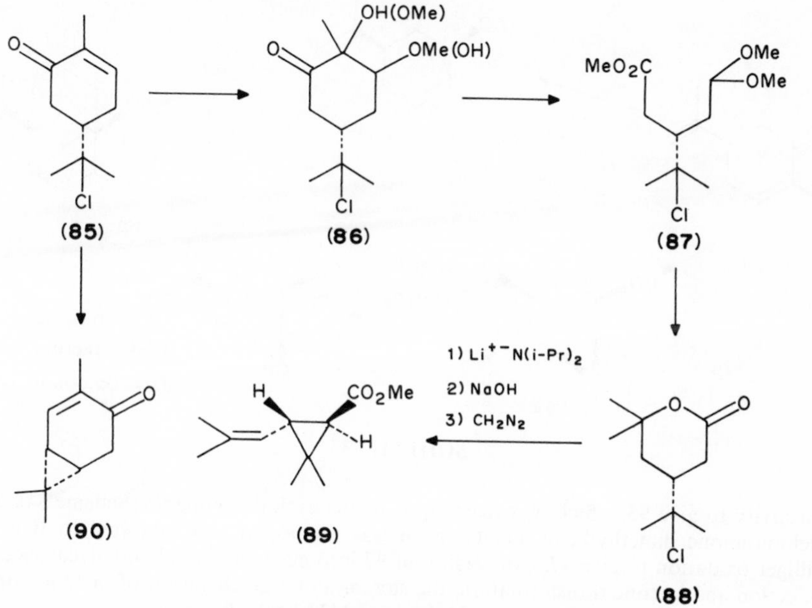

SCHEME 31

(+)-Carvone hydrochloride (85) was shown by Wiemer and coworkers[42] to be readily transformed into (−)-carenone (90) via intramolecular α-alkylation, Wharton rearrangement of the derived hydrazine and oxidation (Scheme 31). Since both enantiomers of 90 are readily available, simply by selecting (+)- or (−)-carvone, these compounds should prove useful in the preparation of a variety of natural products.

The addition of a vinyl nucleophile to the carbonyl of compounds such as 91 followed by Cope rearrangement provides a smooth method for the preparation of macrocyclic, germacrane-like intermediates (Scheme 32). Carvones have been found to be useful precursors to compounds such as 91 and ultimately to natural germacronolides.

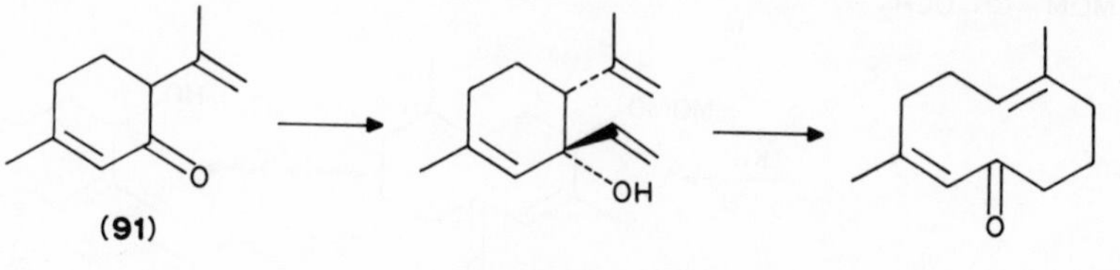

SCHEME 32

The conversion of carvone into enones related to 91 has been accomplished via either a reduction, allylic oxidation sequence (Scheme 33), or a route involving selenide opening of epoxide 92 followed by selenoxide elimination and 1,3-oxidative rearrangement.

The enone 93, derived from (+)-carvone, was employed by Still and coworkers[43] in a synthesis of eucannabinolide which featured an oxy-Cope rearrangement as the key step in the formation of the macrocycle. The cyclobutenyllithium 94 added to 93 with good

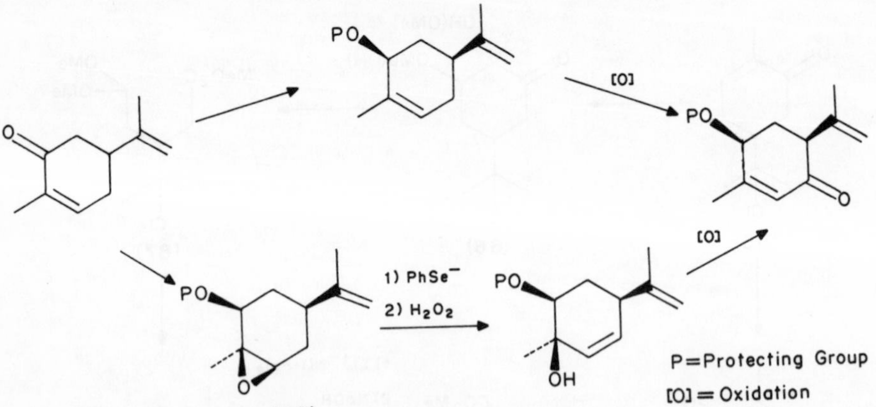

(92)

SCHEME 33

selectivity to give **95**, which was rearranged to the cyclodecenone **96** (Scheme 34). The cyclobutanone dimethylketal moiety of **96** was unmasked and subjected to Baeyer–Villiger oxidation to give **97**. Conversion of **97** into eucannabinolide involved selective reduction and lactone transformation; the stereo- and regiochemistry of these manipulations were effectively predicted on the basis of MM2 calculations.

A similar strategy was employed by Takahashi and coworkers[44] to prepare the heliangolide (**99**) from enone **98** (Scheme 35).

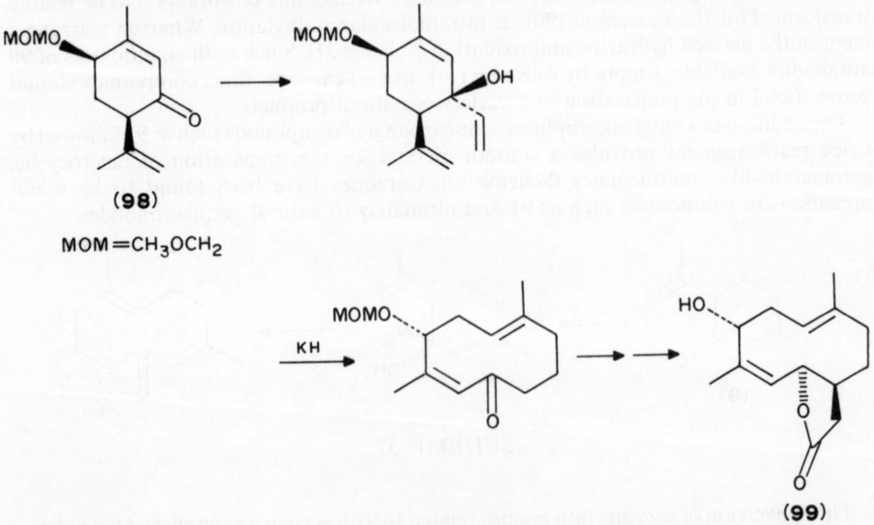

(98)

MOM$=$CH$_3$OCH$_2$

(99)

SCHEME 35

A novel variation of this oxy-Cope macro-expansion methodology was developed by Wender and Holt[45] to prepare 14-membered macrocycles as found in the cembrane series.

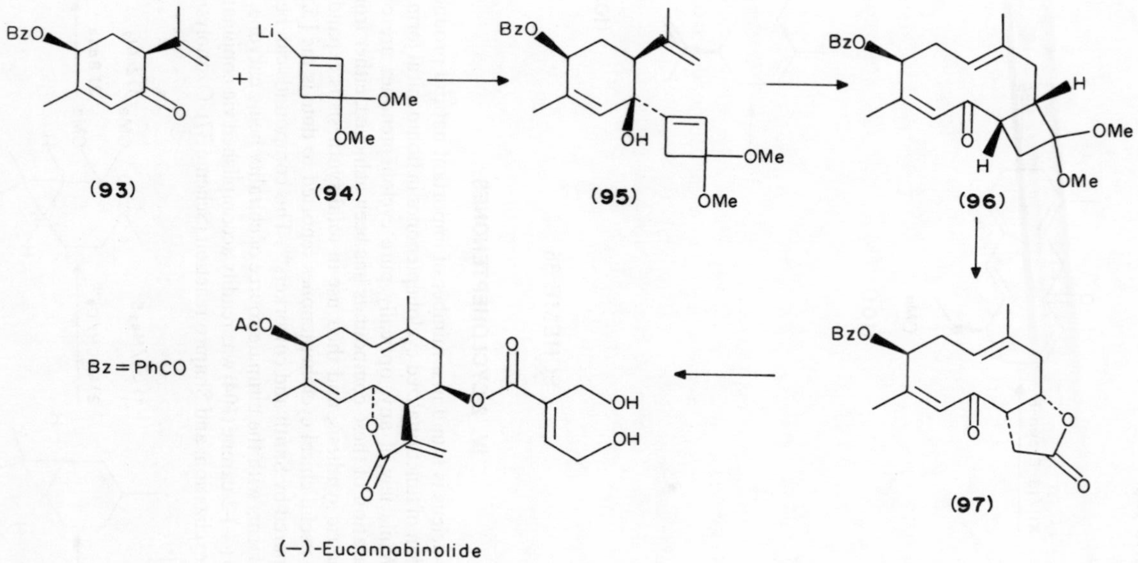

The key step in this approach involves the rearrangement of (100) prepared from (98) via epoxide (99) (Scheme 28) (see text). [reference lines unclear]

(93)

(94)

(95)

(96)

(97)

(−)-Eucannabinolide

Bz = PhCO

SCHEME 34

H. CYCLOHEPTENONES

The cycloheptenones and related types of important terpenoid products indicate that the preparation of medium-sized ring compounds continues to be a particularly desirable goal. Various synthetic routes to cycloheptenones are constructed...

The key step in this approach involves the rearrangement of **100**, prepared from (+)-carvone, to **101** (Scheme 36) which contains all 20 carbons required for elaboration into (−)-(3Z)-cembrane A (**102**). Reductive removal of the carbonyl in **101** was followed by selective hydrogenation of the least substituted double bond and elimination of the methoxy group to give **102**.

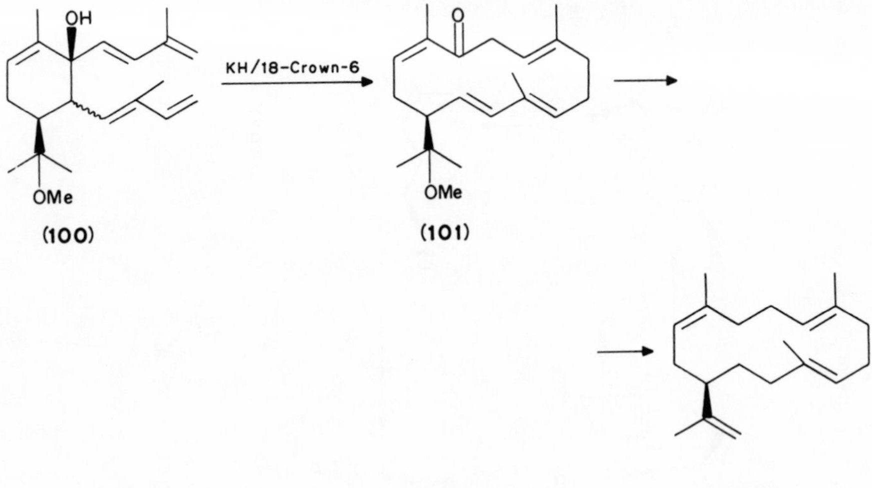

SCHEME 36

IV. 2-CYCLOHEPTENONES

The cycloheptane nucleus is found in a number of important natural products and, as a result, the preparation of functionalized cycloheptenones in homochiral form has become a desirable goal. While few, if any, optically pure cycloheptenones are commercially available, the preparation of these compounds has been achieved, either from the chiral pool or via asymmetric synthesis, and their use in total synthesis is expanding.

One of the more useful chiral cycloheptenones reported to date is the [5.1.0] bicyclic compound **103** prepared by Smith and coworkers[46]. This compounds can be prepared in both enantiomeric forms with the ultimate source of chirality being carvone. Conversion of (+)-carvone into (−)-2-carene (**104**) was readily accomplished via conjugate reduction, hydrochlorination-cyclization, and Shapiro reaction (Scheme 37). Ozonolysis of **104** and

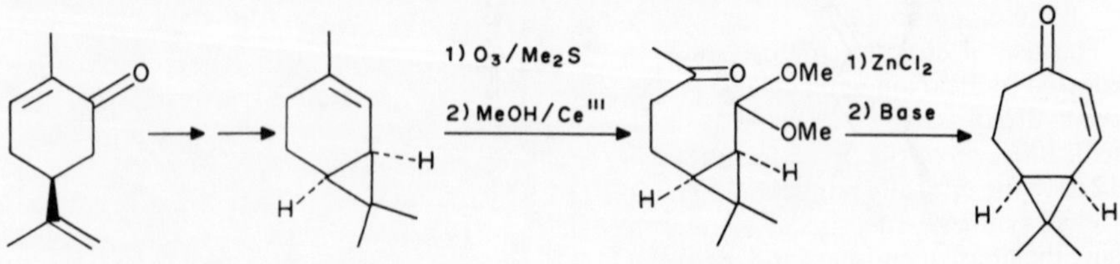

SCHEME 37

selective protection gave **105**, which was cyclized using the Mukaiyama protocol and eliminated to give homochiral **103** in good yield.

The enone functionality of **103** was exploited by Taylor and Smith in a short synthesis of the sesquiterpene (+)-hanegokedial as outlined in Scheme 38[47]. Sequential treatment of **103** with the cuprate prepared from bis(1,1-diethoxy-2-propenyl)lithium and formaldehyde gave **106** along with its C2 epimer, which was subjected to methylenation, oxidation and hydrolysis to give natural (+)-hanegokedial.

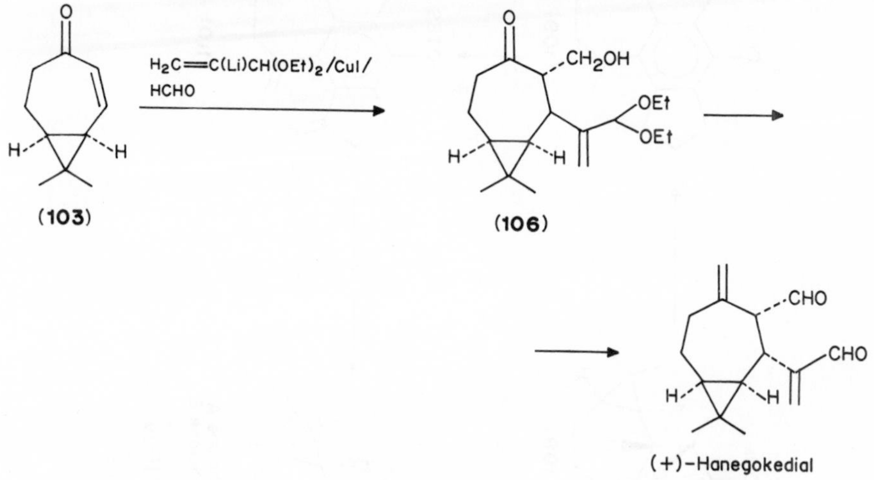

SCHEME 38

The enone **103** was also exploited by Smith and coworkers as a dienophile in a route to the jatropholone skeleton which featured a high-pressure Diels–Alder reactions[48]. The Diels–Alder reaction of **103** and furan **107** at 5 kbar occurred with complete diastereoselectivity and in high yield to give **108**. The ease of this reaction is significant, since cycloalkenones are known to be reluctant partners in Diels–Alder reactions and application of this process to more readily available chiral cyclohexenones and cyclopentenones may prove to be profitable. Aromatization of **108**, followed by methylenation, gave **109** which underwent regioselective oxidation to **110** after protecting group manipulation (Scheme 39). Methylation of **110**, followed by deprotection, gave (+)-jatropholones A and B in homochiral form, which established the absolute stereochemistry of these compounds.

The use of enzymes to prepare optically active cycloheptenones has recently been reported by Pearson and coworkers (Scheme 40)[49]. Enantioselective hydrolysis of **111** to give hydroxy acetate **112** could be achieved using electric eel acetylcholinesterase (39% yield, 100% ee) or the lipase from *Candida cyclindracea* (40% yield, 44% ee). Oxidation of **112** gave the optically pure cycloheptenone **113**, which should prove to be useful in natural product synthesis. In a similar manner the diacetate **114** was enantioselectively hydrolyzed using the lipase mentioned above to give **115** (61% yield, 100% ee). However, the direction of induced chirality was reversed. Oxidation of **115** afforded cycloheptenone **116**, which is related to enones known to be intermediates for the synthesis of the Prelog–Djerassi lactone.

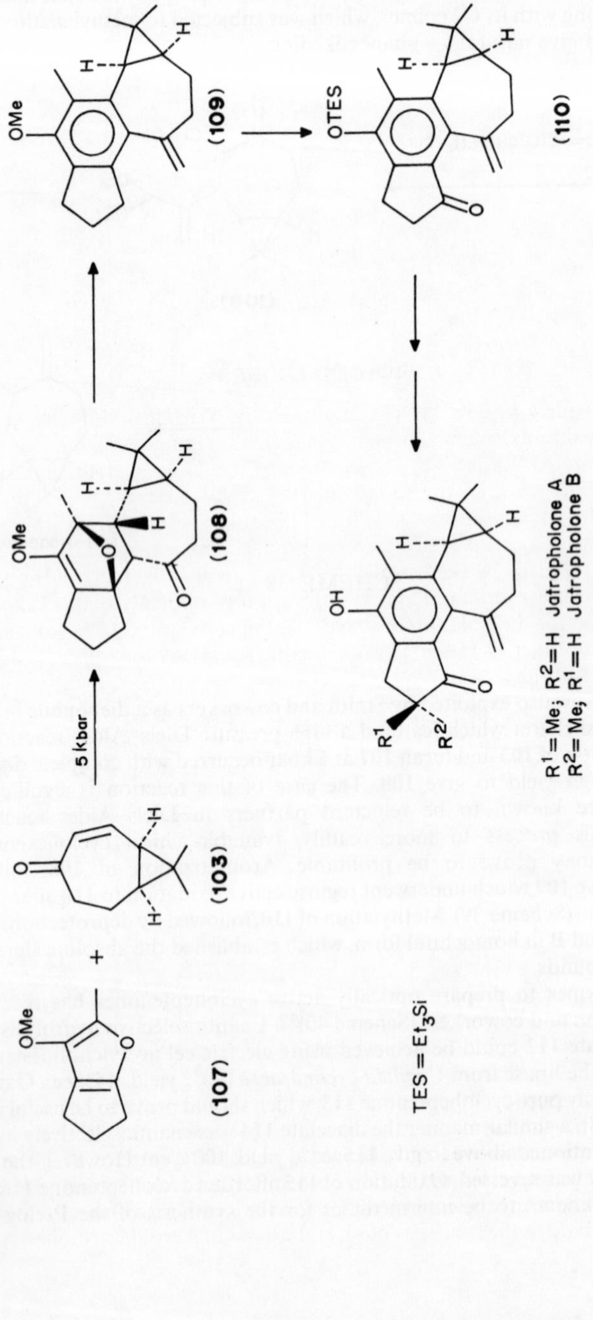

1114

SCHEME 39

(107)

(103)

(108)

5 kbar

(109)

(110)

R¹ = Me; R² = H   Jatropholone A
R² = Me; R¹ = H   Jatropholone B

TES = Et₃Si

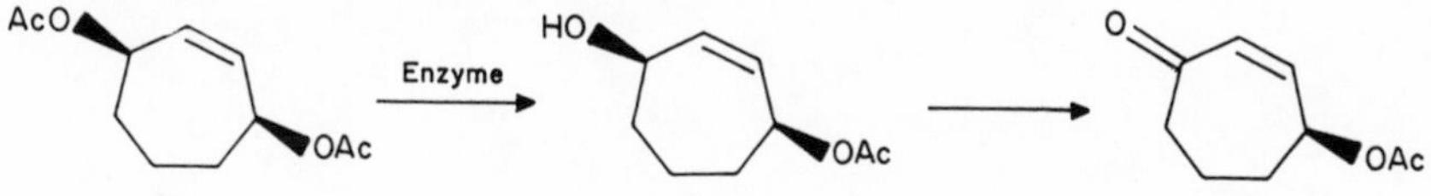

(111)   **(112)**   **(113)**

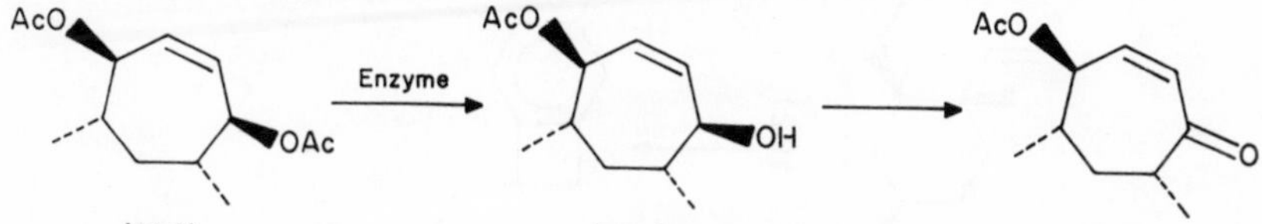

(114)   **(115)**   **(116)**

SCHEME 40

V. BICYCLIC ENONES

One of the most widely known bicyclic enones is the Wieland–Miescher ketone (118) which has been used in its racemic form as a versatile building block for the synthesis of steroids and terpenoids. Enone 120 became readily available in homochiral form as a result of independent work by the groups of Hajos[50] and Eder[51] who found that the intramolecular aldol condensation of triketone 119 could be rendered highly enantioselective through the use of a chiral catalyst, i.e. (S)-proline. Application of this process to the triketone 117 was reported by Furst and coworkers[52] to give the bicyclic enone 118 with 70% optical purity; however, alternate crystallization of optically pure 118 followed by racemic 118 allowed for the isolation of essentially optically pure 118 in reasonable yield. Clearly, either enantiomer of 118 or 120 is available simply by the appropriate choice of catalyst, (S)- or (R)-proline.

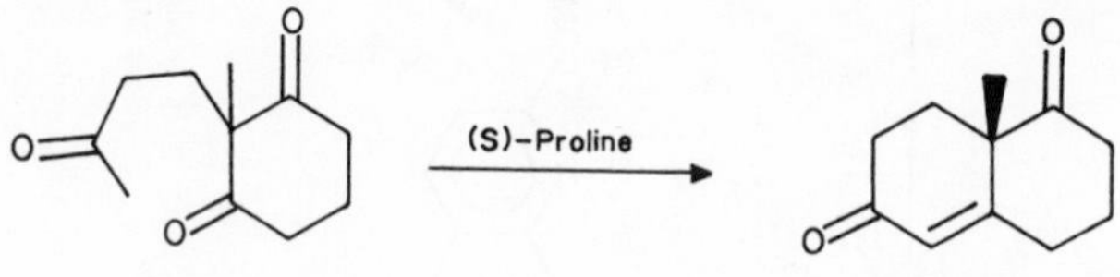

(117)   (S)–Proline   **(118)**

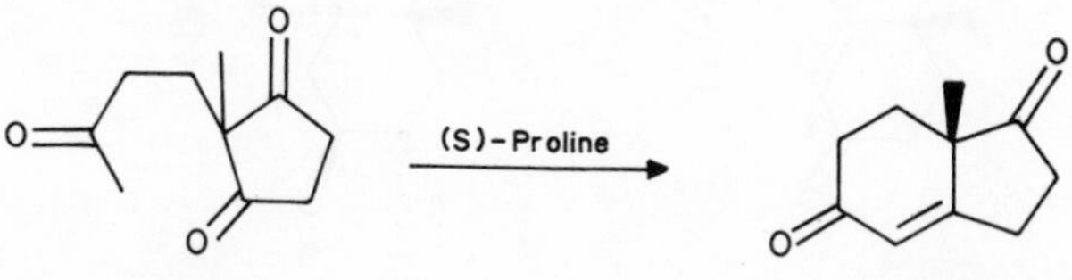

(119)   (S)–Proline   **(120)**

The total synthesis of natural and unnatural steroids continues to be an area of considerable synthetic interest[53] and the stereochemistry and functionality present in 118 makes it an attractive starting material for such endeavors. The (+)-D-homosteroid 124 is an important intermediate in the synthesis of several classes of steroid hormones. An elegant

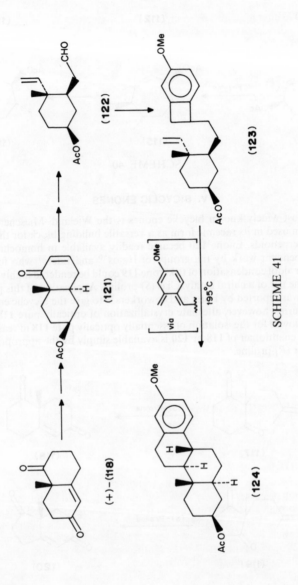

SCHEME 41

1116

approach to **124** was reported by Kametani and coworkers[54] based on the intramolecular Diels–Alder reaction of an *ortho*-quinodimethane (Scheme 41). The enone (+)-**118** was transformed into enone **121** using known chemistry. Enone **121** was epoxidized and cleaved via the Eschenmoser process to an acetylenic ketone which, upon partial hydrogenation, gave **122**, which was condensed with a benzocyclobutene to give the key intermediate **123** after reductive removal of the hydroxy and cyano groups. Thermolysis of **123** effected a completely stereoselective cyclization to give the D-homosteroid **124**.

Enone (+)-**118** can also serve as a synthon for the CD ring system of homosteroids as demonstrated by Furst and coworkers in their synthesis of (+)-D-homoestrone (Scheme 42)[55]. Alkylation of enone **125**, readily prepared from (+)-**118**, with *m*-methoxyphenacyl bromide rapidly assembles all the carbons required for conversion into homoestrone. Hydrogenation of **126** was followed by treatment with acid to effect cyclization to **127**; hydrogenation, deprotection and oxidation gave D-homoestrone.

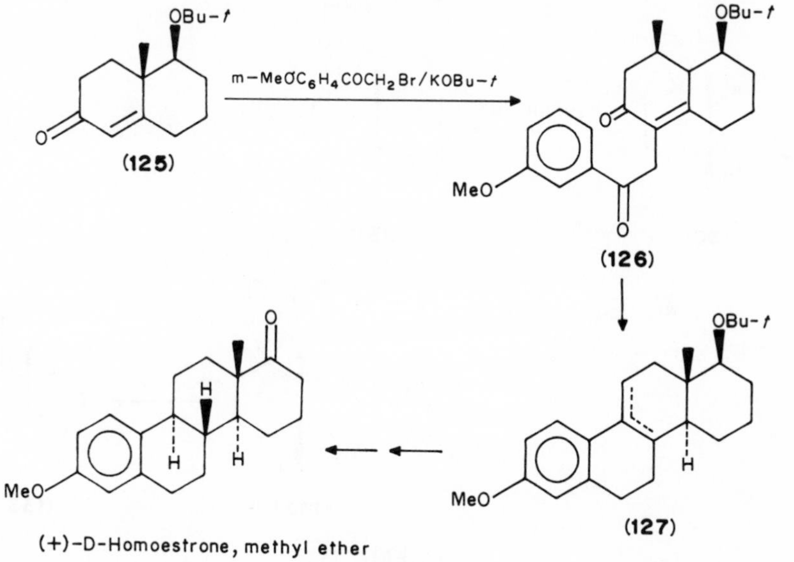

SCHEME 42

Ketone (+)-**118** has served as the starting point in a synthesis of (+)-pallescensin A (**129**), a furanosesquiterpene isolated from a marine sponge, by Smith and Mewshaw[56]. Transketalization of the ethylene ketal of 2-butanone and (+)-**118**, followed by treatment with aqueous formaldehyde, thiophenol and triethylamine, provided intermediate **128**. Conversion of **128** to the target **129** was achieved by a reductive methylation (Li/NH$_3$, then MeI), followed by Wolff–Kishner reduction and elaboration of the fused furan. Intermediate **128** offers possibilities for the synthesis of a variety of architecturally complex natural products containing *trans*-decalin units.

The transformation of (+)-**118** into optically active hydrindenones has been reported by Jung and Hatfield[57] in synthetic efforts directed towards steroid synthesis. Protection and epoxidation of (+)-**118** gave **130**, which underwent Eschenmoser fragmentation to acetylene **131**. Keto acetylene **131** was reductively cyclized to **132** using methodology developed by Stork and this allylic alcohol was rearranged and reduced to the

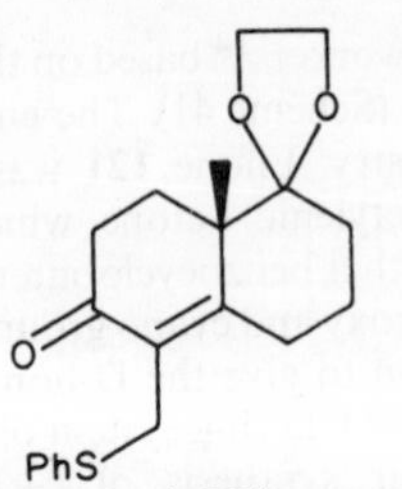

(128) 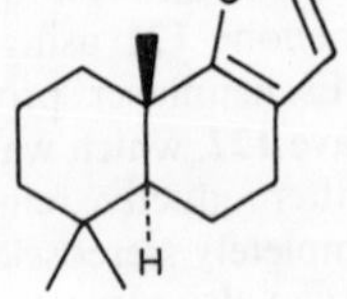   **(129)**

hydrindenone **133** (Scheme 43). This hydrindenone is a synthon for the AB ring portion of steroids via a sequence involving attachment of the C and D rings followed by ozonolysis and cyclization.

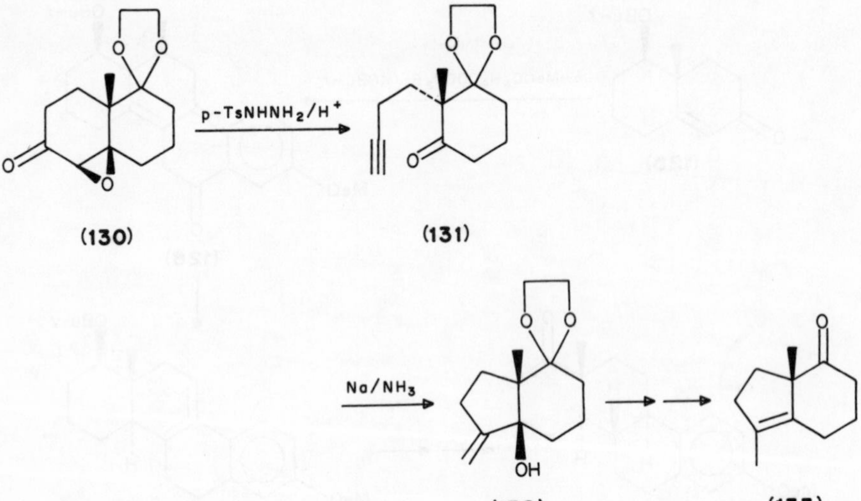

(130)   **(131)**

(132)   **(133)**

SCHEME 43

The preparation of hydrindenone **120** in optically pure form was outlined earlier in this section and this compound has found considerable use as a CD ring synthon in the synthesis of steroids, most notably in the preparation of estrones. The successful use of hydrindenones such as **120** in steroid total synthesis is dependent on their facile conversion

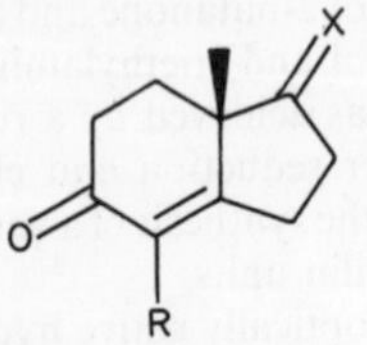

(120) R = H, X = O
(134) R = CO₂H, X = α-H, β-OBu-t
(135) R = CH₂SO₂Ph, X = α-H, β-OBu-t

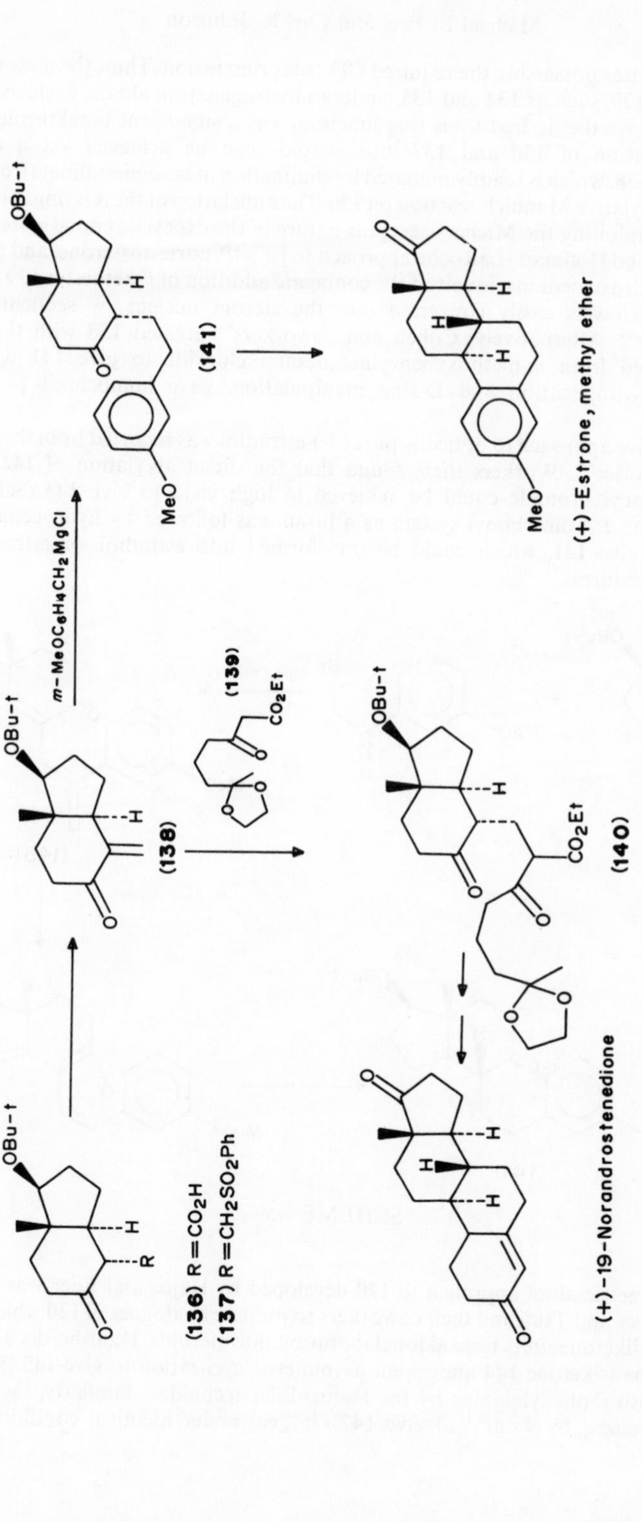

SCHEME 44

into intermediates possessing the required CD-*trans* ring fusion. Thus, the discovery that derivatives of **120**, such as **134** and **135**, undergo hydrogenation almost exclusively from the α face, to give the desired *trans* ring junction, was a significant breakthrough[58].

The elaboration of **136** and **137** into steroids can be achieved via a common intermediate, **138**, which is readily prepared by elimination of benzenesulfinate from **137** or via a decarboxylative Mannich reaction on **136**. The annelation of the AB rings onto **138** is achieved by exploiting the Michael acceptor nature of the exocyclic enone system in **138** (Scheme 44). The Hoffman–LaRoche approach to (+)-19-nortestosterone, and thence to (+)-19-norandrostenedione, involved the conjugate addition of β-ketoester **139** to **138** to give **140**, which was easily converted into the steroid nucleus by sequential aldol condensations[59]. Alternatively, Cohen and coworkers[60] treated **138** with the copper reagent derived from *m*-methoxybenzylmagnesium chloride to give **141** which, on cyclization, hydrogenation and D-ring manipulation, gave homochiral (+)-estrone methyl ether.

An alternative approach to optically pure (+)-estradiol was reported from the Schering A. G. laboratories[61]. Workers there found that the direct alkylation of **142** with *m*-methoxyphenacyl bromide could be achieved in high yield to give **143** (Scheme 45). Masking of the 1,4-dicarbonyl system as a furan was followed by hydrogenation and oxidation to give **141**, which could be transformed into estradiol, or estrone, using standard procedures.

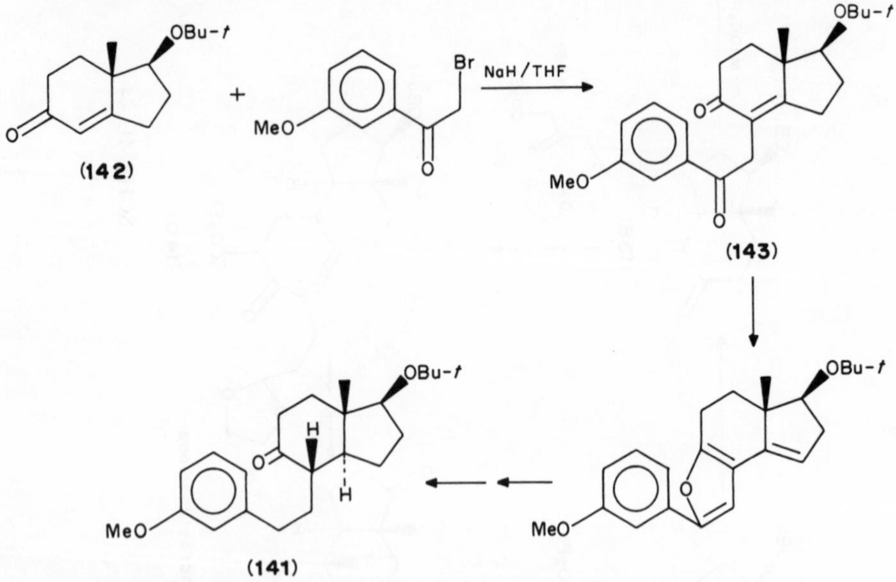

SCHEME 45

The asymmetric aldol approach to **120** developed by Hajos and Eder was employed by Danishefsky and Tsuji and their coworkers to prepare analogues of **120** which contain masked 1,5-diketone units needed for elaboration into steroids. Danishefsky and Cain[62] found that the triketone **144** underwent asymmetric cyclization to give **145** (86% ee) on treatment with L-phenylalanine by the Hajos–Eder technique. Similarly, the triketone (**146**) was cyclized by Tsuji[63] to give **147** (76% ee) under identical conditions. These hydrindenones were further transformed into (+)-estrone and (+)-19-nortestosterone, respectively.

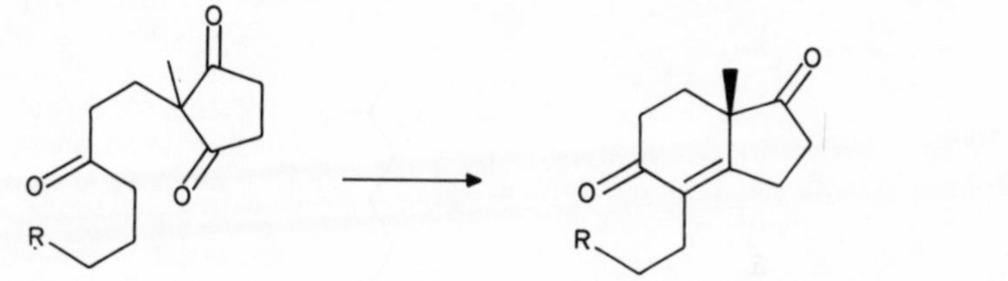

(144) R= 6-Methyl-2-pyridyl   (145) R=6-Methyl-2-pyridyl

(146) R= CH₂=CH   (147) R= CH₂=CH

The structure and stereochemistry of the bicyclic unit present in the vitamin D series and some important metabolites have made them attractive targets for synthesis from hydrindenones derived from **120**. In the synthesis of 1α, 25-dihydroxycholecalciferol reported by Baggiolini and coworkers[39] hydrogenation of **134** served to establish the *trans* hydrindane skeleton and reduction; carboxylate-methyl ketone transformation and elimination gave enone **148** (Scheme 46). Hydrogenation of **148** was followed by Baeyer–Villiger oxidation to give **149**, which was elaborated to the desired Windaus–Grundmann ketone (**80**) using chemistry described by Dauben based on the ene reaction. The Wittig–Horner reaction of **80** with an allylphosphine oxide prepared from (+)-carvone to give 1α, 25-dihydroxycholecalciferol was outlined earlier (Scheme 29).

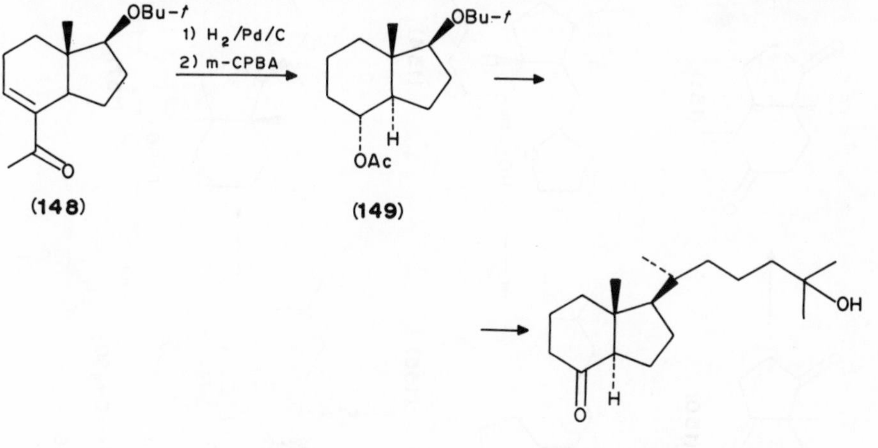

(148)   (149)

(80)

SCHEME 46

A synthesis of vitamin D₃ reported by Fukumoto and coworkers[64] utilized the hydrindenone **150** (the enantiomer of **120**) to establish the required stereochemistry in both the hydrindane skeleton and the side-chain. This synthesis exploits the equatorial nature of the 4-methylpentyl chain in **152**, prepared from **150** via **151**, which ultimately becomes the vitamin D₃ side-chain (Scheme 47). Conversion of **152** into enol acetate **153**

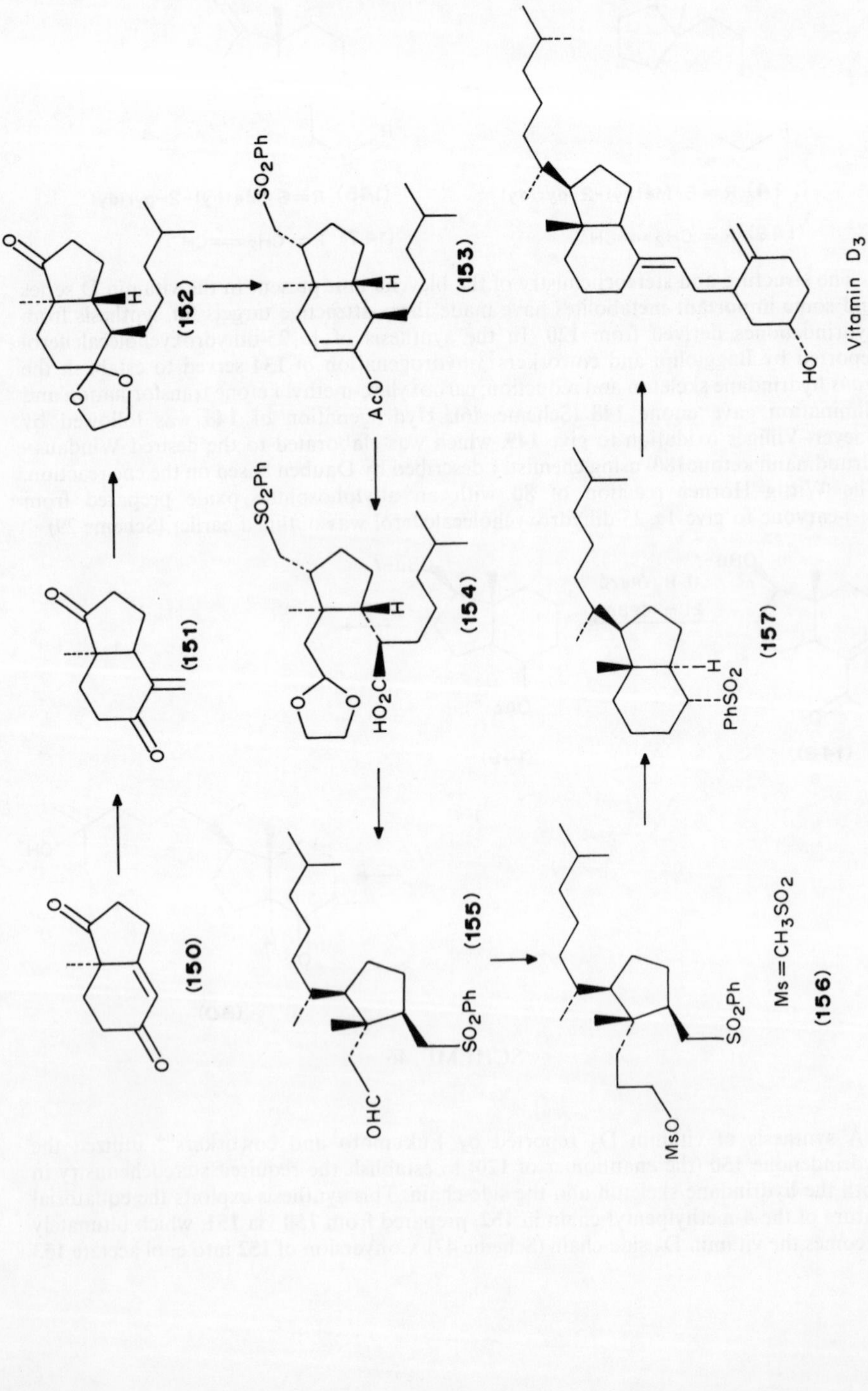

SCHEME 47

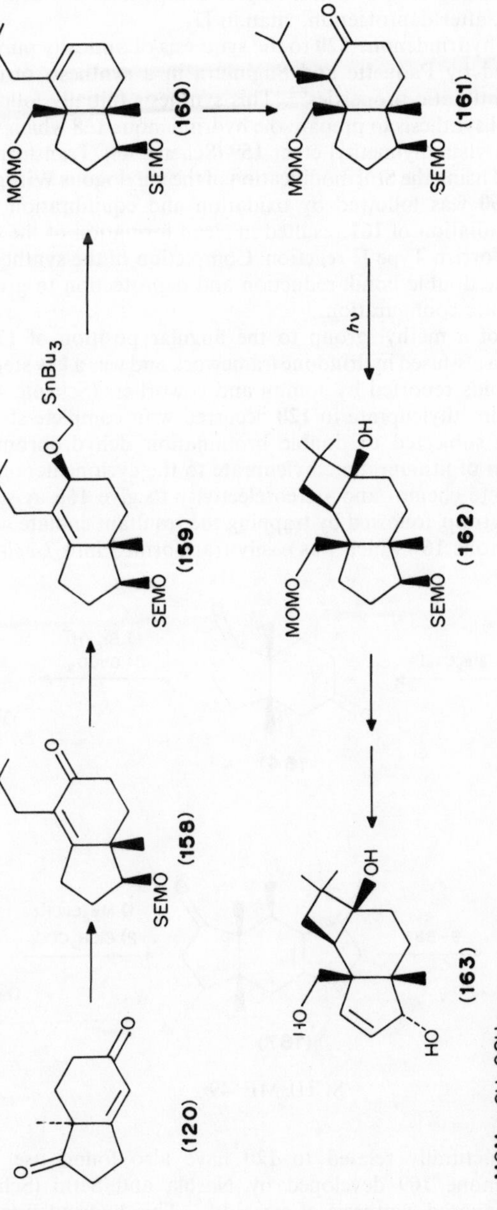

SCHEME 48

MOM = CH_3OCH_2
SEM = $Me_3SiCH_2CH_2OCH_2$

(120)

(158)

(159)

(160)

(161)

(162)

(163)

hν

was accomplished via Wittig reaction, hydrogenation and enol acetylation. Oxidative cleavage of **153** gave, after protection, the ketal acid **154** which was reduced to aldehyde **155**. This aldehyde was elaborated to mesylate **156** which was cyclized to give the hydrindone **157**. Coupling of **157** with a ring A component was achieved using Julia methodology to give, after deprotection, vitamin D$_3$.

The application of hydrindenone **120** to the synthesis of optically pure terpenoids was recently demonstrated by Paquette and Sugimura in a synthesis of (−)-punctatin, a sesquiterpene with antibiotic properties[65]. This synthesis initially follows the protocol established for steroid synthesis to prepare the hydrindenone **158** which was reduced and converted to its tributylstannylmethyl ether **159** (Scheme 48). Transformation of **159** to **160** was accomplished using the Still modification of the vinylogous Wittig rearrangement. Hydroboration of **160** was followed by oxidation and equilibration to give the key intermediate **161**. Irradiation of **161** resulted in clean formation of the cyclobutane **162** as the product of a Norrish Type II reaction. Completion of the synthesis was achieved by introduction of the double bond, reduction and deprotection to give (−)-punctatin (**163**) of known absolute configuration.

The introduction of a methyl group to the angular position of **120** using copper chemistry establishes a *cis*-fused hydrindone framework and was a key step in the synthesis of pinguisane terpenoids reported by Jommi and coworkers (Scheme 49)[66]. Conjugate addition of lithium dimethylcuprate to **120** occurred with complete stereoselectivity to give **164**, which was subjected to double bromination–dehydrobromination to give dienone **165**. Addition of lithium dimethylcuprate to the cyclopentenone moiety of **165** occurred with complete chemo- and stereoselectivity to give **166**. A second conjugate addition of a methyl group, followed by trapping the resultant enolate with chloroacetyl chloride, gave β-furanone **167** which was easily transformed into 7-*epi*-pinguisane.

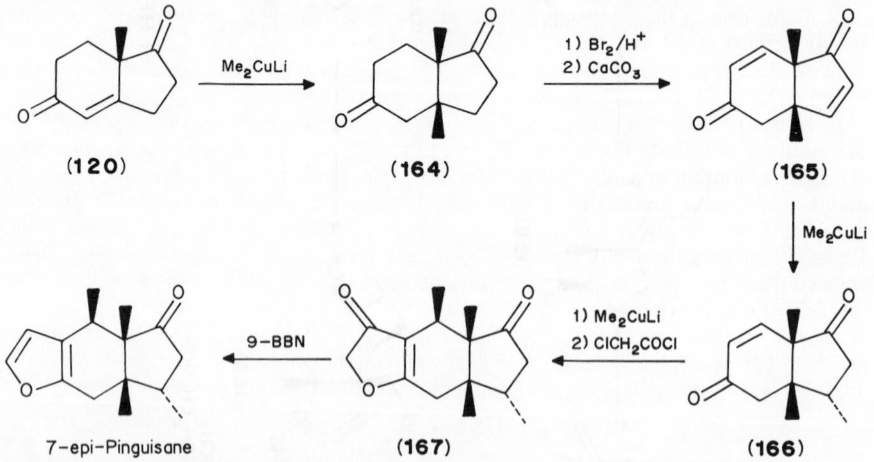

SCHEME 49

Hydrindenones structurally related to **120** have also found use in synthesis as exemplified by the enone **169** developed by Narula and Sethi (Scheme 50) as an intermediate for a proposed synthesis of steroids[67]. This hydrindenone was prepared from the oxime of (−)-π-iodocamphor (**168**) and features an intramolecular S$_N$2 displacement of a neopentylic iodide by an acyl anion equivalent.

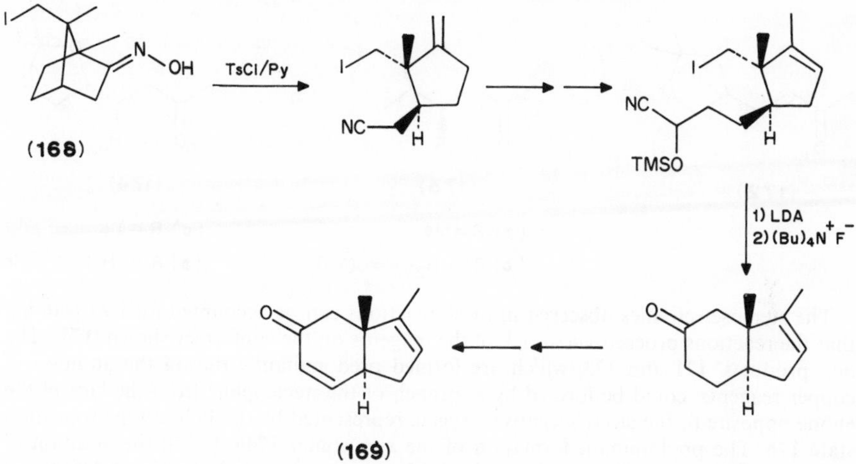

(168)

1) LDA
2) $(Bu)_4N^+F^-$

(169)

SCHEME 50

VI. ACYCLIC ENONES

The diastereoselective addition of organometallic reagents to acyclic α-alkoxycarbonyl compounds is a powerful method in organic synthesis. The vinylogous addition of such reagents to γ-alkoxy-α, β-unsaturated carbonyl systems has received little attention. This is probably due to the relatively flexible nature of such enones, compared to the more rigidly defined cyclic analogues, which makes diastereoface differentiation much more difficult. Recently, however, some examples of conjugate addition to chiral acyclic enones which occur with modest to good selectivity have been reported.

In work directed towards the synthesis of olivin, Roush and Lesur[68] discovered that the addition of lithium divinylcuprate to enone **170** occurred with excellent selectivity (43:1) to give predominantly the *anti* product **171**. Similar results were noted by Cha and Lewis[69], who found that lithium dimethylcuprate added to enone **172** [readily prepared from (R)-glyceraldehyde] to give a 3.8:1 ratio of products **173a** and **174a**. Extensive investigation of enone **172** was carried out by Leonard and Ryan[70], who showed that isopropenylcopper reagents added to **172** to give **173b** in preference to **174b** (8:1). Further investigation by this group revealed a surprising dependence of this conjugate addition upon the counterion. Isopropenyllithium added highly selectively (1:36) to **172** in a 1,4 manner, instead of the expected 1,2 addition, and the direction of addition was opposite to that observed with the corresponding copper reagent.

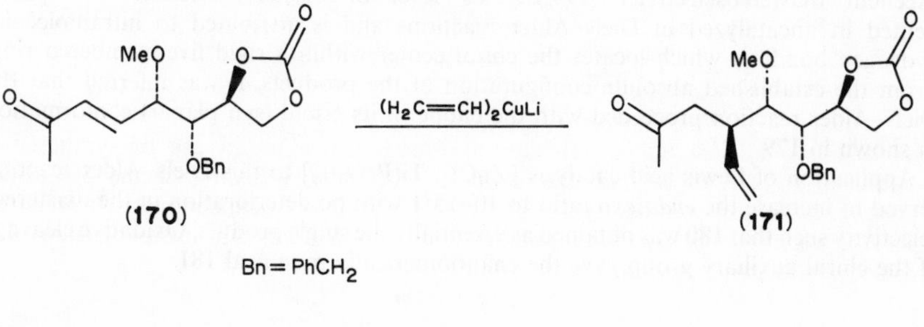

(170) $(H_2C=CH)_2CuLi$ **(171)**

$Bn = PhCH_2$

(172)

(173)

(174)

(a) R = Me

(b) R = H_2C=C(Me)

(a) R = Me

(b) R = H_2C=C(Me)

The stereoselectivities observed in these reactions can be accounted for by assuming that the reactions proceed via attack of the reagents on the conformer shown (175). The *anti* products, 171 and 173, which are formed predominantly during the addition of copper reagents, could be formed by approach of the nucleophile from the face of the enone opposite to the electronegative oxygen, represented by the Felkin-type transition state 176. The predominant formation of the *syn* isomer 174b during the addition of isopropenyllithium to 172 may be explained by assuming chelation assisted delivery of the organometallic reagent to the enone from the same face as the oxygen atom as indicated in 177.

(175)

R—Cu

(176)

R---Li

(177)

VII. ENONES BEARING CHIRAL AUXILIARIES

The use of chiral auxiliaries to effect diastereocontrol in a chemical reaction is an extremely powerful tool in organic synthesis. The Diels–Alder reaction[71], in particular, has proved to be amenable to this process with chiral acrylates, derived from optically pure alcohols, being widely used to prepare optically pure intermediates which are useful in synthesis. With such widespread use of chiral acrylates as partners for the Diels–Alder reaction, the lack of examples of chiral enones in such a process is surprising. This is particularly so in light of the impressive results achieved with the few known chiral enones.

Enone 178, prepared by Masamune and coworkers[72], was found to undergo Diels–Alder reaction with cyclopentadiene with good selectivity (*endo:exo* 8:1), and excellent diastereoselectivity (99%). This level of diastereoselection is unprecedented in uncatalyzed in Diels–Alder reactions and is attributed to intramolecular hydrogen bonding, which locates the chiral center within a rigid five-membered ring. From the established absolute configuration of the products, it was inferred that the Diels–Alder reaction proceeded with the enone in its cisoid (syn planar) conformation as shown in 179.

Application of Lewis acid catalysis [$ZnCl_2$, Ti(PrO-i)$_4$] to this Diels–Alder reaction served to increase the *endo:exo* ratio to 10–15:1 with no deterioration in the diastereoselectivity such that 180 was obtained as essentially the single product. Oxidative cleavage of the chiral auxiliary group gave the enantiomerically pure acid 181.

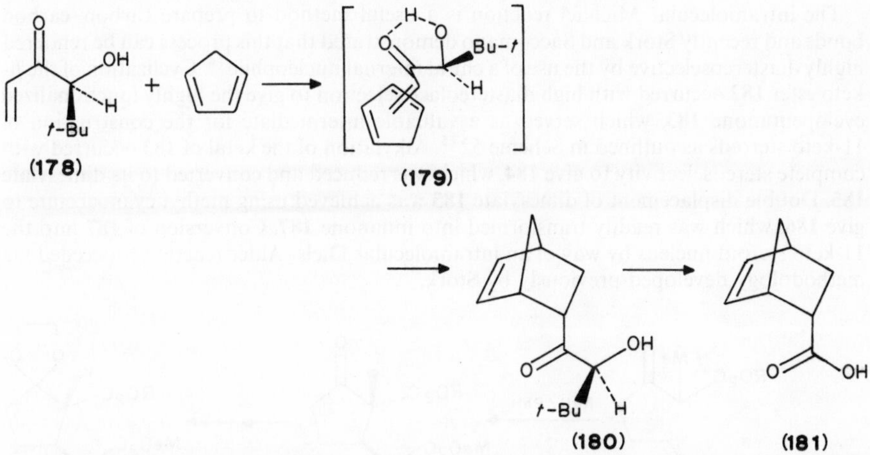

(178)   (179)

(180)   (181)

Masamune and coworkers extrapolated this process to a variety of dienes to give enantiomerically pure intermediates, which were useful for the synthesis of a number of natural products (Scheme 51)[73].

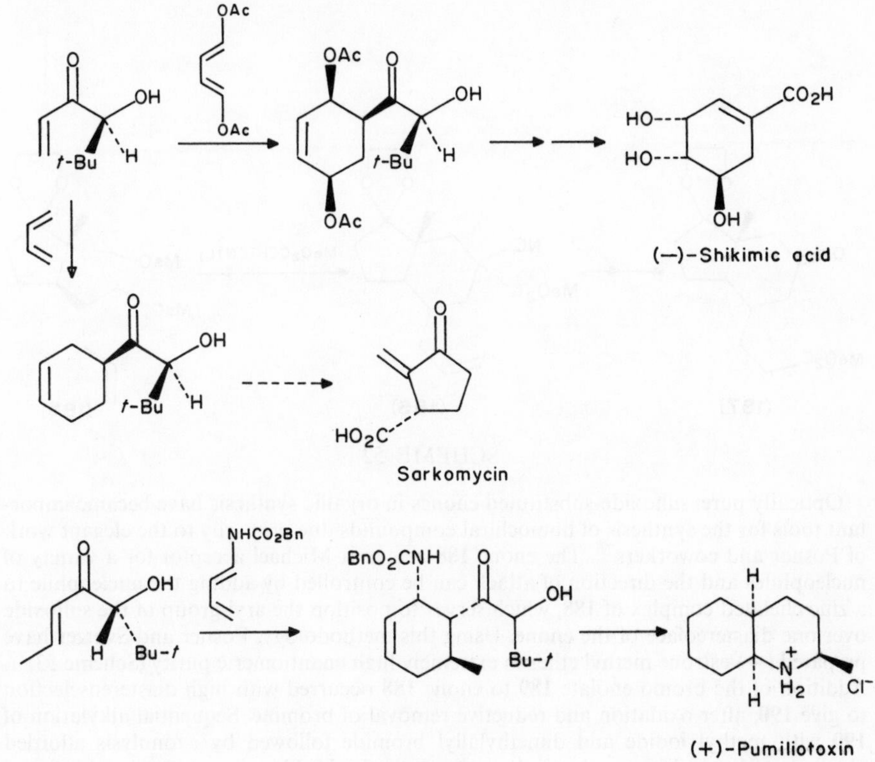

(−)-Shikimic acid

Sarkomycin

(+)-Pumiliotoxin

SCHEME 51

The intramolecular Michael reaction is a useful method to prepare carbon–carbon bonds and recently Stork and Saccomano demonstrated that this process can be rendered highly diastereoselective by the use of a chiral internal nucleophile[74]. Cyclization of the β-keto ester 182 occurred with high diastereoface selection to give the highly functionalized cyclopentanone 183, which served as a valuable intermediate for the construction of 11-keto steroids as outlined in Scheme 52[75]. Alkylation of the ketal of 183 occurred with complete stereoselectivity to give 184, which was reduced and converted to its dimesylate 185. Double displacement of dimesylate 185 was achieved using methyl cyanoacetate to give 186, which was readily transformed into indanone 187. Conversion of 187 into the 11-keto steroid nucleus by way of an intramolecular Diels–Alder reaction proceeded via methodology developed previously by Stork.

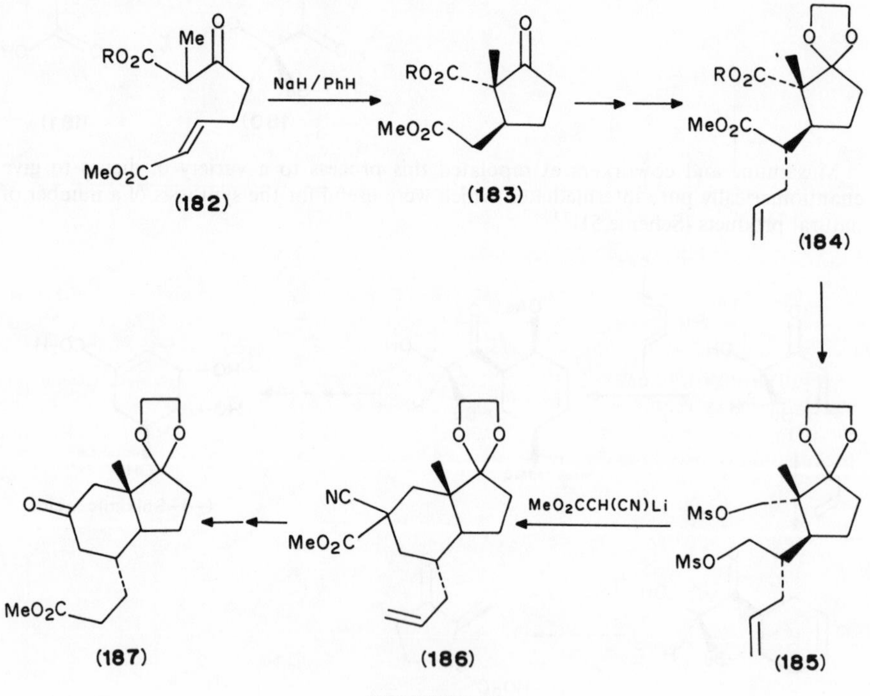

SCHEME 52

Optically pure, sulfoxide-substituted enones in organic synthesis have became impor-tant tools for the synthesis of homochiral compounds due primarily to the elegant work of Posner and coworkers[76]. The enone 188 acts as a Michael acceptor for a variety of nucleophiles and the direction of attack can be controlled by adding the nucleophile to a zinc chelated complex of 188, which serves to position the aryl group of the sulfoxide over one diastereoface of the enone. Using this methodology, Posner and Switzer have prepared (+)-estrone methyl ether in extremely high enantiomeric purity (Scheme 53)[77]. Addition of the bromo enolate 189 to enone 188 occurred with high diastereoselection to give 190, after oxidation and reductive removal of bromine. Sequential alkylation of 190 with methyl iodide and dimethylallyl bromide followed by ozonolysis afforded aldehyde 191, which was reductively cyclized via the McMurry procedure and reduced to give (+)-estrone methyl ether.

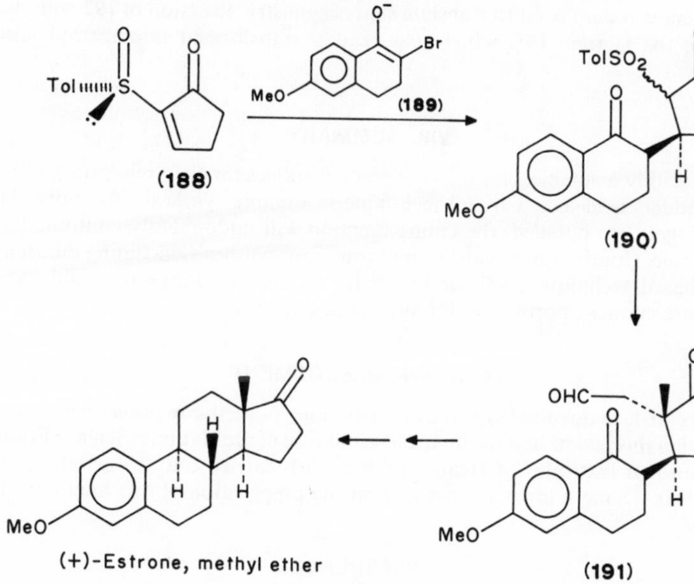

SCHEME 53

Addition of enolate **193** to **192** (the antipode of **188**) served to establish the correct stereochemistry required in **194** for further manipulation into the perfume constituent methyl jasmonate (Scheme 54)[78]. An alternate synthesis of this product from the same precursor was also described by Posner and coworkers in which an additive Pummerer

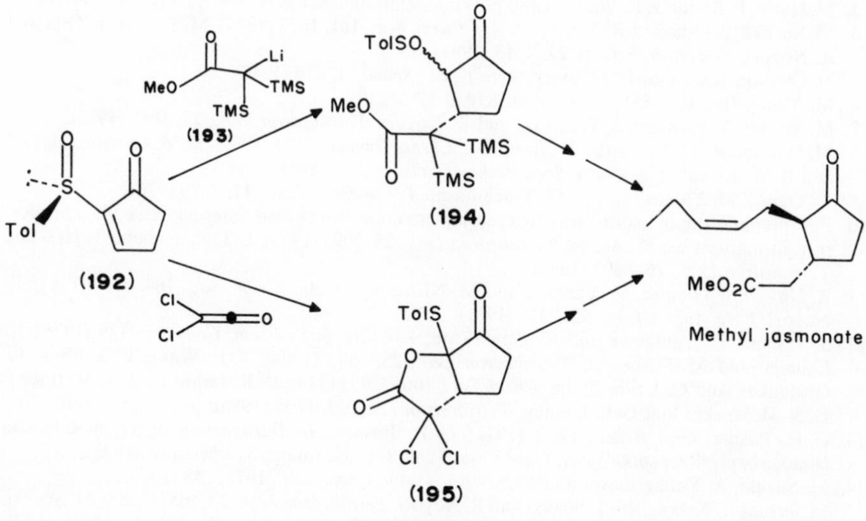

SCHEME 54

rearrangement was employed to translate stereochemistry. Reaction of **192** with dichloro-ketene gave the lactone **195**, which was readily transformed into methyl jasmonate (20% ee)[79].

VIII. SUMMARY

The use of readily available, chiral (non-racemic) enones for the preparation of complex natural products is clearly a useful technique in organic synthesis. As more elaborate synthetic targets are pursued, the enone function will undoubtedly continue to play a prominant role. Continuing advances in asymmetric synthesis, including enzymatic and microbial based techniques, will undoubtedly expand the range of readily available, optically pure enones appropriate for such endeavors.

IX. ACKNOWLEDGEMENT

The authors wish to acknowledge the contributions of their coworkers whose work has been quoted in this review and the financial assistance of the National Science Foundation and the National Institutes of Health for the work carried out in our laboratory. We also thank Ms. Diane Klimas for her help in the preparation of this manuscript.

X. REFERENCES

1. J. S. Bindra and R. Bindra, *Prostaglandin Synthesis*, Academic Press, New York, 1977; A. Mitra, *Synthesis of Prostaglandins*, Wiley–Interscience, New York, 1977; G. A. Garcia, L. A. Maldonado and P. Crabbe, in *Prostaglandin Research*, Chap. 6, Academic Press, New York, 1977; M. P. L. Caton, *Tetrahedron*, **35**, 2705 (1979); K. C. Nicolaou, G. P. Gasic and W. E. Barnette, *Angew. Chem., Int. Ed. Engl.*, **17**, 293 (1978); R. F. Newton and S. M. Roberts, *Tetrahedron*, **36**, 2163 (1980); S. M. Roberts and F. Scheinmann, *New Synthetic Routes to Prostaglandins and Thromboxanes*, Academic Press, New York, 1982; R. F. Newton and S. M. Roberts, *Prostaglandins and Thromboxanes*, Butterworth Scientific, London, 1982.
2. R. Noyori and M. Suzuki, *Angew. Chem., Int. Ed. Engl.*, **23**, 847 (1984).
3. M. Harre, P. Raddatz, R. Walenta and E. Winterfeldt, *Angew. Chem., Int. Ed. Engl.*, **21**, 480 (1982).
4. M. Suzuki, Y. Oda and R. Noyori, *J. Am. Chem. Soc.*, **101**, 1623 (1979); M. Suzuki, Y. Oda and R. Noyori, *Tetrahedron Lett.*, **22**, 4413 (1981).
5. N. Clauson-Kaas and F. Limborg, *Acta Chem. Scand.*, **1**, 619 (1947).
6. M. Minai, Jpn. Pat. 55138505; Japan Kokai 57-62236.
7. M. Suzuki, T. Kawagishi, T. Suzuki and R. Noyori, *Tetrahedron Lett.*, **23**, 4057 (1982).
8. M. Gill and R. W. Rickards, *J. Chem. Soc., Chem. Commun.*, 121 (1979); R. M. Christie, M. Gill and R. W. Rickards, *J. Chem. Soc., Perkin Trans. 1*, 593 (1981).
9. K. Ogura, M. Yamashita and G. Tsuchihashi, *Tetrahedron Lett.*, **17**, 759 (1976).
10. For preparation of chiral 4-hydroxycyclopentenone from *meso* intermediates via chemical transformations see M. Asami, *Tetrahedron Lett.*, **26**, 3099 (1985); L. Duhamel and T. Herman, *Tetrahedron Lett.*, **26**, 5803 (1985).
11. R. Noyori, I. Tomino, M. Yamada and M. Nishizawa, *J. Am. Chem. Soc.*, **106**, 6717 (1984); R. Noyori, *Pure Appl. Chem.*, **53**, 2315 (1981).
12. S. Takano, K. Tanigawa and K. Ogasawara, *J. Chem. Soc., Chem. Commun.*, 189 (1976); K. Laumen and M. Schneider, *Tetrahedron Lett.*, **25**, 5875 (1984); Y. F. Wang, C. S. Chen, G. Girdaukas, and C. J. Sih, *J. Am. Chem. Soc.*, **106**, 3695 (1984); D. R. Deardoff, A. J. Matthews, D. S. McMeekin and C. L. Craney, *Tetrahedron Lett.*, **27**, 1255 (1986).
13. G. H. Posner, *Org. React.*, **19**, 1 (1972); G. H. Posner, *An Introduction to Synthesis Using Organocopper Reagents*, Wiley, New York, 1980; J. F. Normant, *Synthesis*, 63 (1972).
14. M. Suzuki, A. Yanagishawa and R. Noyori, *J. Am. Chem. Soc.*, **107**, 3348 (1985).
15. M. Suzuki, T. Kawagishi, T. Suzuki and R. Noyori, *Tetrahedron Lett.*, **23**, 4057 (1982); M. Suzuki, T. Kawagishi and R. Noyori, *Tetrahedron Lett.*, **23**, 4057 (1982); M. Suzuki, A. Yanagisawa and R. Noyori, *Tetrahedron Lett.*, **25**, 1383 (1984).

16. T. Tanaka, A. Hazato, K. Bannai, N. Okamura, S. Sugiura, K. Manabe, S. Kurozumi, M. Suzuki and R. Noyori, *Tetrahedron Lett.*, **25**, 4947 (1984); T. Tanaka, T. Toru, N. Okamura, A. Hazato, S. Sugiura, K. Manabe, S. Kurozumi, M. Suzuki, T. Kawagishi and R. Noyori, *Tetrahedron Lett.*, **24**, 4103 (1983).

17. T. Tanaka, N. Okamura, K. Bannai, A. Hazato, S. Sugiura, K. Manabe and S. Kurozumi, *Tetrahedron Lett.*, **26**, 5575 (1985).

18. M. Suzuki and R. Noyori, *Tetrahedron Lett.*, **23**, 4817 (1982); S. Sugiura, T. Toru, T. Tanaka, N. Okamura, A. Hazato, K. Bannai, K. Manabe and S. Kurozumi, *Chem. Pharm. Bull.*, **32**, 1248 (1984); T. Tanaka, N. Okamura, K. Bannai, A. Hazato, S. Sugiura, K. Manabe, F. Kamimoto and S. Kurozumi, *Chem. Pharm. Bull.*, **33**, 2359 (1985); A. Hazato, T. Tanaka, K. Watanabe, K. Bannai, T. Toru, N. Okamura, K. Manabe, A. Ohtsu, F. Kamimoto and S. Kurozumi, *Chem. Pharm. Bull.*, **33**, 1815 (1985); J. Nokami, T. Ono, S. Wakabayashi, A. Hazato and S. Kurozumi, *Tetrahedron Lett.*, **26**, 1985 (1985).

19. R., Pappo, P. Collins and C. Jung, *Tetrahedron Lett.*, **14**, 943 (1973).

20. G. Stork and T. Takahashi, *J. Am. Chem. Soc.*, **99**, 1275 (1977).

21. S. Yamada, M. Kitamoto and S. Terashima, *Tetrahedron Lett.*, **17**, 3165 (1976); M. Kitamoto, K. Kameo, S. Terashima and S. Yamada, *Chem. Pharm. Bull.*, **25**, 1273 (1977).

22. (a) S. Kurozumi, T. Toru and S. Ishimoto, *Tetrahedron Lett.*, **14**, 4959 (1973).
    (b) J. B. Heather, R. Sood, P. Price, G. P. Peruzzotti, S. S. Lee, L. F. H. Lee and C. J. Sih, *Tetrahedron Lett.*, **14**, 2313 (1973).
    (c) C. J. Sih, J. B. Heather, G. P. Peruzzotti, P. Price, R. Sood and L. F. H. Lee, *J. Am. Chem. Soc.*, **95**, 1676 (1973).
    (d) C. J. Sih, J. B. Heather, R. Sood, P. Price, G. Peruzzotti, L. F. H. Lee and S. S. Lee, *J. Am. Chem. Soc.*, **97**, 865 (1975).

23. J. G. Miller, W. Kurz, K. G. Untch and G. Stork, *J. Am. Chem. Soc.*, **96**, 6774 (1974).

24. C. R. Johnson and T. D. Penning, *J. Am. Chem. Soc.*, **108**, 5655 (1986).

25. J. R. Medich, K. B. Kunnen and C. R. Johnson, *Tetrahedron Lett.*, **28**, 4131 (1987).

26. M. I. Lim and V. E. Marquez, *Tetrahedron Lett.*, **24**, 5559 (1983); M. I. Lim and V. E. Marquez, *Tetrahedron Lett.*, **26**, 3669 (1985).

27. E. J. Corey, B. C. Pan, D. H. Hua and D. R. Deardoff, *J. Am. Chem. Soc.*, **104**, 6816 (1982); E. J. Corey, D. H. Hua, B. C. Pan and S. P. Seitz, *J. Am. Chem. Soc.*, **104**, 6818 (1982).

28. (a) J. D. White, T. R. Vedananda, K. Kang and S. C. Choudhry, *J. Am. Chem. Soc.*, **108**, 8105 (1986).
    (b) J. D. White, S. Kuo and T. R. Vedananda, *Tetrahedron Lett.*, **28**, 3061 (1987).

29. M. Koreeda and L. Brown, *J. Org. Chem.*, **48**, 2122 (1983).

30. See for example: T. Hudlicky and R. P. Short, *J. Org. Chem.*, **47**, 1522 (1982); D. Solas and J. Wolinsky, *J. Org. Chem.*, **48**, 670 (1983); P. R. Vettel and R. M. Coates, *J. Org. Chem.*, **45**, 5430 (1980).

31. E. J. Corey and H. E. Ensley, *J. Am. Chem. Soc.*, **97**, 6908 (1975).

32. (a) E. J. Corey and R. T. Peterson, *Tetrahedron Lett.*, **26**, 5025 (1985).
    (b) L. A. Paquette, in *Asymmetric Synthesis*, Vol. 3 (Ed. J. D. Morrison), Academic Press, New York, 1984, p. 455.
    (c) D. A. Evans, in *Asymmetric Synthesis*, Vol. 3 (Ed. J. D. Morrison), Academic Press, New York, 1984, p. 94.

33. J. E. Lynch and E. L. Eliel, *J. Am. Chem. Soc.*, **106**, 2943 (1984); E. L. Eliel, in *Asymmetric Synthesis*, Vol. 2 (Ed. J. D. Morrison), Academic Press, New York, 1983, p. 139.

34. M. J. Taschner and A. Shahripour, *J. Am. Chem. Soc.*, **107**, 5570 (1985).

35. J. A. Findlay, D. N. Desai, G. C. Lonergan and P. S. White, *Can. J. Chem.*, **58**, 2827 (1980).

36 F. E. Ziegler, K. J. Hwang, J. F. Kadow, S. I. Klein, U. K. Pati and T. F. Wang, *J. Org. Chem.*, **51**, 4573 (1986).

37. D. Brattesani and C. H. Heathcock, *J. Org. Chem.*, **40**, 2165 (1975).

38. E. J. Corey and H. L. Pearce, *J. Am. Chem. Soc.*, **101**, 5841 (1979).

39. E. G. Baggiolini, J. A. Iacobelli, B. M. Hennesy and M. R. Uskokovic, *J. Am. Chem. Soc.*, **104**, 2945 (1982).

40. F. Kido, T. Abe and A. Yoshikoshi, *J. Chem. Soc., Chem. Commun.*, 590 (1986).

41. S. Torii, T. Inokuchi, and R. Oi, *J. Org. Chem.*, **48**, 1944 (1983).

42. D. D. Maas, M. Blagg and D. F. Wiemer, *J. Org. Chem.*, **49**, 853 (1984).

43. W. C. Still, S. Murata, G. Revial and K. Yoshihara, *J. Am. Chem. Soc.*, **105**, 625 (1983).

44. C. Kuroda, H. Hirota and T. Takahashi, *Chem. Lett.*, 249 (1982).
45. P. A. Wender and D. A. Holt, *J. Am. Chem. Soc.*, **107**, 7771 (1985).
46. M. D. Taylor, G. Minaskanian, K. N. Winzenberg, P. Santone and A. B. Smith, III, *J. Org. Chem.*, **47**, 3960 (1982).
47. M. D. Taylor and A. B. Smith, III, *Tetrahedron Lett.*, **24**, 1867 (1983).
48. A. B. Smith, III, N. J. Liverton, N. J. Hrib, H. Sivaramakrishnan and K. Winzenberg, *J. Am. Chem. Soc.*, **108**, 3040 (1986).
49. A. J. Pearson, H. S. Bansal and Y. S. Lai, *J. Chem. Soc., Chem. Commun.*, 519 (1987).
50. (a) German Offenlegungsschrift (DOS) 2102623, Jan. 21, 1970.
    (b) Z. G. Hajos and D. R. Parrish, *J. Org. Chem.*, **39**, 1615 (1974).
51. (a) German Offenlegungsschrift (DOS) 2014757, March 20, 1970.
    (b) U. Eder, G. Sauer and R. Wiechert, *Angew. Chem.*, **83**, 492 (1971); *Angew. Chem., Int. Ed. Engl.*, **10**, 496 (1971).
52. J. Gutzwiller, P. Buchschacher and A. Furst, *Synthesis*, 167 (1977).
53. D. Taub, in *Total Synthesis of Natural Products*, Vol. 6 (Ed. J. ApSimon), Wiley, New York, 1984, pp. 1–51.
54. T. Kametani, K. Suzuki and H. Nemoto, *J. Chem. Soc., Chem. Commun.*, 1127 (1979); *J. Org. Chem.*, **45**, 2204 (1980).
55. J. Gutzwiller, W. Meier and A. Furst, *Helv. Chim. Acta*, **60**, 2258 (1977).
56. A. B. Smith, III and R. Mewshaw, *J. Org. Chem.*, **49**, 3685 (1984).
57. M. E. Jung and G. L. Hatfield, *Tetrahedron Lett.*, **24**, 3175 (1983).
58. G. Nomine, G. Amiard and V. Torelli, *Bull. Soc. Chim. Fr.*, 3664 (1968); Z. G. Hajos and D. R. Parrish, *J. Org. Chem.*, **38**, 3239 (1973); G. Sauer, U. Eder, G. Haffer, G. Neef and R. Wiechert, *Angew. Chem., Int. Ed. Engl.*, **14**, 417 (1975).
59. R. A. Micheli, Z. G. Hajos, N. Cohen, D. R. Parrish, L. A. Portland, W. Sciamanna, M. A. Scott and P. A. Wehrli, *J. Org. Chem.*, **40**, 675 (1975).
60. N. Cohen, G. L. Banner, W. F. Eichel, D. R. Parrish, G. Saucy, J. M. Cassal, W. Meier and A. Furst, *J. Org. Chem.*, **40**, 681 (1975).
61. U. Eder, H. Gibian, G. Haffer, G. Neef, G. Sauer and R. Wiechert, *Chem. Ber.*, **109**, 2948 (1976); U. Eder, *J. Steroid Biochem.*, **11**, 55 (1979).
62. S. Danishefsky and P. Cain, *J. Org. Chem.*, **39**, 2925 (1974); *J. Am. Chem. Soc.*, **97**, 5282 (1975); *J. Am. Chem. Soc.*, **98**, 4975 (1976).
63. I. Shimizu, Y. Naito and J. Tsuji, *Tetrahedron Lett.*, **21**, 487 (1980).
64. H. Nemoto, H. Kurobe, K. Fukumoto and T. Kametani, *J. Org. Chem.*, **51**, 5311 (1986).
65. L. A. Paquette and T. Sugimura, *J. Am. Chem. Soc.*, **108**, 3841 (1986).
66. S. Bernasconi, M. Ferrari, P. Gariboldi, G. Jommi, M. Sisti and R. Destro, *J. Chem. Soc., Perkin Trans. 1*, 1994 (1981).
67. A. S. Narula and S. P. Sethi, *Tetrahedron Lett.*, **25**, 685 (1984).
68. W. R. Roush and B. M. Lesur, *Tetrahedron Lett.*, **24**, 2231 (1983).
69. J. K. Cha and S. C. Lewis, *Tetrahedron Lett.*, **25**, 5263 (1984).
70. J. Leonard and G. Ryan, *Tetrahedron Lett.*, **28**, 2525 (1987).
71. W. Oppolzer, *Angew. Chem., Int. Ed. Engl.*, **23**, 876 (1984).
72. W. Choy, L. A. Reed, III and S. Masamune, *J. Org. Chem.*, **48**, 1139 (1983).
73. S. Masamune, L. A. Reed, III, J. T. Davis and W. Choy, *J. Org. Chem.*, **48**, 4441 (1983).
74. G. Stork and N. A. Saccomano, *Nouv. J. Chim.*, **10**, 677 (1986).
75. G. Stork and N. A. Saccomano, *Tetrahedron Lett.*, **28**, 2087 (1987).
76. G. H. Posner, *Acc. Chem. Res.*, **20**, 72 (1987). G. H. Posner, in *Asymmetric Synthesis*, Vol. 2 (Ed. J. D. Morrison), Academic Press, New York, 1983, p. 225.
77. G. H. Posner and C. Switzer, *J. Am. Chem. Soc.*, **108**, 1239 (1986).
78. G. H. Posner and E. Asirvatham, *J. Org. Chem.*, **50**, 2589 (1985).
79. G. H. Posner, E. Asirvatham and S. Ali, *J. Chem. Soc., Chem. Commun.*, 542 (1985).

The Chemistry of Enones
Edited by S. Patai and Z. Rappoport
© 1989 John Wiley & Sons Ltd

CHAPTER **22**

Dimerization and polymerization of enones in the fluid and solid states

CHARIS R. THEOCHARIS .

Department of Chemistry, Brunel University, Uxbridge, Middlesex UB8 3PH, UK

I. INTRODUCTION[1]

The commonest polymerizable enone is methyl vinyl ketone (MeCOCH=CH$_2$, **1**) whose polymers have been known for a considerable time. Although the literature in this field is very extensive, recent reviews are not very abundant. Methyl vinyl ketone (MVK, 3-buten-2-one), as well as its polymers or copolymers, and those of its analogues, have been shown to be useful in a variety of practical applications. MVK has also found considerable use in graft polymerization. MVK itself will polymerize spontaneously and, in addition, a variety of catalysts have been used to initiate its polymerization, as well as photochemical means. Radical, anionic and cationic copolymerization is possible. Owing to the applicability of these polymers, many of the recent publications are in a patent form. In many of its applications, MVK appears to be used as a substitute of methyl methacrylate, which is considered, however, to be outside the scope of this review.

A large part of this chapter concentrates on the behaviour of enones upon irradiation with near-UV light, in the crystalline state. These reactions present some interesting features, and are of considerable fundamental and technological interest. A peculiarity of the solid-state reactivity of enones is that the predominant reaction is photochemically induced dimerization, although polymerization does occur and can be of a fairly complex nature. Other reactions such as hydride abstraction, decarboxylation and dehydration are also known, although they are outside the scope of this review. Enones have also been shown to be reversibly photochromic in the solid state.

The chapter has been divided into the following sections. First, a discussion of solid-state reactivity in general, second, sections on the solid-state reactivity of enones, covering both dimerization and polymerization, and finally, sections on fluid-state polymerization, dealing with homopolymerization and copolymerization, separately.

II. TOPOCHEMICAL REACTIONS

'Ein Kristal ist ein chemischer Friedhof'[2]; such was the widely held opinion among chemists during the first two thirds of this century. It was the development of chemical crystallography, and especially of direct methods, that made the systematic study of organic solid-state reactions possible. Several such reactions had been observed and reported, but no interpretation of the mechanism was attempted, or was possible at the time. Organic solid-state chemists consider, justly, that their subject was given birth in 1964 by G. M. J. Schmidt at the Weizmann Institute of Israel[3]. Schmidt, M. D. Cohen and their collaborators studied the solid-state photochemical behaviour of *trans* and *cis* cinnamic acids[4], which was first described by Liebermann in 1889[5]. Part of the reason why the Weizmann group succeeded in providing a consistent and logical interpretation for the behaviour of these solids where others had previously failed, lay in their recognition of the likely role of the crystal structure in the control of the reaction, and the availability of the so-called direct methods of structure solution, which do not require the presence of a 'heavy' atom in the molecule for success.

The Weizmann group observed that different crystalline modifications of *trans* cinnamic acid, obtained by changing the recrystallization solvents, behaved differently towards exposure to near-ultraviolet light (sunlight). They also studied a number of substituted *trans* cinnamic acids[6,7]. It was observed that some of these crystals were unchanged when under prolonged exposure to radiation, whereas others reacted to yield dimers, via opening of the exocyclic double bonds (see Scheme 1). It was found that any crystalline modification, irrespective of substituents, could be classified according to the length of the shortest unit cell axis into three classes, namely α, β and γ. For any substituent, or combination of them, an α crystal would always yield a dimer whose cyclobutane ring had a centre of symmetry, a β crystal would always yield a dimer with a mirror plane, and a

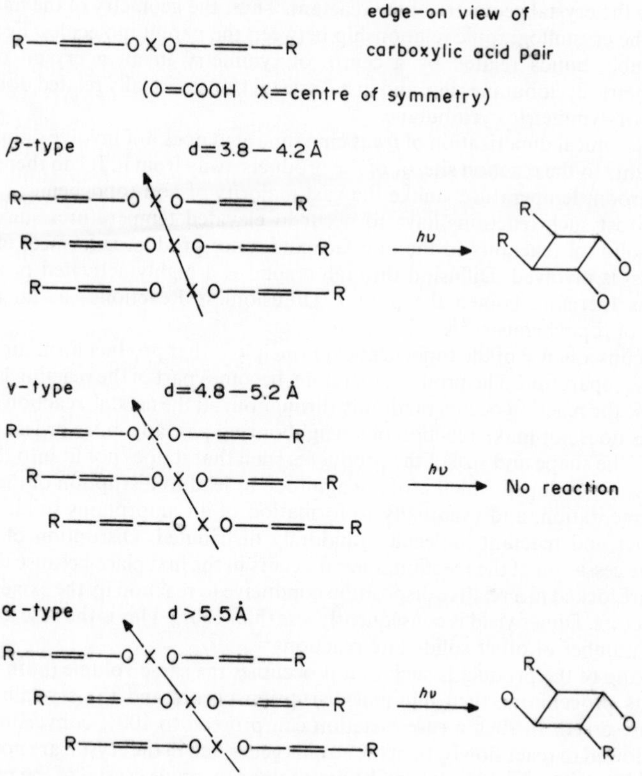

SCHEME 1

γ crystal would always be photostable. No *cis–trans* isomerization, which was known to occur in solution, was ever observed in crystalline *trans* cinnamic acids.

Schmidt and his coworkers determined the X-ray structures of some of these crystals (see for example Reference 8). They found that in an α crystal reactive double bonds were antiparallel, and in a β crystal parallel, to each other. In both cases the centre-to-centre distance of the bonds was between 3.6 and 4.2 Å. In γ crystals, the bond-to-bond distance was in excess of 4.6 Å. Changing the substituent pattern would clearly change the shape of the molecule and its intermolecular interactions, and thus change the crystal structure and hence photochemical behaviour. The ruse usually employed, before the advent of direct methods of introducing a heavy atom (e.g. Br) into the molecule in order to solve its crystal structure, could not be used here. This is because the introduction of an additional atom would change the crystal structure.

The observations on the photochemistry of *trans* cinnamic acid gave rise to the so-called Topochemical Principle, that reactions in the solid state occur with minimum molecular or atomic movement (reviews of this subject include References 4 and 9–20; the list includes mostly articles since 1980 and is not complete). The topochemical principle presupposes that no melting takes place, and no fluid acts as intermediate. The consequences of the principle are far reaching: reaction will only take place if the reactants are in the correct distance and geometry to do so. The nature of the product, if any, is governed by the crystal geometry of the reactant. Thus, the geometry of the final product will reflect the crystallographic relationship between the parent molecules. For example, reactive double bonds related by a centre of symmetry in an α crystal result in a centrosymmetric cyclobutane ring, and in a β crystal translationally related double bonds yield a mirror-symmetric cyclobutane.

The topochemical dimerization of *trans* cinnamic acid does not involve diffusion either of the reactants to the reaction site, or of the products away from it. It can therefore occur at, or near, room temperature, unlike the vast majority of non-topochemical solid-state reactions. Most such reactions have to occur at elevated temperatures since diffusion through a solid (of reactants to the interface and, once product is formed, of reactants through this) is involved. Diffusion through a solid is a highly activated process. J. M. Thomas has therefore coined the phrase 'Diffusionless Reactions', as an alternative description of topochemistry[21].

Another consequence of the topochemical principle is that product formation does not lead to phase separation. The product, therefore, becomes part of the reacting lattice. This is so, because the reaction occurs randomly throughout all the crystal; reaction of one pair of molecules does not make reaction of a neighbouring pair any more or less likely than any other. If the shape and size of the product is such that it does not fit into the reacting lattice, stress is developed which is of a magnitude to lead to disruption of the structure, crystal fragmentation, and eventually to formation of an amorphous solid, containing both product and reactant molecules randomly distributed. Disruption of the lattice results in the cessation of the reaction, since it occurs in the first place because the reactant molecules are locked in a relative disposition conducive to reaction by the exigencies of the crystal structure. Dimer yield is consequently less than 100%. This is the case for cinnamic acid and a number of other solid-state reactions[22].

If the nature of the product is such that it occupies the same volume (both in size and shape) as its progenitors, then minimal disruption occurs and the crystallinity of the system is preserved. In such a case, reaction can proceed to 100% conversion[23]. If this system is allowed to react slowly, then the strains generated in the crystal are not large, and mechanical integrity of the crystal can be maintained. A single crystal of the reactant will therefore yield a single crystal of the product, and crystallography cannot only be used to study the structures of product and reactant and draw conclusions, but can also be used to study the path of the reaction. A whole chemical experiment can therefore be carried out in a single crystal. Examples of this were the reactions of 2-benzyl-5-benzylidenecyclo-pentanone (**2**), and its analogues, to be described in subsequent sections.

Since 1964, a number of systems in addition to the cinnamic acids have been found to be reactive in the solid state, including some which undergo polymerization. Apart from their academic interest, there is potential applicability of such systems in areas ranging from synthesis of chiral or regiospecific polymers to molecular and optoelectronics. Total asymmetric synthesis from achiral precursors has been shown to be possible through solid-state reactions. Probably the most striking use of solid-state reactions is in the preparation of large crystals of regular polymers. In this chapter, a wide interpretation of polymerization will be adopted to include dimerization reactions. Many of the molecules exhibiting solid-state topochemical activity contain conjugated double bonds, often as enones. Their reactivity pattern is very similar to that of the *trans* cinnamic acids.

III. SOLID-STATE CYCLOADDITIONS OF BENZYLIDENE CYCLOPENTANONES

A. Reactions of Benzyl Benzylidene Cyclopentanones

The solid-state reactivity of enones can be exemplified by the behaviour of 2-benzyl-5-benzylidenecyclopentanones[23,24,16] (BBCP, **2**). Crystalline **2** (see Scheme 2) assumes a

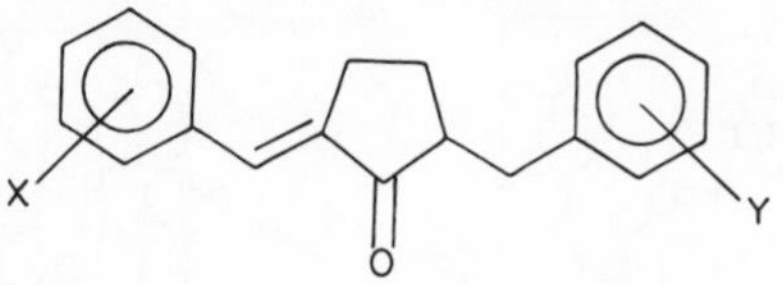

	X	Y
(2)	H	H
(3)	p-Br	H
(4)	H	p-Cl
(5)	H	p-Me
(6)	H	p-Br
(7)	p-Br	p-Cl
(8)	p-Br	p-Me
(9)	m-Br	H
(10)	p-Cl	H
(11)	p-Me	H
(12)	o-Cl	H

SCHEME 2

packing motif in which neighbouring molecules are related by a centre of symmetry, and are situated such that their exocyclic double bonds are antiparallel and separated by 4.1 Å (Figure 1). Further, this packing is conducive to single-crystal to single-crystal reactivity under topochemical control. This is so, because the product dimeric molecule occupies the same volume (Figure 2) and is roughly of the same shape as its two progenitors[25]. The dimer molecule can fit into the monomer lattice, thanks to the presence of the benzyl group. This is a bulky side-group which can change its conformation and relative orientation vis a vis the reacting part of the structure, i.e. the exocyclic double bond. As a result, the position of the benzyl phenyl ring remains unchanged in the dimer and compensates for the movement of other parts of the molecule (Figure 2). The strain produced within the lattice is minimal, and the mechanical integrity of the crystal is maintained throughout the reaction. In crystallographic terms, the change in volume and cell dimensions in going from monomer to dimer is very small. Single-crystal to single-crystal behaviour[26] is shown by molecules 2 to 6. The fact that the product is crystalline means that there is a definite crystallographic relationship between parent and daughter phases. This is the definition of 'topotactic' process. Clearly, in cases such as that for *trans* cinnamic acid where the product is amorphous, no topotactic relationship is possible.

For all compounds described in this section, detection of reactivity was carried out using infrared spectroscopy. Reaction involves the conversion of a C—C double bond to a single bond, which can be observed with the collapse of a peak at $1640 \, cm^{-1}$, which is characteristic of the former group[16].

The mechanism of the reaction is believed to be as follows[27]: on absorption of a photon of light ($\lambda > 360$ nm) one of the molecules in a closest-neighbour pair undergoes an $n \to \pi^*$ transition to an excited singlet state. This crosses over quickly to a vibrationally excited triplet state. This is the species which now reacts with a neighbouring ground-state molecule. The excited triplet state has a conformation which is similar to that of the monomeric residue in the dimer. As a result, the reacting atoms on the two molecules are now closer than the 4.1 Å separating them before photoexcitation. The product results

1138

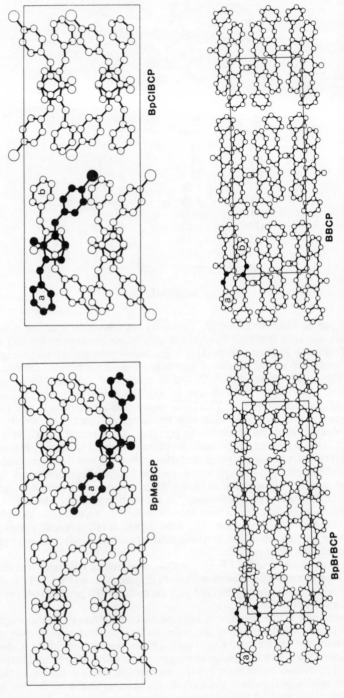

FIGURE 1. Packing diagrams for BpMeBCP (11), BpClBCP (10), BpBrBCP (3) and BBCP (1). Nearest neighbouring molecules in each structure are labeled as a and b.

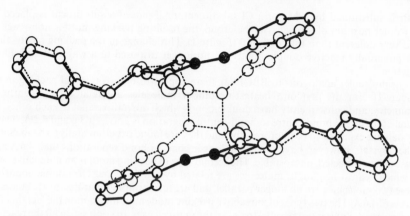

FIGURE 2. Incipient dimer pair and dimer molecule for **2**. The filled circles represent the reacting bonds for the monomer

from the opening-up of the two exocyclic double bonds to yield a centrosymmetric cyclobutane ring (Scheme 3). The source of excitation is usually either a low-pressure Hg lamp (100 or 500 W) with a pyrex filter to exclude low-wavelength radiation, or direct sunlight. The reason for filtering out the radiation with wavelengths shorter than 360 nm is that such UV radiation may cause the cleavage of single C—C bonds.

Substitution of a bromine atom at the *para* position of the benzylidene group (**3**) results in a packing motif which is very similar to that of the parent molecule (Figure 1). The difference in volume of the two cells is almost entirely due to the elongation of the *a* axis, which is necessary in order to incorporate the additional atom[28-30]: In this reactive motif, the long molecular axis is parallel to the long cell axis (i.e. *a*). However,

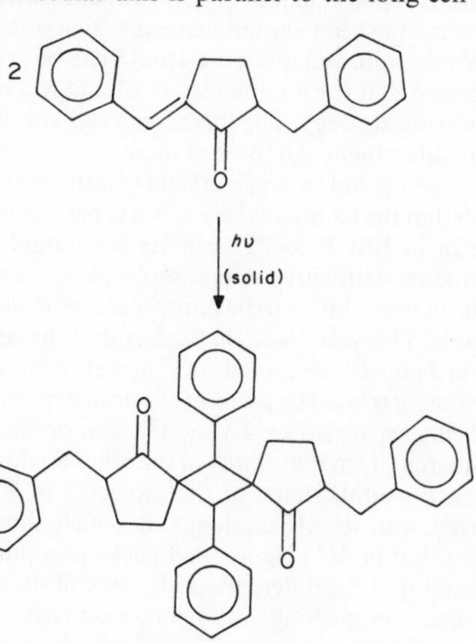

SCHEME 3

if Br is substituted by a Me or a Cl substituent, or if indeed a substituent is placed in any other position of the benzylidene group, the resulting packing motif is non-reactive and very different than that of 2 or 3 (Figure 1). This change in the packing means that the potentially reactive double bonds are no longer disposed in a way that will allow reaction[16,18].

For molecules analogous to BBCP (2), there appear to be two types of packing motif (Figure 1). For the first one, nearest-neighbour molecules are related by a centre of symmetry, and consequently have double bonds which are antiparallel to each other. In some of these, the bond-centre to bond-centre separation is between 3.6 and 4.2 Å. These crystals are expected to be photoreactive and they are found experimentally to be so. Some other crystals, however, have somewhat longer bond-to-bond separations than 4.6 Å and these are, as expected, photostable. The second type of packing motif is an unreactive one, in which nearest-neighbour molecules are related by a glide plane. The double bonds on these two molecules are no longer parallel, and are furthermore separated by distances in excess of 4.6 Å. The two types of unreactive packing mode have in common the fact that the long molecular axis is not parallel to any of the unit-cell axes. In contrast, in all the reactive crystals, the molecular axis and the longest cell axis are parallel. The relative disposition of the nearest-neighbour molecule pair in the reactive as compared with the unreactive packing mode is shown in Figure 3.

Examples of the first type of packing are molecules 2 and 3. We have seen already that a single crystal of 2 can be converted to a single crystal of dimer[25]; the same is true for 3, although the cell changes here are more significant[28]. To achieve a single-crystal to single-crystal change for the irradiation of 3, a slower reaction is needed, which can be achieved by a lower UV dosage. Under these conditions, the mechanical integrity of the crystal can be preserved. Substitution in the benzyl group is also possible. Thus, substitution of Cl[31], Me[16] or Br[32] at the *para* position will yield compounds 4, 5 and 6 respectively, which are isomorphous (i.e. have similar packing modes). The first two have unit cells with roughly equal volumes. This is in agreement with the proposition of Kitaigorodskii[34] that the packing of organic molecules in crystals can be understood as the close packing of spheres of various radii. Therefore, replacement of one substituent with another of similar van der Waals volume at the same position should leave the structure unchanged[33]. Cl and Me have similar van der Waals radii, and therefore 4 and 5 can be expected to be isomorphous. Kitaigorodskii[34] suggested that the interchange of Cl and Me substituents can be used as part of a crystal engineering strategy. The increase in cell volume for the third (6) reflects the larger size of the Br substituent. All three of these compounds pack in a photoreactive motif, and undergo a single-crystal to single-crystal transformation. These structures differ from those of 2 and 3 in that the length of the *a* axis has been halved, because of a change in spacegroup, from PbCa in BBCP to P21/c here, presumably in order to avoid short contacts of the substituents with surrounding molecules.

Since 3 and 4 have both been shown to be photoreactive, it was anticipated that 7 would also be photodimerizable. This was based on the fact that the substitution pattern in 7 is a combination of those in 3 and 4. The crystals of 7, however, were photostable[35], in spite of the nearest neighbours being related by a centre of symmetry. Stability is believed to be due to the double bonds being separated by 4.65 Å. The conformation of the benzyl group in this molecule is very different from that of any of the other analogues of 2. This difference in conformation, and hence overall shape of the molecule of 7 (Figure 4), is made more striking when compared with its Me analogue, 8, which shows a conformation of the benzyl group similar to that in 2[34] (Figure 4). 8 packs in a photoreactive crystal similar to that for 4. It is believed that the difference in the overall shape of the molecules is what gives rise to the differences in packing, and hence reactivity.

The conformational differences between 7 and 8 may be due to the electron-donating nature of the *p*-Me group as distinct from the electron-withdrawing ability of *p*-Cl[16,35].

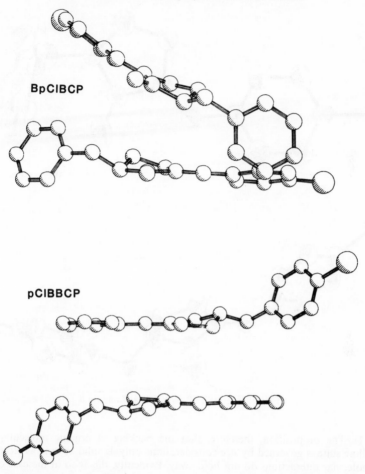

BpCIBCP

pCIBBCP

FIGURE 3. Nearest-neighbour molecules for a reactive crystal **4** (pClBBCP) and a photostable one **10** (BpClBCP)

Arguments solely based on the size of the substituent are insufficient to explain the packing adopted. The presence of the *p*-Br substituent on the benzylidene moiety means that any surplus or deficiency of charge on the carbon atoms of the benzyl group of neighbouring molecules will contribute significantly to the electrostatic interaction involving the *p*-Br atom of the benzylidene group. Me and Cl substituents differ electrostatically in the sense that the former would provide surplus charge to, and the latter extracts charge from, the carbon atoms of the benzyl group to which they are attached. The interaction between these benzyl carbon atoms and the surrounding Br atoms will differ depending on whether the substituent is chloro or methyl. This is reflected in the fact that the shortest contacts between the benzyl phenyl ring carbon atoms and bromine are considerably shorter in **7** than in **8**, as the electron-withdrawing ability of the Cl substituent in **7** will allow the bromine on a neighbouring molecule to form closer contacts (see

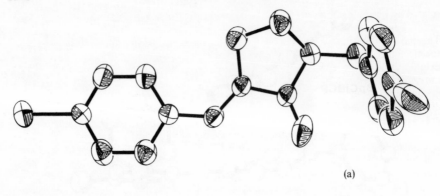

(a)

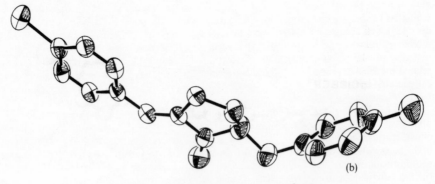

(b)

FIGURE 4. ORTEP plots for (a) **7** and (b) **8**

Table 1). The proposition, therefore, that the packing of organic molecules in the crystalline state is governed by size considerations only, is valid as long as electrostatic intermolecular interactions do not hold sway. Evidently, this is so in the case of **4** and **5**, but not **7** or **8**, where the presence of the polarizable Br substituent makes the electrostatic interactions dominant.

The centrosymmetric unreactive packing motif is represented by the structures of molecules **7** and **9**[18]. The crystal structure of **9** has two molecules in the asymmetric unit

TABLE 1. Shortest Br···C distances in the crystal structures of **7** and **8**[a]

	in molecule	
Br(1) to	**7** (Å)	**8** (Å)
C(7)	3.939	4.695
C(8)	3.868	4.244
C(9)	3.664	5.104
C(10)	3.544	6.228
C(11)	3.634	6.908
C(12)	3.824	5.868

related by a pseudo-centre of symmetry. The nearest double-bond to double-bond contact occurs between molecules related by a crystallographic centre of symmetry, and is 4.36 Å. This may be too long for reaction, but photostability may also be due to steric hindrance: the bromine atoms on each of the reacting benzylidene groups are relatively close to neighbouring benzyl groups; the two groups may clash with each other during the movement necessary for reaction, and thus cause photostability. The limit of 4.2 Å, normally accepted as the longest bond-to-bond distance conductive to [2 + 2] cycloaddition, is not an absolute limit but is based on experimental results. There appears to be a grey area between 4.2 and 4.7 Å, where in some crystals reaction occurs whereas in others it does not. The answer may lie either in the presence of steric hindrance to reaction in some structures and not others, or more detailed geometric requirements for topochemical control than that suggested by the formulation by Schmidt. This aspect of topochemistry will be discussed in later sections.

The packing motif where nearest neighbours are related by a glide plane is represented by 10, 11 and 12[16]. In the crystals of 10, the nearest-neighbour molecules are related by a b glide (Figure 1), with a double-bond centre to double-bond centre separation of 5.40 Å, whilst for 12 they are related by a c glide, with a bond separation of 4.61 Å. In both cases, the double bonds are not parallel and no reaction occurs. The difference between these two molecules lies in the position of the Cl substituent in the benzylidene moiety, and they pack in a very similar packing motif. The closest Cl $\cdots$ Cl distance for 12 was 4.61 Å, for c glide related molecules, whereas for 10 the equivalent distance was 5.00 Å, for centrosymmetrically related molecules. Cl $\cdots$ Cl contacts are believed to have considerable influence on the mode of packing of aromatic compounds.

Crystals of 10 are isomorphous with those of 11 (Figure 1), indicating the interchangeability of chloro and methyl substituents in alkyl or aryl moieties, where volume considerations hold sway[16,18]. The reason why substitution at the *ortho* or *meta* position of the benzylidene group of 2 by Cl, Me or Br groups and by Cl and Me at the *para* position should result in a photostable packing mode, can be explained as follows. Substitution in the flat benzylidene moiety increases its effective size, compared with that it possesses in the unsubstituted 2. This change has to be accommodated either by changing the molecular conformation, or by assuming a different packing. Given the rigidity of the benzylidene group, the only possibility available is the latter. This change of packing is also necessary in order to accommodate the non-bonded interactions in which the substituents take part. In the case of 3, however, size considerations are presumably superseded by the tendency of the bromo substituent, which is a large polarizable atom, to partake in a large number of non-bonded short H $\cdots$ Br contacts, whose number is maximized by retaining the photoreactive motif[33,35]. If, however, substitution is carried out in the benzyl moiety of 2, then a reactive structure is retained, but with a change in spacegroup. Compounds 4, 5 and 6 can retain the same motif as 2 and 3, because the increase in volume of the flexible benzyl moiety can be accommodated by a change in conformation.

In Section II, it was mentioned that the photochemical behaviour of *trans* cinnamic acid in the crystalline state can act as a good guide to the behaviour of enones. It has been seen that the acids can take up three distinct types of packing, namely α, β and γ. Enone 2 and its analogues can assume α- and γ-like packing, but not β. As a consequence, all dimers have a centrosymmetric cyclobutane ring, such as the dimer of 4 shown in Figure 5. It is believed that a molecule such as 2 cannot adopt a β-like structure. This is essentially a non-planar molecule, as the benzyl group always subtends a non-zero dihedral angle with the flat benzylidene moiety, for any substituent. Trying to stack such molecules parallel to each other rather than antiparallel or crossed would lead to a highly open, inefficient, and therefore unlikely crystal structure. Crossing of the molecular axes, such as occurs in some unreactive crystals, leads to non-reactivity, and a type of packing where closest neighbours are related by glide planes, a case not encountered in the cinnamic acid series.

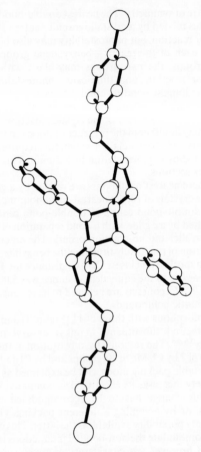

FIGURE 5. Dimer molecule of **4**

B. The Unusual Case of 2,5-Dibenzylidenecyclopentanone

2-Benzyl-5-benzylidenecyclopentanone (**2**) and 2, 5-dibenzylidenecyclopentanone (**13**, in Scheme 4), commonly abbreviated to DBCP, are closely related. The main difference between **2** and **13** is that the C(2)—C(6) single bond (see Figure 6) in **2** has been replaced by a double bond in **13**. DBCP is of interest for several reasons, all consequences of this change. First, the introduction of a second double bond creates additional, potentially reactive centres. Second, comparison of the solid-state photochemical behaviour of **2** and **13** can lead to an understanding of the consequences of rendering the monomer essentially planar and of imparting rigidity to the benzylbenzylidenecyclopentanone backbone. Third, the chiral centre at position 2 (the numbering scheme in Figure 6 is for both **2** and **13**) in the BBCP framework causes all members of this family to crystallize in racemic spacegroups. Racemic spacegroups are those capable of packing molecules of either handedness because they contain mirror planes, or centres of symmetry. The DBCP framework, however, does not contain any chiral centres and can therefore be expected to

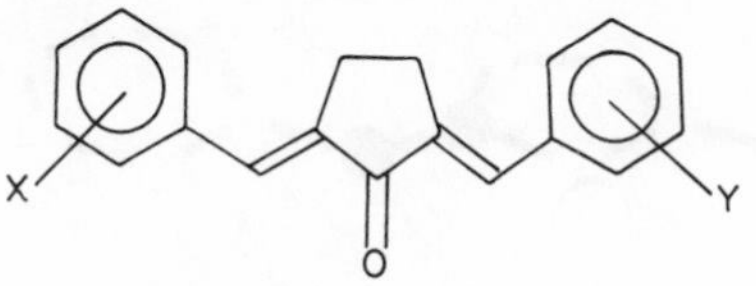

	X	Y
(13)	H	H
(18)	p-Br	p-Br
(19)	p-Br	H
(20)	p-Me	H
(21)	H	p-NO$_2$
(22)	H	p-pyridyl
(25)	p-F	p-F

SCHEME 4

adopt packing arrangements conducive to topochemical reaction which cannot be adopted by either **2** or the cinnamic acids (cinnamic acids also adopt racemic spacegroups because they pack forming hydrogen-bonded centrosymmetric pairs of their carboxylic groups, as shown in Scheme 1). Finally, a racemic mixture of **2** cannot be resolved into optically pure fractions, because the C(2) hydrogen is acidic, and spontaneous racemization occurs in solution via a keto–enol tautomerism mechanism. A new chiral centre can be created at position 3 of the DBCP framework, which is not labile. An additional difference between **2** and **13** is that the former has a low molecular symmetry, whereas the latter can have either a mirror plane or a two-fold axis through its carbonyl, depending upon the substituents present. This symmetry can of course be destroyed by introducing a substituent on only one of the two phenyl groups, or by introducing different substituents.

13 itself has a two fold symmetry and packs in spacegroup C222$_1$, which is a chiral[36,37]. Irradiation with UV light of single crystals of **13** recrystallized from chloroform/methanol

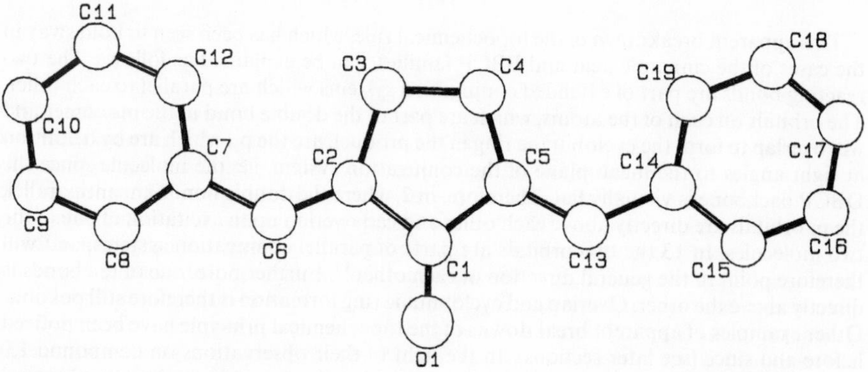

FIGURE 6. Numbering scheme for **13**. Note that the numbering scheme is the same in **2**, but for this latter molecule bond C(2)—C(6) is a single one

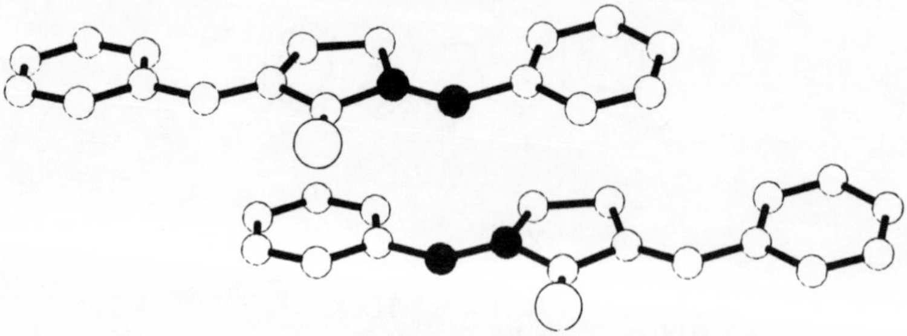

FIGURE 7. Incipient dimer pair for **13**. Filled circles are the two reacting moieties. Note that now they are not totally parallel

in the presence of nitrogen resulted in an amorphous crude product. Using TLC and recrystallization, a number of products were identified. The main product (**14**) is one whose formation can be explained in topochemical terms[38]. Packing in the parent crystal (see Figure 7) is such that nearest-neighbour molecules are parallel, since they are related by translation along the shortest cell axis, *b*. The double bonds on the two molecules closest to each other are shown as filled circles in Figure 7. From that figure, it can be seen that these bonds are in planes which are parallel to each other, but themselves subtend an angle of 56°. This is not a geometry generally considered conductive for a topochemical reaction, although the mean distance separating the potentially reactive centres is 3.71 Å, which is well within the limits previously deduced to be necessary for such reactions. However, comparison of the incipient dimer (Figure 7) and the molecular structure of the dimer (Figure 8) from its crystal structure indicates that **14** is the expected product of a reaction involving the pair in Figure 7, under topochemical control.

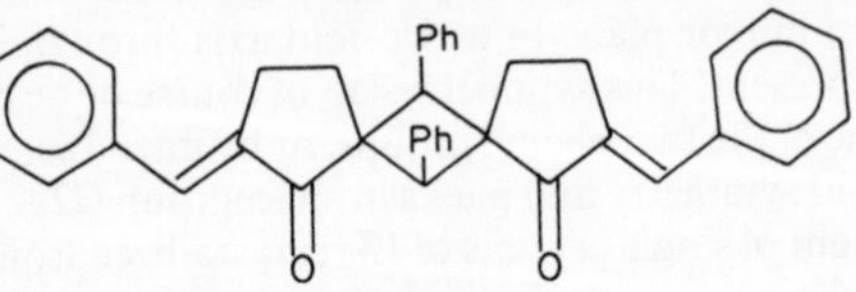

(**14**)

This apparent breakdown of the topochemical rule, which has been seen to hold sway in the cases of the cinnamic acid and BBCP families, can be explained as follows. The two reacting bonds are part of extended conjugation systems which are parallel to each other. The orbitals on each of the atoms, which are part of the double bond in the monomer and will overlap to form the cyclobutane ring in the product, are the p_z, which are by definition at right angles to the mean plane of the conjugation system, i.e. the molecule, since the DBCP backbone is virtually flat. Therefore, in **2**, where the double bonds are antiparallel, the p_z orbitals are directly above each other and can overlap upon excitation of one of the two molecules. In **13** the two orbitals are parts of parallel conjugation systems, and will therefore point in the general direction of each other[37]. Furthermore, one of the bonds is directly above the other. Overlap and cyclobutane ring formation is therefore still possible. Other examples of apparent breakdowns of the topochemical principle have been noticed before and since (see later sections). In the light of their observations on compound **13**, Thomas and coworkers suggested that the prerequisite for reactivity under topochemical control is the ability of the appropriate orbitals to overlap[37].

Compound **15** was identified among the products of the irradiation of **13**. Kaupp and

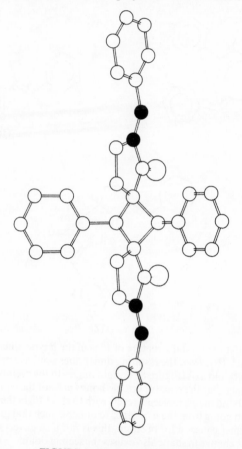

FIGURE 8. Dimer molecule for **13**

Zimmermann[38] suggested that **14** is the product of a reaction between an unreacted molecule of **13**, and the biradical **16**. The formation of this product can be explained in topochemical terms. Kaupp and Zimmermann used a different system than that used by Theocharis and coworkers, namely thin films grown from methylene chloride or methanol solutions. In addition to **14** and **15**, they reported a third product, **17**, no trace of which was detected from reactions in crystals grown from chloroform/methanol. The formation of **17** would not be allowed under topochemical rules from the structure of **13**, identified by Thomas' group. However, duplication of the routine used by Kaupp and Zimmermann did yield this product[39]. Powder XRD studies of **13**, recrystallized from methylene chloride, suggested that more than one phase is obtained, and TLC and NMR suggested that solids obtained by recrystallization from different solvents gave different ratios of products. It is therefore suggested that **13** exhibits polymorphism; one polymorph which is obtainable by recrystallization from chloroform/methanol gives rise upon irradiation to **14** and **15**, whereas other solvent systems yield at least one further polymorph, which is responsible for product **17**. This second polymorph is expected to be a minority component. Efforts to isolate this have so far failed.

(15)

(16)

(17)

It is noteworthy that the crystal structure of **13** is of the β type, since nearest-neighbour molecules are parallel. However, the resulting dimer does not have mirror symmetry, but rather possesses a two-fold axis (Figure 8), at right angles to the cyclobutane ring. This is a consequence of the fact that the reactive double bonds are not the equivalent ones, i.e. the C(5)—C(13) double bond in one molecule reacts with C(2)—C(6) in the second. The change of molecular shape in going from the monomer to dimer is such that growth of the latter in the lattice of the former causes a lot of strain; this is not a single-crystal to single-crystal transformation. This change in shape also causes the second double bond in each molecule to move away from close contact with its neighbour. Therefore, oligomerization is not possible. Irradiation of crystals of **13** leads to an amorphous product. This is caused by the breakdown of the mechanical integrity of the crystals through strain, and the formation of more than one product; however, there is no phase separation. This is a further indication that although this reaction is not topotactic, it is topochemical. The difference in behaviour between **2** and **13** can be traced to the absence of a bulky, flexible anchoring group in **13**, and the rigidity of the whole molecule, caused by the π conjugation system extending over the whole molecule.

Various analogues of **13** have been studied, such as **18** and **19** (Scheme 4). Unlike the benzyl series, where **2** and **3** had very similar packing arrangements, **18**, unlike **13**, is photostable[40]. Crystals of **18** are of the Abm2 spacegroup, whilst the molecule is mirrorsymmetric. Nearest neighbours are related by a glide plane. In contrast, **19** is photoreactive[16], as are molecules **20**, **21** and **22**[41]. This led some workers to suggest that for the dibenzylidene series, photoreactivity is only possible for the parent molecule (**13**) and for non-symmetrically substituted analogues[41]. **20**, **21** and **22** yield dimers as well as oligomers on irradiation. Dimerization results in cyclobutane rings, whereas oligomerization may also involve oxetan formation, and should therefore involve the opening of the carbonyl carbon oxygen double bond in the reaction. This prediction is negated by compounds **23**[42], **24**[42], **25**[43] and **26**[44]. This can, however, be explained as follows. The hydroxy substituents in **23** and **24** will probably be involved in hydrogen bonding; this type of interaction, which is not available in other benzylidenes, is likely to take over as the majority influence, from the $\pi-\pi$ interactions which would normally hold sway. The effect of the fluoro substituents in **25** onto the structure is likely to be complex. The size of the fluoro substituent should not be very different from that of H, but the atom–atom interactions favoured by each would be different. The study of a whole series of fluoro-substituted enones should help in elucidating the relative importance of size and electrostatic considerations, in determining packing patterns. As for **26**, methylenedioxy substitution has been shown to favour strong $\pi-\pi$ interactions, and hence β packing[45] **23** and **24** can act as chelating agents to appropriate transition-metal ions[42]. Coordination polymers (**27**) have been formed with Ni^{2+}, Cu^{2+} and Zn^{2+}. The Ni^{2+} and Cu^{2+} polymers are further photoreactive, but that for Zn^{2+} is photostable. The importance of these observations is that they show that the packing mode, and hence solid-state reactivity of the **13** framework, can be controlled by varying the coordinating metal ion, whilst leaving the substitution pattern intact.

(**23**)

(**24**)

(**26**)

A chiral centre can be created in **13**, by introducing a substituent at position **3** of the cyclopentanone ring. The substituent which can be introduced most easily is Me, and since (+)3-Me cyclopentanone is commercially available, **28** was prepared[37]. These crystals belong to spacegroup $P2_1$, and nearest neighbours are related by the two-fold screw axis. The closest distance separating neighbouring double bonds is 3.87 Å. Although this distance is suitable for $[2+2]$ cycloaddition, the crystal is photostable. This situation arises because the benzylidene groups, and therefore the conjugation systems to which these two bonds belong, are not parallel. This prevents the necessary overlap of potentially reactive orbitals. A malonic acid group can be introduced at position **3**, to yield the enone **29**, which does not dimerize on photoirradiation of its crystals, but undergoes dehydration. **29** can, in common with **23** and **24**, act as a chelating agent[43]. Complexation inhibits the dehydration process, which presumably involves the carboxylic group.

(27)

(28)

(29)

C. Properties of Mixed Crystals

The crystal structures of **4** and **5** are isomorphous, i.e. they have very similar cell dimensions. It is not therefore surprising that single crystals containing both compounds can be obtained from suitable solutions in chloroform/methanol[46]. The two components in such crystals are randomly distributed, forming ideal solid solutions. These crystals yield, upon UV irradiation, a number of dimers: some are the symmetric dimers containing either Cl or Me substituents but not both, as well as dimers which have one Cl and one Me substituent. The mixed dimer has chiral centres at each carbon atom of the cyclobutane ring. This reaction is of the single-crystal to single-crystal type[47]. When the two components **4** and **5** were mixed in varying amounts in a solution which was then allowed to evaporate to dryness, the melting points of the solid residues varied in a linear fashion with composition, between the values for the pure components. This is indicative of ideal-solution behaviour.

Crystal-structure determination on a number of single crystals showed that cell dimensions are intermediate between those of the pure components and dependent upon the Cl:Me ratio[47]. The ratios of dimers obtained upon irradiation was consistent with the Cl:Me ratio for the monomer crystal, as it was determined by crystallographic means. For a given mother solution, different single crystals contained different ratios of the two components, but the structure remained essentially the same, and similar to that of the single component crystals. The range of possible values for the Cl:Me ratio indicates that one can substitute continuously Cl for Me and *vice versa*, and retain the same, reactive packing motif.

When compounds **7** and **8** were dissolved in chloroform/methanol and the solution slowly evaporated, single crystals were obtained with cell dimensions slightly but significantly different from those for **8**. X-ray intensity data were collected for such crystals, and their structure was solved to show that the benzyl benzylidene cyclopentanone framework exhibited a configuration very similar to that for **8**, rather than **7**. Further analysis revealed that both Me and Cl substituents were present, with the former being the majority component, and therefore that mixed crystals were obtained containing both compounds in a statistically averaged fashion[47]. This packing should be conducive to topochemical dimerization, leading to a chiral product. This can be considered as an example of crystal engineering because **8**, which in its native crystal was unreactive, was forced to adopt a different conformation and a reactive packing motif, by incorporating in a lattice (provided by **7**) with those desired attributes. The relative concentrations of **7** and **8** in the solution, and therefore the crystals, was controlled by the low solubility of **8**. Thus, although it should have been possible, in theory, for crystals to be present where **8** was the majority component, thus forcing **7** into a photostable packing mode, none were detected.

Mixed crystals of **25** and **13** have also been studied[44]. The interest in this system is that it enables one to study the influence of the size of the fluoro substituent on the crystal packing: H and F have very similar sizes. Comparison of the rates of solid-state reaction for the two pure phases suggests that **25** reacts much faster than **13**, and that therefore the two crystal structures are likely to be different. The mixed crystals were found to be photoreactive, while mass spectroscopy indicated the presence of mixed dimer, suggesting that the two phases were intermingled. Contrary to the cases reported above, however, the melting points did not vary in a linear fashion with composition, but went through a maximum. This suggests that the solid solution was non-ideal. It is possible that these mixed crystals comprised domains of one compound in a matrix of the other. The presence of appreciable quantities of the mixed dimer is counterindicative to simple coprecipitation. If this had occurred, a mixed dimer would only be possible for reactions at interfaces, and would therefore be present in very small amounts.

D. Other Related Enones

2-Benzylidenecyclopentanone (**30**) has been found to be photostable, in spite of the closest double-bond to double-bond separation being 4.14 Å, for molecules related by a centre of symmetry[48]. This is a geometry which would normally be expected to lead to photoreactivity. However, closer examination of the crystal structure of **30** reveals that the two double bonds are situated in such a way that overlap of the appropriate p_z orbitals upon excitation would not be possible, as the double bonds are not directly above each other. To this extent, **30** is very similar to **7**, where the bonds are also not directly above each other; the presence of the benzyl group in **7**, however, causes the two molecules to be further apart, in which case for that structure the bond-to-bond separation was found to be 4.65 Å. Lactone **31**, however, assumes[49] a photoreactive packing motif, in which the double bonds are separated by 3.67 Å. **30** and **31** are isoelectronic, and might therefore be expected to assume similar packings[48]. It appears, however, that the crystallographic differences arise, at least in part, from the presence of C—H ··· O hydrogen bonds in **31**, but not **30**. What is surprising is that the hydrogen bonds in **31** involve the carbonyl oxygen, not the lactone one. Close examination of the crystal structure of **30** reveals that the six-membered and five-membered rings are not exactly coplanar, as is the case for **31**. This molecular puckering presumably contributes to **30** assuming a photostable packing motif. The presence of hydrogen bonding is reflected in the lower density of **31** and its higher melting point.

(30)   (31)

The dimerization of **31** is not of the single-crystal single-crystal type. In this, it is similar to the case for DBCP (**13**) which, however, poses the additional complication of the generation of side-products. Single crystals of **31** begin to crack very quickly upon photoirradiation. This is due to the generation of strain caused by the mismatch of dimer molecules within the reacting monomer lattice. This behaviour may be traced to the absence of an anchoring group.

2-Benzylidenecyclopentenone[44] (**32**) was studied as a precursor to 3-malonic-2-benzylidene cyclopentanone[44] (**33**). **32** is of interest, because it is a much more rigid molecule than **30** and has a more extensive conjugation system. It has been found to be photoreactive. **33** was not only dimerizable upon irradiation, but also exhibited decarboxylation of the malonic group. Evolution of CO_2 was detected by Fourier-transform infrared spectroscopy of KBr pressed pellets. The CO_2 signal was a single peak, rather than possessing two branches. This would indicate that the product molecules remained trapped within the lattice. This reaction is probably intermolecular. The close chemical similarity of **33** with **29** leads to the assumption that both dehydration for **29** and decarboxylation for **33** are under topochemical control. The malonic acid group can act as a chelating ligand towards metals (e.g. Ni^{2+}). The complex has been shown to be photodimerizable, but the decarboxylation reaction was arrested.

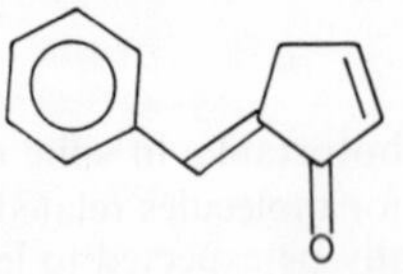

(**32**)

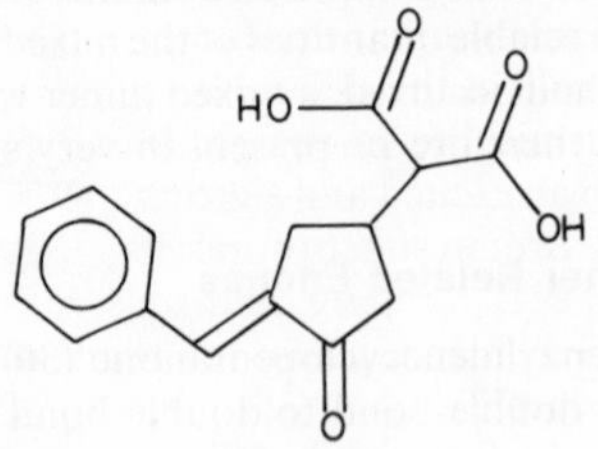

(**33**)

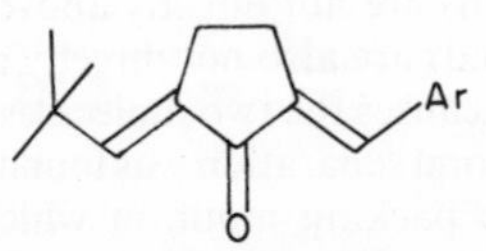

(**34**)

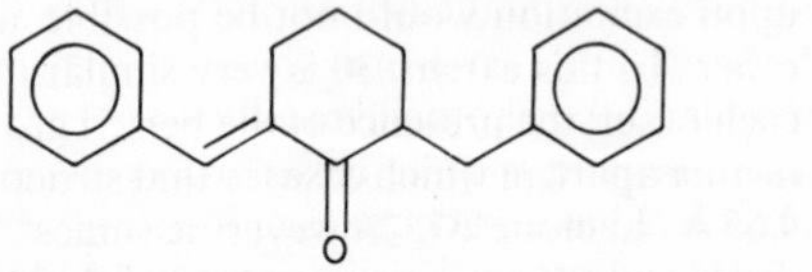

(**35**)

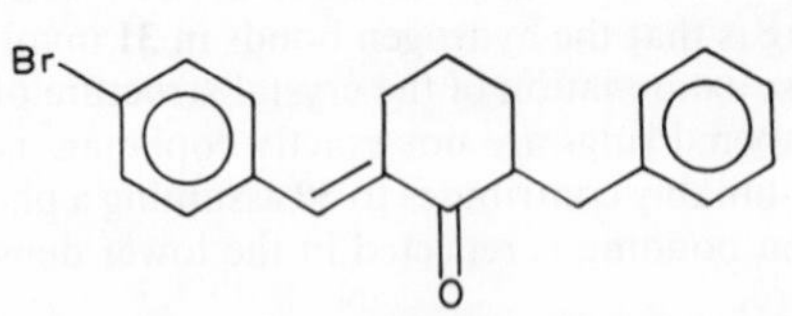

(**36**)

A series of 2-alkylidene-5-arylidenecyclopentanones[41] (**34**) have been studied, with 4-Me, 4-NO_2 or 4-pyridyl substituents on the aryl ring. These were found to be photoreactive, and yield dimers as well as oligomers. The oligomerization reaction appeared to involve the carbonyl group, as well as the exocyclic double bond, leading to oxetan formation.

The solid-state reactivity of the cyclohexanone analogue of 2-benzyl-6-benzylidenecyclohexanone (**35**) was studied, in order to determine the effect of additional molecular volume and flexibility, which is imparted by the extra methylene group[22]. Its

4-Br derivative **36** was also studied[16]. Both were found to be photoreactive. **35** crystallizes in spacegroup PĪ, such that nearest neighbours are related by the centre of symmetry, with a bond-to-bond separation of 3.79 Å. Unlike that of the cyclopentanone analogue, the dimerization of **35** is not single-crystal to single-crystal. In fact, upon partial reaction the crystal melts. This may be due to two facts. First, the short bond separation may not allow the dimer molecule to relax after its formation, and second, the low melting point of **35** (69 °C) will be lowered by the presence of dimer. **36** adopts a packing totally different from that of the unsubstituted cyclohexanone, in spacegroup $P2_1/c$. The steering influence appears to be short Br $\cdots$ Br non-bonded contacts of 3.66 Å, across centres of symmetry. This contact is well short of the sum of the van der Waals radii of the two Br atoms. The shortest double-bond to double-bond separation was found to be 5.26 Å for centrosymmetric pairs. This is probably too long for reaction in the perfect lattice under topochemical control. Reactivity here is thought to arise because of defects: at the defects, molecules are correctly positioned for reaction (cf. the case for anthracenes). The hallmark of reaction at defects is that such reactions are inhomogeneous, i.e. they occur preferentially at some sites and not others. Evidence of inhomogeneity has been found with optical microscopy, where phase separation was observed during photoirradiation. Optical microscopic experiments were carried out under cross-polarized light. The reason for the role of the defects being seen in this reaction and not others may be as follows. Topochemical reactions occur in the perfect lattice when no transfer of energy can occur between an excited and a ground-state molecule, because of the brevity of the excited-state lifetime. Bromo substitution may lengthen the lifetime of the excited state long enough to allow energy hopping, and thus defect-controlled reactivity (see later sections). Defect-controlled reactions have been previously observed for a series of substituted anthracenes[50-52].

IV. THEORETICAL CONSIDERATIONS OF [2 + 2] CYCLOADDITIONS

Molecular-orbital calculations[53] within the MNDO approximation were performed on 1-phenyl-but-1-en-3-one (benzylidene methyl ketone (**37**)). This compound corresponds to the photochemically active portion of the benzyl benzylidene cyclopentanone molecule, and is quite close to those of **2** and **13**, and their analogues. It was therefore considered as an adequate model for the solid-state photodimerization of enones, as the nature and properties of the excited state should be the same, whether the reaction takes place in a fluid or solid environment. Some geometric constraints were imposed, however, on the conformation of the molecule, so as to model more closely the situation that obtains in the solid state. It was initially thought that the theoretical study of solid-state phenomena should involve the consideration of band structures. However, it is nowadays generally accepted that this is not necessary for molecular crystals, as electrons would be largely confined within a given molecule and would not be delocalized.

(37)

The ground state of **37** was found to have a heat of formation of 10.86 kcal mol^{-1}. The maximum electron density for the HOMO was on C(5), and for the LUMO on C(13), the two lobes having the same phase. The geometry, including bond lengths, angles and torsional angles, was close to that found for the benzylidene moiety in the crystal structure of **2**. The lowest excited singlet state was found to have a heat of formation of 42.60 kcal mol^{-1}, with similar disposition of the HOMO and LUMO as the ground state.

The lowest excited state was found to be a triplet state with heat of formation 42.30 kcal mol^{-1}. Maximum electron density for the HOMO was located on C(5), and for the LUMO on C(13), but the two contributions had opposite phases.

The very similar energies of the lowest excited singlet and triplet states mean that transition from the former to the latter is extremely facile. The excited triplet state thus formed will be vibrationally excited. This can be correlated with the so-called 'phonon' assistance of solid-state [2 + 2] cycloaddition reactions previously reported[54]. The molecular-orbital symmetry is such that reaction between two ground-state molecules, or between one ground-state molecule and one in the singlet state, is not allowed. On the other hand, reaction between a ground-state molecule and one in the triplet excited state is allowed. Thus, the facility of energy transfer between states is crucial to the reaction occurring under topochemical control.

The lifetime of the triplet state for **13** as measured from the phosphorescence in emission spectra[55] at 77 K was only 200 μs. The brevity of the lifetime of the excited state means that the excited molecule cannot transfer its energy to a neighbouring one. This process is called energy hopping, and where it occurs the solid-state reaction is not homogeneous, as it is no longer random. Defects in the lattice act as energy traps and therefore such a reaction is more likely to occur at defects. The shapes of the two excited states are very similar to that of the monomeric residues in the dimer. The bond lengths and angles as determined from MNDO for the two excited states of **37** correspond well with those found crystallographically for the dimer of **2**.

The change in shape which accompanies excitation has two consequences: first, it makes energy hopping less likely, since this process is more probable between molecules closely related in structure. Second, this movement probably causes the reactive centres to move closer together, compared to the position they occupy when at the ground state. The speed of reaction is also related to the fact that the transition state is closer in structure to the product than to the reactant. The symmetry of the orbitals in the triplet and ground states indicates that both the head-to-head and head-to-tail reactions are intrinsically possible. Further, the cycloaddition has to be a non-concerted process, since only one pair of orbitals of the two involved are initially of the correct symmetry.

In other sections of this chapter, it will be seen that a number of reactions appear to occur under topochemical control, insofar as the geometry (nature) of the product can be rationalized in terms of the crystal structure of the reactant, yet they occur between double bonds either too far apart, or not totally parallel. A possible explanation for these discrepancies may be that parallel double bonds present the ideal geometry to enable a lobe with correct phase on the ground-state molecule to overlap with one on the excited state. This overlap is clearly possible for orientations other than parallel bonds. Furthermore, since in the present example the phases of the lobes are such that the reaction cannot be concerted, it may be that at the start of the reaction contact has to be favourable for only one atom on each molecule for reaction to be possible, and not for both atoms simultaneously. The term 'minimum movement' probably should only refer to the initial movement of the reacting atoms, and after that the consequent movement for the rest of the molecule may be larger (see, for example, the case for distyrylpyrazine)[91]. This movement will probably cause strain and the breakdown of the mechanical integrity of the crystal, and therefore stop any further reaction, as topochemical control would be lost.

Apparent breakdowns in the topochemical principle, because of separation, are more difficult to explain. There is a grey area consisting of bonds separated by distances between 4.25 and 4.7 Å where molecules, e.g. 4.7 Å apart, react and others separated by 4.3 Å do not, other things being equal. It is sometimes possible to explain stability, because bonds are not parallel (e.g. **27**) or because of steric hindrance to the movement necessary for reaction.

There are cases, however, whether no such clear explanations are possible[56]. It is suggested that in those cases, the reason for stability may be found in the geometric structure of the excited state.

The topochemical principle is a very useful tool for the solid-state chemist, and is capable of application in a variety of situations. It does suffer, however, from the disadvantage that crystallography provides the structure of ground-state molecules when, in the case of photochemical reactions, excited states are involved.

V. SOLID-STATE DIMERIZATION AND POLYMERIZATION OF OTHER ENONES

A. Chalcones

The photochemistry of benzalacetophenone has been studied in solution and in the crystalline state[57,58]. In solution, it undergoes *trans–cis* isomerization. Photoirradiation of crystals leads to formation of both mirror-symmetric and centrosymmetric cyclobutane rings, as well as some resinous byproducts. Irradiation of a solution of *p*-anisal-acetophenone leads only to the formation of a resin. However, in addition to resin, dimers are formed in the solid state, of both the mirror-symmetric and centrosymmetric type.

The photochemistry of chalcones is of interest owing to their occurrence in the form of 4,4′-dioxychalcone functional groups in photo-crosslinkable epoxide resins[59]. In order to mimic their behaviour, the solid- and liquid-state photochemistry of the diglycidyl ether of 4,4′-dihydroxychalcone (38) has been studied[60]. The preferred solvent for the solution studies was acetonitrile. At least in solution, further reaction is preceded by *trans–cis* isomerization. Whether this occurs in the solid state before further reaction takes place is not clear from the paper. Prolonged irradiation with pyrex-filtered UV light (Hg vapour medium pressure lamp) led to 78% dimer and 22% low-molecular-weight polymer in solution, and 63% dimer with 37% polymer in the solid state. Gel permeation chromatography and mass spectroscopy was used to identify the nature of the dimers. It was found that both mirror-symmetric and centrosymmetric cyclobutane rings had been formed. Cleavage of the four-membered rings appears to take place. The olefins that result can either recombine to yield a dimer, or can be converted to a variety of radicals which then polymerize.

(38)

B. 2-Benzyl-5-cinnamylidenecyclopentanone

The enone 39 packs in spacegroup Pbca (Figure 9), with the asymmetric unit comprising two molecules (noted as A and B)[61]. Examination by IR spectroscopy before and after UV irradiation confirmed that reaction had taken place. Examination of the crystal structure indicates that although several double-bond to double-bond short (< 4.3 Å) contacts are present between at least two pairs of molecules (Figure 9), none is for precisely parallel double bonds. ^{13}C NMR spectroscopy indicates that oligomerization has occurred involving both the double bonds and the carbonyl groups. Four-membered rings in the polymer are of both the oxetan and cyclobutane kind. Oxetan formation has been encountered in other oligomerizable systems, such as 34[49], and certain derivatives of 13. In

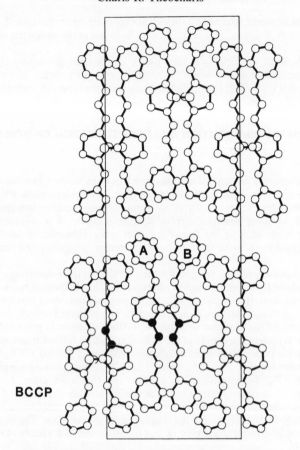

FIGURE 9. Packing diagram for **39**. Labeles A and B refer to
the two independent molecules in the asymmetric unit. Filled
circles indicate the closest bond-to-bond contacts

both **13** and **39**, the double bonds are, presumably, too close to each other in the molecular
framework to allow them to react in the solid state simultaneously. In general, no case has
been found in the solid state where polymerization occurs where only one double bond is
involved. Polymerization occurs only where two widely spaced bonds are present suitably
packed, or where the carbonyl is activated to such an extent that it is able to react and form
an oxetan four-membered ring.

(39)

C. Coumarins

The solid-state photochemistry of a number of 4-, 6- and 7-substituted coumarins (**40**) has been studied[17]. Depending on the substituent, four different types of dimer have been found (Scheme 5). For example, 7-methoxycoumarin crystals[62] yield upon UV irradiation a dimer molecule with a centrosymmetric cyclobutane ring, and 7-chlorocoumarin yields a mirror-symmetric cyclobutane ring[63]. 4-Chlorocoumarin, on the other hand, yields two products, both with cyclobutane rings with two-fold symmetry; in one, the symmetry axis is in the plane of the four-membered ring, and in the other, at right angles (for the structures of these dimers, see Scheme 5).

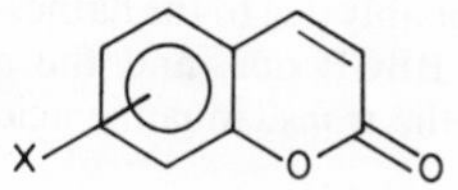

(**40**)

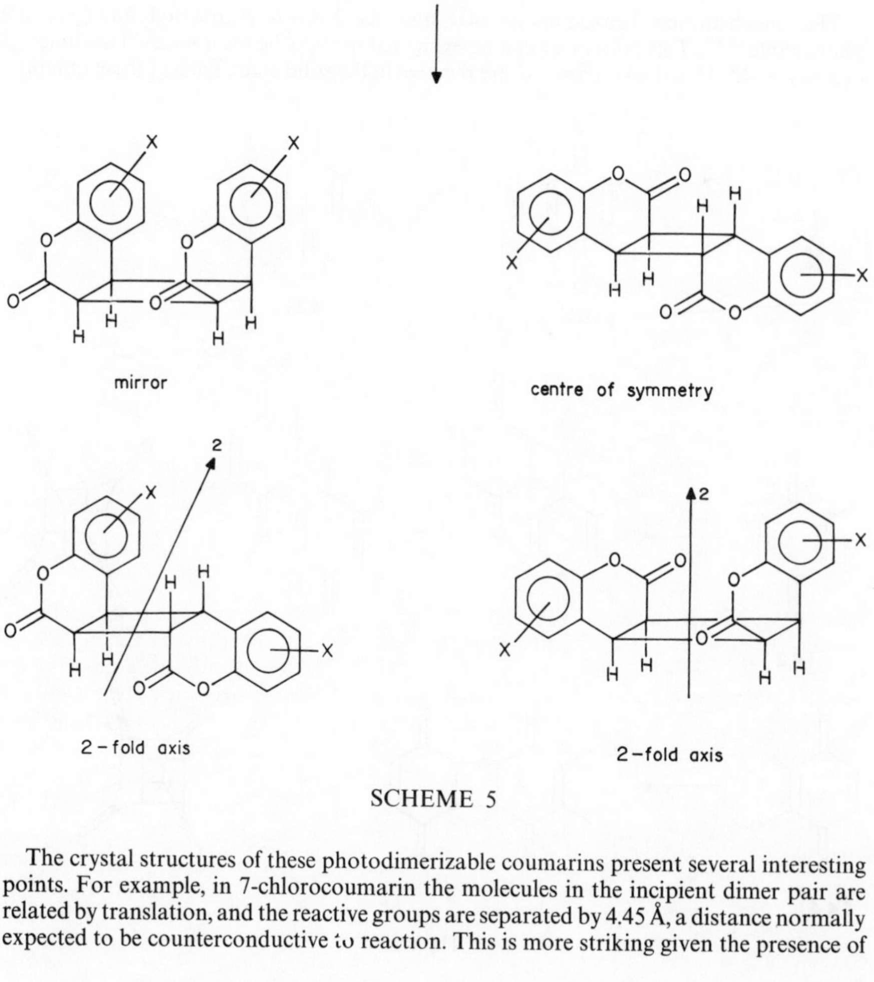

mirror

centre of symmetry

2 – fold axis

2 – fold axis

SCHEME 5

The crystal structures of these photodimerizable coumarins present several interesting points. For example, in 7-chlorocoumarin the molecules in the incipient dimer pair are related by translation, and the reactive groups are separated by 4.45 Å, a distance normally expected to be counterconductive to reaction. This is more striking given the presence of centrosymmetrically positioned double bonds, separated by only 4.12 Å, which would normally be expected to lead to reaction. In 7-methoxycoumarin, the reactive double bonds are separated by 3.8 Å, but are not parallel, and subtend an angle of 65° between them. The explanation proposed for the reactivity of this compound is similar to that put forward for **13**.

It is noteworthy that the four types of dimer yielded by the different coumarins represent all the possible dimers obtainable by [2 + 2] cycloaddition of conjugated *trans* double bonds. Very few chemical systems which exhibit topochemical dimerization have shown such diversity to date: in the *trans* cinnamic acid family only two types of dimer are obtainable, and the same number are possible in the BBCP–DBCP complex. This versatility of the coumarins is probably due to the flatness of the coumarin carbon skeleton as opposed to the non-planar BBCP one, and the absence of the steering effect of hydrogen-bonding operative in the *trans* cinnamic acid system.

D. Quinones

The unsubstituted benzoquinone (**41**) and its 2, 3, 5, 6-tetramethyl analogue are photostable[64,65]. This behaviour can be explained in topochemical terms. The dimethyl derivatives **42**, **43** and **44** (Scheme 6) are reactive in the solid state. Each of these quinones

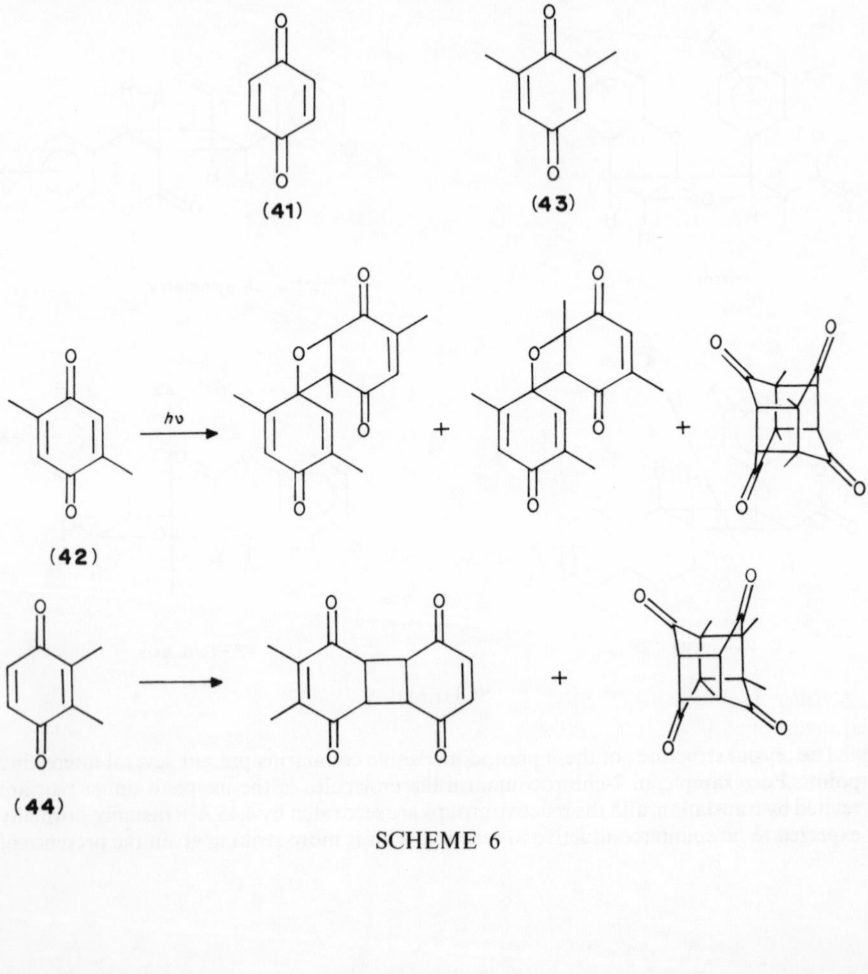

SCHEME 6

yields in general two types of dimer, one cage dimer containing two cyclobutane rings, and an oxetan obtained by the reaction of a carbonyl group on one molecule with a double bond on a neighbour. The solid-state photoreactivity of **42** and **43** can easily be explained in topochemical terms. The crystal structures for these molecules are built up from asymmetric units consisting of two molecules. In both structures, each unique molecule in the asymmetric unit is part of its own stack. Contacts and orbital overlaps are favourable for oxetan formation in one of the two stacks, and for formation of the cage dimer in the other. The symmetry of the oxetan dimer is different for different monomers. In fact, **42** gives two oxetan dimers with different symmetries, whereas **43** yields only one oxetan, in addition to the cage dimer. The solid-state reactivity of **44**, however, cannot be explained easily. The cage dimer which is obtained has a mirror symmetry, whereas nearest neighbours are related by a centre of symmetry. It is possible that reaction in this crystal is controlled by, and occurs at, crystallographic defects. The second product from this crystal is not an oxetan, but contains a cyclobutane ring. Several benzoquinones (e.g. **45**) undergo intramolecular cycloaddition to yield a cage dimer[66,67].

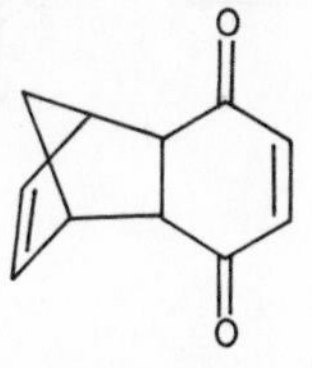

(**45**)

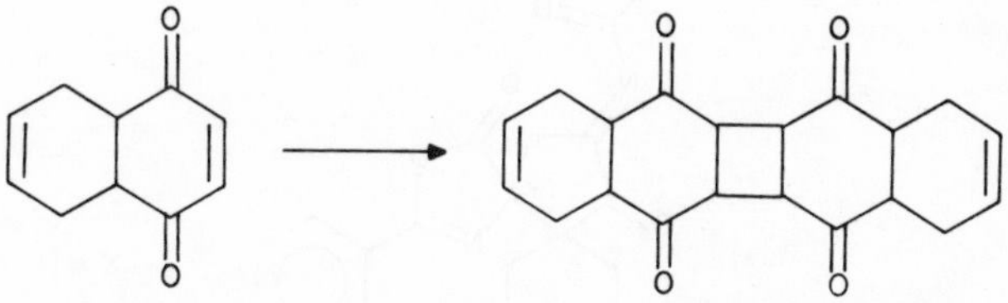

(**46**)   (**47**)

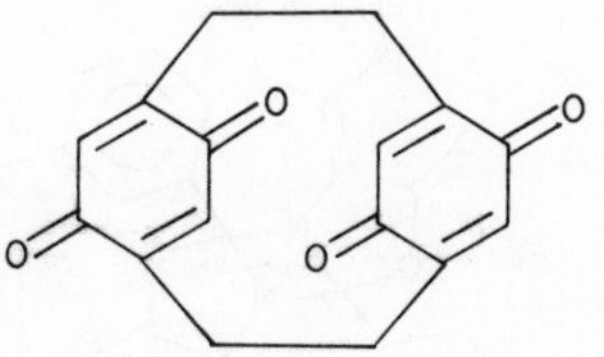

(**48**) 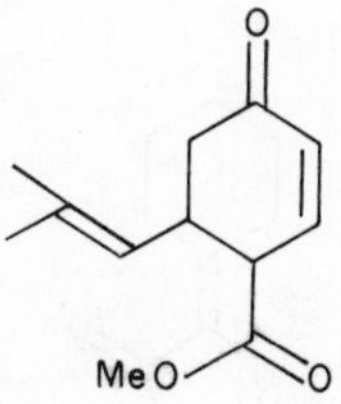   (**49**)

Scheffer, Trotter and other workers have studied the solid-state reactivity of substituted tetrahydronaphthoquinones extensively[11,15,68,69,148], over a number of years. Four different reactivity patterns can be discerned, which are correlated to the disposition of neighbouring molecules and the intermolecular distances. Reactions observed were intermolecular cycloaddition, intramolecular hydrogen abstraction by an oxygen or carbon, and intramolecular oxetan formation[66]. For example, **46** undergoes inter-

Charis R. Theocharis

molecular cyclobutane formation to yield dimer **47**. Scheffer, Trotter and coworkers have shown that the variety of possible reactions is due to the fact that naphthoquinone is frozen into a single conformation in a crystal, irrespective of the substitution pattern. This series of compounds has been extensively reviewed in another volume of this series.

2, 5-Benzoquinonophane (**48**) is polymorphic[70]. In one of its crystal forms, the carbonyl groups of each six-membered ring in the molecule are parallel, and in the second crossed. The latter form is stable, whereas the first undergoes intramolecular cyclization. No intermolecular reactivity is observed. The enone **49** is a natural product, whose crystals are clear and needle-like. Exposure to light quickly changes the crystals into an opaque

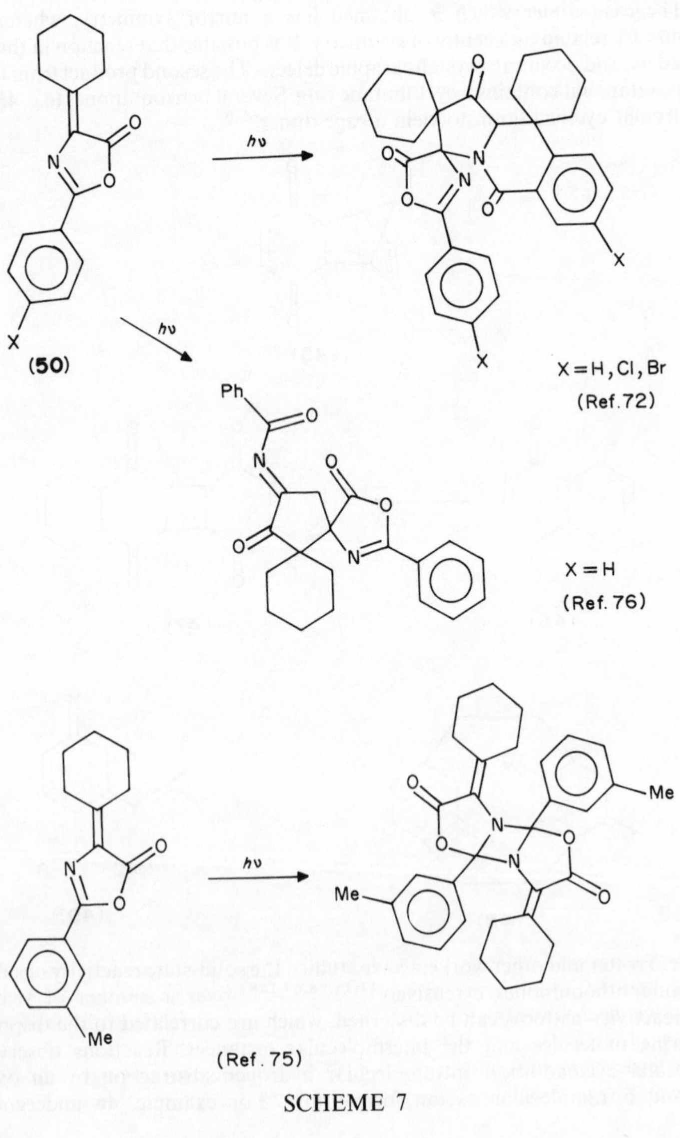

(**50**)

$X = H, Cl, Br$

(Ref.72)

$X = H$

(Ref.76)

(Ref.75)

SCHEME 7

powder[71]. The product has a centrosymmetric cyclobutane ring. In the parent crystal, reactive molecules are antiparallel with a double-bond to double-bond separation of 3.86 Å. However the double bonds, although parallel, are not exactly on top of each other, so that a relatively large movement of ca 2.2 Å is needed from each carbon atom, to react.

E. Heterocyclic Compounds of Enones

4-Alkylidene-oxazol-5(4H)-ones (50) exhibit a variety of light-induced reactions, including asymmetric dimerization with[72] or without[73] H-shift, [2 + 2] dimerizations[74], dimerization reactions involving the C=N bonds[75] (Scheme 7) as well as Diels–Alder dimerizations and Norrish type II processes[75]. Some of these reactions involve opening of one of the lactone ring[76], in addition to ring formation.

3,5-Diphenyl-4-H-thiopyran-4-one-1,1-dioxide (51) undergoes a double Diels–Alder reaction[77] to yield the trimer 52, with an attendant loss of SO_2. Other thiopyranone derivatives have been studied, and of these the 2,6-diphenyl derivative was reactive but the 3,5-dimethyl was photostable. N-Methyl-2-pyridone[78] (53) yields a centrosymmetric cyclobutane compound upon photoirradiation which reverts back to the monomer upon heating, whereas 54[78] and 55[79,80] yield mirror-symmetric cyclobutanes.

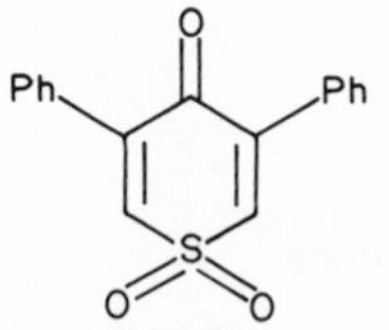

(51)

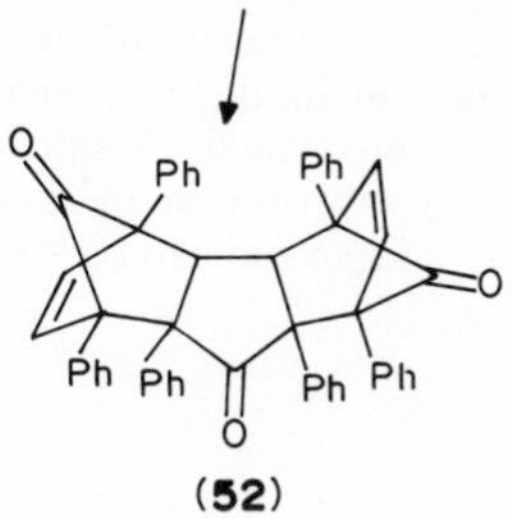

(52)

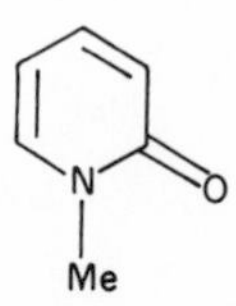

(53) 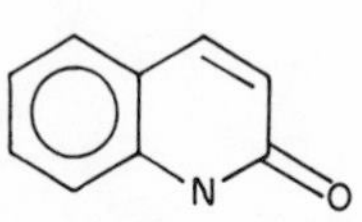   **(54)** 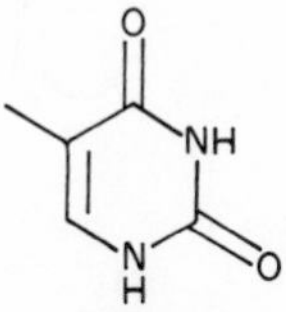   **(55)** 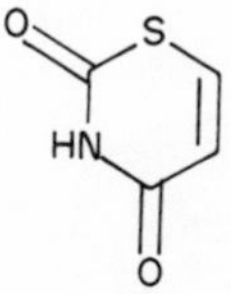   **(56)**

Photoirradiation of crystals of 1-thiouracil[81] (56) yields a dimeric molecule with a puckered, twisted, cyclobutane ring, which has a pseudo two-fold symmetry. Similar reactivity has been observed for uracil itself the dimer of which is obtained by the UV irradiation of RNA[82].

Two heterocycles which undergo [4 + 4] cycloaddition are the α-pyrone[83] (57) and the pyrazinone 58[84], to yield centrosymmetric dimers 59 and 60, respectively. The archetypal

[4 + 4] cycloadditions are those of the anthracenes[51]; a striking difference between **57** and **58** on the one hand and, for instance, 9-cyanoanthracene on the other is that the reactions described here are topochemical, whereas the anthracene one is defect controlled. In fact, the dimer yield for **57** is 100%.

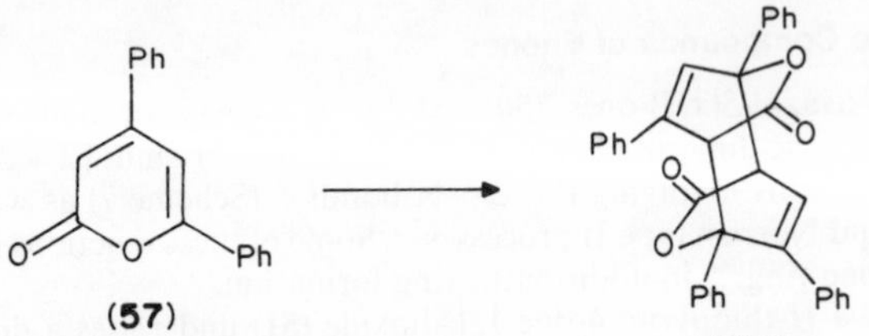

(57)   **(59)**

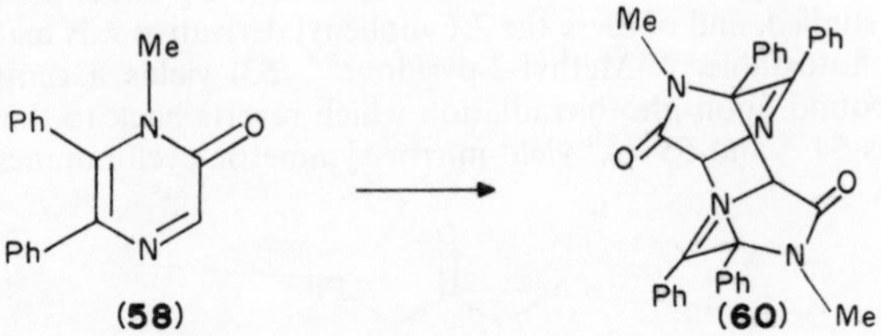

(58)   **(60)**

F. Miscellaneous Other Dimerizations

A number of other enones, such as **61**[85] and **62**[86], dimerize in the solid state to yield cyclobutane rings. Conjugated cyclopentadienones (**63**, R = H, Ph, *t*-Bu, Et, *p*-C$_6$H$_4$Me), on the other hand, undergo cyclization[87] to yield cage dimers (**64**). Dibenzylidene acetone (**65**) is photostable, but its dichloro analogue undergoes facile double cycloaddition to yield compound **66**, which contains two cyclobutane rings. This illustrates the usefulness of chloro substitution in crystal engineering: Cl···Cl close contacts are energetically very favourable and can be maximised by assuming β packing (cf. cinnamic acids).

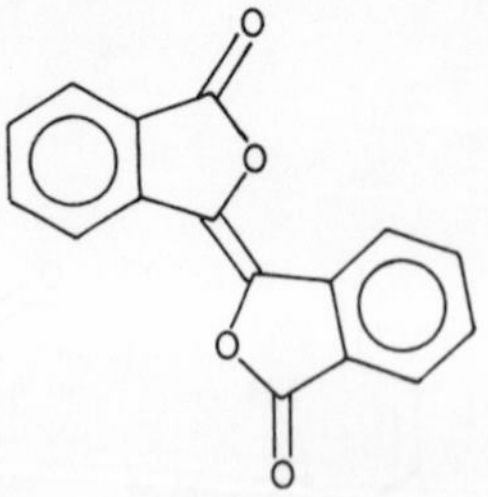

(61) 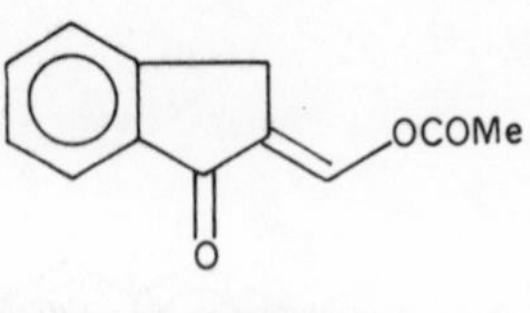   **(62)**

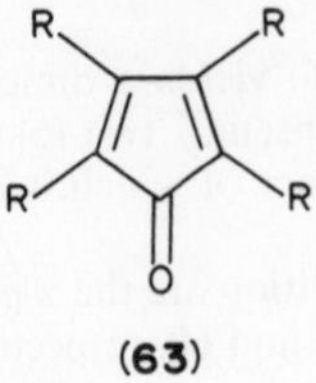

(63) 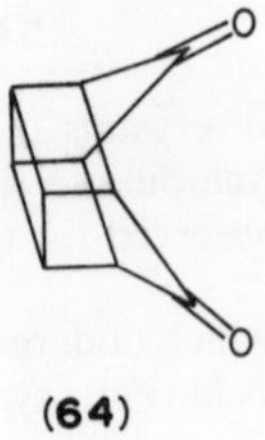   **(64)**

(65)   **(66)**

Mustafa has shown that compounds such as **65** containing[58] extended conjugated π-systems will form complexes with UO_2Cl_2 or $SnCl_4$. In the crystal of the 2:1 adduct of **65** to UO_2Cl_2, the metal ions are related by centres of symmetry with the organic parts of the complex in a packing motif in which they are related by that symmetry. The double bonds are then at the correct orientation and distance for reaction. This crystal-engineering strategy has been used by Moulden and Jones to steer **13** into a packing motif which yielded a centrosymmetric dimer[88].

G. Solid-state Polymerizations

1,7′-Trimethylenebisthymine (**67**) is packed in such a way that reactive double bonds subtend an angle of 4°, and are separated by 3.69 Å. Packing considerations suggest that both intra- and inter-molecular cyclobutane formation is possible, but the reaction actually occurring is the intermolecular one, leading to polymer formation[89].

The archetypal polymerizable dienone is **68**. In its crystals, molecules are arranged such that the reacting pair is skewed, and the intermolecular double-bond separations are 3.98 and 4.09 Å for one incipient cyclobutane, and 3.90 and 3.96 Å for the other[90]. In a single crystal of **68**, dimer **69** is obtained at the initial stages. This reacts further, either intramolecularly to yield the dimer **70** or it yields an oligomer as the minority product, via an intermolecular reaction (**71**).

In general, unsymmetric diolefins (i.e. those unlike **68** which have inequivalent double bonds) can adopt two types of packing conductive to polymerization (Scheme 8): one the so-called hetero-adduct and the other the homo-adduct motif[91]. The former yields chiral cyclobutane rings, and the latter symmetric ones. If a diolefin with a hetero-adduct packing motif crystallizes in a chiral spacegroup (i.e. one which does not contain mirror planes or centres of symmetry), then a single crystal of the monomer will yield a polymer chain of one chirality. If, however, the spacegroup is a racemic one, then polymer strands of

(67)

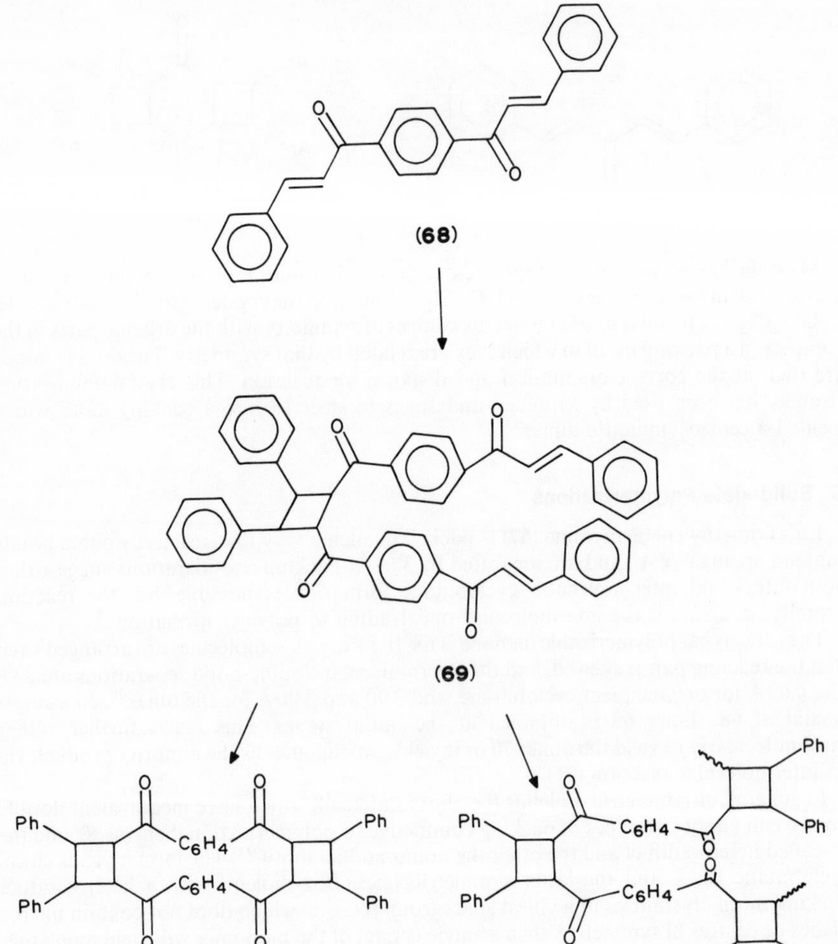

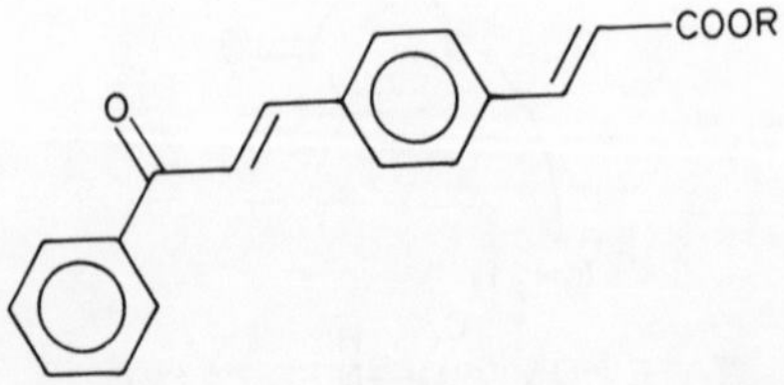

(68)

(69)

(70)

(71)

(72)

both chiralities will be obtained. The enone 72 has been successfully polymerized to yield a chiral polymer. Solid-state polymerization is particularly useful, because it can yield a product of very high crystallinity and relatively large crystals.

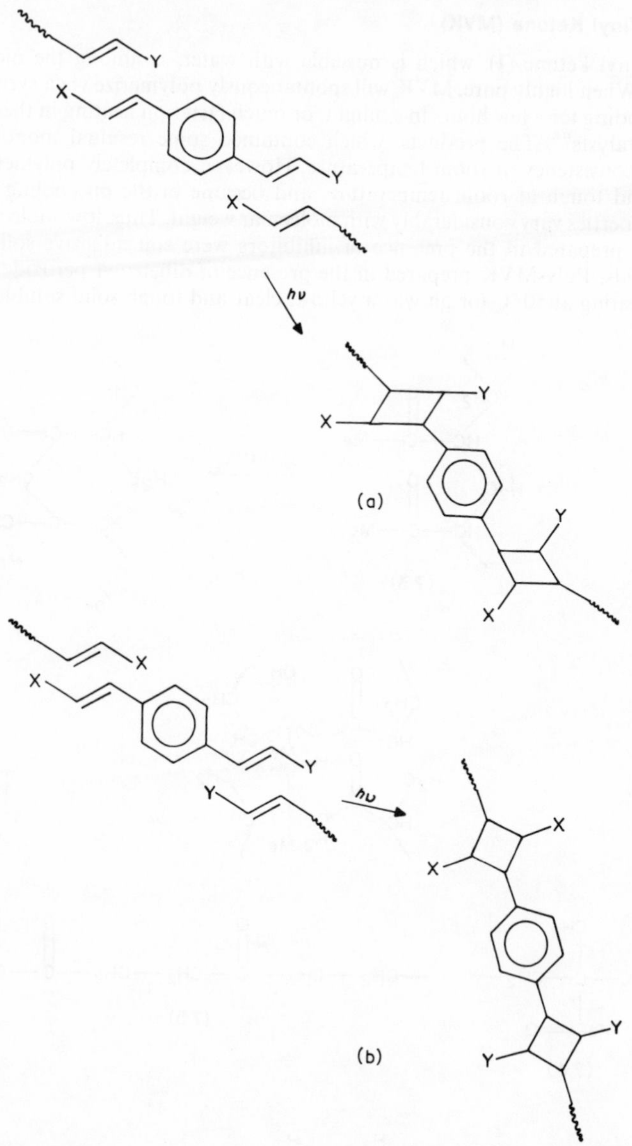

SCHEME 8. (a) Hetero-adduct polymer; (b) homo-adduct polymer

VI. FLUID-STATE HOMOPOLYMERIZATION OF ENONES

α, β-Unsaturated ketones are a particularly interesting class of monomer. At least some of the alkyl vinyl ketones, in addition to being spontaneously polymerizable, are susceptible to various types of initiation, including free radical, anionic and cationic initiation, and photochemical techniques[92,93].

A. Methyl Vinyl Ketone (MVK)

Methyl vinyl ketone (1), which is miscible with water, is among the most reactive monomers. When highly pure, MVK will spontaneously polymerize via a syrup to a solid mass on standing for a few hours in sunlight, or much faster on heating in the presence of peroxide catalysts[94]. The products, which contained some residual monomer, had a rubber-like consistency at room temperature. However, completely polymerized MVK was rigid and tough at room temperature, and became brittle on cooling down. The physical properties vary considerably with molecular weight. Thus, low-molecular-weight poly-MVKs prepared in the presence of inhibitors were soft adhesive solids, or even viscous liquids. Poly-MVK prepared in the presence of dibenzoyl peroxide as catalyst (0.5%) by heating at 50 °C for 5h was a yellow, clear and tough solid soluble in organic

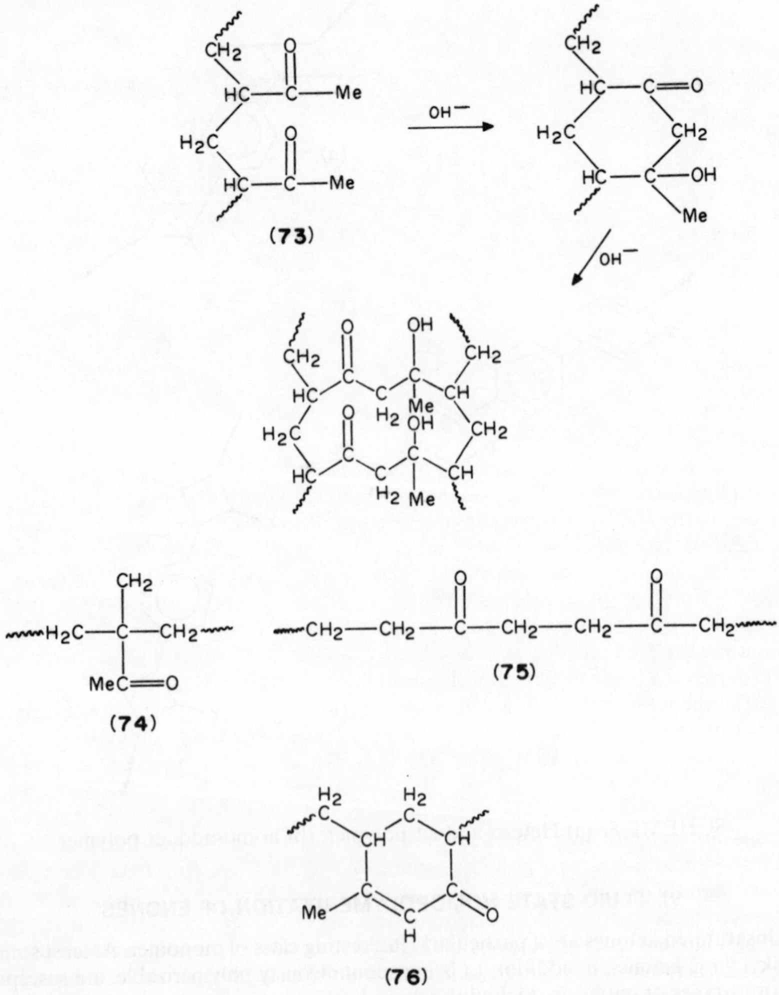

SCHEME 9

solvents such as acetone, acetic acid, dioxan and pyridine[95]. Reaction of this polymer with $ZnCl_2$ in pyridine at 60 °C did not result in dehydration. This was taken to mean that the structure of the polymer was essentially head-to-tail, i.e. it was a 1, 5-diketone (73). Most poly-MVKs are branched to some extent, to give structure 74. This branching may give rise to the observed instability of some of these polymers[92]. Polymerization can also be induced in the gas phase, by UV light irradiation, with CO formed as a by-product[96].

Under certain reaction conditions, MVK undergoes hydrogen transfer polymerization rather than normal vinyl polymerization to yield 73. For example, MVK dissolved in toluene was polymerized in the presence of t-butoxide, to yield a polymer at least in part made up of groups such as 75, obtained via migration of a hydrogen from the methyl group adjacent to a carbonyl, to cause 1, 5 addition[97]. Crystalline, isotactic poly-MVK has been prepared with anionic catalysts, such as Sr or Ca—Zn tetraethyl at 0 °C in toluene[98], and was shown to have a helical structure; some amorphous material was also produced. Use of butyllithium catalyst or sodium naphthalene at − 70 °C yielded a non-crystallizable, red, soluble polymer, which had IR spectra characteristic of structure 76. This is believed to arise from a reaction of 73 with the organometallic catalyst.

(77)

Poly-MVK can also be obtained by γ-ray irradiation of tunnel clathrates of MVK in cyclotriphosphazene (77)[99]. The poly-MVK obtained from this system has a high degree of stereoregularity and has no cross-linking, in contrast to bulk polymerization. Copolymers with random sequences can also be obtained via this route. The technique of group transfer has been used to control the structure of acrylic polymers, including poly-MVK[100]. For example, sequential addition of $Me_2C{=}C(OMe)OSiMe_3$ to 1 can be catalyzed by $(Me_2N)_3S^+$ HF_2^- (or, instead of HF_2^-, CN^-, N_3^- etc. can be used as counterions), or by Lewis acids such as $ZnCl_2$. This leads to poly-MVK with a variety of end groups, via a living polymer mechanism: the silyl group is transferred to the carbonyl oxygen of the monomer. Poly-MVK in common with many polymers containing acidic groups can undergo condensation reactions with mixtures of compounds of groups IIIb, VIIIa, Ia and Vb (e.g. $FeSO_4$). The products can be used as thickeners or retention agents[101].

The softening point of poly-MVK varies between 30 and 50 °C, depending on the mode of preparation. Self-condensation occurs in the presence of mineral bases (Scheme 9), leading to a brittle, insoluble polymer. Amines (e.g. aniline or aniline hydrochloride) react with solutions of poly-MVK to form eventually bright yellow cross-linked polymer, which contains some N function. In acetone solutions, poly-MVK can be reduced to a polymeric secondary alcohol, by reaction with HCHO in the presence of a small amount of mineral acid, which acts as a catalyst[92]. The same effect has been reported from reaction of $LiAlH_4$ with THF solutions of poly-MVK[102]. However, homogeneous reaction of $LiAlH_4$ with poly-MVK prepared by radical polymerization resulted in intramolecular cyclization[103].

Irradiation of poly-MVK or of poly-isopropenyl ketone at room temperature resulted in depolymerization, but at elevated temperatures (80 °C) it resulted in degradation[104]. Heating of poly-MVK in vacuum at 250 °C led to random aldol condensation and a cyclic structure with variable conjugation length. The reaction mechanism is believed to involve $^-CH_2$ groups attacking neighbouring carbonyls[105].

B. Methyl Isopropenyl Ketone

Methyl isopropenyl ketone (**78**, α-methylvinyl methyl ketone) yields polymers with a higher softening temperature and clearer than those of MVK. **78** Polymerizes readily at room temperature, but less so than **1** under similar conditions. Storage of the monomer results in glass-clear polymer, or alternatively polymerization can be brought about by boiling, but only low molecular weights are achieved. Very high molecular weights can be obtained upon exclusion of oxygen. This polymer is believed to be of the head-to-tail type (**79**), and substantially uncross-linked. Coloured polymers can be achieved from aqueous emulsions.

$$ Me-\underset{\underset{O}{\parallel}}{C}-\underset{\underset{Me}{|}}{C}=CH_2 \qquad \sim\!\!\sim\!\!\sim CH_2-\underset{\underset{\underset{Me}{|}}{C=O}}{\overset{\overset{Me}{|}}{C}}-CH_2-\underset{\underset{\underset{Me}{|}}{C=O}}{\overset{\overset{Me}{|}}{C}}-CH_2\!\!\sim\!\!\sim\!\!\sim $$

(**78**)

(**79**)

Catalysts used successfully in polymerizing **78** include dibenzoyl peroxide[106], azodiisobutyronitrile[107], and mixed metal alkyl-transition metal halides (e.g. $AlEt_3$–$FeCl_3$, or $MgEt_2$ with $CoCl_2$ or $MnCl_2$ in ether)[108,109]. Crystalline polymers have been obtained from these catalysts, in a series of hydrocarbon or ether solvents and at temperatures between -60 and 50 °C. For example, in the presence of a $AlEt_3$–$FeCl_3$ catalyst in methylcyclohexane at 18–22 °C, two types of crystalline polymer have been isolated, one isotactic and the other syndiotactic. In these reactions the polymer was precipitated upon addition of water.

Polymerization was also achieved in the presence of phenylmagnesium iodide in EtCl or chloroform solutions. Polymers prepared from radical initiators were not crystalline, whilst those prepared in the presence of BuLi were red in colour[93], the colour being probably due to a structure equivalent to **76**. Analogues of **78**, such as α-ethylvinyl methyl ketone, behave in ways similar to **78**, but propenyl methyl ketone is not polymerizable, presumably owing to its lack of a terminal $CH=CH_2$ group.

C. Uses of Methyl Vinyl Ketone and of Methyl Isopropenyl Ketone

The polymers of both **1** and **78** have been used in photographic or related processes. For example, the use of poly-MVK as an anion-exchange resin component in the manufacture of dye-receptive films has been patented by Kodak[109]. Poly-(methyl isopropenyl ketone) has been used as a component of dry developable resists for Si-wafer manufacture[110]. Poly-MVK obtained from MVK dissolved in dioxane in the presence of 1% Bz_2O_2 was dissolved in a mixture of acetic acid and dioxan with aminoguanidine[111]. Bicarbonate was added slowly under heat and, on addition of water and Zn dust with AcOH, a light amber colour was obtained. On addition of NaOH, **80** was obtained. An equivalent polymer was also obtained from poly-(ethyl vinyl ketone), and poly-(propyl vinyl ketone). **80** can be used in formulations of additives in light-sensitive emulsions for photography, as mordants. A recent patent application describes the use of various enone polymers reacted with cyano dyes as optical laser materials[112].

$$\underset{\text{(80)}}{-(\text{CH}_2-\overset{\overset{\text{Me}}{|}}{\text{CH}}=\text{N}-\text{NH}-\overset{\overset{\text{NH}_2}{|}}{\text{C}}=\text{NH})_{\overline{n}}-}$$

D. Other Alkyl or Aryl Vinyl Ketones

Ethyl vinyl ketone (**81**) polymerizes very readily in sealed tubes at 40 °C in the presence of diacetyl peroxide initiator, to a soft yellowish polymer[92]. Longer periods of reaction time can result in solid polymers. A variety of aryl vinyl ketones, including phenyl, 4-chlorophenyl and napthyl, have been polymerized using dibenzoyl peroxide initiator, yielding polymers of varied hardness[113]. Phenyl vinyl ketone can be polymerized in toluene, in the presence of several organometallic catalysts, at -70 °C[114]. This is not a crystallizable polymer, but a crystalline product has been obtained in the presence of initiators such as lithium dust, sodium hydride, BuLi, etc. The aryl vinyl ketone polymers obtained from this route have higher softening temperatures. Chlorinated monomers can also be used, e.g. **82**, which very readily yield solid polymers at room temperature[115].

$$\underset{\text{(81)}}{\text{CH}_2=\text{CH}-\overset{\overset{}{\underset{\overset{\|}{O}}{}}}{\text{C}}-\text{Et}} \qquad\qquad \underset{\text{(82)}}{\text{CH}_2=\text{CCl}-\overset{\overset{}{\underset{\overset{\|}{O}}{}}}{\text{C}}-\text{Me}}$$

Two types of poly-(t-butyl vinyl ketone) have been produced[116]: the first, made at 25 °C with lithium or organolithium catalysts in hexane or toluene; the second, prepared in THF at 0 °C with lithium biphenyl, or with azobisisobutyronitrile in benzene at 60 °C. The first type is crystalline and much less soluble than the second, and it has been suggested that they are isotactic and moderately syndiotactic, respectively. It was found that with lithium dispersions, BuLi or lithium biphenyl initiators and a mixture of t-butyl vinyl ketone and methyl methacrylate in THF, only homopolymerization of the enone occurred, albeit at twice the rate than in the absence of the methacrylate[117]. Viscosity measurements suggested that chain transfer operated, which was thought to be the reaction of a growing enone chain with the carbonyl group of the methacrylate. The lithium methoxide thus produced would serve to terminate one chain and initiate another. The enhanced rate, however, is probably due to the preferential solvation of growing ion pairs by methyl methacrylate. Also, it is possible that the presence of methacrylate moderates the wasteage of initiator which would otherwise occur, owing to the formation of lithium methoxide, via a reaction of the organolithium compounds with the enone.

The dienone **83** can yield both homopolymers and copolymers[118]. A number of different substituents have been used, e.g., R^1 was cycloalkyl, alkenyl or phenyl, $R^2 = H$, alkyl, phenyl or a halogen, and R^3 or R^4 H, Me or a halogen. Copolymers with **78** have also been formed. For example, a solution of **83**, where $R^1 = Me$, $R^2 = H$, $R^3 = H$ and $R^4 = Me$ in toluene, yielded a *trans*-1, 4 polymer in 1 day at 50 °C in the presence of AlEt$_3$. The product had a molecular weight of approximately 353,000. 2-Hydroxybut-1-en-3-one (**84**) yields brittle polymers[119]. Etherification with MeOH or EtOH of **84** yields a monomer, which can polymerize by heating at 30 °C for 4 days under nitrogen, to a strong transparent product. 2-Methoxymethyl-but-1-en-3-one was polymerized in the absence of oxygen to a clear, hard resin, which was soluble in a variety of organic solvents[120]. Polymerization was initiated by heat, light or peroxides.

(83)   (84)

E. Acrolein

Acrolein, or prop-1-en-3-one (85), was first prepared over 150 years ago[121]. It polymerizes spontaneously to a white non-crystalline polymer. The polymerization reaction is complicated by condensations through the aldehyde group. Clear solid polymers can be obtained in the presence of basic catalysts and buffers. The presence of a little β-naphthol enables 85 to polymerize upon exposure to UV light. The spontaneous polymerization can be inhibited by the presence of hydroquinone.

(85)

α-Methylacrolein (2-methylprop-1-en-3-one) polymerizes almost as readily as 85. Freshly prepared and distilled, it begins polymerizing within a few hours of being left to stand in air, and polymerization may be complete in 4 days to a hard chalky resin. In the presence of hydroquinone, dimerization only occurs. This monomer can also be polymerized in the presence of t-Bu peroxide and $ZnCl_2$ in aqueous solution at room temperature, to an opaque gel. This can be converted to a hard polymer, by oxidation. The ethyl analogue only polymerizes rapidly on heating.

F. Exchange Polymerization

A novel polymerization route has recently been described involving the so-called carbonyl double-bond exchange mechanism. For example, homopolymerization of unsaturated ketones in the presence of WCl_6 yields polyacetylene[122]. Benzylidene acetophenone (86), or 1,3-diphenyl-2-buten-1-one (87) or 1,3,3-triphenyl-2-propen-1-one (88) in the presence of WCl_6 gave poly-phenylacetylene, with molecular weight in the region 1500–3000; increase in the amount of WCl_6 present led to an increase in the degree of polymerization[123]. This polymer was found to be paramagnetic. Reaction of 1,2,3-

(86)   (87)

triphenyl-2-propen-1-one (89) or its 1, 2, 3, 3-tetraphenyl analogue (90) led to poly-diphenylacetylene and 1, 3-dimethyl-2-buten-1-one (91) yielded poly-methylacetylene. Polyacetylenes have generated a lot of excitement in recent years, because they exhibit semiconducting or metal-like conducting behaviour upon p- or n-type doping[124].

(88)

(89)

(90)

(91)

VII. COPOLYMERIZATION AND GRAFT POLYMERIZATION OF ENONES

Enones undergo both copolymerization with a variety of monomers, and grafting on a number of polymers. Both processes have recently received considerable attention. Initiation of these reactions has been carried out by various radical, anionic and cationic catalysts. The usefulness of MVK and of its analogues in copolymerization is a relatively recent development. Copolymers of MVK initially reported tended to be rather water-sensitive, of limited stability and reactive. Products with acid-releasing comonomers tended to be discoloured.

One of the first comonomers that were employed was butadiene[92]. This formed an oil-resistant rubber with MVK which, however, tended to harden upon standing. Initiation was carried out by persulphate emulsions. Better results can be obtained if a small amount of inhibitor is added, which slows down the polymerization of MVK 78 also copolymerizes with butadiene, yielding a product with properties similar to the copolymer of MVK. The 78–butadiene copolymers prepared in an emulsion medium were soluble in aromatic solvents, even at 80% monomer conversion. Terpolymerization of MVK with butadiene and styrene can also be brought about by the same route[125].

Radical mass suspension graft polymerization of methyl vinyl ketone and styrene (92) on polybutadiene results in high impact styrene copolymers with methyl vinyl ketone, which are photodegradable[126]. Copolymerization of 92 and highly pure MVK can also be brought about without the medium of polybutadiene, in the presence of radical initiators[127], such as Bz_2O_2. The reaction is carried out on a water bath in a methyl ethyl ketone solution, and under nitrogen. The polymer is a rubbery mass, which crystallizes to a white powder on stirring with MeOH[128]. In common with homopolymers of MVK, the 92–MVK copolymer can react with $LiAlH_4$, to yield a poly-alcohol. Dienones such as dibenzylideneacetone (65) also form copolymers with 92. These have molecular weights in the range of 20,000 to 30,000, and have thermal stability of form up to 130 °C[129].

Styrene also copolymerizes with a variety of other α, β-unsaturated ketones, including phenyl vinyl ketone, isopropenyl methyl ketone, propenal, 2-methyl propenal, 2-ethyl propenal and methyl methacrylate[127]. These polymers are photodegradable in solution and in the solid state. The reaction that occurs under irradiation is believed to be chain scission[130]. Solid 92–MVK copolymers are susceptible to reaction with aluminium isopropoxide (iso-Pr—O)$_3$Al at 160 °C, which results in evolution of acetone[131]. The reaction results in the elimination of carbonyl groups and the introduction of cross-linking of polymer chains via O—Al—O bridges (93).

(92)   (93)

Enones can also be copolymerized with ethylene. For example, MVK and ethylene react under γ-ray irradiation (Co^{60}) to yield uniquely copolymers; no homopolymerization occurs[132]. Graft copolymers of these two monomers have been used to immobilize Ni^{2+} ions on their surface. Such solids can be used to catalyze isomerization reactions of alkenes, and their dimerization[133]. One of the reactions catalyzed by this solid is ethylene conversion to butadiene. Graft copolymers of MVK on polyethylene can be used to immobolize Ti(IV) compounds, which are present on the polymer surface as clusters[134]. Such a solid is resistant to reduction and can be used as a catalyst.

MVK and other enones can be copolymerized with 2-hydroxymethyl methacrylate in the presence of $(NH_4)_2S_2O_8$ and $Na_2S_2O_3$ as redox catalysts and a small amount of N, N'-methylenediacrylamide, which can act as a cross-linking agent[135]. The product is a network polymer, which can act as an adsorbent of urea. Anionic or cationic copolymerization of MVK with 2, 5, 6-trisubstituted 3, 4-dihydro-2H-pyrans results in head-to-head alternating copolymers[136]. MVK copolymer with 4-vinylpyridine becomes dense and tough when cross-linked with malonyl dihydrazide. This polymer can be made into membranes, which perform well in reverse osmosis with NaCl and $CoCl_2$ containing feeds[137].

A number of vinyl monomers, including MVK, can enter into homogeneous anionic graft copolymerization on Nylon 6. Before reaction with the vinyl monomer, Nylon 6 is metallated in a solution using a variety of alkali metal compounds[138]. Graft copolymerization of MVK onto viscose or cotton fabrics can be carried out by immersing the polymers into an aqueous solution of MVK and irradiating with γ rays[139]. Cellulose can be modified by graft copolymerization of MVK[140]. The thermal stability of poly-(vinyl bromide) is increased if converted to copolymer with MVK. Stability increases with MVK concentration[141]. MVK and vinyl acetate can be copolymerized from ammonia-saturated MeOH solutions by heating at 80 °C for 4h in an autoclave. This polymer can be drawn into a fibre. Copolymerization with butadiene or acrylonitrile leads to fibres with improved dyeability[142].

MVK undergoes radical copolymerization with acrylamide and several of its derivatives[143]. Polymerization is carried out under vacuum at 60 °C, in the presence of dioxan as solvent. Other enone copolymers include those prepared with p-isopropenylphenol and its analogues. These comonomers undergo emulsion copolymerization with MVK at 60–80 °C, in the presence of $- MeC_6H_4SO_3H$ as catalyst[144]. Often, enone copolymers can have their properties changed, by subsequent reactions. For example, poly-MVK or MVK–divinylbenzene copolymers can react with dichlorophosphites (94), to yield poly-(α-OH α-Me-allyl phosphonic acid monoesters)[145]. 2, 4-Dinitrophenylhydrazine has also been shown to react with various MVK copolymers, e.g. with styrene as comonomer[146].

Acrolein can copolymerize with MVK or acrylamide by an anionic mechanism in THF solutions[147] in the presence of imidazole as initiator, at 0 °C. The acrolein–MVK

ROPCl$_2$

R=Me,Et,Pr,Bu

(94)

copolymer was a vinyl polymer with imidazo groups attached to the aldehyde or ketone side-chains. The acrolein–acrylamide copolymer resulted from both 1, 2- and 1, 4-addition polymerization.

VIII. ACKNOWLEDGEMENTS

A large part of this review describes work with which the author has been intimately associated. For that, the financial support of the SERC at Cambridge and of the BRIEF system at Brunel is acknowledged. Thanks are due to Dr W. Jones at Cambridge, and colleagues at Brunel, for useful discussions. The stimulus provided by Professor J. M. Thomas is appreciated.

IX. REEFERENCES AND NOTES

1. This chapter is dedicated to my new-born nephew Constantinos P. Sepetas.
2. Professor Leopold Ruzicka, ETH Zurich, Nobel Prize Laureate.
3. M. D. Cohen and G. M. J. Schmidt, J. Chem. Soc., 1996 (1964).
4. G. M. J. Schmidt, Pure Appl. Chem., 27, 647 (1971).
5. C. Liebermann, Chem. Ber., 22, 782 (1889).
6. M. D. Cohen, G. M. J. Schmidt and F. I. Sonntag, J. Chem. Soc., 2000 (1964).
7. J. Bregmann, G. M. J. Schmidt and F. I. Sonntag, J. Chem. Soc., 2021 (1964).
8. L. Leiserowitz and G. M. J. Schmidt, Acta Crystallogr., 18, 1058 (1965).
9. J. M. Thomas, Phil. Trans. R. Soc., 277, 251 (1974).
10. J. M. Thomas, Pure Appl. Chem., 51, 1065 (1979).
11. J. R. Scheffer, Acc. Chem. Res., 13, 283 (1980).
12. A. Gavezotti and M. Simonetta, Chem. Rev., 82, 1 (1982).
13. M. Hasegawa, Chem. Rev., 83, 507 (1983).
14. J. M. McBride, Acc. Chem. Res., 16, 304 (1983).
15. J. Trotter, Acta Crystallogr., Sect. B, 39, 373 (1983).
16. C. R. Theocharis, PhD Thesis, University of Cambridge, 1982.
17. V. Ramamurthy and K. Venkatesan, Chem. Rev., 87, 433 (1987).
18. C. R. Theocharis and W. Jones, in Organic Solid State Chemistry (Ed. G. R. Desiraju), Elsevier, Amsterdam 1987, pp. 47–68.
19. M. Hasegawa, in Organic Solid State Chemistry, (Ed. G. R. Desiraju), Elsevier, Amsterdam, 1987, pp. 153–178.
20. G. R. Desiraju, Proc. Indian Acad. Sci., 73, 407 (1984).
21. J. M. Thomas, Nature, 289, 633 (1981).
22. H. Nakanishi, C. R. Theocharis and W. Jones, Acta Crystallogr., Sect. B, 37, 758 (1981).
23. W. Jones, H. Nakanishi, C. R. Theocharis and J. M. Thomas, J. Chem. Soc., Chem. Commun., 610 (1980).
24. J. Swiatkiewicz, G. Eisenhardt, P. N. Prasad, J. M. Thomas, W. Jones and C. R. Theocharis, J. Phys. Chem., 86, 1764 (1982).
25. H. Nakanishi, W. Jones, J. M. Thomas, M. B. Hursthouse and M. Motevalli, J. Phys. Chem., 85, 3636 (1981).
26. W. Jones and C. R. Theocharis, J. Cryst. Spec. Res., 14, 447 (1984).
27. E. L. Short, C. R. Theocharis and G. L. Reed, unpublished results.
28. H. Nakanishi, W. Jones, J. M. Thomas, M. B. Hursthouse and M. Motevalli, J. Chem. Soc., Chem. Commun., 611 (1980).
29. D. A. Whiting, J. Chem. Soc. (C), 3396 (1971).

30. G. C. Forward and D. A. Whiting, *J. Chem. Soc. (C)*, 1868 (1969).
31. C. R. Theocharis, H. Nakanishi and W. Jones, *Acta Crystallogr., Sect. B*, **37**, 756 (1981).
32. S. K. Kearsley, *PhD Thesis*, University of Cambridge, 1983.
33. W. Jones, S. Ramdas, C. R. Theocharis, J. M. Thomas and N. W. Thomas, *J. Phys. Chem.*, **85**, 2594 (1981); C. R. Theocharis, W. Jones, M. Motevalli and M. B. Hursthouse, *J. Cryst. Spec. Res.*, **12**, 377 (1982).
34. A. I. Kitaigorodskii, *Molecular Crystals and Molecules*, Academic Press, New York, 1973.
35. H. Nakanishi, W. Jones and J. M. Thomas, *Chem. Phys. Lett.*, **71**, 44 (1980).
36. C. R. Theocharis, W. Jones and J. M. Thomas, *Mol. Cryst. Liq. Cryst.*, **93**, 53 (1983).
37. C. R. Theocharis, W. Jones, J. M. Thomas, M. Motevalli and M. B. Hursthouse, *J. Chem. Soc., Perkin Trans. 2*, 71 (1984).
38. G. Kaupp and I. Zimmermann, *Angew. Chem., Int. Ed. Engl.*, **20**, 1018 (1981).
39. S. E. Hopkin and C. R. Theocharis, in preparation.
40. K. A. Becker, K. Plato and K. Plieth, *Z. Elektrochem.*, **61**, 96 (1957).
41. H. Frey, G. Brehmann and G. Kaupp, *Chem. Ber.*, **120**, 387 (1987).
42. C. R. Theocharis, *J. Chem. Soc., Chem. Commun.*, 80 (1987).
43. C. R. Theocharis, A. M. Clark, S. E. Hopkin, P. Jones, A. C. Perryman and F. Usanga, *Mol. Cryst. Liq. Cryst.*, **156**, 85 (1988).
44. C. R. Theocharis, S. E. Hopkin, A. M. Clark and M. J. Godden, *Solid State Ionics*, in press.
45. G. R. Desiraju, R. Kamala, B. H. Kumani and J. A. R. P. Sarma, *J. Chem. Soc., Perkin Trans. 2*, 187 (1984).
46. W. Jones, C. R. Theocharis, J. M. Thomas and G. R. Desiraju, *J. Chem. Soc., Chem. Commun.*, 1443 (1983).
47. C. R. Theocharis, W. Jones and G. R. Desiraju, *J. Am. Chem. Soc.*, **106**, 3606 (1984).
48. S. K. Kearsley and G. R. Desiraju, *Proc. R. Soc. Ser. A*, **397**, 151 (1985).
49. G. Kaupp, E. Jost Kleigreme and H.-J. Hermann, *Angew. Chem., Int. Ed. Engl.*, **21**, 435 (1982).
50. J. M. Thomas, *Mol. Cryst. Liq. Cryst.*, **52**, 523 (1979).
51. J. M. Thomas and J. O. Williams, *Prog. Solid State Chem.*, **6**, 121 (1971).
52. C. R. Theocharis and W. Jones, *J. Chem. Soc., Faraday Trans. 1*, **81**, 857 (1985).
53. E. L. Short and C. R. Theocharis, submitted.
54. P. A. Prasad, in *Organic Solid State Chemistry* (Ed. G. R. Desiraju), Elsevier, Amsterdam, 1987, pp. 117–151.
55. C. Brauchle and C. R. Theocharis, unpublished results.
56. F. Nakanishi, H. Nakanishi, M. Tsuchiya and M. Hasegawa, *Bull. Chem. Soc. Jpn.*, **49**, 3096 (1976).
57. H. Stobbe and K. Bremer, *J. Prakt. Chem.*, **123**, 1, (1929).
58. A. Mustafa, *Chem. Rev.*, **51**, 1 (1952).
59. N. Bertoniere, W. E. Franklin and S. P. Rowland, *J. Appl. Polym. Sci.*, **15**, 1743 (1971).
60. S. A. Zahir, *J. Appl. Polym. Sci.*, **15**, 1743 (1979).
61. C. R. Theocharis, in preparation.
62. N. Ramasubba, T. N. Guru Row, K. Venkatesan, V. Ramamurthy and C. N. R. Rao, *J. Chem. Soc., Chem. Commun.*, 178 (1982).
63. K. Gnanaguru, N. Ramasubba, K. Venkatesan and V. Ramamurthy, *J. Photochem.*, **27**, 355 (1984).
64. D. Rabinovich and G. M. J. Schmidt, *J. Chem. Soc. (B)*, 144 (1967).
65. R. C. Cookson, D. A. Cox and J. Hudec, *J. Chem. Soc.*, 4499 (1961).
66. R. C. Cookson, E. Crundwell, R. R. Hill and J. Hudec, *J. Chem. Soc.*, 3062 (1964).
67. R. C. Cookson, R. R. Hill and J. Hudec, *J. Chem. Soc.*, 3042 (1964).
68. A. A. Dzakpasu, S. E. V. Philips, J. R. Scheffer and J. Trotter, *J. Am. Chem. Soc.*, **98**, 6049 (1976).
69. J. R. Scheffer and A. A. Dzakpasu, *J. Am. Chem. Soc.*, **101**, 2163 (1979).
70. H. Irngartinger, R. D. Acker, W. Rebafka and H. A. Staab, *Angew. Chem., Int. Ed. Engl.*, **13**, 674 (1974).
71. M. J. Begley, L. Crombie and T. F. W. B. Knupp, *J. Chem. Soc., Perkin Trans. 1*, 976 (1979).
72. S. Mohr, *Tetrahedron Lett.*, 3139 (1979).
73. S. Mohr, *Z. Anal. Chem.*, **304**, 280 (1980).
74. S. Mohr, *Z. Kristallogr.*, **149**, 108 (1979).
75. D. Lawrentz, S. Mohr, and B. Wendlaender, *J. Chem. Soc., Chem. Commun.*, 863 (1984).
76. S. Mohr, *Tetrahedron Lett.*, 2461 (1979).

77. W. Ried, and H. Bopp, *Angew. Chem., Int. Ed. Engl.*, **16**, 653 (1977).
78. E. C. Taylor and W. W. Pundler, *Tetrahedron Lett.*, (issue 25), 1 (1960).
79. S. Y. Wang, *Nature*, **200**, 879 (1963).
80. B. M. Powell and P. Martel, *Photochem. Photobiol.*, **26**, 305 (1977).
81. J. B. Bremner, R. N. Warrener, E. Adman and L. H. Jensen, *J. Am. Chem. Soc.*, **93**, 4574 (1971).
82. G. M. Blackburn and R. J. H. Davies, *Tetrahedron Lett.*, 4471 (1966).
83. R. D. Rieke and R. A. Copenhafer, *Tetrahedron Lett.*, 879 (1971).
84. M. Kaftory, *J. Chem. Soc., Perkin Trans. 1*, 757, (1984).
85. D. B. Chase, R. L. Amey and W. G. Holtje, *Appl. Spectrosc.*, **36**, 155 (1982).
86. E. Waschen, R. Matusch, D. Krampity and K. Hartke, *Liebigs. Ann. Chem.*, 2137 (1978).
87. B. Fucks and M. Pasternak, *J. Chem. Soc., Chem. Commun.*, 537 (1977).
88. N. Moulden, *PhD Thesis*, University of Cambridge, 1985.
89. J. K. Frank and I. C. Paul, *J. Am. Chem. Soc.*, **95**, 2324 (1973).
90. M. Hasegawa, K. Saigo, T. Mori, H. Uno, M. Nomaru and H. Nakanishi, *J. Am. Chem. Soc.*, **107**, 2788 (1985).
91. L. Addadi and M. Lahav, *Pure Appl. Chem.*, **51**, 1269 (1979).
92. C. E. Schildknett, *Vinyl and Related Polymers*, Wiley, New York, 1952.
93. D. M. Wiles, in *Structure and Mechanisms in Vinyl Polymerization* (Eds. T. Tsuruta and K. F. O'Driscoll), M. Dekker, New York, 1969.
94. R. F. Conaway, US Patent 2,088,577.
95. C. S. Marvel and C. L. Levesque, *J. Am. Chem. Soc.*, **60**, 280 (1938).
96. H. V. Melville, T. T. Jones and R. F. Tuckett, *Proc. R. Soc.*, **187**, 19 (1946).
97. S. Iwatsuki, Y. Yamashita and Y. Ishii, *J. Polym. Sci.*, **B1**, 545 (1963).
98. G. Wasai, T. Tsuruta, J. Furukawa and R. Fujio, *Kogyo Kogaku Zasshi*, **66**, 1339 (1963).
99. H. R. Alcock and M. L. Levin, *Macromolecules*, **18**, 1324 (1985).
100. O. W. Webster, W. R. Hertler, D. Y. Sogah, W. B. Farnham and T. V. Rajan-Babu, *J. Am. Chem. Soc.*, **105**, 5706 (1983).
101. J. Plank and A. Aigenberger, Ger. Offen, 3,429,068, (1986).
102. J. A. Blanchette, US Patent 2,862,911.
103. R. C. Schulz, H. Vielhaber and W. Kern, *Kunststoffe*, **50**, 500 (1960).
104. K. F. Wissbrum, *J. Am. Chem. Soc.*, **81**, 58 (1959).
105. J. N. Hay, *Makromol. Chemie*, **67**, 31 (1963).
106. H. Watanabe, R. Kayama, H. Nagai and A. Nishioka, *J. Polym. Sci.*, **62**, 574 (1962).
107. F. Brown, F. Berdinelli, R. J. Kray and L. J. Rosen, *Ind. Eng. Chem.*, **51**, 79 (1959).
108. P. A. Small and D. G. M. Wood, BP 862,862 (1960).
109. Belg. Pat. 553,517 (1959).
110. M. Tsuda, M. Yabuta, K. Yamashita, S. Oikawa, A. Yokoto, H. Nakana, K. Gano and S. Namba, *Nanometer Struct. Electr. Proc. Int. Symp.*, 105 (1984).
111. L. M. Minsk, US Patent 2,882,156 (1959).
112. V. L. Bell, I. J. Ferguson and M. J. Wenderlay, US Patent 4,547,444 (1985).
113. P. R. Thomas, G. J. Tyler, T. E. Edwards, A. T. Radcliffe and R. C. P. Cubbon, *Polymer*, **5**, 525 (1964).
114. T. Tsuruta, R. Fijio and J. Furikawa, *Makromol. Chem.*, **80**, 172 (1964).
115. L. W. Metzer and O. Bayer, US Patent 2,173,066 (1952).
116. C. G. Overberger and A. M. Schiller, *J. Polym. Sci.*, **54**, S30 (1961).
117. C. G. Overberger and A. M. Schiller, *J. Polym. Sci.*, **C1**, 325 (1963).
118. Marazon Petrochemicals, JP 60 76,514 (1985).
119. W. Grimme and J. Woellner, US Patent 2,760,952 (1956).
120. Ger. Offen., 956, 272.
121. J. Redtenbacher, *Ann.*, **47**, 121 (1843).
122. I. Slopov, K. Iossifov and L. Mladenova, *Oesterr. Chem. Z.*, **86**, 208 (1985).
123. I. Slopov, K. Iossifov and L. Mladenova, Springer Ser. Solid State Sci., **63**, 208 (1985).
124. A. G. McDiarmid and A. J. Heeger, *Synth. Metals*, **1**, 101 (1979).
125. V. Yu. Erofeev, N. M. Miranova, A. V. Petuklov and L. N. Korlenko, *Izv. Vyssh. Uchebn. Zaved., Khim. Tekhnol.*, **29**, 119 (1986).
126. A. Stoyanov, G. Nenkov, T. Petrova and V. Kabaivanov, *Dokl. Bolg. Akad. Nauk*, **35**, 929 (1982).
127. O. Seycek, B. Bednav, M. Honska and J. Kalen, *Collect. Czech. Chem. Commun.*, **47**, 785 (1982).
128. K. A. Kun and H. G. Cassidy, *J. Polym. Sci.*, **44**, 383 (1960).

129. Ger. Offen., 1,022,801 (1960).
130. C. R. George and R. R. Gerke, EP 163,496 (1985).
131. R. S. Gregorian and R. V. Bush, *J. Polym. Sci.*, **B2**, 401 (1964).
132. P. Colombo, M. Steinberg and D. Macchia, *J. Polym. Sci.*, **B1**, 483 (1964).
133. A. D. Pomogailo, F. Khrisostomov and F. S. D'yachkovskii, *Kinet. Katal.*, **26**, 1104 (1985).
134. L. Karklino, A. D. Pomogailo, A. P. Lisitskaya and Yu. G. Borod'ko, *Kinet. Katal.*, **24**, 657 (1983).
135. T. Obayashi, N. Yamashita, H. Yuasu and T. Taeshita, *J. Polym. Sci., Polym. Lett. Ed.*, **23**, 593 (1985).
136. I. Chu and J. Y. Lee, *Macromolecules*, **16**, 1245 (1985).
137. E. Oikawa and Y. Igarashi, *J. Appl. Polym. Sci.*, **29**, 1723 (1984).
138. I. Ikeda, K. Suzuki and K. Ishiguro, *Kobushi Rombushu*, **40**, 603 (1983).
139. R. C. Sovish and F. L. Saunders, US Patent 2,998,329 (1962).
140. F. L. Saunders and R. C. Sovish, *J. Appl. Polym. Sci.*, **7**, 357 (1963).
141. M. A. Diab, *Eur. Polym. J.*, **20**, 599 (1984).
142. S. Matsumoto, M. Yano and T. Osugi, JP 9439 (1958).
143. N. Yamashita, K. Ikezawa, S. I. Aynkawa and T. Maeshima, *J. Macromol. Sci. Chem.*, **A21**, 621 (1984).
144. Ger. Offen. 1,153,527 (1963).
145. P. Ho and K. Ye, *Gaodong Xuexiao Huazue Xuebuo*, **3**, 425 (1982).
146. D. R. Burfield, *Polymer*, **23**, 1259 (1982).
147. S. Morita, K. Ikezawa, H. Inone, N. Natsuki and T. Maeshita, *J. Macromol. Sci.*, **A17**, 1495 (1982).
148. J. R. Scheffer and J. Trotter in *The Chemistry of the Quinonoid Compounds Volume 2*, Wiley, (ed. S. Patai and Z. Rappoport) Chichester 1988, p. 1199.

Author index

This author index is designed to enable the reader to locate an author's name and work with the aid of the reference numbers appearing in the text. The page numbers are printed in normal type in ascending numerical order, followed by the reference numbers in parentheses. The numbers in *italics* refer to the pages on which the references are actually listed.

Tam, C.C. 367(79), 368(78, 79), *457*
Tam, W.-C. 183(140), *197*
Tamaru, Y. 371(102), *458*
Tamelen, E.E.van 837(218), *916*
Tamm, C. (183), *916*
Tamura, M. 424, 425(516), *466*
Tamura, T. 423, 424, 426, 428(510), *466*
Tamura, Y. 94(135), *104*, 230(146), *276*, 423(504, 505), *466*
Tan, C.T. 132(21), *149*
Tan, L. 806, 846(132), *915*
Tanabe, M. 321(38), *352*, 932(62), 937(115), *1012, 1013*
Tanaka, H. 311(131), *315*
Tanaka, K. 1007(388), *1021*
Tanaka, T. 1093(16), 1094(17, 18), *1131*
Tancrede, J. 1034(81), *1060*
Tanemura, M. 886(394), *920*
Tang, P.W. 379(151), 381(151, 194, 195), *459, 460*
Tang, Y.S. 324(57), *352*, 571, 576(40), *595*
Tanigawa, K. 1091(12), *1130*
Taniguchi, H. 406(304), *462*
Taniguchi, S. 760(11), 761(12), *778*
Tanis, S.P. 450(661), *469*
Taniyama, E. 497(86), *512*
Tannenbaum, H.P. 182(101), *196*
Tanner, D.D. 497(76), *511*
Tanno, N. 969(260, 261a, 261b), 974(260), *1018*
Tantardini, G. 18(78), *26*
Tantardini, G.F. 18(77), *26*
Taschner, M.J. 1103(34), *1131*
Tashtoush, H. 501(101), *512*
Tatchell, A.R. 970(270–272), *1018*
Tate, D.P. 1034(86), *1060*
Tatsumi, C. 1002(371a), *1021*
Tatsumi, T. 991(337a, 337b), *1020*, 1034(85), *1060*
Tatsuta, K. 436(589), *467*
Taub, D. 212(52), *275*, 1115(53), *1132*
Tauer, E. 1070(29), *1086*
Tautz, W. 255(240), *278*
Tawarayama, Y. 406, 416(307, 308), *462*
Taylor, A.P. 283(10), *312*
Taylor, D.A. 949(175b), *1015*
Taylor, E.C. 147(89), *150*, 247(206), *278*, 553(207, 209, 211), *558*, 1161(78), *1175*
Taylor, G.E. 1050(151), 1052(161), *1062*
Taylor, G.N. 725(201), *755*
Taylor, H.M. 414(399), *464*
Taylor, J.B. 451(674), *469*
Taylor, M.D. 1112(46), 1113(47), *1132*
Taylor, N.J. 1045(123), 1052(176), *1061, 1062*
Taylor, P.J. 65(35), *102*

Taylor, R. 30(3), *53*, 648(45), *751*
Taylor, R.J.K. 379, 393(143), *459*
Taylor, R.T. 512, *512*
Tchir, M. 661(89), 662(90), 664(94), 692, 726(90), *752, 753*
Tchoubar, B. 931(55), *1012*
Teague, P.C. 440(624), 441(624, 629), *468*, 539(143), *557*
Tee, O.S. 342(130), *354*
Telschow, J.E. 293(47d), *313*, 547, 548(188), *558*
Tempel, E. 306(108), *315*
Temple, R.W. 933, 934(68), *1012*
Templeton, J.F. 816(169, 170), 821(170), *915*
Tencer, M. 64(31), 74(58), 80(79), *102, 103*
Teng, J.I. 845, 846, 849(237), *917*
Teng, J.T. 806(129), *915*
Teng, K. 230(140), *276*
Teoh, I. 381(185), *460*
Teoule, R. 770(94), 771(101, 104, 106), 772(117, 120, 121), 773(120), 774(132), *779, 780*
Terada, I. 986(324, 325c), *1019*
Terada, T. 202(15), *274*
Teranishi, A.Y. 218(88), *275*, 545(177), *558*
Terasawa, T. 935(99), *1013*
Terashima, S. 969(260, 261a, 261b), 974(260), *1018*, 1094, 1095(21), *1131*
Terem, B. 605(42), *621*
Terem, R. 612(83), *622*
Terlouw, J.K. 152(3, 11, 14), 153(3, 14), 170(61–63), 171, 173(61, 63), 183(62), 185(3), *194, 195*
Ternai, B. 133(28), *149*
Tero-Kubota, S. 1070(28), *1086*
Terrell, R. 293(48), *313*
Tesarek, J.M. 178, 181(94), *196*
Teschen, H. 1001(366), *1021*
Texier-Boullet, F. 431(545), 432(559), *466, 467*
Tezuka, T. 881(373, 376), 900(455), *919, 921*
Thaler, V. 174(76), *196*
Thanupran, C. 294(57), *313*
Thayer, A.I. 857(279), *918*
Thayer, A.L. 857(275, 277), *918*
Theard, L.M. 768(69), *779*
Thebtaranonth, C. 255(236), *278*, 294(57), *313*
Thebtaranonth, Y. 255(235–237), 257(256, 257), *278, 279*, 294(57), *313*
Theissen, R.J. 215(80), *275*
Theissling, C.B. 173, 174(72), *196*
Theobald, D.W. 934(96), *1013*

Subject index